Research on Plant Genomics and Breeding: 2nd Edition

Research on Plant Genomics and Breeding: 2nd Edition

Guest Editors

Zhiyong Li
Jian Zhang

MDPI

Basel • Beijing • Wuhan • Barcelona • Belgrade • Novi Sad • Cluj • Manchester

Guest Editors
Zhiyong Li
State Key Laboratory of Rice Biology and Breeding
China National Rice Research Institute
Hangzhou
China

Jian Zhang
State Key Laboratory of Rice Biology and Breeding
China National Rice Research Institute
Hangzhou
China

Editorial Office
MDPI AG
Grosspeteranlage 5
4052 Basel, Switzerland

This is a reprint of the Special Issue, published open access by the journal *International Journal of Molecular Sciences* (ISSN 1422-0067), freely accessible at: https://www.mdpi.com/journal/ijms/special_issues/IFNJ2N7VGH.

For citation purposes, cite each article independently as indicated on the article page online and as indicated below:

Lastname, A.A.; Lastname, B.B. Article Title. *Journal Name* **Year**, *Volume Number*, Page Range.

ISBN 978-3-7258-5685-5 (Hbk)
ISBN 978-3-7258-5686-2 (PDF)
https://doi.org/10.3390/books978-3-7258-5686-2

Contents

International Journal of
Molecular Sciences

MDPI

Editorial

Research on Plant Genomics and Breeding 2.0

Long Jin, Zhiyong Li * and Jian Zhang *

State Key Laboratory of Rice Biology and Breeding, China National Rice Research Institute, Hangzhou 311400, China; jinl53719@163.com
* Correspondence: lizhiyong@caas.cn (Z.L.); zhangjian@caas.cn (J.Z.)

Plant genomics and breeding is one among the several highly regarded disciplines in today's field of biological sciences. With the development of modern biotechnology, the research approaches to examining plant gene function have become increasingly diverse. Particularly in recent years, with the accurate determination of reference genome sequences, the popularization of sequencing technologies, and the advancements made in bioinformatics, the study of gene function has become more versatile. Meanwhile, modern biotechnology plays a significant role in enhancing modern agriculture, accelerating biological breeding processes, and ensuring food security. Here, we summarize some of the genomic and breeding research works published in *IJMS* in recent years, hoping to contribute to modern breeding practices.

1. Genome-Wide Association Study (GWAS) Analysis

GWAS is an effective method for studying genetic variations associated with complex traits and diseases [1]. To date, numerous significant genetic loci have been identified through GWAS in both animals and plants. Nitrogen is one of the most crucial elements for plant growth. N. Ahmad et al. identified 16 candidate genes crucial for nitrogen utilization through GWAS, weighted gene co-expression network analysis (WGCNA), and RNA-Seq analysis of 327 rape (*Brassica napus* L.) germplasms, providing a genetic foundation for variety improvement and nitrogen-efficient utilization [2]. Senescence marks the final stage of plant development, accompanied by nutrient assimilation to remobilization. Liu et al. identified a homologous gene, *GhMKK9*, of the Arabidopsis cluster senescence gene *AtMKK9* through GWAS analysis of 355 upland cotton materials. They found that *GhMKK9* positively regulates plant senescence, with gene silencing delaying senescence in cotton and overexpression promoting senescence in Arabidopsis [3]. The sugar content in fresh maize stalks determines their value as green storage feed. Chen et al. conducted GWAS analysis on 188 maize germplasms, including 41 sweet maize inbred lines, 74 glutinous maize inbred lines, and 73 hybrid accessions, identifying 22 candidate genes significantly associated with stalk sugar content, 6 of which exhibited distinct expression differences between high and low sugar accessions [4]. GWAS has extensive application prospects in gene cloning, providing an important genetic basis for breeding improvement and agricultural production.

2. Transcriptome Sequencing Technology

RNA-seq technology enables the detection of the overall transcriptional activity of a specific species at the nucleotide level, thereby comprehensively and rapidly allowing researchers to obtain information on almost all transcripts of the species under certain conditions [5]. Szala conducted transcriptomic analysis on soybean male sterile line SXCMS5A and maintainer line SXCMS5B, revealing significant differences in expression in pathways such as pollen wall development, carbohydrate metabolism, sugar transport, reactive oxygen species (ROS) metabolism, and transcription factors, laying the foundation for understanding the mechanism of male sterility [6]. In previous studies, Zhao et al. found that *AeNAC83* in okra (*Abelmoschus esculentus*) was significantly upregulated under salt stress. Transgenic validation revealed that this gene positively regulates plant salt tolerance. Transcriptomic analysis of *AeNAC83* overexpression lines and wild-type Arabidopsis revealed that the gene enhances plant salt tolerance by affecting photosynthesis, the phenylpropane pathway, and plant hormones [7]. Wang et al. identified a leaf tip wrinkled mutant (*ltr1*) through ethyl methane sulfonate (EMS) mutagenesis in rice and found that the gene may regulate leaf development through hormone signal transduction and improve rice salt tolerance by regulating aquaporins and ion transport proteins through transcriptomic analysis. Fan et al. conducted RNA-Seq on samples treated with varying sucrose concentrations, and gene ontology (GO), Kyoto Encyclopedia of Genes and Genomes (KEGG), and Clusters of Orthologous Groups (COGs) of proteins analyses indicated that sucrose can induce cell wall loosening, extension, hydrolysis, and remodeling, induce changes in plant osmotic pressure, and affect carbohydrate transport and metabolism pathways to promote axillary bud formation [8]. RNA-Seq technology has become increasingly prevalent and essential in gene function research. It provides crucial references for a more thorough understanding of plant biology as well as potential strategies for crop improvement.

3. Genome Research

Complete and accurate genome assembly is crucial for a thorough understanding of the genetics and evolution of organisms. Over the past few years, a large number of genomes from different species have been published. Nurk first reported the first gapless human genome in 2022 [9], followed by an increasing number of gapless genomes of crops such as maize (MOS17) [10], rice (Nipponbare) [11], and soybean (Williams 82) [12]. The release of these genomes has facilitated further understanding of genome structure and function, providing valuable resources for genetics and genomics. Qi et al. summarized the advancements in research on the lotus (*Nelumbo nucifera*) genome and the research value of genomics in breeding. Genomic research is of great significance for studying the origin, classification, and evolution of crops, allowing explorations into the potentially valuable genes in crops and molecular breeding [13]. Xiao et al. revealed the close relationship between synonymous codon usage bias (SCUB) and genetic and epigenetic variations in rice during domestication by analyzing the codon usage bias in wild and cultivated rice [14]. Research on genomes has significantly enhanced our understanding of genome structure and function, providing valuable resources for genetics and genomics. It has made important contributions to exploring the origins, classifications, valuable genes, and molecular breeding of crops.

4. Molecular Marker-Assisted Breeding Research

Molecular markers are essential tools in breeding for multiple desirable traits, including Kompetitive Allele Specific PCR (KASP), Insertion and Deletion (Indel), and Simple Sequence Repeats (SSR), among others. Eltaher et al. conducted an association analysis of two KASP markers, TaDreb-B1 and 1-FEH w3, with drought tolerance in both spring and winter wheat at the seedling stage, revealing that the TaDreb-B1 marker was more effective in selecting for drought tolerance than the 1-FEH w3 allele [15]. Yang et al. employed bulked segregant analysis (BSA) pooling to map a gene controlling grain length in rice at a 96K interval and identified a KASP marker capable of effectively differentiating between grain lengths [16]. Molecular markers play a crucial role in accelerating crop breeding, and identifying molecular markers closely linked to desirable traits is of great significance for achieving genetic improvement in crops.

5. Gene Family Studies

A gene family is a group of genes with a common origin that encode proteins with similar structural characteristics and biochemical functions [17]. Nuclear factor YC (NF-YC) is a class of transcription factors involved in various developmental processes in higher eukaryotes such as nutrient accumulation, flowering regulation, and ABA signal response. In rice, *NF-YC8-12* are homologs with similar seed-specific expression patterns. Xu et al.

demonstrated, through the creation of single mutants and quintuple mutants of *NF-YC8-12*, that these five genes redundantly regulate rice seed quality and seed germination through the abscisic acid (ABA) and gibberellic acid (GA) pathways [18]. The protein arginine methyltransferase (PRMT) family has been shown to be responsible for the methylation of specific arginine residues in proteins in plants, playing important roles in plant growth, development, and abiotic stress responses. In Arabidopsis, *AtPRMT5* plays a key role in regulating flowering time. Ling et al. identified and analyzed the PRMT family in maize, suggesting that ZmPRMT may play a potential role in responding to abiotic stress. The overexpression of the *AtPRMT5* homolog *ZmPRMT1* in Arabidopsis resulted in early flowering and increased heat tolerance [19]. Low-phosphate roots (LPRs) encode proteins containing multicopper oxidase domains. In Arabidopsis, LPR1 regulates root meristem through Fe and Pi signals, while *OsLPR5* in rice demonstrates ferroxidase activity and is necessary for normal rice growth and phosphate homeostasis maintenance. Zhao et al. analyzed the function of *OsLPR5* under salt stress, demonstrating its positive regulation of plant salt tolerance and its enhancement of ROS levels [20]. The natural resistance-associated macrophage protein (NRAMP) family is a widely distributed membrane transport protein family in plants, mainly involved in the transport of divalent metal cations such as Zn, Fe, and Cu, and plays an important role in Mn/Fe homeostasis. Zhou et al. identified the NRAMP family in betel nut, and through bioinformatic analysis of its evolutionary, cis-acting element and sequence features, they demonstrated the important role of the AcNRAMP family in betel nut in response to Zn/Fe stress based on transcriptome data under nutrient-deficient conditions [21]. The basic helix–loop–helix (bHLH) protein is one of the largest transcription factor (TF) families in eukaryotes, involved in regulating the biosynthesis of anthocyanins, flavonoids, and proanthocyanidins. Liu et al. identified 157 bHLH genes in *Ipomoea aquatica* and characterized the features and evolutionary relationships of the bHLH gene family through collinearity analysis, phylogenetic analysis, and sequence feature analysis. The subsequent transcriptome data quantification of bHLH in purple and green watercress and the analysis of 13 differentially expressed genes (DEGs) through cis-acting element analysis indicated that bHLH plays an important role in light response and plant hormone regulation, providing important insights into the role of bHLH in the anthocyanin biosynthesis pathway [22]. Determining the functions of homologous genes based on those of known functional genes can be a way to significantly enhance the study of gene functions. These studies provide important insights the research on gene families, aiding our understanding of their roles in plant growth and adaptation to environmental stress.

6. Conclusions and Perspective

In this review, we examined recent research articles published in *IJMS* pertaining to plant genomics and breeding. Utilizing methods such as genome-wide association analysis, transcriptome sequencing, and gene family studies, several key genes involved in nitrogen utilization, salt tolerance, and anthocyanin biosynthesis pathways were identified across different species. These findings provide important insights for understanding plant biological functions and regulatory networks, and offer valuable information for crop breeding. With the advances in modern biotechnology, numerous new technologies have emerged, including pan-genomics and synthetic biology. It is believed that, with the continuous emergence of new technologies and the maturation of existing ones, we can enrich our understanding of the genetic basis of crops, and breeding methods will become more diverse and streamlined.

Author Contributions: All the authors participated in the editing of this research topic. L.J. wrote the draft, and all the other authors provided comments on the editorial. All authors have read and agreed to the published version of the manuscript.

Funding: This project was funded by the National Key R&D Program of China (2022YFE0125600) and the Zhejiang Provincial Natural Science Foundation of China (LY22C130002 and LD24C130001).

Conflicts of Interest: The authors declare that the research was conducted in the absence of any commercial or financial relationships that could be construed as potential conflicts of interest.

References

1. Risch, N.; Merikangas, K. The future of genetic studies of complex human diseases. *Science* **1996**, *273*, 1516–1517. [CrossRef]
2. Ahmad, N.; Su, B.; Ibrahim, S.; Kuang, L.; Tian, Z.; Wang, X.; Wang, H.; Dun, X. Deciphering the Genetic Basis of Root and Biomass Traits in Rapeseed (*Brassica napus* L.) through the Integration of GWAS and RNA-Seq under Nitrogen Stress. *Int. J. Mol. Sci.* **2022**, *23*, 7958. [CrossRef] [PubMed]
3. Liu, Q.; Feng, Z.; Huang, C.; Wen, J.; Li, L.; Yu, S. Insights into the Genomic Regions and Candidate Genes of Senescence-Related Traits in Upland Cotton via GWAS. *Int. J. Mol. Sci.* **2022**, *23*, 8584. [CrossRef]
4. Chen, J.; Cao, J.; Bian, Y.; Zhang, H.; Li, X.; Wu, Z.; Guo, G.; Lv, G. Identification of Genetic Variations and Candidate Genes Responsible for Stalk Sugar Content and Agronomic Traits in Fresh Corn via GWAS across Multiple Environments. *Int. J. Mol. Sci.* **2022**, *23*, 13490. [CrossRef] [PubMed]
5. Wang, Z.; Gerstein, M.; Snyder, M. RNA-Seq: A revolutionary tool for transcriptomics. *Nat. Rev. Genet.* **2009**, *10*, 57–63. [CrossRef]
6. Bai, Z.; Ding, X.; Zhang, R.; Yang, Y.; Wei, B.; Yang, S.; Gai, J. Transcriptome Analysis Reveals the Genes Related to Pollen Abortion in a Cytoplasmic Male-Sterile Soybean (*Glycine max* (L.) Merr.). *Int. J. Mol. Sci.* **2022**, *23*, 12227. [CrossRef] [PubMed]
7. Zhao, X.; Wu, T.; Guo, S.; Hu, J.; Zhan, Y. Ectopic Expression of AeNAC83, a NAC Transcription Factor from Abelmoschus esculentus, Inhibits Growth and Confers Tolerance to Salt Stress in Arabidopsis. *Int. J. Mol. Sci.* **2022**, *23*, 10182. [CrossRef]
8. Fan, Z.; Huang, G.; Fan, Y.; Yang, J. Sucrose Facilitates Rhizome Development of Perennial Rice (*Oryza longistaminata*). *Int. J. Mol. Sci.* **2022**, *23*, 13396. [CrossRef]
9. Nurk, S.; Koren, S.; Rhie, A.; Rautiainen, M.; Bzikadze, A.V.; Mikheenko, A.; Vollger, M.R.; Altemose, N.; Uralsky, L.; Gershman, A.; et al. The complete sequence of a human genome. *Science* **2022**, *376*, 44–53. [CrossRef]
10. Chen, J.; Wang, Z.; Tan, K.; Huang, W.; Shi, J.; Li, T.; Hu, J.; Wang, K.; Wang, C.; Xin, B.; et al. A complete telomere-to-telomere assembly of the maize genome. *Nat. Genet.* **2023**, *55*, 1221–1231. [CrossRef]
11. Shang, L.; He, W.; Wang, T.; Yang, Y.; Xu, Q.; Zhao, X.; Yang, L.; Zhang, H.; Li, X.; Lv, Y.; et al. A complete assembly of the rice Nipponbare reference genome. *Mol. Plant* **2023**, *16*, 1232–1236. [CrossRef] [PubMed]
12. Wang, L.; Zhang, M.; Li, M.; Jiang, X.; Jiao, W.; Song, Q. A telomere-to-telomere gap-free assembly of soybean genome. *Mol. Plant* **2023**, *16*, 1711–1714. [CrossRef] [PubMed]
13. Qi, H.; Yu, F.; Deng, J.; Yang, P. Studies on Lotus Genomics and the Contribution to Its Breeding. *Int. J. Mol. Sci.* **2022**, *23*, 7270. [CrossRef] [PubMed]
14. Xiao, G.; Zhou, J.; Huo, Z.; Wu, T.; Li, Y.; Li, Y.; Wang, Y.; Wang, M. The Shift in Synonymous Codon Usage Reveals Similar Genomic Variation during Domestication of Asian and African Rice. *Int. J. Mol. Sci.* **2022**, *23*, 12860. [CrossRef] [PubMed]
15. Eltaher, S.; Hashem, M.; Ahmed, A.A.M.; Baenziger, P.S.; Börner, A.; Sallam, A. Effectiveness of TaDreb-B1 and 1-FEH w3 KASP Markers in Spring and Winter Wheat Populations for Marker-Assisted Selection to Improve Drought Tolerance. *Int. J. Mol. Sci.* **2023**, *24*, 8986. [CrossRef] [PubMed]
16. Yang, L.; Li, P.; Wang, J.; Liu, H.; Zheng, H.; Xin, W.; Zou, D. Fine Mapping and Candidate Gene Analysis of Rice Grain Length QTL qGL9.1. *Int. J. Mol. Sci.* **2023**, *24*, 11447. [CrossRef] [PubMed]
17. Li, J.; Yang, S.; Yang, X.; Wu, H.; Tang, H.; Yang, L. PlantGF: An analysis and annotation platform for plant gene families. *Database-J. Biol. Databases Curation* **2022**, *2022*, baab088. [CrossRef] [PubMed]
18. Xu, H.; Li, S.; Kazeem, B.B.; Ajadi, A.A.; Luo, J.; Yin, M.; Liu, X.; Chen, L.; Ying, J.; Tong, X.; et al. Five Rice Seed-Specific NF-YC Genes Redundantly Regulate Grain Quality and Seed Germination via Interfering Gibberellin Pathway. *Int. J. Mol. Sci.* **2022**, *23*, 8382. [CrossRef] [PubMed]
19. Ling, Q.; Liao, J.; Liu, X.; Zhou, Y.; Qian, Y. Genome-Wide Identification of Maize Protein Arginine Methyltransferase Genes and Functional Analysis of ZmPRMT1 Reveal Essential Roles in Arabidopsis Flowering Regulation and Abiotic Stress Tolerance. *Int. J. Mol. Sci.* **2022**, *23*, 12793. [CrossRef]
20. Zhao, J.; Meng, X.; Zhang, Z.; Wang, M.; Nie, F.; Liu, Q. OsLPR5 Encoding Ferroxidase Positively Regulates the Tolerance to Salt Stress in Rice. *Int. J. Mol. Sci.* **2023**, *24*, 8115. [CrossRef]
21. Zhou, G.; An, Q.; Liu, Z.; Wan, Y.; Bao, W. Systematic Analysis of NRAMP Family Genes in Areca catechu and Its Response to Zn/Fe Deficiency Stress. *Int. J. Mol. Sci.* **2023**, *24*, 7383. [CrossRef] [PubMed]
22. Liu, Z.; Fu, X.; Xu, H.; Zhang, Y.; Shi, Z.; Zhou, G.; Bao, W. Comprehensive Analysis of bHLH Transcription Factors in Ipomoea aquatica and Its Response to Anthocyanin Biosynthesis. *Int. J. Mol. Sci.* **2023**, *24*, 5652. [CrossRef] [PubMed]

International Journal of *Molecular Sciences*

MDPI

Article

Characterization and Expression Analysis of the *PvTLP* Gene Family in the Common Bean (*Phaseolus vulgaris*) in Response to Salt and Drought Stresses

Xue Dong [1,2,3], Min Zhao [4], Jia Li [4], Fuyi Qiu [4], Yan Wang [1,2,3], Jiandong Zhao [1,2,3], Jianwu Chang [1,2,3] and Xiaopeng Hao [1,2,3,*]

[1] Center for Agricultural Genetic Resources Research, Shanxi Agricultural University, Taiyuan 030031, China; jodiedong@sxau.edu.cn (X.D.); wangyan204408@sxau.edu.cn (Y.W.); zhaojiandong@sxau.edu.cn (J.Z.); changjianw2005@163.com (J.C.)

[2] Key Laboratory of Crop Gene Resources and Germplasm Enhancement on Loess Plateau, Ministry of Agriculture, Taiyuan 030031, China

[3] Shanxi HouJi Laboratory, Shanxi Agricultural University, Taiyuan 030031, China

[4] College of Agriculture, Shanxi Agricultural University, Taiyuan 030031, China; 20232165@stu.sxau.edu.cn (M.Z.); z20223131@stu.sxau.edu.cn (J.L.); 20233231@stu.sxau.edu.cn (F.Q.)

* Correspondence: haoxiaopeng@sxau.edu.cn

Abstract: Tubby-like proteins (TLPs) are essential multifunctional transcription factors in plants that significantly influence plant growth and development, signal transduction, and adaptation to environmental stress. Despite their importance, there is limited knowledge of the identification and functional roles of the *TLP* gene family in the common bean. In this study, we identified the *PvTLP* gene family, which consists of 10 *PvTLP* genes distributed unevenly across seven chromosomes. Phylogenetic analysis revealed that these genes could be classified into three subfamilies (A, B, and C). All PvTLP proteins contained both conserved tubby and F-box domains, with the exception of PvTLP7, which lacks the F-box domain. Conserved motif analysis revealed that 10 *PvTLP* genes contained motif 1 and motif 3. Cis-acting elements analysis indicated that *PvTLP* genes might be involved in light, hormone, and stress responses. Synteny analysis revealed a closer phylogenetic relationship between the common bean and dicotyledons than monocotyledons. qRT-PCR analysis confirmed the significant differences in the expression of most *PvTLP* genes in both leaves and roots under salt and drought stresses. These findings provide valuable insights for further exploration of the molecular functions of TLPs in plant responses to various stresses and offer key candidate genes for enhancing stress resistance in the common bean through molecular breeding.

Keywords: tubby-like proteins (TLPs); *Phaseolus vulgaris*; salt and drought stresses; expression profile; qRT-PCR

1. Introduction

Tubby-like proteins (TLPs) are multifunctional transcription factors that are widespread across eukaryotic organisms [1]. The *TLP* gene was first observed in obese tubby mice [2]. In mammals, mutations in *TLP* genes may lead to various diseases, including hearing loss, retinal degeneration, and obesity [3]. TLPs are characterized by their conserved tubby domain at the carboxyl terminus [4,5]. This domain demonstrates that a central hydrophobic α-helix traverses through a closed β-barrel composed of 12 antiparallel strands [6]. TLPs are anchored to the plasma membrane through interactions with phosphatidylinositol 4,5-bisphosphate (PI(4,5)P2) [7,8]. Activation of G-protein signaling cascades triggers

https://doi.org/10.3390/ijms26125702

the translocation of TLPs to the nucleus, where they are involved in G-protein-coupled receptor-mediated signal transduction [9,10]. In plants, in addition to the tubby domain at the carboxyl terminus, most TLP proteins harbor a conserved F-box domain in the amino-terminal region that serves as a component of the Skp1-Cullin-F-box (SCF) E3 ubiquitin ligase complex [11]. The F-box domain facilitates participation in ubiquitination pathways through substrate recognition and recruitment to SCF, thereby regulating the SCF-mediated ubiquitin proteolytic pathway [12,13].

More members of the *TLP* gene family have been identified in plants than in animals, including *Arabidopsis thaliana* (11), *Oryza sativa* (14), *Gossypium hirsutum* (12), *Glycine max* (20), *Solanum lycopersicum* (11), and *Malus domestica* (9) [14–19]. TLPs have been suggested to be molecular regulators of plant growth, signal transduction cascades, and adaptive responses to hormone and abiotic stresses. Strawberry *FvTLPs* exhibited dynamic responsiveness to salinity, drought, cold, and hormone treatments [20]. In cassava, *MeTLPs* were found to respond to salt, drought, and low-temperature stresses [21]. *Arabidopsis* AtTLP2 serves as a critical modulator of ABA signaling and dehydration responses, with *AtTLP2*-overexpressing lines exhibiting markedly increased tolerance to drought and salinity during maturation phases. In contrast, the *attlp2* mutant was found to display sensitivity to ionic, osmotic, and oxidative stresses [22]. *AtTLP2* exerted positive regulatory control over UDP-glucose 4-epimerase, modulating homogalacturonan biosynthesis in seed coat mucilage through an ABA-dependent pathway [23]. AtTLP3 and AtTLP9 share high homology. Single mutants of *attlp3* or *attlp9* were found to exhibit tolerance to ABA and osmotic stresses during seed germination and early developmental stages [14,24]. The *attlp3/attlp9* double mutant presented a higher germination frequency than the single mutant under ABA and osmotic stresses, suggesting that *AtTLP3* has a redundant function with *AtTLP9* in *Arabidopsis* [24]. In chickpea, CaTLP1 might interact with protein kinases to regulate ABA-dependent gene transcription and stomatal dynamics, thereby conferring enhanced drought adaptation ability [25,26]. Cucumber CsTLP8 functions as a negative regulator of osmotic stress and ABA signaling by affecting antioxidant enzymatic activity, consequently impairing seed germination and seedling establishment while increasing ABA hypersensitivity [27].

The common bean (*Phaseolus vulgaris* L.) is a globally cultivated legume of critical agricultural importance. It plays a crucial role in maintaining dietary balance and enhancing human nutrition owing to its abundance of plant-based proteins, essential amino acids, dietary fiber, and vitamins [28–30]. However, the growth and yield potential of this legume are often restricted by various abiotic stresses, which severely limit its productivity [31,32]. Consequently, understanding the molecular mechanisms that regulate common bean responses to abiotic stress is essential for developing cultivars that are tolerant to salt and drought stresses, ultimately improving crop yields. Although the *TLP* gene family has been widely characterized in various plants, its role in the common bean remains poorly understood. This research aimed to comprehend the evolutionary history and the abiotic stress response of *PvTLP* genes.

In this study, a comprehensive genome-wide identification of the *PvTLP* gene family in the common bean was performed, and the physicochemical properties, chromosomal distribution and gene duplication, evolutionary relationship, gene structure, conserved motif, synteny, and cis-acting element analyses were determined. Additionally, tissue and abiotic stress-induced expression profiles of *PvTLP* genes were analyzed. The results provide theoretical insights into the stress-responsive molecular mechanisms of *PvTLP* genes and offer a new perspective for resistance breeding in the common bean.

2. Results

2.1. Identification and Characterization of the PvTLP Gene Family

A local BLASTP search was performed using AtTLPs as references, along with HMMER screening targeting the conserved tubby domain (PF01167). This study identified 10 members of the *PvTLP* gene family in the genome of the common bean. These genes were designated *PvTLP1*–*PvTLP10* according to their chromosomal positions. Detailed information on their predicted physiological properties is listed in Table 1. The CDS lengths of *PvTLP* genes ranged from 1083 bp (*PvTLP3*) to 1296 bp (*PvTLP10*). The encoded protein lengths varied from 360 to 431 amino acid residues, with molecular weights (MWs) ranging from 40,469.27 to 48,120.22 Da. The predicted isoelectric points (pI) were between 9.09 and 9.71, with an average of 9.42, indicating that the PvTLP proteins had alkaline properties.

Table 1. Detailed information of the *PvTLP* gene family.

Gene Name	Gene ID	CDS (bp)	Protein (aa [1])	Intron	MW [2] (Da)	pI [3]	Instability Index	Aliphatic Index	GRAVY [4]	Subcellular Localization [5]
PvTLP1	Phvul.001G013500	1290	429	3	47,795.75	9.34	60.34	80.93	−0.341	Nucl
PvTLP2	Phvul.002G031500	1245	414	3	46,047.76	9.46	56.58	80.05	−0.346	Nucl
PvTLP3	Phvul.003G100700	1083	360	4	40,469.27	9.09	57.71	79.67	−0.302	Nucl
PvTLP4	Phvul.003G254900	1242	413	3	46,210.86	9.21	61.47	77.92	−0.334	Chlo
PvTLP5	Phvul.004G095000	1278	425	3	47,566.43	9.56	67.49	76.40	−0.371	Nucl
PvTLP6	Phvul.005G082200	1128	375	3	41,640.27	9.33	53.17	67.09	−0.476	Mito
PvTLP7	Phvul.005G183800	1113	370	8	42,252.31	9.60	37.45	65.62	−0.607	Nucl
PvTLP8	Phvul.006G161100	1152	383	4	43,206.85	9.46	60.29	76.37	−0.327	Mito
PvTLP9	Phvul.006G210500	1284	427	3	47,950.04	9.47	63.82	76.28	−0.363	Nucl
PvTLP10	Phvul.007G079900	1296	431	3	48,120.22	9.71	60.90	75.31	−0.349	Nucl

[1] aa: the number of amino acid residues; [2] MW: molecular weight; [3] pI: theoretical isoelectric point; [4] GRAVY: grand average of hydropathicity; [5] Subcellular Localization: Nucl: nucleus, Chlo: chloroplast, Mito: mitochondria.

In addition to PvTLP7, which exhibited a lower instability index of 37.45, the scores of the other PvTLPs exceeded 50, indicating relative instability. The aliphatic index varied from 65.62 (PvTLP7) to 80.93 (PvTLP1), and all PvTLPs presented negative GRAVY values, confirming their hydrophilic characteristics. Subcellular localization predictions revealed that seven PvTLPs were localized in the nucleus, two in the mitochondria, and one in the chloroplast. These findings provide fundamental data for investigating the evolution and function of the *PvTLP* gene family.

2.2. Chromosomal Localization and Gene Replication Events of the PvTLP Gene Family

Chromosomal mapping analysis revealed that the 10 *PvTLP* genes were distributed on 7 of the 11 chromosomes (Figure 1). Chr03, Chr05, and Chr06 each contained two genes, while Chr01, Chr02, Chr04, and Chr07 each possessed one gene. No *PvTLP* genes were identified on Chr08 to Chr11. This chromosomal distribution pattern reveals the uneven and irregular spatial characteristics of the *PvTLP* gene family.

To elucidate the evolutionary expansion and functional diversification of the *PvTLP* gene family, we examined gene duplication events. The *PvTLP* gene family contained one segmentally duplicated gene pair (*PvTLP2*/*PvTLP4*) and no tandem duplication. Further, the non-synonymous and synonymous substitution ratio (Ka/Ks ratio) was calculated for the segmentally duplicated *PvTLP* gene pair to assess the selection pressure during evolution (Supplementary Table S1). The Ka/Ks ratio was less than 1, indicating that the expansion of *PvTLP* genes has undergone purifying selection pressure during evolution.

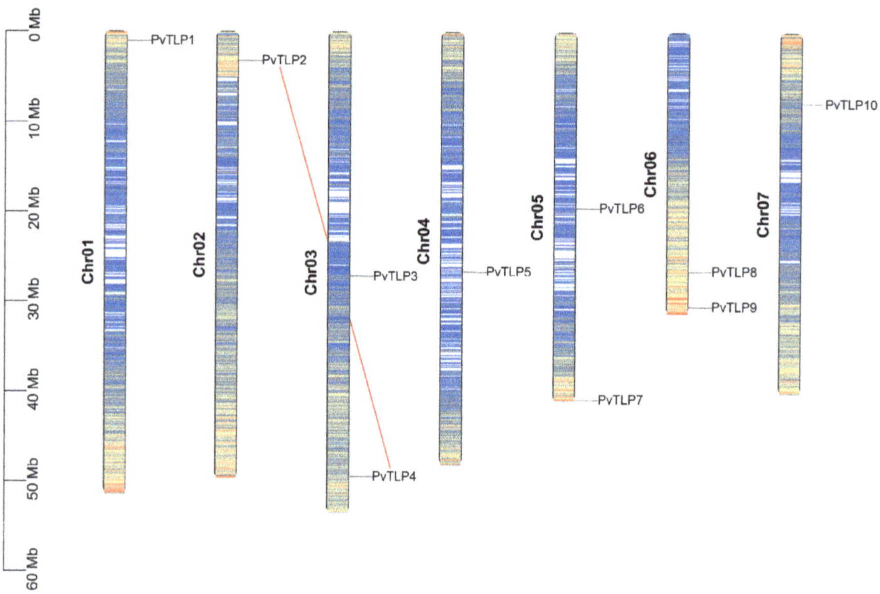

Figure 1. Chromosomal localization and segmental duplication events of *PvTLP* genes. The segmentally duplicated gene pair is marked with a red line.

2.3. Phylogenetic Analysis of the PvTLP Gene Family

To elucidate the evolutionary relationships among TLP proteins across various plant species, multiple sequence alignment of full-length TLP protein sequences from the common bean (10), *Arabidopsis* (11), rice (14), maize (14), and soybean (20) was conducted using Muscle (Supplementary Table S2). Phylogenetic analysis was performed with the MEGA11 software. The neighbor-joining (NJ) phylogenetic tree revealed that 69 TLP proteins were categorized into three subfamilies (A, B, and C, Figure 2). Subfamily A, the largest group, containing 53 TLP members, was further clustered into three distinct subgroups (A1, A2, and A3). Subgroup A1 had 24 TLP proteins, including 3 PvTLPs (PvTLP5, PvTLP9, PvTLP10), 3 AtTLPs, 6 GmTLPs, 6 OsTLPs, and 6 ZmTLPs. Subgroup A2 had 14 TLP proteins, which encompassed 3 PvTLPs (PvTLP1, PvTLP2, PvTLP4), 2 AtTLPs, 6 GmTLPs, 1 OsTLP, and 2 ZmTLPs. Subgroup A3 had 15 TLP proteins, comprising 2 PvTLPs (PvTLP3, PvTLP8), 4 AtTLPs, 5 GmTLPs, 2 OsTLPs, and 2 ZmTLPs. Subfamily B, containing 11 TLP proteins, consisted of 1 PvTLP (PvTLP6), 1 AtTLP, 2 GmTLPs, 4 OsTLPs, and 3 ZmTLPs. Subfamily C was the smallest group, with only one member from each species, including PvTLP7, AtTLP8, GmTLP20, OsTLP4, and ZmTLP13. PvTLPs and GmTLPs exhibited a distinct clustering pattern, with nine PvTLPs showing a 1:2 homologous relationship with GmTLPs. For instance, PvTLP3 paired with GmTLP13/GmTLP19, PvTLP4 with GmTLP12/GmTLP15, and PvTLP6 with GmTLP16/GmTLP17. These findings indicate that the TLPs of the common bean and soybean are significantly more homologous than those of other species.

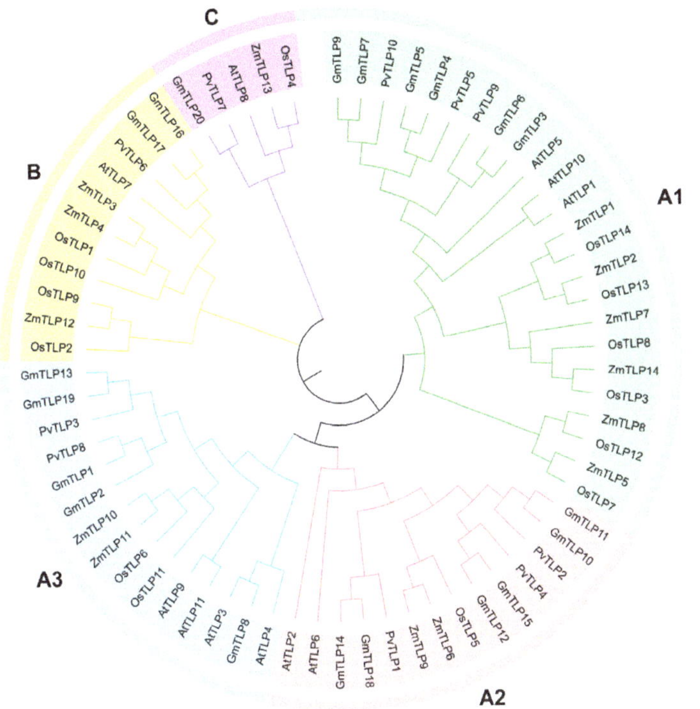

Figure 2. Phylogenic tree of TLP proteins from *Phaseolus vulgaris*, *Arabidopsis thaliana*, *Glycine max*, *Oryza sativa*, and *Zea mays*. A phylogenetic tree was constructed using the neighbor-joining method with 1000 bootstrap replicates. TLP proteins were categorized into three subfamilies (A–C), with subfamily A further clustered into three subgroups (A1–A3).

2.4. Gene Structure, Conserved Domain, and Motif Analysis of the PvTLP Gene Family

Gene structure serves as crucial evidence for elucidating phylogenetic relationships. By utilizing the coding sequences and genomic annotations of *PvTLP* genes, the number and distribution of exons and introns across *PvTLP* genes were predicted. The members of subgroup A1, subgroup A2, and subfamily B contained three introns; subgroup A3 contained four introns; and subfamily C contained eight introns (Figure 3A). These results indicate that *PvTLP* genes present similar gene structures.

Conserved domain analysis revealed that subfamilies A and B contained both conserved tubby and F-box domains, except subfamily C, which had a single tubby domain (Figure 3B). Based on the MEME suite, ten conserved motifs in PvTLPs were identified, and some differences in the distribution of the motifs were observed (Figure 3C). All of the PvTLP proteins contained motif 1 and motif 3, which belong to the tubby domain and likely contribute to the functional conservation of TLP proteins. Subfamilies A and B shared motif 2, motif 4, motif 5, motif 6, and motif 7. Among the subfamily A members, motif 9 was found only in subgroup A1, and motif 8 appeared in both subgroups A1 and A2. These findings suggest that PvTLP proteins in the different subfamilies contain various conserved motifs, which result in functional diversity.

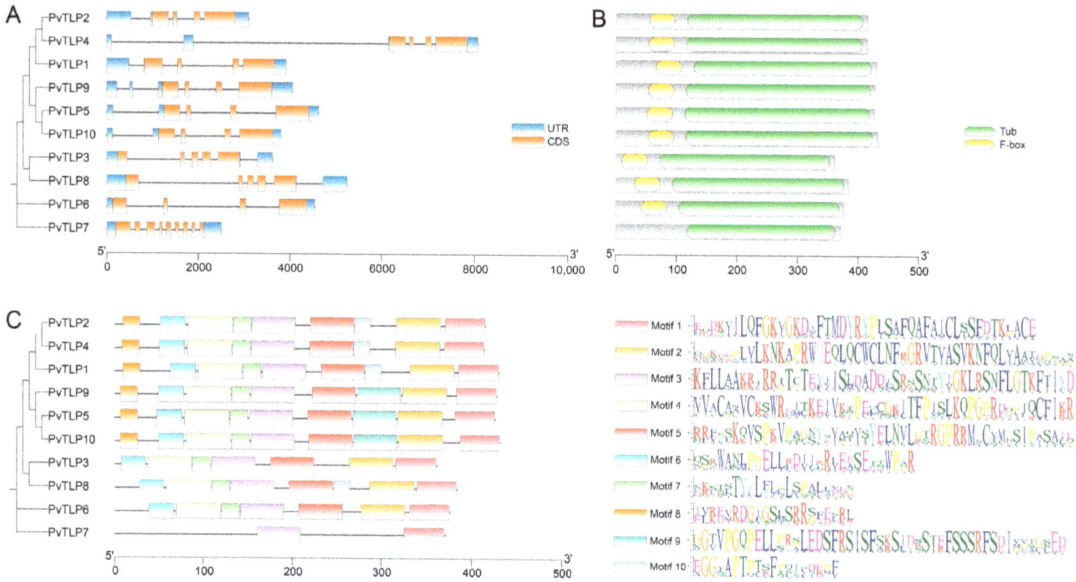

Figure 3. Gene structure, conserved domain, and motif analysis of the *PvTLP* gene family: (**A**) gene structure of *PvTLP* genes; (**B**) conserved domains of PvTLP proteins; (**C**) conserved motifs of PvTLP proteins and the amino acid consensus sequence logos of these motifs.

2.5. Synteny Analysis of TLP Genes

To clarify the evolutionary relationships and phylogenetic characteristics of *TLP* genes, this study constructed synteny diagrams comparing the common bean with other plants, including two dicotyledonous plants (*Arabidopsis* and soybean) and two monocotyledonous plants (rice and maize, Figure 4). The synteny patterns revealed the following syntenic gene pairs: 38 pairs between 10 *PvTLP* genes and 19 *GmTLP* genes, 9 pairs between 8 *PvTLP* genes and 6 *AtTLP* genes, 8 pairs between 5 *PvTLP* genes and 7 *OsTLP* genes, and only 5 pairs between 4 *PvTLP* genes and 4 *ZmTLP* genes. Compared with monocotyledonous plants, the common bean presented stronger synteny with dicotyledonous plants. The highest level of synteny was observed between the common bean and soybean. These findings indicate that *PvTLPs* and *GmTLPs* exhibit a more closed evolutionary conservation than other plants do.

2.6. Analysis of Cis-Acting Elements in the PvTLP Promoter

To elucidate the potential functions and environmental response mechanisms of the *PvTLP* gene family in the common bean, we extracted 2000 bp upstream sequences from the start codons of 10 *PvTLP* genes and analyzed the cis-acting elements using the PlantCARE database. A total of 41 distinct cis-acting elements were identified and classified into four functional categories: light-, hormone-, stress-, and development-responsive elements (Figure 5). All *PvTLP* genes, except for *PvTLP3* and *PvTLP9*, contained the four categories of functional elements. Light-responsive elements, including Box 4, G-Box, and GT1-motif, are widely distributed across all *PvTLP* promoters. There are five types of hormone-responsive elements: abscisic acid-responsive elements (ABRE) in seven genes, MeJA-responsive elements (CGTCA-motif and TGACG-motif) in four genes, gibberellin-responsive elements (GARE-motif and P-box) in six genes, salicylic acid-responsive elements (TCA-element) in seven genes, and auxin-responsive elements (AuxRR-core and TGA-element) in three genes.

Additionally, abundant stress-responsive elements, such as anaerobic induction element (ARE), low-temperature response element (LTR), drought response element (MBS), and defence and stress-responsive element (TC-rich repeats), were identified in the promoter regions. With the exception of *PvTLP9*, one to four development-responsive elements were detected in all *PvTLP* promoters. These included elements related to the cell cycle (MSA-like), palisade mesophyll cells (HD-Zip 1), and Zein metabolism regulation element (O2-site). These results indicate that *PvTLP* genes may play a major role in regulatory pathways and biological processes through complex transcriptional networks.

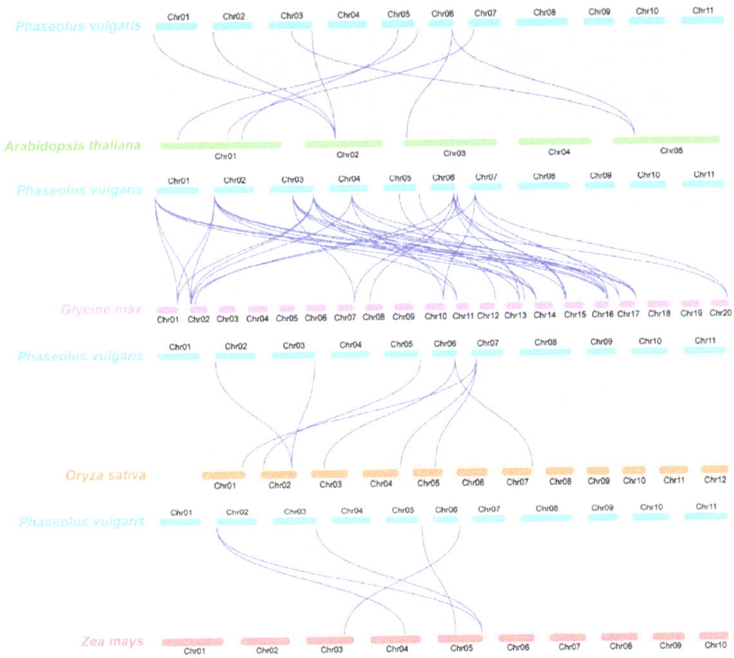

Figure 4. Synteny analysis of *TLP* genes in *Phaseolus vulgaris*, *Arabidopsis thaliana*, *Glycine max*, *Oryza sativa*, and *Zea mays*. Gray lines indicate the collinear blocks between two species, and blue lines indicate syntenic *TLP* gene pairs.

2.7. Expression Patterns of PvTLP Genes in Different Tissues

Using transcriptome sequencing data from the Phytozome database (Supplementary Table S3), we analyzed the expression levels of *PvTLP* genes in various tissues, including leaves, roots, stems, flower buds, flowers, green mature pods, and nodules. The analysis revealed significant differences in the expression patterns of *PvTLP* genes in these tissues (Figure 6). The expression profiles of *PvTLP* genes were categorized into two groups: Group I, which included seven genes (*PvTLP2*, *PvTLP4*, *PvTLP5*, *PvTLP6*, *PvTLP8*, *PvTLP9*, and *PvTLP10*), exhibited widespread expression in all tissues. In contrast, Group II, comprising three genes (*PvTLP1*, *PvTLP3*, and *PvTLP7*), presented specific expression in certain tissues. Notably, *PvTLP6* was expressed at higher levels in flowers and green mature pods than in other tissues, whereas *PvTLP1* was expressed at a relatively higher level in stems compared with other tissues. Additionally, the segmentally duplicated gene pair *PvTLP2* and *PvTLP4* presented similar expression levels in above-ground tissues. These results suggest that *PvTLP* genes are involved in tissue development and growth in the common bean.

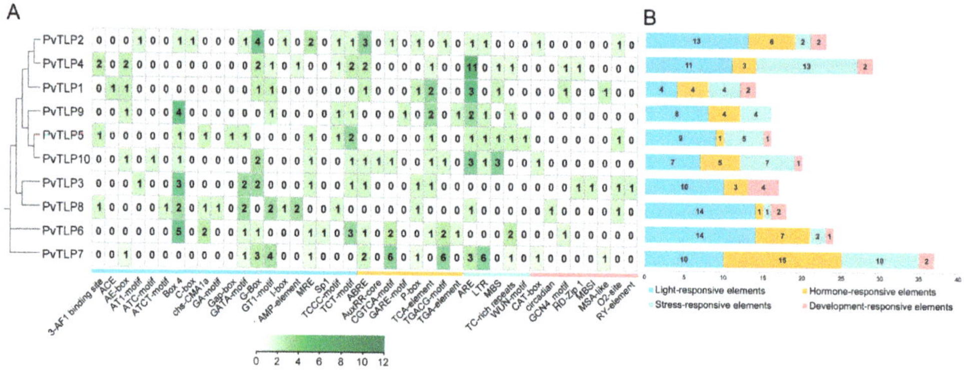

Figure 5. The number of cis-acting elements in the putative promoters of *PvTLP* genes: (**A**) heatmap of the number of cis-acting elements. The color scale represents the abundance of the putative cis-acting elements ranging from white to dark green. The larger the number, the darker the green color. The count of cis-acting elements is marked in the box; (**B**) the stacked bar chart visualizes the statistics of cis-acting elements of each gene in four types. The colors blue, orange, green, and pink indicate the light-, hormone-, stress-, and development-responsive elements, respectively.

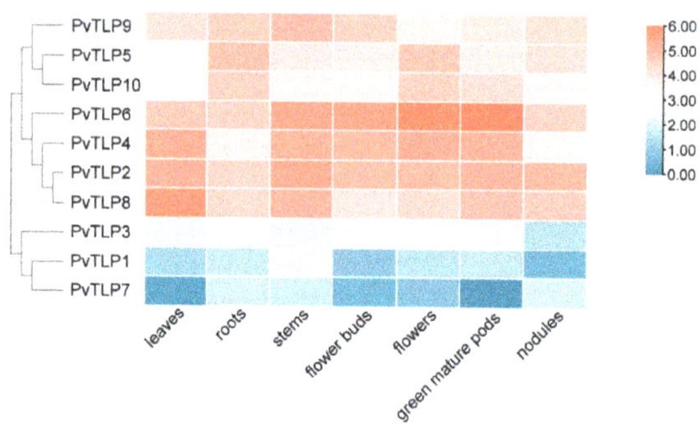

Figure 6. Heatmap of *PvTLP* gene expression profiles in various tissues, including leaves, roots, stems, flower buds, flowers, green mature pods, and nodules. The color scale represents the FPKM values normalized by log2. Orange indicates high expression levels and blue indicates lower expression levels. FPKM, Fragments Per Kilobase of exon model per Million mapped fragments.

2.8. Expression Patterns of PvTLP Genes Under Salt and Drought Stresses

Previous studies have confirmed that *TLP* genes respond to various abiotic stresses. To further validate this mechanism, we analyzed the expression patterns of 10 *PvTLP* genes in the leaves and roots of the common bean under salt and drought stresses using qRT-PCR (Figure 7). Our findings revealed various changes in the expression levels of these genes in response to both types of stresses. Under salt stress, the expression levels of eight genes (*PvTLP2, PvTLP3, PvTLP4, PvTLP5, PvTLP6, PvTLP7, PvTLP8,* and *PvTLP10*) in the leaves peaked at 6 h after treatment, with *PvTLP5* showing the greatest induction (81-fold greater than that of the control), followed by gradual downregulation. In the roots, the expression levels of three genes (*PvTLP7, PvTLP8,* and *PvTLP10*) exhibited a highly significant increase at 12 h, whereas the other three genes (*PvTLP2, PvTLP3,* and *PvTLP6*) peaked at 24 h. Under drought stress, the expression levels of eight genes (*PvTLP1, PvTLP2, PvTLP4, PvTLP6, PvTLP7, PvTLP8, PvTLP9,* and *PvTLP10*) in the leaves increased to a peak at 24 h, whereas five genes (*PvTLP3, PvTLP5, PvTLP6, PvTLP7,* and *PvTLP10*) in the roots reached their peak significantly at 12 h. Additionally, most *PvTLP* genes tended to exhibit low/high/low expression patterns under drought stress. These findings imply that *PvTLP* genes may be essential for the response of the common bean to salt and drought stresses.

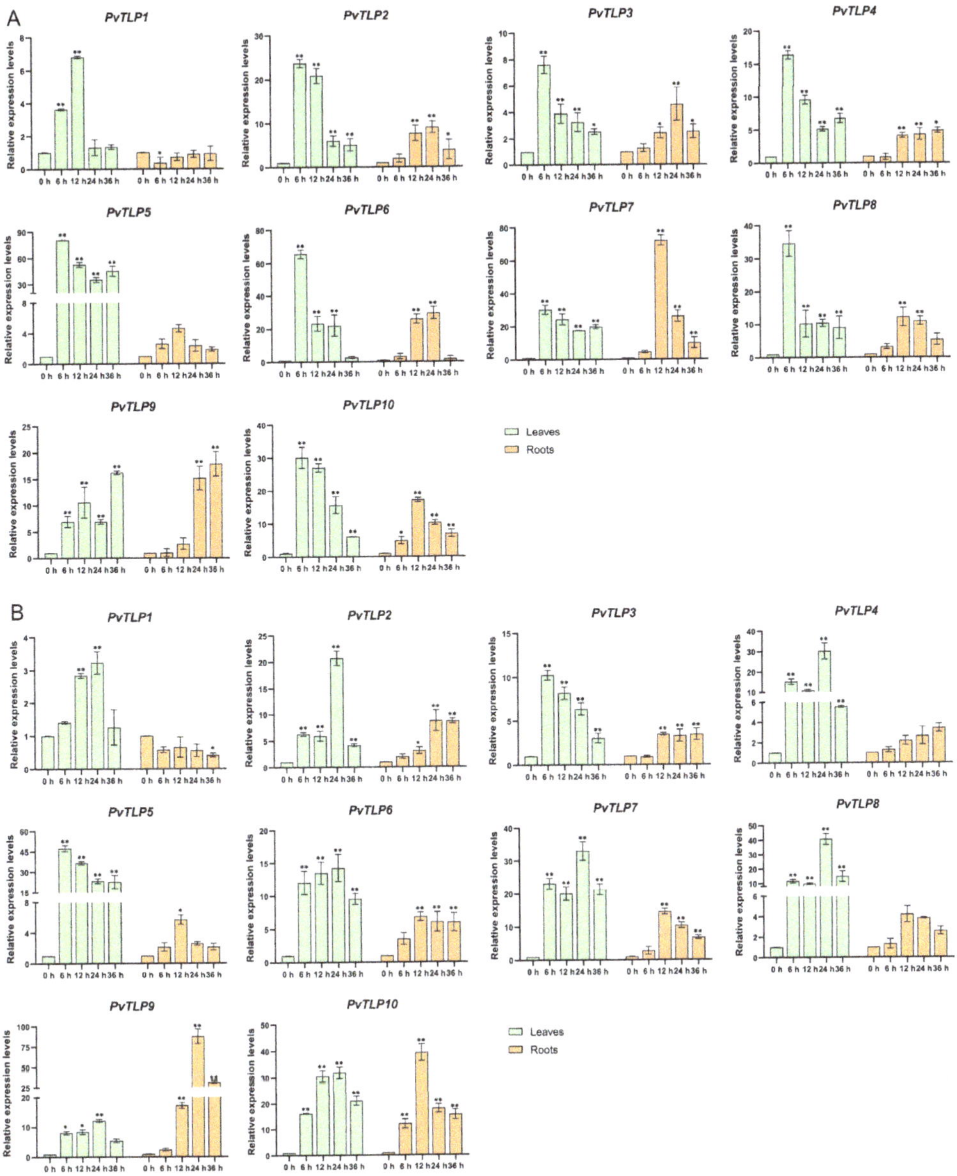

Figure 7. Expression patterns of *PvTLP* genes in the leaves and roots under salt and drought stresses by qRT-PCR analysis: (**A**) expression patterns of *PvTLP* genes under 200 mmol L^{-1} NaCl treatment; (**B**) expression patterns of *PvTLP* genes under 20% PEG6000 treatment. Data were normalized using the common bean *actin* gene. Error bars represent ±SD of three biological replicates. Asterisks indicate statistically significant differences (*, $p < 0.05$; **, $p < 0.01$), determined by the two-way ANOVA statistical test.

3. Discussion

TLP genes play critical roles in plant biological processes, including growth and development, signal transduction, and stress responses. In this study, we identified 10 *PvTLP* genes in the common bean through a comprehensive genome-wide analysis. The number of *PvTLP* genes is significantly lower than that in *Triticum aestivum* (40) and *Brassica napus* (29), and slightly lower than that in *Salvia miltiorrhiza* (12), *Brachypodium distachyon* (12), *Cucumis sativus* (11), and *Arabidopsis thaliana* (11), but greater than that in *Malus domestica* (9) and *Fragaria vesca* (8) [14,19,20,33–37]. The subcellular localization of plant TLPs is diverse. In tomato, 8 out of 11 SlTLP-GFP fusion proteins were found in the nucleus, 2 were detected in both the nucleus and plasma membrane, and 1 was located in the plasma membrane [18]. In addition, the AtTLP5-GFP fusion protein was observed in both the cytosol and plasma membrane [24]. The results of computational predictions concerning the subcellular localization suggested diverse distribution patterns of 10 PvTLP proteins. This diversity may be attributed to different domains or regulatory mechanisms. Therefore, it is essential to conduct further subcellular localization experiments involving PvTLPs fused with GFPs to confirm these findings and gain a deeper understanding of their roles.

Phylogenetic analysis has clarified evolutionary relationships within the *TLP* gene family in the common bean and other plants. To investigate these relationships, a phylogenetic tree of 69 full-length TLP proteins from the common bean, *Arabidopsis*, soybean, rice, and maize was constructed. The results demonstrated their classification into three subfamilies (A, B, and C, Figure 2). The distribution of TLPs across all subfamilies in the five species suggested that the TLPs might have derived from an ancient common ancestor before the divergence of monocotyledons and dicotyledons. The number of PvTLPs in subfamily A was much greater than that in subfamilies B and C, which was consistent with findings related to poplar and wheat [15,33]. Subfamily A was further divided into three subgroups (A1–A3), suggesting the importance of the functional diversification of TLPs. Subfamilies A and B contained both tubby and F-box domains, whereas subfamily C lacked the F-box domain (Figure 3B), likely because the F-box domain was lost during the long-term evolutionary process. Furthermore, the differences in intron number among eukaryotic species are closely associated with the functional diversity of genes [38,39]. Intron/exon analysis revealed that subfamilies A and B of *PvTLP* genes contained three to four introns (Figure 3A), which is similar to the findings of previous studies in rice and strawberry [15,20]. In contrast, subfamily C presented a notable increase in the number of introns, such as *AtTLP8*, *OsTLP4*, *MdTLP8*, *MdTLP9*, and *FvTLP8*, which contained eight, seven, seven, nine, and eight introns, respectively [14,15,19,20]. In the common bean, *PvTLP7* had eight introns (Figure 3A). The high number of introns in subfamily C suggests that these genes may be involved in specific biological functions. Notably, at least one member belonging to subfamily C has been identified in the *TLP* gene family of different plants, indicating that subfamily C is a relatively unique cluster that may perform critical biological functions in plant development. Significant structural and domain-based distinction among subfamilies verifies the validity of the phylogenetic classification of the *TLP* gene family. Furthermore, the distribution of the motifs identified by the MEME suite demonstrates conservation among PvTLP proteins. According to the phylogenetic tree, gene structure, and motif analysis, members of the same subfamily presented similar gene structures and conserved motifs.

Gene duplication events significantly contributed to the rapid expansion of plant gene families. One consequence of gene duplication is the generation of genetic redundancy [40]. In the *Arabidopsis* genome, gene duplication events are widespread. Most mutants with a single loss-of-function mutation do not produce obvious phenotypic effects, which may be due to functional compensation resulting from gene redundancy [41]. Previous reports have shown that *WRKY54* and *WRKY70*, as a pair of duplicated genes, both act as negative regulators of leaf senescence [42]. The genes presented similar expression patterns during leaf development. Mutants of *wrky54* or *wrky70* displayed no obvious or only a weak phenotype of premature senescence, respectively. However, compared to the single mutants, the *wrky54/wrky70* double mutants presented a pronounced early senescence phenotype. These findings indicate that *WRKY54* and *WRKY70* present functional redundancy in regulating leaf senescence [43]. In the common bean, a segmentally duplicated gene pair (*PvTLP2/PvTLP4*) has been identified. The identity of these proteins was 78.74%, and they exhibited similar expression patterns in above-ground tissues. This finding suggests their potential functional redundancy, and they may cooperate to participate in similar regulatory pathways related to plant growth.

Analysis of cis-acting elements in the promoters of different gene families can elucidate their potential biological functions [44]. This study revealed that the promoters of *PvTLP* genes contained four types of cis-acting elements: light-responsive, hormone-responsive, stress-responsive, and development-responsive elements (Figure 5). Light-responsive elements, such as Box 4, GATA-motif, and G-box, accounted for 46.73% of the total cis-acting elements in the *PvTLP* promoters. In *Mahonia bealei*, the GATA-motif in the *ndhF* promoter might suppress benzylisoquinoline alkaloid biosynthesis during the dark/light transitions by downregulating *ndh* expression [45]. Moreover, the G-box interacts with G-box binding factor (GBF1) to regulate blue light-induced photomorphogenic growth in cotyledons in *Arabidopsis* [46]. Light-regulated GBFs translocated from the cytoplasm to the nucleus, suggesting their importance in coordinating light transduction pathways with light-dependent gene expression [47]. The presence of multiple light-responsive elements in all *PvTLP* promoters indicates that the *PvTLP* gene family may play a significant role in light response regulation. Furthermore, the promoters contain various hormone- and stress-responsive elements. AREBs and AREs were widespread in most *PvTLP* genes, which is consistent with research on wheat [33]. These findings suggest that *PvTLP* genes may regulate hormone and abiotic stress responses in the common bean. Notably, the *PvTLP7* promoter had the greatest number of cis-acting elements, including 15 hormone-, 10 light-, 10 stress-, and 2 development-responsive elements. In addition, considering that PvTLP7 has a unique gene structure and conserved motifs, it suggests that PvTLP7 may play a critical role in plant development and stress adaptation, but more experimental evidence is needed.

Plant growth and development are regulated by complex gene regulatory networks and signaling pathways. Expression profiles have revealed the dynamic regulatory mechanisms of *TLP* genes. In tomato, *SlTLP1* and *SlTLP2* are highly exhibited throughout fruit development, with experimental evidence suggesting their involvement in ethylene-dependent fruit ripening pathways [48]. *AtTLP6* and *AtTLP7* presented pollen-specific expression, where the *attlp6* mutants produced 5–10% aborted pollen, 15% pollen with aberrant male germ units, and 5% bicellular pollen. Similarly, *attlp7* mutants displayed 20–25% pollen with disorganized male germ units and 5–10% bicellular pollen due to cytoskeleton dysfunction [24,49]. Heatmap analysis revealed that 10 *PvTLP* genes presented constitutive or tissue-specific expression patterns (Figure 6). For example, *PvTLP6* was significantly expressed in flowers and green mature pods, indicating its regulatory role in the process of floral-to-pod development. These findings provide a theoretical foundation for the functional characterization of the *PvTLP* gene family.

Abiotic stress seriously threatens crop yields by impairing essential physiological and biochemical processes. Research has verified that *TLP* genes play multiple functional roles in enhancing crop tolerance to abiotic stress such as drought and salinity. In foxtail millet, most *SiTLP* genes were regulated by salt and high-temperature conditions [50].

The expression patterns of apple *MdTLP* genes significantly differed under various abiotic stresses and hormone treatments, with *MdTLP4* and *MdTLP7* consistently upregulated in response to all stresses [19,51,52]. In *Arabidopsis*, *MdTLP7*-overexpressing seedlings presented improved tolerance to cold, heat, osmotic, and salt stresses [53]. Through qRT-PCR analysis, the expression levels of most *PvTLP* genes exhibited significant changes in both leaves and roots after salt and drought stresses (Figure 7). In the leaves, the expression of eight *PvTLP* genes, excluding *PvTLP1* and *PvTLP9*, peaked at 6 h under salt stress. Nevertheless, eight *PvTLP* genes reached their maximum expression levels at 24 h under drought stress, suggesting that *PvTLP* genes in the leaves respond more rapidly to salt stress. In the roots, most *PvTLP* genes exhibited peak expression at 12–24 h after both treatments. Notably, *PvTLP5* expression levels in the leaves rapidly increased at 6 h under both stresses, with expression levels reaching 81-fold greater than that of the control after salt treatment and 48-fold greater than that of the control after drought treatment. The presence of multiple stress-responsive elements in the promoter regions of *PvTLP5* further indicates that *PvTLP* genes may be essential for adapting to abiotic stress in the common bean. Therefore, *PvTLP5* may be a promising candidate gene for genetic improvement to increase stress resistance. The results suggest that *PvTLP* genes respond to both salt and drought stresses. However, these findings are based on gene expression levels, and the exact mechanism by which *PvTLP* genes operate in the regulatory pathway remains unclear. Based on the excellent genes identified in this study, molecular biology techniques will be further employed for plant transformation (such as gene overexpression or knockout experiments) and functional verification, to elucidate the specific functions of *PvTLP* genes and the mechanisms involved in response to environmental stress. This study provides an important theoretical foundation for understanding the functions of *PvTLP* genes and developing breeding strategies to increase stress tolerance in the common bean.

4. Materials and Methods

4.1. Genome-Wide Identification of PvTLP Genes

The genome data and gene annotation files of the common bean (v2.1) were downloaded from the Phytozome database (http://phytozome.jgi.doe.gov/, accessed on 18 February 2025), and *Arabidopsis* AtTLP protein sequences were obtained from the TAIR database (https://www.arabidopsis.org/, accessed on 18 February 2025). To identify the candidate TLP proteins, AtTLP protein sequences were used as queries in a local BLASTP alignment [54]. For further validation, the Hidden Markov Model (HMM) of the tubby domain (PF01167) was retrieved from the Pfam database (http://pfam.xfam.org/, accessed on 20 February 2025) and analyzed using HMMER 3.3 with the default parameters [55,56]. All proteins of the putative TLP proteins were subsequently subjected to batch CD-search (https://www.ncbi.nlm.nih.gov/cdd, accessed on 20 February 2025) to confirm the presence of complete tubby domains [57]. The physicochemical properties of the TLPs were predicted using the ExPASy online server (http://web.expasy.org/protparam/, accessed on 22 February 2025) [58]. Subcellular localization analysis was conducted using the WoLF PSORT online server (https://www.genscript.com/wolf-psort.html/, accessed on 24 February 2025) [59].

4.2. Chromosomal Localization and Gene Replication Events

According to the common bean genome data and annotation information, TBtools-II (version 2.156) was employed to visualize the chromosomal localization map for 10 *PvTLP* genes [60]. The synteny analysis of *TLP* genes was conducted using the MCScanX software [61]. The ratio of the non-synonymous substitution rate (Ka) to the synonymous substitution rate (Ks) for the duplicated *PvTLP* gene pairs was calculated using the TBtools-

II software. A Ka/Ks ratio greater than 1 indicates positive selection, a ratio of 1 indicates neutral selection, and a ratio less than 1 indicates purifying selection or negative selection.

4.3. Multiple Sequence Alignment and Phylogenetic Analysis

Full-length protein sequences of 10 PvTLPs, 11 AtTLPs, 20 GmTLPs, 14 OsTLPs, and 14 ZmTLPs were retrieved from previous studies (Supplementary Table S2) [14,15,17,62]. Previous research identified 15 *ZmTLP* genes in maize; however, we found that *ZmTLP15* is a non-translating gene. Multiple sequence alignment wa s performed on the 69 TLP protein sequences using Muscle. The neighbor-joining (NJ) method was employed to construct the phylogenetic tree in the MEGA11 software with 1000 bootstrap replicates [63,64]. iTOl (https://itol.embl.de/, accessed on 9 March 2025) was used for visualization and beautification [65].

4.4. Gene Structure, Conserved Domain, and Motif Analysis

A batch CD-search (https://www.ncbi.nlm.nih.gov/cdd, accessed on 22 February 2025) was performed to identify the conserved domains of PvTLP proteins. The PvTLP protein sequences were submitted to the MEME online software (https://meme-suite.org/meme/, accessed on 22 February 2025) to predict the conserved motifs, with the maximum number of motifs configured to 10 [66]. Gene structure, conserved domain, and motif were integrated and visualized using TBtools-II.

4.5. Synteny Analysis

The genomic sequences and annotation information for *Arabidopsis* and soybean were downloaded from the TAIR (https://www.arabidopsis.org/, accessed on 18 February 2025) and Phytozome (http://phytozome.jgi.doe.gov/, accessed on 26 February 2025) databases, respectively. Genomic resources and annotation files for rice and maize were retrieved from Ensembl Plants (https://plants.ensembl.org/index.html, accessed on 26 February 2025). We analyzed the inter-species synteny using MCScanX and visualized the results using TBtools-II.

4.6. Promoter Cis-Acting Elements Analysis

The 2000 bp upstream sequences of 10 *PvTLP* genes in the common bean were extracted as promoter regions. Subsequently, these sequences were uploaded to the PlantCARE database (https://bioinformatics.psb.ugent.be/webtools/plantcare/html/, accessed on 28 February 2025) to cis-acting elements prediction [67]. After screening and classification, the main cis-acting elements related to light, hormone, stress, and development were identified. Data visualization was performed with TBtools-II and Excel 2021.

4.7. Heatmap Analysis

The transcriptome sequencing data for *PvTLP* genes in various tissues were retrieved from the Phytozome database (http://phytozome.jgi.doe.gov/, accessed on 25 February 2025, Supplementary Table S3). Expression levels were determined using FPKM (Fragments Per Kilobase of exon model per Million mapped fragments) values and normalized using the log2 (Fold Change) transformation. Expression profiles were generated using the heatmap function from OmicShare tools (https://www.omicshare.com/tools/, accessed on 1 March 2025) [68].

4.8. Plant Material Source and Treatments

The common bean cultivar Pinjinyun 4 is a bred variety from Shanxi, China. The seedlings were grown until the trifoliate leaves had fully expanded in an artificial climate chamber at 25 °C, with a 16 h light/8 h dark photoperiod and 70% relative humidity. For salt treatment, the seedlings were cultured in Hoagland solution supplemented with 0 or 200 mmol L^{-1} NaCl. For drought treatment, the seedlings were cultured in Hoagland solution supplemented with 0 or 20% PEG6000. Leaves and roots were collected at 0, 6, 12, 24, and 36 h after NaCl and PEG6000 treatments, frozen quickly in liquid nitrogen, and stored at −80 °C.

4.9. RNA Extraction and qRT-PCR Analysis

Total RNA was extracted from the leaves and roots of the common bean using the RNA Easy Fast Plant Tissue Kit (Tiangen, Beijing, China). The concentration and purity of the RNA were assessed using a NanoDrop 2000 spectrophotometer (Thermo Fisher Scientific, Waltham, MA, USA), and the quality of the RNA was confirmed by 1% agarose gel electrophoresis. Total RNA was reverse-transcribed into cDNA using the PrimeScript Fast RT reagent Kit with gDNA Eraser (Takara, Otsu, Japan). qRT-PCR was conducted on an Applied Biosystems QuantStudio 6 Flex cycler (Applied Biosystems, Waltham, MA, USA) following the protocol of the SYBR Premix Ex Taq Kit (Takara Bio, Otsu, Japan): initial denaturation at 95 °C for 30 s, followed by 40 cycles of 95 °C for 5 s, and 60 °C for 34 s. Each experiment included three technical replicates and three biological replicates. The common bean *actin* gene (Phvul.011G064500) was used as the internal control. Relative quantification was performed using the $2^{-\Delta\Delta Ct}$ method [69]. Statistically significant differences were determined by two-way ANOVA followed by Dunnett's multiple comparisons test and data visualization, which were performed using GraphPad Prism version 8.4.0 (Boston, MA, USA, www.graphpad.com). Primer sequences are provided in Supplementary Table S4.

Supplementary Materials: The following supporting information can be downloaded at https://www.mdpi.com/article/10.3390/ijms26125702/s1.

Author Contributions: Conceptualization, X.D. and X.H.; Data curation, M.Z.; Funding acquisition, X.D. and X.H.; Resources, Y.W.; Software, F.Q.; Validation, J.L.; Visualization, J.Z.; Writing—original draft, X.D., M.Z., J.L., and F.Q.; Writing—review and editing, X.D., J.C., and X.H. All authors have read and agreed to the published version of the manuscript.

Funding: This research was funded by the China Agriculture Research System of MOF and MARA-Food Legumes, grant number CARS-08; the National Key Research and Development Program of China, grant number 2021YFD1600601; Hou Ji Laboratory in Shanxi Province, grant number 202304010930003; the Major Special Science and Technology Plan in Shanxi Province, grant number 202301140601016 and 202101140601027; and the Biological Breeding Engineering Project of Shanxi Agricultural University, grant number YZGC148.

Institutional Review Board Statement: Not applicable.

Informed Consent Statement: Not applicable.

Data Availability Statement: All the data that support the findings of this study are available in the paper and its Supplementary Materials published online.

Conflicts of Interest: The authors declare no conflicts of interest.

References

1. Bano, N.; Aalam, S.; Bag, S.K. Tubby-Like Proteins (TLPs) Transcription Factor in Different Regulatory Mechanism in Plants: A Review. *Plant Mol. Biol.* **2022**, *110*, 455–468. [CrossRef] [PubMed]
2. Kleyn, P.W.; Fan, W.; Kovats, S.G.; Lee, J.J.; Pulido, J.C.; Wu, Y.; Berkemeier, L.R.; Misumi, D.J.; Holmgren, L.; Charlat, O.; et al. Identification and Characterization of the Mouse Obesity Gene *tubby*: A Member of a Novel Gene Family. *Cell* **1996**, *85*, 281–290. [CrossRef] [PubMed]
3. Hagstrom, S.A.; North, M.A.; Nishina, P.L.; Berson, E.L.; Dryja, T.P. Recessive Mutations in the Gene Encoding the Tubby-Like Protein TULP1 in Patients with Retinitis Pigmentosa. *Nat. Genet.* **1998**, *18*, 174–176. [CrossRef] [PubMed]

4. Ikeda, A.; Nishina, P.M.; Naggert, J.K. The Tubby-Like Proteins, a Family with Roles in Neuronal Development and Function. *J. Cell Sci.* **2002**, *115*, 9–14. [CrossRef]
5. Mukhopadhyay, S.; Jackson, P.K. The Tubby Family Proteins. *Genome Biol.* **2011**, *12*, 225. [CrossRef]
6. Wang, M.; Xu, Z.; Kong, Y. The Tubby-Like Proteins Kingdom in Animals and Plants. *Gene* **2018**, *642*, 16–25. [CrossRef]
7. Boggon, T.J.; Shan, W.S.; Santagata, S.; Myers, S.C.; Shapiro, L. Implication of Tubby Proteins as Transcription Factors by Structure-Based Functional Analysis. *Science* **1999**, *286*, 2119–2125. [CrossRef]
8. Thallmair, V.; Schultz, L.; Zhao, W.; Marrink, S.J.; Oliver, D.; Thallmair, S. Two Cooperative Binding Sites Sensitize PI(4,5)P_2 Recognition by the Tubby Domain. *Sci. Adv.* **2022**, *8*, eabp9471. [CrossRef]
9. Reitz, M.U.; Bissue, J.K.; Zocher, K.; Attard, A.; Hückelhoven, R.; Becker, K.; Imani, J.; Eichmann, R.; Schäfer, P. The Subcellular Localization of Tubby-Like Proteins and Participation in Stress Signaling and Root Colonization by the Mutualist *Piriformospora indica*. *Plant Physiol.* **2012**, *160*, 349–364. [CrossRef]
10. Santagata, S.; Boggon, T.J.; Baird, C.L.; Gomez, C.A.; Zhao, J.; Shan, W.S.; Myszka, D.G.; Shapiro, L. G-Protein Signaling Through Tubby Proteins. *Science* **2001**, *292*, 2041–2050. [CrossRef]
11. Thulasi Devendrakumar, K.; Copeland, C.; Adamchek, C.; Zhong, X.; Huang, X.; Gendron, J.M.; Li, X. Arabidopsis Tubby Domain-Containing F-Box Proteins Positively Regulate Immunity by Modulating PI4Kβ Protein Levels. *New Phytol.* **2023**, *240*, 354–371. [CrossRef] [PubMed]
12. Xu, K.; Wu, N.; Yao, W.; Li, X.; Zhou, Y.; Li, H. The Biological Function and Roles in Phytohormone Signaling of the F-Box Protein in Plants. *Agronomy* **2021**, *11*, 2360. [CrossRef]
13. Xiao, W.; Jang, J. F-Box Proteins in *Arabidopsis*. *Trends Plant Sci.* **2000**, *5*, 454–457. [CrossRef] [PubMed]
14. Lai, C.P.; Lee, C.L.; Chen, P.H.; Wu, S.H.; Yang, C.C.; Shaw, J.F. Molecular Analyses of the Arabidopsis TUBBY-Like Protein Gene Gamily. *Plant Physiol.* **2004**, *134*, 1586–1597. [CrossRef]
15. Yang, Z.; Zhou, Y.; Wang, X.; Gu, S.; Yu, J.; Liang, G.; Yan, C.; Xu, C. Genomewide Comparative Phylogenetic and Molecular Evolutionary Analysis of Tubby-Like Protein Family in *Arabidopsis*, Rice, and Poplar. *Genomics* **2008**, *92*, 246–253. [CrossRef]
16. Bano, N.; Fakhrah, S.; Mohanty, C.S.; Bag, S.K. Genome-Wide Identification and Evolutionary Analysis of *Gossypium* Tubby-Like Protein (TLP) Gene Family and Expression Analyses during Salt and Drought Stress. *Front. Plant Sci.* **2021**, *12*, 667929. [CrossRef]
17. Zeng, Y.; Wen, J.; Fu, J.; Geng, H.; Dan, Z.; Zhao, W.; Xu, W.; Huang, W. Genome-Wide Identification and Comprehensive Analysis of Tubby-Like Protein Gene Family in Multiple Crops. *Front. Plant Sci.* **2022**, *13*, 1093944. [CrossRef]
18. Zhang, J.; Wang, X.; Dong, X.; Wang, F.; Cao, L.; Li, S.; Liu, Z.; Zhang, X.; Guo, Y.D.; Zhao, B.; et al. Expression Analysis and Functional Characterization of Tomato Tubby-Like Protein Family. *Plant Sci.* **2022**, *324*, 111454. [CrossRef]
19. Xu, J.N.; Xing, S.S.; Zhang, Z.R.; Chen, X.S.; Wang, X.Y. Genome-Wide Identification and Expression Analysis of the Tubby-Like Protein Family in the *Malus Domestica* Genome. *Front. Plant Sci.* **2016**, *7*, 1693. [CrossRef]
20. Li, S.; Wang, G.; Chang, L.; Sun, R.; Wu, R.; Zhong, C.; Gao, Y.; Zhang, H.; Wei, L.; Wei, Y.; et al. Genome-Wide Identification, Characterization, and Expression Analysis of Tubby-Like Protein (TLP) Gene Family Members in Woodland Strawberry (*Fragaria vesca*). *Int. J. Mol. Sci.* **2022**, *23*, 11961. [CrossRef]
21. Dong, M.Y.; Fan, X.W.; Pang, X.Y.; Li, Y.Z. Decrypting Tubby-Like Protein Gene Family of Multiple Functions in Starch Root Crop Cassava. *AoB Plants* **2019**, *11*, plz075. [CrossRef] [PubMed]
22. Jain, N.; Khurana, P.; Khurana, J.P. AtTLP2, a Tubby-Like Protein, Plays Intricate Roles in Abiotic Stress Signaling. *Plant Cell Rep.* **2023**, *42*, 235–252. [CrossRef]
23. Wang, M.; Xu, Z.; Ahmed, R.I.; Wang, Y.; Hu, R.; Zhou, G.; Kong, Y. *Tubby-Like Protein 2* Regulates Homogalacturonan Biosynthesis in *Arabidopsis* Seed Coat Mucilage. *Plant Mol. Biol.* **2019**, *99*, 421–436. [CrossRef] [PubMed]
24. Bao, Y.; Song, W.M.; Jin, Y.L.; Jiang, C.M.; Yang, Y.; Li, B.; Huang, W.J.; Liu, H.; Zhang, H.X. Characterization of *Arabidopsis* Tubby-Like Proteins and Redundant Function of AtTLP3 and AtTLP9 in Plant Response to ABA and Osmotic Stress. *Plant Mol. Biol.* **2014**, *86*, 471–483. [CrossRef] [PubMed]
25. Wardhan, V.; Jahan, K.; Gupta, S.; Chennareddy, S.; Datta, A.; Chakraborty, S.; Chakraborty, N. Overexpression of CaTLP1, A Putative Transcription Factor in Chickpea (*Cicer arietinum* L.), Promotes Stress Tolerance. *Plant Mol. Biol.* **2012**, *79*, 479–493. [CrossRef]
26. Wardhan, V.; Pandey, A.; Chakraborty, S.; Chakraborty, N. Chickpea Transcription Factor CaTLP1 Interacts with Protein Kinases, Modulates ROS Accumulation and Promotes ABA-Mediated Stomatal Closure. *Sci. Rep.* **2016**, *6*, 38121. [CrossRef]
27. Li, S.; Wang, Z.; Wang, F.; Lv, H.; Cao, M.; Zhang, N.; Li, F.; Wang, H.; Li, X.; Yuan, X.; et al. A Tubby-Like Protein CsTLP8 Acts in the ABA Signaling Pathway and Negatively Regulates Osmotic Stresses Tolerance during Seed Germination. *BMC Plant Biol.* **2021**, *21*, 340. [CrossRef]
28. Lisciani, S.; Marconi, S.; Le Donne, C.; Camilli, E.; Aguzzi, A.; Gabrielli, P.; Gambelli, L.; Kunert, K.; Marais, D.; Vorster, B.J.; et al. Legumes and Common Beans in Sustainable Diets: Nutritional Quality, Environmental Benefits, Spread and Use in Food Preparations. *Front. Nutr.* **2024**, *11*, 1385232. [CrossRef]

29. Alcázar-Valle, M.; Lugo-Cervantes, E.; Mojica, L.; Morales-Hernández, N.; Reyes-Ramírez, H.; Enríquez-Vara, J.N.; García-Morales, S. Bioactive Compounds, Antioxidant Activity, and Antinutritional Content of Legumes: A Comparison between Four *Phaseolus* Species. *Molecules* **2020**, *25*, 3528. [CrossRef]
30. Losa, A.; Vorster, J.; Cominelli, E.; Sparvoli, F.; Paolo, D.; Sala, T.; Ferrari, M.; Carbonaro, M.; Marconi, S.; Camilli, E.; et al. Drought and Heat Affect Common Bean Minerals and Human Diet—What We Know and Where to Go. *Food Energy Secur.* **2022**, *11*, e351. [CrossRef]
31. Dutta, T.; Neelapu, N.R.R.; Wani, S.H.; Challa, S. Response of Pulses to Drought and Salinity Stress Response: A Physiological Perspective. In *Pulse Improvement*; Wani, S.H., Jain, M., Eds.; Springer: Cham, Switzerland, 2018; Volume 4, pp. 77–98. [CrossRef]
32. Uebersax, M.A.; Cichy, K.A.; Gomez, F.E.; Porch, T.G.; Heitholt, J.; Osorno, J.M.; Kamfwa, K.; Snapp, S.S.; Bales, S. Dry Beans (*Phaseolus vulgaris* L.) as a Vital Component of Sustainable Agriculture and Food Security–A Review. *Legume Sci.* **2022**, *5*, e155. [CrossRef]
33. Altaf, A.; Zada, A.; Hussain, S.; Gull, S.; Ding, Y.; Tao, R.; Zhu, M.; Zhu, X. Genome-Wide Identification, Characterization, and Expression Analysis of TUBBY Gene Family in Wheat (*Triticum aestivum* L.) under Biotic and Abiotic Stresses. *Agronomy* **2022**, *12*, 1121. [CrossRef]
34. Wang, T.; Hu, J.; Ma, X.; Li, C.; Yang, Q.; Feng, S.; Li, M.; Li, N.; Song, X. Identification, Evolution and Expression Analyses of Whole Genome-Wide *TLP* Gene Family in *Brassica napus*. *BMC Genom.* **2020**, *21*, 264. [CrossRef] [PubMed]
35. Wang, K.; Cheng, Y.; Yi, L.; He, H.; Zhan, S.; Yang, P. Genome-Wide Identification of the Tubby-Like Protein (TLPs) Family in Medicinal Model Plant *Salvia miltiorrhiza*. *PeerJ* **2021**, *9*, e11403. [CrossRef] [PubMed]
36. Mejia, S.; Santos, J.L.B.; Noutsos, C. Comprehensive Genome-Wide Natural Variation and Expression Analysis of Tubby-Like Proteins Gene Family in *Brachypodium distachyon*. *Plants* **2024**, *13*, 987. [CrossRef]
37. Ullah, I.; Uzair, M.; Rehman, O.U.; Attacha, S.; Attia, K.A.; Ahmad, S.; Salman, M.; Ali, H.; Rashid, M.A.R.; Abushady, A.M.; et al. Characterization of the Tubby-Like Proteins (TLPs) Gene Family in *Cucumis sativus* L.: Insights into the Evolution, Structure, and Expression Patterns under Salt Stress. *Genet. Resour. Crop Evol.* **2023**, *71*, 2019–2036. [CrossRef]
38. Xu, G.; Guo, C.; Shan, H.; Kong, H. Divergence of Duplicate Genes in Exon–Intron Structure. *Proc. Natl. Acad. Sci. USA* **2012**, *109*, 1187–1192. [CrossRef]
39. Rogozin, I.B.; Wolf, Y.I.; Sorokin, A.V.; Mirkin, B.G.; Koonin, E.V. Remarkable Interkingdom Conservation of Intron Positions and Massive, Lineage-Specific Intron Loss and Gain in Eukaryotic Evolution. *Curr. Biol.* **2003**, *13*, 1512–1517. [CrossRef]
40. Wagner, A. Genetic Redundancy Caused by Gene Duplications and Its Evolution in Networks of Transcriptional Regulators. *Biol. Cybern.* **1996**, *74*, 557–567. [CrossRef]
41. Chapman, B.A.; Bowers, J.E.; Feltus, F.A.; Paterson, A.H. Buffering of Crucial Functions by Paleologous Duplicated Genes May Contribute Cyclicality to Angiosperm Genome Duplication. *Proc. Natl. Acad. Sci. USA* **2006**, *103*, 2730–2735. [CrossRef]
42. Li, Z.; Woo, H.R.; Guo, H. Genetic Redundancy of Senescence-Associated Transcription Factors in Arabidopsis. *J. Exp. Bot.* **2018**, *69*, 811–823. [CrossRef] [PubMed]
43. Besseau, S.; Li, J.; Palva, E.T. WRKY54 and WRKY70 Co-Operate as Negative Regulators of Leaf Senescence in *Arabidopsis thaliana*. *J. Exp. Bot.* **2012**, *63*, 2667–2679. [CrossRef]
44. Baillo, E.H.; Kimotho, R.N.; Zhang, Z.; Xu, P. Transcription Factors Associated with Abiotic and Biotic Stress Tolerance and Their Potential for Crops Improvement. *Genes* **2019**, *10*, 771. [CrossRef] [PubMed]
45. Chen, J.; Jiang, T.; Jiang, J.; Deng, L.; Liu, Y.; Zhong, Z.; Fu, H.; Yang, B.; Zhang, L. The Chloroplast GATA-Motif of *Mahonia bealei* Participates in Alkaloid-Mediated Photosystem Inhibition during Dark to Light Transition. *J. Plant Physiol.* **2023**, *280*, 153894. [CrossRef]
46. Mallappa, C.; Yadav, V.; Negi, P.; Chattopadhyay, S. A Basic Leucine Zipper Transcription Factor, G-Box-Binding Factor 1, Regulates Blue Light-Mediated Photomorphogenic Growth in *Arabidopsis*. *J. Biol. Chem.* **2006**, *281*, 22190–22199. [CrossRef] [PubMed]
47. Harter, K.; Kircher, S.; Frohnmeyer, H.; Krenz, M.; Nagy, F.; Schäfer, E. Light-Regulated Modification and Nuclear Translocation of Cytosolic G-Box Binding Factors in Parsley. *Plant Cell* **1994**, *6*, 545–559. [CrossRef]
48. Zhang, Y.; He, X.; Su, D.; Feng, Y.; Zhao, H.; Deng, H.; Liu, M. Comprehensive Profiling of Tubby-Like Protein Expression Uncovers Ripening-Related *TLP* Genes in Tomato (*Solanum lycopersicum*). *Int. J. Mol. Sci.* **2020**, *21*, 1000. [CrossRef] [PubMed]
49. Reňák, D.; Dupl'áková, N.; Honys, D. Wide-Scale Screening of T-DNA Lines for Transcription Factor Genes Affecting Male Gametophyte Development in *Arabidopsis*. *Sex. Plant Reprod.* **2012**, *25*, 39–60. [CrossRef]
50. Jiang, Z.; Chen, X.; Ruan, L.; Xu, Y.; Li, K. Molecular Analyses of the Tubby-Like Protein Gene Family and Their Response to Salt and High Temperature in the Foxtail Millet (*Setaria italica*). *Funct. Integr. Genom.* **2024**, *24*, 170. [CrossRef]
51. Du, F.; Xu, J.N.; Zhan, C.Y.; Yu, Z.B.; Wang, X.Y. An Obesity-Like Gene *MdTLP7* from Apple (*Malus* × *Domestica*) Enhances Abiotic Stress Tolerance. *Biochem. Bioph. Res. Co.* **2014**, *445*, 394–397. [CrossRef]
52. Du, F.; Xu, J.N.; Li, D.; Wang, X.Y. The Identification of Novel and Differentially Expressed Apple-Tree Genes under Low-Temperature Stress Using High-Throughput Illumina Sequencing. *Mol. Biol. Rep.* **2015**, *42*, 569–580. [CrossRef]

53. Xu, J.; Xing, S.; Sun, Q.; Zhan, C.; Liu, X.; Zhang, S.; Wang, X. The Expression of a Tubby-Like Protein from *Malus domestica* (*Md*TLP7) Enhances Abiotic Stress Tolerance in *Arabidopsis*. *BMC Plant Biol.* **2019**, *19*, 60. [CrossRef] [PubMed]
54. Altschul, S.F.; Gish, W.; Miller, W.; Myers, E.W.; Lipman, D.J. Basic Local Alignment Search Tool. *J. Mol. Biol.* **1990**, *215*, 403–410. [CrossRef]
55. Eddy, S.R. Accelerated Profile HMM Searches. *PLoS Comput. Biol.* **2011**, *7*, e1002195. [CrossRef]
56. Mistry, J.; Chuguransky, S.; Williams, L.; Qureshi, M.; Salazar, G.A.; Sonnhammer, E.L.L.; Tosatto, S.C.E.; Paladin, L.; Raj, S.; Richardson, L.J.; et al. Pfam: The Protein Families Database in 2021. *Nucleic Acids Res.* **2021**, *49*, D412–D419. [CrossRef]
57. Wang, J.; Chitsaz, F.; Derbyshire, M.K.; Gonzales, N.R.; Gwadz, M.; Lu, S.; Marchler, G.H.; Song, J.S.; Thanki, N.; Yamashita, R.A.; et al. The Conserved Domain Database in 2023. *Nucleic Acids Res.* **2023**, *51*, D384–D388. [CrossRef] [PubMed]
58. Artimo, P.; Jonnalagedda, M.; Arnold, K.; Baratin, D.; Csardi, G.; de Castro, E.; Duvaud, S.; Flegel, V.; Fortier, A.; Gasteiger, E.; et al. ExPASy: SIB Bioinformatics Resource Portal. *Nucleic Acids Res.* **2012**, *40*, W597–W603. [CrossRef]
59. Horton, P.; Park, K.J.; Obayashi, T.; Fujita, N.; Harada, H.; Adams-Collier, C.J.; Nakai, K. WoLF PSORT: Protein Localization Predictor. *Nucleic Acids Res.* **2007**, *35*, W585–W587. [CrossRef] [PubMed]
60. Chen, C.; Wu, Y.; Li, J.; Wang, X.; Zeng, Z.; Xu, J.; Liu, Y.; Feng, J.; Chen, H.; He, Y.; et al. TBtools-II: A "One for All, All for One" Bioinformatics Platform for Biological Big-Data Mining. *Mol. Plant.* **2023**, *16*, 1733–1742. [CrossRef]
61. Wang, Y.; Tang, H.; Debarry, J.D.; Tan, X.; Li, J.; Wang, X.; Lee, T.H.; Jin, H.; Marler, B.; Guo, H.; et al. *MCScanX*: A Toolkit for Detection and Evolutionary Analysis of Gene Synteny and Collinearity. *Nucleic Acids Res.* **2012**, *40*, e49. [CrossRef] [PubMed]
62. Chen, Y.; Wei, D.; Sun, B.; Zhao, Y.; Ma, Q. Genome-Wide Identification and Comparative Analysis of the TUBBY-Like Protein Gene Family in Maize. *Genes Genom.* **2015**, *38*, 25–36. [CrossRef]
63. Saitou, N.; Nei, M. The Neighbor-Joining Method: A New Method for Reconstructing Phylogenetic Trees. *Mol. Biol. Evol.* **1987**, *4*, 406–425. [CrossRef]
64. Tamura, K.; Stecher, G.; Kumar, S. MEGA11: Molecular Evolutionary Genetics Analysis Version 11. *Mol. Biol. Evol.* **2021**, *38*, 3022–3027. [CrossRef] [PubMed]
65. Letunic, I.; Bork, P. Interactive Tree of Life (iTOL) V6: Recent Updates to the Phylogenetic Tree Display and Annotation Tool. *Nucleic Acids Res.* **2024**, *52*, W78–W82. [CrossRef] [PubMed]
66. Bailey, T.L.; Johnson, J.; Grant, C.E.; Noble, W.S. The MEME Suite. *Nucleic Acids Res.* **2015**, *43*, W39–W49. [CrossRef] [PubMed]
67. Lescot, M.; Déhais, P.; Thijs, G.; Marchal, K.; Moreau, Y.; Van de Peer, Y.; Rouzé, P.; Rombauts, S. PlantCARE, a Database of Plant Cis-Acting Regulatory Elements and a Portal to Tools for in Silico Analysis of Promoter Sequences. *Nucleic Acids Res.* **2002**, *30*, 325–327. [CrossRef]
68. Mu, H.; Chen, J.; Huang, W.; Huang, G.; Deng, M.; Hong, S.; Ai, P.; Gao, C.; Zhou, H. OmicShare Tools: A Zero-Code Interactive Online Platform for Biological Data Analysis and Visualization. *iMeta* **2024**, *3*, e228. [CrossRef]
69. Livak, K.J.; Schmittgen, T.D. Analysis of Relative Gene Expression Data Using Real-Time Quantitative PCR and the $2^{-\Delta\Delta Ct}$ Method. *Methods* **2001**, *25*, 402–408. [CrossRef]

International Journal of *Molecular Sciences*

MDPI

Article

Genome-Wide Identification of the ARF Gene Family in Safflower (*Carthamus tinctorius* L.) and Their Response Patterns to Exogenous Hormone Treatments

Shuwei Qin [1,†], Xinrong Wen [1,†], Mengyuan Ma [1], Jiaxing Wang [1], Jianhang Zhang [1], Meihui Huang [1], Kexin Sun [1], Ya Zhao [1], Meng Zhao [1], Asigul Ismayil [1], Min Liu [2,*] and Aiping Cao [1,2,*]

1 College of Life Sciences, Shihezi University, Shihezi 832003, China; qsw2623201516@163.com (S.Q.); 15299941807@163.com (X.W.); mengyuanma2022@163.com (M.M.); 18119249542@163.com (J.W.); zjh17630412656@163.com (J.Z.); 18503174612@163.com (M.H.); 13223072825@163.com (K.S.); 18160362098@163.com (Y.Z.); zm022362@163.com (M.Z.); asgli12@163.com (A.I.)

2 Key Laboratory of Xinjiang Phytomedicine Resource and Utilization, Institute for Safflower Industry Research, Pharmacy College, Shihezi University, Ministry of Education, Shihezi 832003, China

* Correspondence: liumin@shzu.edu.cn (M.L.); caoaiping@shzu.edu.cn (A.C.)

† These authors contributed equally to this work.

Abstract: Auxin response factors (ARFs) are a class of transcription factors widely present in plants. As an important economic crop, research on the effects of safflower ARFs on endogenous auxin and effective components is relatively limited. In this study, a total of 23 ARF genes were identified from the safflower genome. Sequence alignment and domain analysis indicated the presence of conserved B3 and Auxin_resp domains in these ARFs. Phylogenetic analysis indicated that CtARF could be classified into five subfamilies, a conclusion also supported by gene structure, consensus motifs, and domain compositions. Transcriptome data showed that ARFs are expressed in all flower colors, but the expression levels of ARF family members vary among different flower colors. *CtARF19* had relatively higher expression in deep red flowers, *CtARF3* had higher expression in white flowers, *CtARF2/12* had higher expression in yellow flowers, and *CtARF21/22* had higher expression in light red flowers. Protein–protein interaction network analysis indicated that ARF family members (CtARF2/3/4/5/15/18/19/22) are located within the interaction network. Cis-acting element analysis suggested that *CtARF* genes may be regulated by hormone treatment (AuxRR-core) and abiotic stress, and the results of qRT-PCR also confirmed this. Additionally, the content of endogenous auxin and active components in safflower with different flower colors significantly changed upon treatment with hormones that affect IAA content. In summary, our study provides valuable insights into the biological functions of *CtARF* genes under exogenous hormone conditions and their effects on active components.

Keywords: ARF; safflower; *Aux/IAA*; hormone treatment; active components; flower color

1. Introduction

Safflower (*Carthamus tinctorius* L.) has been cultivated in China for a period exceeding 2100 years. Its distribution is primarily concentrated in Xinjiang, Yunnan, Henan, Inner Mongolia, and several other regions. Remarkably, the cultivation area of safflower in Xinjiang constitutes over 80% of the total cultivation area across China. Moreover, safflower is cultivated in numerous regions across the globe, including countries such as the United States, India, and Canada. Safflower is a versatile economic crop, integrating medicinal [1], dyeing [2], oil-yielding [3], and feed [4] functions, and thus holds broad prospects for development and application. Hydroxysafflor yellow A (HSYA), extracted from the inflorescences of safflower, is composed of one C-glucosylquinochalcone and has been widely applied in the treatment of coronary heart disease, cerebral infarction, and tumors [5–8]. Therefore, investigating the regulatory mechanisms of safflower development and screening for potential genes that regulate HSYA biosynthesis are important approaches to improve the agronomic traits of safflower and increase the content of active components.

Auxin (*Aux/IAA*) is a class of plant hormones that plays a crucial role in plant growth and development [9]. It not only influences processes, such as plant embryo development and cell elongation, but also affects apical bud growth, root growth, and fruit maturation [10]. Due to the extensive functional repertoire of auxin, studying its biological functions through auxin-deficient mutants presents significant challenges. However, this difficulty can be overcome by employing auxin inhibitors to reduce auxin activity [11]. Different auxin inhibitors exert distinct inhibitory effects. For instance, N-1-naphthylphthalamic acid (NPA) is a chemical that alters IAA transport and thus affects the responses to indigenous IAA in planta. Scientifically modulating the ratio of these two in plants can regulate plant growth processes by affecting endogenous hormone levels, and can also achieve effective weed control in agricultural management [12].

The Aux/IAA and ARF families are core components of the auxin signaling pathway [13]. The Aux/IAA proteins can be divided into three functional domains: the N-terminal DNA-binding inhibitory domain, Domain II, and the C-terminal domain [14]. The DNA-binding inhibitory domain can recruit the TOPLESS (TPL) co-repressor, and the specific binding between them mediates the regulation of biological processes by auxin [15]. Domain II can interact with the TIR1/AFB co-receptor. The C-terminal domain, also known as the PB1 domain, consists of the Phox and Bem1 modules, which can be referred to as Domain III and Domain IV [16]. ARFs can also be divided into three functional domains: the DNA-binding domain (DBD), the middle region (MR), and the C-terminal dimerization domain (CTD) [17,18], which are involved in the regulation of gene activation or repression. The ARF family exhibits species-specific characteristics. Initially, 23 members of the ARF family were identified in *Arabidopsis thaliana*. Subsequently, 25, 23, 25, 36, and 24 members were identified in *Oryza sativa*, *Triticum aestivum*, *Sorghum bicolor*, *Setaria viridis*, and *Sesamum indicum*, respectively [19].

The ARF family is not only involved in the regulation of plant growth and development but also plays a crucial role in plant responses to abiotic stress and in mitigating heavy metal stress [20–22]. Previous studies have revealed that in *Arabidopsis thaliana*, *AtARF1* and *AtARF2* are involved in the regulation of flower organ development and senescence [23]. *AtARF6* and *AtARF8* promote the maturation of flower organs [24], while *AtARF17* facilitates anther development and pollen tube wall formation [25]. Concurrently, studies have demonstrated that ARF genes in apples can regulate anthocyanin biosynthesis pathways, thereby influencing flower color [26]. Therefore, ARF genes also possess potential regulatory functions in flower color formation. Flower color, as an important trait of plants, not only provides ornamental value, but also serves as a crucial signal to attract insects for pollination [27]. Previous studies have shown that the flower color of safflower is controlled by two dominant epistatic loci, each with two alleles, and these loci follow Mendelian inheritance laws [28,29]. Studies have shown that there are significant differences in HSYA content among different genotypes of safflower [30], and that HSYA content is highly correlated with flower color. Studies have indicated that genes encoding cinnamate-4-hydroxylase (C4H), chalcone synthase (CHS), C-glycosyltransferase (CGT), and cytochrome P450 (P450) may be involved in the biosynthetic pathway of HSYA [30]. Additionally, kaempferol (KAE) is synthesized through the catalytic actions of enzymes, such as cinnamate-4-hydroxylase (C4H), chalcone synthase (CHS), chalcone isomerase

(CHI), and flavanone 3-hydroxylase (F3H). The biosynthetic pathways of KAE and HSYA indeed intersect, with the biosynthetic route of naringenin chalcone being conserved [30]. However, there is a lack of experimental evidence linking different flower colors of safflower to HSYA biosynthesis. Additionally, the mechanisms by which ARF family genes influence plant growth, development, and HSYA content in different flower colors are worth further investigation.

In this study, bioinformatics techniques were employed to identify *CtARF* genes from the whole genome of safflower. A comprehensive analysis of this gene family was conducted in terms of physicochemical properties, conserved sequences, gene structure, phylogenetics, and gene expression. Additionally, the effects of exogenous hormone treatment on safflower ARF genes, endogenous IAA, and active components were investigated. This study provides a theoretical basis for further exploring the functions of *CtARF* family genes and their involvement in the auxin pathway, which affects the growth, development, and synthesis of HSYA in safflower.

2. Results

2.1. Identification and Physicochemical Property Analysis of ARF Transcription Factor Family Members in Safflower

With the protein sequences of the ARF family in *Arabidopsis thaliana* and *Helianthus annuus* as references, 89 homologous genes (preliminary screen of the safflower genome) were obtained from the safflower genome through BLAST alignment. The identified proteins were successively submitted to the Pfam website for a domain search, and the presence of the NBD and CTD domains was confirmed. In summary, A total of 23 ARF genes were identified in the safflower genome, and they were named *CtARF1* to *CtARF23* according to their chromosomal locations (Figure S6). The protein lengths of CtARF family members ranged from 548 to 1119 aa (Tables S3 and S4), with molecular weights of 61.87–125.48 kD and aliphatic indices of 63.41–76.91. The isoelectric points of the encoded proteins of CtARF family genes ranged from 5.13 to 7.64, suggesting that they may function in weakly acidic or neutral subcellular environments. The instability indices of all encoded proteins of CtARF family genes were greater than 40, indicating that they are unstable proteins. The grand average of hydropathicity values of all proteins was negative, indicating that they are hydrophilic proteins. Subcellular localization prediction showed that all proteins are located in the nucleus (Table 1).

2.2. Chromosomal Distribution of ARF Family Members in Safflower

Based on the genome annotation information of safflower, the chromosomal localization of ARF genes was conducted (Figure S6). ARF genes were distributed across all 12 chromosomes of safflower. Specifically, chromosomes 2, 3, 4, 6, 7, and 9 each harbored one ARF gene, chromosomes 1, 10, and 12 each had two ARF genes, chromosomes 5 and 8 each possessed four ARF genes, and chromosome 11 contained three ARF genes. These findings indicated that ARF genes in safflower were dispersed throughout the chromosomal set.

2.3. Phylogenetic Analysis of ARF Proteins in Safflower

To further investigate the evolutionary relationships of *CtARF* genes and explore their potential functions, a phylogenetic tree was constructed using 85 members of the ARF family from safflower, *Arabidopsis thaliana*, and *Helianthus annuus* (Figure 1). Based on the phylogenetic tree and the classification of ARF genes in *Arabidopsis* and *Helianthus*, the *CtARF* family was divided into five subfamilies (I, II, III, IV, and V). The results indicated that safflower ARF genes shared similarities with those in *Arabidopsis* and *Helianthus*.

Subfamily I included *CtARF3/4/10/16/19/22* (six members), subfamily II included *CtARF1/14* (two members), subfamily III included *CtARF6/7/8/9/15/18/23* (seven members), subfamily IV included *CtARF2/5/12/20* (four members), and subfamily V included *CtARF11/13/17/21* (four members).

Table 1. Physiochemical properties of the *CtARF* gene family in safflower.

Gene Name	Locus ID	Protein Length (aa)	Molecular Weight (KDa)	PI	Instability Indices	Aliphatic Indices	Hydrophobicity	Subcellular Localization
CtARF1	CtAH01T0069600	813	90.53341	5.85	60.62	73.14	−0.468	nucleus
CtARF2	CtAH01T0211500	614	67.8474	8.38	53.06	71.04	−0.436	nucleus
CtARF3	CtAH02T0026700	641	71.89815	6.41	56.42	70.83	−0.582	nucleus
CtARF4	CtAH03T0268400	651	73.22105	5.51	55.27	73.01	−0.437	nucleus
CtARF5	CtAH04T0143600	604	67.71435	6.66	53.22	68.08	−0.483	nucleus
CtARF6	CtAH05T0001600	1119	125.48255	6.92	73.1	69.63	−0.638	nucleus
CtARF7	CtAH05T0047600	1059	117.82049	6.16	59.13	72.83	−0.551	nucleus
CtARF8	CtAH05T0213800	1054	116.50864	5.86	60.98	68.55	−0.575	nucleus
CtARF9	CtAH05T0279800	1049	117.9647	6.36	75.38	65.73	−0.765	nucleus
CtARF10	CtAH06T0272100	707	79.08847	6.08	68.44	63.41	−0.688	nucleus
CtARF11	CtAH07T0101600	648	72.28166	6.82	52.99	69.95	−0.44	nucleus
CtARF12	CtAH08T0107100	787	86.37022	5.94	56.73	70.32	−0.433	nucleus
CtARF13	CtAH08T0109300	665	73.11407	5.9	47.25	74.6	−0.362	nucleus
CtARF14	CtAH08T0176800	854	95.41481	5.81	63.57	70.2	−0.557	nucleus
CtARF15	CtAH08T0303600	855	94.24119	5.71	55.24	68.87	−0.521	nucleus
CtARF16	CtAH09T0132300	716	79.16385	6.84	58.87	65.21	−0.6	nucleus
CtARF17	CtAH10T0191300	557	62.06441	8.21	51.29	71.92	−0.447	nucleus
CtARF18	CtAH10T0242900	573	63.1186	5.13	57.16	71.61	−0.43	nucleus
CtARF19	CtAH11T0158500	548	61.86642	6.11	45.67	75.71	−0.451	nucleus
CtARF20	CtAH11T0163400	785	85.89191	6.06	54.01	69.87	−0.414	nucleus
CtARF21	CtAH11T0194100	664	72.94625	7.64	46.99	66.49	−0.462	nucleus
CtARF22	CtAH12T0040800	679	76.07008	5.75	57.36	76.91	−0.462	nucleus

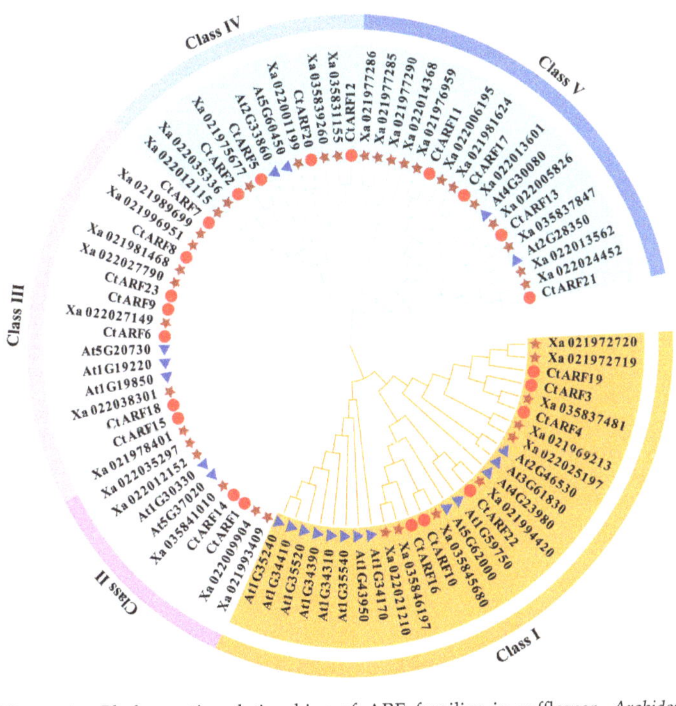

Figure 1. Phylogenetic relationships of ARF families in safflower, *Arabidopsis thaliana*, and *Helianthus annuus*. Colors represent different subfamilies, and shapes represent different species: circles for safflower, triangles for *Arabidopsis thaliana*, and stars for *Helianthus annuus*.

2.4. Gene Structure Analysis of ARF Proteins in Safflower

The gene structure analysis of *CtARF* revealed that the ARF family genes in safflower contained 2–15 exons (Figure S2). Specifically, subfamily I genes had 12–15 exons, subfamily II genes had 14 exons, subfamily III genes had 8–14 exons, subfamily IV genes had 10–13 exons, and subfamily V genes had 2–5 exons. These results indicated that genes within the same subfamily were identical or similar, and the gene structure of subfamily V was significantly different from the other four subfamilies.

To elucidate the function and mechanism of ARF proteins in safflower, we conducted secondary and tertiary structure predictions of safflower ARF proteins. The secondary structure analysis revealed that the content of random coils in safflower ARF proteins was the highest, exceeding 60% in all proteins (except for ARF19, which had 59.67%). In contrast, the content of α-helices was over 10%, and the content of extended strands ranged from 8.77% to 14.96%. The β-sheet content was the lowest, around 2%. These results indicated that random coils are the predominant secondary structure component of ARF proteins and may play a role in the formation of the tertiary structure (Table S9). We further modeled the tertiary structure of CtARF proteins, and the results showed successful modeling for all proteins (Figure S3). There were significant differences in the proportions of α-helices, extended strands, β-turns, and random coils among proteins from different subfamilies. The structures of CtARF4 and CtARF19 were highly similar, and ARF genes that clustered together in the phylogenetic tree encoded proteins with similar tertiary structures.

2.5. Analysis of Conserved Domains and Motifs in Safflower ARF Proteins

The analysis of conserved domains revealed that 13 safflower ARF proteins possessed B3, Auxin_resp, and AUX_IAA superfamily domains, including 6 members of subfamily I, 2 members of subfamily II, 4 members of subfamily III, and 1 member of subfamily IV. Additionally, 9 ARF proteins contained only B3 and Auxin_resp domains, including 2 members of subfamily III, 3 members of subfamily IV, and 4 members of subfamily V. Notably, CtARF18 in subfamily III lacked the B3 domain but contained Auxin_resp and AUX_IAA superfamily domains (Figure 2B). The conserved domains of safflower ARF proteins were predicted using the online software MEME (Version 5.5.7) (Figure 2C). The results showed that closely related members exhibited similar conserved motif characteristics. Subfamily I contained 8–10 conserved motifs, subfamily II contained 10 conserved motifs, subfamily III contained 6–10 conserved motifs, subfamily IV contained 7–10 conserved motifs, and subfamily V contained 8–10 conserved motifs. Specifically, CtARF2 had 7 conserved motifs, CtARF5/10/16/17 had 8 conserved motifs, CtARF3/11 had 9 conserved motifs, CtARF18 had 6 conserved motifs, and the other sequences all had 10 conserved motifs. All safflower ARF proteins possessed Motif6 and Motif9. These findings indicated that members of the same subfamily have identical or similar conserved structures, whereas CtARF18 was significantly different from the other sequences.

2.6. Analysis of Synteny and Interactions of Safflower ARF Proteins

Intragenomic synteny analysis revealed that there were evident duplication relationships of *CtARF* genes in the safflower genome. Among the 23 *CtARF* genes, 10 segmental duplication events were identified (Figure 3A), which were distributed across all chromosomes except chromosome 6. It is speculated that these genes may have similar functions. Synteny analysis between *Arabidopsis* and safflower (Figure 3C) revealed 14 pairs of homologous genes, indicating the conservation of ARF genes across different species.

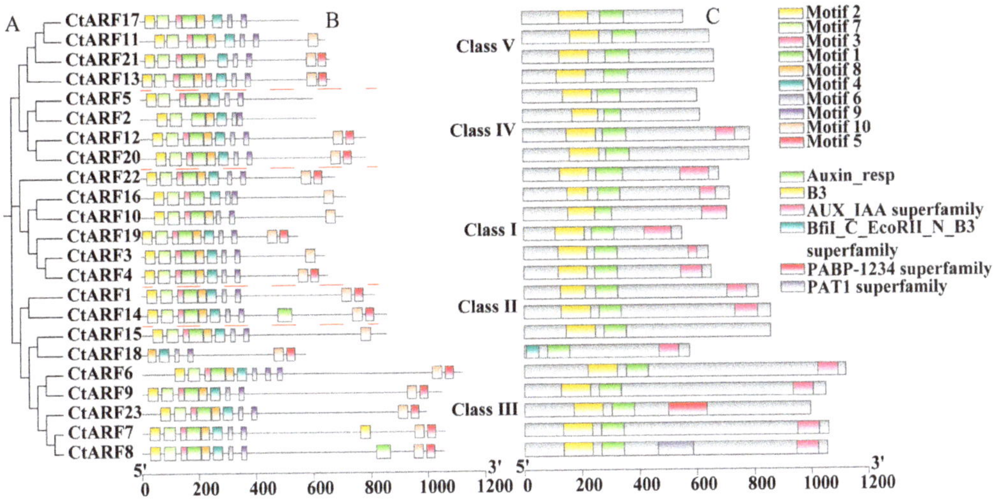

Figure 2. Analysis of conserved domains and conserved motifs in safflower ARF proteins. (**A**) Phylogenetic tree of the ARF gene family. (**B**) Conserved motifs of the ARF gene family. (**C**) Conserved domains of the ARF gene family.

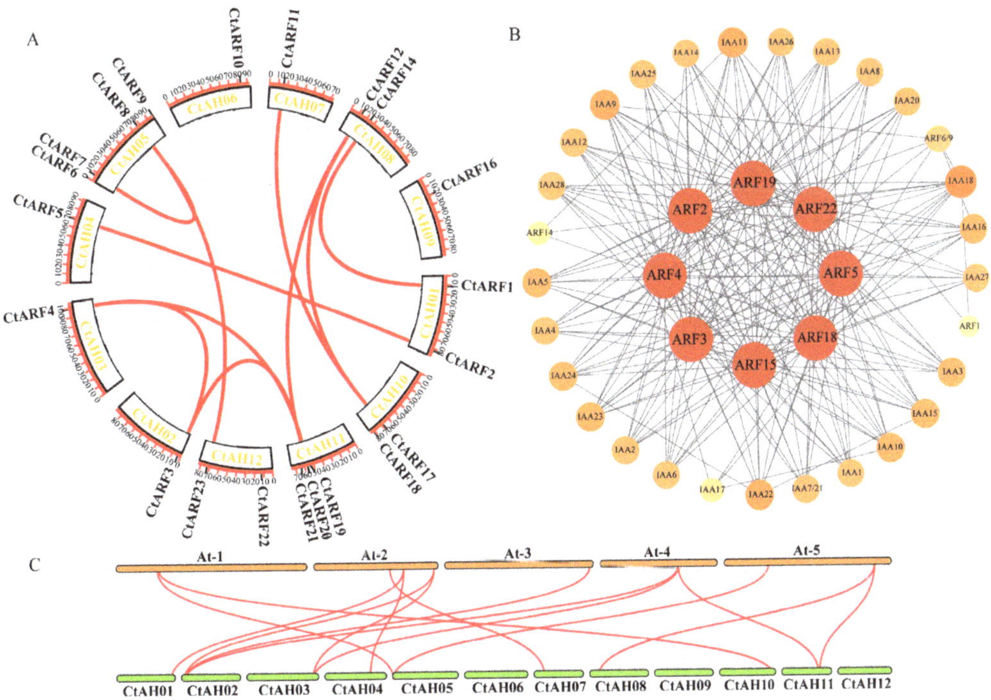

Figure 3. Intraspecific synteny analysis (**A**), interspecific synteny analysis (**C**), and protein–protein interaction analysis (**B**) of safflower ARF proteins.

To analyze the potential interactions among CtARF proteins, we compared the protein sequences of all ARF and IAA gene families in safflower using the STRING website to identify the highest-scoring homologous proteins, and the results were visualized using Cytoscape (Version 3.10.0) (Figure 3B). By aligning the ARF and IAA protein sequences with the STRING database, we identified a protein interaction network comprising 216 edges and 37 nodes. Among them, eight ARF family members (CtARF2/3/4/5/15/18/19/22) were located within the interaction network, indicating that they may be core members of the CtARF gene family regulating the growth and development of safflower. The remaining 29 members were located outside the interaction network, and some members were found to have no interactions. Notably, all core members were from the ARF family, suggesting that ARF family members have more diverse functions. Additionally, the GO enrichment results indicated that safflower ARF family members have a wide range of biological functions and primarily act as transcriptional activators to regulate gene expression (Figure S1).

2.7. Analysis of Cis-Acting Elements in the Promoter Regions of Safflower ARF Genes

The promoters (2000 bp upstream of the translation initiation codon) of the ARF family genes in safflower were analyzed (Figure S7). In addition to the core promoter elements (TATA-box) and upstream promoter elements (CAAT-box), 31 cis-acting elements were identified, which could be mainly classified into two categories: light-responsive elements and hormone-responsive elements (Table S5). Among the 23 ARF genes, auxin-responsive elements (TGA-element and AuxRR-core) were detected, indicating that the ARF proteins likely played important roles in regulating IAA-related events. Moreover, nine types of light-responsive elements were found in the promoters of these genes, with each gene containing one or more light-responsive element. This suggests that light has a significant impact on the growth and development of safflower. Additionally, a variety of hormone-responsive elements, such as those for abscisic acid, methyl jasmonate, and auxin, were identified in the promoters of the ARF family genes, indicating that these genes may be involved in hormone regulation to cope with environmental changes. Further analysis revealed that different ARF members had distinct cis-acting elements (Figure S4). For example, *CtARF6* contained a large number of ABRE (10) and G-box (11) elements, suggesting its involvement in ABA regulation and light response pathways. *CtARF9* had a high number of CGTCA-motif (8) elements, which may be related to the expression of methyl jasmonate. *CtARF10* contained a large number of ARE (8) and 3-AF1 binding site (8) elements, which may be related to hormone expression, environmental stress, and transcriptional regulation. *CtARF12* and *CtARF20* contained a large number of ARE (9/8) elements, while *CtARF13* had a high number of ABRE (7), G-box (8), and Box II (8) elements. The ATCT-motif element was abundant in *CtARF15* and *CtARF17*. These results indicated that different genes have distinct cis-acting elements, and genes within the same subfamily share similar cis-acting elements.

2.8. Expression Analysis of Safflower ARF Genes in Different Flower Colors

Flower color, as a significant trait of safflower, not only provides ornamental value but also attracts pollinating insects. Hydroxysafflor yellow A (HSYA), composed of one C-glucosylquinochalcone and responsible for the yellow pigments in the red color of safflower, has been widely used in the treatment of diseases, such as coronary heart disease, hypertension, and cerebral thrombosis. The expression of the 23 ARF genes in different flower colors of safflower (deep red, light red, yellow, and white) was analyzed (Figure 4D). The transcriptome data were derived from the safflower (*Carthamus tinctorius* L.) transcriptome database constructed by Wang et al., with sequencing materials obtained from safflower inflorescences (Table S6) [30]. Subsequently, the transcriptome data were analyzed using the "pheatmap" package in R to generate expression heatmaps. The results showed that all 23 ARF genes were expressed in different flower colors, with genes from the same subfamily exhibiting similar expression patterns. However, there were significant differences in the expression levels of ARF genes from different subfamilies across different flower colors. *CtARF7/8/23* had high expression in all flower colors, while *CtARF4/5/15/17/18* had relatively low expression in all flower colors. *CtARF19* had relatively higher expression in deep red flowers, *CtARF3* had higher expression in white flowers, *CtARF2/12* had higher expression in yellow flowers, and *CtARF21/22* had higher expression in light red flowers.

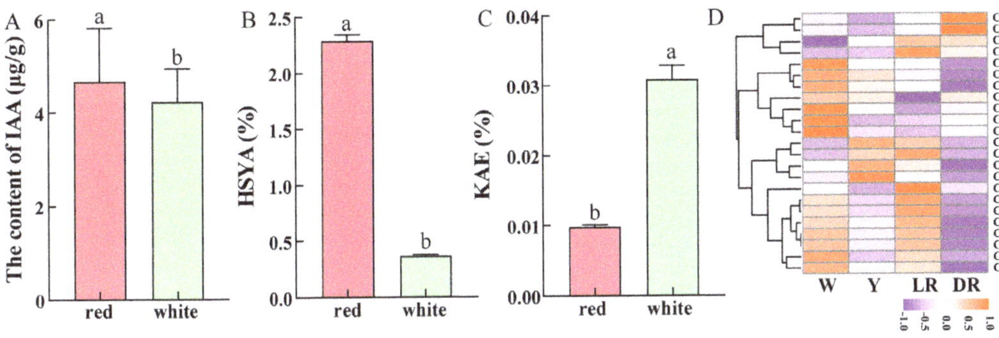

Figure 4. Content of endogenous IAA (**A**) and active components (HSYA (**B**) and KAE (**C**)) in safflower with different flower colors. HSYA represents hydroxysafflor yellow A, and KAE represents kaempferol. Different colors represent different safflower varieties, and different lowercase letters indicate significant differences. Expression analysis of the ARF gene family in safflower with different flower colors (**D**). W represents white safflower, Y represents yellow safflower, LR represents light red safflower, and DR represents dark red safflower.

2.9. Content of Endogenous IAA and Active Components in Safflower with Different Flower Colors

The endogenous IAA content in safflower with different flower colors was measured (Figure 4A). The results showed that there was a significant difference in endogenous IAA content between RS (4.66 μg/g) and WS (4.22 μg/g). KAE is a flavonoid compound, whose biosynthetic pathway partially overlaps with that of HSYA, and it exhibits pharmacological functions, such as anticancer activity and antioxidant capacity. The content of two antioxidants, including HSYA and KAE, was also determined in safflower with different flower colors (Figure 4B,C). There were significant differences in the content of active components among safflower with different flower colors. The content of HSYA was higher in RS (2.29%) and lower in WS (0.37%), while the content of KAE was higher in WS (0.03%) and lower in RS (0.01%), showing an opposite trend.

2.10. Effects of Different Exogenous Plant Hormone Treatments on Agronomic Traits of Safflower

Safflower plants were treated with exogenous IAA and an inhibitor of IAA transport (NPA, Figure 5). For both red safflower (RS) and white safflower (WS), the plant height and stem diameter increased to some extent after IAA application (Figure 5F,H). With the increasing IAA concentration, the plant height and stem diameter first increased and then decreased. The highest plant height was observed with IAA75 treatment, with RS reaching 40.6 cm and WS reaching 20.73 cm. The maximum stem diameter was achieved at IAA50 for RS (0.69 cm) and IAA25 for WS (0.64 cm). Compared with CK, plant growth status improved after IAA application. However, the high IAA concentration (IAA100) showed some inhibitory effects compared with moderate IAA treatments. After NPA application, both RS and WS exhibited some degree of deformity, with more pronounced deformity in RS (Figure 5B,D). Specific manifestations included stem bending and leaf curling. The degree of malformation in RS was more severe, with a more pronounced degree of bending. Interestingly, plant height significantly increased with the increasing NPA concentration, reaching the maximum with NPA70 treatment, with values of 48.23 cm for RS and 28.8 cm for WS. Unlike plant height, stem diameter initially increased and then decreased with the increasing NPA concentration, reaching the minimum with NPA70 treatment, with values of 0.48 cm for RS and 0.49 cm for WS. The deformity also became more severe at higher NPA concentrations.

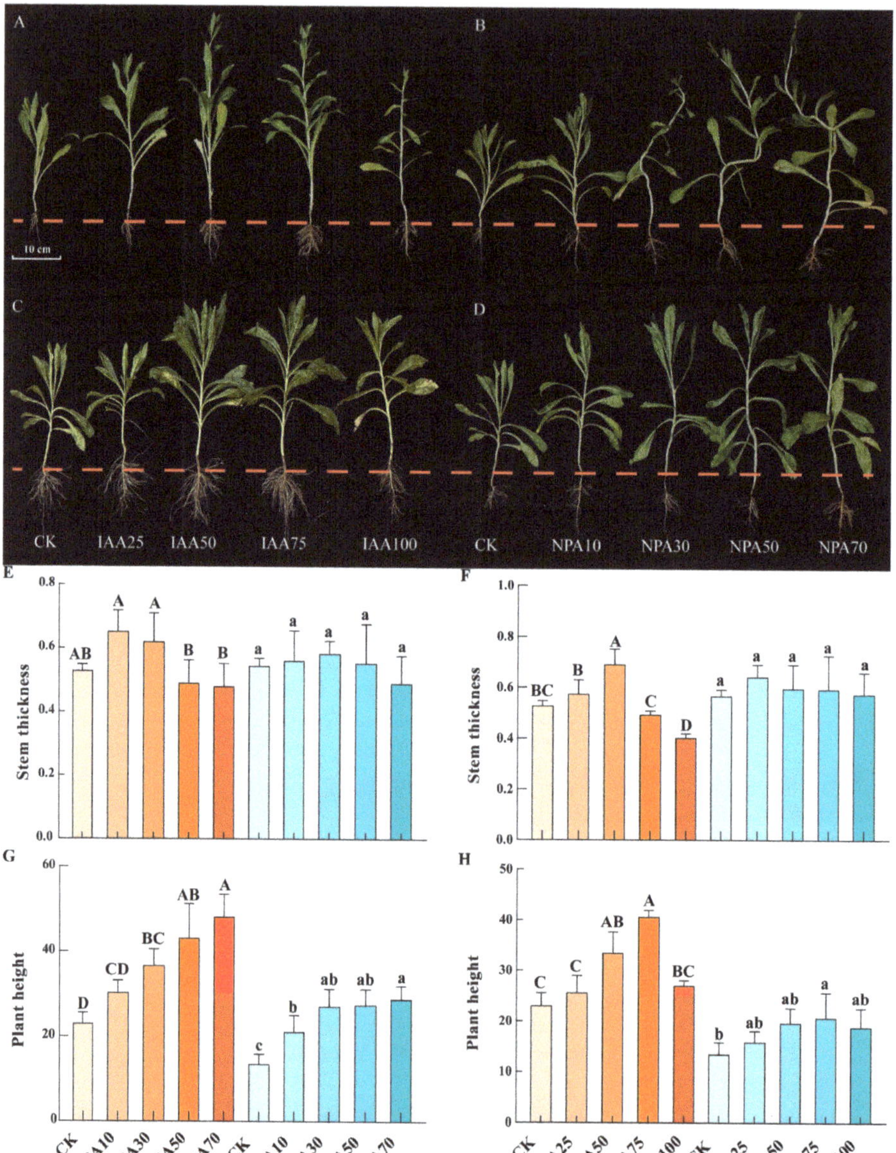

Figure 5. Growth status of safflower after exogenous hormone treatment (IAA and NPA). (**A**,**B**) Red safflower (RS) and (**C**,**D**) white safflower (WS). Measurement of agronomic traits of safflower after exogenous hormone treatment (IAA and NPA). (**E**,**G**) NPA treatment and (**F**,**H**) IAA treatment. Brown indicates red safflower (RS), and blue indicates white safflower (WS). The axes indicate the specific treatments, with CK representing the control group. Different letters represent significant differences ($p < 0.05$) among three plants, with three biological replicates each.

2.11. Expression Analysis of CtARF Genes and Determination of Endogenous IAA and Active Component Content After IAA and NPA Treatment

We selected treatments with more pronounced agronomic traits (IAA75 and NPA50) for subsequent experiments. By measuring the endogenous auxin content in safflower inflorescences after hormone treatment, we found that safflower with different flower colors exhibited different responses to exogenous hormones (Figure 6). Compared with CK, the endogenous IAA content in RS was highest when treated with NPA and then endogenous IAA, while WS exhibited the opposite trend. The auxin signaling pathway or auxin biosynthesis-related genes in RS may be more sensitive to exogenous IAA and NPA, thereby being activated and promoting the synthesis of endogenous auxin. In contrast, WS may be more sensitive to feedback regulation, and exogenous IAA treatment may inhibit the synthesis of endogenous hormones.

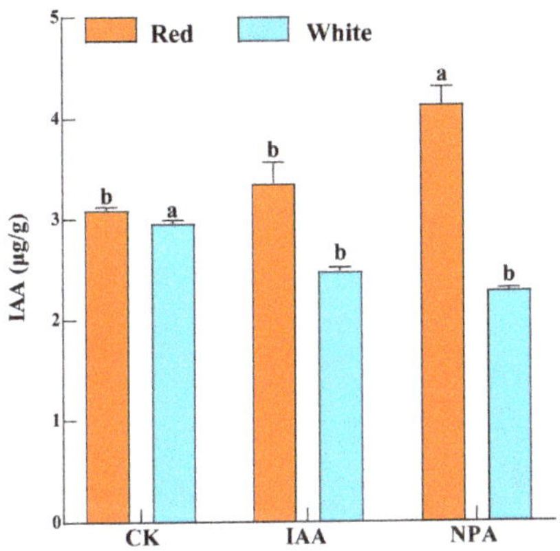

Figure 6. Determination of endogenous auxin content in safflower after exogenous hormone treatment (IAA and NPA). Brown represents red safflower (RS), and blue represents white safflower (WS). Different lowercase letters indicate significant differences ($p < 0.05$) among three plants, with three biological replicates each.

The expression of the ARF gene family in different organs of safflower was detected after spraying IAA and NPA (Figures 7 and S5). The results showed that different ARF genes exhibited different expression patterns after exogenous application of auxin and auxin inhibitor. In the roots of RS, the expression trends of *ARF2/3/5/19/22* were consistent after exogenous IAA application, with increased expression levels, while NPA treatment reduced the expression levels significantly below those of CK. In contrast, the expression trends of *ARF4/15* were consistent, showing a downward trend compared with CK, and the expression levels after NPA treatment were significantly lower than those of CK and IAA treatments. In the roots of WS, a completely different trend was observed. Except for *ARF2*, the expression levels of the other family members were NPA > IAA > CK, with extremely significant differences between each treatment group. Compared with CK, the expression levels of the *ARF2* gene increased significantly after IAA and NPA treatments, but there was no significant difference between the two treatments. In the stems of RS, the expression trends of ARF gene family members were consistent, with significant increases after treatment, the highest increase being observed after IAA treatment. However, there were no significant differences in the expression levels of *ARF3/4/19/22* between IAA and NPA treatments. In the stems of WS, the responses of ARF family members to IAA and NPA were more complex. NPA treatment significantly reduced the expression levels of *ARF2/3*, while IAA treatment significantly increased the expression levels of *ARF4/19/22*. The expression trends of *ARF5* and *ARF15* were consistent, both showing a downward trend and reaching the lowest point after NPA treatment. In the leaves of RS, the expression levels of *ARF2/3/5/15/19/22* genes all increased, reaching the maximum after NPA treatment. The *ARF4* member showed a different trend, with significant increases after treatment, reaching the highest level after IAA treatment. Interestingly, in the leaves of WS, the expression trends of ARF family members were generally consistent. Compared with CK, gene expression levels showed some differences after IAA treatment, but the differences were not significant. However, a highly significant increase was observed after NPA treatment. In the flowers of RS, the expression trends of ARF family members were mainly divided into two patterns. Compared with CK, the expression levels of *ARF2/3/4/22* were NPA > IAA > CK, while those of *ARF5/15/19* were IAA > NPA > CK, with a highly significant increase after IAA treatment. In the flowers of WS, ARF family members exhibited different trends. The expression levels of *ARF2/3/4/5/15* gradually decreased and reached the lowest point after NPA treatment. In contrast, *ARF22* showed an opposite trend, with the highest expression level after NPA treatment.

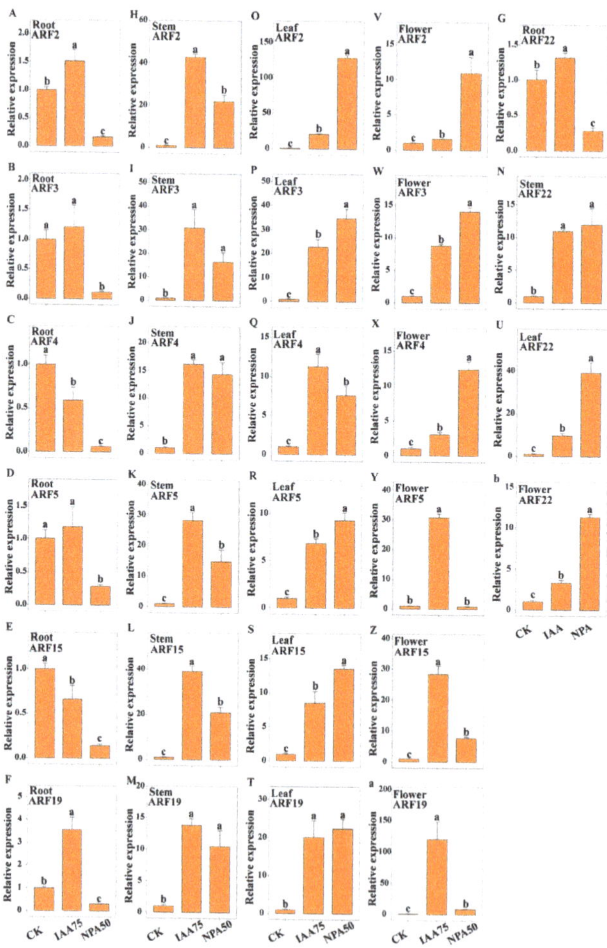

Figure 7. Determination of the relative expression levels of some ARF family members in different organs of safflower (RS) after exogenous hormone treatment (IAA and NPA). Root: **A/B/C/D/E/F/G**, Stem: **H/I/J/K/L/M/N**, Leaf: **O/P/Q/R/S/T/U**, Flower: **V/W/X/Y/Z/a/b**. Different lowercase letters indicate significant differences.

HSYA is a flavonoid compound unique to the inflorescence of safflower. In safflower, the content of HSYA was relatively high, while the content of KAE was low. In contrast, in white flowers, the content of KAE was higher, whereas the content of HSYA was lower. Safflower with different flower colors exhibited distinct responses to treatments with IAA or NPA (Figure 8). For RS, the content of HSYA decreased to varying degrees after IAA treatment (except for a 10.88% increase in IAA100 and a 3.06% increase in NPA30). The most significant decreases were observed in the plants treated with IAA75 and NPA10, with reductions of 44.22% and 26.87%, respectively. In contrast to the changes in HSYA content, the content of KAE increased to some extent after exogenous hormone treatment. Compared with CK, IAA75 (150%) and IAA100 (150%) showed significant increases, as did NPA30 (154.99%) and NPA50 (150%). After treatment with high concentrations (IAA100 and NPA70), the HSYA content in WS was extremely low, at around 0.01%. Except for the NPA50 treatment, which showed a significant increase in HSYA content by 41.03%, all other treatments reduced the HSYA content in WS. The content of KAE in WS responded differently to different exogenous hormones. IAA treatment significantly increased the KAE content in WS, with the highest increase of 85% observed after WIAA75 treatment. In contrast, NPA treatment resulted in a significant decrease in KAE content in WS, with no significant differences in KAE content among different NPA concentrations.

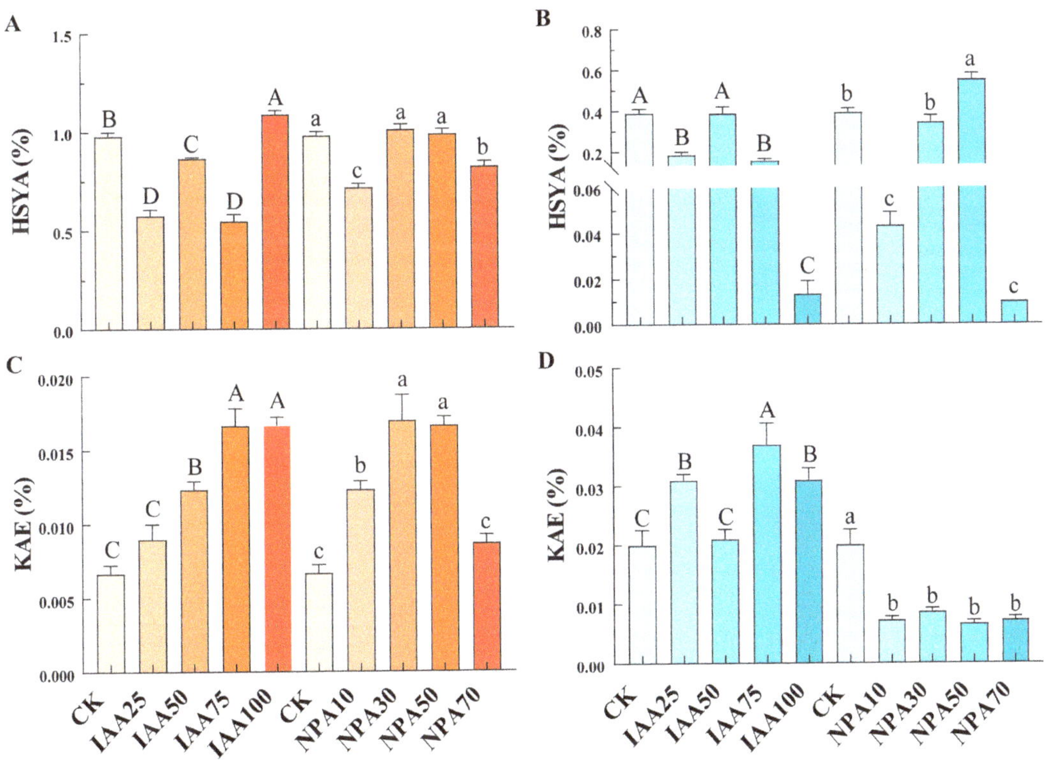

Figure 8. Determination of HSYA and KAE content in red safflower (RS; **A**,**C**) and white safflower (WS; **B**,**D**) after exogenous hormone treatment (IAA and NPA). Uppercase letters indicate IAA treatments, while lowercase letters indicate NPA treatments. Different letters represent significant differences ($p < 0.05$) among three plants, with three biological replicates each.

3. Discussion

3.1. Identification and Structural Analysis of the CtARF

The ARF transcription factors play a crucial role in the auxin signaling pathway, thereby influencing plant growth and development [12,31–33]. Different ARF proteins can act as transcriptional activators or repressors, and whether the ARF functions as a transcriptional activator or repressor depends on the sequence and corresponding structure of the middle region of the protein [34,35]. In addition, auxins and auxin inhibitors can be used as plant growth regulators. Exogenous application can influence the endogenous hormone levels within plants, modulate the content of secondary metabolites, and thereby alter plant quality [12]. We performed a bioinformatics analysis of the whole genome of safflower and identified a total of 23 *CtARF* genes (Table 1). The results showed that these genes were dispersed across the 12 chromosomes of safflower, with chromosome 3 being the longest but having fewer ARF genes [36–38]. Through phylogenetic analysis, it was found that safflower, sunflower, and *Arabidopsis* are closely related in terms of evolution (Figure 1). The *CtARF* gene family could be divided into five subfamilies. Genes within the same subfamily were highly similar, but the gene structure of subfamily V differed significantly from the other four subfamilies, which may be related to their distinct functions. In this study, *CtARF12* and *CtARF13* were found to be tandemly duplicated (Figure 3). Collinearity analysis revealed ten segmental duplication events for *CtARF* genes, all of which were formed through random duplication events. Intergroup collinearity analysis identified 14 pairs of homologous genes, indicating the conservation of ARF genes across different species and their similar functions. From an evolutionary perspective, gene duplication leading to the expansion of gene families often results in genes with identical or similar structures and functions. However, in some cases, gene duplication events can also lead to the acquisition of new functions by homologous genes [39], thereby enhancing the adaptability of plants.

3.2. Structural Analysis of CtARF Protein

By predicting the secondary and tertiary structures of ARF proteins in safflower, it was found that the proportions of α-helices, extended strands, β-turns, and random coils in proteins of different subfamilies varied significantly (Figure S3 and Table S9). Among them, CtARF18 lacked the B3 domain compared with other sequences, while the ARF proteins in sunflower did not have this deficiency. The B3 domain is a component of the DNA-binding domain (DBD) at the N-terminus of ARF proteins. In *Arabidopsis thaliana* ARF1 and ARF5, it also includes a dimerization domain (DD), which is located on both sides of the B3 domain, as well as an auxiliary Tudor-like domain. The DD can mediate the homodimerization of AtARF1 and AtARF5, which is essential for the cooperative binding of AtARF5 to target DNA [40]. The absence of the B3 domain is also observed in other species, such as *Eucalyptus grandis* [41] and *Physcomitrella patens* [42]. Therefore, different ARF members have diverged during the evolutionary process and may also exhibit functional differences.

3.3. Cis-Acting Element and Interaction Network Analysis of CtARFs

By analyzing the 2000 bp promoter regions upstream of the ARF family genes, we identified not only the core promoter TATA-box and the upstream promoter element CAAT-box but also light-responsive and hormone-responsive cis-acting elements (Figures S4 and S7). Each *CtARF* gene contained light-responsive cis-acting elements, suggesting that *CtARF* genes may play important roles in light-signaling regulation. Additionally, auxin-responsive elements (TGA-element and AuxRR-core) were found upstream of *CtARF3*, *CtARF4*, *CtARF14*, *CtARF15*, *CtARF16*, *CtARF18*, *CtARF20*, *CtARF21*, and *CtARF23*, indicating that these genes may be involved in auxin signaling. The types and numbers of cis-acting elements in the upstream regions of different genes varied, which may confer specificity and diversity to the functions of these genes.

In the phylogenetic tree, *AtARF7* and *AtARF19*, which are located on the same branch, exhibit identical functions and are involved in auxin signal transduction in roots [43,44]. The genes *CtARF1* and *CtARF14* are located on the same branch in the phylogenetic tree and may also have similar functions. ARFs, as key auxin regulators, can interact with other biological factors to jointly regulate plant growth and development. Research has shown that *Arabidopsis* has a total of 29 Aux/IAA genes and 23 ARF genes [45]. *AtARF4* can regulate branch regeneration through synergistic action with *AtARF5* and *AtIAA12* [46]. *IAA14* can interact with *ARF7/ARF19* to affect the rate of lateral root development [47]. The protein sequences of ARF and IAA from safflower were compared with the STRING database, revealing a protein–protein interaction network with 37 nodes and 216 edges (Figure 3). Eight members of the ARF family (*CtARF2*, *CtARF3*, *CtARF4*, *CtARF5*, *CtARF15*, *CtARF18*, *CtARF19*, and *CtARF22*) were identified within this interaction network, suggesting that they may be core members of the *CtARF* gene family involved in the regulation of safflower growth and development. Notably, all core members belonged to the ARF family, indicating that ARF family members are involved in a broader range of functions, which is consistent with the GO enrichment results of safflower. Transcriptome data analysis revealed significant differences in the expression of ARF genes among safflower with different flower colors. Additionally, it has been reported that *NtARF8* in tobacco can synergistically regulate vegetative growth and seed production with *TTG2*, and can control flower color by regulating anthocyanin biosynthesis genes [48]. Therefore, we speculate that the ARF family genes in safflower may also have similar functions.

3.4. The Response of Endogenous IAA and Active Ingredient Content in Safflower with Different Flower Colors to Exogenous Hormone Treatment

Flower color is one of the important traits of safflower. The expression patterns and active ingredient content of safflower with different flower colors showed significant differences (Figure 4). Studies have shown that the differences in auxin biosynthesis and transport pathways as well as ARF expression in safflower with different flower colors may affect flower color formation and maintenance [49,50]. It has also been found that *CtUGT9* may play an important role in the color transition of safflower and the biosynthesis of flavonoid glycosides [51]. In addition, our experiments found that the endogenous IAA, HSYA, and KAE content in safflower with different flower colors showed significant differences. The endogenous IAA and HSYA contents in RS were significantly higher than those in WS, while the KAE content was the opposite, which may be related to the growth environment and genetic background of the plants [50]. However, there is a lack of research evidence on whether auxin can regulate the content of flavonoids in safflower with different flower colors.

Plant growth regulators can influence plant growth by modulating the endogenous hormone levels within plants [52]. We treated safflower with different flower colors using various concentrations of plant growth regulators (IAA and NPA). The results showed that treatment with IAA75 and NPA50 significantly increased the IAA content in RS, while decreasing the IAA content in WS, with the effects of NPA treatment being more pronounced (Figure 6). This may be related to the influence of exogenous auxin on the processes of IAA biosynthesis and transport, as well as auxin signaling pathways [53]. Further research revealed that the application of exogenous plant hormones significantly increased plant height and had a pronounced effect on stem thickness (Figure 5). Auxin (IAA) promoted plant cell expansion and elongation, thereby increasing plant height and stem thickness. In contrast, NPA primarily interfered with the polar transport of auxin, thereby affecting plant growth. This effect is closely related to the concentration of exogenous application [54,55]. Furthermore, it has been found that high concentrations of NPA can inhibit the polar transport of auxin, leading to abnormal development of the hypocotyls in safflower [55]. The stigmas of safflower contain a variety of flavonoids, among which HSYA (hydroxysafflor yellow A) is the major bioactive compound. Stigmas with different flower colors have different contents of flavonoids [56]. Relatively speaking, RS contains more HSYA, while WS contains more KAE (Figure 8). It has been found that treatment with IAA75 significantly reduced the content of HSYA and increased the content of KAE. Treatment with NPA50 significantly increased the content of HSYA (in WS) and KAE (in RS), while decreasing the content of KAE in WS. This indicated that the content of active ingredients in safflower may be regulated by endogenous hormone levels. Furthermore, the results of real-time quantitative PCR showed that in RS, the expression levels of *ARF2/3/4/22* were in the order of NPA > IAA > CK (control), while the expression levels of *ARF15/19* were in the order of IAA > NPA > CK, with significant differences observed (Figure 7 and S5). However, in WS, the members of the ARF family showed different trends. The expression levels of *ARF2/3/4/5/15* gradually decreased, in the order of CK > IAA > NPA. In contrast, *ARF22* showed an opposite trend, with the highest expression level under NPA treatment. This suggested that *ARF2/3/4/5/15/19/22* respond to exogenous hormone treatment and are core members regulating the content of endogenous IAA and active ingredients.

4. Materials and Methods

4.1. Materials

In this study, two safflower cultivars, "Yumin spineless (red, RS)" and "Xin Honghua No. 7 (white, WS)", were used as experimental materials. Safflower seeds were directly sown in pots (substrate composed of loam, vermiculite, and perlite in a volume ratio of 3:1:1), with 10 plants per pot and 3 pots per treatments, and placed outdoors for cultivation. When the safflower plants reached the jointing stage (with the stem elongating to approximately 2 cm), plants with inconsistent growth were removed (one safflower plant was retained per pot), and the treatments were initiated. Additionally, safflower plants grown in the experimental field were treated when they reached the blooming stage (with approximately 60% of the flowers open). The plants were treated with IAA and NPA (Table S2), with applications occurring every five days a total of three times, ensuring even spraying. Fifteen days after treatment, root, stem, leaf, and flower samples were collected from the safflower plants, flash-frozen in liquid nitrogen, and stored at −80 °C for subsequent use.

All data obtained from the experiments were organized using Excel software. Analysis of variance (ANOVA) and graphing were performed using R (Version 4.4.1) and GraphPad Prism 9.5 software. The figures were refined and enhanced using Adobe Illustrator (Version 24.0.1) software.

4.2. Identification and Physicochemical Properties and Structural Analysis of Safflower ARF Transcription Factor Family Members

Our laboratory obtained the complete genome and annotation information of safflower. The protein sequences of all ARF members from *Arabidopsis thaliana* were downloaded from the TAIR (https://www.arabidopsis.org/, accessed on 18 October 2024) [57], and those from *Helianthus annuus* were obtained from the NCBI database (https://www.ncbi.nlm.nih.gov/, accessed on 27 October 2024) [58]. These sequences were aligned (e-value 1×10^{-5}) against the safflower genome in our laboratory to identify homologous sequences, which were retained as candidate ARF family protein sequences. The Pfam (http://Pfam.sanger.ac.uk/, accessed on 20 November 2024) [59] and SMART (https://smart.embl.de/smart/

set_mode.cgi?GENOMIC=1, accessed on 20 November 2024) [60] online analysis tools were used to annotate the domains of the candidate sequences. Sequences containing the PF06507 domain were ultimately identified as the candidate ARF sequences in safflower.

The basic physicochemical properties of ARF proteins were analyzed using the Prot Param online tool (https://web.expasy.org/protparam/, accessed on 28 November 2024) [61]. Subcellular localization was predicted using the Plant-mPLoc website (http://www.csbio.sjtu.edu.cn/bioinf/plant-multi/, accessed on 28 November 2024) [62]. The secondary structure of the proteins was predicted via the SPOMA website (https://npsa.lyon.inserm.fr/cgi-bin/npsa_automat.pl?page=/NPSA/npsa_sopma.html, accessed on 28 November 2024) [63]. The tertiary structure of the proteins was predicted using the SWISS-MODEL website (https://swissmodel.expasy.org/, accessed on 28 November 2024) [64].

4.3. Multiple Sequence Alignment and Phylogenetic Analysis

The ARF protein sequences obtained from safflower were aligned with those of *Arabidopsis thaliana* and *Helianthus annuus* using the ClustalW tool in MEGA7.0 [65] software. A phylogenetic tree was constructed using the neighbor-joining (NJ) method with a bootstrap value of 1000 (Table S8). Finally, the tree was visually enhanced using the iTOL online tool (https://itol.embl.de/, accessed on 20 November 2024) [66].

4.4. Chromosomal Distribution, Gene Structure, and Collinearity Analysis

The chromosomal distribution map was constructed based on the chromosomal location information of the genes using TBtools (Version 2.210) [67]. The gene structure of the ARF family was visualized using the gff3 file in TBtools [67]. The gene duplication and collinearity analysis was completed using the Advanced Circos and Dual Synteny Plot MCscanX modules in TBtools [67].

4.5. Conserved Domain and Conserved Motif Analysis

The conserved domains of ARF proteins were analyzed using the online tool CD-search from NCBI (https://www.ncbi.nlm.nih.gov/Structure/cdd/wrps b.cgi, accessed on 23 November 2024) [68]. The conserved motifs of ARF proteins were analyzed using the online database MEME (https://meme-suite.org/meme/tools/meme, accessed on 23 November 2024) [69]. Finally, the results were visualized using TBtools [67].

4.6. Cis-Acting Element Analysis and Protein–Protein Interaction Prediction

The promoter sequences (2000 bp upstream of the start codon) of safflower ARF genes were extracted using TBtools (Table S7) [67], and then input into the PlantCare website (https://bioinformatics.psb.ugent.be/webtools/plantcare/html/, accessed on 24 November 2024) [70] to predict cis-acting elements in the promoter regions. The distribution map of cis-acting elements and the heatmap were generated using TBtools [67]. The interactions between safflower ARF and IAA families were analyzed using the STRING website (https://cn.string-db.org/, accessed on 15 December 2024) [71]. The protein sequence analysis confidence was set to 0.4, and the output results were visualized using Cytoscape [72]. GO enrichment analysis was performed using TBtools [67].

4.7. Expression Pattern Analysis of ARF Genes in Different Flower Colors

Based on the safflower transcriptome database constructed by Wang et al. [30], the expression data of safflower with different flower colors (white—W, yellow—Y, light red—LR, and deep red—DR) were extracted. The expression patterns of the 23 ARF genes in different tissues were analyzed using TBtools [67] to generate a heatmap.

4.8. Determination of Endogenous IAA and Active Component Content

The endogenous IAA content in safflower was determined using a plant IAA content determination kit (Plant Indole-3-Acetic Acid (IAA) ELISA kit, Jingmei, Yancheng, China). The specific preparation method was referred to the product manual [73].

The content of active components (HSYA and KAE) in safflower flowers was determined by high-performance liquid chromatography (HPLC) [74]. The chromatographic column used was an Agilent EC-C-18 column (150 mm × 4.6 mm, 4 μm). The detection wavelengths were set at 403 nm and 367 nm. The mobile phase consisted of A: water, B: methanol, C: acetonitrile, and D: 0.4% phosphoric acid solution. The elution conditions were as follows: 0–60 min, 5% B and 95% D; 60–65 min, 95% B and 5% D; 65–70 min, 5% B and 95% D. The column temperature was maintained at 30 °C, with an injection volume of 10 μL. The standard concentrations were 0.424 mg/mL and 0.432 mg/mL.

4.9. RNA Extraction and Quantitative Real-Time Reverse Transcription PCR (qRT-PCR) Analysis

To detect the response of safflower ARF genes to exogenous hormones (IAA and NPA), roots, stems, leaves, and flowers of treated safflower were collected. The samples were ground in liquid nitrogen and RNA was extracted using the FastPure Plant Total RNA Isolation Kit (Vazyme, NanJing, China). The integrity of the extracted RNA was assessed by agarose gel electrophoresis, and its concentration was measured using a NanoDrop spectrophotometer (Thermo Fisher, Waltham, MA, USA). Subsequently, 0.5 μg of RNA was reverse transcribed into cDNA using the HiScript III RT SuperMix for qPCR (Vazyme). The diluted cDNA was used for quantitative real-time PCR to measure the expression levels of ARF genes. The qPCR was performed using the ChamQ SYBR Color qPCR Master Mix (Vazyme), with *Ct60s* as the reference gene [75]. The primers used are listed in Table S1. Each sample was analyzed in triplicate, and the data were analyzed using the $2^{-\Delta\Delta CT}$ method [76].

5. Conclusions

This study represents the first genome-wide analysis of ARFs in safflower. A total of 23 *CtARFs* were identified across the 12 chromosomes of safflower. These genes were analyzed in terms of conserved motifs, conserved domains, gene structure, secondary and tertiary protein structures, and phylogenetic evolution. These findings revealed the high conservation and evolutionary relationships among *CtARF* genes. Collinearity analysis revealed that tandem duplication is the primary driving force for the expansion of the *CtARF* gene family. The identification of cis-acting elements related to hormone response in the promoter regions of *CtARF* genes facilitated a better understanding of hormone response pathways in safflower. Through protein interaction network analysis, we identified the core members of the *CtARF* family (*CtARF2/3/4/5/15/18/19/22*) and studied their expression patterns in different organs of safflower under hormone treatments. Additionally, the study found that different hormone treatments led to significant changes in the endogenous auxin and bioactive compound contents of safflower with different flower colors, which may be related to the expression of ARF family members, especially the core members. The findings of this study will provide a reference for further research on the involvement of safflower ARF genes in plant hormone pathways and secondary metabolite synthesis pathways and will be of certain value for the breeding of safflower cultivars.

Supplementary Materials: The following supporting information can be downloaded at: https://www.mdpi.com/article/10.3390/ijms26083773/s1.

Author Contributions: Conceptualization, A.I. and A.C.; methodology, M.L.; investigation, S.Q., X.W., M.M., J.Z. and M.H.; resources, A.C.; data curation, S.Q.; writing—original draft preparation, S.Q.,

J.W., K.S., Y.Z. and M.Z.; writing—review and editing, A.C.; supervision, A.C.; project administration, M.L.; funding acquisition, A.C. All authors have read and agreed to the published version of the manuscript.

Funding: This project was financially supported by the "Science and Technology Project (2024D05l)" of the Bingtuan, the "Science and Technology Project (YZZX202301)" of the Shihezi University, the "Guiding Technology Program Projects (2022ZD049)" of the Bingtuan, and the "Xinjiang safflower industry development fund (2024D08C)".

Institutional Review Board Statement: Not applicable.

Informed Consent Statement: Not applicable.

Data Availability Statement: The original contributions presented in this study are included in the article/Supplementary Materials. Further inquiries can be directed to the corresponding authors.

Conflicts of Interest: The authors declare no conflicts of interest.

References

1. Orgah, J.O.; He, S.; Wang, Y.; Jiang, M.; Wang, Y.; Orgah, E.A.; Duan, Y.; Zhao, B.; Zhang, B.; Han, J. Pharmacological potential of the combination of Salvia miltiorrhiza (Danshen) and *Carthamus tinctorius* (Honghua) for diabetes mellitus and its cardiovascular complications. *Pharmacol. Res.* **2020**, *153*, 104654. [CrossRef] [PubMed]
2. Baerle, A.; Savcenco, A.; Tatarov, P.; Fetea, F.; Ivanova, R.; Radu, O. Stability limits of a red Carthamin–cellulose complex as a potential food colourant. *Food Funct.* **2021**, *12*, 8037–8043. [CrossRef]
3. Zanetti, F.; Angelini, L.G.; Berzuini, S.; Foschi, L.; Clemente, C.; Ferioli, F.; Vecchi, A.; Rossi, A.; Monti, A.; Tavarini, S. Safflower (*Carthamus tinctorius* L.) a winter multipurpose oilseed crop for the Mediterranean region: Lesson learnt from on-farm trials. *Ind. Crops Prod.* **2022**, *184*, 115042. [CrossRef]
4. Khalid, N.; Khan, R.S.; Hussain, M.I.; Farooq, M.; Ahmad, A.; Ahmed, I. A comprehensive characterisation of safflower oil for its potential applications as a bioactive food ingredient—A review. *Trends Food Sci. Technol.* **2017**, *66*, 176–186. [CrossRef]
5. Chen, G.; Li, C.; Zhang, L.; Yang, J.; Meng, H.; Wan, H.; He, Y. Hydroxysafflor yellow A and anhydrosafflor yellow B alleviate ferroptosis and parthanatos in PC12 cells injured by OGD/R. *Free Radic. Biol. Med.* **2022**, *179*, 1–10. [CrossRef] [PubMed]
6. Feng, X.; Du, M.; Li, S.; Zhang, Y.; Ding, J.; Wang, J.; Wang, Y.; Liu, P. Hydroxysafflor yellow A regulates lymphangiogenesis and inflammation via the inhibition of PI3K on regulating AKT/mTOR and NF-κB pathway in macrophages to reduce atherosclerosis in ApoE-/-mice. *Phytomedicine* **2023**, *112*, 154684. [CrossRef] [PubMed]
7. Wang, W.; Liu, M.; Fu, X.; Qi, M.; Zhu, F.; Fan, F.; Wang, Y.; Zhang, K.; Chu, S. Hydroxysafflor yellow A ameliorates alcohol-induced liver injury through PI3K/Akt and STAT3/NF-κB signaling pathways. *Phytomedicine* **2024**, *132*, 155814. [CrossRef]
8. Yu, H.-C.; Bai, Q.-R.; Guo, J.-J.; Chen, M.-Y.; Wang, L.; Tong, F.-C.; Zhang, S.-L.; Wu, J. Elucidating hydroxysafflor yellow A's multi-target mechanisms against alcoholic liver disease through integrative pharmacology. *Phytomedicine* **2024**, *134*, 155956. [CrossRef]
9. Bao, D.; Chang, S.; Li, X.; Qi, Y. Advances in the study of auxin early response genes: Aux/IAA, GH3, and SAUR. *Crop J.* **2024**, *12*, 964–978. [CrossRef]
10. Wang, J.; Hu, D.; Shi, X.; Luo, J.; Ren, G.; Dai, Z.; Qi, S.; Du, D. Different Responses of Invasive Weed Alternanthera philoxeroide s and Oryza sativa to Plant Growth Regulators. *Life* **2022**, *12*, 1069. [CrossRef]
11. Kakei, Y.; Yamazaki, C.; Suzuki, M.; Nakamura, A.; Sato, A.; Ishida, Y.; Kikuchi, R.; Higashi, S.; Kokudo, Y.; Ishii, T. Small-molecule auxin inhibitors that target YUCCA are powerful tools for studying auxin function. *Plant J.* **2015**, *84*, 827–837. [CrossRef] [PubMed]
12. Guo, F.; Huang, Y.; Qi, P.; Lian, G.; Hu, X.; Han, N.; Wang, J.; Zhu, M.; Qian, Q.; Bian, H. Functional analysis of auxin receptor OsTIR1/OsAFB family members in rice grain yield, tillering, plant height, root system, germination, and auxinic herbicide resistance. *New Phytol.* **2021**, *229*, 2676–2692. [CrossRef]
13. Guilfoyle, T.J.; Hagen, G. Auxin response factors. *Curr. Opin. Plant Biol.* **2007**, *10*, 453–460. [CrossRef] [PubMed]
14. Lavy, M.; Prigge, M.J.; Tao, S.; Shain, S.; Kuo, A.; Kirchsteiger, K.; Estelle, M. Constitutive auxin response in Physcomitrella reveals complex interactions between Aux/IAA and ARF proteins. *Elife* **2016**, *5*, e13325. [CrossRef]
15. Szemenyei, H.; Hannon, M.; Long, J.A. TOPLESS mediates auxin-dependent transcriptional repression during Arabidopsis embryogenesis. *Science* **2008**, *319*, 1384–1386. [CrossRef]
16. Guilfoyle, T.J.; Hagen, G. Getting a grasp on domain III/IV responsible for Auxin Response Factor–IAA protein interactions. *Plant Sci. Amst. Neth.* **2012**, *190*, 82–88. [CrossRef] [PubMed]
17. Finet, C.; Berne-Dedieu, A.; Scutt, C.P.; Marlétaz, F. Evolution of the ARF gene family in land plants: Old domains, new tricks. *Mol. Biol. Evol.* **2013**, *30*, 45–56. [CrossRef]

18. Roosjen, M.; Paque, S.; Weijers, D. Auxin response factors: Output control in auxin biology. *J. Exp. Bot.* **2018**, *69*, 179–188. [CrossRef]
19. Li, Y.; Han, S.; Qi, Y. Advances in structure and function of auxin response factor in plants. *J. Integr. Plant Biol.* **2023**, *65*, 617–632. [CrossRef]
20. Hu, W.; Zuo, J.; Hou, X.; Yan, Y.; Wei, Y.; Liu, J.; Li, M.; Xu, B.; Jin, Z. The auxin response factor gene family in banana: Genome-wide identification and expression analyses during development, ripening, and abiotic stress. *Front. Plant Sci.* **2015**, *6*, 742. [CrossRef]
21. Bouzroud, S.; Gasparini, K.; Hu, G.; Barbosa, M.A.M.; Rosa, B.L.; Fahr, M.; Bendaou, N.; Bouzayen, M.; Zsögön, A.; Smouni, A. Down regulation and loss of auxin response factor 4 function using CRISPR/Cas9 alters plant growth, stomatal function and improves tomato tolerance to salinity and osmotic stress. *Genes* **2020**, *11*, 272. [CrossRef] [PubMed]
22. Liu, N.; Dong, L.; Deng, X.; Liu, D.; Liu, Y.; Li, M.; Hu, Y.; Yan, Y. Genome-wide identification, molecular evolution, and expression analysis of auxin response factor (ARF) gene family in *Brachypodium distachyon* L. *BMC Plant Biol.* **2018**, *18*, 336. [CrossRef] [PubMed]
23. Ellis, C.M.; Nagpal, P.; Young, J.C.; Hagen, G.; Guilfoyle, T.J.; Reed, J.W. Auxin Response Factor1 and Auxin Response Factor2 regulate senescence and floral organ abscission in *Arabidopsis thaliana*. *Development* **2005**, *132*, 4563–4574. [CrossRef]
24. Nagpal, P.; Ellis, C.M.; Weber, H.; Ploense, S.E.; Barkawi, L.S.; Guilfoyle, T.J.; Hagen, G.; Alonso, J.M.; Cohen, J.D.; Farmer, E.E. Auxin response factors ARF6 and ARF8 promote jasmonic acid production and flower maturation. *Development* **2005**, *132*, 4107–4118. [CrossRef]
25. Wang, B.; Xue, J.-S.; Yu, Y.-H.; Liu, S.-Q.; Zhang, J.-X.; Yao, X.-Z.; Liu, Z.-X.; Xu, X.-F.; Yang, Z.-N. Fine regulation of ARF17 for anther development and pollen formation. *BMC Plant Biol.* **2017**, *17*, 243. [CrossRef]
26. Wang, C.-K.; Han, P.-L.; Zhao, Y.-W.; Yu, J.-Q.; You, C.-X.; Hu, D.-G.; Hao, Y.-J. Genome-wide analysis of auxin response factor (ARF) genes and functional identification of MdARF2 reveals the involvement in the regulation of anthocyanin accumulation in apple. *N. Z. J. Crop Hortic. Sci.* **2021**, *49*, 78–91. [CrossRef]
27. Wan, W.; Jia, F.; Liu, Z.; Sun, W.; Zhang, X.; Su, J.; Guan, Z.; Chen, F.; Zhang, F.; Fang, W. Quantitative evaluation and genome-wide association studies of chrysanthemum flower color. *Sci. Hortic.* **2024**, *338*, 113561. [CrossRef]
28. Pahlavani, M.; Mirlohi, A.; Saeidi, G. Inheritance of flower color and spininess in safflower (*Carthamus tinctorius* L.). *J. Hered.* **2004**, *95*, 265–267. [CrossRef]
29. Durbin, M.L.; Lundy, K.E.; Morrell, P.L.; Torres-Martinez, C.L.; Clegg, M.T. Genes that determine flower color: The role of regulatory changes in the evolution of phenotypic adaptations. *Mol. Phylogenet. Evol.* **2003**, *29*, 507–518. [CrossRef]
30. Wang, R.; Ren, C.; Dong, S.; Chen, C.; Xian, B.; Wu, Q.; Wang, J.; Pei, J.; Chen, J. Integrated metabolomics and transcriptome analysis of flavonoid biosynthesis in safflower (*Carthamus tinctorius* L.) with different colors. *Front. Plant Sci.* **2021**, *12*, 712038. [CrossRef]
31. Feng, L.; Li, G.; He, Z.; Han, W.; Sun, J.; Huang, F.; Di, J.; Chen, Y. The ARF, GH3, and Aux/IAA gene families in castor bean (*Ricinus communis* L.): Genome-wide identification and expression profiles in high-stalk and dwarf strains. *Ind. Crops Prod.* **2019**, *141*, 111804. [CrossRef]
32. Wilmoth, J.C.; Wang, S.; Tiwari, S.B.; Joshi, A.D.; Hagen, G.; Guilfoyle, T.J.; Alonso, J.M.; Ecker, J.R.; Reed, J.W. NPH4/ARF7 and ARF19 promote leaf expansion and auxin-induced lateral root formation. *Plant J.* **2005**, *43*, 118–130. [CrossRef] [PubMed]
33. Zhao, S.; Zhang, M.-L.; Ma, T.-L.; Wang, Y. Phosphorylation of ARF2 relieves its repression of transcription of the K+ transporter gene HAK5 in response to low potassium stress. *Plant Cell* **2016**, *28*, 3005–3019. [CrossRef] [PubMed]
34. Ulmasov, T.; Hagen, G.; Guilfoyle, T.J. ARF1, a transcription factor that binds to auxin response elements. *Science* **1997**, *276*, 1865–1868. [CrossRef] [PubMed]
35. Ulmasov, T.; Hagen, G.; Guilfoyle, T.J. Dimerization and DNA binding of auxin response factors. *Plant J.* **1999**, *19*, 309–319. [CrossRef]
36. Zhang, Y.; Wu, W.; Shen, H.; Yang, L. Genome-wide identification and expression analysis of ARF gene family in embryonic development of Korean pine (*Pinus koraiensis*). *BMC Plant Biol.* **2024**, *24*, 267. [CrossRef]
37. Liu, Y.; Wang, R.; Yu, J.; Huang, S.; Zhang, Y.; Wei, H.; Wei, Z. Genome-wide identification and characterization of auxin response factor (ARF) gene family involved in wood formation and response to exogenous hormone treatment in Populus trichocarpa. *Int. J. Mol. Sci.* **2023**, *24*, 740. [CrossRef]
38. Hu, Z.; Zong, D.; Zhang, Q.; Zhang, X.; Lu, Y.; He, C. PyuARF16/33 are involved in the regulation of lignin synthesis and rapid growth in Populus yunnanensis. *Genes* **2023**, *14*, 278. [CrossRef]
39. Jiao, Y. Double the genome, double the fun: Genome duplications in angiosperms. *Mol. Plant* **2018**, *11*, 357–358. [CrossRef]
40. Boer, D.R.; Freire-Rios, A.; Van Den Berg, W.A.; Saaki, T.; Manfield, I.W.; Kepinski, S.; López-Vidrieo, I.; Franco-Zorrilla, J.M.; de Vries, S.C.; Solano, R. Structural basis for DNA binding specificity by the auxin-dependent ARF transcription factors. *Cell* **2014**, *156*, 577–589. [CrossRef]

41. Yu, H.; Soler, M.; Mila, I.; San Clemente, H.; Savelli, B.; Dunand, C.; Paiva, J.A.; Myburg, A.A.; Bouzayen, M.; Grima-Pettenati, J. Genome-wide characterization and expression profiling of the AUXIN RESPONSE FACTOR (ARF) gene family in *Eucalyptus grandis*. *PLoS ONE* **2014**, *9*, e108906. [CrossRef] [PubMed]
42. Paponov, I.A.; Teale, W.; Lang, D.; Paponov, M.; Reski, R.; Rensing, S.A.; Palme, K. The evolution of nuclear auxin signalling. *BMC Evol. Biol.* **2009**, *9*, 126. [CrossRef]
43. Tian, C.e.; Muto, H.; Higuchi, K.; Matamura, T.; Tatematsu, K.; Koshiba, T.; Yamamoto, K.T. Disruption and overexpression of auxin response factor 8 gene of Arabidopsis affect hypocotyl elongation and root growth habit, indicating its possible involvement in auxin homeostasis in light condition. *Plant J.* **2004**, *40*, 333–343. [CrossRef] [PubMed]
44. Okushima, Y.; Fukaki, H.; Onoda, M.; Theologis, A.; Tasaka, M. ARF7 and ARF19 regulate lateral root formation via direct activation of LBD/ASL genes in Arabidopsis. *Plant Cell* **2007**, *19*, 118–130. [CrossRef]
45. Feng, S.; Li, N.; Chen, H.; Liu, Z.; Li, C.; Zhou, R.; Zhang, Y.; Cao, R.; Ma, X.; Song, X. Large-scale analysis of the ARF and Aux/IAA gene families in 406 horticultural and other plants. *Mol. Hortic.* **2024**, *4*, 13. [CrossRef]
46. Hakman, I.; Hallberg, H.; Palovaara, J. The polar auxin transport inhibitor NPA impairs embryo morphology and increases the expression of an auxin efflux facilitator protein PIN during *Picea abies* somatic embryo development. *Tree Physiol.* **2009**, *29*, 483–496. [CrossRef] [PubMed]
47. Guseman, J.M.; Hellmuth, A.; Lanctot, A.; Feldman, T.P.; Moss, B.L.; Klavins, E.; Calderón Villalobos, L.I.A.; Nemhauser, J.L. Auxin-induced degradation dynamics set the pace for lateral root development. *Development* **2015**, *142*, 905–909.
48. Li, P.; Chen, X.; Sun, F.; Dong, H. Tobacco TTG 2 and ARF 8 function concomitantly to control flower colouring by regulating anthocyanin synthesis genes. *Plant Biol.* **2017**, *19*, 525–532. [CrossRef]
49. Solanki, M.; Shukla, L.I. Recent advances in auxin biosynthesis and homeostasis. *3 Biotech* **2023**, *13*, 290. [CrossRef]
50. Huang, R.; Zhang, X.; Luo, K.; Tembrock, L.R.; Li, S.; Wu, Z. The Identification of Auxin Response Factors and Expression Analyses of Different Floral Development Stages in Roses. *Genes* **2025**, *16*, 41. [CrossRef]
51. Ren, C.; Chen, C.; Dong, S.; Wang, R.; Xian, B.; Liu, T.; Xi, Z.; Pei, J.; Chen, J. Integrated metabolomics and transcriptome analysis on flavonoid biosynthesis in flowers of safflower (*Carthamus tinctorius* L.) during colour-transition. *PeerJ* **2022**, *10*, e13591. [PubMed]
52. Tian, Y.; Chen, Z.; Jiang, Z.; Huang, X.; Zhang, L.; Zhang, Z.; Sun, P. Effects of plant growth regulators on flower abscission and growth of tea plant *Camellia sinensis* (L.) O. Kuntze. *J. Plant Growth Regul.* **2022**, *41*, 1161–1173.
53. Hou, M.; Zhang, Y.; Xu, X.; Ai, H. Advances in auxin synthesis, transport, and signaling in rice: Implications for stress resilience and crop improvement. *Front. Plant Sci.* **2025**, *15*, 1516884.
54. Wang, J.; Jin, D.; Deng, Z.; Zheng, L.; Guo, P.; Ji, Y.; Song, Z.; Zeng, H.Y.; Kinoshita, T.; Liao, Z. The apoplastic pH is a key determinant in the hypocotyl growth response to auxin dosage and light. *Nat. Plants* **2025**, *11*, 279–294. [PubMed]
55. Kong, M.; Liu, X.; Sun, L.; Tan, S. Molecular mechanisms of N-1-naphthylphthalamic acid, a chemical tool in plant biology and agriculture. *Mol. Hortic.* **2022**, *2*, 22.
56. Xian, B.; Wang, R.; Jiang, H.; Zhou, Y.; Yan, J.; Huang, X.; Chen, J.; Wu, Q.; Chen, C.; Xi, Z. Comprehensive review of two groups of flavonoids in *Carthamus tinctorius* L. *Biomed. Pharmacother.* **2022**, *153*, 113462.
57. Lamesch, P.; Berardini, T.Z.; Li, D.; Swarbreck, D.; Wilks, C.; Sasidharan, R.; Muller, R.; Dreher, K.; Alexander, D.L.; Garcia-Hernandez, M. The Arabidopsis Information Resource (TAIR): Improved gene annotation and new tools. *Nucleic Acids Res.* **2012**, *40*, D1202–D1210.
58. Pruitt, K.D.; Tatusova, T.; Maglott, D.R. NCBI Reference Sequence (RefSeq): A curated non-redundant sequence database of genomes, transcripts and proteins. *Nucleic Acids Res.* **2005**, *33*, D501–D504.
59. Finn, R.D.; Coggill, P.; Eberhardt, R.Y.; Eddy, S.R.; Mistry, J.; Mitchell, A.L.; Potter, S.C.; Punta, M.; Qureshi, M.; Sangrador-Vegas, A. The Pfam protein families database: Towards a more sustainable future. *Nucleic Acids Res.* **2016**, *44*, D279–D285
60. Letunic, I.; Goodstadt, L.; Dickens, N.J.; Doerks, T.; Schultz, J.; Mott, R.; Ciccarelli, F.; Copley, R.R.; Ponting, C.P.; Bork, P. Recent improvements to the SMART domain-based sequence annotation resource. *Nucleic Acids Res.* **2002**, *30*, 242–244.
61. Duvaud, S.; Gabella, C.; Lisacek, F.; Stockinger, H.; Ioannidis, V.; Durinx, C. Expasy, the Swiss Bioinformatics Resource Portal, as designed by its users. *Nucleic Acids Res.* **2021**, *49*, W216–W227. [CrossRef] [PubMed]
62. Chou, K.-C.; Shen, H.-B. Plant-mPLoc: A top-down strategy to augment the power for predicting plant protein subcellular localization. *PLoS ONE* **2010**, *5*, e11335.
63. Geourjon, C.; Deleage, G. SOPMA: Significant improvements in protein secondary structure prediction by consensus prediction from multiple alignments. *Bioinformatics* **1995**, *11*, 681–684.
64. Guex, N.; Peitsch, M.C. SWISS-MODEL and the Swiss-Pdb Viewer: An environment for comparative protein modeling. *Electrophoresis* **1997**, *18*, 2714–2723. [PubMed]
65. Kumar, S.; Stecher, G.; Tamura, K. MEGA7: Molecular evolutionary genetics analysis version 7.0 for bigger datasets. *Mol. Biol. Evol.* **2016**, *33*, 1870–1874. [PubMed]

66. Letunic, I.; Bork, P. Interactive Tree Of Life (iTOL): An online tool for phylogenetic tree display and annotation. *Bioinformatics* **2007**, *23*, 127–128.
67. Chen, C.; Wu, Y.; Li, J.; Wang, X.; Zeng, Z.; Xu, J.; Liu, Y.; Feng, J.; Chen, H.; He, Y. TBtools-II: A "one for all, all for one" bioinformatics platform for biological big-data mining. *Mol. Plant* **2023**, *16*, 1733–1742.
68. Marchler-Bauer, A.; Anderson, J.B.; DeWeese-Scott, C.; Fedorova, N.D.; Geer, L.Y.; He, S.; Hurwitz, D.I.; Jackson, J.D.; Jacobs, A.R.; Lanczycki, C.J. CDD: A curated Entrez database of conserved domain alignments. *Nucleic Acids Res.* **2003**, *31*, 383–387.
69. Bailey, T.L.; Boden, M.; Buske, F.A.; Frith, M.; Grant, C.E.; Clementi, L.; Ren, J.; Li, W.W.; Noble, W.S. MEME SUITE: Tools for motif discovery and searching. *Nucleic Acids Res.* **2009**, *37*, W202–W208.
70. Lescot, M.; Déhais, P.; Thijs, G.; Marchal, K.; Moreau, Y.; Van de Peer, Y.; Rouzé, P.; Rombauts, S. PlantCARE, a database of plant cis-acting regulatory elements and a portal to tools for in silico analysis of promoter sequences. *Nucleic Acids Res.* **2002**, *30*, 325–327.
71. Szklarczyk, D.; Franceschini, A.; Kuhn, M.; Simonovic, M.; Roth, A.; Minguez, P.; Doerks, T.; Stark, M.; Muller, J.; Bork, P. The STRING database in 2011: Functional interaction networks of proteins, globally integrated and scored. *Nucleic Acids Res.* **2010**, *39*, D561–D568. [PubMed]
72. Shannon, P.; Markiel, A.; Ozier, O.; Baliga, N.S.; Wang, J.T.; Ramage, D.; Amin, N.; Schwikowski, B.; Ideker, T. Cytoscape: A software environment for integrated models of biomolecular interaction networks. *Genome Res.* **2003**, *13*, 2498–2504. [PubMed]
73. Mao, W.; Bao, C.; Cheng, Q.; Liang, N.; Wang, L.; Yang, H. All-year high IAA and ABA contents in rhizome buds may contribute to natural four-season shooting in woody bamboo cephalostachyum pingbianense. *Plants* **2024**, *13*, 410. [CrossRef] [PubMed]
74. Zhang, Y.; Yu, L.; Jin, W.; Li, C.; Wang, Y.; Wan, H.; Yang, J. Simultaneous optimization of the ultrasonic extraction method and determination of the antioxidant activities of hydroxysafflor yellow A and anhydrosafflor yellow B from safflower using a response surface methodology. *Molecules* **2020**, *25*, 1226. [CrossRef]
75. Tan, Z.W.; Lu, D.D.; Yu, Y.L.; Li, L.; Xu, L.J.; Dong, W.; Li, C.M.; Yang, Q.; Liang, H.Z. Genome-Wide Analysis of the APETALA2/Ethylene-Responsive Factor Gene Family in *Carthamus tinctorius* L. *Plant Direct* **2025**, *9*, e70032.
76. Livak, K.J.; Schmittgen, T.D. Analysis of relative gene expression data using real-time quantitative PCR and the $2^{-\Delta\Delta CT}$ method. *Methods* **2001**, *25*, 402–408.

International Journal of *Molecular Sciences*

MDPI

Article

Multiomics Studies on the Effects of High-Temperature Stress on Male Sterility in *Gossypium barbadense*

Jiangbo Li †, Xiaojuan Deng, Man Gao, Tao Lv, Yongsheng Cai, Yanying Qu, Quanjia Chen and Kai Zheng *

Xinjiang Key Laboratory of Crop Biology Breeding, College of Agriculture, Xinjiang Agricultural University, Urumqi 830052, China; 18699160581@163.com (J.L.); dengxj007@163.com (X.D.); 13209972635@163.com (M.G.); 13909447464@163.com (T.L.); cys0620@126.com (Y.C.); xjyyq5322@126.com (Y.Q.); chqjia@126.com (Q.C.)

* Correspondence: zhengkai555@xjau.edu.cn

† This author is the first author.

Abstract: High-temperature (HT) stress has been recognized as one of the main factors restricting the normal growth and development of cotton and severely affects fiber quality and yield. To elucidate the regulatory mechanism of male sterility-related hormones in *Gossypium barbadense* under HT stress, we explored candidate genes closely related to male sterility in *G. barbadense*. We studied the expression profiles of hormones and genes in the anthers of *G. barbadense* GB150 under HT stress by combining transcriptomic and metabolomic analyses. Through a combined analysis of the transcriptional metabolism of the anthers of *G. barbadense* GB150, we determined the contents of ABA, JA, SA, IAA, tZR, and GA_{20} and the expression of genes related to biosynthetic pathways and signal transduction pathways. The results revealed that the ABA and JA contents significantly increased after HT; the IAA, tZR, and GA_{20} contents significantly decreased; and the SA content did not significantly change after HT. We then used weighted gene coexpression network analysis (WGCNA) to further analyze the interactions among hormones, transcription factors, and core genes and constructed hormone coexpression networks and genome-wide coexpression networks. Through these network analyses, we ultimately identified 10 candidate genes closely related to male sterility in *G. barbadense*. Using qRT-PCR, resequencing data from 221 *G. barbadense* materials revealed that ALA4 (*Arabidopsis thaliana* has been proven to be associated with male fertility) and SBP1 (two stop gains in the gene structure) may play important roles in the process of male sterility in *G. barbadense*. The results of this study provide a theoretical basis for the molecular mechanism of male sterility in *G. barbadense*.

Keywords: *Gossypium barbadense*; transcriptome; metabolome; anther; male sterility

1. Introduction

High-temperature (HT) stress is defined as irreversible damage to the growth and development of plants caused by a temperature exceeding a certain threshold level within a certain period. When the temperature increases frequently to 10–15 °C above the ambient temperature in a short period, it is called HT shock or HT stress [1]. Cotton is an annual or perennial herb of the *Malvaceae* family, and its optimum growth temperature is 20–30 °C [2]. *Gossypium barbadense* is very sensitive to HT stress, and HT stress inhibits anther dehiscence, affecting pollen vitality and resulting in boll fall and reduced yield. Moreover, continuous HT stress affects the development of fiber quality, resulting in a decrease in the fiber strength and fiber uniformity [3]. In addition, HT stress can affect plant reproductive organs and vegetative organs, but HT has a greater effect on reproductive organs than on vegetative organs [4].

The decrease in the cotton yield is due mainly to HT, which increases the possibility of male sterility [5,6]. Previous studies have shown that HT affects several key stages of cotton reproductive growth [7]. During the reproductive organ development stage, HT causes the pollen quantity and stamen vitality to decrease significantly, resulting in the formation of deformed pollen grains, weakening pollination function, and eventually leading to male sterility and abortion. Moreover, the pistil is also negatively affected by HT: the female gametophyte is degraded, the style and ovary are deformed, and the structural integrity of the pistil and the function of the ovule are reduced. In the process of pollination to fertilization, HT also has adverse effects, such as reduced pollen adhesion and germination, hindered pollen tube growth, and reduced ovule viability and lifespan, which ultimately affect normal gamete union.

Under HT stress, the above factors hinder improvements in cotton yield. If no mitigation measures are taken as the climate warms, the global average surface temperature will increase (2.6–4.8 °C) by the end of the century, with every 1 °C increase in temperature affecting crop yields by up to 17% during the growing season [8]. HT stress has become an important limiting factor affecting cotton production. To reduce the impact of HT on the cotton industry, it is necessary to accelerate the collection and identification of HT-resistant cotton resources, analyze the molecular mechanism of HT resistance, and create new HT-resistant materials.

Combined transcriptomic and metabolomic analysis is conducive to obtaining a more comprehensive understanding of biological processes and their regulatory networks inside plants and is widely used in various organisms [9–14]. Joint analyses can not only identify novel metabolic pathways and key metabolites, which is highly important for the study of secondary plant metabolism and the production of chemical substances, but also break traditional boundaries and promote crossover and fusion between different research fields [15].

Under HT stress, dynamic changes in hormones directly affect the normal growth and development of plants [16]. Abscisic acid (ABA), an important hormone that regulates the water state of plants, plays an important role in HT stress. By regulating transpiration and water absorption, ABA can meet the demand of atmospheric CO_2 for photosynthesis [17]. *Arabidopsis* mutants that are not sensitive to ABA present significant defects in HT resistance [18]. As a stress-related hormone, abnormal jasmonic acid content also leads to pollen inactivation and anther dehiscence. Auxin, gibberellin, and cytokinin, key hormones in plant growth and development, have also been confirmed to play roles in male plant sterility [19].

The aim of this study was to explore the regulatory mechanism by which HT stress affects medicinal anther dehiscence in *G. barbadense*. First, 36 cotton materials were selected from 534 *G. barbadense* resources according to the extent of four flower traits (style length, stigma length, single anther length, and stamen length). Among these materials, the GB150 material was selected as a key research object because of its long anther length. To assess the heat resistance of these materials, 36 resources were identified at the seedling and adult stages. The results revealed that GB150 was highly sensitive to HT stress during different periods (Figure S1). Moreover, in the field warming stress experiment, we observed that the GB150 material presented a nondehiscent anther phenotype, whereas the other materials did not. To date, many studies have investigated the molecular mechanism of male sterility in cotton [20,21], but the relationship between male sterility and hormone-related regulatory networks in *G. barbadense* under HT stress is still relatively uncharacterized. To analyze the response mechanism of *G. barbadense* to male sterility and HT stress, it is important to study the regulatory network of hormone metabolism and signal transduction. Therefore, we conducted in-depth hormone and transcriptomic analyses of GB150 to analyze the interaction between male sterility and hormone metabolism in *G. barbadense* under HT stress. Based on a detailed analysis of the interactions between hormones and hormone signaling pathways, we constructed a transcriptional regulation model of male sterility in *G. barbadense* under HT stress and identified 10 related candidate genes. These findings provide a foundation for further analysis of the molecular mechanism of male sterility in *G. barbadense*.

2. Results and Analysis

2.1. Quality Assessment of the Transcriptomic and Metabolomic Sequencing Data

We constructed an RNA-seq library using six anther samples of the *G. barbadense* resource GB150 under HT stress and normal conditions. A total of 61.74 Gb of clean data was obtained, and more than 6.43 Gb of clean data was obtained for each sample. The total reads of each sample were subsequently compared with the reference genome of *G. barbadense*. The efficiency of the comparison was 93.82–95.87%, the percentage of Q30 bases was no less than 94.28%, and the GC content was no less than 43.76% (Table 1). These results confirmed that the RNA-seq data in this study were of high quality and could be used for subsequent analysis.

Table 1. Statistical results of the sequencing data.

Sample	Raw Reads	Reads Mapped	Clean Reads	Clean Base (G)	Error Rate (%)	Q30 (%)	GC Content (%)
CK-1	47,820,474	40,368,138 (94.21%)	42,847,734	6.43	0.01	95.2	44.46
CK-2	74,758,782	65,763,208 (93.82%)	70,097,786	10.51	0.02	94.28	44.44
CK-3	88,691,976	77,567,651 (94.88%)	81,755,950	12.26	0.01	96.38	44.29
HOT-1	99,523,708	87,674,236 (95.02%)	92,264,390	13.84	0.01	97.19	43.76
HOT-2	87,482,908	77,555,852 (95.87%)	80,895,212	12.13	0.01	97.12	43.86
HOT-3	49,351,338	41,977,255 (95.83%)	43,803,982	6.57	0.01	96.2	44.14

We conducted a principal component analysis of the expressed genes (Figure 1A) and found that samples in the same group were clustered into one class, and samples in different groups with anther dehiscence (CK group) or anther nondehiscence (HOT group) were far apart, indicating that the expression levels of anther-related genes under the two treatments greatly differed. The principal component results indirectly demonstrated the reliability of the transcriptomic data.

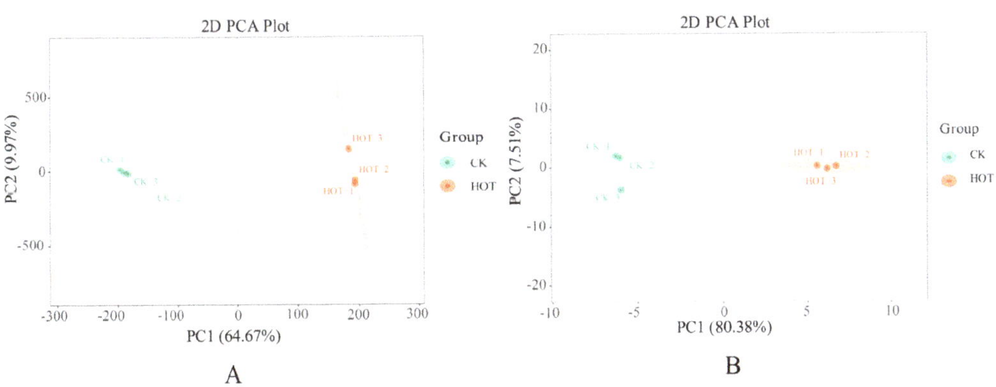

Figure 1. Principal component analyses of transcriptomic and metabolomic samples. (**A**) PCA of transcriptomic samples. (**B**) PCA of metabolomic samples. HT: High-temperature, CK: control check, the same as below.

The abundances of anther-related metabolites of GB150 subjected to different treatments were determined, and the differences in metabolite abundances among the different treatment groups were analyzed. The metabolite PCA under different treatments revealed that the contribution rate of PC1 was 80.38%, and the contribution rate of PC2 was 7.51% (Figure 1B). The data in the two groups had good reproducibility, and there were significant differences in metabolite abundances between the groups.

2.2. *Effects of High-Temperature Stress on Floral Organ Development and Hormonal Regulation in G. barbadense GB150*

Many studies have shown that HT stress has significant effects on many aspects of plant morphology, photosynthesis, and the antioxidant system. In this study, we found that GB150 resources caused no anther dehiscence and decreased pollen viability following HT stress (Figure 2A,B). To explore the effects of HT stress on the reproductive development of *G. barbadense*, we investigated the relationships among the four flower traits of GB150 (Figure 2C). The results revealed a significant correlation between the stylar length and stigma length among the four flower traits but no significant correlation between the other flower traits. We found that of the four flower traits, only the stigma length was greater than that of the normal control group under HT stress, and the values of the other three flower traits (style length, stamen length, and single anther length) were shorter than those of the control group following HT stress. We systematically analyzed the correlations between the levels of six hormones and four flower traits and found that the levels of tZR, IAA, and GA_{20}, three hormones that are involved mainly in plant growth and development, were significantly correlated with four flower traits, indicating that these hormones may play important roles in plant growth and development, especially flower formation and development. The tZR levels were significantly correlated with two indicators (style length and stamen length), the IAA levels were significantly correlated with two indicators (style length and stamen length), and the GA_{20} levels were significantly correlated with one indicator (style length). The levels of the hormones ABA, JA, and SA, which are involved in the plant stress response and defense response, were not significantly correlated with the four flower traits, which does not mean that they have no role in plant growth and development; they may not directly participate in or affect the development of flower traits.

2.3. *Changes in ABA Content and Expression Levels in G. barbadense Anthers Under HT Stress*

Under HT stress, changes in the ABA content in plants play a core role. Through in-depth studies of the biosynthesis and catabolic pathways of ABA in many plants, we elucidated its regulatory mechanism. In this study, the ABA contents of anthers subjected to HT stress were determined (Figure 3A). Compared with normal conditions, HT stress significantly increased the endogenous ABA content in anthers (Figure 3B). For the RNAseq expression analysis, we examined the expression patterns of differentially expressed genes (DEGs) involved in the ABA biosynthesis pathway (Figure 3C,D). Among these genes, VDE1 (Gbar_A02G014800, Gbar_A11G025550) and NCED5 (Gbar_A06G014870) presented upregulated expression, and the ABA2 (1), NCED1 (2), CYP707A1 (1), and CYP707A2 (2) genes presented downregulated expression. By analyzing ABA signal transduction pathway-related DEGs, we found that ABF2 (Gbar_A05G025600) and ABF3 (Gbar_A12G002250) expression levels were upregulated and that ABF2 (Gbar_D05G020810) expression was downregulated.

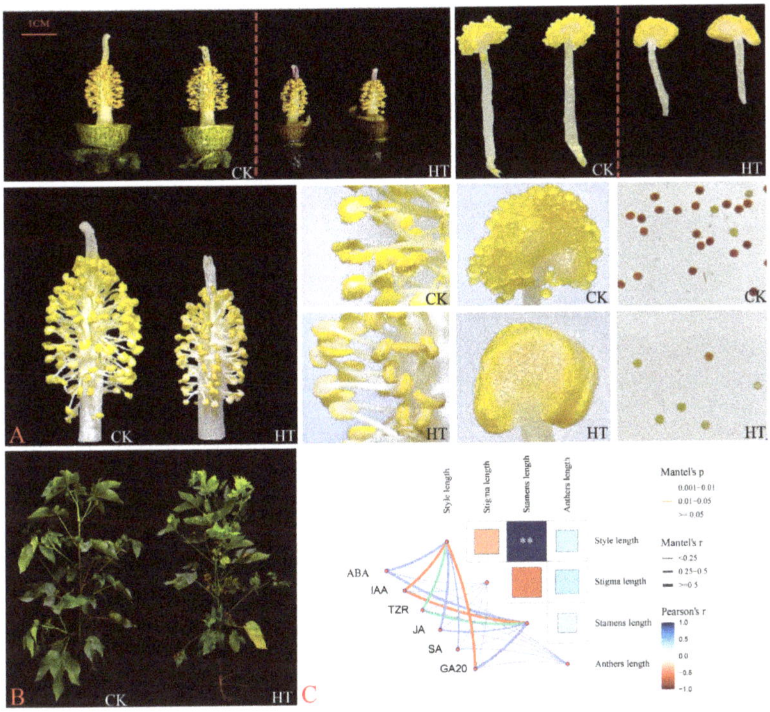

Figure 2. Response of GB150 to HT stress. (**A**) Flower phenotypes of the GB150 germplasm after HT treatment. (**B**) Phenotypes GB150 germplasm HT-treated plants. (**C**) Correlation network of 4 floral organ indices and the levels of 6 major hormones in the GB150 germplasm. ABA: Abscisic Acid, JA: Jasmonic Acid, SA: Salicylic Acid, GA_{20}: Gibberellic Acid 20, tZR: trans-Zeatin Riboside, IAA: 3-Indoleacetic acid, the same as below ** $p < 0.01$.

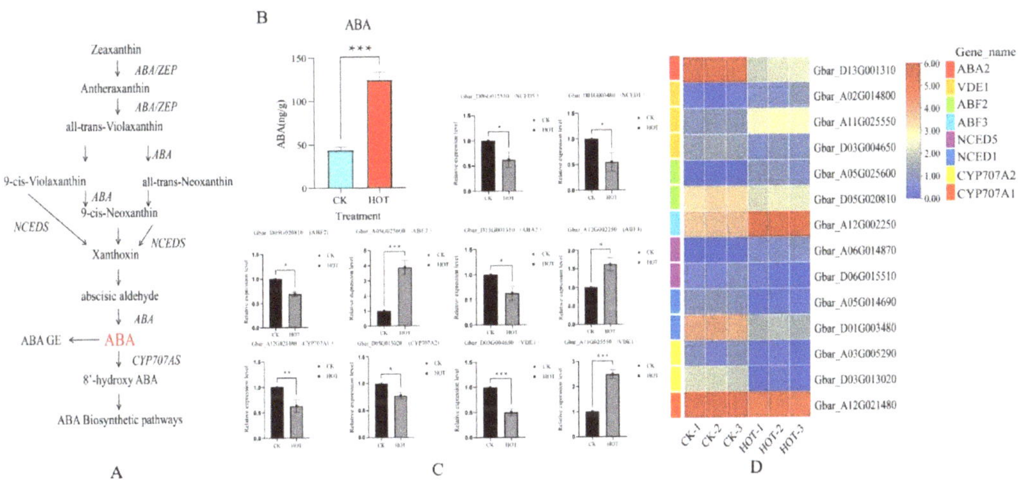

Figure 3. Effects of HT stress on ABA biosynthesis and signal transduction. (**A**) ABA biosynthesis pathway. (**B**) Changes in ABA content under HT stress. (**C**) Expression patterns of 10 ABA pathway genes. (**D**) Expression profiles of genes involved in ABA biosynthesis and signal transduction after HT stress. * $p < 0.05$; ** $p < 0.01$; *** $p < 0.001$.

qRT-PCR was performed on eight core genes (10) of the ABA biosynthesis and signal transduction pathways (Figure 3C). The expression levels of ABF2 (Gbar_A05G025600), VDE1 (Gbar_A11G025550), and ABF3 (Gbar_A12G002250) were upregulated, and the expression levels of the other seven genes were downregulated, consistent with the RNA-seq results. Further analysis revealed that 5 of the 14 ABA-related DEGs were upregulated and that 9 were downregulated. Based on these results, we speculated that the core genes closely related to the ABA content in anthers mainly negatively regulate ABA biosynthesis and signal transduction pathways under HT stress.

2.4. Effects of HT Stress on JA and SA Biosynthesis and Signal Transduction

JA and SA, as regulators of HT stress, also play important roles in the response to HT. The results revealed that the JA content significantly increased under HT stress (Figure 4A). Among the four types of genes encoding JA biosynthesis-related proteins (AOS3 (one), LOX5 (one), OPR3 (one), and CYP94B3 (three), for a total of six genes) (Figure 4C), the expression levels of two genes, LOX5 (Gbar_D09G022770) and OPR3 (Gbar_A06G007940), were upregulated, and the expression levels of the other four genes were downregulated. However, among the two classes of genes in the JA signal transduction pathway (MYC2 (one) and JAR1 (three), for a total of four genes), only one gene presented upregulated expression (MYC2 (Gbar_D09G020720)), whereas the other three genes presented down-regulated expression (Figure 4E).

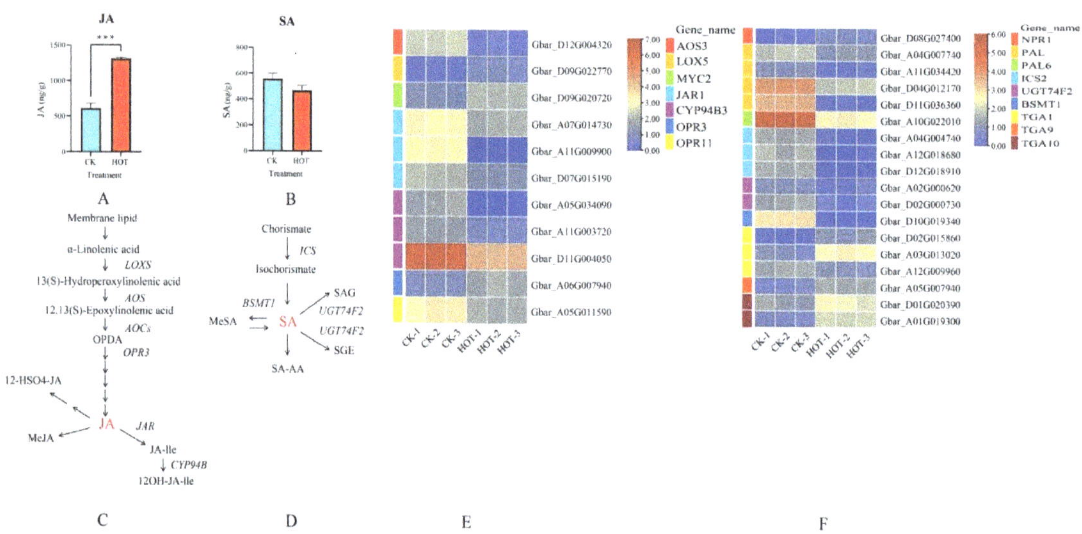

Figure 4. Effects of HT stress on JA and SA biosynthesis and signal transduction. (**A**) Changes in JA content under HT stress. (**B**) Changes in SA content under HT stress. (**C**) JA biosynthesis pathway. (**D**) SA biosynthesis pathway. (**E**) Expression profiles of JA biosynthesis- and signal transduction-related genes after HT stress. (**F**) Expression profiles of genes involved in SA biosynthesis and signal transduction after HT stress.*** $p < 0.001$.

However, under HT stress, the SA content decreased, but the change was not obvious (Figure 4B). Therefore, we thoroughly analyzed the expression levels of core genes in the SA biosynthetic and signaling pathways. In the biosynthetic pathway (Figure 4D), the expression levels of three types of genes (for a total of six genes), ICS2 (three), UGT74F2 (two) and BSMT1 (one), tended to be downregulated. The changes in expression were consistent with the decrease in SA content, indicating that these SA biosynthesis-related genes expressed in anthers positively regulate SA synthesis after HT stress. In the signal transduction pathway, we observed expression changes in three types of core genes (twelve genes in total): TGA (six), NPR (one), and PAL (five). The results revealed six genes with upregulated expression and six genes with downregulated expression, among which five TGA genes (of six) presented upregulated expression, and only one TGA gene (Gbar_A12G009960) was downregulated (Figure 4F). Notably, the expression levels of SA-related core genes did change, but the SA content did not fluctuate significantly. Based on these findings, we hypothesized that although SA is involved in anther regulation after HT stress, it is not the main regulatory hormone.

2.5. Effects of HT Stress on IAA, GA_{20} and tZR Biosynthesis and Signal Transduction

The change in the IAA content under HT stress was very complicated. We found that the IAA content decreased significantly after HT stress (Figure 5A). However, the expression levels of both the CYP83B1 genes (Gbar_A10G020760 and Gbar_D10G021010) involved in the IAA biosynthesis pathway (Figure 5B) were upregulated. Moreover, among the four types of core genes (47 in total) in the signaling pathway, ARF (1), SAUR (21), AUX (3), and IAA (22), 12 genes (ARF (1), IAA (5), and SAUR (6)) presented upregulated expression, and the expression levels of 35 genes (AUX (3), IAA (17), SAUR (15)) were downregulated (Figure 5C). These results suggest that the expression levels of most signal transduction-related genes in anthers are downregulated under HT stress, which may lead to the inhibition of IAA expression and thus affect anther dehiscence.

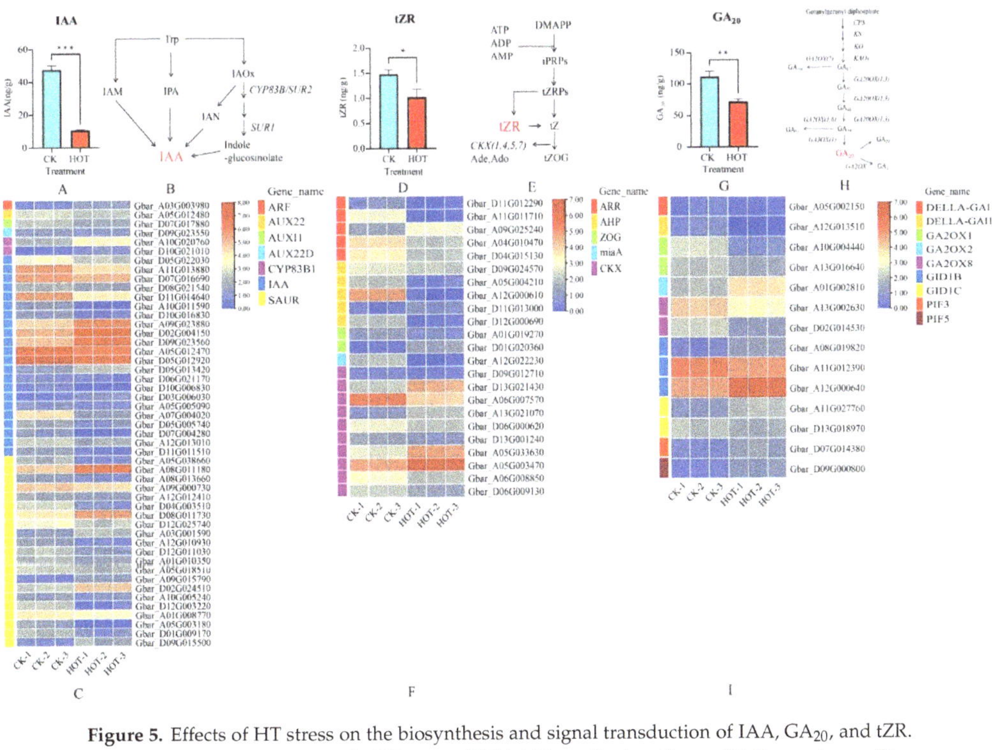

Figure 5. Effects of HT stress on the biosynthesis and signal transduction of IAA, GA_{20}, and tZR. (**A**) Changes in IAA content under HT stress. (**B**) IAA biosynthesis pathway. (**C**) Expression profiles of genes involved in IAA biosynthesis and signal transduction after HT stress. (**D**) Changes in TZR content under HT stress. (**E**) TZR biosynthetic pathway. (**F**) Expression profiles of tZR biosynthesis and signal transduction genes after HT stress. (**G**) Changes in the GA_{20} content under HT stress. (**H**) GA20 biosynthetic pathway. (**I**) Expression profiles of genes involved in GA_{20} biosynthesis and signal transduction after HT stress. * $p < 0.05$; ** $p < 0.01$; *** $p < 0.001$.

The tZR content decreased significantly under HT stress (Figure 5D). To explore the molecular mechanisms underlying this phenomenon, we analyzed the expression patterns of genes involved in tZR biosynthesis (Figure 5E) and signal transduction. In the tZR biosynthesis pathway, we focused on four classes of genes (18 in total), including AHP (5), miaA (1), ZOG (2), and CKX (10). The expression levels of 6 genes (AHP (1), ZOG (1), and CKX (4)) were increased, whereas the expression levels of the other 12 genes (AHP (4), miaA (1), ZOG (1), and CKX (6)) were decreased. Moreover, in the tZR signal transduction pathway, we focused on the expression levels of ARR genes (5). The results revealed that the expression levels of four ARR genes were decreased, whereas that of only one ARR gene (Gbar_A09G025240) was increased (Figure 5F). Therefore, we concluded that the significant reduction in the tZR content under HT stress may be closely related to changes in the expression patterns of genes related to biosynthesis and signal transduction.

Under HT stress, the GA_{20} content also tended to decrease (Figure 5G). GA2OX plays a key role in the biosynthesis of GA_{20} (Figure 5H). However, we observed that only one GA2OX-related gene (Gbar_A01G002810) was increased under HT stress, whereas the expression levels of the remaining four GA2OX genes were decreased. We also focused on nine genes involved in the GA_{20} signaling pathway, including DELLA (two), GID (five), and PIF (two). Among these genes, the expression levels of five genes (GID (three) and PIF (two)) were increased, whereas the expression levels of the other four genes (DELLA (two), GID (two)) were decreased (Figure 5I). In summary, the decrease in the GA_{20} content under HT stress may be related to the downregulation of the expression of the GA20 biosynthesis-related gene GA2OX and the complex regulation of the related genes DELLA, GID, and PIF in the signal transduction pathway.

2.6. Changes in the Whole Transcriptome Under HT Stress

We conducted an in-depth analysis of the transcriptomic data, and through comparative analysis, we identified 14,857 DEGs (Figure 6A) that significantly differed between the two conditions. Among these DEGs, 5131 genes presented upregulated expression, and 9726 genes presented downregulated expression (Figure 6B). Through a hierarchical cluster analysis, we identified nine statistically significant gene expression clusters (Figure 6C) and successfully generated multiple pulse expression profiles. These clusters show the expression patterns of different genes under HT stress. Clusters 1, 3, 4, 5, 6, and 8 presented trends toward downregulated expression, and the cluster with the greatest number of DEGs in the downregulated expression profile was cluster 1, with 5494 genes, which may suggest that this cluster plays an important role in the response to HT stress. The cluster with the fewest DEGs in the downregulated expression profiles was cluster 3, with 480 genes. Although the number of genes involved is small, these genes may also play key roles in specific pathways or processes. Conversely, clusters 2, 7, and 9 tended to show upregulated gene expression under different treatments, and the group with the greatest number of DEGs in the upregulated expression profile was cluster 7, with 2720 genes, which may indicate that these genes were inhibited under HT stress, thus affecting related biological functions. In the upregulated expression profile, the nine classes with the fewest DEGs included 1071 genes. These results revealed the complex response mechanism of *G. barbadense* to HT stress. Changes in different gene expression patterns may regulate the tolerance of *G. barbadense* to HT stress.

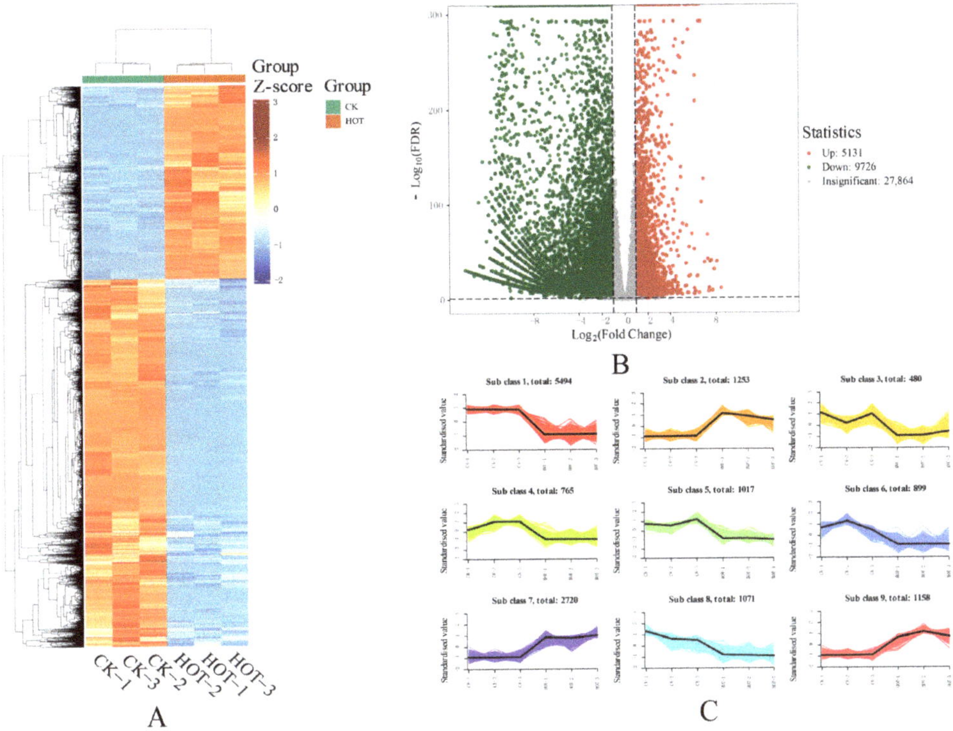

Figure 6. Changes in the whole transcriptome under HT stress. (**A**) Expression patterns of differentially expressed genes under HT stress. (**B**) Volcano map of differentially expressed genes. (**C**) Cluster and line graphs showing the trends of gene expression in the clusters.

2.7. Construction and TF Analysis of the Hormone-Related Coexpression Network

To reveal the functions of the regulatory networks, we used a weighted gene coexpression network analysis (WGCNA) of the HT stress-related coexpression networks in *G. barbadense* anthers (Figure 7A,C). The WGCNA results of the DEGs revealed four coexpressed gene modules (Figure 7B). We further calculated the correlations between modules and hormone contents and found that the gene expression profiles in the blue module were highly correlated with the ABA, IAA, and JA levels and that the gene expression profiles in the turquoise module were highly correlated with the IAA, SA, and GA_{20} levels. The blue module was composed of 5 TFs, 5 hub genes of 5 different types (AP2/ERF-AP2, MYB, C3H, C2H2, and bZIP), and the top 100 genes with the highest connectivity, among which the top hub gene was Gbar_A09G027130 (Figure 7D). The turquoise module was composed of 5 TFs, 5 hub genes of 5 different types (AP2/ERF-AP2, NAC, HB HD ZIP, and bHLH), and the top 100 genes with the highest connectivity, among which the top hub gene was Gbar_A04G010820 (Figure 7E). To further investigate the transcription factor (TF) expression patterns, we used RNA-seq to analyze the expression patterns of eight TF genes (Figure S2). Our findings identified potential candidate genes for further exploration of the genes related to anther dehiscence in *G. barbadense*.

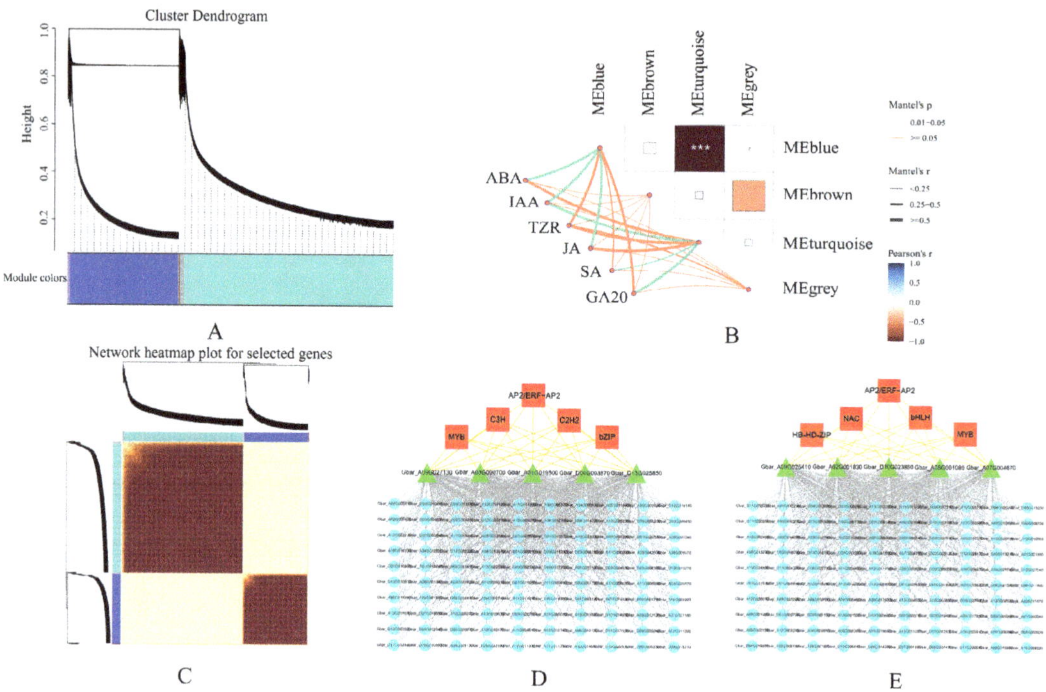

Figure 7. Correlation and coexpression networks between different gene modules and hormones. (**A**) Division of different gene modules. (**B**) Correlations between hormones and modules. (**C**) Modular gene clustering heatmap. (**D**,**E**) Gene coexpression network. *** $p < 0.001$.

2.8. Influence of HT Stress on TF Expression and Construction of a Genome-Wide Coexpression Network

A total of 1114 TFs were differentially expressed under different treatments, and their expression profiles were clearly distinguished between the upregulated and downregulated groups (Figure 8A–C). We found that the genes of five TF families, including AP2/ERF-AP2, MYB, HB-HD-ZIP, bHLH, and NAC, presented upregulated expression. Conversely, the expression levels of five TF family genes, AP2/ERF-AP2, C2H2, MYB, bZIP, and C3H, were decreased. Because TFs can mediate the inhibition of target gene expression, there should be a high correlation between their expression and the expression profiles of the target genes. We identified correlations between four modules previously detected by WGCNA and various processes. We selected the blue and turquoise modules, which were significantly and highly correlated with different treatments (Figure 8D), and constructed a whole-genome coexpression network of *G. barbadense* using the genes with the top-ranked TFs and connectivity. A total of eight classes of TFs and the top 100 genes with the highest connectivity in each module were identified (Figure 8E). In conclusion, not only were the specific expression patterns of TFs in *G. barbadense* under HT stress revealed, but a genome-wide coexpression network based on hormones, TFs, and key node genes was also successfully constructed. These findings provide a basis for further studies on the molecular mechanism of HT resistance in *G. barbadense*.

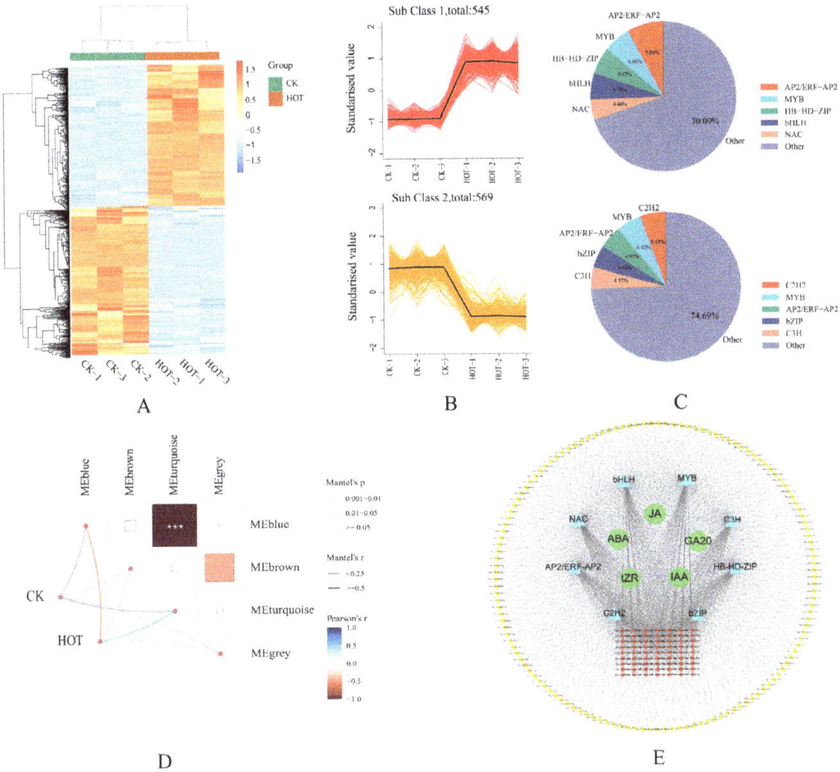

Figure 8. Expression patterns of differentially expressed TFs under HT stress, the correlations between modules and stages, and the whole-genome coexpression network. (**A**,**B**) Heatmaps and line maps of differentially expressed TFs. (**C**) Ratios of TFs whose expression levels were increased to those whose expression levels were decreased. (**D**) Heatmap of the relationships between modules and time points. (**E**) Genome-wide coexpression network. *** $p < 0.001$.

2.9. Expression Validation and Genetic Variation Analysis of Candidate Genes

Gene expression patterns are usually closely related to gene function. To better understand the expression patterns of the candidate genes, we conducted qRT-PCR experiments on 10 candidate genes in anthers (Figure 9). The results of the qRT-PCR analysis of 10 candidate genes were consistent with the RNA-seq results. The expression levels of Gbar_A09G027130 (ALA4), Gbar_D04G003870 (DUS3), and Gbar_D13G025850 (PI4KG1) significantly decreased under the different treatments.

To further study the 10 candidate genes associated with male sterility in *G. barbadense*, we used the resequencing data of 221 *G. barbadense* materials to type these candidate genes (Figure 10). The results revealed that 9 of the 10 candidate genes presented single-nucleotide polymorphisms (SNPs) in their upstream, downstream, exon, and intron regions. Notably, the Gbar_A07G004670 (APS1) gene had no SNPs at any location, whereas the Gbar_A02G001830 (SBP1) gene had the highest number of SNPS (up to 28). Mutations were identified at positions 21444092 and 21448044 of this gene that resulted in the stop gain of the amino acid sequence. These findings strongly suggest that the Gbar_A02G001830 (SBP1) gene may play a crucial role in the male sterility of *G. barbadense*.

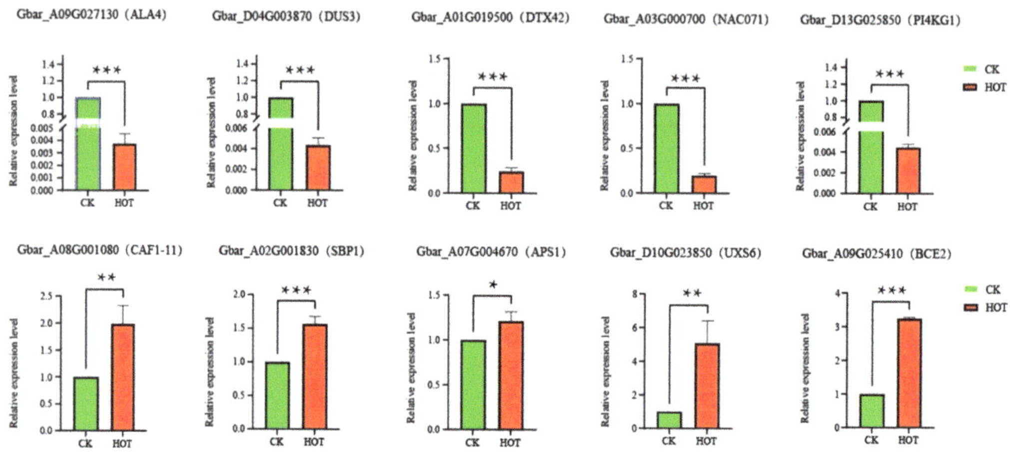

Figure 9. Analysis of the specific expression profiles of 10 candidate genes in anther tissues. * $p < 0.05$; ** $p < 0.01$; *** $p < 0.001$.

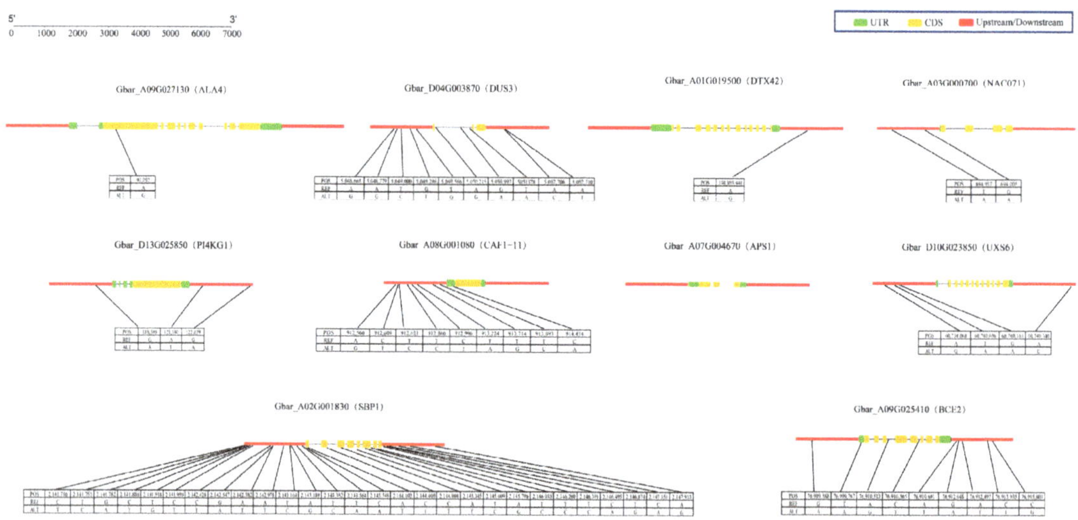

Figure 10. Information on SNP variations in 10 candidate genes in 221 *G. barbadense* accessions.

3. Discussion

The appropriate temperature is crucial for the entire growth period of cotton [22]. Once the ambient temperature exceeds the threshold that cotton growth can withstand, the plant will be destroyed by heat damage. From the perspective of the plant phenotype, HT can lead to the excessive elongation of plant stems and leaves, increase the leaf area, and affect many apparent traits, such as the fresh and dry weights of leaves, the length and width of the main stem, and the plant height [23]. Moreover, HT stress can also cause the accumulation of reactive oxygen species inside plants, which leads to a significant increase in the level of intracellular oxidative stress [24]. When the temperature is too high, it may even trigger lipid peroxidation of the cell membrane, resulting in rupture and damage to the cell membrane and severely affecting the normal physiological functions of the plant [25]. In response to this oxidative damage, plants activate a series of protective mechanisms, including the activation of antioxidants and the induction of antioxidant enzyme activity [26], to maintain the stability of cell membranes and ensure that plants can continue to grow and develop in HT environments. In this study, among the four flower traits investigated (style length, stigma length, stamen length, and single anther length), only the style length was significantly correlated with stigma length, whereas no significant relationships were found among the other traits. These findings suggest that these flower traits may involve relatively independent regulatory mechanisms during growth and development. Under HT stress, only the stigma length increased, whereas the style length, stamen length, and individual anther length decreased. These results indicate that HT stress has a selective effect on the development of reproductive organs in *G. barbadense*, in which stigma length may have a special adaptive effect, whereas other floral traits are sensitive to HT. The reproductive capacity or adaptability of *G. barbadense* may be enhanced by altering the development of several flower traits (such as increasing stigma length) under HT stress. The tZR, IAA, and GA_{20} levels, which are closely related to plant growth and development, are significantly correlated with many indices of flower traits. These findings suggest that these metabolites may play important roles in flower formation and development, especially in the regulation of style and stamen length. The levels of three hormones, ABA, JA, and SA, which are related to the plant stress response and defense response, were not significantly correlated with flower traits. This does not mean that these hormones have no role in plant growth and development or may not directly participate in or mainly affect the development of flower traits; rather, they may participate more in the plant stress response under adverse conditions.

Male sterility, especially abnormal anther dehiscence, is usually the result of multiple metabolic pathways regulated by the expression of multiple genes (including TFs) [27]. Among these complex regulatory mechanisms, plant hormones play crucial roles, especially ABA, which plays an irreplaceable role in the development of plant reproductive organs [28]. The accumulation of ABA in cotton often leads to male sterility [29]. In tomato, high concentrations of ABA may also lead to abnormal anther development [30]. The ABA content increased significantly after HT stress. By screening differentially expressed genes (DEGs), we identified 14 key pathway genes in eight classes. Most of the genes (nine) tended to have downregulated expression, whereas a few genes (five) tended to have upregulated expression. The results also revealed that in response to HT stress, the accumulation of ABA was negatively regulated by the expression of ABA response element binding factor (ABF) and violaxanthin decycloxygenase (VDE). Moreover, the expression of CYP707A (ABA 8′-hydroxylase) and NCED (9-cis-epoxide carotenoid dioxygenase) was induced to positively regulate ABA accumulation.

Jasmonic acid (JA) is an essential plant hormone regulator that plays a crucial role in the HT stress response [31]. In the biosynthetic pathway of JA, genes such as those encoding acyl-CoA oxidase (ACO) and propylene oxide cyclase (AOC) are considered of special importance because they are involved in the production of sterile male plants during HT stress [32]. Histone H3K27me3 also affects the male fertility of cotton by inhibiting JA synthesis and signaling [33]. Aamir Hamid Khan et al. [34] reported that a decrease in the JA content resulted in abnormal microspore development, pollen inactivation, and anther nondehiscence. In this study, the JA content significantly increased after HT stress. However, the expression levels of most of the pathway genes involved in JA biosynthesis and signaling were decreased (11 genes, with 8 genes downregulated). These results indicate that most of the genes that may regulate JA content are negative regulatory genes that play important roles under stress conditions.

Auxin, a plant growth hormone, plays crucial roles in the growth and development of plants and in coping with stress conditions [35]. Two TFs, ARF6 and ARF8, regulate the expression patterns of auxin response genes by binding to their promoter elements.

However, in double loss-of-function mutants of ARF6 and ARF8, plants exhibit flower stagnation, stamen shortening, and anther nondehiscence and are thus unable to release pollen grains [36]. In our study, we observed a significant decrease in the IAA content in the plants after HT stress. Further analysis revealed that most of the genes related to IAA biosynthesis and signal transduction presented downregulated expression (49 genes involved, 35 of which presented significantly downregulated expression), and the plants in the present study also presented male sterility and no anther dehiscence. However, in contrast to previous studies, the changes in the IAA content under HT stress observed in this study were not consistent. We speculate that this may be because the interaction and regulatory mechanisms between different hormones in plants responding to HT stress may be more complex and need to be further explored.

The role of cytokinin is often neglected in studies of the mechanism of male sterility caused by HT stress. However, in this study, we observed that tZR (zeosin riboside, a major cytokinin) levels were significantly downregulated under different treatments, and the expression levels of the genes related to the intrinsic regulation of tZR also significantly differed. These findings suggest that tZR may play an important role in the phenomenon of male sterility under HT stress. Therefore, the effect of tZR on male sterility under HT stress is worthy of further study and confirmation.

GA20, an endogenous regulator [37], has significant effects on multiple developmental processes in plants, including the development of seeds, anthers, and pollen and the response of plants to various abiotic stresses [38,39]. In a previous study, researchers reported that a lack of GA (gibberellin) can significantly affect the normal development of stamens and that the overactivation of GA signals can also adversely affect the fertility of plants [40]. Under high-temperature stress, the DELLA protein encoded by the core gene of the germ plays a key regulatory role [41], which effectively increases the survival rate of plants under abiotic stress by regulating the scavenging activity of reactive oxygen species (ROS). We found that the content of GA_{20} (a key intermediate in the gibberellin synthesis pathway) was significantly decreased after HT stress. Further analysis revealed that the core regulatory genes DELLA and GA2OX encoding GA_{20} were also significantly downregulated, which usually implies the inhibition of gibberellin synthesis and signal transduction pathways. However, simultaneously, the expression levels of GID (gibberellin-insensitive dwarfing protein) and PIF (photochromic interaction factor) were significantly increased. The synergistic effect of the expression of these genes led to a decrease in the GA_{20} content, which affected the male fertility of plants and produced male-sterile plants.

Under high-temperature stress, various hormones such as ABA, JA, IAA, tZR, and GA_{20} exhibit complex interactions and synergistic regulatory mechanisms in *G. barbadense* [42]. These mechanisms collectively influence anther dehiscence, pollen viability, and male fertility. As a stress-responsive hormone, ABA accumulates under high temperatures and works synergistically with other hormones like JA to regulate the plant's stress response. Meanwhile, the coordinated changes in IAA, GA_{20}, and tZR directly affect anther dehiscence and pollen release, with their reduced levels potentially being a key factor in male sterility. Under high-temperature stress, these hormonal interactions and synergistic regulations are disrupted, further exacerbating male sterility in *G. barbadense*.

WGCNA is an effective technique for classifying transcriptome data into coexpression modules to reduce the number of potential candidate genes [43–46]. In this study, we identified four modules, among which the blue and turquoise modules were strongly correlated with the male sterility of *G. barbadense*. Through transcriptomic and hormone profile analysis of *G. barbadense* tissue, we established a coexpression network of hormones, TFs, and their target genes under HT stress. The synergistic relationship between hormone signals under HT stress was revealed, providing valuable insights into the molecular mechanism of male sterility in *G. barbadense*.

By integrating transcriptomic and metabolomic data, researchers can gain a deeper understanding of biological processes such as plant growth, development, and stress responses [47]. This comprehensive analysis not only helps to reveal the regulatory networks between genes and metabolism but also further explores the complex biological mechanisms that occur within plants. In this experiment, we selected two modules (blue and turquoise) that were significantly correlated with hormones at different times and identified the top five genes in each module as candidate genes (Table S1). Among these genes, the ALA4 (Gbar_A09G027130) gene (ranked first in the blue module) has been reported to be associated with male fertility in *Arabidopsis thaliana* [48]. Two mutations in the Gbar_A02G001830 (SBP1) gene lead to the acquisition of stop codons in the amino acid sequence, and these two mutations can be used as candidate gene markers.

To gain a more comprehensive understanding of the molecular mechanisms underlying male sterility in *G. barbadense* under high-temperature stress, future research should delve deeper into the interactions [49] and synergistic regulatory mechanisms [50] among these hormones. On one hand, focusing on candidate genes closely associated with male sterility (such as ALA4 and SBP1), functional studies using gene editing and expression pattern analysis [51] could elucidate their specific roles under high-temperature stress, revealing how they regulate anther dehiscence and pollen viability. On the other hand, multi-omics approaches, including transcriptomics and metabolomics, should be employed to construct and refine hormone regulatory network models [52]. This would help analyze the interactions and coordinated regulation of key hormones (such as ABA, JA, IAA, tZR, and GA20) under high-temperature stress, clarifying how these hormones influence male fertility in *G. barbadense* through complex signal transduction pathways [53].

This study not only summarizes the interaction network of hormones, transcription factors, and genes related to HT stress in *G. barbadense* but also provides a basis for analyzing the molecular mechanism of male sterility in *G. barbadense*.

4. Materials and Methods

4.1. Plant Materials and Growth Conditions

In this study, *G. barbadense* GB150 was used as the test material (this material is an HT-sensitive material, and the anther does not dehisce following HT stress). The test was conducted in 2023 in Awati County (80°39′ E, 40°39′ N). A warming shed was built to stress GB150 in the field, and GB150 materials in the natural state of the field were used as normal controls. In accordance with the weather conditions in Awati County and the growth of *G. barbadense*, field warming treatment was performed for 10 days during the cotton blooming period (9 July 2023–18 July 2023). The temperature was increased every day from 11:00 to 20:00. Thermometers were hung 30 cm above the canopy inside and outside the greenhouses. The average temperature increase during the stress period was 3.645 °C (Figure S3).

4.2. Determination of Flower Organ Indices

The stylar length, stigma length, stamen length, and single anther length were measured after HT stress.

4.3. Determination of Plant Hormone Contents

Anther samples of *G. barbadense* GB150 stored at ultralow temperature were removed and ground (30 Hz, 1 min) with a grinder until powdered. Fifty milligrams of ground sample was weighed, 10 μL of internal standard mixed solution with a concentration of

100 ng/mL, and 1 mL of methanol/water/formic acid (15:4:1, v/v/v) extractant were added, and the mixture was mixed well. The mixture was vortexed for 10 min and then centrifuged at 4 °C and 12,000 r/min for 5 min, after which the supernatant was transferred to a new centrifuge tube for concentration. After concentration, the mixture was redissolved in 100 μL of 80% methanol/aqueous solution, and the concentration was filtered through a 0.22 μm filter. The contents of ABA, JA, SA, tZR, IAA, and GA_{20} were determined by liquid chromatography–tandem mass spectrometry (LC-MS/MS) analysis.

4.4. Transcriptomic Sequencing

Three biological replicates were used to construct RNA-seq libraries of 6 samples from GB150 anther tissue of *G. barbadense* under HT stress compared with a normal control, and sequencing was performed using an Illumina platform. Strict quality control was performed using fastp, and high-quality data were used for subsequent analysis. The *G. barbadense* 3-79 genome (https://cottonfgd.net (*G. barbadense*, HAU)) (accessed on 10 December 2024) [54] was used as a reference genome. Clean reads were sequentially compared with the reference genome in HISAT2 to obtain their position information on the reference genome or gene. Finally, the reads were assembled into transcripts using StringTie.

4.5. Coexpression Network Construction

We used the pickSoftThreshold function in the WGCNA package in R to calculate the optimal power value (the RsquaredCut value was set to 0.85), which identified 4 different expression modules. Finally, Cytoscape (v3.10.0) software was used to visualize the coexpression network.

4.6. qRT-PCR

Primer Premier 5 software was used for primer design, and primer specificity was detected using DNAMAN (9.0) software. qRT-PCR analysis was performed using an Applied Biosystems 7500 Fast real-time quantitative PCR instrument (Applied Biosystems, Foster City, CA, USA) and a Thermo SYBR qPCR Master Mix kit (Thermo Fisher Scientific (China) Co., Ltd., Shanghai, China) according to the instructions. Each template was set up for 3 biological replicates, and the relative gene expression was calculated using the $2^{-\Delta\Delta Ct}$ method.

5. Conclusions

Through integrated transcriptomic and metabolomic analysis, this study investigated the molecular mechanisms underlying high-temperature-induced male sterility in *Gossypium barbadens*. Our findings demonstrate that elevated temperatures significantly altered the phytohormone profiles, with marked increases in the ABA and JA levels accompanied by significant decreases in the IAA, tZR, and GA20 concentrations, while the SA content remained stable. Our comprehensive analysis identified multiple differentially expressed genes associated with hormone biosynthesis and signaling pathways. Utilizing weighted gene coexpression network analysis (WGCNA), we established hormone interaction networks and genome-wide coexpression patterns, revealing 10 candidate genes potentially involved in male sterility. The subsequent validation through qRT-PCR and polymorphism analysis highlighted the functional significance of the ALA4 and SBP1 genes in the sterility phenotype. These results provide novel insights into the molecular basis of thermosensitive male sterility in *Gossypium barbadens* and offer valuable genetic resources for developing heat-tolerant *Gossypium barbadens* cultivars through molecular breeding approaches.

Supplementary Materials: The following supporting information can be downloaded at https://www.mdpi.com/article/10.3390/ijms26083693/s1.

Author Contributions: Writing—original draft, J.L.; drafting and revising the manuscript, M.G., T.L. and Y.C.; data analysis, Y.Q. and Q.C.; funding acquisition, X.D.; conceptualization, K.Z. All authors have read and agreed to the published version of the manuscript.

Funding: This work was supported by the Key Research and Development Project in the Autonomous Region (2024A0003-2), The projects sponsored by the development fund for Xinjiang talents XL, Xinjiang Long-staple Cotton Breeding Joint Research Project (2022MH-01), Tianshan Talents Training Program (2023TSYCLJ0012).

Data Availability Statement: The original contributions presented in the study are included in the article/Supplementary Materials. Further inquiries can be directed to the corresponding author.

Conflicts of Interest: The authors declare that they have no known competing financial interests or personal relationships that could have appeared to influence the work reported in this paper.

Abbreviations

The following abbreviations are used in this manuscript:

HT	High-temperature
G. Barbadense	*Gossypium barbadense*
qRT-PCR	Quantitative real-time polymerase chain reaction
DEGs	Differentially expressed genes
PCA	Principal components analysis
WGCNA	weighted gene co-expression network analysis
ABA	Abscisic Acid
JA	Jasmonic Acid
SA	Salicylic Acid
GA20	Gibberellic Acid 20
tZR	trans-Zeatin Riboside
IAA	3-Indoleacetic acid
VED	Violaxanthin de-epoxidase
ABF	Abscisic Acid (ABA) Responsive Element Binding Factors
NCED	9-cis-epoxycarotenoid dioxygenase gene
CYP	Cytochrome P450
AOS	Allene Oxide Synthase
LOX	Lysyl oxidase
MYC	Transcription factor MYC
JAR	Jasmonate Responsive
OPR	12-oxo-phytodienoic acid reductase
NPR	Non-expressor of pathogenesis-related
PAL	Phenylalanine ammonia-lyase
ICS	Isochorismate synthase
UGT	UDP-glucuronosyltransferase
BSMT	Benzoic acid/salicylic acid carboxyl methyltransferase
TGA	TGACG motif binding factor
ARF	Auxin Response Factors
AUX	Auxin/Indole-3-Acetic Acid
SAUR	Small Auxin Up RNA
ARR	Arabidopsis response regulators
AHP	Histidine-containing Phosphotransfer Protein
ZOG	Zeatin O-glucosyltransferase
miaA	tRNA dimethylallyltransferase
CKX	Cytokinin Oxidase/Dehydrogenase

DELLA	DELLA protein
GA2OX	Gibberellin 2-oxidase
GID	Gibberellin receptor
PIF	Phytochrome-interacting Factors

References

1. Cao, Z.; Tang, H.; Cai, Y.; Zeng, B.; Zhao, J.; Tang, X.; Lu, M.; Wang, H.; Zhu, X.; Wu, X.; et al. Natural variation of *HTH5* from wild rice, *Oryza rufipogon* Griff., is involved in conferring high-temperature tolerance at the heading stage. *Plant Biotechnol. J.* **2022**, *20*, 1591–1605. [CrossRef] [PubMed]
2. Reddy, K.R.; Hodges, H.F.; McCarty, W.H.; McKinion, J.M. *Weather and Cotton Growth: Present and Future*; MSU-MAFES: Starkville, MS, USA, 1996.
3. Pettigrew, W.T. The effect of higher temperatures on cotton lint yield production and fiber quality. *Crop Sci.* **2008**, *48*, 278–285. [CrossRef]
4. Achard, P.; Renou, J.P.; Berthomé, R.; Harberd, N.P.; Genschik, P. Plant DELLAs restrain growth and promote survival of adversity by reducing the levels of reactive oxygen species. *Curr. Biol.* **2008**, *18*, 656–660. [CrossRef]
5. Hedhly, A.; Hormaza, J.I.; Herrero, M. Global warming and sexual plant reproduction. *Trends Plant Sci.* **2009**, *14*, 30–36. [CrossRef]
6. Zinn, K.E.; Tunc-Ozdemir, M.; Harper, J.F. Temperature stress and plant sexual reproduction: Uncovering the weakest links. *J. Exp. Bot.* **2010**, *61*, 1959–1968. [CrossRef]
7. Zahid, K.R.; Ali, F.; Shah, F.; Younas, M.; Shah, T.; Shahwar, D.; Hassan, W.; Ahmad, Z.; Qi, C.; Lu, Y.; et al. Response and tolerance mechanism of cotton *Gossypium hirsutum* L. to elevated temperature stress: A review. *Front. Plant Sci.* **2016**, *7*, 937. [CrossRef]
8. Lobell, D.B.; Asner, G.P. Climate and management contributions to recent trends in US agricultural yields. *Science* **2003**, *299*, 1032. [CrossRef]
9. Pan, X.; Du, L.; Tao, J.; Jiang, S.; Qian, D.; Duan, J. Dynamic changes of flavonoids in *Abelmoschus manihot* different organs at different growth periods by UPLC-MS/MS. *J. Chromatogr. B* **2017**, *1059*, 21–26. [CrossRef]
10. Yi, D.; Zhang, H.; Lai, B.; Liu, L.; Pan, X.; Ma, Z.; Wang, Y.; Xie, J.; Shi, S.; Wei, Y. Integrative analysis of the coloring mechanism of red longan pericarp through metabolome and transcriptome analyses. *J. Agric. Food Chem.* **2020**, *69*, 1806–1815. [CrossRef]
11. Wang, S.; Liu, L.; Mi, X.; Zhao, S.; An, Y.; Xia, X.; Guo, R.; Wei, C. Multi-omics analysis to visualize the dynamic roles of defense genes in the response of tea plants to gray blight. *Plant J.* **2021**, *106*, 862–875. [CrossRef]
12. Li, S.; Deng, B.; Tian, S.; Guo, M.; Liu, H.; Zhao, X. Metabolic and transcriptomic analyses reveal different metabolite biosynthesis profiles between leaf buds and mature leaves in *Ziziphus jujuba* mill. *Food Chem.* **2021**, *347*, 129005. [CrossRef]
13. Li, C.; Xin, M.; Li, L.; He, X.; Yi, P.; Tang, Y.; Li, J.; Zheng, F.; Liu, G.; Sheng, J.; et al. Characterization of the aromatic profile of purple passion fruit (*Passiflora edulis* Sims) during ripening by HS-SPME-GC/MS and RNA sequencing. *Food Chem.* **2021**, *355*, 129685. [CrossRef] [PubMed]
14. Zhu, Y.; Wang, Q.; Wang, Y.; Xu, Y.; Li, J.; Zhao, S.; Wang, D.; Ma, Z.; Yan, F.; Liu, Y. Combined transcriptomic and metabolomic analysis reveals the role of phenylpropanoid biosynthesis pathway in the salt tolerance process of *Sophora alopecuroides*. *Int. J. Mol. Sci.* **2021**, *22*, 2389–2399. [CrossRef]
15. Wang, X.; Shan, X.; Wu, Y.; Su, S.; Li, S.; Liu, H.; Han, J.; Xue, C.; Yuan, Y. iTRAQ-based quantitative proteomic analysis reveals new metabolic pathways responding to chilling stress in maize seedlings. *J. Proteom.* **2016**, *146*, 14–24. [CrossRef]
16. Yu, Z.; Duan, X.; Luo, L.; Dai, S.; Ding, Z.; Xia, G. How plant hormones mediate salt stress responses. *Trends Plant Sci.* **2020**, *25*, 1117–1130. [CrossRef]
17. Yoshida, T.; Mogami, J.; Yamaguchi-Shinozaki, K. ABA-dependent and ABA-independent signaling in response to osmotic stress in plants. *Curr. Opin. Plant Biol.* **2014**, *21*, 133–139. [CrossRef] [PubMed]
18. Tamura, N.; Yoshida, T.; Tanaka, A.; Sasaki, R.; Bando, A.; Toh, S.; Lepiniec, L.; Kawakami, N. Isolation and characterization of high temperature-resistant germination mutants of *Arabidopsis thaliana*. *Plant Cell Physiol.* **2006**, *47*, 1081–1094. [CrossRef]
19. Nie, H.; Cheng, C.; Kong, J.; Li, H.; Hua, J. Plant non-coding RNAs function in pollen development and male sterility. *Front. Plant Sci.* **2023**, *14*, 1109941. [CrossRef]
20. You, J.; Li, M.; Li, H.; Bai, Y.; Zhu, X.; Kong, X.; Chen, X.; Zhou, R. Integrated methylome and transcriptome analysis widen the knowledge of cytoplasmic male sterility in cotton (*Gossypium barbadense* L.). *Front. Plant Sci.* **2022**, *13*, 770098. [CrossRef]
21. Zhang, J.; Wu, P.; Li, N.; Xu, X.; Wang, S.; Chang, S.; Zhang, Y.; Wang, X.; Liu, W.; Ma, Y.; et al. A male-sterile mutant with necrosis-like dark spots on anthers was generated in cotton. *Front. Plant Sci.* **2022**, *13*, 1102196. [CrossRef]
22. Ma, X.; Zheng, J.; Zhang, X.; Hu, Q.; Qian, R. Salicylic acid alleviates the adverse effects of salt stress on *Dianthus superbus* (Caryophyllaceae) by activating photosynthesis, protecting morphological structure, and enhancing the antioxidant system. *Front. Plant Sci.* **2017**, *8*, 600. [CrossRef] [PubMed]
23. Ali, S.; Rizwan, M.; Arif, M.S.; Ahmad, R.; Hasanuzzaman, M.; Ali, B.; Hussain, A. Approaches in enhancing thermotolerance in plants: An updated review. *J. Plant Growth Regul.* **2020**, *39*, 456–480. [CrossRef]

24. Suzuki, N.; Mittler, R. Reactive oxygen species and temperature stresses: A delicate balance between signaling and destruction. *Physiol. Plant.* **2006**, *126*, 45–51. [CrossRef]
25. Suzuki, N.; Koussevitzky, S.; Mittler, R.O.N.; Miller, G.A.D. ROS and redox signalling in the response of plants to abiotic stress. *Plant Cell Environ.* **2012**, *35*, 259–270. [CrossRef]
26. Mittler, R. Oxidative stress, antioxidants and stress tolerance. *Trends Plant Sci.* **2002**, *7*, 405–410. [CrossRef] [PubMed]
27. Oliver, S.N.; Dennis, E.S.; Dolferus, R. ABA regulates apoplastic sugar transport and is a potential signal for cold-induced pollen sterility in rice. *Plant Cell Physiol.* **2007**, *48*, 1319–1330. [CrossRef]
28. Qin, G.; Zhao, N.; Wang, W.; Wang, M.; Zhu, J.; Yang, J.; Lin, F.; Huang, X.; Zhang, Y.; Min, L.; et al. Glyphosate-Induced Abscisic Acid Accumulation Causes Male Sterility in Sea Island Cotton. *Plants* **2023**, *12*, 1046–1058. [CrossRef] [PubMed]
29. Zhang, M.; Liu, J.; Ma, Q.; Qin, Y.; Wang, H.; Chen, P.; Ma, L.; Fu, X.; Zhu, L.; Wei, H.; et al. Deficiencies in the formation and regulation of anther cuticle and tryphine contribute to male sterility in cotton PGMS line. *BMC Genom.* **2020**, *21*, 825. [CrossRef]
30. Gan, Z.; Feng, Y.; Wu, T.; Wang, Y.; Xu, X.; Zhang, X.; Han, Z. Downregulation of the auxin transporter gene SlPIN8 results in pollen abortion in tomato. *Plant Mol. Biol.* **2019**, *99*, 561–573. [CrossRef]
31. Chen, C.; Pang, Y.; Pan, X.; Zhang, L. Impacts of climate change on cotton yield in China from 1961 to 2010 based on provincial data. *J. Meteorol. Res.* **2015**, *29*, 515–524. [CrossRef]
32. Fragkostefanakis, S.; Mesihovic, A.; Simm, S.; Paupière, M.J.; Hu, Y.; Paul, P.; Mishra, S.K.; Tschiersch, B.; Theres, K.; Bovy, A.; et al. HsfA2 controls the activity of developmentally and stress-regulated heat stress protection mechanisms in tomato male reproductive tissues. *Plant Physiol.* **2016**, *170*, 2461–2477. [CrossRef] [PubMed]
33. Li, Y.; Chen, M.; Khan, A.H.; Ma, Y.; He, X.; Yang, J.; Zhang, R.; Ma, H.; Zuo, C.; Li, Y.; et al. Histone H3 lysine 27 trimethylation suppresses jasmonate biosynthesis and signaling to affect male fertility under high temperature in cotton. *Plant Commun.* **2023**, *4*, 100660. [CrossRef] [PubMed]
34. Khan, A.H.; Ma, Y.; Wu, Y.; Akbar, A.; Shaban, M.; Ullah, A.; Deng, J.; Khan, A.S.; Chi, H.; Zhu, L.; et al. High-temperature stress suppresses allene oxide cyclase 2 and causes male sterility in cotton by disrupting jasmonic acid signaling. *Crop J.* **2023**, *11*, 33–45. [CrossRef]
35. Luo, P.; Di, D.; Wu, L.; Yang, J.; Lu, Y.; Shi, W. MicroRNAs are involved in regulating plant development and stress response through fine-tuning of TIR1/AFB-dependent auxin signaling. *Int. J. Mol. Sci.* **2022**, *23*, 510. [CrossRef]
36. Nagpal, P.; Ellis, C.M.; Weber, H.; Ploense, S.E.; Barkawi, L.S.; Guilfoyle, T.J.; Hagen, G.; Alonso, J.M.; Cohen, J.D.; Farmer, E.E.; et al. Auxin response factors ARF6 and ARF8 promote jasmonic acid production and flower maturation. *Development* **2005**, *132*, 4107–4118. [CrossRef]
37. Hu, Y.; Zhou, L.; Huang, M.; He, X.; Yang, Y.; Liu, X.; Li, Y.; Hou, X. Gibberellins play an essential role in late embryogenesis of Arabidopsis. *Nat. Plants* **2018**, *4*, 289–298. [CrossRef]
38. Kawai, K.; Takehara, S.; Kashio, T.; Morii, M.; Sugihara, A.; Yoshimura, H.; Ito, A.; Hattori, M.; Toda, Y.; Kojima, M.; et al. Evolutionary alterations in gene expression and enzymatic activities of gibberellin 3-oxidase 1 in Oryza. *Commun. Biol.* **2022**, *5*, 67. [CrossRef]
39. Liu, H.; Shang, H.; Yang, H.; Liu, W.; Tsugama, D.; Nonomura, K.-I.; Zhou, A.; Wu, W.; Takano, T.; Liu, S. RNA-Binding Protein MAC5A Is Required for Gibberellin-Regulated Stamen Development. *Int. J. Mol. Sci.* **2022**, *23*, 2009. [CrossRef]
40. Plackett, A.R.G.; Ferguson, A.C.; Powers, S.J.; Wanchoo-Kohli, A.; Phillips, A.L.; Wilson, Z.A.; Hedden, P.; Thomas, S.G. DELLA activity is required for successful pollen development in the Columbia ecotype of Arabidopsis. *New Phytol.* **2014**, *201*, 825–836. [CrossRef]
41. Yang, D.-L.; Yao, J.; Mei, C.-S.; Tong, X.-H.; Zeng, L.-J.; Li, Q.; Xiao, L.-T.; Sun, T.-P.; Li, J.; Deng, X.-W.; et al. Plant hormone jasmonate prioritizes defense over growth by interfering with gibberellin signaling cascade. *Proc. Natl. Acad. Sci. USA* **2012**, *109*, E1192–E1200. [CrossRef]
42. Khan, A.H.; Min, L.; Ma, Y.; Zeeshan, M.; Jin, S.; Zhang, X. High-temperature stress in crops: Male sterility, yield loss and potential remedy approaches. *Plant Biotechnol. J.* **2023**, *21*, 680–697. [CrossRef] [PubMed]
43. Hollender, C.A.; Kang, C.; Darwish, O.; Geretz, A.; Matthews, B.F.; Slovin, J.; Alkharouf, N.; Liu, Z. Floral transcriptomes in woodland strawberry uncover developing receptacle and anther gene networks. *Plant Physiol.* **2014**, *165*, 1062–1075. [CrossRef]
44. Greenham, K.; Guadagno, C.R.; Gehan, M.A.; Mockler, T.C.; Weinig, C.; Ewers, B.E.; McClung, C.R. Temporal network analysis identifies early physiological and transcriptomic indicators of mild drought in *Brassica rapa*. *elife* **2017**, *6*, e29655. [CrossRef] [PubMed]
45. Schaefer, R.J.; Michno, J.-M.; Jeffers, J.; Hoekenga, O.; Dilkes, B.; Baxter, I.; Myers, C.L. Integrating coexpression networks with GWAS to prioritize causal genes in maize. *Plant Cell* **2018**, *30*, 2922–2942. [CrossRef]
46. Azam, M.; Zhang, S.; Huai, Y.; Abdelghany, A.M.; Shaibu, A.S.; Qi, J.; Feng, Y.; Liu, Y.; Li, J.; Qiu, L.; et al. Identification of genes for seed isoflavones based on bulk segregant analysis sequencing in soybean natural population. *Theor. Appl. Genet.* **2023**, *136*, 13. [CrossRef]

47. Zou, J.; Yang, L.; Li, Y.; Piao, M.; Li, Y.; Yao, N.; Zhang, X.; Zhang, Q.; Hu, G.; Yang, D.; et al. Comparative Proteomics Combined with Morphophysiological Analysis Revealed Chilling Response Patterns in Two Contrasting Maize Genotypes. *Cells* **2022**, *11*, 1321. [CrossRef] [PubMed]
48. Davis, J.A.; Pares, R.B.; Bernstein, T.; McDowell, S.C.; Brown, E.; Stubrich, J.; Rosenberg, A.; Cahoon, E.B.; Cahoon, R.E.; Poulsen, L.R.; et al. The lipid flippases ALA4 and ALA5 play critical roles in cell expansion and plant growth. *Plant Physiol.* **2020**, *182*, 2111–2125. [CrossRef]
49. Raza, A.; Salehi, H.; Rahman, A.; Zahid, Z.; Haghjou, M.M.; Najafi-Kakavand, S.; Charagh, S.; Osman, H.S.; Albaqami, M.; Zhuang, Y.; et al. Plant hormones and neurotransmitter interactions mediate antioxidant defensesunder induced oxidative stress in plants. *Front. Plant Sci.* **2022**, *13*, 961872. [CrossRef]
50. Waadt, R.; Seller, C.A.; Hsu, P.-K.; Takahashi, Y.; Munemasa, S.; Schroeder, J.I. Plant hormone regulation of abiotic stress responses. *Nat. Rev. Mol. Cell Biol.* **2022**, *23*, 680–694. [CrossRef]
51. Wang, J.Y.; Doudna, J.A. CRISPR technology: A decade of genome editing is only the beginning. *Science* **2023**, *379*, eadd8643. [CrossRef]
52. Singh, S.; Praveen, A.; Dudha, N.; Bhadrecha, P. Integrating physiological and multi-omics methods to elucidate heat stress tolerance for sustainable rice production. *Physiol. Olecularbiol. Plants* **2024**, *30*, 1185–1208. [CrossRef] [PubMed]
53. Lu, F.; Sun, X.; Dai, X.; Zhang, P.; Ma, Y.; Xu, Y.; Wang, L.; Zhang, J. Integrated Multi-Omics Analysis to Investigate the Molecular Mechanisms Underlying the Response of *Auricularia heimuer* to High-Temperature Stress. *J. Fungi* **2025**, *11*, 167. [CrossRef] [PubMed]
54. Wang, M.; Tu, L.; Yuan, D.; Zhu, D.; Shen, C.; Li, J.; Liu, F.; Pei, L.; Wang, P.; Zhao, G.; et al. Reference genome sequences of two cultivated allotetraploid cottons, *Gossypium hirsutum* and *Gossypium barbadense*. *Nat. Genet.* **2019**, *51*, 224–229. [CrossRef] [PubMed]

International Journal of
Molecular Sciences

MDPI

Article

Genome-Wide Association Study and Candidate Gene Mining of Husk Number Trait in Maize

Yancui Wang, Shukai Wang, Dusheng Lu, Ming Chen, Baokun Li, Zhenhong Li, Haixiao Su, Jing Sun, Pingping Xu and Cuixia Chen *

College of Agronomy, Shandong Agricultural University, Tai'an 271018, China; wyc5515@163.com (Y.W.); sdauwangshukai@163.com (S.W.); ldusheng@163.com (D.L.); chming_11@126.com (M.C.); b1774562702@163.com (B.L.); lzh980525@163.com (Z.L.); suhaixiao2019@163.com (H.S.); sunjing010525@163.com (J.S.); 18706388706@163.com (P.X.)

* Correspondence: cxchen@sdau.edu.cn

Abstract: Husk number (HN) trait is an important factor affecting maize kernel dehydration rate after the physiological maturity stage. In general, a reasonable reduction in HN is a key target sought for breeding maize varieties that are suitable for mechanized harvesting. In this study, the HN of a maize natural population panel containing 232 inbred lines was analyzed, and the results showed a broad-sense heritability of 0.89, along with a wide range of phenotypic variation. With the best linear unbiased prediction (BLUP) values across the three environments, a genome-wide association study (GWAS) was conducted using 995,106 single-nucleotide polymorphism (SNP) markers. A total of 16 SNPs significantly associated with HN were identified by the mixed linear model and general linear model using the TASSEL 5.0 software program. A local linkage disequilibrium (LD) study was performed to infer the candidate interval around the lead SNPs. A total of 19 functionally annotated genes were identified. The candidate genes were divided into multiple functional types, including transcriptional regulation, signal transduction, and metabolic and cellular transport. These results provide hints for the understanding of the genetic basis of the HN trait and for the breeding of maize varieties with fewer HN and faster dehydration rate.

Keywords: maize (*Zea mays* L.); husk number; GWAS; favorable allele; dehydration rate

1. Introduction

Maize (*Zea mays* L.) is one of the most widely cultivated crops in the world and serves as an important source of food, feed, and industrial material [1]. Maize husk is a natural protective barrier enwrapping the ear, originating from the husk primordium at the internodes of the ear stalk [2]. The husks play a crucial role in maintaining optimal temperature and humidity levels for kernel growth, preventing poor grain filling from heat loss or water deficiency [3–5], and act as a barrier that not only reduces the risk of pathogen infection but also prevents pests and birds from damaging the ear [6–8]. In addition, the husks can temporarily store nutrients coming from other organs and transport them to the ear, and they are also capable of both C3 and C4 photosynthesis, therefore accumulating photosynthetic products by themselves [9,10].

However, excessive husk coverage can hinder moisture dissipation from the kernel after physiological maturity, therefore slowing down the kernel dehydration. Husk-related traits in maize such as husk number (HN), length, width, thickness, and tightness, are important factors influencing the kernel dehydration rate [9,11]. Among these, HN is a key

https://doi.org/10.3390/ijms26073437

trait that negatively correlates with the kernel dehydration rate [12]. The HN in modern maize varieties has decreased to 9 compared with 14 in old varieties in America [13]. Li et al. even investigated the effects of removing husks on the grain yield and dehydrating rate, and the results indicated that as more husks were removed, the kernel dehydration became faster [12]. Previous studies examined 508 maize inbred lines, and the results showed that the HN ranged from 5 to 20, suggesting significant phenotypic variation in HN and great potential for the genetic improvement of the HN trait [11,14].

The HN is a complex quantitative trait controlled by multiple quantitative trait loci (QTL) with minor effects in maize. A number of studies have used genome-wide association studies (GWAS) and QTL mapping to identify QTL and single-nucleotide polymorphisms (SNPs) related to HN. Zhang et al. identified a total of 7 QTLs for HN on chromosomes 1, 2, 3, 4, 6, and 9 [15]. Among these, the physical interval of *qHN4* was 0.89 Mb and contained 7 candidate genes, which was detected across three different environments. Zhang et al. studied a BC_2F_8 population derived from the maize inbred line Mo17 and teosinte X26-4, and found that two HN-associated QTLs (*qHN-1-1* and *qHN-1-2*) each explained 8.9% of the phenotypic variation [16]. Zhou et al. performed a QTL analysis on 204 recombinant inbred lines derived from DH4866 and T877 and identified one QTL on each of chromosomes 1, 3, 6, and 7, among which *qHN7* showed the largest effect, explaining 19.6% of phenotypic variation [17]. Through fine mapping, they narrowed down *qHN7* to a physical interval of 721.1 kb and predicted four candidate genes. With the development of high-throughput sequencing technology, GWAS based on linkage disequilibrium (LD) has emerged as a powerful approach for dissecting the genetic bases of complex traits [18–20]. Zhou et al. used 253 maize inbred lines and the Maize SNP3 K Bead-chip to conduct a GWAS for HN. They identified 24 SNPs that were significantly associated with HN ($-\log_{10}(P) > 3$), among which 8 SNPs were consistently detected across three different environments [14]. Cui et al. utilized 508 maize inbred lines and 543,641 SNPs to detect associated SNPs for husk traits [11]. They identified 5 and 9 SNPs significantly associated with HN using mixed linear models (MLMs) and general linear models (GLMs), respectively. Wang et al. revealed that a 10 bp natural variant in the 3′UTR of *Zea methyltransferase2* (*ZMET2*) was associated with HN in maize [21]. *ZMET2* encodes a DNA methyltransferase, and the HN of *zmet2* mutants significantly increased.

In this study, GWAS was performed for the HN trait in the association panel of 232 maize inbred lines with 995,106 SNPs. Genes within the same LD block of the significantly associated SNPs were further screened to identify potential candidate genes affecting HN. This study aimed to identify HN-related candidate genes and provide new resources for understanding the genetic basis of husk development.

2. Results

2.1. Phenotypic Variations in Husk Number (HN)

The association panel in this study was planted in three environments in Sanya and Taian from 2015 to 2017 (SY2015, TA2016, and TA2017). The HN ranged from 4.80 to 19.00, 6.00 to 18.33, and 5.50 to 16.75, with mean values of 8.56, 9.85, and 9.70 in the three environments, respectively (Table 1), indicating abundant phenotypic variation in the HN trait. The variation in HN follows a normal distribution (Figure 1), with the skewness ranging from 0.94 to 1.15 and kurtosis ranging from 0.63 to 2.01 in different environments, and this is consistent with the characteristics of quantitative traits. High broad-sense heritability (H^2 = 88.62%) was observed for the HN, suggesting that the HN trait is mainly controlled by genetic factors and suitable for further association analysis.

Table 1. Descriptive statistical analysis of husk number traits in maize.

Environment	Range	Mean ± SD	CV (%)	Skewness	Kurtosis	H^2 (%) [a]
SY2015	4.80–19.00	8.56 ± 2.14	25.35	1.15	2.01	
TA2016	6.00–18.33	9.85 ± 2.25	22.84	1.04	1.33	88.62
TA2017	5.50–16.75	9.70 ± 2.22	22.89	0.94	0.63	

SY2015: Sanya in 2015; TA2016: Taian in 2016; TA2017: Taian in 2017; SD: standard deviation; CV: coefficient of variation; [a] broad-sense heritability estimated across three environments.

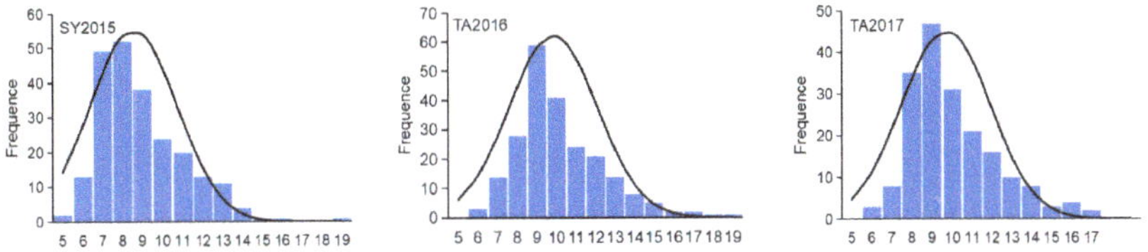

Figure 1. Frequency distribution of husk number. SY2015: Sanya in 2015; TA2016: Taian in 2016; TA2017: Taian in 2017.

Adopting the classification by Yang et al. [22], the association panel was categorized into four subpopulations: TST, NSS, SS, and Mixed (Table S1). SS and NSS germplasms originate from temperate regions, and TST germplasms from tropical and subtropical areas, with the remaining inbred lines being the Mixed subpopulation. To investigate the impact of population structure on the HN trait, we used phenotypic best linear unbiased prediction (BLUP) values to analyze the phenotypic variation among these four subpopulations. The results indicate that the average husk number in the TST subpopulation was higher than those in NSS and SS (Figure S1), suggesting that maize inbred lines from tropical or subtropical regions tend to have higher husk number.

2.2. Genome-Wide Association Analysis (GWAS) and Mining the Favorable Alleles

To reduce the effect of environmental variation, BLUP values across three environments (SY2015, TA2016, and TA2017) were used for the association study. The mixed linear model (MLM) was used to conduct GWAS for mining the significant SNPs associated with HN. Both population structure (Q matrix) and kinship relationship (K matrix) were incorporated to avoid false associations. With a threshold of $-\log_{10}(P) > 5.12$ ($P < 7.6 \times 10^{-6}$), a total of 16 significantly associated SNP sites were detected, which were distributed on chromosomes 1, 4, 6, 7, and 10 (Figure 2a, Table S2). These significant SNPs explained between 9.6% to 11.9% of the observed phenotypic variation. Furthermore, five SNPs were detected on chromosomes 4 and 6 at the threshold of $-\log_{10}(P) > 6.42$ ($P < 3.8 \times 10^{-7}$) by the general linear model (GLM), which could explain 7.8–10.4% of the phenotypic variation (Figure 2b, Table S2). A total of 16 SNPs were identified by the GLM and MLM, of which five SNPs were present in both models. When multiple significant SNPs were highly linked within a linkage disequilibrium (LD) block, they were merged into one genomic locus. We selected the peak SNP within each LD block for a subsequent analysis.

In this study, SNP alleles with a lower HN were defined as favorable alleles, whereas those with a higher HN were defined as unfavorable alleles. The allelic effects of different haplotypes of each significant SNPs on the HN trait were analyzed (Figure 3). The average HN for allele T of chr1.S_88001442 was 9.16, which was significantly smaller than that for allele C (10.39, $p \leq 0.001$). Similarly, other favorable alleles for the HN included allele A of chr1.S_281597633, allele C of chr4.S_11389760, allele G of chr6.S_162115837, allele A of chr6.S_162882875, allele C of chr7.S_9364384, and allele C of chr10.S_141001231. The HN for lines carrying these alleles was 1.85, 1.50, 2.56, 2.08, 1.89, and 2.28 lower than those carrying unfavorable alleles, respectively.

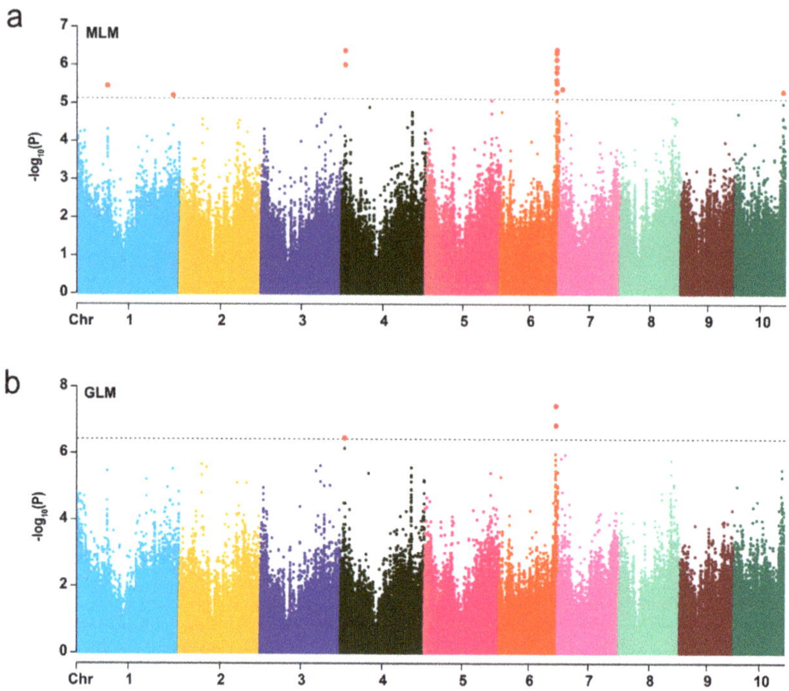

Figure 2. Manhattan plot for genome-wide association analysis of husk number. The SNPs are represented by dots. The horizontal dashed line represents the $-\log_{10}(P) > 5.12$ ($P < 7.6 \times 10^{-6}$) (**a**) and $-\log_{10}(P) > 6.42$ ($P < 3.8 \times 10^{-7}$) (**b**) significant thresholds, respectively. Red dots above the dashed line indicated SNPs significantly associated with husk number. MLM: mixed linear model; GLM: general linear model.

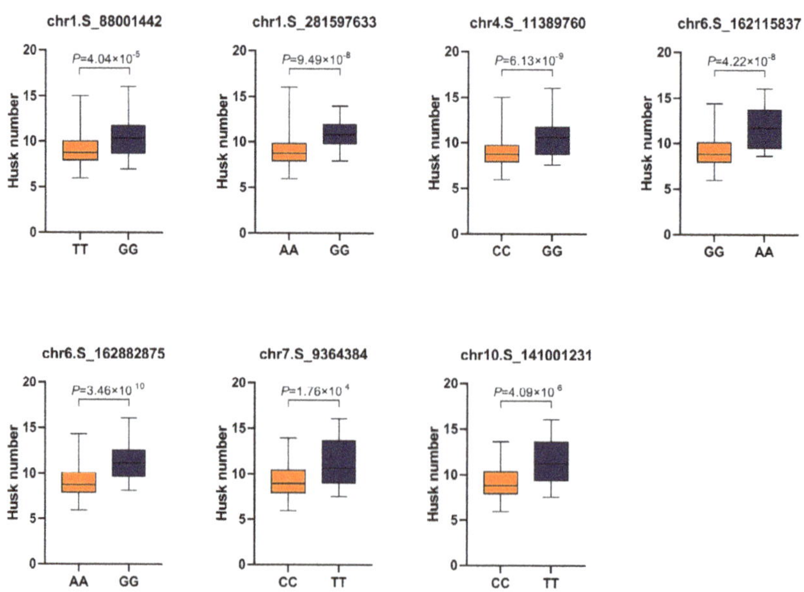

Figure 3. Allelic effects of the significant sites associated with husk number.

Further, we examined the allele frequency of significantly associated SNP sites in different subpopulations. Interestingly, the temperate maize inbred lines (SS and NSS) had higher allele frequencies for low-HN alleles for the most of significant SNPs, whereas tropical maize inbred lines (TST subpopulation) tended to favor the high-HN alleles (Figure 4). For example, for chr6.S_162115837, the high-HN allele (AA) is found in the TST subpopulation at a frequency of 87.80%, while it was zero in the SS and NSS subpopulations with the remaining 23.33% found in the Mixed population. Similarly, for chr4.S_11389760, the high-HN allele (GG) was observed with a frequency of 70.21% in the TST subpopulation, but its frequency was 11.34% in the SS and NSS subpopulations (Figure 4). These results suggest that the low-HN trait is favored in the temperate germplasms and may play a role during maize domestication and adaptation, leading to a rapid rise in the low-HN alleles within temperate maize germplasm.

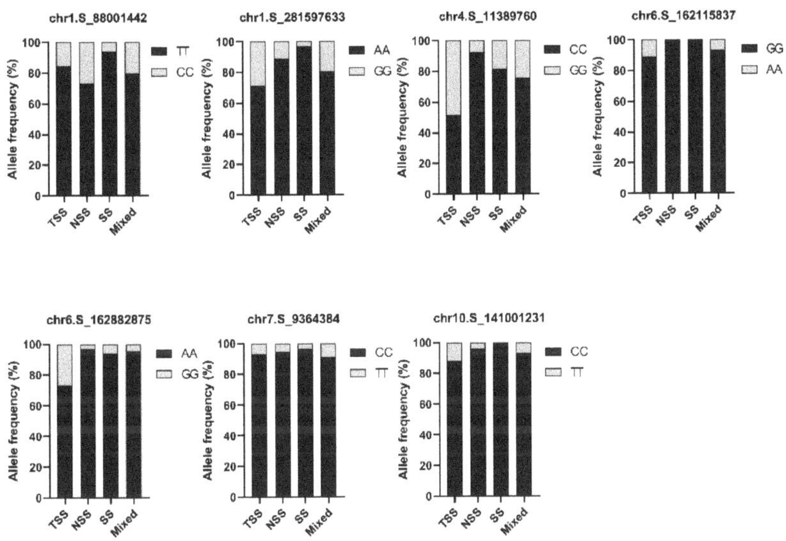

Figure 4. Allele frequency of SNPs significantly associated with husk number in different subpopulations. Dark-colored columns represent less-HN allele, and light-colored columns represent high-HN allele. TST, tropical/subtropical; SS, stiff-stalk; NSS, non-stiff-stalk; Mixed, the remaining inbred lines being Mixed subpopulation.

2.3. Candidate Genes Related to the HN

Among the 16 SNPs that were detected by the MLM and GLM, 14 were present in the genic regions (Table S3). The SNP chr1.S_88001442 is located in the 5′UTR of *GRMZM2G071172*, and chr4.S_11389625 and chr4.S_11389760 are located in the fourth exon of *GRMZM2G008259*. chr6.S_162115732, chr6.S_162115778, chr6.S_162115837, chr6.S_162115848 and chr6.S_162115911 are located in the 5′UTR of *GRMZM2G175089*. chr6.S_162882875, chr6.S_162882956 and chr6.S_162883019 are located in the first exon of *GRMZM2G038032*, and chr6.S_162883376 is located in the intron of this gene. chr7.S_9364384 is located in the 18th intron of *GRMZM2G456059*. chr1.S_281597633 is located in the intergenic region, and the nearest gene is *GRMZM2G028151*. In addition, we calculated the LD between the associated SNP and neighboring SNPs within a 200 kb region (100 kb each upstream and downstream) to substantiate the evidence for identifying the candidate region (Figure S2). The genes located within the LD block around significant SNPs were considered as candidate genes. Finally, nineteen candidate genes have been located on chromosomes 1, 4, 6, 7, and 10 (3, 1, 10, 4, and 1 gene, respectively).

The candidate genes were divided into six functional types, including transcriptional regulation, signal transduction, metabolic, cellular transport, cell cycle regulation, and programmed cell death (Table S3). Based on published transcriptome data, the expression patterns of candidate genes were analyzed at different stages (including the primordium initiation period) of husk and leaf development (Figure 5). Except for *GRMZM2G476357*, whose expression is extremely low, the other genes are expressed in both the husk and leaf tissue. Specifically, *GRMZM2G038032*, *GRMZM2G174990*, and *GRMZM2G008259* are highly expressed in both tissues. The expression of *GRMZM2G038032* was higher in the early stage of primordium development, suggesting that this gene may regulate husk development at the critical early stage. In addition, nucleotide diversity analysis was performed on the *GRMZM2G038032* gene region and 5 kb upstream and downstream of the coding region using the third-generation Zea mays haplotype map (HapMap 3) data including teosintes, landraces, and modern maize varieties (Figure S3). The results showed that the nucleotide diversity of this gene's promoter region in modern maize varieties was lower than that in teosinte and landraces, suggesting that the promoter/regulatory region of *GRMZM2G038032* was affected by selection during maize domestication. *GRMZM2G153754*, *GRMZM2G008259*, *GRMZM2G079723*, *GRMZM2G079617*, *GRMZM2G028151*, and *GRMZM2G174949* had similar temporal expression patterns to that of *GRMZM2G038032*: the expression levels of these genes were all higher at early stage of primordium development (Figure 5).

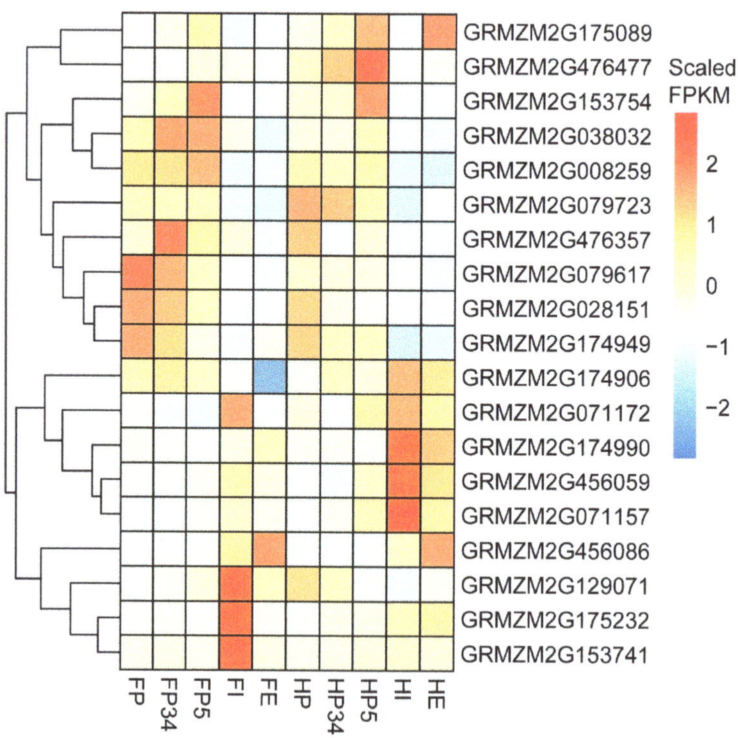

Figure 5. Heat-map of expression levels of the candidate genes identified by GWAS. The expression data are from leaf and husk leaf, and the description of the tissue is detailed in the study of Wang et al. [2]. Foliar primordia (FP), Foliar immature (FI), and foliar expanded (FE) leaf blades; husk primordia (HP), the outermost leaf on the ear (HE), and the third leaf in from the outside (HI). The color scale represents scaled FPKM values normalized using the z-score approach.

3. Discussion

3.1. Phenotypic Analysis of Husk Number (HN)

The maize husks wrap around the ear to help the normal growth and development of the kernel. However, excessive husk wrapping can significantly hinder grain dehydration after physiological maturity. The HN is a quantitative trait controlled by multiple genes or loci [11,23]. In this study, the HN of 232 maize inbred lines exhibited wide phenotypic variation and normal distribution. Maize originated in the tropics and subsequently adapted to the high latitudes of the temperate regions [24]. Comparing the HN among different subpopulations, our results indicate that maize from tropical/subtropical regions tends to have a higher husk number (Figure S1). The results of our study are consistent with those presented by Cui et al., who also showed that the rapid accumulation of fewer, narrower, and thinner husks in temperate maize germplasm as maize cultivation spreads from tropical to temperate regions [11].

3.2. SNP Sites Associated with HN via GWAS

Compared with linkage analysis, association analysis includes a wider range of natural variants and has the advantage of mining for more favorable allelic variants at higher resolution in a shorter period [25,26]. Maize is a model plant for association analysis because of its rich phenotypic variation and genetic diversity [20,27]. Currently, association analysis has become an efficient and powerful method to analyze the genetic mechanism of maize complex traits. In this study, we identified 16 SNPs significantly associated with the HN trait via GWAS. The favorable allelic variation in the identified SNPs had a significant regulatory effect on the HN (Figure 3), indicating that these SNPs had great potential application. These novel SNPs provide novel genetic loci for further analysis of the regulation mechanism of the HN in maize.

Previous studies have analyzed the correlations between husk number and other agronomic traits. The results showed that husk number was significantly correlated with the number of leaves above the ear, number of tassel branches, and flowering time-related traits, as well as some other traits [11]. In this study, to explore whether SNPs related to HN affect other agronomic traits, we analyzed the correlation between specific loci and eight traits using data sets from previous studies [28] (Figure S4). *tasselsheath1* (*tsh1*) and *tsh4* have been proven to establish developmental boundaries and maintain meristems, and mutants of these genes showed a reduced tassel branch and increased HN [29,30]. In this study, the SNPs chr4.S_11389760 and chr6.S_162882875 significantly affected tassel branch number (TBN) (Figure S4d), suggesting that HN and TBN may be subject to similar genetic regulation mechanisms, although these two traits were positively correlated. The leaf number above ear (LNE) at chr6.S_162115837 and chr6.S_162882875 was also positively correlated with the HN (Figure S4e). In addition, lines carrying low-HN alleles flowered significantly earlier than those with high-HN alleles (Figure S4a–c). Previous research has found that photoperiodic response factor *ZmCLΓ3.1* regulates gene expression of tsh4 by forming a protein complex with RAMOSA2 (RA2) [31]. These HN-related SNPs did not affect yield traits, including panicle length, row number, and 100-grain weight (Figure S4f–h). The above evidence shows the correlation between the HN and other traits. It is of great significance to understand the potential similar genetic mechanism between different traits and consider breeding for multiple traits.

3.3. Putative HN-Related Genes

The genes related to ear development and their regulatory pathways have been extensively studied. However, the molecular mechanism of the development of husk, which is a protective organ for healthy ear growth, is poorly understood. At present, only one gene, *ZMET2*, which regulates the HN, has been cloned [21]. Most of the existing studies remain at the preliminary stage, including large QTL intervals and SNP loci identified by GWAS and linkage analysis, and key genes at these loci have not been verified. In this study, several novel SNP sites associated with HN were identified via GWAS. We extracted genes within the LD block around peak SNPs and identified 19 candidate genes with functional annotations. According to the functional annotation, these genes are primarily categorized into several functional groups: transcriptional regulation, signal transduction, and metabolic. These processes are essential for plant growth and organ development.

Transcription factors are a class of important regulatory proteins, which is involved in multiple processes of plant growth and development. In this study, we identified five transcription factors *GRMZM2G028151*, *GRMZM2G008259*, *GRMZM2G175232*, *GRMZM2G476357*, and *GRMZM2G153754*. Specifically, *GRMZM2G028151* encodes AP2/EREBP protein. AP2/EREBP family regulates many biological processes in plants, such as plant morphogenesis, hormone signal transduction, and the regulation of the metabolic process. The loss of AP2 function has been shown to cause the homeotic transition of sepals to carpels, and petals to stamens, as *AP2-5*, *AP2-6*, and *AP2-7* mutants in Arabidopsis [32,33]. The maize *branched silkess1* (*bd1*) of the AP2/ERF family was specifically expressed in the ear, which affects flower organ development [34]. The AP2-like genes played an important role in modifying spike characteristics in barley and wheat [35]. Thus, it is speculated that the husk developed from the axillary meristem may also be regulated by the AP2 family gene. The MYB family is one of the largest transcription factors families in plants. *GRMZM2G175232* encodes MYB99 in maize. In Arabidopsis, several MYB transcription factors such as *AtMYB37/RA1*, *AtMYB38/RAX2/BIT1*, *AtMYB84/RAX3*, *AtMYB105/LOF2*, and *AtMYB117/LOF1* have been shown to be regulators of axillary meristem formation [36–38]. In addition, *AtMYB59* and *AtMYB77* have been confirmed to regulate root development and lateral root formation, respectively [39,40]. These studies indicate that MYB transcription factors play an important role in plant organ development.

Proper plant growth and development require various metabolites such as hormones, lipids, vitamins, and amino acids, and the synthesis and regulation of these substances are integral components of the plant metabolic pathway [41,42]. Plant metabolic pathways are complex and usually involve multiple enzymes. We found that candidate genes include several related enzymes in different metabolic pathways, including glucosaminly transferase (*GRMZM2G071172*), Polygalacturonase (*GRMZM2G079617*), and ribose-5-phosphate isomerase (*GRMZM2G456086*). Among them, Polygalacturonase is involved in cell expansion and division, and influences cell wall dynamics by regulating the structure of the cell wall [43].

How plants regulate and determine their growth, development, and morphogenesis is inseparable from complex signal transduction processes. Eight genes related to cell signal transduction were identified in this study. We identified a Rho of plants (ROP) guanine-nucleotide exchange factor (GEF), *GRMZM2G071157*. ROP is a key protein for polar signal transduction and controls auxin-dependent polar cell elongation during plant development [44]. *ROPGEF1* was shown to regulate the polarization of the auxin influx carrier AUX1 and the accumulation of efflux carriers PIN7 and PIN2 [45]. Mutations in *ROPGEF1* and *ROPGEF7* result in embryonic defects and abnormal cotyledon development [45,46]. Protein kinases control a series of cell processes, including metabolism, transcription, and cell cycle [47]. *GRMZM2G456059* encodes a serine/threonine-protein kinase, which is one of the major protein kinases.

GRMZM2G038032 encodes a receptor for activated C kinase 1 (RACK1). As a scaffold protein, RACK1 regulates plant growth and development by connecting various plant hormone signaling pathways [48], and its expression can be induced by auxin and cytokinin [49]. RACK1 has been demonstrated to interact with *GIF1* (GRF-interacting factor1), a key gene that regulates hormone biosynthesis and meristem determination and controls the structure and morphological development of maize ear [50]. *GIF1* mutants frequently develop axillary shoots in the axils of husk leaves and have an increased husk number [51]. In Arabidopsis, *GIF1/AN3* is necessary for the maintenance of shoot apical meristem, and is involved in the regulation of cell proliferation, growth, and lateral organ growth [52–54]. As *GRMZM2G038032* was highly expressed in the husk primordium (Figure 5), we postulate that this gene may be involved in the regulation of husk primordia developed from the axillary meristem through interactions with GIF1 in maize. In addition, through a nucleotide polymorphism analysis, we found that the sequence diversity of the promoter region of *GRMZM2G038032* in modern maize is low, indicating that this region was selected during domestication (Figure S3). Hence, it is reasonable to speculate that *GRMZM2G038032* is the most likely candidate gene on chromosome 6 for the regulation of HN in maize.

In summary, by analyzing the functional annotation and expression level of candidate genes, we speculated the possible pathways of some candidate genes controlling the HN trait. The other candidate genes that have not been discussed may also be involved in husk development in currently unknown regulatory pathways.

4. Materials and Methods

4.1. Plant Materials and Field Experiment

GWAS was conducted on a diverse panel comprising 232 maize inbred lines, which included tropical/subtropical (TST), temperate (SS, NSS), and mixed germplasms, as detailed in Table S1. This association panel was cultivated in three environments in China: Sanya, Hainan Province (SY, 18.22° N, 109.01° E) in November 2015, and Taian, Shandong Province (TA, 36.09° N, 117.09° E) in May 2016, and in May 2017. The husk number (HN) of 230, 222, and 188 inbred lines were counted in the three environments, respectively. Each inbred line was planted with seven plants per row, the and row spacing and plant spacing were about 0.6 and 0.2 m, respectively. Field management during the entire growing period was the same as local field management practices.

4.2. Phenotyping and Statistical Analysis

After the ears matured, the plants with grow well and uniformly in each inbred line were harvested to count the HN. Phenotypic data were represented by the mean values of the selected three to five ears. The Microsoft Excel 2010 software package was used to organize phenotypic data and conduct descriptive statistical analysis. Phenotypic variation in the husk number trait was performed using the lme4 package of the R 4.0.1 software program. The ANOVA model is as follows: $y_{ij} = \mu + g_i + e_j + \varepsilon_{ij}$, where y_{ij} is the phenotypic value of the husk number, μ is the mean value across environments, g_i is the effect of the genotype, e_j is the effect of the environment, and ε_{ij} is the residual error. The broad-sense heritability (H^2) was calculated as follows: $H^2 = \frac{\sigma_g^2}{\sigma_g^2 + (\sigma_{gy}^2/n)}$ [55], where σ_g^2 is genetic variance, σ_{gy}^2 is the residual variance, and n is the number of environments. To minimize environmental effects, the best linear unbiased prediction (BLUP) values for the husk number across all environments were estimated for each line using the same ANOVA model.

4.3. Genome-Wide Association Analysis

The genotype data used in this association panel were detailed in a prior study by Liu et al. [56]. The genotype data included 1.25 million SNPs with minor allele frequency (MAF) $\geq$ 0.05 and consisted of four sets (MaizeSNP50 BeadChip, 600K SNP array, RNA-seq, and genotyping by sequencing (GBS)), and the data can be downloaded at http://www.maizego.org/Resources.html (accessed on 26 March 2025). In addition, we retained only bi-allelic SNPs with a missing data rate less than 20% within this association panel. A total of 995,106 SNPs were ultimately retained to conduct GWAS. The GWAS for HN was performed using TASSEL (v5.0) [57] (Bradbury et al., 2007) under the mixed linear model (MLM) and the general linear model (GLM). The MLM was conducted using population structure (Q matrix) and kinship relationship (K matrix) to avoid spurious associations [58,59]. The SNPs in the GWAS are not independent due to the linkage disequilibrium (LD) among them. We therefore used the effective number of independent markers for the adjustment of multiple markers to obtain the P-value thresholds. A total of 131,432 markers in approximate linkage equilibrium (the LD R^2 threshold is 0.2) were found by PLINK [60]. We used the uniform Bonferroni-corrected thresholds at $\alpha = 1$ for the MLM and $\alpha = 0.05$ for the GLM as the cut-offs as applied in previous studies [28,61,62]. Therefore, the suggestive *p*-value 7.61×10^{-6} and 3.8×10^{-7} were established for the MLM and GLM, respectively, to identify significant SNPs with the HN.

4.4. Annotation of Candidate Genes

We used the PLINK 1.9 package to calculate the LD between the associated SNP and neighboring SNPs within a 200 kb region (100 kb upstream and downstream) to identify the candidate region. Genes in the same LD block ($r^2 > 0.2$) with significant SNPs were considered candidate genes. Functional annotation and prediction of candidate genes were performed through the NCBI (http://www.ncbi.nlm.nih.gov/) website and MaizeGDB database (http://www.maizegdb.org/). Transcriptome data were obtained from a published study by Wang et al. [2]. The third-generation *Zea mays* haplotype map (HapMap 3) data were downloaded from https://www.panzea.org/ and were used to examine the nucleotide diversity around *GRMZM2G038032* in maize and its wild relative, teosinte.

5. Conclusions

Based on the GWAS of the MLM and GLM, a total of 16 SNPs were associated with HN. The discovered candidate genes may co-regulate husk development in different pathways, among which *GRMZM2G038032* located on chromosome 6 is likely the most relevant gene, though further verification is needed. The identified SNPs related to the HN trait provide a reference set not only for marker-assisted selective breeding for low husk number and faster grain dehydration, but also for the prediction of candidate genes for unveiling the regulatory mechanism of husk development.

Supplementary Materials: The following supporting information can be downloaded at: https://www.mdpi.com/article/10.3390/ijms26073437/s1.

Author Contributions: Conceptualization, Y.W. and C.C.; methodology, Y.W., S.W., D.L. and B.L.; software, Y.W. and S.W.; investigation, M.C., Z.L. and H.S.; data curation, Y.W. and J.S. writing—original draft preparation, Y.W.; writing—review and editing, Y.W., C.C. and P.X. All authors have read and agreed to the published version of the manuscript.

Funding: This research was funded by the National Key R&D Program of China (2023YFD1200501), and the National Natural Science Foundation of China (32370672).

Institutional Review Board Statement: Not applicable.

Informed Consent Statement: Not applicable.

Data Availability Statement: The original contributions presented in this study are included in the article/Supplementary Materials. Further inquiries can be directed to the corresponding authors.

Conflicts of Interest: The authors declare no conflicts of interest.

References

1. Duvick, D.N. Genetic progress in yield of United States maize (*Zea mays* L.). *Maydica* **2005**, *50*, 193–202.
2. Wang, P.; Kelly, S.; Fouracre, J.P.; Langdale, J.A. Genome-wide transcript analysis of early maize leaf development reveals gene cohorts associated with the differentiation of C4 Kranz anatomy. *Plant J.* **2013**, *75*, 656–670. [CrossRef]
3. Kang, M.S.; Zuber, M.S. Combining Ability for Grain Moisture, Husk Moisture, and Maturity in Maize with Yellow and White Endosperms. *Crop Sci.* **1989**, *29*, 689–692. [CrossRef]
4. Liu, S.; Liu, S.; Ma, Y.; Liu, Y.; Li, Z.; Jian, S. Dehydration characteristics of spring maize husk leaves in eastern Jilin Province. *Soils Crops* **2023**, *12*, 283–292. [CrossRef]
5. Westgate, M.E.; Grant, D.L. Water deficits and reproduction in maize: Response of the reproductive tissue to water deficits at anthesis and mid-grain fill. *Plant Physiol.* **1989**, *91*, 862–867. [CrossRef] [PubMed]
6. Betrán, F.J.; Isakeit, T. Aflatoxin accumulation in maize hybrids of different maturities. *Agron. J.* **2004**, *96*, 565–570. [CrossRef]
7. Cao, A.; Santiago, R.; Ramos, A.J.; Souto, X.C.; Aguín, O.; Malvar, R.A.; Butrón, A. Critical environmental and genotypic factors for Fusarium verticillioides infection, fungal growth and fumonisin contamination in maize grown in northwestern Spain. *Int. J. Food Microbiol.* **2014**, *177*, 63–71. [CrossRef]
8. Widstrom, N.W.; Butrón, A.; Guo, B.; Wilson, D.M.; Snook, M.E.; Cleveland, T.E.; Lynch, R.E. Control of preharvest aflatoxin contamination in maize by pyramiding QTL involved in resistance to ear-feeding insects and invasion by *Aspergillus* spp. *Eur. J. Agron.* **2003**, *19*, 563–572.
9. Ji, H.C.; Yamakawa, T. Inheritance of Long Husk Leaves of Maize in Recombinant Inbred lines (RILs). *J. Fac. Agr. Kyushu U.* **2008**, *53*, 379–384. [CrossRef]
10. Pengelly, J.J.; Kwasny, S.; Bala, S.; Evans, J.R.; Voznesenskaya, E.V.; Koteyeva, N.K.; Edwards, G.E.; Furbank, R.T.; von Caemmerer, S. Functional analysis of corn husk photosynthesis. *Plant Physiol.* **2011**, *156*, 503–513. [CrossRef]
11. Cui, Z.; Luo, J.; Qi, C.; Ruan, Y.; Li, J.; Zhang, A.; Yang, X.; He, Y. Genome-wide association study (GWAS) reveals the genetic architecture of four husk traits in maize. *BMC Genom.* **2016**, *17*, 946. [CrossRef]
12. Li, L.; Xie, R.; Wang, K.; Ming, B.; Hou, P.; Li, S. Effects of peeling Husk on Grain Dehydration of Maize. *Crops* **2018**, *34*, 114–117. [CrossRef]
13. Cavalieri, A.; Smith, O.S. Grain Filling and Field Drying of a Set of Maize Hybrids Released From 1930[1]. *Crop Sci.* **1985**, *25*, 856–860. [CrossRef]
14. Zhou, G.; Hao, D.; Chen, G.; Lu, H.; Shi, M.; Mao, Y.; Zhang, Z.; Huang, X.; Xue, L. Genome-wide association study of the husk number and weight in maize (*Zea mays* L.). *Euphytica* **2016**, *210*, 195–205. [CrossRef]
15. Zhang, J.; Zhang, F.; Tian, L.; Ding, Y.; Qi, J.; Zhang, H.; Mu, X.; Ma, Z.; Xia, L.; Tang, B. Molecular mapping of quantitative trait loci for 3 husk traits using genotyping by sequencing in maize (*Zea mays* L.). *G3* **2022**, *12*, jkac198. [CrossRef]
16. Zhang, X.; Lu, M.; Xia, A.; Xu, T.; Cui, Z.; Zhang, R.; Liu, W.; He, Y. Genetic analysis of three maize husk traits by QTL mapping in a maize-teosinte population. *BMC Genom.* **2021**, *22*, 386. [CrossRef]
17. Zhou, G.; Mao, Y.; Xue, L.; Chen, G.; Lu, H.; Shi, M.; Zhang, Z.; Huang, X.; Xudong, S.; Hao, D. Genetic dissection of husk number and length across multiple environments and fine-mapping of a major-effect QTL for husk number in maize (*Zea mays* L.). *Crop J.* **2020**, *8*, 1071–1080. [CrossRef]
18. Atwell, S.; Huang, Y.S.; Vilhjálmsson, B.J.; Willems, G.; Horton, M.; Li, Y.; Meng, D.; Platt, A.; Tarone, A.M.; Hu, T.T.; et al. Genome-wide association study of 107 phenotypes in *Arabidopsis thaliana* inbred lines. *Nature* **2010**, *465*, 627–631. [CrossRef]
19. Huang, X.; Zhao, Y.; Wei, X.; Li, C.; Wang, A.; Zhao, Q.; Li, W.; Guo, Y.; Deng, L.; Zhu, C.; et al. Genome-wide association study of flowering time and grain yield traits in a worldwide collection of rice germplasm. *Nat. Genet.* **2012**, *44*, 32–39. [CrossRef]
20. Xiao, Y.; Liu, H.; Wu, L.; Warburton, M.; Yan, J. Genome-wide Association Studies in Maize: Praise and Stargaze. *Mol. Plant* **2017**, *10*, 359–374. [CrossRef]
21. Wang, Z.; Xia, A.; Wang, Q.; Cui, Z.; Lu, M.; Ye, Y.; Wang, Y.; He, Y. Natural polymorphisms in *ZMET2* encoding a DNA methyltransferase modulate the number of husk layers in maize. *Plant Physiol.* **2024**, *195*, 2129–2142. [CrossRef] [PubMed]
22. Yang, X.; Gao, S.; Xu, S.; Zhang, Z.; Prasanna, B.M.; Li, L.; Li, J.; Yan, J. Characterization of a global germplasm collection and its potential utilization for analysis of complex quantitative traits in maize. *Mol. Breed.* **2011**, *28*, 511–526. [CrossRef]
23. Liang, Z.; Xi, N.; Liu, T.; Li, M.; Sang, M.; Zou, C.; Chen, Z.; Yuan, G.; Pan, G.; Ma, L.; et al. A combination of QTL mapping and genome-wide association study revealed the key gene for husk number in maize. *Theor. Appl. Genet.* **2024**, *137*, 112. [CrossRef] [PubMed]
24. Chen, L.; Luo, J.; Jin, M.; Yang, N.; Liu, X.; Peng, Y.; Li, W.; Phillips, A.; Cameron, B.; Bernal, J.S.; et al. Genome sequencing reveals evidence of adaptive variation in the genus Zea. *Nat. Genet.* **2022**, *54*, 1736–1745. [CrossRef]
25. Flint-Garcia, S.A.; Thuillet, A.C.; Yu, J.; Pressoir, G.; Romero, S.M.; Mitchell, S.E.; Doebley, J.; Kresovich, S.; Goodman, M.M.; Buckler, E.S. Maize association population: A high-resolution platform for quantitative trait locus dissection. *Plant J.* **2005**, *44*, 1054–1064. [CrossRef]

26. Shikha, K.; Shahi, J.P.; Vinayan, M.T.; Zaidi, P.H.; Singh, A.K.; Sinha, B. Genome-wide association mapping in maize: Status and prospects. *3 Biotech* **2021**, *11*, 244. [CrossRef]
27. Huang, X.; Han, B. Natural variations and genome-wide association studies in crop plants. *Annu. Rev. Plant Biol.* **2014**, *65*, 531–551. [CrossRef]
28. Yang, N.; Lu, Y.; Yang, X.; Huang, J.; Zhou, Y.; Ali, F.; Wen, W.; Liu, J.; Li, J.; Yan, J. Genome Wide Association Studies Using a New Nonparametric Model Reveal the Genetic Architecture of 17 Agronomic Traits in an Enlarged Maize Association Panel. *PLoS Genet.* **2014**, *10*, e1004573. [CrossRef]
29. Dong, Z.; Hu, G.; Chen, Q.; Shemyakina, E.A.; Chau, G.; Whipple, C.J.; Fletcher, J.C.; Chuck, G. A regulatory network controlling developmental boundaries and meristem fates contributed to maize domestication. *Nat. Genet.* **2024**, *56*, 2528–2537. [CrossRef]
30. Xiao, Y.; Guo, J.; Dong, Z.; Richardson, A.; Patterson, E.; Mangrum, S.; Bybee, S.; Bertolini, E.; Bartlett, M.; Chuck, G.; et al. Boundary domain genes were recruited to suppress bract growth and promote branching in maize. *Sci. Adv.* **2022**, *8*, eabm6835. [CrossRef]
31. Xie, Y.; Zhao, Y.; Chen, L.; Wang, Y.; Xue, W.; Kong, D.; Li, C.; Zhou, L.; Li, H.; Zhao, Y.; et al. *ZmELF3.1* integrates the *RA2-TSH4* module to repress maize tassel branching. *New Phytol.* **2024**, *241*, 490–503. [CrossRef] [PubMed]
32. Causier, B.; Schwarz-Sommer, Z.; Davies, B. Floral organ identity: 20 years of ABCs. *Semin. Cell Dev. Biol.* **2010**, *21*, 73–79. [CrossRef] [PubMed]
33. Kunst, L.; Klenz, J.E.; Martínez-Zapater, J.M.; Haughn, G.W. AP2 Gene Determines the Identity of Perianth Organs in Flowers of Arabidopsis thaliana. *Plant Cell.* **1989**, *1*, 1195–1208. [PubMed]
34. Chuck, G.; Muszynski, M.; Kellogg, E.; Hake, S.; Schmidt, R.J. The control of spikelet meristem identity by the branched silkless1 gene in maize. *Science* **2002**, *298*, 1238–1241. [CrossRef]
35. Gil-Humanes, J.; Pistón, F.; Martín, A.; Barro, F. Comparative genomic analysis and expression of the APETALA2-like genes from barley, wheat, and barley-wheat amphiploids. *BMC Plant Biol.* **2009**, *9*, 66. [CrossRef]
36. Keller, T.; Abbott, J.; Moritz, T.; Doerner, P. Arabidopsis Regulator of Axillary Meristems1 controls a leaf axil stem cell niche and modulates vegetative development. *Plant Cell* **2006**, *18*, 598–611. [CrossRef]
37. Lee, D.K.; Geisler, M.; Springer, P.S. Lateral Organ Fusion1 and Lateral Organ Fusion2 function in lateral organ separation and axillary meristem formation in Arabidopsis. *Development* **2009**, *136*, 2423–2432. [CrossRef]
38. Müller, D.; Schmitz, G.; Theres, K. Blind homologous R2R3 Myb genes control the pattern of lateral meristem initiation in Arabidopsis. *Plant Cell* **2006**, *18*, 586–597. [CrossRef]
39. Mu, R.-L.; Cao, Y.-R.; Liu, Y.-F.; Lei, G.; Zou, H.-F.; Liao, Y.; Wang, H.-W.; Zhang, W.-K.; Ma, B.; Du, J.-Z.; et al. An R2R3-type transcription factor gene *AtMYB59* regulates root growth and cell cycle progression in Arabidopsis. *Cell Res.* **2009**, *19*, 1291–1304. [CrossRef]
40. Shin, R.; Burch, A.Y.; Huppert, K.A.; Tiwari, S.B.; Murphy, A.S.; Guilfoyle, T.J.; Schachtman, D.P. The Arabidopsis transcription factor MYB77 modulates auxin signal transduction. *Plant Cell* **2007**, *19*, 2440–2453. [CrossRef]
41. Keurentjes, J.J.; Fu, J.; de Vos, C.H.; Lommen, A.; Hall, R.D.; Bino, R.J.; van der Plas, L.H.; Jansen, R.C.; Vreugdenhil, D.; Koornneef, M. The genetics of plant metabolism. *Nat. Genet.* **2006**, *38*, 842–849. [CrossRef] [PubMed]
42. Sulpice, R.; McKeown, P.C. Moving toward a comprehensive map of central plant metabolism. *Annu. Rev. Plant Biol.* **2015**, *66*, 187–210. [CrossRef] [PubMed]
43. Mellerowicz, E.J.; Sundberg, B. Wood cell walls: Biosynthesis, developmental dynamics and their implications for wood properties. *Curr. Opin. Plant Biol.* **2008**, *11*, 293–300. [CrossRef] [PubMed]
44. Denninger, P. Rho of Plants signalling and the activating Rop Guanine Nucleotide Exchange Factors: Specificity in cellular signal transduction in plants. *J. Exp. Bot.* **2024**, *75*, 3685–3699. [CrossRef]
45. Liu, Y.; Dong, Q.; Kita, D.; Huang, J.B.; Liu, G.; Wu, X.; Zhu, X.; Cheung, A.Y.; Wu, H.M.; Tao, L.Z. RopGEF1 Plays a Critical Role in Polar Auxin Transport in Early Development. *Plant Physiol.* **2017**, *175*, 157–171. [CrossRef]
46. Chen, M.; Liu, H.; Kong, J.; Yang, Y.; Zhang, N.; Li, R.; Yue, J.; Huang, J.; Li, C.; Cheung, A.Y.; et al. RopGEF7 regulates PLETHORA-dependent maintenance of the root stem cell niche in Arabidopsis. *Plant Cell* **2011**, *23*, 2880–2894. [CrossRef]
47. Nakashima, K.; Yamaguchi-Shinozaki, K. ABA signaling in stress-response and seed development. *Plant Cell Rep.* **2013**, *32*, 959–970. [CrossRef]
48. Masood, J.; Zhu, W.; Fu, Y.; Li, Z.; Zhou, Y.; Zhang, D.; Han, H.; Yan, Y.; Wen, X.; Guo, H.; et al. Scaffold protein RACK1A positively regulates leaf senescence by coordinating the EIN3-miR164-ORE1 transcriptional cascade in Arabidopsis. *J. Integr. Plant Biol.* **2023**, *65*, 1703–1716. [CrossRef]
49. Chen, J.G.; Ullah, H.; Temple, B.; Liang, J.; Guo, J.; Alonso, J.M.; Ecker, J.R.; Jones, A.M. RACK1 mediates multiple hormone responsiveness and developmental processes in Arabidopsis. *J. Exp. Bot.* **2006**, *57*, 2697–2708. [CrossRef]
50. Li, M.; Zheng, Y.; Cui, D.; Du, Y.; Zhang, D.; Sun, W.; Du, H.; Zhang, Z. GIF1 controls ear inflorescence architecture and floral development by regulating key genes in hormone biosynthesis and meristem determinacy in maize. *BMC Plant Biol.* **2022**, *22*, 127. [CrossRef]

51. Zhang, D.; Sun, W.; Singh, R.; Zheng, Y.; Cao, Z.; Li, M.; Lunde, C.; Hake, S.; Zhang, Z. GRF-interacting factor1 Regulates Shoot Architecture and Meristem Determinacy in Maize. *Plant Cell* **2018**, *30*, 360–374. [CrossRef]
52. Horiguchi, G.; Kim, G.T.; Tsukaya, H. The transcription factor AtGRF5 and the transcription coactivator AN3 regulate cell proliferation in leaf primordia of Arabidopsis thaliana. *Plant J.* **2005**, *43*, 68–78. [CrossRef] [PubMed]
53. Lee, B.H.; Ko, J.H.; Lee, S.; Lee, Y.; Pak, J.H.; Kim, J.H. The Arabidopsis GRF-INTERACTING FACTOR gene family performs an overlapping function in determining organ size as well as multiple developmental properties. *Plant Physiol.* **2009**, *151*, 655–668. [CrossRef]
54. Lee, B.H.; Wynn, A.N.; Franks, R.G.; Hwang, Y.S.; Lim, J.; Kim, J.H. The Arabidopsis thaliana GRF-INTERACTING FACTOR gene family plays an essential role in control of male and female reproductive development. *Dev. Biol.* **2014**, *386*, 12–24. [CrossRef] [PubMed]
55. Knapp, S.J.; Stroup, W.W.; Ross, W.M. Exact Confidence Intervals for Heritability on a Progeny Mean Basis. *Crop Sci.* **1985**, *25*, 192–194. [CrossRef]
56. Liu, H.; Luo, X.; Niu, L.; Xiao, Y.; Chen, L.; Liu, J.; Wang, X.; Jin, M.; Li, W.; Zhang, Q.; et al. Distant eQTLs and Non-coding Sequences Play Critical Roles in Regulating Gene Expression and Quantitative Trait Variation in Maize. *Mol. Plant* **2017**, *10*, 414–426. [CrossRef] [PubMed]
57. Bradbury, P.J.; Zhang, Z.; Kroon, D.E.; Casstevens, T.M.; Ramdoss, Y.; Buckler, E.S. TASSEL: Software for association mapping of complex traits in diverse samples. *Bioinformatics* **2007**, *23*, 2633–2635. [CrossRef]
58. Yu, J.; Pressoir, G.; Briggs, W.H.; Vroh Bi, I.; Yamasaki, M.; Doebley, J.F.; McMullen, M.D.; Gaut, B.S.; Nielsen, D.M.; Holland, J.B.; et al. A unified mixed-model method for association mapping that accounts for multiple levels of relatedness. *Nat. Genet.* **2006**, *38*, 203–208. [CrossRef]
59. Zhang, Z.; Ersoz, E.; Lai, C.Q.; Todhunter, R.J.; Tiwari, H.K.; Gore, M.A.; Bradbury, P.J.; Yu, J.; Arnett, D.K.; Ordovas, J.M.; et al. Mixed linear model approach adapted for genome-wide association studies. *Nat. Genet.* **2010**, *42*, 355–360. [CrossRef]
60. Purcell, S.; Neale, B.; Todd-Brown, K.; Thomas, L.; Ferreira, M.A.; Bender, D.; Maller, J.; Sklar, P.; de Bakker, P.I.; Daly, M.J.; et al. PLINK: A tool set for whole-genome association and population-based linkage analyses. *Am. J. Hum. Genet.* **2007**, *81*, 559–575. [CrossRef]
61. Mao, H.; Wang, H.; Liu, S.; Li, Z.; Yang, X.; Yan, J.; Li, J.; Tran, L.-S.P.; Qin, F. A transposable element in a NAC gene is associated with drought tolerance in maize seedlings. *Nat. Commun.* **2015**, *6*, 8326. [CrossRef] [PubMed]
62. Wang, M.; Yan, J.; Zhao, J.; Song, W.; Zhang, X.; Xiao, Y.; Zheng, Y. Genome-wide association study (GWAS) of resistance to head smut in maize. *Plant Sci.* **2012**, *196*, 125–131. [CrossRef] [PubMed]

International Journal of
Molecular Sciences

MDPI

Article

Breeding D1-Type Hybrid *Japonica* Rice in Diverse Upland Rainfed Environments

Chunli Wang [1,†], Juan Li [1,†], Qian Zhu [1,2,3], Junjie Li [1], Cui Zhang [1], Ruke Hong [1], Dajun Huang [1], Zhonglin Zhang [1], Jin Xu [1], Dandan Li [1], Jiancheng Wen [1], Chengyun Li [2], Youyong Zhu [2], Dongsun Lee [2,*] and Lijuan Chen [1,3,*]

1 Rice Research Institute, Yunnan Agricultural University, Kunming 650201, China; wchunli1989@163.com (C.W.); lijuan1661@126.com (J.L.); gabriel731@ynau.edu.cn (Q.Z.); ad9979li@163.com (J.L.); 15096694734@139.com (C.Z.); hlshrk@163.com (R.H.); hdj668@163.com (D.H.); zzl681128@163.com (Z.Z.); genexu@126.com (J.X.); lidanzaizhe@163.com (D.L.); jcwen1117@163.com (J.W.)
2 State Key Laboratory for Conservation and Utilization of Bio-Resources in Yunnan, Yunnan Agricultural University, Kunming 650201, China; lichengyun@ynau.edu.cn (C.L.); yyzhu@ynau.edu.cn (Y.Z.)
3 The Key Laboratory for Crop Production and Smart Agriculture of Yunnan Province, Yunnan Agricultural University, Kunming 650201, China
* Correspondence: dongeast@naver.com (D.L.); chenlijuan@ynau.edu.cn (L.C.)
† These authors contributed equally to this work.

Abstract: 'Dianheyou615' (DHY615) is an elite Dian (D1)-type hybrid *japonica* rice variety, renowned for its high yield, exceptional grain quality, and unique adaptability to both irrigated and rainfed conditions in the Yungui Plateau of southwestern China. However, the genetic mechanisms underlying the agronomic performance of the D1-type hybrid *japonica* rice remain unclear. In this study, a comprehensive analysis of 'DHY615"s agronomic performance, genetic genealogy, and molecular genetic foundation was conducted to dissect its desirable traits for upland rainfed cultivation across diverse ecological environments. The main findings indicate that 'DHY615' possesses 6432 heterozygous SNPs, with 57.48% and 14.43% located in the promoter and coding regions, respectively, potentially affecting key phenotypic traits. High-impact SNPs variants and numerous well-known functional genes were identified, such as *OsAAP6*, *GS3*, *Sd1*, *Rf1*, *BADH2*, *BPh14*, *Rymv1*, *OsFRO1*, *NRT1.1B*, *SKC1*, *OsNCED2*, and *qUVR-10*, which are likely linked to traits including plant architecture, grain yield, grain quality, and resistance to various biotic and abiotic stresses (e.g., disease, cold, drought, salt, high iron, and high UV radiation). Notably, 'Nan615' harbors a greater number of functional allele variants compared to 'H479A', which potentially explaining its superior grain yield and remarkable adaptability. This study offers novel and valuable insights into the molecular genetic foundation of the plateau D1-type hybrid *japonica* rice, underscoring its potential for sustainable rice production across diverse ecological zones, especially with its unparalleled high-altitude adaptability to rainfed upland planting.

Keywords: hybrid *japonica* rice; unique adaptability; rainfed upland planting; diverse ecological zones

1. Introduction

Rice (*Oryza sativa* L.) is a typical monoecious and self-pollinated crop. The successful development of the three-line hybrid rice system (cytoplasmic male sterility or CMS (A-), maintainer (B-), and restorer (R-) lines) was the core genetic tool used to harness heterosis. Currently, the Wild–Abortive (WA) type and Hong Lian (HL) type are the primary CMS sources used in the production of the three-line hybrid *xian* (*O. sativa* ssp. *indica*), while the Boro–Taichung (BT) type and Dian I (D1) type of CMS are frequently employed in the breeding of the three-line hybrid *geng* (*O. sativa* ssp. *japonica*). Notably, the D1-type CMS has been a key source for developing 'Yuza', 'Dianza', 'Dianheyou', 'Chunyou', and 'Yongyou' series hybrid *japonica* or *indica–japonica* rice combinations in China [1–3].

In 1965, the Chinese scientist Li Zhengyou identified a male sterile (MS) plant from the natural hybridization progeny in a high-altitude *indica* rice landrace and a low-altitude *japonica* rice cultivar 'Taipei 8' in Baoshan, Yunnan Province. Following eight generations of backcrossing with the MS plant, the first D1-CMS line was bred in 1969, characterized by gametophytic sterility, and named *japonica* rice 'Hongmaoying A' [4,5]. The pollen abortion in D1-CMS occurred at the trinucleate stage, leading to significant starch accumulation, with I_2-KI staining appearing blue-black. Genetic analysis revealed that the sterile genes (mitochondrial chimeric gene *atp6-orf79*) and restorer genes (*Rf1a* and *Rf1b*) of the D1-CMS were nearly identical to those of the BT-CMS [6–9]. In 1973, the successful breeding of the first D1-type *japonica* rice R-line 698 marked the establishment of D1-type hybrid rice breeding technology. The first D1-type super-high-yield hybrid rice, 'Yuza 29', was a significant breakthrough and set a world record for *japonica* rice yield at 16.63 t/hm^2 in Binchuan, Yunnan, in 1994 [4]. The synergistic innovation of breeding strategies and germplasm utilization has enabled the effective pyramiding of elite traits. These D1-type varieties of A-, B-, and R-lines, which contain *indica* kinship, not only exhibit rich genetic background and robust multi-resistance but also superior rice quality and high yields, particularly in the southwestern provinces of China. As a result, D1-type hybrid rice holds significant value for functional genomics and genetic improvement in rice [10].

'DHY615' is a D1-type three-line hybrid *japonica* rice variety that exhibits high adaptability. It can be cultivated as both irrigated paddy rice and rainfed upland rice under direct seeding, making it well-suited for high-altitude regions up to 2200 m above sea level (masl). This rice variety excels in the diverse ecological zones of the Yunnan–Guizhou (Yungui) Plateau and southwestern China, becoming a predominant variety in these areas. Over the past five years, the total planting area of 'DHY615' has exceeded 66,666.67 hectares (hm^2). Rice cultivation in the southwest region of China, including in the Yungui Plateau and the Qinghai–Tibet Plateau, accounts for approximately 8% of the national rice cultivation area. The region is divided into three sub-areas: the eastern Guizhou and western Hunan Plateau Mountains (single and double cropping rice), the Yunnan–Sichuan Plateau Ridges and Valleys (single cropping rice and double harvest), and the Qinghai–Tibet Plateau cold river valleys (single cropping rice). The accumulated temperature of $\geq$10 °C ranges from 2900 to 8000 °C, with sunshine hours varying from 800–1500 h and a significant variation in rice distribution based on altitude. In the low-altitude areas, *indica* rice is primarily cultivated, whereas *japonica* rice dominates the high-altitude regions, with a mixed distribution of both rice types in the intermediate zones. The rice grain yield averages between 7.5 and 9.0 t/hm^2, varying by variety. The rice production season lasts from 180 to 260 days, and annual precipitation ranges from 500 to 1400 mm [11]. This unique plateau *japonica* rice region [12,13] is characterized by its high altitudes, intense solar and ultraviolet (UV) radiation, infertile soil, and frequent natural disasters, necessitating special traits in cultivated rice varieties. These include cold tolerance, strong resistance to UV radiation, efficient fertilizer utilization, and broad adaptability to other harsh environments.

Studies on the structural genomics, functional genomics, and pan-genomics of rice have progressed rapidly in the recent two decades [14]. The applications of molecular markers and SNPs chips have enabled the development of high-yield rice and green super rice varieties [15–17]. Several genomic breeding chips have been developed globally for rice, such as Rice6K, Rice60K, Rice90K, and C6AIR [18,19]. Notably, Rice6K incorporates markers of genes such as *GS3* (grain length), *GS5* (grain width), *NRT1.1B* (nutrient-use efficiency), *Badh* (fragrance), *IPA1* (plant architecture), and *Sd-1* (plant height), applying in the genotyping of germplasm resources, gene fingerprint identification of varieties, QTL mapping, marker-assisted backcross breeding, and pedigree verification of breeding lines [20,21]. As quantitative genetics, population genetics, and comparative genomics continue to develop, the genetic mechanisms underlying heterosis in hybrid *indica* rice are becoming increasingly clearer. A few heterosis-associated loci from female parents explained a large proportion of the yield advantage of hybrids over their male parents [22]. The dominance, overdominance, and epistasis underlying functional complementation are indispensable contributors to yield heterosis [23,24]. However, the genotyping and genetic mechanisms underlying the heterosis of D1-type hybrid *japonica* rice remain unclear.

The objectives of this study are to (1) analyze the genealogy and phenotypic characterization of 'DHY615', (2) evaluate and validate the aggregation of the excellent agronomic traits in 'DHY615' for breeding the plateau hybrid *japonica* rice, and (3) annotate and predict the effects of genetic variation on gene function in 'DHY615', and identify genes likely associated with biotic and abiotic stress tolerance (including disease, drought, cold, high iron content in red soils, and high UV radiation), as well as nitrogen use efficiency, productivity, quality, and other important traits related to the wide adaptability of 'DHY615'. These findings provide a better understanding of the agronomic performance and genetic basis of hybrid *japonica* rice, particularly its adaptability to high-altitude regions, and contribute to agricultural sustainability and food security in diverse rice-growing zones.

2. Results

2.1. Ecological Adaptation and Field Performance of 'DHY615'

To validate the effectiveness of the aggregation of the numerous outstanding agronomic traits in 'DHY615' for breeding purposes, we compiled data on the distribution and adaptability, grain yield and quality, biotic/abiotic resistance exhibited by 'DHY615' throughout its participation in China's National and Yunnan Provincial Variety Approval and Production Trials. Available online: https://www.ricedata.cn/variety/varis/618086.htm (accessed on 6 November 2024).

These authoritative results indicate that 'DHY615' is predominantly distributed and well-adapted across five provinces in China, encompassing the mid-season *japonica* rice region in southwestern China (Yunnan, Guizhou, and Sichuan provinces, and Chongqing Municipality) along the Upper Yangtze River, as well as the *indica–japonica* rice mixed cultivation area in Shaanxi Province. Geographically, this spans from 97°79′ E (Dehong Prefecture, Yunnan) to 107°98′ E (Qiannan Prefecture, Guizhou) longitudinally and from 21°52′ N (Mengla Country, Yunnan) to 33°06′ N (Hanzhong Country, Shaanxi) latitudinally, encompassing 10.19 degrees of latitude and 11.54 degrees of longitude, specifically for irrigated paddy fields (Figure 1A).

In addition to irrigated cultivation, 'DHY615' is suitable for rainfed conditions in the high-altitude regions of the Yunnan, Guizhou, and Sichuan provinces. Rainfed upland cultivation has expanded to over 60 counties across 14 prefectures/cities (Figure 1B), with altitudes ranging from 826 to 2200 masl in Yunnan Province. The optimal regions for natural rainfed upland planting are characterized by hot and rainy climates, with average altitudes of 800–1600 masl and annual rainfall exceeding 1200 mm. These regions include the Lincang, Honghe, Wenshan, Puer, and Nujiang prefectures. Regions with altitudes between 1600 and 2200 masl, such as Qujing, Yuxi, Dali, Chuxiong, Baoshan, and Kunming, are considered moderately suitable. From 2021 to 2024, 'DHY615' exhibited robust growth and a compact plant architecture under direct seeding conditions in rainfed upland fields, as observed in Yunshan Village, Zhutang Township, and Lancang County (Figure 1C). Its development has been recognized as a milestone in the renewal of D1-type hybrid rice, highlighting its exceptional adaptability to high-altitude environments in the Plateau.

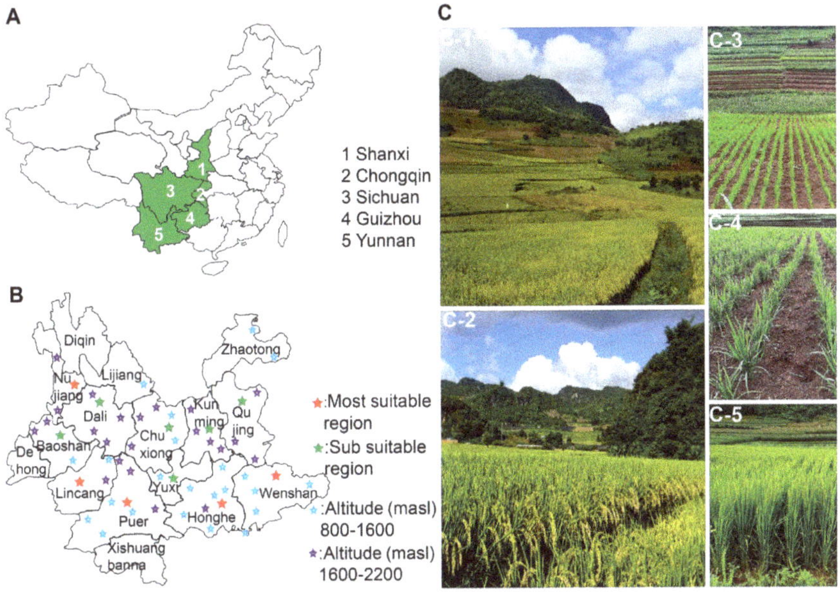

Figure 1. The geographic distribution of the 'DHY615' plant area and the growth periods of 'DHY615' under rainfed upland conditions. (**A**) The provinces where 'DHY615' is significantly planted in southwestern China are indicated in green. (**B**) Considering altitude and annual rainfall, 'DHY615' has been found suitable for cultivation under rainfed (non-irrigated direct seeding) conditions across 13 prefectures (encompassing 52 counties) in Yunnan Province. Five prefectures/cities (Nujiang, Lincang, Puer, Honghe, and Wenshan) have been identified as the most suitable cultivation zones (★), while six prefectures/cities (Dali, Chuxiong, Yuxi, Kunming, Qujing, and Baoshan) have been classified as sub-optimal zones (★). (**C**) The various growth stages of 'DHY615' in the rainfed upland fields in Lancang County (altitude: 1300–1700 masl, annual rainfall: 1600 mm) in Yunnan Province are as follows: (**C-1**,**C-2**) represent the maturity stage; (**C-3**) is the seedling stage; (**C-4**) is the tillering stage; and (**C-5**) is the heading stage.

In 2013, we conducted varietal comparison trials in Fumin paddy fields, and the results demonstrated significant yield differences among the parental and hybrid F_1 varieties. 'DHY615' achieved the highest average yield at 11.19 t/hm^2, followed by the maintainer line 'H479B' at 8.37 t/hm^2 and the restorer line 'Nan 615' at 7.38 t/hm^2. 'DHY615' exhibited significant heterosis, with yield advantages of 25.1% over 'H479B' and 34.1% over 'Nan 615'.

The yield performance of 'DHY615' in paddy fields and rainfed upland planting areas was comparable from 2020 to 2024 (Table 1). Across 22 irrigated paddy field locations (934–2200 masl), the grain yield ranged from 6.97 to 14.27 t/hm^2, with an average yield of 11.09 t/hm^2 and a coefficient of variation of 13.00%. Similarly, on-farm demonstration yields in 22 rainfed upland cultivation areas (1075–2132 masl) ranged from 5.64 to 12.22 t/hm^2, with an average yield of 8.13 t/hm^2 and a coefficient of variation of 10.43%. The average highest yield of 'DHY615' in the most suitable region (Langcang, Puer) reached 9.99 t/hm^2 under rainfed upland cultivation, showing no significant difference compared to its yield under paddy cultivation. For irrigated paddy fields, among 22 locations, only four (Maguan, Mangshi, Yunxian in Yunnan, and Bijie) showed yields statistically indistinguishable from the lowest-yielding locations (Fuping, Shaanxi). The remaining 17 locations revealed significantly higher yields than the Fuping locations. However, the yield statistical analysis revealed no significant yield differences among the 22 rainfed upland cultivation locations compared to the lowest-yielding location (Anshun, Guizhou), indicating consistent performance across diverse rainfed environments.

According to the Chinese Agricultural Industry Standard NY/T 593-2013 [25], the quality of 'DHY615' was evaluated in four irrigated paddy field ecological regions—Lufeng, Yunnan; Lueyang, Shanxi; Huishui, Guizhou; Anshi, Hubei—and in rainfed upland cultivation areas (Yanshang, Yunnan). The results indicated that 'DHY615' has a high brown rice percentage (82.50–84.10%) and milled rice percentage (63.00–68.8%), intermediate amylose content (16.4–18.4%), a gel consistency exceeding 64.0 mm (Table S1-1), and transparent, polished grains (Figure S1). The grain quality met the II- or III-class standards. Additionally, the grain quality traits of the D1-type hybrid *japonica* 'DHY34' (H479A/Nan34) were evaluated under both paddy and rainfed upland field cultivation in Yunnan Province, and it also met the II- or III-class standards. Notably, the grains exhibited enriched micronutrients, including iron (Fe), zinc (Zn), selenium (Se), magnesium (Mg), and manganese (Mn), while remaining free of heavy metals (Tables S1-2 and S1-3). 'DHY615' also showed resistance to multiple major rice diseases, such as leaf blast, spike blast, bacterial blight, and sheath blight, and exhibited high resistance to rice false smut. Furthermore, it demonstrated combined tolerance to cold and drought stress (Table S2).

In conclusion, DHY615' demonstrated a clear advantage in yield and quality, along with robust resistance or tolerance to multiple rice diseases, as well as resistance to cold and drought. 'DHY615' is distinguished by its exceptional yield potential, superior quality, multi-disease resistance, and adaptability to diverse rainfed upland environments, making it a valuable resource for rice production in challenging environments.

Table 1. The yield performance of 'DHY615' in paddy fields and rainfed upland planting representative areas.

Paddy Field Locations	Altitude (masl)	Yield (t/hm^2)	Rainfed Upland Locations	Altitude (masl)	Yield (t/hm^2)
YN, Lincang, Yunxian	1580	8.02 ± 0.78	YN, Wenshan, Maguan	1260	7.18 ± 0.56
YN, Qujing, Qilin	2000	10.85 ± 1.21 *	YN, Wenshan, Guangnan	1280	7.90 ± 0.25
YN, Qujing, Luliang	1800	12.75 ± 0.45 ***	YN, Kunming, Yiliang	1801	7.20 ± 1.23
YN, Dali, Xiagua	1970	12.15 ± 0.75 ***	YN, Kunming, Xishan	1905	7.56 ± 0.34
YN, Dali, Heqing	2100	11.32 ± 1.66 **	YN, Kunming, Luquan	1900	7.63 ± 0.67
YN, Dali, Yongping	1690	10.73 ± 0.76 *	YN, Kunming, Shilin	1870	7.64 ± 0.99
YN, Dali, Jianchuan	2200	11.50 ± 0.88 **	YN, Nujiang, Fugong	1800	8.78 ± 1.26
YN, Kunming, Xundian	1553	11.64 ± 0.92 **	YN, Nujiang, Lushui	1637	8.97 ± 0.94
YN, Kunming, Yiliang	1540	12.07 ± 1.33 ***	YN, Chuxiong, Yuanmou	1445	7.68 ± 0.92
YN, Wenshan, Maguang	1337	10.28 ± 1.07	YN, Chuxiong, Lufeng	1584	9.09 ± 1.33
YN, Dehong, Mangshi	1516	9.46 ± 1.23	YN, Lincang, Linxiang	1813	7.91 ± 0.45

Table 1. *Cont.*

Paddy Field Locations	Altitude (masl)	Yield (t/hm^2)	Rainfed Upland Locations	Altitude (masl)	Yield (t/hm^2)
YN, Chuxiong, Lufeng	1548	12.79 ± 1.33 ***	YN, Lincang, Yunxian	1580	8.77 ± 0.72
YN, Hehong, Jianshui	1700	12.27 ± 0.56 ***	YN, Honghe, Hekou	1075	7.71 ± 1.35
YN, Lijiang	2195	10.84 ± 2.1 *	YN, Honghe, Luchun	1306	7.51 ± 2.55
YN, Yuxi, Eshan	1845	11.84 ± 0.12 ***	YN, Honghe, Yuanyang	1700	7.69 ± 1.54
YN, Zhaotong, Ludian	1850	12.23 ± 2.04 ***	YN, Honghe, Kaiyuan	1450	9.65 ± 1.72
YN, Baoshan, Longyang	1650	12.55 ± 0.33 ***	YN, Honghe, Gejiu	1499	8.93 ± 3.29
GZ, Bijie	1520	8.93 ± 0.54	YN, Puer, Mojiang	1700	7.53 ± 0.36
GZ, Qiannan, Huish	1400	11.13 ± 1.95 **	YN, Puer, Ninger	1600	9.20 ± 1.79
SC, Liangshan, Xichang	1480	10.86 ± 1.34 *	YN, Puer, Lancang	1460	9.99 ± 1.81
SC, Liangshan, Mianning	1620	12.39 ± 0.76 ***	GZ, Anshun	2132	7.01 ± 0.38
SX, Weinan, Fuping	934	7.47 ± 0.50	SC, Jingchuan	1300	7.37 ± 1.47
Mean		11.09			8.13
SD		1.44			0.85
Coefficient variation (%)		13.05			10.43

Note: YN: Yunnan Province, GZ: Guizhou Province, SC: Sichuan Province, SX: Shaanxi Province locations (province, prefecture or county, and city). The data was compiled derived from the National/Provincial field yield trials and on-farm demonstrations in 2020–2024. The yield measurement was conducted in accordance with the Chinese Agricultural Industry Standard NY/T 789-2004 [26]. *: yield given in average ± SD. The asterisks indicate significant differences: * $p < 0.05$, ** $p < 0.01$, and *** $p < 0.001$. Grain moisture derived using a PM-8188-A grain moisture meter.

2.2. Genealogy and Phenotypic Characterization of 'DHY615'

'DHY615' was developed from the female parent (A-line), 'H479A', and the male parent (R-line), 'Nan615'. Its genealogy (Figure 2) reveals a rich and diverse genetic heritage, integrating Yunnan local rice resources with elite genetic materials from *indica* and *japonica* rice subspecies and ecotypes across Asia (e.g., China, Japan, Korea, and Philippines–IRRI), Europe (Italy), and America (USA)). This highlights the successful application of advanced breeding technologies in developing D1-CMS lines and hybrid *japonica* rice in China.

The morphology of 'DHY615' (Figure S1) was conducted alongside evaluations of yield-related agronomic traits with 'H479A' and 'Nan615'. The results showed that 'DHY615' exhibits significantly greater plant height, effective spike number, and grain number per panicle compared to 'Nan615'. Its effective spike number also surpassed that of 'H479A'. These findings indicate that 'DHY615' inherits dominant traits from both parents, including enhanced spike productivity, grain number per panicle, moderate plant height, appropriate thousand-seed weight, and a stable seed-setting rate. The total growth duration of 'DHY615' was approximately 150 days under rainfed upland conditions and 180 days in paddy fields, varying with altitude (Table S3).

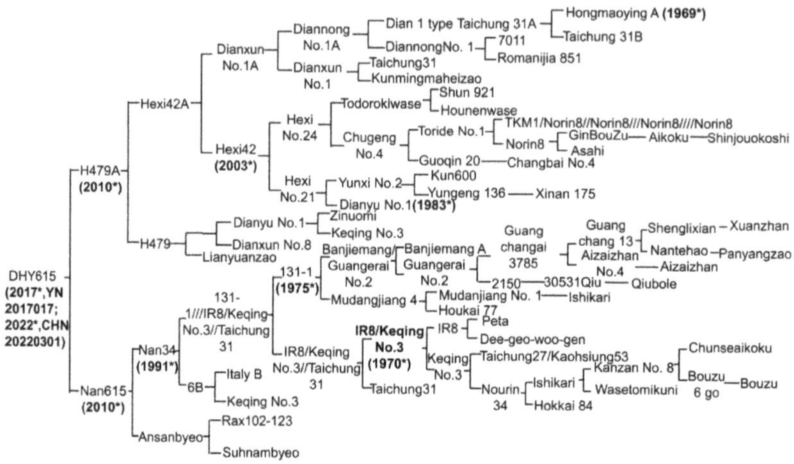

Figure 2. Genealogy of the D1-type hybrid *japonica* rice, 'DHY615'. * Indicates the year the variety was developed. Over the course of approximately five decades, from 1969 to 2017, through systematic selection and breeding, two elite lines, 'Nan 615' and 'H479A', were successfully developed. The R-line, 'Nan615', was bred by crossing 'Nan34' (which includes the *indica* 'IR8' and a series of Japanese *japonica* kinships) with Ansanbyeo from Korea. The D1-type CMS line, 'H479A', was developed by crossing 'Hongmaoying A' with plateau-adapted B-lines ('Diannong No.1', 'Dianxun No.1', 'Dianyu No.1', 'Hexi42', and 'H479').

2.3. *Indica–Japonica* Component and Genomic Differences

The SNP profiles of 'DHY615' and its parental lines were compared with the reference genomes of 'Nipponbare' and '9311'. Chromosomal analyses revealed that 'H479A' (Figure 3A), which carries the *indica* variety 'Lianyuanzao' in its pedigree, exhibited an *indica* component of 1.47% and a *japonica* component of 94.13%, aligning most closely with the 'Nipponbare' genome. In contrast, 'Nan615' (Figure 3B), an *indica–japonica* bridging R-line, displayed an *indica* component of 16.27% and a *japonica* component of 74.27%. The genomic composition of 'DHY615' (Figure 3C) was intermediary, with 87.60% *japonica* and 3.60% *indica* components.

Analysis using the rice GSR40k SNPs chip identified 32,887 high-quality SNPs, of which 32,607 were mapped to specific genomic coordinates. These loci were distributed across the 12 rice chromosomes (Figure 3D–F). 'H479A' had five heterozygous SNPs loci, with a homozygosity rate of 99.98% (Figure 3D), and 'Nan615' exhibited 24 heterozygous loci, with a homozygosity rate of 99.93% (Figure 3E). Both parental lines displayed stable genotypes, without phenotypic segregation. Conversely, the homozygosity rate of 'DHY615' was 80.19%, with 6432 heterozygous SNPs (Figure 3F). These loci were unevenly distributed, with chromosome 1 containing the most heterozygous loci and chromosome 5 containing the fewest heterozygous loci (Figure 4A). A total of 25,629 homozygous SNPs loci (78.60% of all SNPs markers) were identical across 'DHY615' and its parents, while 19.72% of markers were heterozygous, indicating that the heterozygous advantage of 'DHY615' likely resides in these genomic regions. The detailed analysis of the 6432 heterozygous loci (Figure 4B) showed that 57.48% were in promoter regions, 22.06% in introns, 2.83% in 3′UTRs, 3.20% in 5′UTRs, and 14.43% in coding sequences.

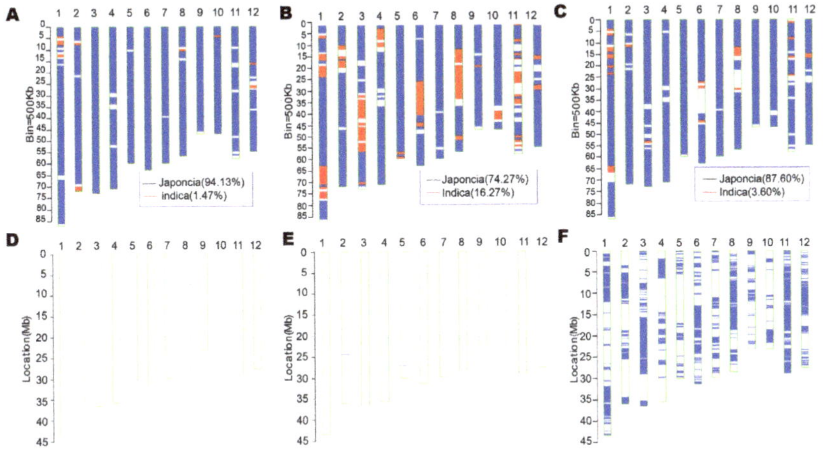

Figure 3. Genetic background screening using the GSR40K array. (**A**) The *indica–japonica* genetic background of 'H479A'. (**B**) The *indica–japonica* genetic background of 'Nan615'. (**C**) The *indica–japonica* genetic background of 'DHY615', with the blue color representing the *japonica* component and the red color indicating the *indica* component. (**D**) The detailed genotyping map of the plant 'H479A'. (**E**) The detailed genotyping map of the plant 'Nan615'. (**F**) The detailed genotyping map of the plant 'DHY615'. The twelve chromosomes of rice are labeled from 1 to 12. The reference genome is Nipponbare (rice MSU7.0). The blue short line (**D–F**) on each chromosome indicates the position of the single nucleotide polymorphisms (SNPs) with heterozygous genotypes, where genomic fragments of the parents ('H479A' and 'Nan615') were introgressed, and the genotypes of the remaining genomic regions were the same as the parent.

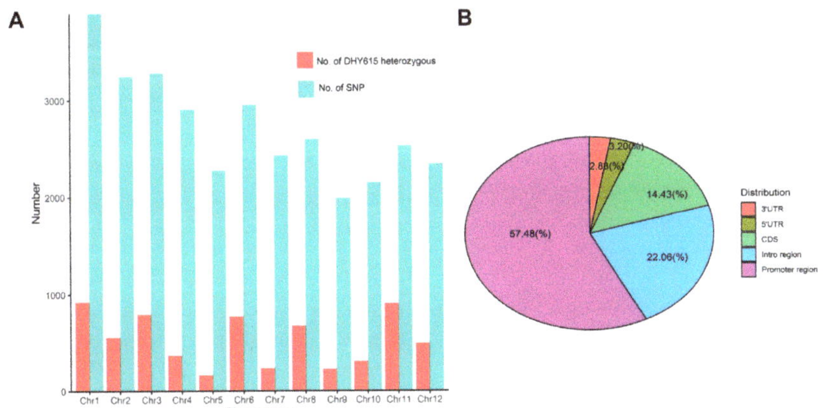

Figure 4. Number and distribution of the heterozygous loci in 'DHY615'. (**A**) The number of SNPs and heterozygous loci across 12 chromosomes. (**B**) The differential distribution proportions of the heterozygous loci for 'DHY615' within the gene structure.

2.4. Functional Impact of SNPs Variations

The functional effects of SNPs variations were predicted using the ANNOVAR program (ANNOVAR version 2023-06-08). Variants were classified into four impact categories: high, moderate, low, and modifier. The chromosomal distribution of variant effects across the four groups revealed significant disparities (Figure 5A), particularly with the number of heterozygous loci in 'DHY615'. Chromosome 1 exhibited the highest variation density, whereas chromosome 9 had the lowest. The high-impact group, consisting of 218 variants (0.35%), included splice–site, stop–gain, stop–loss, and start–loss mutations.

Moderate-impact variants (3488, 5.63%) were predominantly missense mutations. Low-impact variants (1791, 2.89%) comprised synonymous and splice–region changes. The majority (91.12%) were modifier variants (56,403), encompassing intron and intergenic variants. The stop–gain mutations dominated the high-impact group, while missense variants were the most common in the moderate-impact group (Figure 5B).

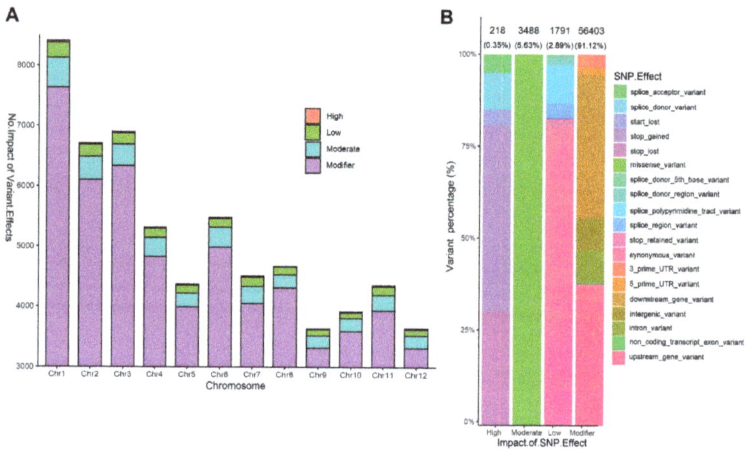

Figure 5. Classification of the variants by their impact on gene function and by the effects of these impacts. (**A**) The distribution of the variant impact effects (high, low, moderate, and modifier group) across the 12 chromosomes. (**B**) The classification of the variants based on the impact effects on gene function.

Additional details on high- and moderate-impact variants are provided in Table S4. Among 64 high-impact variants referenced in the RAP-DB database. Available online: https://rapdb.dna.affrc.go.jp (accessed on 6 November 2024) (Table S5), genes such as *OsLPR4*, *OsSCP42*, *OsGELP26*, and *OsWAK47* were likely implicated in their biochemical processes, seed development, and morphological traits. Stress-response genes like *OsSPX6*, *OsNCED2*, and *PR10* were also identified, along with the resistance genes Pi3, *OsWRKY123*, *OsLOX11*, and *Bph14*.

Of the 428 heterozygous SNPs loci in the coding regions, 80 were identified as high- and moderate-impact variants (Table S6). Genes including *OsMC6*, *OsWRKY22*, and *OsACBP3* were likely involved in both abiotic and biotic stress responses. Additional genes, such as *OsPUB58*, *OsMYB1R1*, and *OsSPX1*, were likely linked to tolerance against salt, cold, and drought stresses (Figure 6). Disease-resistance genes like *Pi64*, *CYP76M6*, *OsCIPK05*, and *OsBSR-K1* were implicated in resistance to leaf blight, insect infestation, and blast disease (Table 2). These findings suggest that most of these genes may have contributed to the enhanced stress and disease resistance observed in 'DHY615'.

Table 2. Summary of SNPs loci variations in abiotic stress and biotic stress genes.

Gene Name	Gene ID	Trait	No. of Variation Sites	Reference
OsMC6	LOC_Os01g58580	abiotic stress and biotic stress	1	[27]
OsWRKY22	LOC_Os01g60490	abiotic stress and biotic stress	1	[28]

Table 2. *Cont.*

Gene Name	Gene ID	Trait	No. of Variation Sites	Reference
OsACBP3	LOC_Os03g37960	cold tolerance and blast disease	1	[29]
OsPUB58	LOC_Os10g40120	drought and cold tolerance	1	[30]
OsMYB1R1	LOC_Os04g49450	drought resistance	1	[31]
OsSPX1	LOC_Os06g40120	cold tolerance	1	[32]
Pi64	LOC_Os01g57280	leaf blast disease resistance	1	[33]
CYP76M6	LOC_Os02g36280	insect resistance	1	[34]
OsCIPK05	LOC_Os01g10890	blast disease resistance	1	[35]
OsBSR-K1	Os10g0548200	blast disease	1	[36]

Figure 6. 'DHY615' water-saving planting in the arid region with poor soil of Yuanmou County (1445 msal, annual rainfall less than 300 mm) in Yunnan (2021). At the maturity stage, 'DHY615' exhibits drought-adaptive features (**A-1**–**A-3**). The root structure and density of 'DHY615' (**B-1**,**B-2**).

2.5. Functional Locus Analysis of 'H479A', 'Nan615', and 'DHY615'

We identified key genes likely associated with essential agronomic traits in 'H479A', 'Nan615', and 'DHY615. The results reveal that 'DHY615' inherits alleles from 'Nan615,' including *SKC1*, which enhances root sodium content for improved salt tolerance [37]; *qUVR-10*, which increases UV–B tolerance [38]; *NRT1.1B*, associated with enhanced nitrogen use efficiency; *STV11*, conferring resistance to rice stripe virus; *OsAAP6*, promoting high protein content; *GS3*, increasing stigma exertion rate and potentially grain size [39]; *Hd17*, delaying heading and flowering [40]; and *Os11g08410*, also delaying heading. Notably, alleles shared by both parents, such as *Waxy* and *ALK*, responsible for taste quality [41,42]; *Hd3a* and *Os01g62780*, controlling heading date [43]; *SaF*, conferring wide *japonica* affinity [44]; *sh4*, associated with the seed shattering trait [45]; *OsFRO1*, imparting iron tolerance [46]; *Rymv1*, associated with resistance to yellow mottle virus disease [47]; and *Sdt97*, contributing to semi-dwarfism [48], were also inherited by 'DHY615. However, genes such as *Sd-1* and *qSW5/GW5* from 'Nan615' were lost during breeding, while *SUB1A* was newly acquired in 'DHY615'. These findings indicate that 'Nan615', as the R-line, carries more functional alleles than the A-line 'H479A'. Consequently, 'DHY615' exhibits trait complementarity and promotes favorable phenotypes (Table 3).

Table 3. Summary of genes associated with important agronomic traits in 'H479A', 'Nan615', and 'DHY615'.

Trait	Gene	Probe Type	Chr	Representative Variety	Phenotype	H479A	Nan615	DHY615
TQ	*BADH2 (fgr)*	SNP	8	Suyunuo	Fragrance	√		√o
ABS	*SKC1*	SNP	1	Nona Bokra	Salt resistant		√	√o

Table 3. *Cont.*

Trait	Gene	Probe Type	Chr	Representative Variety	Phenotype	H479A	Nan615	DHY615
ABS	*qUVR-10*	SNP	10	Sasanishiki	Enhanced phytoremediation activity		√	√o
ABS	*NTR1.1B*	SNP	10	9311	Enhanced nitrogen uptake		√	√o
BS	*STV11*	SNP	11	Kasalath	Persistent resistance to rice stripe virus		√	√o
TQ	*OsAAP6*	INDEL	1	9311	High protein		√	√o
TQ	*GS3*	SNP	3	Minghui 63	Long grain		√	√o
HD	*Hd17*	SNP	6	Koshihikari	Delayed flowering		√	√o
HD	*Os11g08410*	SNP	11	Haplotype C	Delayed heading		√	√o
TQ	Waxy	SNP	6	Nipponbare	Increase the branched starch content	√	√	√
TQ	*ALK*	SNP	6	Minghui 63	Long branched chain starch and higher pasting temperature	√	√	√
HD	*Os01g62780*	SNP	1	Haplotype B	Delayed heading	√	√	√
HD	*Hd3a*	SNP	6	Nipponbare	Photoperiod-sensitive genes	√	√	√
FE	*SaF*	SNP	1	Nipponbare	Wide affinity	√	√	√
OT	*sh4*	SNP	4	Nipponbare	Non-shattering	√	√	√
ABS	*OsFRO1*	SNP	4	KDML105	Iron tolerance	√	√	√
BS	*Rymv1*	SNP	4	Nipponbare	Resistance yellow mottle virus disease	√	√	√
PA	*Sdt97*	SNP	6	Y98149	Semidwarf	√	√	√
PA	*Sd-1*	INDEL	1	DGWG-type	Semidwarf		√	
TQ	*qSW5/GW5*	INDEL	5	Nipponbare	Increase grain width		√	
ABS	*SUB1A*	SNP	9	FR13A	Flood-resistant	o	o	√o

Note: The "√" symbol indicates the genotypic consistent with the representative variety and does not signify whether mutations have occurred; "√o" indicates heterozygous; the "o" symbol indicates the material has no allele. The genes highlighted with yellow and blue are inherited by 'DHY615' from 'H479A' and 'Nan615', respectively. The genes highlighted with green are inherited by 'DHY615' from both 'H479A' and 'Nan615'. The unhighlighted genes are unique to 'DHY615' and 'Nan615', respectively. HD, heading date; FE, fertility; TQ, taste quality; OT, others; ABS, abiotic stress; BS, biotic stress; PA, plant architecture.

To validate the functional loci, gene amplification analysis of *Rf1a* and *OsBADH2* was conducted in 'DHY615' and its parents (Figure S2). The results confirm that the R-line 'Nan615' carries the *Rf1a* gene, while the A-line 'H479A', B-line 'H479B', and hybrid 'DHY615' harbor the *OsBADH2* gene. These findings affirm that the aggregation of functional genes in 'DHY615' positively impacts the breeding of D1-type *japonica* hybrid rice for practical production.

3. Discussion

The variety 'DHY615' represents the culmination of 50 years of D1-type *japonica* hybrid rice breeding. Possessing a highly diverse genetic background, 'DHY615' demonstrates exceptional ecological adaptability to the high-altitude regions of the Yungui Plateau in southwestern China (Figures 1 and 6). Its successful cultivation exemplifies the breeding value and molecular mechanisms underlying plateau *japonica* hybrids, serving as a vital reference for future research and development. The genealogical tracing of the D1-type hybrid *japonica* rice, 'DHY615' (Figure 2), reveals a rich integration of superior *indica* and *japonica* genetic resources from both domestic and international origins [4]. This lineage was achieved by creating an R-line through an "*indica–japonica* bridging" strategy, particularly involving the crossbreeding of IR8 and Keqing No.3. This approach facilitated extensive genetic exchange between subspecies. We analyzed the genetic polymorphism and structure of *japonica* rice populations in the Yungui Plateau, showing that *japonica* rice resources predominantly consist of the 'Dianza', 'Dianheyou' (DHY), and 'Hexi' series [49]. The genetic composition of 'Nan615' and 'DHY615' reflects a high proportion of *indica* components (Figure 4), demonstrating the historical application of a three-line breeding strategy for D1-type *japonica* hybrids through *indica–japonica* integration.

In this study, by utilizing high-impact SNP variants, we identified hundreds of functional genes in 'DHY615' and its parental lines. These encompass known genes involved in plant architecture, grain yield, quality, and resistance to both biotic and abiotic stresses, such as *NRT1.1B*, *SKC1*, *qUVR-10*, *GS3*, *Waxy*, *ALK*, *Hd3a*, and *OsFRO1* (Table 3). These genes may have contributed to key traits, including adaptability to cold, drought, and poor soil conditions, as well as resistance to high UV and iron levels (Figures 1 and 6). This underscores the molecular and genetic basis of 'DHY615''s high-altitude adaptability in the Yungui Plateau.

We further identified 32,607 SNPs in 'DHY615' and its parents, of which 6432 were completely heterozygous markers. This extensive variation likely underpins the superior phenotypic traits of 'DHY615', including improved plant architecture, disease resistance, grain quality, and tolerance to environmental stresses. Using the ANNOVAR program, we predicted the functional impacts of these variants. Among them, 218 variants (0.35%) were classified as high impact, involving changes such as stop codon gain or loss, start codon loss, and splice–site alterations. These variants are strongly associated with phenotypic variation [50]. Additionally, 53 high-impact SNPs were annotated to agronomically important genes, while 3488 (5.63%) missense mutations with moderate functional effects may also influence phenotypic traits. Even variants predicted to have low or modifier effects could contribute to observed phenotypic diversity, as evidenced by the 428 heterozygous SNP loci in the coding regions predicted to have these impacts. Thus, 'DHY615' likely exhibited complementarity in key traits and promoted favorable phenotypes. It was particularly significant that scholars confirmed through field experiments that *japonica* varieties carrying the *NRT1.1B indica* allele showed a substantial increase in seed yield and nitrogen use efficiency compared to those without the *NRT1.1B* allele [51], and the genotyping of 'Nan615' for *NRT1.1B* was consistent with the representative variety. The findings of this study indicated that 'Nan615' also harbors more significant genes, potentially conferring desirable traits, which align with the outcomes previously reported [22]. Consequently, the hybrid loci contributing to the heterotic advantage are primarily derived from 'Nan615'. Building on this, selecting beneficial traits from the parent 'H479A' could be harnessed to further enhance rice quality. The fragrance gene from 'DHY34' also inherits alleles from 'H479A' [52]. Coupled with the results of the fragrance gene detection, we can affirm the reliability of functional gene analysis. Furthermore, we tested the field drought resistance of 198 *japonica* rice variety resources [53], demonstrating that 'DHY615' exhibited high resistance to drought (Table S2). This is consistent with the potential of 'DHY615' to be cultivated as upland rice in high-altitude areas under rainfed conditions.

The ecological adaptation and field performance of 'DHY615' were assessed by collecting phenotypic data related to on-farm yield, disease resistance, and grain quality (Tables 1, S1-1 and S2). Field trials confirmed the yield stability of 'DHY615' under rainfed cultivation across diverse environments in Yunnan, Guizhou, and other regions in southwest China, with no statistically significant yield difference between the optimal and marginal growing zones (Figure 6). Furthermore, the yield variability observed in irrigated locations could potentially be attributable to different management practices and environmental pressures. Additionally, 'DHY615''s capability for rainfed upland cultivation was compared with other water-saving and drought-resistant rice (WDR) varieties.

Li et al. analyzed the yield of 12 approved varieties in Hubei Province and found that the yield of current WDR varieties was 7.5 t/hm^2~9.8 t/hm^2, still lack significant advantages in yield compared to irrigated paddy field rice varieties and 'DHY615' (rainfed upland planting) [54]. However, the yield performance of 'DHY615' in paddy fields and rainfed upland planting areas was comparable, suggesting robust eco-adaptation, particularly valuable for climate-resilient cultivation in montane areas.

Notably, 'DHY615' could be effectively utilized as a rainfed upland rice variety (Figure 6), showcasing suitable ecological characteristics for water-saving and drought-resistant rice (WDR) [55]. It has transformed traditional seedling raising, plowing, and transplanting practices, breaking away from the conventional paddy rice cultivation that yielded low productivity, high cost, and inefficient use of resources (and, in some instances, labor). The success of 'DHY615' has laid the foundation for further innovation in sustainable ecological agriculture. Its cultivation alongside other crops offers opportunities to optimize land use and minimize pest and disease pressure. In this study, we have utilized the GSR40K SNP chip for a preliminary analysis of the genetic background and broad adaptability of 'DHY615'. Our ongoing research aims to integrate the haplotype-resolved telomere-to-telomere (T2T) and high-throughput chromosome conformation capture (Hi-C) reference genome, multi-omics, pan-omics, and advanced biotechnology for the design and genomic selection (GS) of DHY hybrid *japonica* rice. Furthermore, we are conducting a long-term environmental impact assessment of D1-type hybrid upland rice cultivation, focusing on water and land use efficiency and environmentally sustainable agriculture, which includes reducing CH_4 gas emissions, total nutrient runoff loss, and total pesticide loss. Future breeding efforts will aim to expand the DHY series of hybrid *japonica* rice varieties, ensuring their suitability for rainfed upland agriculture while maintaining high yield, quality, and resilience.

4. Materials and Methods

4.1. Research Materials

The rice variety 'DHY615', approved by both the Crop Variety Committee of Yunnan (approval No.2017017) and China (approval No. CHN20220301), is an exceptional D1-type three-line *japonica* hybrid rice combination. It is characterized by the CMS line (A-line) 'H479A' as the female parent and the restorer line (R-line) 'Nan615' as the male parent. In 2019, this variety was bestowed with the Gold Award in the Japonica Rice Category at the 2nd National Quality Rice Tasting Evaluation. In 2023, 'DHY615' received the "Breakthrough New Rice Variety" accolade from the Department of Agriculture and Rural Affairs of Yunnan. All rice cultivars used in this study were developed by the Rice Research Institute of Yunnan Agricultural University.

4.2. Breeding of 'DHY615'

The maintainer line (B-line) 'H479B' was developed by another culture of F_1 (Dianyu No.1/Dianxun No.8//Lianyuanzao), and the D1-CMS A-line 'H479A' was developed by the continuous backcrossing of 'H479B' with 'Hexi 42A'. Therefore, 'H479A' inherited the high-yield and ideal plant type from 'Dianyu No.1', as well as its good taste and fragrance from 'Dianxun No.8'. The R-line 'Nan615' was bred by crossbreeding, using an elite core R-line 'Nan34' as the female parent, with 'Ansanbyeo' introduced from South Korea as the male parent for hybridization to obtain F_1 in 2003. From 2003 to 2009, continuous self-crossed breeding and hybrid offspring selection were conducted in Kunming and Hainan based on dominant agronomic characteristics. Until 2010, the R-line 'Nan615' was self-pollinated until the F_{13} generation, with the dominant agronomic traits being completely stable and uniform.

4.3. Phenotyping and Assessment of Field Performance

Assessment of the phenotypic and field performance of 'DHY615' under both paddy and upland cultivation of 'DHY615' was carried out in areas below 2200 masl (Yunnan *japonica* growing region) that are suitable for upland rice cultivation from 2020 to 2024. The agronomic practices for 'DHY615' were implemented in accordance with the Technical Specification for Cultivation of Upland High-quality Rice (DB53/T 1318.1-2024) [56] of the Yunnan Province Standard, including seed treatment, irrigation methods, fertilization, pest management, planting density, etc. The comprehensive characteristics of 'DHY615', including the varietal feature of DUS (distinctiveness, uniformity, and stability), biological traits (plant height, effective spikes, panicle length, grain number per panicle, seed-setting rate, thousand seed weight, etc.), yield, quality, disease resistance, cold tolerance, drought tolerance, and ecological adaptability, were evaluated according to the National and Provincial field trial standards (DB42/T 1404-2018 and NY/T 2863-2015) [57,58]. The grain yield of 'DHY615' in more than 40 core on-farm demonstration locations was measured in the field by experts organized by National or Provincial Science and Technology Departments, following the "Rice Yield Measurement Method" of the Ministry of Agriculture and Rural Affairs of China.

4.4. DNA Extraction and High-Density Genome-Wide SNP Microarray Analysis

'H479A', 'H479B', 'Nan615', and 'DHY615' (F_1) were grown in a greenhouse at the Rice Research Institute of Yunnan Agricultural University. The fresh young seedling leaflets of 3–5 cm in length were harvested and ground, after freezing with liquid nitrogen, and the genomic DNA of 'H479A', 'H479B', 'Nan615', and 'DHY615' (F_1) was extracted following the CTAB protocol. The DNA quality was assessed by 1–1.5% agarose gel electrophoresis. The DNA samples with high quality (>10-kb fragments) and an appropriate concentration (10–50 ng/uL) were analyzed using the whole genome SNP array GSR40K containing 44263 variations, which was developed by Wuhan Shuanglvyuan Chuangxin Technology Research Institute Co. LTD, Wuhan, Hubei Province). The overall steps involved DNA amplification, fragmentation, chip hybridization, single base extension, staining, and scanning, all of which were performed according to the Infinium assay standard protocol (Infinium HD Assay Ultra Protocol Guide. Available online: http://www.illumina.com/ (accessed on 6 November 2024)). A HiScan scanner (Illumina Inc., San Diego, CA, USA) was used for the SNP microarray scanning, and GenomeStudio software version v2.0.5 was used for raw SNP genotyping analysis. The SNP markers were filtered using the following criteria: (1) a call rate $\geq$80% (missing data < 20%), and (2) a MAF (minor allele frequency) $\geq$5%. The resulting high-quality SNPs were used for downstream analyses. The R platform (R version 4.4.2) was employed for genotype identification, comparison, and map drawing [59]. The unique chromosome frames were constructed based on the 'Nipponbare' 7.0 Genome. Based on the genomic sequence of *japonica* 'Nipponbare', *indica* '9311' was used to analyze the *indica–japonica* components and homozygosity of 'DHY615' and its parents. Here, the SNP genotypes of 'DHY615' and its two parents were compared and classified as homozygous and heterozygous (male parent genotype AA, maternal genotype BB, and hybrid genotype AB).

Based on the gene sequences of *OsBADH2* and *Rf1a* in 'Nipponbare', three pairs of primers were designed in Table S7. The DNA from the A-line 'H479A', R-line 'Nan615', B-line 'H479B', and hybrids of 'DHY615' were used as templates, and the target genes were amplified by Taq DNA polymerase (Thermo Fisher Scientific, Waltham, MA, USA). The total 15 µL PCR reaction system included 0.3 µL of template DNA, 1 µL of primers, and 13.7 µL of mix. The amplification reaction conditions were as follows: pre-denaturation at 94 °C for 5 min, 94 °C denaturation for 30 s, 55 °C/56 °C annealing for 10 s, 72 °C extension for 30 s, 35 cycles, and a final extension at 72 °C for 7 min. Finally, the PCR products were electrophoresed on 1% agarose gel.

4.5. Functional Locus and Statistical Analysis

The reference genome IRGSP-1.0 and the annotation of genomes with GFF files from the Rice Genome Annotation Project. Available online: https://rice.uga.edu/ (accessed on 6 November 2024) were used to annotate and extract the genotypes, positions of varieties, and variants with high, moderate, low, and modifier impact effects using ANNOVAR software with default parameters [60]. To identify SNP variations in well-known, agronomically important genes, we referred to the genes listed in the RAP database RAP-DB. Available online: https://rapdb.dna.affrc.go.jp (accessed on 6 November 2024) [61].

The differences in the yield and yield-related characteristics of 'DHY615' were analyzed using GraphPad Prism 8.4.2. The average, standard deviation and coefficient of the variations were calculated. The graphs were created using Adobe Photoshop CC 2019 and Adobe Illustrator CC 2017.

5. Conclusions

In this study, we conducted a comprehensive analysis of the genealogy and phenotypic characterization of 'DHY615', and identified genes likely associated with biotic and abiotic stress tolerance based on genomic annotation and predicting the effects of genetic variation on gene function in 'DHY615'. Our findings provide valuable insights into the genetic underpinnings of the elite traits and agronomic performance of the D1-type *japonica* hybrid rice, particularly its exceptional adaptability to the diverse high-altitude environments of the Yungui Plateau in southwestern China. We should prioritize the application of 'DHY615' in the most suitable regions of five prefectures/cities, while the suboptimal cultivation conditions will require complementary water regulation technologies to achieve the yield potential. Moreover, the yield demonstrated stable performance across diverse rainfed environments, showcasing its stress-resilient yield stability. It also offers valuable resources for developing breeding strategies for marker-assisted selection (MAS) and genomic selection (GS), presenting new opportunities for breeding and unlocking molecular mechanisms in plateau rice varieties. This work lays a foundation for the continued improvements in rice varieties suited to high-altitude environments and highlights the broader potential of 'DHY615' in promoting sustainable agricultural practices. The large-scale adoption of 'DHY615' via rainfed upland cultivation systems demonstrates significant potential for advancing research in water-efficient agriculture, optimized land utilization, and ecological sustainability in high-altitude plateau regions.

Supplementary Materials: The following supporting information can be downloaded at: https://www.mdpi.com/article/10.3390/ijms26073246/s1.

Author Contributions: Conceptualization, J.W. and C.L.; data curation, R.H., D.H. and Z.Z.; investigation, R.H., D.H. and Z.Z.; methodology, C.W., J.L. (Juan Li), J.X. and D.L. (Dandan Li); project administration, L.C. and D.L. (Dongsun Lee); supervision, Y.Z.; writing—original draft, C.W., J.L. (Juan Li) and Q.Z.; Writing—review and editing, Q.Z., J.L. (Junjie Li) and C.Z. All authors have read and agreed to the published version of the manuscript.

Funding: This study was supported by grants from the Major Science and Technology Projects of Yunnan (Grant No. 202202AE09002102 and 202402AE090026), the Central Leading Local Science and Technology Development Project (Grant No. 202207AA110010), and the National Natural Science Foundation of China (Grant No. 31860108).

Institutional Review Board Statement: Not applicable.

Informed Consent Statement: Not applicable.

Data Availability Statement: The data presented in this study are available upon request from the first author, Chunli Wang. Due to privacy restrictions, the data are not publicly available.

Acknowledgments: The authors would like to express their gratitude to all the anonymous reviewers for their valuable comments and suggestions. We are thankful to Xu Yang (Yunnan Agricultural Technology Extension Station, Kunming) and Fan Luo, as well as Xiaoli Zhou (Xichang University), for their assistance with data collection and analysis. We also wish to extend our thanks to our collaborators at Yunnan Yuanfang Agricultural Science and Technology Limited Company, including Shihuang Pu, Zilin Duan, Fei Gao, Dingchang Chen, Zibing Ban, Youwei Wang, Zeng Wang, and Zhi Wang, for their diligent work in breeding and applying 'DHY615'. Additionally, we would like to thank the senior researchers Quanheng Li (Seed Management Station of Yunnan Province, Kunming) and Xianhan Xu (Qiannan Academy of Agricultural Sciences, Duyun) for their administrative support throughout the project.

Conflicts of Interest: The authors declare no competing financial interests or personal relationships that could have influenced the work reported in this paper.

References

1. Xu, J.; Li, Q. Developing Dian-type *japonica* hybrid rice. *China Seed Ind.* **2019**, *5*, 48–49. [CrossRef]
2. Hua, Y.; Song, X.; Lin, J.; Wu, M. Breeding and application of Yunnan late *Japonica* rice male sterile line Chunjiang 29A. *Hybrid Rice* **2019**, *34*, 12.
3. Ma, R.; Wang, X.; Lu, Y.; Zhou, H.; Cai, K.; Li, X.; Zhang, Z. Breeding and application of late *japonica* male sterile line Yongjing 2A and its late *indica-japonica* hybrid rice combinations. *Hybrid Rice* **2010**, 185–189. [CrossRef]
4. Wang, Z.; Shen, M. Historical development of Li Zhengyou's creation of Dian-type hybrid rice. *Food Saf. Guide* **2017**, 76.
5. Hu, X.; Qian, Q. Revisiting and thinking on research of *Japonica* hybrid rice in China. *China Rice* **2021**, *27*, 9–11.
6. Xu, Y.; Yu, D.; Chen, J.; Duan, M. A review of rice male sterility types and their sterility mechanisms. *Heliyon* **2023**, *9*, e18204.
7. He, H. Studies on fertility inheritance and sterility mechanism of Dian-type and BT hybrid rice. *J. Yunnan Agric. Univ.* **1988**, *1*, 154–168. [CrossRef]
8. Tan, X.; Tan, Y.; Zhao, Y.; Zhang, X.; Hong, R.; Jin, S.; Liu, X.; Huang, D. Identification of the Rf gene conferring fertility restoration of the CMS Dian-type 1 in rice by using simple sequence repeat markers and advanced inbred lines of restorer and maintainer. *Plant Breed.* **2004**, *123*, 338–341.
9. Luan, J.; Liu, T.; Luo, W.; Liu, W.; Peng, M.; Li, W.; Dai, X.; Liang, M.; Chen, L. Mitochondrial DNA genetic polymorphism in thirteen rice cytoplasmic male sterile lines. *Plant Cell Rep.* **2013**, *32*, 545–554.
10. Zheng, X.; Wei, F.; Cheng, C.; Qian, Q. A historical review of hybrid rice breeding. *J. Integr. Plant Biol.* **2024**, *66*, 532–545.
11. Mei, F.; Wu, X.; Yao, C.; Li, L.; Wang, L.; Chen, Q. Rice Cropping Regionalization in China. *Chin. J. Rice Sci.* **1988**, *2*, 97–110.
12. Cheng, K. Cheng Kansheng's Selected Papers on Rice Farming. Yunnan Academy of Agricultural Sciences Press: Kunming, China, 1987.
13. Jiang, Z. *Rice Farming in Yunnan*; Yunnan Science and Technology Press: Kunming, China, 1995; p. 266.
14. Chen, R.; Deng, Y.; Ding, Y.; Guo, J.; Qiu, J.; Wang, B.; Wang, C.; Xie, Y.; Zhang, Z.; Chen, J. Rice functional genomics: Decades' efforts and roads ahead. *Sci. China Life Sci.* **2022**, *65*, 33–92. [PubMed]
15. Xiong, L.; Uga, Y.; Li, Y. Rice functional genomics: Theories and practical applications. *Mol. Breed.* **2020**, *40*, 72.
16. Yu, S.; Ali, J.; Zhou, S.; Ren, G.; Xie, H.; Xu, J.; Yu, X.; Zhou, F.; Peng, S.; Ma, L. From green super rice to green agriculture: Reaping the promise of functional genomic research. *Mol. Plant* **2022**, *15*, 9–26.
17. Zhang, Q. Strategies for developing green super rice. *Proc. Natl. Acad. Sci. USA* **2007**, *104*, 16402–16409.
18. Zhang, Q.; Li, J.; Xue, Y.; Han, B.; Deng, X. Rice 2020: A call for an international coordinated effort in rice functional genomics. *Mol. Plant* **2008**, *1*, 715–719.
19. Thomson, M.J.; Singh, N.; Dwiyanti, M.S.; Wang, D.R.; Wright, M.H.; Perez, F.A.; DeClerck, G.; Chin, J.H.; Malitic-Layaoen, G.A.; Juanillas, V.M. Large-scale deployment of a rice 6K SNP array for genetics and breeding applications. *Rice* **2017**, *10*, 40.
20. Yu, H.; Xie, W.; Li, J.; Zhou, F.; Zhang, Q. A whole-genome SNP array (RICE 6K) for genomic breeding in rice. *Plant Biotechnol.* **2014**, *12*, 28–37.
21. Chen, H.; Xie, W.; He, H.; Yu, H.; Chen, W.; Li, J.; Yu, R.; Yao, Y.; Zhang, W.; He, Y. A high-density SNP genotyping array for rice biology and molecular breeding. *Mol. Plant* **2014**, *7*, 541–553.
22. Huang, X.; Yang, S.; Gong, J.; Zhao, Q.; Feng, Q.; Zhan, Q.; Zhao, Y.; Li, W.; Cheng, B.; Xia, J. Genomic architecture of heterosis for yield traits in rice. *Nature* **2015**, *537*, 629–633.

23. Wang, C.; Wang, Z.; Cai, Y.; Zhu, Z.; Yu, D.; Hong, L.; Wang, Y.; Lv, W.; Zhao, Q.; Si, L.; et al. A higher-yield hybrid rice is achieved by assimilating a dominant heterotic gene in inbred parental lines. *Plant Biotechnol.* **2024**, *22*, 1669–1680. [CrossRef] [PubMed]
24. Gu, Z.; Han, B. Unlocking the mystery of heterosis opens the era of intelligent rice breeding. *Plant Physiol.* **2024**, *196*, 735–744. [CrossRef] [PubMed]
25. *NY/T 593-2013*; Quality of Edible Rice Varieties. China Agriculture Press: Beijing, China, 2013.
26. *NY/T 789-2004*; Technical Specification for Rice Variety Trials. China Agriculture Press: Beijing, China, 2004.
27. Huang, L.; Zhang, H.; Hong, Y.; Liu, S.; Li, D.; Song, F. Stress-responsive expression, subcellular localization and protein-protein interactions of the rice Metacaspase family. *Int. J. Mol. Sci.* **2015**, *16*, 16216–16241. [CrossRef]
28. Li, G.; Wang, Z.; Yokosho, K.; Ding, B.; Fan, W.; Gong, Q.; Li, G.; Wu, Y.; Yang, J.; Ma, J. Transcription factor *WRKY22* promotes aluminum tolerance via activation of *OsFRDL4* expression and enhancement of citrate secretion in rice (*Oryza sativa*). *New Phytol.* **2018**, *219*, 149–162. [CrossRef]
29. Meng, W.; Hsiao, A.-S.; Gao, C.; Jiang, L.; Chye, A.-L. Subcellular localization of rice acyl Coa-binding proteins (ACBPs) indicates that OsACBP6: GFP is targeted to the peroxisomes. *New Phytol.* **2014**, *203*, 469–482. [CrossRef]
30. Tran, T.T.T.; Le, L.H.M.; Nguyen, T.T.; Nguyen, T.C.; Hoang, T.T.H.; Do, P.T.; To, H.T.M. QTL-seq identifies genomic region associated with the crown root development under jasmonic acid response. *Funct. Integr. Genom.* **2024**, *24*, 141. [CrossRef]
31. Peng, Y.; Tang, N.; Zou, J.; Ran, J.; Chen, X. Rice MYB transcription factor *OsMYB1R1* negatively regulates drought resistance. *Plant Growth Regul.* **2023**, *99*, 515–525. [CrossRef]
32. Wang, Z.; Hu, H.; Huang, H.; Duan, K.; Wu, Z.; Wu, P. Regulation of *OsSPX1* and *OsSPX3* on expression of *OsSPX* domain genes and Pi-starvation signaling in rice. *J. Integr. Plant Biol.* **2009**, *51*, 663–674. [CrossRef]
33. Ma, J.; Lei, C.; Xu, X.; Hao, K.; Wang, J.; Cheng, Z.; Ma, X.; Ma, J.; Zhou, K.; Zhang, X. *Pi64*, encoding a novel CC-NBS-LRR protein, confers resistance to leaf and neck blast in rice. *Mol. Plant Microbe Interact.* **2015**, *28*, 558–568. [CrossRef]
34. Wu, Y.; Wang, Q.; Hillwig, M.L.; Peters, R.J. Picking sides: Distinct roles for CYP76M6 and CYP76M8 in rice oryzalexin biosynthesis. *Biochem. J.* **2013**, *454*, 209–216.
35. Yu, S.; Luo, X.; Lian, L.; Xu, H.; Chen, L.; Wei, Y.; Cai, Q.; Xie, H.; Zhang, J. Identification of CIPK family in rice and qRT-PCR analysis on *OsCIPK5* induced by *Magnaporthe oryzae*. *Fujian J. Agric. Sci.* **2019**, *34*, 1237–1245.
36. Zhou, X.; Liao, H.; Chern, M.; Yin, J.; Chen, Y.; Wang, J.; Zhu, X.; Chen, Z.; Yuan, C.; Zhao, W.; et al. Loss of function of a rice TPR-domain RNA-binding protein confers broad-spectrum disease resistance. *Proc. Natl. Acad. Sci. USA* **2018**, *115*, 3174–3179. [PubMed]
37. Raghuvanshi, R.; Srivastava, A.K.; Verulkar, S.; Suprasanna, P. Unlocking allelic diversity for sustainable development of salinity stress tolerance in rice. *Curr. Genom.* **2021**, *22*, 393–403. [CrossRef] [PubMed]
38. Ueda, T.; Sato, T.; Hidema, J.; Hirouchi, T.; Yamamoto, K.; Kumagai, T.; Yano, M. *qUVR10*, a major quantitative trait locus for Ultraviolet-B resistance in rice, encodes cyclobutane pyrimidine dimer photolyase. *Genetics* **2005**, *171*, 1941–1950. [CrossRef]
39. Zhou, H.; Li, P.; Xie, W.; Hussain, S.; Li, Y.; Xia, D.; Zhao, H.; Sun, S.; Chen, J.; Ye, H. Genome-wide association analyses reveal the genetic basis of stigma exertion in rice. *Mol. Plant* **2017**, *10*, 634–644.
40. Matsubara, K.; Ogiso-Tanaka, E.; Hori, K.; Ebana, K.; Ando, T.; Yano, M. Natural variation in *Hd17*, a homolog of Arabidopsis *ELF3* that is involved in rice photoperiodic flowering. *Plant Cell Physiol.* **2012**, *53*, 709–716. [CrossRef]
41. Adegoke, T.V.; Wang, Y.; Chen, L.; Wang, H.; Liu, W.; Liu, X.; Cheng, Y.-C.; Tong, X.; Ying, J.; Zhang, J. Posttranslational modification of waxy to genetically improve starch quality in rice grain. *Int. J. Mol. Sci.* **2021**, *22*, 4845. [CrossRef]
42. Hori, K.; Suzuki, K.; Ishikawa, H.; Nonoue, Y.; Nagata, K.; Fukuoka, S.; Tanaka, J. Genomic regions involved in differences in eating and cooking quality other than *Wx* and *Alk* genes between *indica* and *japonica* Rice cultivars. *Rice* **2021**, *14*, 8. [CrossRef]
43. Tamaki, S.; Matsuo, S.; Wong, H.; YoKol, S.; Shimamoto, K. Hd3a protein is a mobile flowering signal in Rice. *Science* **2007**, *316*, 1033–1036.
44. Xie, Y.; Niu, B.; Long, Y.; Li, G.; Tang, J.; Zhang, Y.; Ren, D.; Liu, Y.-G.; Chen, L. Suppression or knockout of *SaF/SaM* overcomes the *Sa*-mediated hybrid male sterility in rice. *J. Integr. Plant Biol.* **2017**, *59*, 669–679. [CrossRef]
45. Zhu, Y.; Ellstrand, N.C.; Lu, B. Sequence polymorphisms in wild, weedy, and cultivated rice suggest seed-shattering locus *sh4* played a minor role in Asian rice domestication. *Ecol. Evol.* **2012**, *2*, 2106–2113. [PubMed]
46. Li, L.; Ye, L.; Kong, Q.; Shou, H. A vacuolar membrane *Ferric-Chelate Reductase, OsFRO1*, alleviates Fe toxicity in rice (*Oryza sativa* L.). *Front. Plant Sci.* **2019**, *10*, 700.
47. Pidon, H.; Chéron, S.; Ghesquière, A.; Albar, L. Allele mining unlocks the identification of RYMV resistance genes and alleles in African cultivated rice. *BMC Plant Biol.* **2020**, *20*, 222.
48. Tong, J.; Han, Z.; Han, A.; Liu, X.; Zhang, S.; Fu, B.; Hu, J.; Su, J.; Li, S.; Wang, S.; et al. *Sdt97*: A point mutation in the 5′ untranslated region confers Semi-dwarfism in rice. *G3-Genes Genom. Gene.* **2016**, *6*, 1491–1502.
49. Wang, T.; Li, J.; Luo, D.; Zhu, Q.; Wang, C.; Liu, F.; Li, X.; Li, J.; Su, P.; He, Y.; et al. Population genetic structure analysis of *japonica* rice in Yunnan Plateau based on 56K SNP chip. *Mol. Plant Breed.* **2024**, *22*, 5630–5639.

50. Ji, H.; Shin, Y.; Lee, C.; Oh, H.; Yoon, I.S.; Baek, J.; Cha, Y.-S.; Lee, G.-S.; Kim, S.L.; Kim, K.-H. Genomic variation in Korean *japonica* rice varieties. *Genes* **2021**, *12*, 1749. [CrossRef]
51. Hu, B.; Wang, W.; Ou, S.; Tang, J.; Li, H.; Che, R.; Zhang, Z.; Chai, X.; Wang, H.; Wang, Y. Variation in *NRT1.1B* contributes to nitrate-use divergence between rice subspecies. *Nat. Genet.* **2015**, *47*, 834–838.
52. Liu, C.; Wang, Z.; Yu, J.; Zhang, Y.; Zhu, S.; Li, M.; Tan, Y.; Xu, j. The molecular basis of rice fragrance in the high quality *Japonica* hybrid rice variety Dianheyou34. *J. Yunnan Univ. Nat. Sci. Ed.* **2020**, *35*, 371–376.
53. Luo, D.; Wang, T.; Li, J.; Zhu, Q.; He, Y.; Li, J.; Jiang, C.; Huang, D.; Chen, L.; Lee, D. Identification of drought resistance and difference analysis of related gene loci of *japonica* rice in Yunnan Plateau. *Mol. Plant Breed.* **2024**, *22*, 6033–6045.
54. Li, X.; Zhang, S.; Wang, H.; Cai, X.; Zhou, Q.; Zhou, N. Effects of different planting methods on yield, appearance quality and economic benefit of water-saving and drought-resistant rice. *Jiangsu Agric. Sci.* **2021**, *49*, 78–82.
55. Luo, L.; Mei, H.; Yu, X.; Xia, H.; Chen, L.; Liu, H.; Zhang, A.; Xu, K.; Wei, H.; Liu, G. Water-saving and drought-resistance rice: From the concept to practice and theory. *Mol. Breed.* **2019**, *39*, 145.
56. *DB53/T 1318.1-2024*; Technical specifications for highland japonica rice variety testing—Part 1: Regional trials. Yunnan Provincial Standard: Kunming, China, 2024.
57. *DB42/T 1404-2018*; Technical specifications for rice variety field trials. Standard Press of Hubei: Wuhan, China, 2018.
58. *NY/T 2863-2015*; Technical code of practice for regional field trials of rice varieties. China Agriculture Press: Beijing, China, 2015.
59. R Development Core Team. *R: A Language and Environment for Statistical Computing*; R Foundation for Statistical Computing: Vienna, Austria, 2011.
60. Yang, H.; Wang, K. Genomic variant annotation and prioritization with ANNOVAR and wANNOVAR. *Nat. Protoc.* **2015**, *10*, 1556–1566. [PubMed]
61. Sakai, H.; Lee, S.S.; Tanaka, T.; Numa, H.; Kim, J.; Kawahara, Y.; Wakimoto, H.; Yang, C.C.; Iwamoto, M.; Abe, T.; et al. Rice Annotation Project Database (RAP-DB): An integrative and interactive data-base for rice genomics. *Plant Cell Physiol.* **2013**, *54*, e6.

International Journal of
Molecular Sciences

MDPI

Article

Fine Mapping of the QTL *qRLP12* That Controls Root Length Under Polyethylene glycol-Induced Drought Stress During the Early Seedling Stage of Sesame

Junchao Liang [1], Yanxin Deng [1,2], Xiaowen Yan [1], Zhiqi Wang [1], Pan Zeng [1], Meiwang Le [1], Hongying Zhou [1] and Jian Sun [1,*]

[1] Crop Research Institute, Jiangxi Academy of Agricultural Sciences, Jiangxi Province Key Laboratory for Genetic Improvement of Oilcrops, Nanchang 330200, China; njljc@163.com (J.L.); dengyanxin0717.stu@yangtzeu.edu.cn (Y.D.); yanxiaowen1983@126.com (X.Y.); 18679145645@126.com (Z.W.); 15271750715@163.com (P.Z.); mwyuecarl@163.com (M.L.); zhouhongying99@163.com (H.Z.)

[2] College of Agriculture, Yangtze University, Jingzhou 434025, China

* Correspondence: sunjian@jxaas.cn

Abstract: A deeper root system can improve the efficiency of water and nutrient absorption from soil; therefore, genetic improvements to the root length of crops are essential for yield stability under drought stress. We previously identified a stable quantitative trait locus (QTL) *qRLP12* for root length under polyethylene glycol (PEG)-induced drought stress in a Jinhuangma (JHM, sensitive)/Zhushanbai (ZSB, tolerant) recombinant inbred line (RIL) population. To validate and fine map this QTL, in this study, a secondary F_2 population was constructed, and the genetic effect of the target QTL was validated by comparing the phenotype data of different genotypes. Using newly developed markers, 14 genotypes of recombinant F_2 individuals were obtained. A phenotypic analysis of homozygous recombinant progeny lines narrowed *qRLP12* to a 91 kb region. Seven putative predicted genes were identified in the target region, among which *LOC105165547*, a callose synthase gene, was the only one containing nonsynonymous variations in the coding region between two parents. Quantitative real-time PCR analysis revealed that *LOC105165547* was significantly induced by PEG stress in the *qRLP12*+ line. These indicated that *LOC105165547* might be the candidate gene for *qRLP12*, which is responsible for root length subjected to PEG stress. Our results provide a favored gene resource for improving root length under drought stress in sesame.

Keywords: sesame; root length; *qRLP12*; drought stress; fine mapping

1. Introduction

Drought is recognized as one of the most challenging abiotic stresses affecting crop productivity worldwide. The percentage of the planet affected by drought has more than doubled in the last 40 years, and during the same period, droughts have impacted more people worldwide than any other natural hazard [1]. This alarming trend underscores the urgent need for effective strategies to enhance drought tolerance in crops. Deep root systems in terrestrial crops allow them to avoid drought stress by absorbing water from subsoil. Improving the root length under water shortage conditions is therefore a critical strategy to enhance drought tolerance in crops [2]. Research has shown that deeper rooting can enhance grain yield under drought conditions in crops such as maize [3], sorghum [4], wheat [5], and rice [6]. Therefore, understanding the genetic and molecular mechanisms underlying root development is essential for improving agricultural sustainability and addressing food security challenges.

Sesame (*Sesamum indicum* L., 2n = 26) is one of the world's most important oilseed crops, often referred to as the "queen of oilseeds" due to its rich nutrient profile, including tocopherols, lignans, proteins, and plenty of various mineral elements [7,8]. Although sesame is considered a typical "survivor crop" since it has relatively strong tolerance to water deficits, it is particularly vulnerable to drought stress at the germination and flowering stages, primarily due to its shallow root characterization [9,10]. Moreover, the increasing frequency and intensity of drought events, exacerbated by climate change, have further heightened the risk to sesame production over recent decades [11,12]. Major sesame-producing areas worldwide, such as West and Central Africa [13], India [14], and China, frequently experience drought stress during sesame growth season. In China, flash droughts and seasonal droughts frequently coincide with the sesame sowing period in most parts of the main sesame-producing areas, especially in southern China; this significantly hampers sesame seedling growth and production potential [11,15,16]. This highlights the pressing need to enhance the drought tolerance of sesame to ensure yield stability.

Despite the existing knowledge surrounding drought tolerance, significant gaps remain in our understanding of the genetic mechanisms regulating root growth under drought stress in sesame. So far, several studies using transcriptome analysis have revealed multiple transcription factors (TFs), such as ERF [17], MYB [18], bZIP [19], NAC [20], and HD-ZIP [21], participating in the response to drought stress in sesame. Some of these TFs, including *SiERF5* [22], *SiNAC104* [22], *SiMYB75* [23], and *SiMYB77* [24], have been found to improve root elongation and enhance tolerance to drought stress in transgenic *Arabidopsis thaliana* and tobacco plants. Besides these, some other genes, such as the osmotin-like proteins gene (*SindOLP*) and S-adenosylmethionine synthetase gene (*SiSAM*), regulate drought tolerance through various pathways [25,26]. Overexpression of *SindOLP* in sesame could provide longer roots and enhanced tolerance against drought, salinity, and the charcoal rot pathogen [25], while *SiSAM* was found to be related to capsule number and stem length under drought stress in a genome-wide association (GWAS) study examining drought tolerance-related traits at the flowering stage [26]. The latter gene may improve drought tolerance by modulating the polyamine level and ROS homeostasis [26]. Another sesame orphan gene, '*Big Root biomass*', improved yield parameters under normal growth conditions and increased drought stress sensitivity when overexpressed in *Arabidopsis thaliana* plants [27]. In addition, transcriptome profiling of root and leaf samples has revealed hundreds of genes and microRNAs participating in drought response in sesame [28–32]. However, while some of these identified TFs or genes were found to be able to improve root growth under drought stress in transgenic *Arabidopsis thaliana* and tobacco experiments, major QTLs or candidate genes for root growth under drought stress remain largely unexplored in sesame. To our knowledge, only three stable major QTLs for root length under drought stress were identified by far, which was performed by our lab [33]. The genetic mechanism and underlying functional genes that govern root growth under drought stress in sesame are still unknown, representing a critical knowledge gap in the field. Validation, fine mapping, and cloning of major QTLs are essential for understanding their function and applying them in sesame improvement.

We previously identified a major QTL for the main root length under polyethylene glycol (PEG)-induced drought stress (RLP) at the early seedling stage using an RIL population derived from Jinhuangma (JHM, sensitive) and Zhushanbai (ZSB, tolerant) [33]. This QTL explained 11.85–14.46% of the root length under PEG stress variations across different experiments, with the landrace ZSB conferring the favorable allele for longer roots [33]. Here, the drought-tolerant line RA11 and drought-sensitive line RA171 selected from the

JHM/ZSB RIL population were used to construct a large F_2 mapping population to fine map *qRLP12*, develop robust markers for breeding, and identify the underlying gene for root length under drought stress in sesame.

2. Results

2.1. Validation of *qRLP12* for Root Length Under PEG Stress

qRLP12 was initially mapped to a 4.2 cM genetic interval in the RIL population derived from JHM and ZSB [33]. To further verify the genetic effect of *qRLP12*, we constructed a secondary F_2 population by using two RILs: RA11 and RA171. RA11 harbored only one QTL (*qRLP12*) for root length under PEG stress, while RA171 contained no such QTL. All the F_2 plants were planted in nylon net houses to obtain $F_{2:3}$ lines. Under PEG stress, RA11 had a significantly longer root length than RA171, while the root length under water control conditions showed no difference between these two lines (Figure 1, Table S1). Two flanking markers, *Z9* and *Z10*, were developed based on variations from our previous whole-genome resequencing data. After genotyping with *Z9* and *Z10*, one hundred and six $F_{2:3}$ lines showing no cross-over between these two markers were used for phenotyping. According to the marker genotypes of *Z9* and *Z10*, the $F_{2:3}$ lines could be divided into three groups, including Jinhuangma type (JJ), Zhushanbai type (ZZ), and heterozygous type (ZJ). Analysis of variance components for the phenotypic data revealed that the RLP mean of the ZZ group was significantly larger than that of the JJ group ($p < 0.01$) (Figure 2, Tables 1 and S2). No significant variation was detected among different replicates (Table S3). The distribution of RLP in the heterozygous group overlapped with those of both the JJ and the ZZ groups. These results suggest that *qRLP12* was segregated in this F_2 population with the enhancing allele derived from ZSB. Therefore, *qRLP12* was identified as a stable locus and suitable for further fine mapping and cloning.

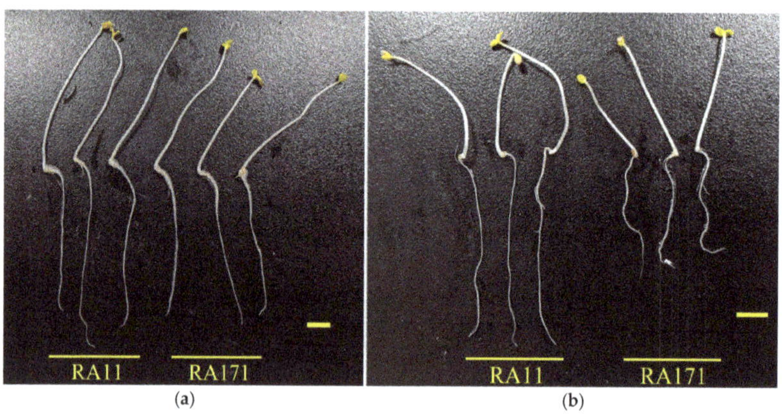

Figure 1. Root phenotypes of the mapping parents RA11 and RA171. (**a**): water condition; (**b**): 15% PEG treatment. Scale bars represent 1 cm.

Table 1. Statistical analysis for root length under PEG stress (RLP) of $F_{2:3}$ population.

Genotype [1]	No. [2]	RLP [3] (cm)
JJ	23	3.80 ± 0.68 a
ZJ	56	5.26 ± 1.05 b
ZZ	27	6.05 ± 0.87 c

[1] Genotypes identified using flanking markers *Z9* and *Z10*; ZZ, JJ and ZJ represent ZSB type, JHM type and heterozygous type. [2] Number of lines. [3] Data for RLP are shown as mean ± standard deviation. Letters represent significant differences at $p = 0.01$.

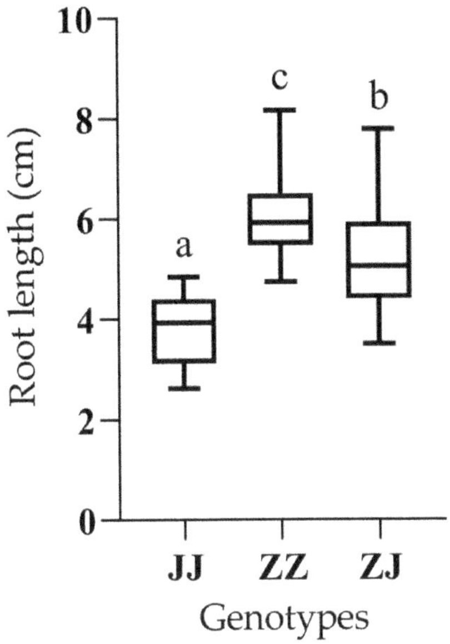

Figure 2. Root length under PEG-stress of $F_{2:3}$ lines grouped according to their genotypes of flanking markers *Z9* and *Z10*. ZZ: homozygous ZSB genotype; JJ: homozygous JHM genotype; ZJ: heterozygous genotype. Different letters above indicate significant differences at $p = 0.01$ between genotypes.

2.2. Fine Mapping of *qRLP12*

To refine the position of *qRLP12*, we screened 1025 F_2 plants derived from RA11 and RA171 for recombinants between markers *Z9* and *Z10*. This led to the identification of 27 plants with cross-over events between these two markers. To saturate this target region, seven additional markers, *Z65*, *Z61*, *Z66*, *Z63*, *Z67*, *Z62*, and *Z64*, were developed based on variations between JHM and ZSB according to previous resequencing data (Table S4, Figure 2). Among the 27 recombinants, we identified 14 genotypes by surveying with these markers in the target interval. No cross-over event was observed between the two markers *Z66* and *Z63* in our mapping population. By means of linkage analysis using the genotype data of recombinants, a high-resolution map was constructed (Figure 3). These markers were mapped to a 1.32 cM interval. Homozygous F_3 progenies of recombinants were singled out by using these markers and planted in a net house to obtain two groups of homozygous $F_{3:4}$ lines (including recombinant and non-recombinant types) for phenotyping.

After comparisons and a two-way ANOVA of the phenotype data among different homozygous progeny $F_{3:4}$ lines of recombinants, significant differences ($p < 0.01$) in RLP were detected only between different groups of progeny lines of seven recombinants (R3, R6, R7, R8, R9, R10, and R13) that had heterozygous regions delimited by the markers *Z61* and *Z67*. The effect of different replications on phenotype was not significant (Table S5). Among these progenies, the lines harboring homozygous ZSB alleles in the *Z61*–*Z67* interval exhibited longer RLP. On the other hand, the remaining recombinant genotypes (R1, R2, R4, R5, R11, and R14) carried homozygous segments in the same interval, and no significant differences in RLP were detected between their progeny groups (Figure 4). Consequently, *qRLP12* was finally mapped to a 91 kb region between markers *Z61* and *Z67* based on the location of critical recombination events (Figure 4).

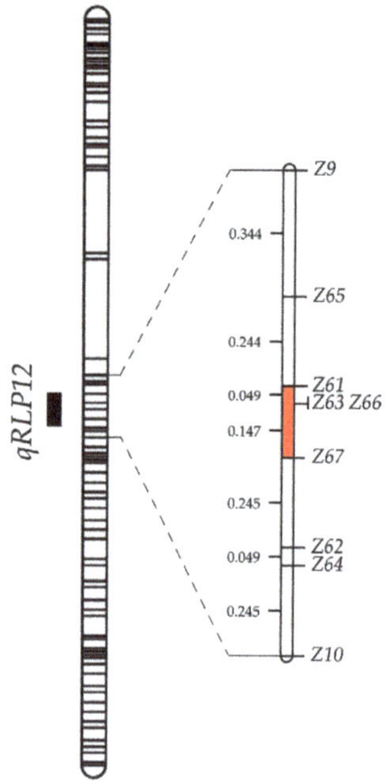

Figure 3. High-resolution linkage map of the *qRLP12* interval. The QTL interval shown to the left of the map was excerpted from Liang et al. (2021) [33]. The red rectangle shown in the map represents the co-segregation region of *qRLP12*.

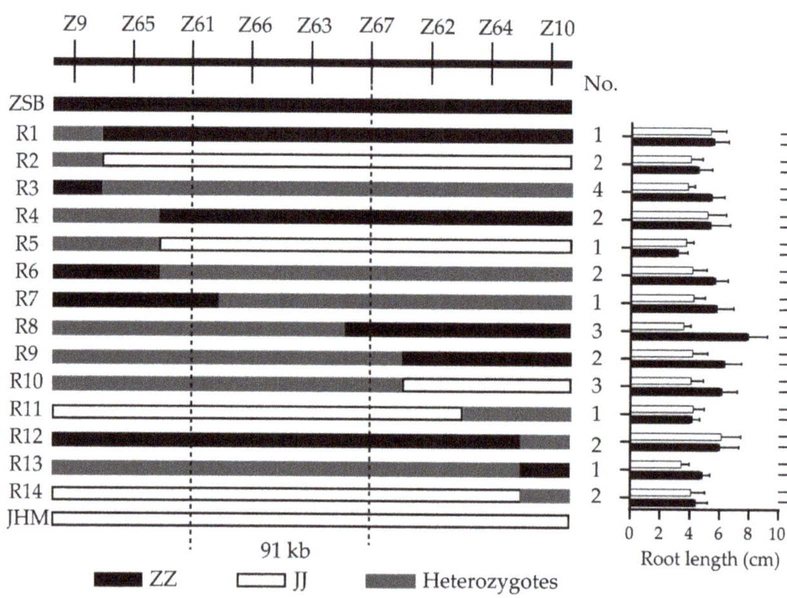

Figure 4. Genotypes (**left panel**) and the corresponding phenotypic assays (**right panel**) of *qRLP12* recombinants. Dark, open, and gray rectangles represent the ZSB genotypes, JHM genotypes, and heterozygotes, respectively. The vertical dashed lines define the interval harboring *qRLP12*. Statistical analyses of phenotypes between different genotypes from self-pollinated progenies of each recombinant are shown with **, ***, and ns, significant at $p < 0.01$ and $p < 0.001$, and non-significant, respectively.

2.3. Candidate Genes in qRLP12 Region

By using the sesame reference genome v1.0 database, a total of seven predicted genes were identified in the targeted 91 kb [34] (https://www.ncbi.nlm.nih.gov/datasets/genome/GCF_000512975.1/, accessed on 5 March 2024). These genes were annotated to encode callose synthase 7 (*LOC105165547*), conserved oligomeric Golgi complex subunit 5 (*LOC105165546*), GDSL esterase/lipase (*LOC105165545*), protein VACUOLELESS1 (*LOC105165544*), 3-hydroxyisobutyryl-CoA hydrolase-like protein 5 (*LOC105165543*), heavy metal-associated isoprenylated plant protein (*LOC105165542*) and subtilisin-like protease (*LOC105165541*). Based on the previous whole-genome resequencing data for the two parents, we discovered 14 SNPs and four insertions and deletions (InDels) in the 91 kb candidate region. Among these seven genes, only the callose synthase 7 (*LOC105165547*) contained nonsynonymous variations between the two parents. The subtilisin-like protease gene (*LOC105165541*) harbored one SNP in its fifth intron. The flanking marker Z67 was developed based on the corresponding SNP in *LOC105165541*, while *LOC105165547* had 10 variations, including two InDels and eight SNPs (Tables S6 and S7). Two nonsynonymous substitutions (A to G and T to C) in *LOC105165547* resulted in amino acid changes (Thr to Ala and Val to Ala) between ZSB and JHM. Two insertions of 35 bp and 37 bp were detected in the last exon of *LOC105165547* in JHM, leading to a premature stop codon (Table 2, Figure S1). The insertion sites of these two fragments were only 5 bp apart. Notably, the cleaved amplified polymorphic sequences (CAPS) marker *Z66* and InDel marker *Z63* in the *qRLP12* co-segregation region were designed based on the missense SNP (A to G) and the two insertions in *LOC105165547*. These suggested that the callose synthase-encoding *LOC105165547* was the most likely causal locus for *qRLP12*.

Table 2. Predicted candidate genes in the 91 kb target region of *qRLP12*.

Gene ID [1]	Annotation	Variation [2]
LOC105165541	subtilisin-like protease SBT5.3	NA
LOC105165542	heavy metal-associated isoprenylated plant protein 16-like	NA
LOC105165543	3-hydroxyisobutyryl-CoA hydrolase-like protein 5	NA
LOC105165544	protein VACUOLELESS1	NA
LOC105165545	GDSL esterase/lipase	NA
LOC105165546	conserved oligomeric Golgi complex subunit 5	NA
LOC105165547	callose synthase 7	Exon14: A1558G (Thr520Ala); Exon42: 5679indel35bp, 5684indel37bp; Exon42: T5696C (Val1910Ala)

[1] Gene accession numbers were retrieved from the annotation of sesame reference genome (https://www.ncbi.nlm.nih.gov/datasets/gene/GCF_000512975.1/, accessed on 5 March 2024). [2] Numbers represent the positions of polymorphic sites in the coding region of candidate genes between JHM and ZSB, the location of start codon was set as 1, "NA" indicates no sequence variation in the coding region of annotated genes.

2.4. Expression Analysis of Candidate Genes

To further identify the candidate gene for *qRLP12*, quantitative real-time PCR (qRT-PCR) was used to detect the expression levels of all seven predicted genes of the target region in RA11 and RA171 subjected to PEG stress (Figure 5). Comparison with water control treatment, *LOC105165546* was significantly induced in the roots of both RA11 and RA171 under PEG stress by similar degrees ($p < 0.01$). Two genes, *LOC105165544* and *LOC105165547*, were significantly up-regulated only in RA11 ($p < 0.01$). *LOC105165541* was significantly down-regulated in both RA11 and RA171 ($p < 0.01$ for RA11, $p < 0.05$ for RA171). The expression level of *LOC105165547* was significantly suppressed in RA171 ($p < 0.05$). There was no difference in the expression of the other three genes (*LOC105165545*, *LOC105165543*, and *LOC 105165542*) between PEG stress and water conditions in the root of both RA11 and RA171. These results further supported our conclusion that *LOC105165547* might be a good candidate gene for PEG-induced drought stress tolerance at the *qRLP12* locus.

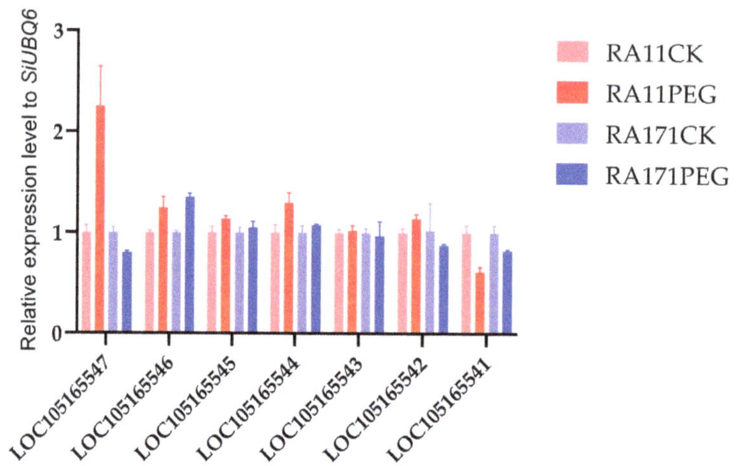

Figure 5. Expression levels of the seven candidate genes in the roots of RA11 (*qRLP12+*) and RA171 (*qRLP12-*) subjected to PEG-stress or water control. *SiUBQ6* was used as an internal control.

3. Discussion

Deep roots are often favored for plant seedling establishment and growth, especially during periods of water scarcity [35–38]. As roots are the hidden part of plants, their traits are relatively more difficult to investigate than those of above-ground parts. To date, few genes are associated with root development and root-related traits for drought tolerance in sesame. Phenotypic investigations using hydroponic systems and field experiments have revealed that sesame roots have great morphological and anatomical diversity [39,40]. This allows the identification of major loci/genes for root traits through genetic analysis approaches, such as linkage mapping and GWAS. By using GWAS, an orphan gene, *BRB*, associated with root number and root dry weight, was identified on LG15 [27]. *BRB*-overexpressing *Arabidopsis* lines displayed reduced total lateral root length and fresh root weight [27]. Besides this gene, six QTLs associated with root length in hydroponic growth conditions were also detected on LG1, LG4, LG5, LG7, and LG10 using the same diversity panel [27]. However, no major genes for root traits under drought stress were cloned in sesame so far, and the genetic mechanism of root growth under water deficits in sesame remained unclear.

In our previous study, we identified three stable QTLs associated with root length under PEG stress in a JHM × ZSB recombinant inbred lines population [33]. Among them, *qRLP12* located on LG6 had the highest LOD value and strongest effect, meaning that it could be an important gene for better drought tolerance at the early seedling stage in sesame. The ZSB allele of *qRLP12* conferred a longer root length. Here in this study, using 27 recombinants from the secondary F_2 population, we fine mapped the *qRLP12* locus to an interval between the markers *Z61* and *Z67*, spanning a 91 kb region on LG6 according to the sesame reference genome sequence v1.0. To our knowledge, this is the first fine map of a root length QTL under drought conditions in sesame. Our results provide the first understanding of the genetic control of root length response to drought stress in sesame. The genome location of the *qRLP12* interval is different from that of any other reported root trait QTL or drought tolerance-related QTL in sesame [26,27,41], suggesting a potential unique genetic pathway influencing root development and drought response in sesame. Further studies on the interaction of *qRLP12* with other drought tolerance QTLs may reveal the genetic mechanism of water deprivation response in sesame.

There are seven annotated genes in the target region of *qRLP12* according to the sesame reference genome database v1.0. Three of them have been highlighted in association with root development or growth. In particular, *LOC105165545* encodes GDSL esterase/lipase (GELP), which have been linked to root development through hormone-mediated signaling pathways in rice [42], soybean [43], and *Arabidopsis* [44]. *LOC105165541* encodes subtilisin-like protease potentially involved in regulating root growth or drought tolerance [45,46]. However, our genomic resequencing data revealed no functional polymorphism in *LOC105165545* between the two parent lines used in this study. Although *LOC105165541* harbors one missense mutation, this SNP was converted into the flanking marker *Z67* and mapped at the boundary of the *qRLP12* interval (Figures 3 and 4). The key recombinant, R8, carrying a homozygous ZSB type in the marker *Z67* locus, showed segregating phenotypes among the progeny lines (Figure 4). These results indicated that *LOC105165545* and *LOC105165541* could be excluded as the *qRLP12* candidate. The other gene, *LOC105165547* encoding callose synthase 7 gene (here designated as *SiCalS7*), contains two InDels and two nonsynonymous substitutions in its open reading frame (Tables S6 and S7). The 35 bp insertion of *SiCalS7* in JHM caused a premature termination. Moreover, a gene expression analysis revealed that *SiCalS7* was significantly induced under PEG treatment in the *qRLP12+* line RA11 but down- regulated in the *qRLP12-* line RA171 (Figure 5). A comparison of the 2 kb promoter region of *SiCalS7* revealed an SNP and a 2-bp InDel between two parents (Table S6), leading to a change in the CCAAT-box element, which is a typical stress-responsive element [47], indicating that these variations may be the cause of the differential expression of *SiCalS7*. Therefore, *SiCalS7* is strongly suggested as the candidate gene for *qPLP12*.

SiCalS7 is homologous to *Arabidopsis CALLOSE SYNTHASE 7/GLUCAN SYNTHASE-LIKE7 (CALS7/GSL7, At1g06490)*, which belongs to glycosyltransferase family 48 (GT48) [48]. This enzyme is required for callose biosynthesis during phloem development [49]. It has been proposed that callose may serve as an induced defense mechanism in response to both biotic and abiotic stresses [50]. Callose deposition and trafficking in the cell wall play a crucial role in tissue development and responses to environmental stresses, while callose synthase genes regulate the level of callose in plants [51,52]. During root development, CALS7 is reported to be highly and specifically enriched in phloem sieve elements, which possess numerous symplastic connections to adjacent cells [53,54]. Loss-of-function mutations in *CALS7* result in complete loss of the callose lining of sieve plate pores between phloem sieve elements in roots [55]. Callose synthase genes can be activated by various biotic and abiotic stresses, and the accumulation of callose has been found to contribute to enhanced root cell walls and offer a physical barrier against drought stress [51]. For example, in soybean, callose accumulation in seedlings was a response to water deficit in roots [55]; in barley, PEG treatment could lead to drought-resistant genotypes accumulating higher levels of callose in their root systems compared to drought-sensitive genotypes [56], and high tolerance to drought and salt stress in Tibetan wild barley is closely related to enhanced callose [57]. Elevated callose content may impede the plant's defense mechanisms against drought by enhancing its water retention capabilities, which could be associated with improved water use efficiency. Further mutagenesis studies or complementation experiments are needed to examine the function of *SiCalS7* and how it regulates root growth under drought stress. Recently, the clustered regularly interspaced short palindromic repeats (CRISPR)/CRISPR-associated protein 9 (Cas9) system had been successfully applied in sesame through hairy root transformation [58]. The biological functions of *SiCalS7* and its associated genomic variations could be confirmed using gene editing or high-efficiency overexpression transformation systems in future studies.

4. Materials and Methods

4.1. Plant Materials

RA11 is a recombinant inbred line (RIL) that carries *qRLP12*, and RA171 is a PEG stress-sensitive RIL containing no QTL for root length under drought stress. RA11 and RA171 were selected from an RIL population developed in our previous study from two parents, Jinghuangma (JHM) and Zhushanbai (ZSB), with contrasting drought tolerance [33]. An F_2 population was developed by crossing the PEG-induced drought stress-resistant RIL RA11 and the susceptible RIL RA171 to fine map *qRLP12*. To verify the initial *qRLP12* mapping results, a total of 106 $F_{2:3}$ lines were phenotyped and genotyped. All the plants were propagated in a nylon net house to prevent cross-pollination caused by insects.

4.2. Marker Development and Recombinant Screening

A total of 1025 F_2 plants from the crossing of RA11 and RA171 were screened to identify recombinants in the *qRLP12* interval. Nine markers, including one SSR marker, four InDel (insertion and deletion) markers, one derived cleaved amplified polymorphic sequences (dCAPS) marker, and three CAPS markers, within the physical region delimiting *qRLP12* were developed based on a comparison of re-sequencing data for JHM and ZSB. All primers were designed with Primer Premier 6 (https://www.premierbiosoft.com/). The polymorphic primer pair sequences are shown in Table S4. Total genomic DNA was extracted from young, fresh leaves based on the CTAB method [59]. PCR was run on a Bio-Rad T100 thermal cycler (Bio-Rad Laboratories, Hercules, CA, USA) using the following protocol: 94 °C for 3 min, followed by 35 cycles of 94 °C for 30 s, 50–60 °C (depending on the specific primers) for 40 s, and 72 °C for 50 s, and a final extension at 72 °C for 5 min. PCR products were digested with corresponding restriction enzymes when needed, separated on agarose gel or 8% non-denaturing polyacrylamide gels (acrylamide/bis-acrylamide = 19:1 or 39:1) with 1 × TBE buffer, and visualized by means of silver staining. Recombinants were identified via genotyping with two flanking markers, *Z9* and *Z10*. The identified recombinants were then genotyped with seven additional markers to further differentiate recombinant event types. A linkage analysis was carried out with JoinMap® v4 [60], and the genetic map was displayed with MapChart v2.32 [61]. The $F_{3:4}$ lines of recombinants were also genotyped with newly developed markers to single out the homozygous ones for phenotyping. At least two homozygous lines for each genotype were used for phenotyping.

4.3. Phenotype Evaluation

The root length under polyethylene glycol (PEG)-simulated drought stress (RLP) was evaluated according to the methods used in our previous study [33]. Thirty-five mature seeds of each line were surface sterilized with 75% ethanol for 5 min, rinsed five times with sterile water, and dried at 25 °C. Sterilized seeds were first germinated on two layers of filter paper in a plastic container (10 cm × 10 cm × 5 cm) with a lid. Thirty hours after germination in darkness at 28 °C, when the roots of the seedling had reached 1–2 cm, twelve similar seedlings from each line were transferred to new containers with 30 mL of 15% *w*/*v* PEG6000/water solution or ddH_2O. For the $F_{2:3}$ population and the recombinants' progenies, three replicates were performed for each line, and three replications were performed for each $F_{3:4}$ line under PEG stress. After four days of incubation in darkness at 28 °C (five days after germination), the root length of the twelve seedlings for each line was measured with a ruler. Root samples of RA11 and RA171 after five days of PEG treatment or water treatment were collected for gene expression analysis.

4.4. Statistical Analysis

An analysis of variance (ANOVA) and multiple comparisons (Fisher's unprotected LSD) were carried out based on a general linear model to test for phenotypic differences between the different genotypes. Statistical analyses were performed, and graphs were drawn using GraphPad Prism 9 (GraphPad Software, LLC, Boston, MA, USA).

4.5. RNA Extraction and Quantitative PCR with Reverse Transcription

Total RNA was isolated using a FastPure Universal Plant Total RNA Isolation Kit (RC411, Vazyme, Nanjing, China) according to the manufacturer's instructions. cDNA was synthesized using a HiScript III All-in-one RT SuperMix Perfect kit for qPCR (RC333, Vazyme, Nanjing, China). Quantitative real-time PCR (qRT-PCR) was conducted on a Roche Lightcycler® 96 (Roche Molecular Systems, Mannheim, Germany) instrument using ChamQ Universal SYBR qPCR Master Mix (Q711, Vazyme). The qRT-PCR volume was 20 μL, consisting of 10 μL of 2× ChamQ Universal SYBR qPCR Master Mix, 0.4 μL of each primer (10 μM), 2 μL of cDNA, and 7.2 μL of ddH_2O. The amplification conditions were as follows: 95 °C for 30 s, followed by 40 cycles of 95 °C for 10 s and 60 °C for 30 s. Three biological replicates of each sample were tested for all tested genes. The raw qPCR data was processed using LightCycler® 96 software v1.1 (Roche Molecular Systems, Mannheim, Germany). The relative gene expression levels were calculated using the $2^{-\Delta\Delta CT}$ method using *SiUBQ6* as an internal control [62]. The primer pairs used for quantitative PCR experiments are listed in Table S4.

5. Conclusions

In this study, we fine mapped the novel QTL *qRLP12* for root length under drought stress to a physical distance of 91 kb in sesame, establishing it as a key locus for root length enhancement under drought conditions. Among the annotated genes within the target region, *LOC105165547* encoding a callose synthase 7 protein is the most likely candidate gene of *qRLP12* based on their putative functions and sequence variations between two parents and expression analysis. To our knowledge, this *qRLP12* locus is the first report to refine a QTL associated with root length under drought stress in sesame. Our study paves the way for future investigations into the molecular mechanisms governing root growth under drought stress in sesame, while the linked markers developed herein can facilitate marker-assisted selection for root improvement under water deficits in sesame breeding efforts.

Supplementary Materials: Supporting information can be downloaded at https://www.mdpi.com/article/10.3390/ijms26072886/s1.

Author Contributions: Conceptualization, J.L. and J.S.; methodology, J.L. and M.L.; validation, J.L., Y.D. and X.Y.; investigation, J.L., Y.D., Z.W., H.Z., P.Z. and X.Y.; resources, J.L., J.S., H.Z., Z.W., P.Z. and X.Y.; writing—original draft preparation, J.L.; writing—review and editing, J.S. and M.L.; supervision, J.S. and M.L.; funding acquisition, J.L. and J.S. All authors have read and agreed to the published version of the manuscript.

Funding: This research was funded by the National Natural Science Foundation of China (32160482) and the China Agriculture Research System (CARS-14).

Informed Consent Statement: Not applicable.

Data Availability Statement: The data used in this study are available within the article and its accompanying Supplementary Materials.

Acknowledgments: We thank Chunhua Hu and Yunyan Zhao for their support on data processing.

Conflicts of Interest: The authors declare no conflicts of interest.

Abbreviations

The following abbreviations are used in this manuscript:

QTL	quantitative trait locus
JHM	jinhuangma
ZSB	zhushanbai
RIL	recombinant inbred line
PEG	polyethylene glycol
RLP	root length under PEG-induced drought stress
TFs	transcription factors
qRT-PCR	quantitative real-time PCR
GWAS	genome-wide association
GELP	GDSL esterase/lipase
InDels	insertions and deletions
dCAPS	derived cleaved amplified polymorphic sequences
CAPS	cleaved amplified polymorphic sequences
CALS7	CALLOSE SYNTHASE 7

References

1. Orimoloye, I.R.; Belle, J.A.; Orimoloye, Y.M.; Olusola, A.O.; Ololade, O.O. Drought: A Common Environmental Disaster. *Atmosphere* **2022**, *13*, 111. [CrossRef]
2. Frolov, A.; Bilova, T.; Paudel, G.; Berger, R.; Balcke, G.U.; Birkemeyer, C.; Wessjohann, L.A. Early responses of mature Arabidopsis thaliana plants to reduced water potential in the agar-based polyethylene glycol infusion drought model. *J. Plant Physiol.* **2017**, *208*, 70–83. [CrossRef] [PubMed]
3. Feng, X.; Jia, L.; Cai, Y.; Guan, H.; Zheng, D.; Zhang, W.; Xiong, H.; Zhou, H.; Wen, Y.; Hu, Y. ABA-inducible DEEPER ROOTING 1 improves adaptation of maize to water deficiency. *Plant Biotechnol. J.* **2022**, *20*, 2077–2088. [PubMed]
4. Chen, X.; Wu, Q.; Gao, Y.; Zhang, J.; Wang, Y.; Zhang, R.; Zhou, Y.; Xiao, M.; Xu, W.; Huang, R. The role of deep roots in sorghum yield production under drought conditions. *Agronomy* **2020**, *10*, 611. [CrossRef]
5. Bacher, H.; Montagu, A.; Herrmann, I.; Walia, H.; Schwartz, N.; Peleg, Z. Stress-induced deeper rooting introgression enhances wheat yield under terminal drought. *J. Exp. Bot.* **2023**, *74*, 4862–4874.
6. Uga, Y.; Sugimoto, K.; Ogawa, S.; Rane, J.; Ishitani, M.; Hara, N.; Kitomi, Y.; Inukai, Y.; Ono, K.; Kanno, N. Control of root system architecture by DEEPER ROOTING 1 increases rice yield under drought conditions. *Nat. Genet.* **2013**, *45*, 1097–1102.
7. Yadav, R.; Kalia, S.; Rangan, P.; Pradheep, K.; Rao, G.P.; Kaur, V.; Pandey, R.; Rai, V.; Vasimalla, C.C.; Langyan, S.; et al. Current Research Trends and Prospects for Yield and Quality Improvement in Sesame, an Important Oilseed Crop. *Front. Plant Sci.* **2022**, *13*, 863521. [CrossRef]
8. Teboul, N.; Gadri, Y.; Berkovich, Z.; Reifen, R.; Peleg, Z. Genetic Architecture Underpinning Yield Components and Seed Mineral–Nutrients in Sesame. *Genes* **2020**, *11*, 1221. [CrossRef]
9. Jeyaraj, S.; Beevy, S.S. Insights into the Drought Stress Tolerance Mechanisms of Sesame: The Queen of Oilseeds. *J. Plant Growth Regul.* **2024**, *43*, 3370–3391. [CrossRef]
10. Dossa, K.; Wei, X.; Li, D.; Fonceka, D.; Zhang, Y.; Wang, L.; Yu, J.; Boshou, L.; Diouf, D.; Cissé, N.; et al. Insight into the AP2/ERF transcription factor superfamily in sesame and expression profiling of DREB subfamily under drought stress. *BMC Plant Biol.* **2016**, *16*, 171. [CrossRef]
11. Chen, S.; Gong, B. Response and adaptation of agriculture to climate change: Evidence from China. *J. Dev. Econ.* **2021**, *148*, 102557. [CrossRef]
12. Hermans, K.; McLeman, R. Climate change, drought, land degradation and migration: Exploring the linkages. *Curr. Opin. Environ. Sustain.* **2021**, *50*, 236–244.
13. Boureima, S.; Diouf, S.; Amoukou, M.; Van Damme, P. Screening for sources of tolerance to drought in sesame induced mutants: Assessment of indirect selection criteria for seed yield. *Int. J. Pure Appl. Biosci.* **2016**, *4*, 45–60.
14. Pandey, B.B.; Ratnakumar, P.; Usha Kiran, B.; Dudhe, M.Y.; Lakshmi, G.S.; Ramesh, K.; Guhey, A. Identifying traits associated with terminal drought tolerance in sesame (*Sesamum indicum* L.) genotypes. *Front. Plant Sci.* **2021**, *12*, 739896.
15. Wang, L.; Yuan, X. Two types of flash drought and their connections with seasonal drought. *Adv. Atmos. Sci.* **2018**, *35*, 1478–1490.
16. Deng, Y.; Wang, X.; Lu, T.; Du, H.; Ciais, P.; Lin, X. Divergent seasonal responses of carbon fluxes to extreme droughts over China. *Agric. For. Meteorol.* **2023**, *328*, 109253.

17. Su, R.; Dossou, S.S.K.; Dossa, K.; Zhou, R.; Liu, A.; Zhong, Y.; Fang, S.; Zhang, X.; Wu, Z.; You, J. Genome-wide characterization and identification of candidate ERF genes involved in various abiotic stress responses in sesame (*Sesamum indicum* L.). *BMC Plant Biol.* **2022**, *22*, 256.
18. Mmadi, M.A.; Dossa, K.; Wang, L.; Zhou, R.; Wang, Y.; Cisse, N.; Sy, M.O.; Zhang, X. Functional characterization of the versatile MYB gene family uncovered their important roles in plant development and responses to drought and waterlogging in sesame. *Genes* **2017**, *8*, 362. [CrossRef]
19. Wang, Y.; Zhang, Y.; Zhou, R.; Dossa, K.; Yu, J.; Li, D.; Liu, A.; Mmadi, M.A.; Zhang, X.; You, J. Identification and characterization of the bZIP transcription factor family and its expression in response to abiotic stresses in sesame. *PLoS ONE* **2018**, *13*, e0200850.
20. Zhang, Y.; Li, D.; Wang, Y.; Zhou, R.; Wang, L.; Zhang, Y.; Yu, J.; Gong, H.; You, J.; Zhang, X. Genome-wide identification and comprehensive analysis of the NAC transcription factor family in *Sesamum indicum*. *PLoS ONE* **2018**, *13*, e0199262.
21. Wei, M.; Liu, A.; Zhang, Y.; Zhou, Y.; Li, D.; Dossa, K.; Zhou, R.; Zhang, X.; You, J. Genome-wide characterization and expression analysis of the HD-Zip gene family in response to drought and salinity stresses in sesame. *BMC Genom.* **2019**, *20*, 748. [CrossRef] [PubMed]
22. Dossa, K.; Mmadi, M.A.; Zhou, R.; Zhang, T.; Su, R.; Zhang, Y.; Wang, L.; You, J.; Zhang, X. Depicting the core transcriptome modulating multiple abiotic stresses responses in sesame (*Sesamum indicum* L.). *Int. J. Mol. Sci.* **2019**, *20*, 3930. [CrossRef] [PubMed]
23. Dossa, K.; Mmadi, M.A.; Zhou, R.; Liu, A.; Yang, Y.; Diouf, D.; You, J.; Zhang, X. Ectopic expression of the sesame MYB transcription factor SiMYB305 promotes root growth and modulates ABA-mediated tolerance to drought and salt stresses in Arabidopsis. *AoB Plants* **2020**, *12*, plz081. [CrossRef]
24. Rajput, P.; Agarwal, P.; Agarwal, P.K. A *Sesamum indicum* SiMYB77 Transcription Factor, Enhances Drought and Salt Tolerance in Transgenic Tobacco Via Maintaining Higher Osmolytes and ROS Homeostasis. *J. Plant Growth Regul.* 2024, *in press*. [CrossRef]
25. Chowdhury, S.; Basu, A.; Kundu, S. Overexpression of a new osmotin-like protein gene (SindOLP) confers tolerance against biotic and abiotic stresses in sesame. *Front. Plant Sci.* **2017**, *8*, 410. [CrossRef]
26. Dossa, K.; Li, D.; Zhou, R.; Yu, J.; Wang, L.; Zhang, Y.; You, J.; Liu, A.; Mmadi, M.A.; Fonceka, D. The genetic basis of drought tolerance in the high oil crop *Sesamum indicum*. *Plant Biotechnol. J.* **2019**, *17*, 1788–1803. [CrossRef]
27. Dossa, K.; Zhou, R.; Li, D.; Liu, A.; Qin, L.; Mmadi, M.A.; Su, R.; Zhang, Y.; Wang, J.; Gao, Y. A novel motif in the 5′-UTR of an orphan gene 'Big Root Biomass' modulates root biomass in sesame. *Plant Biotechnol. J.* **2021**, *19*, 1065–1079. [CrossRef]
28. Baghery, M.A.; Kazemitabar, S.K.; Dehestani, A.; Mehrabanjoubani, P.; Naghizadeh, M.M.; Masoudi-Nejad, A. Transcription Factors and microRNA Genes Regulatory Network Construction Under Drought Stress in Sesame (*Sesamum indicum* L.). *Res. Sq.* **2020**. [CrossRef]
29. Baghery, M.A.; Kazemitabar, S.K.; Dehestani, A.; Mehrabanjoubani, P.; Naghizadeh, M.M.; Masoudi-Nejad, A. Insight into gene regulatory networks involved in sesame (*Sesamum indicum* L.) drought response. *Biologia* **2022**, *77*, 1181–1196. [CrossRef]
30. Song, Q.; Joshi, M.; Wang, S.; Johnson, C.D.; Joshi, V. Comparative analysis of root transcriptome profiles of sesame (*Sesamum indicum* L.) in response to osmotic stress. *J. Plant Growth Regul.* **2021**, *40*, 1787–1801. [CrossRef]
31. Dossa, K.; Li, D.; Wang, L.; Zheng, X.; Liu, A.; Yu, J.; Wei, X.; Zhou, R.; Fonceka, D.; Diouf, D. Transcriptomic, biochemical and physio-anatomical investigations shed more light on responses to drought stress in two contrasting sesame genotypes. *Sci. Rep.* **2017**, *7*, 8755.
32. You, J.; Zhang, Y.; Liu, A.; Li, D.; Wang, X.; Dossa, K.; Zhou, R.; Yu, J.; Zhang, Y.; Wang, L. Transcriptomic and metabolomic profiling of drought-tolerant and susceptible sesame genotypes in response to drought stress. *BMC Plant Biol.* **2019**, *19*, 267.
33. Liang, J.; Sun, J.; Ye, Y.; Yan, X.; Yan, T.; Rao, Y.; Zhou, H.; Le, M. QTL mapping of PEG-induced drought tolerance at the early seedling stage in sesame using whole genome re-sequencing. *PLoS ONE* **2021**, *16*, e0247681.
34. Wang, L.; Yu, S.; Tong, C.; Zhao, Y.; Liu, Y.; Song, C.; Zhang, Y.; Zhang, X.; Wang, Y.; Hua, W. Genome sequencing of the high oil crop sesame provides insight into oil biosynthesis. *Genome Biol.* **2014**, *15*, R39.
35. Siddiqui, M.N.; Léon, J.; Naz, A.A.; Ballvora, A. Genetics and genomics of root system variation in adaptation to drought stress in cereal crops. *J. Exp. Bot.* **2021**, *72*, 1007–1019. [PubMed]
36. Courtois, B.; Ahmadi, N.; Khowaja, F.; Price, A.H.; Rami, J.-F.; Frouin, J.; Hamelin, C.; Ruiz, M. Rice root genetic architecture: Meta-analysis from a drought QTL database. *Rice* **2009**, *2*, 115–128. [CrossRef]
37. Comas, L.H.; Becker, S.R.; Cruz, V.M.V.; Byrne, P.F.; Dierig, D.A. Root traits contributing to plant productivity under drought. *Front. Plant Sci.* **2013**, *4*, 442.
38. Meister, R.; Rajani, M.; Ruzicka, D.; Schachtman, D.P. Challenges of modifying root traits in crops for agriculture. *Trends Plant Sci.* **2014**, *19*, 779–788.
39. Su, R.; Zhou, R.; Mmadi, M.A.; Li, D.; Qin, L.; Liu, A.; Wang, J.; Gao, Y.; Wei, M.; Shi, L. Root diversity in sesame (*Sesamum indicum* L.): Insights into the morphological, anatomical and gene expression profiles. *Planta* **2019**, *250*, 1461–1474.
40. Hamedani, N.G.; Gholamhoseini, M.; Bazrafshan, F.; Amiri, B.; Habibzadeh, F. Variability of root traits in sesame genotypes under different irrigation regimes. *Rhizosphere* **2020**, *13*, 100190.

41. Li, D.; Dossa, K.; Zhang, Y.; Wei, X.; Wang, L.; Zhang, Y.; Liu, A.; Zhou, R.; Zhang, X. GWAS uncovers differential genetic bases for drought and salt tolerances in sesame at the germination stage. *Genes* **2018**, *9*, 87. [CrossRef] [PubMed]
42. Zhao, H.; Ma, B.; Duan, K.-X.; Li, X.-K.; Lu, X.; Yin, C.-C.; Tao, J.-J.; Wei, W.; Zhang, W.-K.; Xin, P.-Y. The GDSL lipase MHZ11 modulates ethylene signaling in rice roots. *Plant Cell* **2020**, *32*, 1626–1643. [CrossRef] [PubMed]
43. Su, H.-G.; Zhang, X.-H.; Wang, T.-T.; Wei, W.-L.; Wang, Y.-X.; Chen, J.; Zhou, Y.-B.; Chen, M.; Ma, Y.-Z.; Xu, Z.-S. Genome-wide identification, evolution, and expression of GDSL-type esterase/lipase gene family in soybean. *Front. Plant Sci.* **2020**, *11*, 726.
44. Liu, D.; Zhao, X.; Liu, Y.; Tian, M.; Zhao, J.; Bai, N.; Huang, R.; Li, M.; Sui, S. The GDSL lipase CpGLIP1 from Chimonanthus praecox improves drought and cold tolerance in Arabidopsis and poplar. *Ind. Crop. Prod.* **2024**, *215*, 118636.
45. Stührwohldt, N.; Bühler, E.; Sauter, M.; Schaller, A. Phytosulfokine (PSK) precursor processing by subtilase SBT3. 8 and PSK signaling improve drought stress tolerance in Arabidopsis. *J. Exp. Bot.* **2021**, *72*, 3427–3440.
46. Zhao, X.; Goher, F.; Chen, L.; Song, J.; Zhao, J. Genome-wide identification, phylogeny and expression analysis of subtilisin (SBT) gene family under wheat biotic and abiotic stress. *Plants* **2023**, *12*, 3065. [CrossRef]
47. Laloum, T.; De Mita, S.; Gamas, P.; Baudin, M.; Niebel, A. CCAAT-box binding transcription factors in plants: Y so many? *Trends Plant Sci.* **2013**, *18*, 157–166.
48. Shi, X.; Han, X.; Lu, T.-g. Callose synthesis during reproductive development in monocotyledonous and dicotyledonous plants. *Plant Signal. Behav.* **2016**, *11*, e1062196.
49. Ušák, D.; Haluška, S.; Pleskot, R. Callose synthesis at the center point of plant development—An evolutionary insight. *Plant Physiol.* **2023**, *193*, 54–69. [CrossRef]
50. Nishimura, M.T.; Stein, M.; Hou, B.-H.; Vogel, J.P.; Edwards, H.; Somerville, S.C. Loss of a callose synthase results in salicylic acid-dependent disease resistance. *Science* **2003**, *301*, 969–972. [CrossRef]
51. Li, N.; Lin, Z.; Yu, P.; Zeng, Y.; Du, S.; Huang, L.-J. The multifarious role of callose and callose synthase in plant development and environment interactions. *Front. Plant Sci.* **2023**, *14*, 1183402. [CrossRef] [PubMed]
52. Diana, S.-G.; Kamila, G.-J.; Ewa, K.; Małgorzata, K.-K.; Monika, T.; Emilia, G.; Kaja, S.; Magdalena, R.; Karolina, U.; Monika, K. The effect of silicon supplementation and drought stress on the deposition of callose and chemical components in the cell walls of the Brassica napus roots. *BMC Plant Biol.* **2024**, *24*, 1249. [CrossRef]
53. Vatén, A.; Dettmer, J.; Wu, S.; Stierhof, Y.-D.; Miyashima, S.; Yadav, S.R.; Roberts, C.J.; Campilho, A.; Bulone, V.; Lichtenberger, R. Callose biosynthesis regulates symplastic trafficking during root development. *Dev. cell* **2011**, *21*, 1144–1155. [CrossRef]
54. Kalmbach, L.; Bourdon, M.; Belevich, I.; Safran, J.; Lemaire, A.; Heo, J.-o.; Otero, S.; Blob, B.; Pelloux, J.; Jokitalo, E. Putative pectate lyase PLL12 and callose deposition through polar CALS7 are necessary for long-distance phloem transport in Arabidopsis. *Curr. Biol.* **2023**, *33*, 926–939.e9. [CrossRef]
55. Ehrhardt-Brocardo, N.C.M.; Coelho, C.M.M.; Souza, C.A.; Mathias, V. Callose accumulation in roots of soybean seedlings under water deficit. *Theor. Exp. Plant Physiol.* **2019**, *31*, 475–481. [CrossRef]
56. Ahmed, I.M.; Nadira, U.A.; Cao, F.; He, X.; Zhang, G.; Wu, F. Physiological and molecular analysis on root growth associated with the tolerance to aluminum and drought individual and combined in Tibetan wild and cultivated barley. *Planta* **2016**, *243*, 973–985. [CrossRef] [PubMed]
57. Ahmed, I.M.; Nadira, U.A.; Bibi, N.; Cao, F.; He, X.; Zhang, G.; Wu, F. Secondary metabolism and antioxidants are involved in the tolerance to drought and salinity, separately and combined, in Tibetan wild barley. *Environ. Exp. Bot.* **2015**, *111*, 1–12. [CrossRef]
58. You, J.; Li, D.; Yang, L.; Dossou, S.S.K.; Zhou, R.; Zhang, Y.; Wang, L. CRISPR/Cas9-mediated efficient targeted mutagenesis in sesame (*Sesamum indicum* L.). *Front. Plant Sci.* **2022**, *13*, 935825. [CrossRef]
59. Porebski, S.; Bailey, L.G.; Baum, B.R. Modification of a CTAB DNA extraction protocol for plants containing high polysaccharide and polyphenol components. *Plant Mol. Biol. Report.* **1997**, *15*, 8–15. [CrossRef]
60. Van Ooijen, J. *JoinMap 4 ® Software for the Calculation of Genetic Linkage Maps in Experimental Populations*; Kyazma B.V.: Wageningen, The Netherlands, 2006.
61. Voorrips, R. MapChart: Software for the graphical presentation of linkage maps and QTLs. *J. Hered.* **2002**, *93*, 77–78. [CrossRef]
62. Wei, L.; Miao, H.; Zhao, R.; Han, X.; Zhang, T.; Zhang, H. Identification and testing of reference genes for Sesame gene expression analysis by quantitative real-time PCR. *Planta* **2013**, *237*, 873–889. [PubMed]

International Journal of *Molecular Sciences*

MDPI

Article

Integrated Transcriptome and Metabolome Analyses Reveal Candidate Genes Associate with Phenolic Compound Biosynthesis in Different Varieties of *Perilla frutescens*

Hong Ye [1,2,3,4,†], Jiaxin Mu [2,†], Tong Yang [2], Qi Shen [5,*] and Yukun Wang [1,2,3,4,*]

1 Guangdong Provincial Key Laboratory of Utilization and Conservation of Food and Medicinal Resources in Northern Region, Shaoguan University, Shaoguan 512005, China; 19881212hong@163.com
2 College of Biology and Agriculture, Shaoguan University, Shaoguan 512005, China; muuixyy@163.com (J.M.); sguanxueyuan@163.com (T.Y.)
3 Engineering and Technology Research Center of Shaoguan Horticulture in Shaoguan University, Shaoguan 512005, China
4 Guangdong Provincial Engineering and Technology Research Center of Special Fruit and Vegetables in Northern Region, Engineering and Technology Research Center of Shaoguan Horticulture in Shaoguan University, Shaoguan 512005, China
5 School of Pharmaceutical Sciences, Institute of Medical Plant Physiology and Ecology, Guangzhou University of Chinese Medicine, Guangzhou 510006, China
* Correspondence: shenqi@gzucm.edu.cn (Q.S.); wangyu_kun1@163.com (Y.W.)
† These authors contributed equally to this work.

Abstract: *Perilla frutescens* (Perilla) has great potential for utilization in food and medicine due to the abundance of secondary metabolites, especially phenolic compounds. However, the molecular mechanism underlying phenolic compound synthesis in perilla remains poorly understood. By targeted metabolome analysis, we found nine differentially accumulated metabolites among QS2, QS6, and QO7 leaves and nine unique metabolites which only exist in QS6. Furthermore, transcriptome analysis showed the differential expression genes (DEGs) were significantly enriched into phenylpropanoid- and flavonoid-related pathways. Moreover, the integrated transcriptome and metabolome analyses indicated 14 candidates highly correlated with phenolic compound biosynthesis. In addition, phenylpropanoid- and flavonoid-biosynthesis-related DEGs, including one PAL, one CYP73A, one 4CL, two CHI, two F3H, one FLS, three CHS, two CYP75B1, one ANS, and two DFR, were isolated. The results in this study provide useful information for the metabolic regulation of phenolic compounds and serve as essential gene resources for future breeding programs.

Keywords: *Perilla frutescens*; transcriptome; metabolome; phenolic compounds; biosynthesis

1. Introduction

Perilla is an annual herb belonging to the *Labiaceae* family and is broadly planted in East Asian regions, especially in China, Japan, Korea, and so on [1]. In China, perilla has been cultivated for more than 2000 years and is known as an important medicinal and edible plant [2]. Leaves of perilla can be eaten with vegetables. As a kind of food, perilla leaves can be cooked as a side dish, boiled into soup, or used as a seasoning for cooking meat to add flavor [2]. As a kind of traditional medicine, perilla has several properties such as antibacterial, antitumor, antioxidative, anti-inflammatory, and antitussive abilities [3–7]. These effects can be mainly attributed to the abundant primary and secondary components in perilla tissues. Specifically, phenolic acids, flavonoids, and volatile oils serve as effective constituents in perilla leaves, stems, and seeds [8–10]. Due to these, perilla is utilized as a crucial source of natural pigments, preservatives, and sweeteners [11].

In plants, phenolic compounds are regarded as a complex class of secondary metabolites. At present, over 8000 phenolic compounds have been found [12]. Phenolic compounds can be categorized as simple phenolics, polyphenols, and other phenolic compounds. Simple phenolics include phenolic acids (hydroxybenzoic and hydroxycinnamic acids) and coumarins. Polyphenols are composed of flavonoids and tannins, while other phenolic compounds are made up of stilbenes, lignans, and lignins [13]. Simple phenolic acid contains a single aromatic ring and can be further divided into hydroxybenzoic acids and hydroxycinnamic acids [14]. The biosynthesis of simple phenolic acids derived from cinnamic acid occurs with the participation of shikimate and phenylpropanoid pathways. Based on related studies, simple phenolic acids are found in almost all foods [12]. For instance, simple phenolic acids can be found and consumed in fruits, legumes, oats, cereals, oils, and their byproducts [12,15,16]. However, studies of the biological mechanism of phenolic compounds in perilla varieties are still rare.

As a group of polyphenols, flavonoids are an essential class of secondary metabolites that widely existed in plants. They are thought of as plant growth and development regulators and prominently used in food and medicine [17]. It is well known that flavonoids are derived from the phenylpropanoid metabolic pathway and can be separated into 12 substances such as chalcones, stilbenes, aurones, flavanones, flavones, isoflavones, and so on [17–19]. In detail, flavonoids are generated from phenylalanine, which is synthesized through the shikimate pathway [20]. In the general phenylpropanoid pathway, phenylalanine is converted to *p*-coumaroyl-CoA by the catalysis of phenylalanine ammonia lyase (PAL), cinnamic acid 4-hydroxylase (C4H), and 4-coumarate: CoA ligase (4CL) [17]. Then, the downstream flavonoids are formed via the catalysis of corresponding enzymes. Some perilla varieties are rich in several kinds of flavonoids and thus are always used as the materials to study the regulatory mechanism of flavonoid biosynthesis.

In this study, we performed a global survey of phenolic acids in three different perilla varieties using targeted metabolic analysis, and the type and number of phenolic compounds were uncovered, respectively. Through comparative analysis, the differential phenolic compounds, including the unique and common components in three perilla varieties, were screened. Furthermore, comparative transcriptome analysis was carried out, and the corresponding results highlighted differences in three different perilla varieties at transcriptional level. Subsequently, the integrated transcriptome and metabolome analysis not only provided candidate genes but also revealed valuable information for the metabolic engineering of phenolic compound biosynthesis in the leaves of perilla.

2. Results

2.1. Differences in Phenolic Compounds in Three Perilla Varieties

Three perilla varieties QS2, QS6, and QO7, which have different leaf colors, were chose as the materials in this study. Both sides of the leaves of QS2 and QO7 were purple and green, respectively, while the front was green and the back was purple in QS6 leaves (Figure 1A), indicating that these varieties have different numbers of pigments, especially anthocyanin. To further understand the reasons for these differences, we performed the targeted metabolome analysis to detect 128 types of phenolic compounds such as anthocyanins, benzoic acid derivatives, flavanones, flavones, isoflavones, phenylpropanoids, terpenoids, and so on. In QS2 leaves, 48 subclasses of metabolites, which belonged to 17 subclasses of phenolic compounds, were detected. Flavones (n = 8), flavonols (n = 7), and benzoic acid derivatives (n = 6) were the top three subclasses of metabolites (Figure 1B). A total of 56 classes of metabolites, which mainly belonged to flavonols (n = 10), flavones (n = 9), benzoic acid derivatives (n = 6), and phenylpropanoids (n = 6), were found in QS6 leaves (Figure 1C). There were also 17 subclasses of phenolic compounds found in QO7 leaves, but the number of metabolite classes (45) was lower than that of QS2 and QS6. Similarly to QS2 and QS6, flavonols (n = 7), flavones (n = 7), benzoic acid derivatives (n = 5), and phenylpropanoids (n = 5) were the main metabolites in QO7 leaves (Figure 1D).

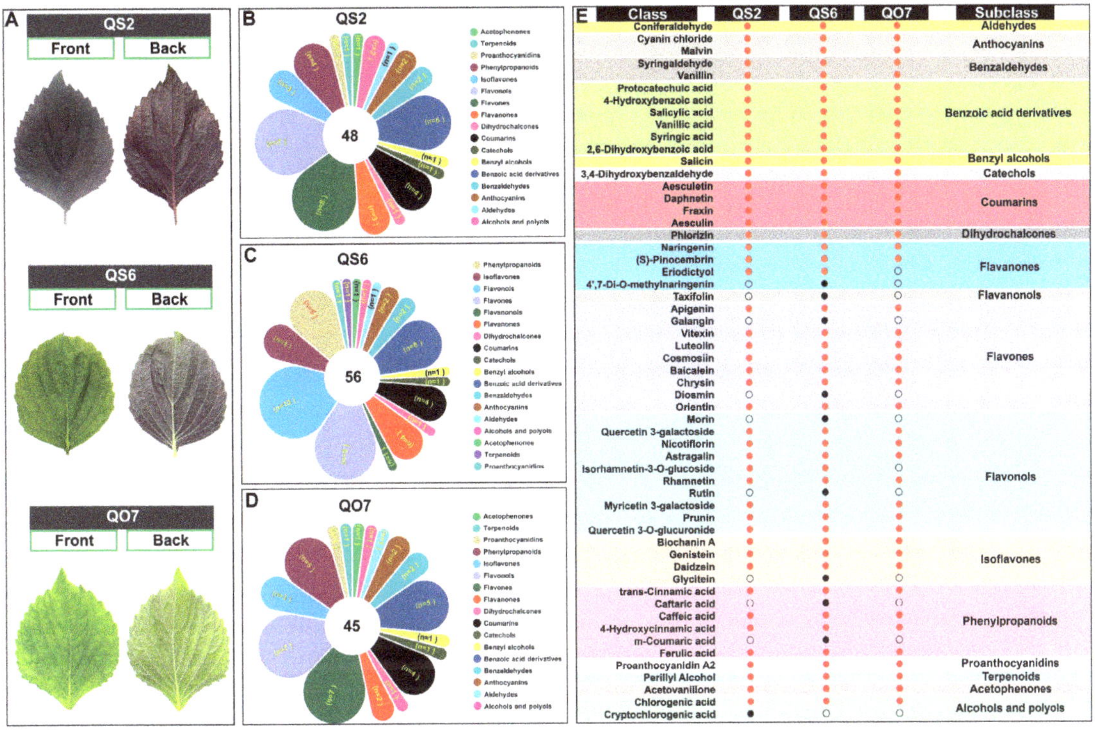

Figure 1. Differences in the species and quantity of phenolic compounds in three perilla varieties. (**A**) Phenotypic characteristics of perilla QS2, QS6, and QO7 leaves. (**B–D**) Targeted metabolome analysis of phenolic compounds in three perilla leaves. The total numbers are shown in the center of the figures. (**E**) Details of the species and quantity of phenolic compounds in the three perilla leaves. Red dots indicate phenolic compounds that were found in all the three perilla leaves. Black dots indicate phenolic compounds that were found in only one perilla. Black circles indicate phenolic compounds that were not found in any of the perilla leaves.

The targeted metabolome analysis can provide absolute quantification data of each metabolite. Next, we compared metabolites in each perilla variety. As shown in Figure 1E, most of the metabolites in QS2, QS6, and QO7 leaves were coincident. Specifically, among 17 subclasses of metabolites, all the perilla leaves contained aldehydes (n = 1), anthocyanins (n = 2), benzaldehydes (n = 2), benzyl alcohols (n = 6), catechol (n = 1), coumarins (n = 4), dihydrochalcones (n = 1), proanthocyanidins (n = 1), terpenoids (n = 1), and acetophenones (n = 1) (Figure 1E). There were clear differences in the components of flavanones (n = 4), flavones (n = 9), flavonols (n = 10), phenylpropanoids (n = 6), isoflavones (n = 4), and alcohols and polyols (n = 2). QO7 leaves had no eriodictyol and isorhamnetin-3-O-glucoside, which were found in QS2 and QS6 leaves. Cryptochlorogenic acid was only detected in QS2 leaves. Moreover, 4′,7-Di-O-methylnaringenin, taxifolin, galangin, diosmin, morin, rutin, glycitein, caftaric acid and m-coumaric acid were only detected in QS6 leaves (Figure 1E). These results suggest that different perilla varieties has different species and quantities of phenolic compounds, especially in flavonoids, and these differences might be an important reason for phenotypic differences between the three kinds of perilla leaves.

2.2. Isolation and Enrichment Analysis of Differentially Accumulated Phenolic Compounds

Having determined the unique metabolites in QS2 and QS6 (Figure 1E), we then carried out the isolation of differentially accumulated phenolic compounds among the common metabolites. Finally, nine (one up- and eight downregulated) and eight (seven up- and one downregulated) DAMs were screened out from comparable groups 'QS2 vs. QS6' and 'QS6 vs. QO7', respectively (Figure 2A). The level of caffeic acid in QS6 leaves was the lowest (Figure 2B,C). In addition, levels of protocatechuic acid, quercetin 3-galactoside, apigenin, (*S*)-pinocembrin, luteolin, cyanin chloride, astragalin, and nicotiflorin were the highest in QS6 leaves than in QS2 and QO7 leaves (Figure 2B,C). KEGG enrichment of these DAMs showed that they mainly belonged to 'flavone and flavonol biosynthesis', 'degradation of flavonoids', and 'isoflavonoid biosynthesis' pathways (Figure 3A,B). These results indicated that the activity of the flavonoid biosynthesis pathway in QS2, QS6, and QO7 was inconsonant and the differences in levels of flavonoids might be another reason for phenotypic differences in the three types of perilla leaves.

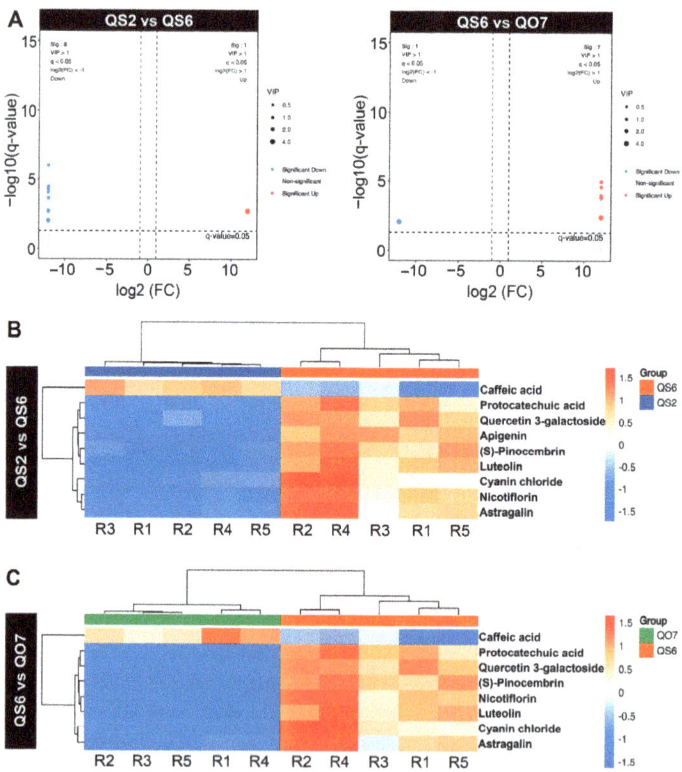

Figure 2. Isolation of differentially accumulated phenolic compounds. (**A**) Volcano plot of differentially accumulated phenolic compounds in comparative groups 'QS2 vs. QS6' and 'QS6 vs. QO7'. (**B**) Heatmap of differentially accumulated phenolic compounds in comparative groups 'QS2 vs. QS6'. (**C**) Heatmap of differentially accumulated phenolic compounds in comparative groups 'QS6 vs. QO7'.

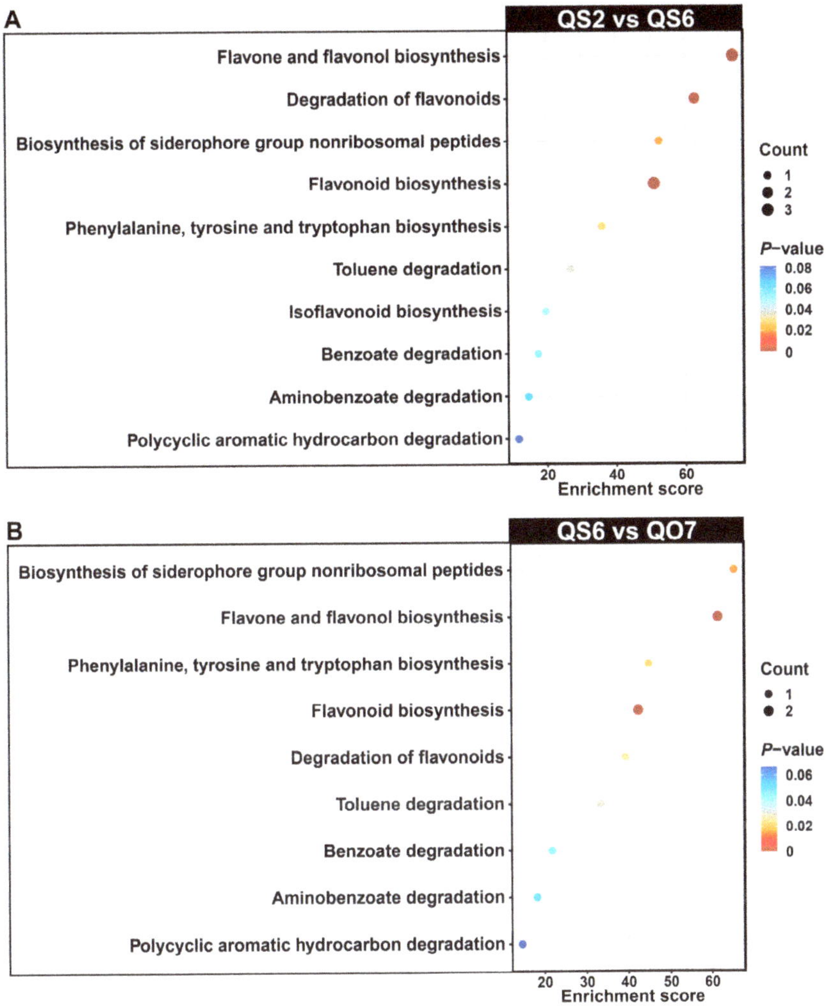

Figure 3. KEGG enrichment analysis of differentially accumulated phenolic compounds in comparative groups (**A**) 'QS2 vs. QS6' and (**B**) 'QS6 vs. QO7'.

2.3. Comparative Transcriptome Analysis

In order to study more information at the transcriptional level, we performed a comparative transcriptome analysis among QS2, QS6, and QO7 leaves. PCA of gene expression profiles of nine analyzed samples was performed. It showed that the clear distinction between QS6 and QO7 samples was largely accounted for by PC1 (55.15%), while the difference between QS2 and QS6 samples was largely accounted for by PC2 (23.31%) (Figure 4A). In the comparative group 'QS2 vs. QS6', 2626 DEGs, including 933 up- and 1693 downregulated genes, were screened out (Figure 4B). In the comparative group 'QS6 vs. QO7', 1967 upregulated and 1685 downregulated genes were isolated, respectively (Figure 4C). The expression heatmaps of DEGs in 'QS2 vs. QS6' and 'QS6 vs. QO7' indicated that each DEG was expressed relatively steadily in each replication and the DEGs in two comparative groups were clustered into two modules displaying clear differences in expression patterns (Figure 4D,E).

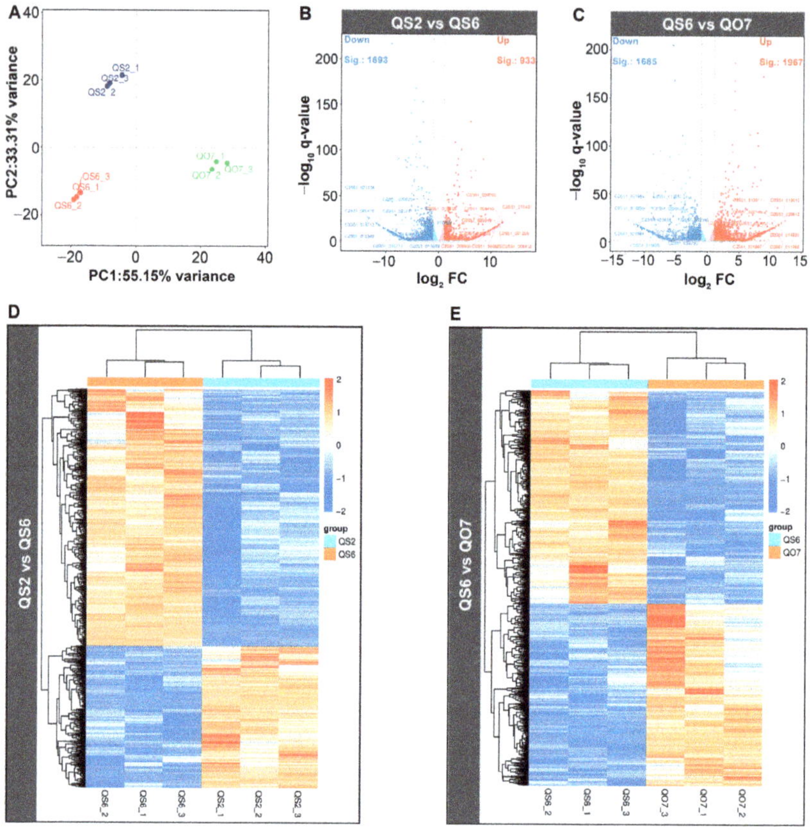

Figure 4. Comparative transcriptome analysis of DEGs. (**A**) PCA of samples used for transcriptome analysis. (**B**) Volcano plot of the DEGs in comparative group 'QS2 vs. QS6'. (**C**) Volcano plot of the DEGs in comparative group 'QS6 vs. QO7'. (**D**) Heatmap of the DEGs in comparative group 'QS2 vs. QS6'. (**E**) Heatmap of the DEGs in comparative group 'QS6 vs. QO7'.

Then, GO enrichment analysis of DEGs was carried out to study the possible functions of DEGs. In the comparative group 'QS2 vs. QS6', 52 non-redundant DEGs were involved in pigmentation (GO:0043473; nine genes), flavonoid biosynthetic process (GO:0009813; 17 genes), anthocyanin-containing compound biosynthetic process (GO:0009718; 20 genes), monoterpenoid biosynthetic process (GO:0016099; five genes), phenylpropanoid metabolic process (GO:0009698; six genes), terpene metabolic process (GO:0042214; three genes), and triterpenoid biosynthetic process (GO:0016104; four genes) (Figure 5A). In the comparative group 'QS6 vs. QO7', 48 non-redundant DEGs were enriched into flavonoid 3′-monooxygenase activity (GO:0016711; five genes), pigment biosynthetic process (GO:0046148; eight genes), anthocyanin-containing compound biosynthetic process (GO:0009718; 19 genes), anthocyanin 5-O-glucoside 6‴-O-malonyltransferase activity (GO:0033810; four genes), pigmentation (GO:0043473; seven genes), flavonoid biosynthetic process (GO:0009813; 16 genes), and anthocyanin accumulation in tissues in response to UV light (GO:0043481; five genes) (Figure 5B). The results of KEGG pathway analysis demonstrated that several DEGs were clustered into phenylpropanoid biosynthesis (ko00940), flavonoid biosynthesis (ko00941), and anthocyanin biosynthesis (ko00942) pathways. In the comparative group 'QS2 vs. QS6', 29, 26, and 5 DEGs were involved in phenylpropanoid biosynthesis, flavonoid biosynthesis, and anthocyanin biosynthesis pathways, respectively (Figure 5C). For the comparative group 'QS6 vs. QO7', 37, 27, and 5 DEGs were involved in phenylpropanoid biosynthesis, flavonoid biosynthesis, and anthocyanin biosynthesis pathways, respectively (Figure 5D).

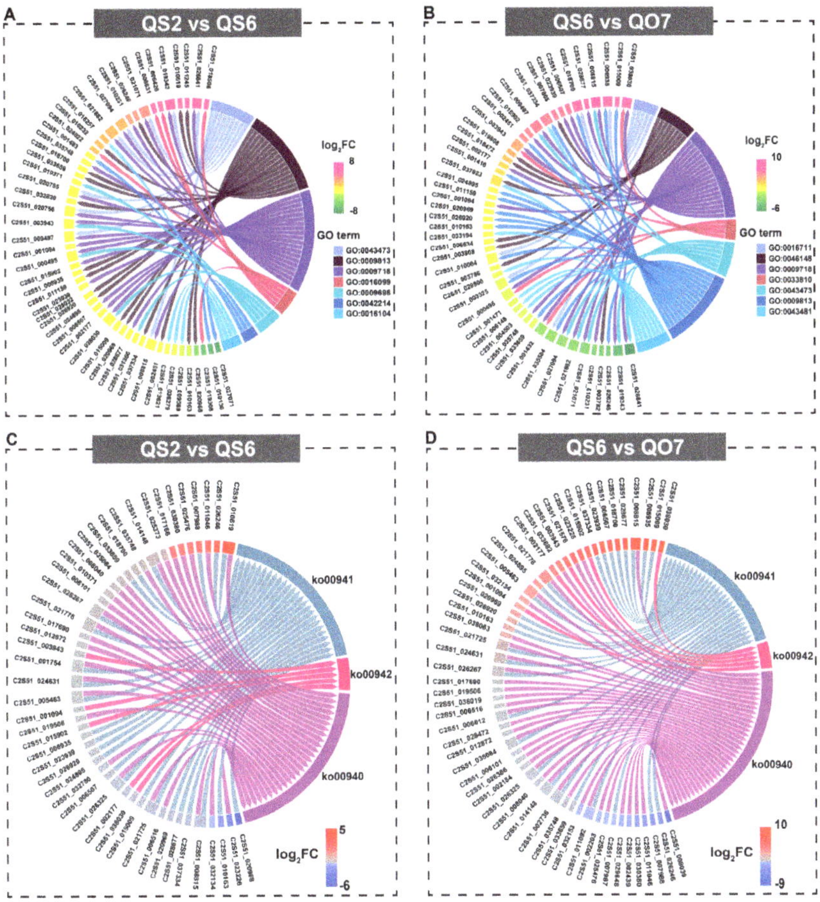

Figure 5. GO and KEGG enrichment analysis of DEGs. (**A**) GO analysis of DEGs in comparative group 'QS2 vs. QS6'. (**B**) GO analysis of DEGs in comparative group 'QS6 vs. QO7'. (**C**) KEGG analysis of DEGs in comparative group 'QS2 vs. QS6'. (**D**) KEGG analysis of DEGs in comparative group 'QS6 vs. QO7'.

2.4. qRT-PCR Validation of DEGs

Based on the results of the GO and KEGG analyses, we combined the DEGs in GO terms and KEGG pathways and finally obtained 104 candidates. We surveyed the expression abundance of each DEG and built a heatmap using the FPKM values. As shown in Figure 6A, 70 candidates were clearly separated into two clades, of which 31 DEGs displayed distinct differences in QS2, QS6, and QO7 leaves, and 39 DEGs showed relatively low FPKM values in QS2, QS6, and QO7 leaves. In order to validate the quality of RNA-seq in this study, we chose 10 DEGs and performed a qRT-PCR assay. The results revealed that all the selected genes showed similar expression patterns in qRT-PCR analysis as that in RNA-seq data (Figure 6B), suggesting that the expression data obtained by RNA-seq of this study were highly reliable.

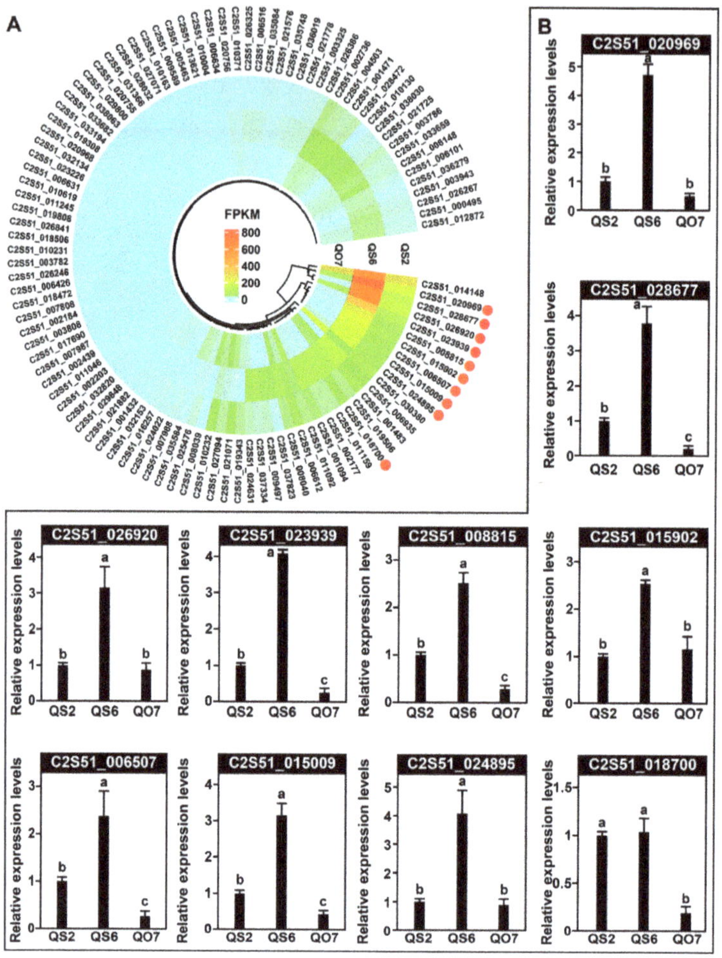

Figure 6. Validation of selected DEGs. (**A**) Expression heatmap of DEGs based on the GO and KEGG enrichment analysis results. Red dots indicate 10 selected DEGs for validation. (**B**) qRT-PCR validation of 10 selected DEGs. Data are presented as means ± standard deviations (n = 3). Lowercase letters are used to indicate differences. Identical letters mean that the differences are not significant, while different letters indicate significant differences.

2.5. Integrated Analysis of Transcriptome and Metabolome Data

To further investigate the relationship between DEGs and DAMs, the coexpression network analysis of the transcriptome and metabolome was conducted. We analyzed the correlations between 50 DEGs (top 50) and 15 phenolic compounds (including 8 DAMs, 6 unique metabolites in QS6, and 1 unique metabolite in QS2). Among these 50 DEGs, 5 DEGs (C2S51_010231, C2S51_002203, C2S51_025476, C2S51_011046, and C2S51_007988) were negatively correlated with the relative phenolic compounds, while 4 DEGs (C2S51_026325, C2S51_010232, C2S51_013621, and C2S51_009589) were positively and negatively correlated with the relative phenolic compounds. The rest of 41 DEGs were all positively correlated with the relative phenolic compounds. In addition, each DEG was correlated with at least six kinds of phenolic metabolites (Figure 7A). These results not only indicated that one DEG might be involved in several phenolic metabolisms, but also suggested that one phenolic metabolite might be regulated by multiple factors.

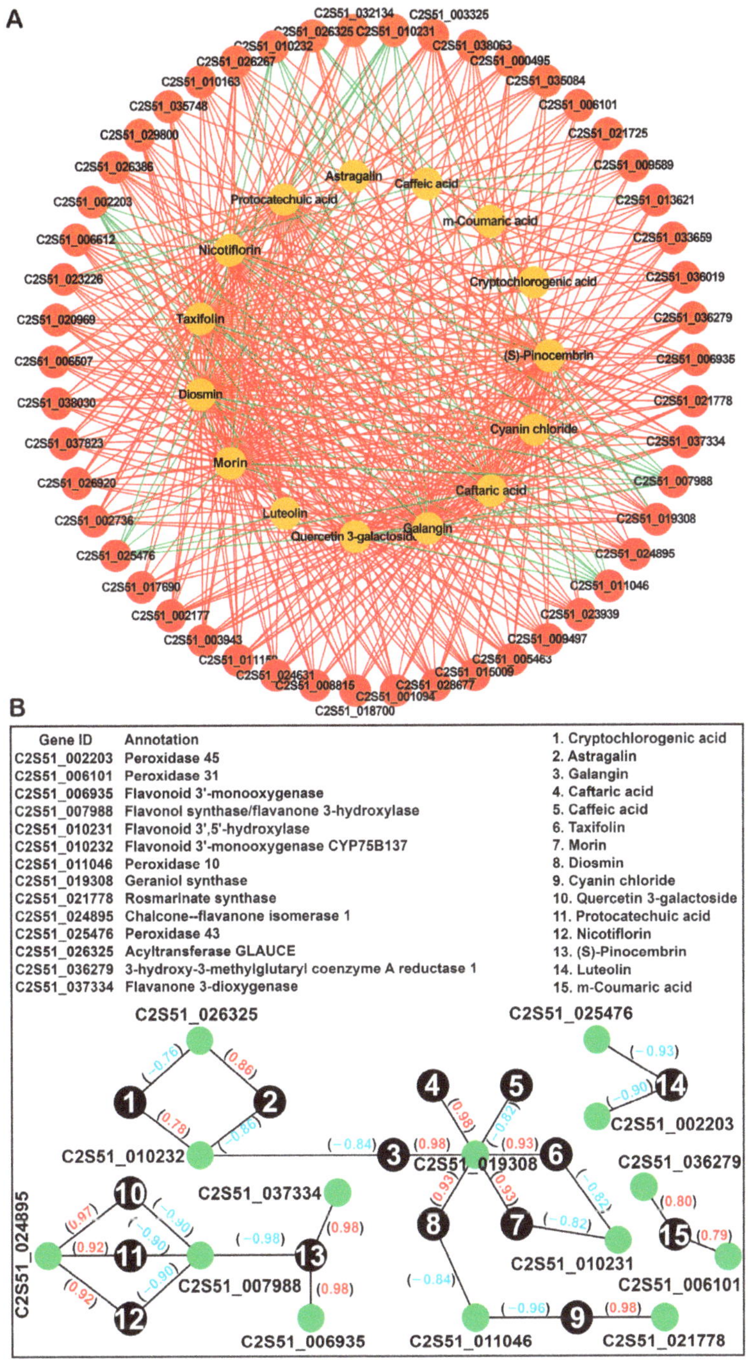

Figure 7. Integrated analysis of the transcriptome and metabolome data. (**A**) Correlation network of DEGs and DAMs. Data showed the top 50 DEGs. (**B**) Correlation network of DEG and DAM pairs that had the highest correlation, including positive (red) and negative (blue) correlations. The annotation of each DEG was shown as well.

Based on the Pearson correlation coefficient, we chose the top two DEGs with the highest or lowest Pearson correlation coefficient for each phenolic metabolite. The DEG (C2S51_010232) that encoded a flavonoid 3′-monooxygenase CYP75B protein showed a positive correlation (r = 0.78) with cryptochlorogenic acid, but displayed a negative correlation (r = −0.86) with astragalin. In addition, this gene was also negatively correlated (r = −0.84) with galangin (Figure 7B). C2S51_026325 encoded an acyltransferase GLAUCE and was positively correlated (r = −0.76) with astragalin but negatively correlated (r = 0.86) with cryptochlorogenic acid (Figure 7B). C2S51_019308, which encoded a geraniol synthase, was positively correlative with galangin (r = 0.98), caftaric acid (r = 0.98), taxifolin (r = 0.93), morin (r = 0.93), diosmin (r = 0.93), while negatively correlative with caffeic acid (r = −0.82) (Figure 7B). C2S51_011046 encoded a peroxidase 10 and was negatively correlative with diosmin (r = −0.84) and cyanin chloride (r = −0.96) (Figure 7B). For cyanin chloride, the rosmarinate synthase encoding gene C2S51_021778 showed a positive correlation (r = 0.98) (Figure 7B). The flavonol synthase encoding gene C2S51_007988 was negatively correlative with quercetin 3-galactoside (r = −0.90), protocatechuic acid (r = −0.90), nicotiflorin (r = −0.90), and (S)-Pinocembrin (r = −0.98) (Figure 7B). For (S)-Pinocembrin, a flavonoid 3′-monooxygenase (C2S51_006935) and a flavanone 3-dioxygenase (C2S51_037334) might be the positive regulators (Figure 7B). Chalcone-flavanone isomerase (C2S51_024895) was a positive regulator for quercetin 3-galactoside (r = 0.97), protocatechuic acid (r = 0.92), and nicotiflorin (r = 0.92). Furthermore, two peroxidase-encoding genes C2S51_002203 and C2S51_025476 were negatively correlative with luteolin, and the Pearson correlation coefficients were −0.90 and −0.93, respectively (Figure 7B). Finally, peroxidase-encoding gene C2S51_006101 and 3-hydroxy-3-methylglutaryl CoA reductase encoding gene C2S51_037334 were positively correlative with *m*-coumaric acid, and the Pearson correlation coefficients were 0.79 and 0.80, respectively (Figure 7B). These results provided several potential candidates for phenolic metabolisms in perilla, but their precise functions should be studied in the future.

2.6. Profiles of Key Genes and Metabolites Associated with Phenolic Metabolisms in Perilla

Phenolic compounds are known as the most prevalent secondary metabolites, synthesized in all plant cells [21,22]. In line with this, our abovementioned results also showed that many kinds of phenolic metabolisms, such as flavanols, flavanones, flavanonols, flavones, flavonols, isoflavones, phenylpropanoids, and anthocyanins, were synthesized in perilla leaves (Figures 1–3). The biosynthesis of simple phenolic acids derived from cinnamic acid needs the combined action of shikimate and phenylpropanoid pathways. The shikimate pathway is the metabolic way for the synthesis of phenylalanine, tyrosine, and tryptophan [23,24]. For the synthesis of benzoic acids, shikimate and phenylpropanoid pathway intermediates are used as precursors [25]. As a kind of benzoic acid derivative, protocatechuic acid is formed from dehydration and hydroxylation of 3-dehydroshikimate [26]. In this study, we found that the content of DAM protocatechuic acid was the highest in QS6 leaves (Figures 2B,C and 8). However, we found that the expression levels of 3-Dehydroquinate dehydratase (DHQD) encoding gene (C2S51_002722) did not reach a significant level (Figure 8). In addition, the direct regulatory factors of the DAMs caffeic acid, apigenin, cyanin chloride, astragalin, nicotiflorin, and quercetin 3-galactoside were not detected (Figure 8), indicating that the biosynthesis of these metabolites might be regulated by key enzymes in upstream pathways. A similar situation was observed in the biosynthesis of unique metabolites (only detected in QS6), such as *m*-coumaric acid, caftaric acid, rutin, glycitein, and 4′,7-Di-O-methylnaringenin (Figure 8). Given these, it was worth confirming the functions of the DEGs that showed high correlation with metabolites mentioned above (Figures 7 and 8). Moreover, we found that most of the DAMs and unique metabolites (only detected in QS6) were involved in phenylpropanoid biosynthesis and flavonoid biosynthesis pathways (Figure 8). A phenylalanine ammonia lyase (PAL) encoding gene C2S51_028472, which was highly expressed in QS6 leaves, was screened out as the initiator enzyme of phenylpropanoid biosynthesis pathway (Figure 8). PAL catalyzes the deamination of phenylalanine to cinnamic acid; thus, as the derivative of cinnamic acid, *m*-coumaric acid might be controlled by PAL. In addition, caffeic acid and caftaric acid, which were derived from cinnamic acid in phenylpropanoid biosynthesis pathway, might be directly regulated by trans-cinnamate 4-monooxygenase (CYP73A) encoded gene (C2S51_024631) (Figure 8).

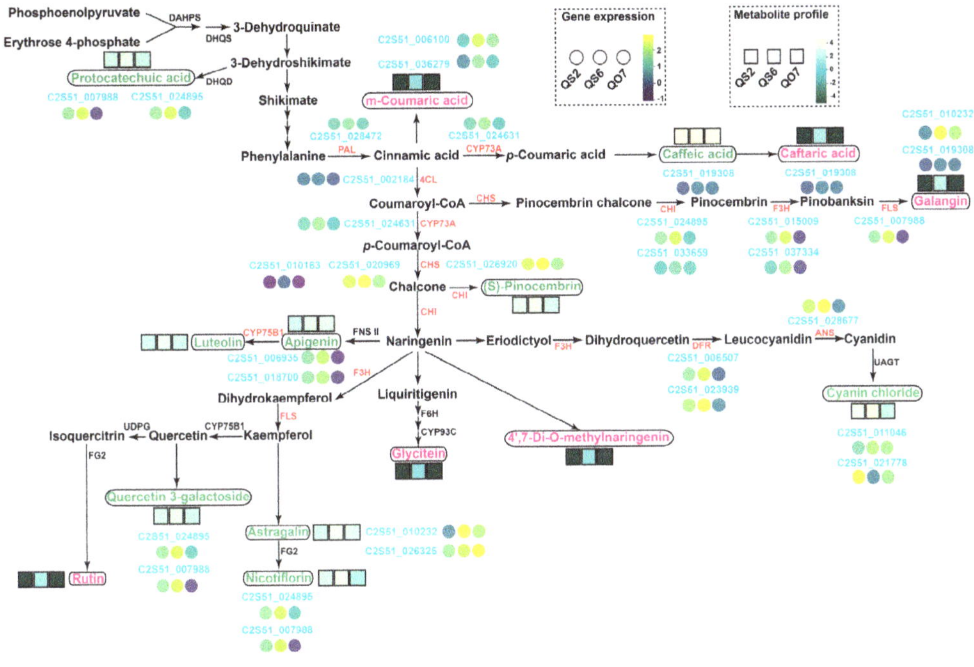

Figure 8. Expression profiles of genes and metabolites involved in shikimate, phenylpropanoid, and flavonoid biosynthesis pathways. The circular patterns represent the expression levels of genes in QS2, QS6, and QO7 leaves. The rectangle patterns with different colors represent the accumulations of metabolites. Metabolites labeled by green color indicate DAMs. Metabolites labeled by pink color represent the unique one only in QS6.

By performing integrative analysis of transcriptome and metabolome (Figure 7), we concluded that flavonoid metabolism, especially flavones and flavonols, showed different patterns in QS2, QS6, and QO7 leaves. Therefore, we isolated several essential genes involved in flavonoid biosynthesis pathway (Figure 8). Based on the KEGG pathway database, a brief flavonoid biosynthetic pathway was established in this study (Figure 8). The enzymes involved in the flavonoid pathway were isolated as follows: 4-coumarate-CoA-ligase (4CL, C2S51_002184), chalcone synthase (CHS, C2S51_010163, C2S51_020969, C2S51_026920), chalcone isomerase (CHI, C2S51_024895, C2S51_033659), naringenin 3-dioxygenase (F3H, C2S51_015009, C2S51_037334), flavonol synthase (FLS, C2S51_037334), bifunctional dihydroflavonol 4-reductase/flavanone 4-reductase enzymes (DFR, C2S51_006507, C2S51_023939), anthocyanidin synthase (ANS, C2S51_028677), and flavonoid 3′-monooxygenase (CYP75B1, C2S51_006935, C2S51_018700) (Figure 8). Moreover, we noticed that these flavonoid biosynthesis genes had different expression patterns among the three perilla leaves. The expression levels of these structural genes were the highest in QS6 leaves compared to the QS2 and QO7 leaves (Figure 8).

3. Discussion

3.1. New Insights into the Constitution of Phenolic Compounds in Perilla Leaves

As a kind of edible and medical herb, the nutritive value and effective constituents of perilla have been studied widely. So far, 271 types of phytochemical compounds have been isolated and reported in perilla tissues [3]. Among these phytochemical compounds, phenolic compounds occupy a certain proportion. By using capillary electrophoresis, Peng et al. reported that catechin, ferulic acid, apigenin, luteolin, rosmarinic acid, and caffeic acid are major important active ingredients in perilla leaves and seeds [10]. In addition, a considerable number of phenolic compounds such as gallic acid, isovanillic acid, chlorogenic acid, sinapic acid, and quercetin have been detected in aqueous and organic solvent extracts of perilla [27,28].

In this study, we chose three perilla varieties QS2, QS6, and QO7, which have phenotypical differences in leaf colors (Figure 1A). We aimed to find differences in phenolic compounds among these perilla varieties by using targeted metabolome analysis, which is an efficient way to identify qualitative and quantitative information of metabolites. The results showed that QS6 leaves contained the greatest number of phenolic compounds comparing with QS2 and QO7 (Figure 1B–D). In line with the previous study, ferulic acid, apigenin, luteolin, caffeic acid, and two quercetin derivatives quercetin 3-galactoside and quercetin 3-O-glucuronide, which are known as major metabolites, were detected in QS2, QS6, and QO7 leaves as well (Figure 1E). Moreover, caffeic acid, apigenein, luteolin, and quercetin 3-galactoside were DAMs in comparable groups 'QS2 vs. QS6' and 'QS6 vs. QO7' (Figure 2B,C). These results indicated that the major compounds in different perilla varieties were almost consistent. Interestingly, we found several flavonoids, such as 4′,7-Di-O-methylnaringenin, taxifolin, galangin, diosmin, morin, rutin, and glycitein, were uniquely isolated in QS6 leaves. In addition, cryptochlorogenic acid, which is a kind of phenolic acid, was specifically detected in QS2 leaves (Figure 1E). This study is the first to report these above-mentioned phenolic compounds in perilla. These results suggest that different perilla varieties have different compositions of phenolic compounds, and some phenolic compounds may be unique to some perilla varieties. Since the metabolites reported in this study have clear pharmaceutical effects on antiphlogosis and antioxidant, antitumor, and other properties [3,29,30], QS6 might be a good variety of perilla for utilization in pharmaceutics. Based on these, our data provided new insights into composition of phenolic compounds in perilla and proved that targeted metabolome analysis is an efficient way to identify good perilla varieties with high potential for use in food and medicine.

3.2. Candidate Genes Involved in Phenolic Compound Biosynthesis in Perilla

In plants, phenolic compounds are thought to be the most widely distributed secondary metabolites [31]. To date, the biosynthesis of plant phenolics is almost clear, and there is no doubt that different phenolic compounds are generated via several pathways. The Shikimate pathway, phenylpropanoid pathway, and flavonoid branch pathway are three main pathways resulting in the formation of phenolic compounds [31,32]. Shikimic acid is the core of the shikimate pathway, and several aromatic amino acids, including L-phenylalanine, L-tyrosine, and L-tryptophan, and phenolic compounds originate from shikimic acid [33,34]. Subsequently, shikimic acid can generate several hydroxybenzoic acids, such as *p*-hydroxybenzoic acid, protocatechuic acid, and gallic acid [14]. At the beginning of the phenylpropanoid pathway, L-phenylalanine is catalyzed by the key phenolic metabolism enzyme PAL and results in the formation of trans-cinnamic acid, which is a bridge between the metabolism of aromatic amino acids and phenolic compounds [35,36]. In higher plants, the condensation of *p*-Coumaroyl-CoA and three molecules of malonyl-CoA leads to the formation of chalcone. Then, the reaction of biosynthesis of flavonoids is catalyzed by CHS [17,37]. So far, flavonoid-formation-related enzymes have been widely studied in plants [31,38,39]. In perilla, most of the enzyme encoding genes involved in flavonoid biosynthetic pathway have been isolated and identified [2]. However, factors that potentially participate in phenolic compounds are still unclear.

In this study, we performed the targeted metabolome analysis of phenolic compounds in three perilla varieties and found that several compounds belonging to productions of the shikimate pathway, phenylpropanoid pathway, and flavonoid branch pathway were differentially accumulated or unique in QS2 or QS6 leaves (Figures 1 and 8). Based on the integrated transcriptome and metabolome analyses, we screened out 14 genes, which could be divided into modules of the shikimate pathway, phenylpropanoid pathway, and flavonoid pathway. These genes displayed high positive or negative correlation with 15 phenolic compounds (Figure 7). Protocatechuic acid is a type of phenolic acid, which has a wide range of pharmacological activities including antioxidant, anti-inflammatory, neuroprotective, antibacterial, antiviral, anticancer, and so on [40]. In this study, we found chalcone-flavanone isomerase 1 (C2S51_024895) and flavonol synthase (C2S51_007988) encoding genes were positively and negatively correlated with protocatechuic acid biosynthesis (Figures 7 and 8). Although chalcone-flavanone isomerase 1 and flavonol synthase are key enzymes in the flavonoid branch pathway, they were still highly correlative with the phenylpropanoid pathway, indicating that the shikimate pathway and flavonoid pathway were not independent and there might be a feedback loop between these two pathways. In addition, we found that several key enzymes in the flavonoid pathway, for instance, one 4CL, two CHI, two F3H, one FLS, two DFR, and one ANS, were isolated. They all showed different expression profiles in QS2, QS6, and QO7 leaves (Figure 8). In perilla, some studies have reported enzymes in the flavonoid biosynthesis metabolic pathway [41,42]. However, these studies did not use the newest reference genome of perilla and this is inconvenient for gene cloning. Additionally, their results did not show the complete flavonoid biosynthesis metabolic pathway. Our results were based on the newest reference genome of perilla, and it was easy to clone the relative genes. Moreover, our results further covered the gaps seen in previous studies.

In addition, we found several peroxidase-encoding genes, including peroxidase 10, 31, 43, and 45, were highly correlative with luteolin, *m*-coumaric acid, diosmin, and cyanin chloride (Figure 7B), indicating that peroxidase displayed a close relationship with the biosynthesis of phenolic compounds. In plants, a series of secondary metabolites, especially phenols, may act as reductant substrates of class III peroxidases [43–45]. Peroxidases are thought as ubiquitous enzymes in plant cell walls and vacuoles, which are also the target compartments for the accumulation of most secondary metabolites [46–48]. Given these, it was highly possible that luteolin, *m*-coumaric acid, diosmin, and cyanin chloride were substrates of peroxidases.

4. Materials and Methods

4.1. Plant Materials, Growth Conditions, and Sample Collection

Perilla varieties QS2, QS6, and QO7 were planted in an experimental field in Shaoguan University (the research center for engineering and technology of aromatic plants, 24°46′ N 113°40′ E). All plants were given a unified and coincident treatment. Samples in the experiments were collected from fresh leaves at the maturity stage of branches at the same height, and each sample set three biological replicates. The leaves of each variety were divided into two parts, one for targeted metabolic analysis and the other for the transcriptome sequencing. The leaves for targeted metabolic analysis were frozen in liquid nitrogen and ground to a powder, then stored at −80 °C. Samples selected for transcriptome sequencing were cleaned after harvesting and immediately frozen using liquid nitrogen and stored at −80 °C.

4.2. RNA Isolation and Library Preparation

Total RNA was extracted from perilla leaves using the TRIzol reagent (Invitrogen, Waltham, CA, USA) according to the manufacturer's protocol. RNA purity and quantification were evaluated using the NanoDrop 2000 spectrophotometer (Thermo Scientific, Waltham, CA, USA). RNA integrity was assessed using the Agilent 2100 Bioanalyzer (Agilent Technologies, Santa Clara, CA, USA). cDNA libraries were constructed using VAHTS Universal V6 RNA-seq Library Prep Kit according to the manufacturer's instructions.

4.3. RNA Sequencing and Differentially Expressed Genes Analysis

The libraries were sequenced on an Ilumina Novaseq 6000 platform (Industrial Blvd., Hayward, CA, USA), and 150 bp paired-end reads were generated. Raw reads of fastq format were firstly processed using fastp [49], and the low-quality reads were removed to obtain the clean reads. The clean reads were mapped to the reference genome of perilla (https://www.ncbi.nlm.nih.gov/genome/?term=PRJNA431002 (accessed on 18 March 2025)) using HISAT2 [50]. FPKM [51] of each gene was calculated and the read counts of each gene were obtained by HTSeq-count. Differential expression analysis was performed using the DESeq2 [52]. Q-value < 0.05 and Fold change > 2 or Fold change < 0.5 were set as the thresholds for significantly differentially expression genes (DEGs). Hierarchical cluster analysis of DEGs was performed using R (v 3.2.0) to demonstrate the expression pattern of genes in different groups and samples. Based on the hypergeometric distribution, GO and KEGG pathway enrichment analyses of DEGs were performed to screen the significant enriched term using R (v 3.2.0). Pearson's correlations of samples were performed on variance-stabilized transformed values, while principal component analysis (PCA) was performed on regularized log values of read counts.

4.4. Quantitative Real-Time PCR Assay

RNA was extracted using the RNAprep Pure Plant Plus kit DP441 (TIANGEN BIOTECH (BEIJING) CO., LTD, Beijing, China), and the extracted RNA was reverse-transcribed using FastKing gDNA Dispelling RT SuperMix kit KR118 (TIANGEN BIOTECH (BEIJING) CO., LTD) after being qualified by NanoDrop detection. The primers were designed using primer 5.0 software, and perilla actin gene was employed as an internal control (Table S1). qRT–PCR was performed using 2 × SYBR Green qPCR Master Mix reagent (1708891, Bio–Rad, Hercules, CA, USA) using an iCyeler iQTM/Clooo system (Bio–Rad, USA). The qRT-PCR cycling conditions were as follows: predenaturation at 95 °C for 30 s, followed by 40 cycles of denaturation at 95 °C for 5 s and annealing/extension at 60 °C for 30 s. The final step included a melting curve analysis at 95 °C for 5 s, followed by 60 °C for 1 min. Each gene was amplified in triplicate. The relative expression levels of each gene were computed using the $2^{-\Delta\Delta Ct}$ method.

4.5. Targeted Metabolome Analysis

In this study, the targeted metabolome analysis of phenolic compounds in perilla leaves was performed. The sampling method was the same as the transcriptome analysis experiment, and five biological replicates were set up. For the methods of targeted metabolome analysis, 600 μL of ice-cold MeOH-water (2:1, v/v, containing interior standard) was added into the freeze-dried perilla leaves. Then, the mixture was crushed using a mixer mill (Wonbio-E, WONBIO BIOTECHNOLOGY, Shanghai, China) with a zirconia bead for 2 min at 60 Hz. Afterward, the whole samples were ultrasonically extracted for 20 min in ice-water bath, and then centrifuged at 4 °C (13,000 rpm) for 10 min prior to decanting of 200 μL of supernatant to sample vials. Next, 400 μL ice-cold MeOH-water (2:1, *v*/*v*, containing interior standard) was added into the residual sample. The samples were ultrasonically extracted for 20 min in an ice-water bath, and then centrifuged at 4 °C (13,000 rpm) for 10 min prior to decanting 200 μL of supernatant into sample vials. The two supernatants were combined and mixed well. Then, the mixed supernatant (400 μL) was dried under a nitrogen stream and re-dissolved in 200 μL of MeOH-water (7:18, *v*/*v*, containing interior standard), extracted by ultrasonic for 5 min in ice-water bath, and then filtered through a 0.22 μm organic phase pinhole filter for subsequent UPLC-MS/MS (AB Sciex Qtrap 5500/Nexera UHPLC LC-30A) analysis. Quality control (QC) samples were prepared by mixing all samples to determine the reproducibility of the obtained results. Liquid chromatography was performed on a Nexera UHPLC LC-30A (SHIMADZU, Kyoto, Japan). A Waters ACQUITY UPLC HSS T3 column (2.1 × 100 mm, 1.8 μm) was used for analysis. Injection volume was 5 μL. The mobile phase A was water, containing 0.1% formic acid, and the mobile phase B was ACN. Gradient conditions were as follows with 0.3 mL/min flow rate: 0–2 min, 0 B; 2–30 min, 0–50% B; 30–32 min, 50–95% B; 32–34 min, 95% B; 34–34.1 min, 0–100% B; 34.1–35.5 min, 0 B. All the samples were kept at 4 °C during the analysis, and the column temperature was set at 40 °C.

Mass spectrometry was performed on the AB SCIEX Selex ION Triple Quad™ 5500 System, with an electrospray ionization (ESI) source, operating in both positive and negative ion mode. Nitrogen was employed as the collision gas. Additional instrumental parameters were as follows: positive ion mode: CUR: 35Psi; EP: 10 V; IS: 5500 V; CXP: 10 V; TEM: 500 °C; Gas1: 60 Psi; Gas2: 50 Psi; negative ion mode: CUR: 35 Psi; EP: −10 V; IS: −4500 V; CXP: −20 V; TEM: 500 °C; Gas1: 60 Psi; Gas2: 50 Psi. Targeted metabolites were analyzed in schedule multiple reaction monitoring (SRM) mode. The MRM pairs, decluttering potentials (DPs), and collision energies (CEs) were optimized for each analyte. Data acquisitions and further analysis were conducted using Analyst software (https://sciex.com/products/software/analyst-software (accessed on 18 March 2025)). SCIEX OS-MQ software (https://sciex.com/products/software/sciex-os-software (accessed on 18 March 2025)) was used to quantify all metabolites.

Differentially accumulated metabolites (DAMs) were isolated based on the variable importance of the projection (VIP) ≥ 1 and |log2 (fold change)| ≥ 1 (*p*-value < 0.05). The functions of DAMs were further annotated using the KEGG compound database to determine the metabolic pathways.

5. Conclusions

This research provided new information on the types and amounts of phenolic compounds in different perilla leaves. The results extended our understanding of the biosynthesis of phenolic compounds and isolated 14 candidate genes that highly correlated with 9 DAMs in QS2, QS6, and QO7 leaves and 9 phenolic compounds only in QS6 leaves. In addition, we focused on key enzymes in shikimate pathway, phenylpropanoid pathway, and flavonoid branch pathway and screened out 16 enzyme encoding genes, which further enhanced the possibilities of gene cloning from perilla. In conclusion, our results provide valuable information on phenolic compound biosynthesis and the genetic resources for molecular breeding in *Perilla frutescens*.

Supplementary Materials: The supporting information can be downloaded at: https://www.mdpi.com/article/10.3390/ijms26072841/s1.

Author Contributions: Conceptualization, H.Y. and Y.W.; methodology, J.M.; software, Q.S.; validation, J.M. and Q.S.; data curation, T.Y.; writing—original draft preparation, H.Y. and J.M.; writing—review and editing, Y.W.; visualization, J.M. and Y.W.; supervision, H.Y. and Y.W.; funding acquisition, Y.W. All authors have read and agreed to the published version of the manuscript.

Funding: This research was funded by the Start-up Fund of Shaoguan University (to H.Y. and Y.W.) and the National College Students' Innovation and Entrepreneurship Training Program (202410576007).

Institutional Review Board Statement: Not applicable.

Informed Consent Statement: Not applicable.

Data Availability Statement: The transcriptome data were released in National Genomics Data Center (https://ngdc.cncb.ac.cn/?lang=en (accessed on 13 October 2024)) with the accession number CRA020281 and the targeted metabolome data were released in National Genomics Data Center with the accession number OMIX007760.

Conflicts of Interest: The authors declare no conflicts of interest.

Abbreviations

The following abbreviations are used in this manuscript:

4CL	4-coumarate-CoAligase
ANR	anthocyanidin reductase
ANS	anthocyanidin synthase
C4H	cinnamate 4- hydroxylase
cDNA	complementary DNA
CHI	chalcone isomerase
CHS	chalcone synthase
CYP73A	trans-cinnamate 4-monooxygenase
CYP75B1	flavonoid 3′- monooxygenase
CYP93C	cytochrome P450 93C
DAHPS	3-Deoxy-D-arabino-heptulosonate 7-phosphate synthase
DAM	differentially accumulated metabolite
DEGs	differentially expressed genes
DFR	dihydrofavonol-4-reductase
DHQD	3-Dehydroquinate dehydratase
DHQS	3-Dehydroquinate synthase
F3H	naringenin 3-dioxygenase
F6H	feruloyl-CoA 6-hydroxylase
FG2	flavonol-3-O-glucoside L-rhamnosyltransferase
FLS	flavonol synthase
FNS I	flavone synthase I
FPKM	fragments per kilobase per transcript per million mapped reads
GO	gene ontology
KEGG	Kyoto encyclopedia of genes and genomes
PAL	phenylalanine ammonia-lyase
PCA	principal component analysis
QC	quality control
RT-qPCR	Real-time quantitative polymerase chain reaction
RNA-Seq	RNA sequencing
UAGT	Anthocyanidin 3-O-glucosyltransferase
UDPG	UDP-glucosyltransferase
UGT	UDP-glycosyltransferases
UPLC-MS/MS	ultra-performance liquid chromatography/tandem mass spectrometry
VIP	variable importance of the projection

References

1. Hou, T.; Netala, V.R.; Zhang, H.; Xing, Y.; Li, H.; Zhang, Z. *Perilla frutescens*: A Rich Source of Pharmacological Active Compounds. *Molecules* **2022**, *27*, 3578. [CrossRef]
2. Wu, X.; Dong, S.; Chen, H.; Guo, M.; Sun, Z.; Luo, H. *Perilla frutescens*: A traditional medicine and food homologous plant. *Chin. Herb. Med.* **2023**, *15*, 369–375. [CrossRef] [PubMed]
3. Ahmed, H.M. Ethnomedicinal, phytochemical and pharmacological investigations of *Perilla frutescens* (L.) Britt. *Molecules* **2018**, *24*, 102. [CrossRef] [PubMed]
4. Huang, W.Y.; Wang, J.; Liu, Y.M.; Zheng, Q.S.; Li, C.Y. Inhibitory effect of Malvidin on TNF-alpha-induced inflammatory response in endothelial cells. *Eur. J. Pharmacol.* **2014**, *723*, 67–72. [CrossRef] [PubMed]
5. Lin, C.S.; Kuo, C.L.; Wang, J.P.; Cheng, J.S.; Huang, Z.W.; Chen, C.F. Growth inhibitory and apoptosis inducing effect of Perilla frutescens extract on human hepatoma HepG2 cells. *J. Ethnopharmacol.* **2007**, *112*, 557–567. [CrossRef]
6. Saita, E.; Kishimoto, Y.; Tani, M.; Iizuka, M.; Toyozaki, M.; Sugihara, N.; Kondo, K. Antioxidant activities of Perilla frutescens against low density lipoprotein oxidation in vitro and in human subjects. *J. Oleo Sci.* **2012**, *61*, 113–120. [CrossRef]
7. Yamamoto, H.; Ogawa, T. Antimicrobial activity of perilla seed polyphenols against oral pathogenic bacteria. *Biosci. Biotechnol. Biochem.* **2002**, *66*, 921–924. [CrossRef]
8. Ghimire, B.K.; Yoo, J.H.; Yu, C.Y.; Chung, I.M. GC-MS analysis of volatile compounds of Perilla frutescens Britton var. Japonica accessions: Morphological and seasonal variability. *Asian Pac. J. Trop. Med.* **2017**, *10*, 643–651. [CrossRef]
9. Peng, Y.; Ye, J.; Kong, J. Determination of phenolic compounds in Perilla frutescens L. by capillary electrophoresis with electrochemical detection. *J. Agric. Food Chem.* **2005**, *53*, 8141–8147. [CrossRef]
10. Xie, G.; Zou, X.; Liang, Z.; Zhang, K.; Wu, D.; Jin, H.; Wang, H.; Shen, Q. GBF family member PfGBF3 and NAC family member PfNAC2 regulate rosmarinic acid biosynthesis under high light. *Plant Physiol.* **2004**, *195*, 1728–1744. [CrossRef]
11. da Silva, A.P.G.; Sganzerla, W.G.; John, O.D.; Marchiosi, R. A comprehensive review of the classification, sources, biosynthesis, and biological properties of hydroxybenzoic and hydroxycinnamic acids. *Phytochem. Rev.* **2023**. [CrossRef]
12. Al Manari, H.H. *Phenolic Compounds: Classification, Chemistry, and Updated Techniques of Analysis and Synthesis*; IntechOpen: London, UK, 2022. [CrossRef]
13. Marchiosi, R.; dos Santos, W.D.; Constantin, R.P.; de Lima, R.B.; Soares, A.R.; Finger-Teixeira, A.; Maota, T.R.; de Oliveira, D.M.; Foletto-Felipe, M.P.; Abrahão, J.; et al. Biosynthesis and metabolic actions of simple phenolic acids in plants. *Phytochem. Rev.* **2020**, *19*, 865–906. [CrossRef]
14. Albuquerque, B.R.; Heleno, S.A.; Oliveira, M.B.P.P.; Barros, L.; Ferreira, I.C.F.R. Function phenolic compounds: Current industrial applications, limitations and future challenges. *Food Funct.* **2021**, *12*, 14–29. [CrossRef] [PubMed]
15. Rashmi, H.B.; Negi, P.S. Phenolic acids from vegetables: A review on processing stability and health benefits. *Food Res. Int.* **2020**, *136*, 109298. [CrossRef]
16. Liu, W.; Feng, Y.; Yu, S.; Fan, Z.; Li, X.; Li, J.; Yin, H. The Flavonoid Biosynthesis Network in Plants. *Int. J. Mol. Sci.* **2021**, *22*, 12824. [CrossRef]
17. Sasaki, N.; Nakayama, T. Achievements and Perspectives in Biochemistry Concerning Anthocyanin Modification for Blue Flower Coloration. *Plant Cell Physiol.* **2015**, *56*, 28–40. [CrossRef]
18. Winkel-Shirley, B. Flavonoid Biosynthesis. A Colorful Model for Genetics, Biochemistry, Cell Biology, and Biotechnology. *Plant Physiol.* **2001**, *126*, 485–493. [CrossRef]
19. Wang, Z.L.; Wang, S.; Kuang, Y.; Hu, Z.M.; Qiao, X.; Ye, M. A comprehensive review on phytochemistry, pharmacology, and flavonoid biosynthesis of *Scutellaria baicalensis*. *Pharm. Biol.* **2018**, *56*, 465–484. [CrossRef]
20. Li, Y.; Kong, D.; Fu, Y.; Sussman, M.R.; Wu, H. The effect of developmental and environmental factors on secondary metabolites in medicinal plants. *Plant Physiol. Biochem.* **2020**, *148*, 80–89. [CrossRef]
21. Qaderi, M.M.; Martel, A.B.; Strugnell, C.A. Environmental factors regulate plant secondary metabolites. *Plants* **2023**, *12*, 447. [CrossRef]
22. Herrmann, K.M.; Weaver, L.M. The shikimate pathway. *Annu. Rev. Plant Physiol. Plant Mol. Biol.* **1999**, *50*, 473–503. [CrossRef] [PubMed]
23. Tzin, V.; Galili, G. New insights into the shikimate and aromatic amino acids biosynthesis pathways in plants. *Mol. Plant* **2010**, *3*, 956–972. [CrossRef] [PubMed]
24. Guo, J.; Carrington, Y.; Alber, A.; Ehlting, J. Molecular characterization of quinate and shikimate metabolism in populus trichocarpa. *J. Biol. Chem.* **2014**, *289*, 23846–23858. [CrossRef] [PubMed]
25. Muir, R.M.; Ibáñez, A.M.; Uratsu, S.L.; Ingham, E.S.; Leslie, C.A.; McGranahan, G.H.; Batra, N.; Goyal, S.; Joseph, J.; Jemmis, E.D.; et al. Mechanism of gallic acid biosynthesis in bacteria (*Escherichia coli*) and walnut (*Juglans regia*). *Plant Mol. Biol.* **2011**, *75*, 555–565. [CrossRef]
26. Hong, E.; Park, K.H.; Kim, G.H. Phenolic-enriched fractions from perilla frutescens var. acuta: Determinating rosmarinic acid and antioxidant activity. *J. Food Biochem.* **2011**, *35*, 1637–1645. [CrossRef]

27. Zhao, Y.; Kong, H.; Zhang, X.; Hu, X.; Wang, M. The effect of Perilla (*Perilla frutescens*) leaf extracts on the quality of surimi fish balls. *Food Sci. Nutr.* **2019**, *7*, 2083–2090. [CrossRef]
28. Ghorbani, A. Mechanisms of antidiabetic effects of flavonoid rutin. *Biomed. Pharmacother.* **2017**, *96*, 305–312. [CrossRef]
29. Thuan, N.H.; Shrestha, A.; Trung, N.T.; Tatipamula, V.B.; Van Cuong, D.; Canh, N.X.; Van Giang, N.; Kim, T.S.; Sohng, J.K.; Dhakal, D. Advances in biochemistry and the biotechnological production of taxifolin and its derivatives. *Biotechnol. Appl. Biochem.* **2022**, *69*, 848–861. [CrossRef]
30. Wang, D.; Chen, J.; Pu, L.; Yu, L.; Xiong, F.; Sun, L.; Yu, Q.; Cao, X.; Chen, Y.; Peng, F.; et al. Galangin: A food-derived flavonoid with therapeutic potential against a wide spectrum of diseases. *Phytother. Res.* **2023**, *37*, 5700–5723. [CrossRef]
31. Cheynier, V.; Comte, G.; Davies, K.M.; Lattanzio, V.; Martens, S. Plant phenolics: Recent advances on their biosynthesis, genetics, and ecophysiology. *Plant Physiol. Biochem.* **2013**, *72*, 1–20. [CrossRef]
32. Quideau, S.; Deffieux, D.; Douat-Casassus, C.; Pouységu, L. Plant polyphenols: Chemical properties, biological activities, and synthesis. *Angew. Chem. Int. Ed. Engl.* **2011**, *50*, 586–621. [CrossRef] [PubMed]
33. Herrmann, K.M. The shikimate pathway as an entry to aromatic secondary metabolism. *Plant Physiol.* **1995**, *107*, 7–12. [CrossRef] [PubMed]
34. Maeda, H.; Dudareva, N. The shikimate pathway and aromatic amino acid biosynthesis in plants. *Annu. Rev. Plant Biol.* **2012**, *63*, 73–105. [CrossRef] [PubMed]
35. Yadav, V.; Wang, Z.; Wei, C.; Amo, A.; Ahmed, B.; Yang, X.; Zhang, X. Phenylpropanoid Pathway Engineering: An Emerging Approach towards Plant Defense. *Pathogens* **2020**, *9*, 312. [CrossRef]
36. Kołton, A.; Długosz-Grochowska, O.; Wojciechowska, R.; Czaja, M. Biosynthesis regulation of folates and phenols in plants. *Sci. Hortic.* **2022**, *291*, 110561. [CrossRef]
37. Nabavi, S.M.; Šamec, D.; Tomczyk, M.; Milella, L.; Russo, D.; Habtemariam, S.; Suntar, I.; Rastrelli, L.; Daglia, M.; Xiao, J.; et al. Flavonoid biosynthetic pathways in plants: Versatile targets for metabolic engineering. *Biotechnol. Adv.* **2020**, *38*, 107316. [CrossRef]
38. Shen, N.; Wang, T.; Gan, Q.; Liu, S.; Wang, L.; Jin, B. Plant flavonoids: Classification, distribution, biosynthesis, and antioxidant activity. *Food Chem.* **2022**, *383*, 132531. [CrossRef]
39. Saigo, T.; Wang, T.; Watanabe, M.; Tohge, T. Diversity of anthocyanin and proanthocyanin biosynthesis in land plants. *Curr. Opin. Plant Biol.* **2020**, *55*, 93–99. [CrossRef]
40. Song, J.; He, Y.; Luo, C.; Feng, B.; Ran, F.; Xu, H.; Ci, Z.; Xu, R.; Han, L.; Zhang, D. New progress in the pharmacology of protocatechuic acid: A compound ingested in daily foods and herbs frequently and heavily. *Pharmacol. Res.* **2020**, *161*, 105109. [CrossRef]
41. Jiang, T.; Guo, K.; Liu, L.; Tian, W.; Xie, X.; Wen, S.; Wen, C. Integrated transcriptomic and metabolomic data reveal the flavonoid biosynthesis metabolic pathway in *Perilla frutescens* (L.) leaves. *Sci. Rep.* **2020**, *10*, 16207. [CrossRef]
42. Xie, G.; Zou, X.; Liang, Z.; Wu, D.; He, J.; Xie, K.; Jin, H.; Wang, H.; Shen, Q. Integrated metabolomic and transcriptomic analyses reveal molecular response of anthocyanins biosynthesis in perilla to light intensity. *Front. Plant Sci.* **2022**, *13*, 976449. [CrossRef]
43. Ferreres, F.; Figueiredo, R.; Bettencourt, S.; Carqueijeiro, I.; Oliveira, J.; Gil-Izquierdo, A.; Pereira, D.M.; Valentão, P.; Andrade, P.B.; Duarte, P.; et al. Identification of phenolic compounds in isolated vacuoles of the medicinal plant Catharanthus roseus and their interaction with vacuolar class III peroxidase: An H_2O_2 affair? *J. Exp. Bot.* **2011**, *62*, 2841–2854. [CrossRef] [PubMed]
44. Veitch, N.C. Horseradish peroxidase: A modern view of a classic enzyme. *Phytochemistry* **2004**, *65*, 249. [CrossRef]
45. Passardi, F.; Cosio, C.; Penel, C.; Dunand, C. Peroxidases have more functions than a Swiss army knife. *Plant Cell Rep.* **2005**, *24*, 255–265. [CrossRef] [PubMed]
46. Tognolli, M.; Penel, C.; Greppin, H.; Simon, P. Analysis and expression of the class III peroxidase large gene family in *Arabidopsis thaliana*. *Gene* **2002**, *288*, 129–138. [CrossRef]
47. Welinder, K.G.; Justesen, A.F.; Kjaersgård, I.V.; Jensen, R.B.; Rasmussen, S.K.; Jespersen, H.M.; Duroux, L. Structural diversity and transcription of class III peroxidases from *Arabidopsis thaliana*. *Eur. J. Biochem.* **2002**, *269*, 6063–6081. [CrossRef]
48. Sirikantaramas, S.; Yamazaki, M.; Saito, K. Mechanisms of resistance to self-produced toxic secondary metabolites in plants. *Phytochem. Rev.* **2008**, *7*, 467–477. [CrossRef]
49. Chen, S.; Zhou, Y.; Chen, Y.; Gu, J. fastp: An ultra-fast all-in-one FASTQ preprocessor. *Bioinformatics* **2018**, *34*, i884–i890. [CrossRef]
50. Kim, D.; Langmead, B.; Salzberg, S.L. HISAT: A fast spliced aligner with low memory requirements. *Nat. Methods* **2015**, *12*, 357–360. [CrossRef]
51. Roberts, A.; Trapnell, C.; Donaghey, J.; Rinn, J.L.; Pachter, L. Improving RNA-Seq expression estimates by correcting for fragment bias. *Genome Biol.* **2011**, *12*, R22. [CrossRef]
52. Love, M.I.; Huber, W.; Anders, S. Moderated estimation of fold change and dispersion for RNA-seq data with DESeq2. *Genome Biol.* **2014**, *15*, 550. [CrossRef]

International Journal of *Molecular Sciences*

MDPI

Article

NAC047/052/104 Synergistically Regulate the Dark-Induced Leaf Senescence in Non-Heading Chinese Cabbage

Bing Yang [1,2,†], Dingyu Zhang [2,†], Zitong Meng [1,2], Yijiang Yin [2], Xiao Yang [2], Mengqin Cao [1,2], Ruixin Li [2], Yishan Song [1,*] and Hongfang Zhu [2,*]

1 College of Food Science and Technology, Shanghai Ocean University, Shanghai 201306, China; y17739190411@163.com (B.Y.); 17797622297@163.com (Z.M.); 18536495112@163.com (M.C.)
2 Shanghai Key Laboratory of Protected Horticultural Technology, Horticultural Research Institute, Shanghai Academy of Agricultural Sciences, Shanghai 201403, China; zhangdy1225@126.com (D.Z.); 13577597931@163.com (Y.Y.); yangxiao2113@163.com (X.Y.); 113368613997@outlook.com (R.L.)
* Correspondence: yssong@shou.edu.cn (Y.S.); zhuhongfang@saas.sh.cn (H.Z.)
† These authors contributed equally to this work.

Abstract: Non-heading Chinese cabbage (NHCC) is an important vegetable, and its leaves are harvested for consumption. Thus, the initiation and progression of leaf senescence in NHCC directly impact its yield and quality. In multiple plant species, NAC transcription factors are known to act as critical regulators of leaf senescence. However, in NHCC, the NAC transcription factors contributing to leaf senescence regulation remain to be identified, and the mechanisms underlying dark-induced leaf senescence remain unclear. To explore the molecular mechanisms of leaf senescence in NHCC, we stored NHCC away from light and subsequently examined dark-induced transcriptional alterations via RNA sequencing. Interestingly, three *NAC* transcription factors, *BrNAC047*, *BrNAC052*, and *BrNAC104*, were found to be potently activated by darkness. Subsequently, the virus-induced gene silencing of *BrNAC047*, *BrNAC052*, and *BrNAC104* demonstrated that these three *NACs* act as positive regulators of dark-induced leaf senescence in NHCC. Dual-luciferase assays further confirmed that BrNAC047, BrNAC052, and BrNAC104 directly activate the promoters of certain senescence-associated genes. This study uncovers the molecular signaling pathways governing dark-induced leaf senescence in NHCC, highlighting the role of three key regulators and offering valuable molecular targets for delaying leaf senescence in NHCC.

Keywords: NHCC; dark; NACs; leaf senescence

1. Introduction

Non-heading Chinese cabbage (NHCC; *Brassica rapa*) is extensively cultivated throughout Asia [1,2]. The leaves of this plant act as distinctive autotrophic apparatuses, converting light energy into energy substrates such as sugar via photosynthesis [3,4]. However, when the leaves enter the senescence phase, the disintegration of chloroplasts triggers the efflux of nutrients into the nascent parts of the plant, causing the leaves to wither and undergo de-greening [5,6]. Moreover, leaf senescence is not only a developmental phenomenon but also a direct response to external stress in these plants [7].

Leaf senescence is regulated by a series of senescence-associated genes (SAGs). The transcriptional activity of SAGs increases leaf age and is strictly controlled by multi-level regulatory mechanisms [8–10]. Leaf senescence is thus a consequence of the combined regulatory effects of both endogenous and exogenous environmental elements [11,12].

Notably, light is the primary external factor modulating leaf senescence. Light is not only essential for energy production via photosynthesis, facilitating plant growth, but also operates as a signal to regulate the development and life cycle of plants [13]. During post-harvest storage, NHCC is inevitably exposed to dark conditions for prolonged periods, which induces rapid leaf decay. Studies have demonstrated that dark conditions trigger rapid senescence in plants, and dark treatment is thus a common approach for studying leaf senescence across multiple plants [14,15]. Hence, investigations into the molecular machinery regulating dark-induced senescence in NHCC could be valuable for resolving the quality deterioration issues associated with storage.

Transcription factors regulate plant growth, development, and stress responses by modulating downstream functional genes [16]. The NAC family is among the largest and most important families of transcription factors in plants [17]. NAC proteins contain a highly conserved NAC domain, which functions as a transcriptional regulator [18]. In model plants, such as *Arabidopsis thaliana* and rice, the expression of NAC transcription factors is correlated with stress responses and senescence regulation [19–21]. Specifically, in *A. thaliana*, the germination capacity of *AtNAC092* knockout and overexpression lines appears to be augmented and diminished under salt stress, respectively [22]. Meanwhile, the up-regulation of *ATNAP* can accelerate leaf senescence in this species [23]. Similarly, *TaNAC29* expression enhances the antioxidant potency of wheat plants, diminishes the accumulation of hydrogen peroxide (H_2O_2), and mitigates membrane impairment, thereby improving the resistance of wheat to salt and drought stresses [24,25]. Additionally, NAC transcription factors also regulate phytohormone signals to regulate leaf senescence. *ANAC019*, *ANAC055*, and *ANAC072* positively regulate abscisic acid (ABA)-induced leaf senescence in *A. thaliana*. In addition, ATNAP interacts with SAG113, further mediating and regulating ABA-triggered leaf senescence [26–28]. These studies highlight the pivotal functions of NAC transcription factors in leaf senescence and stress responses. Nevertheless, which NAC transcription factors modulate the progression of leaf senescence in NHCC remains unclear, and their potential mechanisms of action are yet to be delineated.

In this study, we investigated the mechanism of leaf senescence in NHCC by incubating detached leaves under dark conditions. Through transcriptomic analysis, we identified some NAC transcription factors that act as cardinal responsive genes during dark induction in NHCC. Subsequently, virus-induced gene silencing (VIGS) assays revealed that *BrNAC047*, *BrNAC052*, and *BrNAC104* play a positive role in regulating dark-induced leaf senescence in NHCC. Molecular assays confirmed that BrNAC047, BrNAC052, and BrNAC104 can directly activate the promoters of several SAGs, thereby accelerating leaf senescence in NHCC. The present study not only uncovers some novel NAC transcription factors involved in modulating leaf senescence but also delineates the molecular mechanisms underlying dark-induced leaf senescence in NHCC.

2. Results

2.1. Phenotypic and Physiological Analysis of Dark-Induced Senescence in NHCC

To uncover the regulatory mechanism underlying dark-induced leaf senescence in NHCC, we incubated NHCC leaves in the dark. As shown in Figure 1, significant differences in leaf color emerged in NHCC after 3 days of dark treatment. Specifically, the leaves started to turn yellow due to chlorophyll degradation (Figure 1A). DAB and NBT staining assays indicated that, compared with the control group, the treatment group exposed to darkness contained higher accumulated levels of ROS (Figure 1B). Moreover, both the photosynthetic function and chlorophyll content of the leaves displayed a marked downward trend following dark treatment (Figure 1C,D). Expression analyses of the key enzymes involved in the chlorophyll degradation and ROS biosynthesis pathways revealed that the chlorophyll catabolism genes *NYC1* (*BrC04g009380*), *SAG12* (*BrC02g034440*) and *NYE1* (*BrC03g052200*), along with the ROS-producing enzyme genes *RBOH* (*BrC02g041610*) and *RBOH* (*BrC03g016000*), were significantly upregulated in the dark treatment group (Figure 1E–I). Collectively, these results demonstrated that dark treatment promotes the degradation of chlorophyll in NHCC leaves, accelerates the biosynthesis of ROS, and thus accelerates the leaf senescence process.

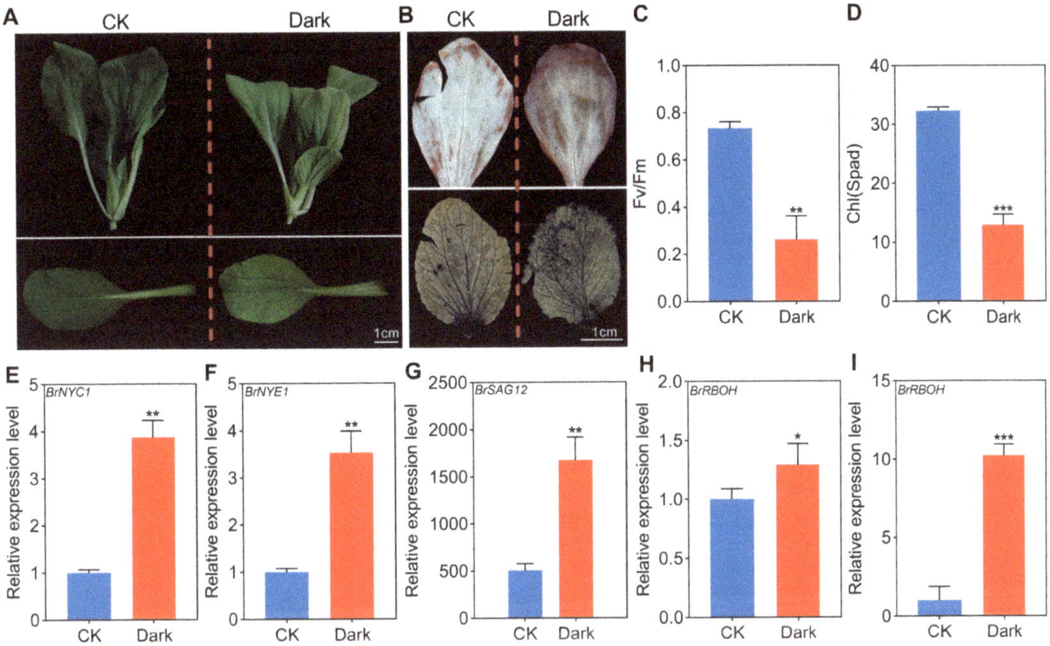

Figure 1. Phenotypes of dark-treated NHCC leaves. (**A**) Phenotypes of NHCC leaves after dark treatment; (**B**) DAB and NBT staining of NHCC leaves after dark treatment; (**C**) F_V/F_m of NHCC leaves after dark treatment; (**D**) Chlorophyll content of NHCC leaves after dark treatment; (**E–I**) Expression levels of *BrNYC1*, *BrNYE1*, *BrSAG12* and *BrRBOHs* in NHCC leaves after dark treatment. Processing time: 3 days. Data represent the mean ± SD (n = 3 biological replicates), *** $p < 0.001$, ** $p < 0.01$, and * $p < 0.05$ (t-test).

2.2. Analysis of Dark-Induced Transcriptomic Alterations in NHCC Leaves

To explore the molecular signaling pathways involved in dark-induced leaf senescence, we performed transcriptomic sequencing analysis on NHCC leaves following dark treatment. The data revealed that dark treatment triggered drastic transcriptional changes in NHCC leaves. Specifically, after dark treatment, 2605 genes were significantly upregulated, while 2782 genes were significantly downregulated (Figure 2A,B). Using Gene Ontology (GO) analysis, we found that the upregulated genes were associated with the abscisic acid pathway, organic acid catabolism pathway, and response to abiotic stimuli in NHCC. Meanwhile, the downregulated genes were associated with pathways such as photosynthesis and glycolysis (Figure 2C,E). Through the Kyoto Encyclopedia of Genes and Genomes (KEGG) analysis of the DEGs, we found that the upregulated genes were mainly enriched in signaling pathways such as amino acid metabolism, fatty acid degradation, and carotenoid biosynthesis. In contrast, the downregulated genes were primarily linked to light signaling, including photosynthesis antenna proteins and photosynthesis (Figure 2D,F). Collectively, these findings offer compelling proof of the significant molecular differences between NHCC leaves receiving light and those stored under dark conditions.

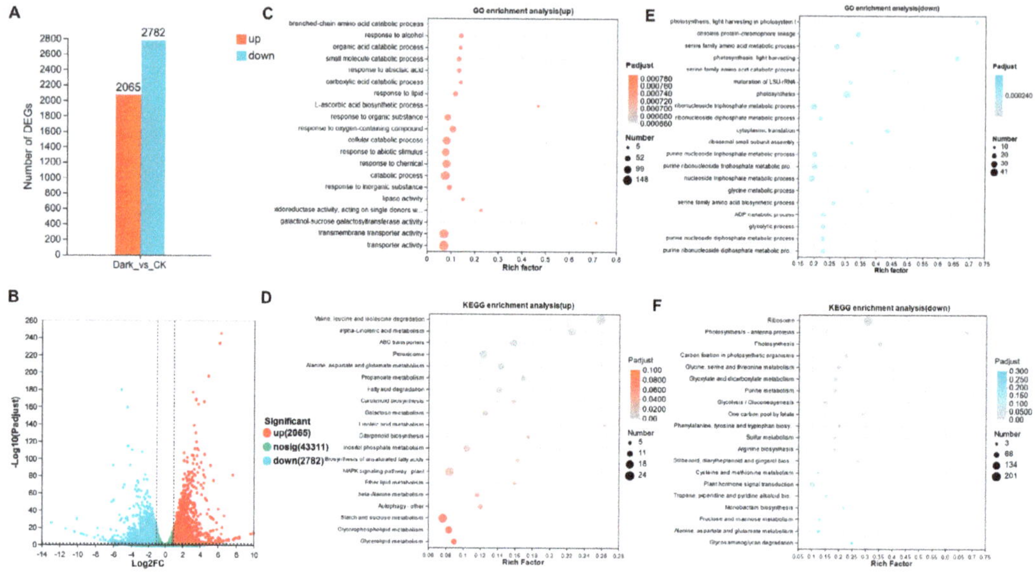

Figure 2. Overview of transcriptome data and analysis of signaling pathways triggered by dark treatment in NHCC: (**A**) Statistical analysis of differentially expressed genes (DEGs) in NHCC leaves after dark treatment; (**B**) volcano plot showing the DEGs in NHCC leaves after dark treatment; (**C**,**D**) GO and KEGG analyses of the upregulated DEGs in NHCC leaves; and (**E**,**F**) GO and KEGG analyses of the downregulated DEGs in NHCC leaves. Data represent the mean ± SD (n = 3 biological replicates). Full name of the pathway in the diagram: oxidoreductase activity, acting on single donors with incorporation of molecular oxygen, incorporation of two atoms of oxygen; photosynthesis, light harvesting in photosystem II; purine ribonucleotide triphosphate metabolic process; purine ribonucleotide metabolic process; phenylanine, tyrosine and gingerol biosynthesis; Stilbenoid, diarylheptanoid, and gingerol biosynthesis; Tropane, Piperidine, and Pyridine Alkaloid Biosynthesis.

2.3. Darkness Primarily Activates NAC Transcription Factors in NHCC

Given that dark treatment induced considerable transcriptomic variations in NHCC, we further analyzed the response of various transcription factor families to darkness in this species. The *NAC*, *bHLH*, *MYB*, *WRKY*, and *AP2/ERF* families are the most important transcription factor families in plants. By examining the expression patterns of these transcription factors, we observed that the vast majority of *NAC* transcription factors were significantly upregulated in NHCC leaves after dark treatment. Similarly, most *MYB* transcription factors also showed a trend of upregulation following dark treatment. In contrast, the distribution of upregulated and downregulated *bHLH*, *AP2/ERF*, and *WRKY* transcription factors was comparable. Collectively, these findings suggested that *NAC* transcription factors are likely to play a decisive role in dark-induced leaf senescence (Figure 3A,B).

Subsequently, a Venn diagram analysis was conducted by comparing the DEGs altered by dark treatment with those in the leaf senescence gene database. Notably, of the dark-induced SAGs, 806 genes exhibited an upregulated expression profile and 463 exhibited a downregulated expression profile (Figure 3C). Notably, 13 *NAC* genes were prominently activated and showed significant transcriptional elevation following dark treatment (Figure S1). The expression analysis of these genes revealed that most of them exhibited elevated expression levels 3 days after dark treatment. Three genes in particular—*BraC01g048160* (*BrNAC047*), *BraC01g045260* (*BrNAC052*), and *BraC06g034160* (*BrNAC104*)—were exceptionally sensitive to darkness and remained in an activated state under dark conditions (Figure 3D). These findings strongly implied that these three *NAC* genes may play a cardinal role in dark-induced leaf senescence in NHCC.

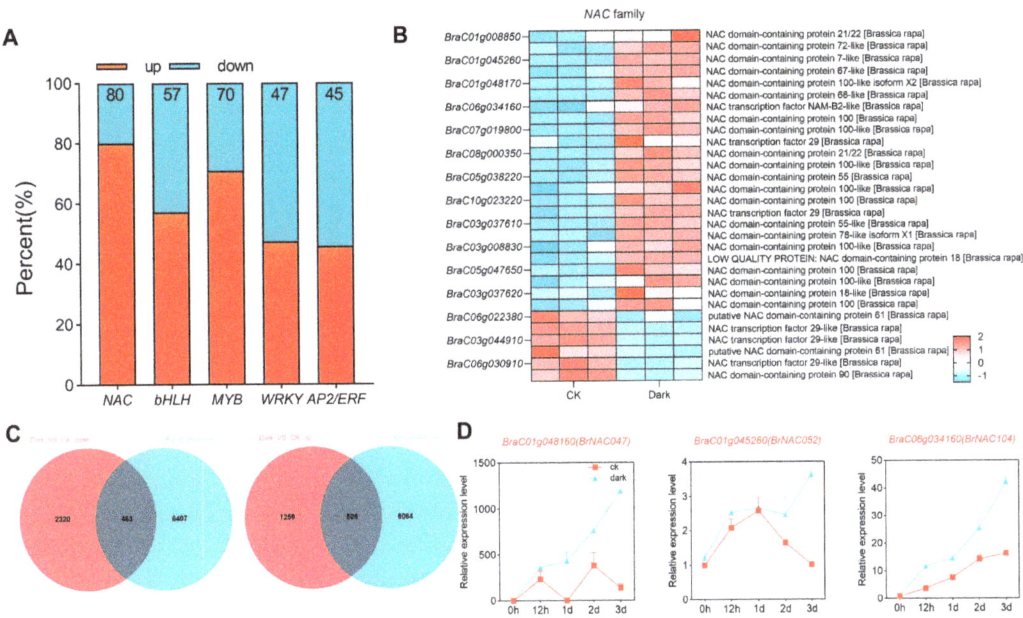

Figure 3. Dark-induced signal transduction involves *NAC* transcription factors. (**A**) A 100% stacked bar chart, showing the dark-induced transcription factors in NHCC leaves. (**B**) Heatmap of dark-induced *NAC* family genes in NHCC leaves. (**C**) Venn analysis comparing differentially expressed genes with those from the leaf senescence gene database. (**D**) Expression profiles of *BrNAC047*, *BrNAC052*, and *BrNAC104* in the control and dark treatment groups. DEGs were defined based on the following thresholds: Log2 (foldchange) ≥ 1, FDR < 0.05 ($n = 2$ biological replicates). Data represent the mean $\pm$ SD ($n = 3$ biological replicates).

2.4. Molecular Characterization of BrNAC047, BrNAC052, and BrNAC104

BrNAC047, BrNAC052, and BrNAC104 appeared to be positive regulators of leaf senescence in NHCC. Hence, we conducted an in-depth comparative analysis of their molecular characteristics and upstream regulators. Amino acid sequence analysis shows that all these proteins contain highly conserved NAC DNA binding domains and other specific protein domains (Figure 4A). Subsequently, phylogenetic trees also confirmed that BrNAC047, BrNAC052, and BrNAC104 were closely related to ATNAC47, ATNAC052, and ATNAC1 (Figure 4B), respectively. Using the plantCARE and TBtools 2.136 software, a predictive analysis of the 1000 bp promoter sequences upstream of these genes was performed. The promoters of these three genes were found to contain several cis-regulatory elements that are potentially involved in responses to MeJA, ABA, and auxin (Figure 4C). Thus, we examined the induction of these three genes in response to different phytohormones via qPCR. Upon the exogenous application of ABA for 2 h, all three of these *NACs* showed pronounced upregulation (Figure 4D). This implied that ABA may play a crucial role in the dark-induced expression of *BrNAC047*, *BrNAC052*, and *BrNAC104* in NHCC.

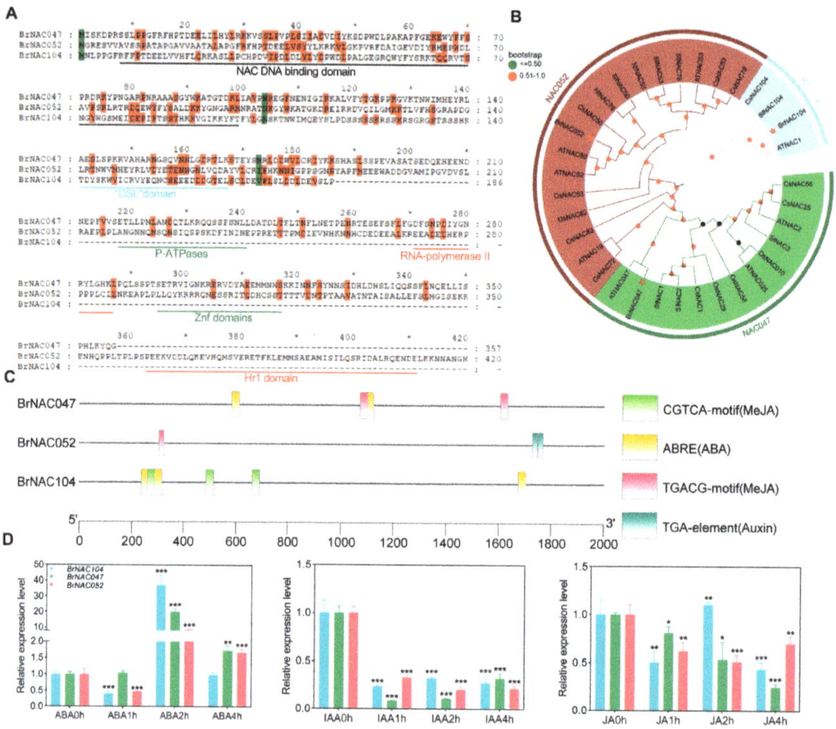

Figure 4. Protein and DNA sequence analyses of BrNAC047, BrNAC052, and BrNAC104. (**A**) Protein structure prediction of BrNAC047, BrNAC052, and BrNAC104. * Number is an omission of the number of amino acid sequences. The green color indicates the same amino acids, while the red color represents unique amino acids. (**B**) Phylogenetic analysis of BrNAC047, BrNAC052, and BrNAC104 in different species. Specially marked with a red asterisk. Multiple alignment was performed CLUSTALW, and the phylogenetic tree was constructed with MEGA6.0 using a bootstrap test of phylogeny with the neighbor-joining test and default parameters. (**C**) Analysis of the promoter sequences of *BrNAC047*, *BrNAC052*, and *BrNAC104*. (**D**) qPCR analysis of the responses of *BrNAC047*, *BrNAC052*, and *BrNAC104* to different phytohormones. Data represent the mean ± SD (n = 3 biological replicates). *** $p < 0.001$, ** $p < 0.01$, and * $p < 0.05$ (t-test).

2.5. *BraNAC047/BrNAC052/BrNAC104 Positively Regulate Dark-Induced Leaf Senescence in NHCC*

To investigate whether *BrNAC047*, *BrNAC052*, and *BrNAC104* participate in the regulation of dark-induced leaf senescence in NHCC, we conducted VIGS experiments to inhibit *BrNAC047*, *BrNAC052*, and *BrNAC104* expression in NHCC. qPCR assays confirmed the knockdown of these *NAC* genes after VIGS (Figure 5D). After incubation in the dark for 3 days, the *BrNAC047*-, *BrNAC052*-, and *BrNAC104*-silenced plants and leaves exhibited delayed senescence when compared with control samples (Figure 5A–C). The gene-silenced leaves also displayed relatively higher chlorophyll levels and more robust photosynthesis capacities (Figure 5E,I). Enzyme assays demonstrated that following dark treatment, the activities of POD, CAT, and SOD in *BrNAC047*-, *BrNAC052*-, and *BrNAC104*-silenced leaves were significantly higher than those in the control leaves (Figure 5F–H). In addition, the proline and soluble sugar contents were significantly higher in the *BrNAC047*-, *BrNAC052*- and *BrNAC104*-silenced plants, while the MDA contents were lower (Figure 5J–L). Collectively, these findings demonstrated that BrNAC047, BrNAC052, and BrNAC104 positively regulate the senescence of NHCC by targeting multiple metabolic pathways.

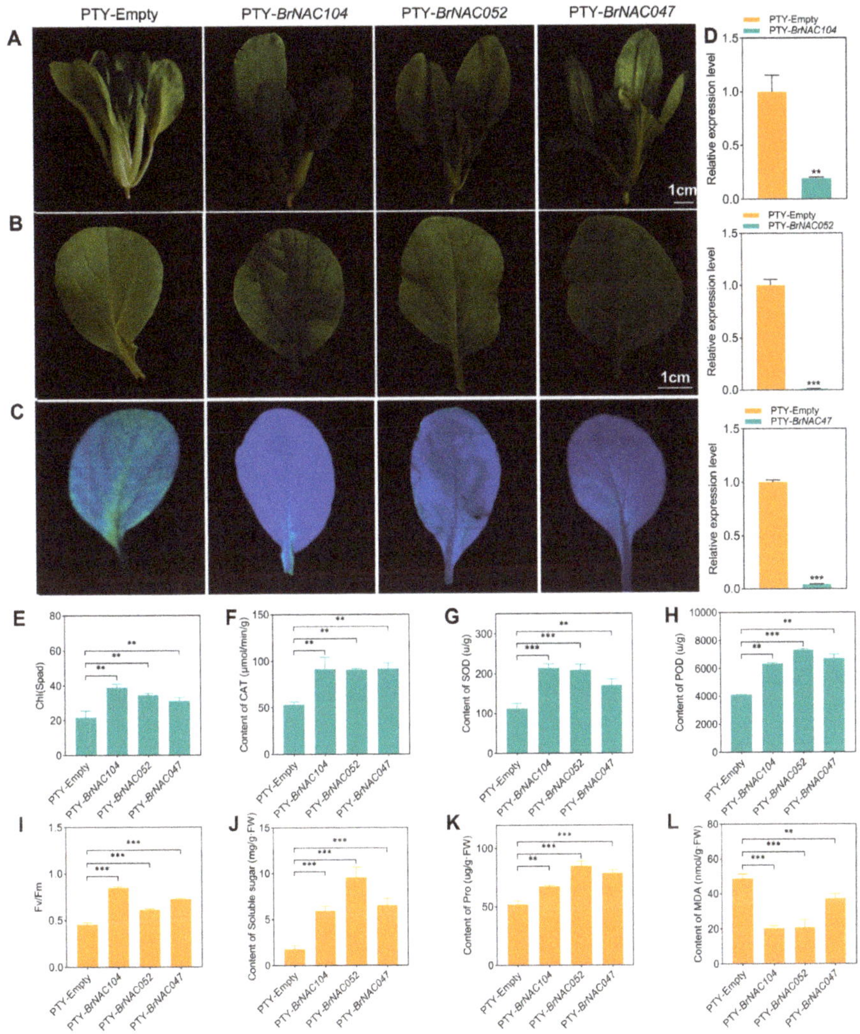

Figure 5. Functional validation of BrNAC047, BrNAC052, and BrNAC104 during dark-induced leaf senescence in NHCC. (**A**) Phenotypes of gene-silenced NHCC after dark treatment. (**B**) Phenotype of the first true leave from gene-silenced NHCC after dark treatment. (**C**) Chlorophyll fluorescence in the first true leave from the gene-silenced NHCC after dark treatment. (**D**) Gene expression levels of *NACs* in the gene-silenced NHCC plants. (**E**) Chlorophyll content in the gene-silenced NHCC leaves after dark treatment. (**F–H**) CAT, SOD, and POD contents in the gene-silenced NHCC leaves after dark treatment. (**I**) Photosynthetic capacity of the gene-silenced NHCC leaves after dark treatment. (**J–L**) Soluble sugar, proline, and MDA contents in the gene-silenced NHCC leaves after dark treatment. Data represent the mean ± SD (n = 3 biological replicates). *** $p < 0.001$, ** $p < 0.01$ (t-test).

2.6. BrNAC047, BrNAC052, and BrNAC104 Activate SAG Expression in NHCC

To explore whether BrNAC047, BrNAC052, and BrNAC104 directly regulate the expression of SAGs, we performed dual-luciferase assays in a tobacco model system to assess the regulatory effects of these NACs on the promoters of certain SAGs, namely, *CRK3*, *proDH2*, *NAS3*, *NAC18*, and *NAC55*. The promoters of *CRK3*, *proDH2*, *NAS3*, *NAC18*, and

NAC55 were cloned into pGreenII 0800-LUC vectors. Subsequently, these constructs were co-transfected along with *BrNAC047*, *BrNAC052*, and *BrNAC104* overexpression vectors into tobacco leaf tissues via *Agrobacterium tumefaciens*-mediated gene transfer (Figure 6A). As shown in Figure 6B, the overexpression of *BrNAC047*, *BrNAC052*, and *BrNAC104* significantly activated the promoters of the target SAGs. These confirmed that BrNAC047, BrNAC052, and BrNAC104 act as positive regulators of leaf senescence by directly controlling the expression of various SAGs.

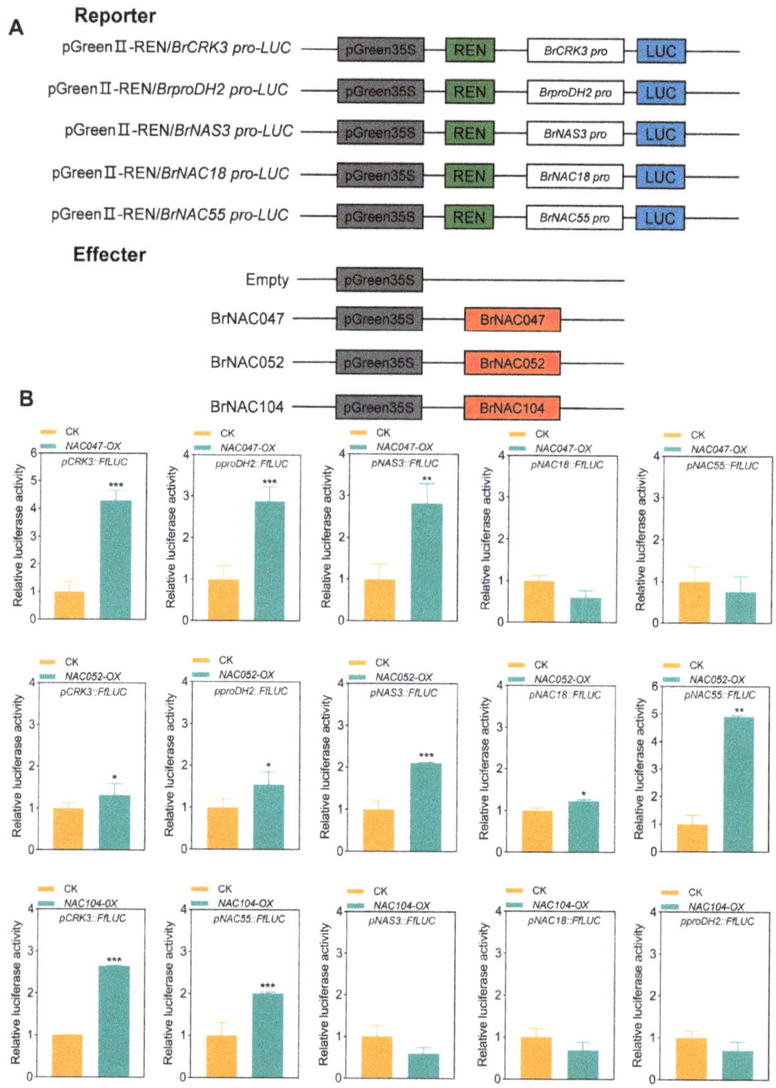

Figure 6. Dual-luciferase assay showing the effects of BrNAC047, BrNAC052, and BrNAC104 on the promoters of SAGs. (**A**) Schematics of the reporter and effector vectors. (**B**) Dual-luciferase transient expression assay in NHCC leaves showing that *BrNAC047*, *BrNAC052*, and *BrNAC104* activate the transcription of *pCRK3*, *pproDH2*, *pNAS3*, *pNAC18*, and *pNAC55*. The reporter and effector vectors are illustrated in the top panel. Data represent the mean ± SD (n = 3 biological replicates). *** $p < 0.001$, ** $p < 0.01$, and * $p < 0.05$ (t-test).

3. Discussion

Darkness is an important factor inducing the deterioration of green leafy vegetables during the post-harvest stage. In model plants, such as *A. thaliana* and rice, the mechanism of dark-induced leaf senescence has been studied extensively. Nevertheless, how dark environments induce leaf senescence in NHCC has so far remained unclear [29–31]. In the present study, NHCC was exposed to the dark for a prolonged period. The results showed that dark treatment accelerates the de-greening of NHCC leaves. In addition, the expression levels of chlorophyll catabolism genes also exhibited remarkable upregulation (Figure 1A–F). Furthermore, ROS levels in NHCC leaves were substantially elevated after dark treatment (Figure 1B), and the genes encoding the ROS-generating enzyme *RBOH* were activated during the senescence process (Figure 1G–I). These findings suggested that darkness was indeed a key initiator of rapid leaf decay and quality loss in NHCC at the post-harvest stage.

Several studies have demonstrated that light is a pivotal environmental determinant governing the progression of leaf senescence [32]. Phytochrome-interacting factors (PIFs), which are key regulatory molecules, orchestrate the cellular responsiveness to light and temperature oscillations in plants [33]. Furthermore, phytochrome B (phyB) plays a critical role in plant growth and development, exerting a significant regulatory influence on the function of PIF4/5 [34,35]. Under conditions of appropriate illumination, phyB remains in an activated state and retains its capacity to repress the transcriptional function of PIF4/5, thereby sustaining the regular growth cadence of plants and maintaining morphogenetic processes. This regulatory mechanism has a very important position within the intricate network of plant photomorphogenesis and environmental acclimatization [36]. However, we did not observe any conspicuous upward trend in *PIF* expression under dark conditions in the present study, and the expression of *PIFs* appeared to be unresponsive to dark treatment (Figure S2). These results implied that the responsiveness of PIFs to light mainly occurs at the post-transcriptional level.

The *NAC* family is one of the largest transcription factor families in plants and plays a pivotal role in the senescence program in many plant species, including *A. thaliana*, wheat, and rice [37]. For instance, *ANAC016*, *ANAC002*, and *ANAC032* facilitate senescence, whereas *ANAC083*, *ANAC041*, and *JUB1* inhibit senescence in *A. thaliana* [27,38–40]. In rice, *OsNAP*, *OsNAC2*, and *ONAC011* have been implicated in the modulation of aging- and dark-induced leaf senescence [41–43]. In the study of poplar, through RNA sequencing and gene co-expression network analysis, it was found that the alternative splicing variant of PtRD26, PtRD26IR, can delay leaf senescence. In the study of Arabidopsis thaliana, by using molecular biological and metabolomic methods, it was confirmed that RD26 is a key regulator of metabolic reprogramming during dark-induced senescence [44,45]. In the present study, through comprehensive bioinformatic screening, we found that *BrNAC104*, *BrNAC052*, and *BrNAC047* demonstrate remarkable upregulation under dark-induced senescence (Figures 3D and S4). We also demonstrated that *BrNAC047*, *BrNAC052*, and *BrNAC104* actively modulate the senescence of NHCC leaves by influencing multiple metabolic processes (Figures 4 and 6). Given that these candidate genes could synergistically induce the promoter activity of various SAGs, we examined their direct protein interactions using a yeast hybrid assay. However, no direct protein interactions were detected among BrNAC047, BrNAC052, and BrNAC104 (Figure S3). The results suggested that BrNAC047, BrNAC052, and BrNAC104 synergistically regulate leaf senescence in a parallel manner. In future agriculture production, various technical measures, such as gene-editing with CRISPR/Cas9 system, can be developed to delay leaf senescence by targeting these three genes. This contributes to cultivating new NHCC varieties with a delayed leaf senescence trait. On the other hand, inhibitors targeting these genes can also be developed to precisely suppress the expression of *BrNAC047*, *BrNAC052*, and *BrNAC104*, thus inhibiting the progression of leaf senescence in NHCC.

Light plays a crucial role in regulating plant growth and development by interacting with various phytohormones. Light influences the movement and distribution of auxin as well as the synthesis and signal transduction processes of cytokinin. Additionally, it modulates the inhibition of gibberellin biosynthesis and maintains the synthesis and efficacy of ethylene [46,47]. Among these intricate phytohormonal interaction networks, the association between ABA and light is unique and is crucial in the context of leaf senescence [48]. ABA has been shown to modulate the expression of SAGs via multiple signaling cascades [49–51]. In the present study, during the promoter element analysis of *BrNAC047*, *BrNAC052*, and *BrNAC104*, several phytohormone-responsive elements such as ABA-, auxin-, and JA-responsive elements were detected (Figure 4C). GO analysis of dark-induced DEGs further demonstrated that the dark-induced genes were conspicuously enriched in the ABA response pathway (Figure 2C). This result was consistent with our finding that *BrNAC047*, *BrNAC052*, and *BrNAC104* are induced by the ABA. Taken together, these results demonstrate that dark-induced leaf senescence in NHCC is mediated by ABA signaling.

4. Materials and Methods

4.1. Plant Materials and Growth Conditions

The NHCC variety used in the present study was previously developed by our laboratory and was stored at the Horticultural Research Institute of the Shanghai Academy of Agricultural Sciences. The plant materials were cultivated in incubators under a 16:8 h light/dark photoperiod, with a light intensity of approximately 60 μmol m^{-2} s^{-1} and a humidity level of 60%.

4.2. Dark Treatment

NHCC that had reached the five-true-leaf stage was subjected to treatment. Detached leaves were stored in the dark for 3 days, with light treatment serving as the control. The experiment was repeated three times.

4.3. Total RNA Extraction, Library Construction, and Sequencing

RNA extraction, library construction, and subsequent transcriptome sequencing were outsourced to Shanghai Majorbio Bio-Pharm Technology Co., Ltd. (Shanghai, China). The chinensis (Cultivar NHCC001) v1.0 Genome (Download Detail (njau.edu.cn)) was adopted as the reference genome. DESeq2 was employed for differential expression analysis. Genes showing a log2 (fold change) value ≥ 1 and adjusted p-value < 0.05 were designated as differentially expressed genes (DEGs).

4.4. Measurements of Chlorophyll Content and the F_V/F_m Ratio

Positionally matched leaves from the control and experimental groups were examined to measure the maximum photochemical efficiency (F_V/F_m) of Photosystem II (PSII). Briefly, the leaves were placed in darkness for approximately half an hour, and subsequently, F_V/F_m was determined using a calibrated chlorophyll fluorescence imaging system (WALZ IMAG—MAX/L, Effeltrich, Germany). The chlorophyll content was examined with a chlorophyll meter.

4.5. DAB and NBT Staining

NHCC leaves were stained for detecting reactive oxygen species (ROS) accumulation, as described previously [52]. The leaves were incubated in a 3,3′-diaminobenzidine (DAB)

(1 mg/mL [pH 3.8]) and NBT (nitro blue tetrazolium) staining solution (0.5 mg/mL [pH 7.8]) for 6 h away from light. Subsequently, the leaves were immersed in 95% ethanol to remove chlorophyll, preserved in 95% ethanol, and photographed using a camera.

4.6. Virus-Induced Gene Silencing (VIGS)

First, 40-bp sequences from the coding region of *BrNAC047*, *BrNAC052*, and *BrNAC104* were selected and reverse-complemented to obtain 80-bp palindrome sequences. The sequences used for gene silencing are listed in Supplementary Table S1. The *BrNAC047*-silencing vector (PTY-*BrNAC047*), *BrNAC052*-silencing vector (PTY-*BrNAC052*), and *BrNAC104*-silencing vector (PTY-*BrNAC104*) were synthesized and constructed using GeneScript. Two-week-old NHCC seedlings were used for the VIGS assay. PTY-*BrNAC047*, PTY-*BrNAC052*, PTY-*BrNAC104*, and PTY (control vector) (Kingsray Biotechnology Co., Ltd., Nanjing, China) were coated with gold powder and inserted into NHCC leaves using a GJ-1000 gene-gun (Zhejiang Xinzhi, Taizhou, China), as described previously [53]. One month later, qPCR was conducted to examine the gene silencing efficacy of these constructs in NHCC leaves.

4.7. Determination of Antioxidant Activity and Soluble Sugar Content

The soluble sugar and proline content of NHCC, as well as the malondialdehyde (MDA), peroxidase (POD), catalase (CAT), and superoxide dismutase (SOD) activities, were determined using commercial kits (Comin, Item No. POD-1-Y, CAT-1-W, SOD-1-W, BCAP-1-W, Suzhou, China) according to the manufacturer's instructions.

4.8. Dual-Luciferase Assay

The promoter sequences of SAGs were cloned into the pGreenII 0800-LUC vector. Meanwhile, the coding sequences of the upstream transcription factors were inserted into the PCHF3 vector [54]. Tobacco leaves were infected with a transformed *Agrobacterium* strain (GV3101: pSoup) using a 9:1 v/v (effector:reporter) mixture. Empty PCHF3 plasmids were utilized as the negative control. Following 2 days of incubation, the Dual-Luciferase Reporter Assay System (TransGen, Beijing, China) was used to examine Firefly and Renilla luciferase activities. Fluorescence was monitored using a Synergy two multi-mode microplate reader (Bio-Tek, Winooski, VT, USA) according to the manufacturer's instructions.

4.9. Yeast Two-Hybrid Assay

Full-length genes were amplified with TaKaRa Primer STAR Max DNA polymerase. *BrNAC047* and *BrNAC052* were added into a pGBKT7 vector, whereas *BrNAC104* was added into a pGADT7 vector. The plasmids were co-transformed into the yeast strain AH109, and single colonies were resuspended and grown on SD/−Leu/−Trp and SD/−Leu/−Trp/−Ade/−His medium for further analysis.

4.10. qRT-PCR Analysis

RNA was reverse transcribed into cDNA using the PrimeScript RT Master Mix Kit (TaKaRa, Kyoto, Japan). A 20 μL qPCR reaction mixture containing 1 μL of cDNA (250 ng/mL), 1 μL each of upstream and downstream primers (10 μm), and 10 μL of TransStart Top Green qPCR SuperMix was prepared, and the final volume was brought up to 20 μL using ddH_2O. Primers were synthesized by Qingke Biotechnology Co., Ltd. (Beijing, China) and are listed in Table S1.

5. Conclusions

In this study, BrNAC047, BrNAC052, and BrNAC104 were identified as key promoters of dark-induced senescence in NHCC. These transcription factors were found to directly target the promoters of certain SAGs. With regard to upstream signals, our findings suggested that ABA may be involved in the dark-induced leaf senescence of NHCC by regulating the expression of *BrNAC047*, *BrNAC052*, and *BrNAC104*. Hence, this study highlights the role of these three key regulators and provides valuable molecular targets for delaying leaf senescence in NHCC. In the future, more *BrNAC047*, *BrNAC052*, and *BrNAC104* constructs should be generated to obtain genetic data for elucidating the mechanism of dark-induced leaf senescence in NHCC.

Supplementary Materials: The supporting information can be downloaded at: https://www.mdpi.com/article/10.3390/ijms26052340/s1.

Author Contributions: D.Z. designed the experiments. B.Y. performed the experiments. Z.M., Y.Y., X.Y., M.C. and R.L. analyzed the data. B.Y. and D.Z. wrote the manuscript. D.Z. and Y.S. discussed the results and commented on the manuscript. H.Z. provided the funds. All authors contributed to the article and approved the submitted version. All authors have read and agreed to the published version of the manuscript.

Funding: This work was supported by the Shanghai Agricultural Science and Technology Innovation Program (Grant numbers: T2024309 to Dingyu Zhang).

Institutional Review Board Statement: Not applicable.

Informed Consent Statement: Not applicable.

Data Availability Statement: The datasets presented in this study can be found in online repositories. The names of the repository/repositories and accession number(s) can be found below: http://www.ncbi.nlm.nih.gov/bioproject/1181662, PRJNA1181662, accessed on 4 November 2024.

Conflicts of Interest: The authors declare no conflicts of interest.

References

1. Xiao, D.; Liu, S.T.; Wei, Y.P.; Zhou, D.Y.; Hou, X.L.; Li, Y.; Hu, C.M. cDNA-AFLP analysis reveals differential gene expression in incompatible interaction between infected non-heading Chinese cabbage and *Hyaloperonospora parasitica*. *Hortic. Res.* **2016**, *3*, 16034. [CrossRef] [PubMed]
2. Jeon, J.; Lim, C.J.; Kim, J.K.; Park, S.U. Comparative Metabolic Profiling of Green and Purple Pakchoi (*Brassica Rapa* Subsp. *Chinensis*). *Molecules* **2018**, *23*, 1613. [CrossRef] [PubMed]
3. Zhang, W.; Tang, S.; Li, X.; Chen, Y.; Li, J.; Wang, Y.; Bian, R.; Jin, Y.; Zhu, X.; Zhang, K. *Arabidopsis* WRKY1 promotes monocarpic senescence by integrative regulation of flowering, leaf senescence and nitrogen remobilization. *Mol. Plant* **2024**, *17*, 1289–1306. [CrossRef]
4. Zhu, Y.G.; Peng, J.; Chen, C.; Xiong, C.; Li, S.; Ge, A.; Wang, E.; Liesack, W. Harnessing biological nitrogen fixation in plant leaves. *Trends Plant Sci.* **2023**, *28*, 1391–1405. [CrossRef] [PubMed]
5. Zohner, C.M.; Mirzagholi, L.; Renner, S.S.; Mo, L.; Rebindaine, D.; Bucher, R.; Palouš, D.; Vitasse, Y.; Fu, Y.H.; Stocker, B.D.; et al. Effect of climate warming on the timing of autumn leaf senescence reverses after the summer solstice. *Science* **2023**, *381*, eadf5098. [CrossRef]
6. Wingler, A.; Marès, M.; Pourtau, N. Spatial patterns and metabolic regulation of photosynthetic parameters during leaf senescence. *New Phytol.* **2004**, *161*, 781–789. [CrossRef]
7. Guo, Y.; Ren, G.; Zhang, K.; Li, Z.; Miao, Y.; Guo, H. Leaf senescence: Progression, regulation, and application. *Mol. Hortic.* **2021**, *1*, 5. [CrossRef]
8. Guo, Y.; Gan, S. Leaf senescence: Signals, execution, and regulation. *Curr. Top. Dev. Biol.* **2005**, *71*, 83–112. [CrossRef]
9. Sun, L.; Xu, H.; Song, J.; Yang, X.; Wang, X.; Liu, H.; Pang, M.; Hu, Y.; Yang, Q.; Ning, X.; et al. OsNAC103, a NAC Transcription Factor, Positively Regulates Leaf Senescence and Plant Architecture in Rice. *Rice* **2024**, *17*, 15. [CrossRef]
10. Woo, H.R.; Kim, H.J.; Nam, H.G.; Lim, P.O. Plant leaf senescence and death—Regulation by multiple layers of control and implications for aging in general. *J. Cell Sci.* **2013**, *126*, 4823–4833. [CrossRef]
11. Ma, L.; Han, R.; Yang, Y.; Liu, X.; Li, H.; Zhao, X.; Li, J.; Fu, H.; Huo, Y.; Sun, L.; et al. Phytochromes enhance SOS2-mediated PIF1 and PIF3 phosphorylation and degradation to promote Arabidopsis salt tolerance. *Plant Cell* **2023**, *35*, 2997–3020. [CrossRef] [PubMed]

12. Tachibana, R.; Abe, S.; Marugami, M.; Yamagami, A.; Akema, R.; Ohashi, T.; Nishida, K.; Nosaki, S.; Miyakawa, T.; Tanokura, M.; et al. BPG4 regulates chloroplast development and homeostasis by suppressing GLK transcription factors and involving light and brassinosteroid signaling. *Nat. Commun.* **2024**, *15*, 370. [CrossRef] [PubMed]
13. Iqbal, Z.; Iqbal, M.S.; Hashem, A.; Abd Allah, E.F.; Ansari, M.I. Plant Defense Responses to Biotic Stress and Its Interplay with Fluctuating Dark/Light Conditions. *Front. Plant Sci.* **2021**, *12*, 631810. [CrossRef]
14. Buchanan-Wollaston, V.; Page, T.; Harrison, E.; Breeze, E.; Lim, P.O.; Nam, H.G.; Lin, J.F.; Wu, S.H.; Swidzinski, J.; Ishizaki, K.; et al. Comparative transcriptome analysis reveals significant differences in gene expression and signalling pathways between developmental and dark/starvation-induced senescence in Arabidopsis. *Plant J.* **2005**, *42*, 567–585. [CrossRef] [PubMed]
15. Song, X.; Zhao, W.; Cui, S.; Su, X.; Yu, J.; Guo, L.; Song, K. Deciphering the dual role of persistent luminescence materials: Toxicity and photoreception effects on rice development. *Sci. Total Environ.* **2024**, *947*, 174542. [CrossRef]
16. Liu, X.W.; Wang, T.; Bartholomew, E.; Black, K.; Dong, M.M.; Zhang, Y.Q.; Yang, S.; Cai, Y.L.; Xue, S.D.; Weng, Y.Q.; et al. Comprehensive analysis of NAC transcription factors and their expression during fruit spine development in cucumber (*Cucumis sativus* L.). *Hortic. Res.* **2018**, *5*, 31. [CrossRef]
17. Aida, M.; Ishida, T.; Fukaki, H.; Fujisawa, H.; Tasaka, M. Genes involved in organ separation in Arabidopsis: An analysis of the cup-shaped cotyledon mutant. *Plant Cell* **1997**, *9*, 841–857. [CrossRef]
18. Hegedus, D.; Yu, M.; Baldwin, D.; Gruber, M.; Sharpe, A.; Parkin, I.; Whitwill, S.; Lydiate, D. Molecular characterization of *Brassica napus* NAC domain transcriptional activators induced in response to biotic and abiotic stress. *Plant Mol. Biol.* **2003**, *53*, 383–397. [CrossRef]
19. Lei, L.; Wu, D.; Cui, C.; Gao, X.; Yao, Y.; Dong, J.; Xu, L.; Yang, M. Transcriptome Analysis of Early Senescence in the Post-Anthesis Flag Leaf of Wheat (*Triticum aestivum* L.). *Plants* **2022**, *11*, 2593. [CrossRef]
20. Parrott, D.L.; McInnerney, K.; Feller, U.; Fischer, A.M. Steam-girdling of barley (*Hordeum vulgare*) leaves leads to carbohydrate accumulation and accelerated leaf senescence, facilitating transcriptomic analysis of senescence-associated genes. *New Phytol.* **2007**, *176*, 56–69. [CrossRef]
21. Liu, L.; Zhou, Y.; Zhou, G.; Ye, R.; Zhao, L.; Li, X.; Lin, Y. Identification of early senescence-associated genes in rice flag leaves. *Plant Mol. Biol.* **2008**, *67*, 37–55. [CrossRef] [PubMed]
22. Balazadeh, S.; Siddiqui, H.; Allu, A.D.; Matallana-Ramirez, L.P.; Caldana, C.; Mehrnia, M.; Zanor, M.I.; Köhler, B.; Mueller-Roeber, B. A gene regulatory network controlled by the NAC transcription factor ANAC092/AtNAC2/ORE1 during salt-promoted senescence. *Plant J.* **2010**, *62*, 250–264. [CrossRef] [PubMed]
23. Guo, Y.; Gan, S. AtNAP, a NAC family transcription factor, has an important role in leaf senescence. *Plant J.* **2006**, *46*, 601–612. [CrossRef] [PubMed]
24. Huang, Q.J.; Wang, Y.; Li, B.; Chang, J.L.; Chen, M.J.; Li, K.X.; Yang, G.X.; He, G.Y. TaNAC29, a NAC transcription factor from wheat, enhances salt and drought tolerance in transgenic *Arabidopsis*. *BMC Plant Biol.* **2015**, *15*, 268. [CrossRef]
25. Xu, Z.Y.; Gongbuzhaxi; Wang, C.Y.; Xue, F.; Zhang, H.; Ji, W.Q. Wheat NAC transcription factor TaNAC29 is involved in response to salt stress. *Plant Physiol. Biochem.* **2015**, *96*, 356–363. [CrossRef]
26. Yang, S.D.; Seo, P.J.; Yoon, H.K.; Park, C.M. The *Arabidopsis* NAC transcription factor VNI2 integrates abscisic acid signals into leaf senescence via the COR/RD genes. *Plant Cell* **2011**, *23*, 2155–2168. [CrossRef]
27. Takasaki, H.; Maruyama, K.; Takahashi, F.; Fujita, M.; Yoshida, T.; Nakashima, K.; Myouga, F.; Toyooka, K.; Yamaguchi-Shinozaki, K.; Shinozaki, K. SNAC-As, stress-responsive NAC transcription factors, mediate ABA-inducible leaf senescence. *Plant J.* **2015**, *84*, 1114–1123. [CrossRef]
28. Zhang, K.; Gan, S.S. An abscisic acid-AtNAP transcription factor-SAG113 protein phosphatase 2C regulatory chain for controlling dehydration in senescing Arabidopsis leaves. *Plant Physiol.* **2012**, *158*, 961–969. [CrossRef]
29. Zareen, S.; Ali, A.; Lim, C.J.; Khan, H.A.; Park, J.; Xu, Z.Y.; Yun, D.J. The Transcriptional Corepressor HOS15 Mediates Dark-Induced Leaf Senescence in Arabidopsis. *Front. Plant Sci.* **2022**, *13*, 828264. [CrossRef]
30. Chen, P.Y.; Nguyen, T.T.T.; Lee, R.H.; Hsu, T.W.; Kao, M.H.; Gojobori, T.; Chiang, T.Y.; Huang, C.L. Genome-wide expression analysis of vegetative organs during developmental and herbicide-induced whole plant senescence in *Arabidopsis thaliana*. *BMC Genom.* **2024**, *25*, 621. [CrossRef]
31. Song, G.; Kwon, C.T.; Kim, S.H.; Shim, Y.; Lim, C.; Koh, H.J.; An, G.; Kang, K.; Paek, N.C. The Rice SPOTTED LEAF4 (SPL4) Encodes a Plant Spastin That Inhibits ROS Accumulation in Leaf Development and Functions in Leaf Senescence. *Front. Plant Sci.* **2018**, *9*, 1925. [CrossRef]
32. Lee, J.; Kang, M.H.; Kim, J.Y.; Lim, P.O. The Role of Light and Circadian Clock in Regulation of Leaf Senescence. *Front. Plant Sci.* **2021**, *12*, 669170. [CrossRef] [PubMed]
33. Rosado, D.; Trench, B.; Bianchetti, R.; Zuccarelli, R.; Rodrigues Alves, F.R.; Purgatto, E.; Segal Floh, E.I.; Silveira Nogueira, F.T.; Freschi, L.; Rossi, M. Downregulation of PHYTOCHROME-INTERACTING FACTOR 4 Influences Plant Development and Fruit Production. *Plant Physiol.* **2019**, *181*, 1360–1370. [CrossRef] [PubMed]

34. Zhang, Y.; Liu, Z.; Chen, Y.; He, J.X.; Bi, Y. PHYTOCHROME-INTERACTING FACTOR 5 (PIF5) positively regulates dark-induced senescence and chlorophyll degradation in Arabidopsis. *Plant Sci.* **2015**, *237*, 57–68. [CrossRef]
35. Song, Y.; Yang, C.; Gao, S.; Zhang, W.; Li, L.; Kuai, B. Age-triggered and dark-induced leaf senescence require the bHLH transcription factors PIF3, 4, and 5. *Mol. Plant* **2014**, *7*, 1776–1787. [CrossRef] [PubMed]
36. Thiele, A.; Herold, M.; Lenk, I.; Quail, P.H.; Gatz, C. Heterologous expression of Arabidopsis phytochrome B in transgenic potato influences photosynthetic performance and tuber development. *Plant Physiol.* **1999**, *120*, 73–82. [CrossRef]
37. Chun, I.; Kim, H.J.; Hong, S.; Kim, Y.G.; Kim, M.S. Structural basis of DNA binding by the NAC transcription factor ORE1, a master regulator of plant senescence. *Plant Commun.* **2023**, *4*, 100510. [CrossRef]
38. Garapati, P.; Xue, G.P.; Munné-Bosch, S.; Balazadeh, S. Transcription Factor ATAF1 in Arabidopsis Promotes Senescence by Direct Regulation of Key Chloroplast Maintenance and Senescence Transcriptional Cascades. *Plant Physiol.* **2015**, *168*, 1122–1139. [CrossRef]
39. Kim, H.J.; Nam, H.G.; Lim, P.O. Regulatory network of NAC transcription factors in leaf senescence. *Curr. Opin. Plant Biol.* **2016**, *33*, 48–56. [CrossRef]
40. Oda-Yamamizo, C.; Mitsuda, N.; Sakamoto, S.; Ogawa, D.; Ohme-Takagi, M.; Ohmiya, A. The NAC transcription factor ANAC046 is a positive regulator of chlorophyll degradation and senescence in Arabidopsis leaves. *Sci. Rep.* **2016**, *6*, 23609. [CrossRef]
41. Liang, C.; Wang, Y.; Zhu, Y.; Tang, J.; Hu, B.; Liu, L.; Ou, S.; Wu, H.; Sun, X.; Chu, J.; et al. OsNAP connects abscisic acid and leaf senescence by fine-tuning abscisic acid biosynthesis and directly targeting senescence-associated genes in rice. *Proc. Natl. Acad. Sci. USA* **2014**, *111*, 10013–10018. [CrossRef] [PubMed]
42. Mao, C.; Lu, S.; Lv, B.; Zhang, B.; Shen, J.; He, J.; Luo, L.; Xi, D.; Chen, X.; Ming, F. A Rice NAC Transcription Factor Promotes Leaf Senescence via ABA Biosynthesis. *Plant Physiol.* **2017**, *174*, 1747–1763. [CrossRef]
43. El Mannai, Y.; Akabane, K.; Hiratsu, K.; Satoh-Nagasawa, N.; Wabiko, H. The NAC Transcription Factor Gene OsY37 (ONAC011) Promotes Leaf Senescence and Accelerates Heading Time in Rice. *Int. J. Mol. Sci.* **2017**, *18*, 2165. [CrossRef] [PubMed]
44. Kamranfar, I.; Xue, G.P.; Tohge, T.; Sedaghatmehr, M.; Fernie, A.R.; Balazadeh, S.; Mueller-Roeber, B. Transcription factor RD26 is a key regulator of metabolic reprogramming during dark-induced senescence. *New Phytol.* **2018**, *218*, 1543–1557. [CrossRef]
45. Wang, H.L.; Zhang, Y.; Wang, T.; Yang, Q.; Yang, Y.; Li, Z.; Li, B.; Wen, X.; Li, W.; Yin, W.; et al. An alternative splicing variant of PtRD26 delays leaf senescence by regulating multiple NAC transcription factors in Populus. *Plant Cell* **2021**, *33*, 1594–1614. [CrossRef]
46. Xu, Y.; Zhu, Z. PIF4 and PIF4-Interacting Proteins: At the Nexus of Plant Light, Temperature and Hormone Signal Integrations. *Int. J. Mol. Sci.* **2021**, *22*, 10304. [CrossRef]
47. Cackett, L.; Luginbuehl, L.H.; Schreier, T.B.; Lopez-Juez, E.; Hibberd, J.M. Chloroplast development in green plant tissues: The interplay between light, hormone, and transcriptional regulation. *New Phytol.* **2021**, *233*, 2000–2016. [CrossRef]
48. Mahapatra, K.; Dwivedi, S.; Mukherjee, A.; Pradhan, A.A.; Rao, K.V.; Singh, D.; Bhagavatula, L.; Datta, S. Interplay of Light and ABA signaling to modulate plant development. *J. Exp. Bot.* **2024**, *76*, 730–745. [CrossRef] [PubMed]
49. Liebsch, D.; Keech, O. Dark-induced leaf senescence: New insights into a complex light-dependent regulatory pathway. *New Phytol.* **2016**, *212*, 563–570. [CrossRef]
50. Deepika; Ankit; Sagar, S.; Singh, A. Dark-Induced Hormonal Regulation of Plant Growth and Development. *Front. Plant Sci.* **2020**, *11*, 581666. [CrossRef]
51. Zheng, Q.; Teng, Z.; Zhang, J.; Ye, N. ABA Inhibits Rice Seed Aging by Reducing H_2O_2 Accumulation in the Radicle of Seeds. *Plants* **2024**, *13*, 809. [CrossRef]
52. Zhang, D.Y.; Wu, S.D.; Li, N.; Gao, J.; Liu, S.H.; Zhu, S.; Li, Z.L.; Ren, G.D.; Kuai, B.K. Chemical induction of leaf senescence and powdery mildew resistance involves ethylene-mediated chlorophyll degradation and ROS metabolism in cucumber. *Hortic. Res.* **2022**, *9*, uhac101. [CrossRef] [PubMed]
53. Huang, F.Y.; Liu, T.K.; Tang, J.; Duan, W.K.; Hou, X.L. BcMAF2 activates BcTEM1 and represses flowering in Pak-choi (*Brassica rapa* ssp. *chinensis*). *Plant Mol. Biol.* **2019**, *100*, 19–32. [CrossRef] [PubMed]
54. Zhang, D.; Zhu, Z.; Gao, J.; Zhou, X.; Zhu, S.; Wang, X.; Wang, X.; Ren, G.; Kuai, B. The NPR1-WRKY46-WRKY6 signaling cascade mediates probenazole/salicylic acid-elicited leaf senescence in *Arabidopsis thaliana*. *J. Integr. Plant Biol.* **2021**, *63*, 924–936. [CrossRef] [PubMed]

International Journal of *Molecular Sciences*

MDPI

Article

Functional Study of *GbSMXL8*-Mediated Strigolactone Signaling Pathway in Regulating Cotton Fiber Elongation and Plant Growth

Lingyu Chen, Wennuo Xu, Lingyu Zhang, Qin Chen, Yongsheng Cai, Quanjia Chen *,† and Kai Zheng *,†

Xinjiang Key Laboratory of Crop Biology Breeding, College of Agriculture, Xinjiang Agricultural University, Ürümqi 830052, China; ingyu0904@126.com (L.C.); xu_wen_nuo@163.com (W.X.); zlyyuuu@163.com (L.Z.); cqq0777@163.com (Q.C.); cys0620@126.com (Y.C.)

* Correspondence: cqj@xjau.edu.cn (Q.C.); zhengkai555@xjau.edu.cn (K.Z.)

† These authors contributed equally to this work.

Abstract: The novel plant hormone strigolactones (SL) are involved significantly in plant growth and development. Its key members *SMXL6, 7, 8* can modulate SL signal reception and response negatively and can regulate plant branching remarkably. There are relatively scarce studies of cotton SMXL gene family, and this study was carried out to clarify the role of *GbSMXL8* in cotton fiber development. Phylogenetic analysis identified 48 cotton SMXL genes, which were divided into SMXL-I (*SMXL 1, 2*), SMXL-II (*SMXL 3*) and SMXL-III (*SMXL6, 7, 8*) groups. The results of the cis-element analysis indicated that the SMXL gene could respond to hormones and the environment to modulate cotton growth process. A candidate gene *GbSMXL8* was screened out based on the expression difference in extreme varieties of *Gossypium barbadense*. Tissue-specific analysis indicated that *GbSMXL8* was mainly expressed in roots, 20D, 25D, and 35D and was involved in SL signaling pathways. In vitro ovule culture experiments showed that exogenous SLs (GR24) could promote the fiber elongation of *G. barbadense*, and *GbSMXL8* expression was increased after GR24 treatment, indicating that *GbSMXL8* was specifically responsive to GR24 in regulating fiber growth. *GbSMXL8* knockout resulted in creased length and number of epidermal hairs and the length of fiber, indicating the interference role of *GbSMXL8* gene with the development of cotton fiber. The *GbSMXL8* transgenic plant was detected with a higher chlorophyll content and photosynthetic rate than those of the control plant, producing a direct impact on plant growth, yield, and biomass accumulation. *GbSMXL8* gene knockout could increase plant height, accelerate growth rate, and lengthen fiber length. Intervening *GbSMXL8* may mediate cotton growth, plant type formation and fiber elongation. In conclusion, the present study uncovers the function of *GbSMXL8*-mediated SL signal in cotton, providing theoretical insight for future breeding of new cotton varieties.

Keywords: cotton; *GbSMXL8*; functional analysis; tissue expression

1. Introduction

Cotton is an important cash crop (a fiber and oil source); it is a grain crop containing high protein, a raw material for textiles and fine chemicals, and an important strategic material [1]. Cotton occupies a major position in the raw material supply of textile industry and is also an important strategic reserve resource in China. *Gossypium hirsutum* (*G. hirsutum*) and *Gossypium barbadense* (*G. barbadense*) are tetraploid cultivated cotton species domesticated from the same ancestor. Specifically, *G. hirsutum* has high yield and good tolerance to stress, but ordinary fiber quality, while *G. barbadense* has small bolls and relatively low yield, but is an important source of more slender and strong high-quality fibers [2]. Therefore, mining fiber development genes in *G. barbadense* may be a potent path to improve the fiber quality of *G. hirsutum*, which is convenient for large-scale popularization and planting [3]. It is challenging for cotton breeding owing to the multifactorial process related to fiber varieties (both genetic and environmental). Mining and functional analysis of cotton fiber-related genes are important means to achieve this goal [4,5].

Strigolactones (SLs) are a novel class of plant endogenous hormones that coordinate plant growth and development through multi-layered regulatory mechanisms. Their core functions include: (1) inhibiting lateral bud elongation to control branching formation [6]; (2) optimizing root architecture by promoting root hair elongation while suppressing lateral root formation [7]; (3) negatively regulating rice mesocotyl elongation under dark conditions [8]; (4) delaying leaf senescence [9]; and (5) acting as allelopathic signals to induce the germination of root parasitic weed seeds [10]. Studies have shown that SLs interact with hormones such as auxin, cytokinin, and gibberellins (GAs) to form a cross-regulatory network, coordinating plant development through the following mechanisms [11]. Synergy with auxin: SL signaling acts upstream of auxin to promote primary root elongation and inhibit lateral root formation, establishing the molecular basis for root plasticity regulation [12]. Interaction with cytokinin: SLs regulate the expression of cytokinin biosynthesis genes (*IPT3*), influencing downstream signaling pathways to control rice tillering and panicle development [13]. Integration with gibberellins: the SL signaling component D53 interacts with the GA signaling repressor *SLR1* [14].

Studies of the SLs signaling pathway have shown that similar to mechanisms of auxin, jasmonic acid, and GA signaling, SL signaling is based on hormone-activated proteolysis [15,16]. The SL receptor complex consists of α/β folding hydrolase *DWAERF14* (*D14*), leucine-rich repeat F-box protein *DWARF3* (*D3*), and the Clp protease family *DWARF53* (SMXLs) [17]. *D14* and *D3* can form SCF complex during signal transduction in witchweed, SCF can be used to accommodate witchweed lactone molecules and hydrolyze them into a covalently linked inter mediator molecule as well as promote the Cp protease family degradation of *D53/SMXLs*, which can exert a regulatory role in the branching of plants [18].

SMXL-like proteins represented by repressors *SMXL6*, *SMXL7*, *SMXL8*, and *D53* mainly regulate SL-related traits in rice and are targeted by AtD14-SCFMAX2 after SL signal perception [19]. *D53* undergoes rapid ubiquitination modification and degradation upon rac-G R24 treatment in Arabidopsis, depending on endogenous active forms of *D14* and *D3*, which is involved in SL signal transduction [20]. The dominant mutation of *D53* gene resulted in a gain-of-function mutant d53, which was insensitive to SL, and showed a dwarf and multi-tillering phenotype. SL can promote *D14* to interact with *D53* and F-box protein *D3*, induce *D53* ubiquitination modification and degradation, and inhibit plant branching. *D53* contains the EAR motif, allowing *D53* to recruit a TPL co-repressor that binds to the transcription factor IPA1 protein and inhibits its transcriptional activation function [18,21].

D53/SMXL6,7,8, as core repressors in the strigolactone (SL) signaling pathway, function similarly to DELLA proteins in the gibberellin (GA) signaling pathway, AUX/IAA proteins in the auxin signaling pathway, and JAZ proteins in the jasmonate (JA) signaling pathway. These repressors directly bind downstream transcription factors, inhibiting their transcriptional activity and thereby suppressing the expression of hormone-responsive genes [20]. Although the regulatory mechanisms of the SL signaling pathway in plant branching, root development, and environmental adaptation have been extensively studied, its role in cotton fiber development remains poorly understood. Therefore, elucidating the specific mechanisms of D53/SMXL6,7,8 in cotton growth and development will not only reveal the unique functions of the SL signaling pathway in fiber development but also provide new molecular targets for improving cotton fiber quality, highlighting its significant theoretical and practical implications.

2. Results

2.1. Gene Member Identification of SMXL Family in Cotton

In order to study the evolutionary relationship of SMXL families in different G. species, this study analyzed SMXL gene families in *Oryza sativa* L., *Arabidopsis thaliana*, *G. hirsutum*, *G. barbadense*, *G. hirsutum*, and *G. raymond*. Among them, 16 GbSMXL genes, 16 GbSMXL genes, 8 GaSMXL genes, and 8 GrSMXL genes were identified in *G. hirsutum*, *G. barbadense*, *G. raymond*, and *G. raymond*, respectively (Table 1).

Table 1. Members of genes from the SMXL family in cotton.

Gene Name	Gene ID	Chromosomal Location	Gene Length (bp)	Number of Amino Acids (aa)	Chromosomal Assignment	Isoelectric Point (IP)	Protein Molecular Weight (kD)
GaSMXL3-1	Ga04G1902	Chr04	2583	860	95242458–95245014	7.692	87.256
GaSMAX1	Ga05G0071	Chr05	3174	1057	748814–752504	8.021	115.686
GaSMXL7	Ga05G3318	Chr05	3294	1097	43831373–43835849	6.859	120.874
GaSMXL2	Ga08G2769	Chr08	3366	1121	127820527–127828153	6.425	123.610
GaSMXL3-2	Ga09G1542	Chr09	2583	860	72503416–72506219	6.833	95.317
GaSMXL6	Ga10G0058	Chr10	3015	1004	617219–621650	8.049	111.242
GaSMXL3-3	Ga13G0098	Chr13	2574	857	977357–979930	6.817	88.262
GaSMXL4	Ga13G2559	Chr10	2769	922	617219–621650	7.633	103.048
GbSMAX1-D	GB_D05G0073	D05	3168	1055	702090–705771	8.024	115.392
GbSMAX2-A	GB_A08G2797	A08	2388	795	119163347–119165732	6.689	89.16
GbSMAX2-D	GB_D08G2782	D08	3168	1055	66175674–66178844	8.024	115.392
GbSMAX1-A	GB_A05G0073	A05	3168	1055	812909–816590	8.024	115.392
GbSMXL3-1-A	GB_A05G4239	A05	2583	860	104306914–104309470	6.833	95.365
GbSMXL3-2-A	GB_A09G1634	A09	2346	781	67281364–67284167	7.939	87.135
GbSMXL3-3-A	GB_A13G0093	A13	2583	795	888559–891131	6.817	88.262
GbSMXL3-1-D	GB_D04G0224	D04	2346	781	2897103–2899659	7.96	87.437
GbSMXL3-2-D	GB_D09G1478	D09	2583	860	42975932–42978737	6.962	95.336
GbSMXL3-3-D	GB_D13G0095	D13	2769	795	744946–747524	6.833	95.365
GbSMXL4-A	GB_A13G2506	A13	2769	922	110467899–110471049	8.124	103.066
GbSMXL4-D	GB_D13G2442	D13	2388	795	58526390–58529536	6.728	89.203
GbSMXL6-A	GB_A10G2794	A10	3018	1005	110924219–110928660	8.286	111.286
GbSMXL6-D	GB_D10G2753	D10	3006	1001	66148734–66153240	8.334	110.726
GbSMXL7-A	GB_A05G3145	A05	3351	1116	42056154–42060637	6.784	122.845
GbSMXL7-D	GB_D05G3116	D05	3351	1116	33847628–33852136	6.618	122.798
GbSMXL8-A	GB_A05G0893	A05	3369	1122	8339616–8343711	6.281	123.759
GbSMXL8-D	GB_D05G0872	D05	3369	1122	7368329–7372426	6.395	123.767
GhSMAX1	Gh_A05G005600	A05	3168	1055	874949–879694	8.024	115.441
GhSMXL2-A	Gh_A08G270100	A08	2583	860	123972508–123980097	6.962	95.162
GhSMXL2-D	Gh_D08G260600	D08	2583	860	67120320–67123606	6.962	95.162

Table 1. *Cont.*

Gene Name	Gene ID	Chromosomal Location	Gene Length (bp)	Number of Amino Acids (aa)	Chromosomal Assignment	Isoelectric Point (IP)	Protein Molecular Weight (kD)
GhSMXL3-1-A	Gh_A05G402300	A05	2346	781	106202251–106205180	7.961	87.156
GhSMXL3-2-A	Gh_A09G154300	A09	2583	860	71317475–71320277	6.833	95.33
GhSMXL3-3-A	Gh_A13G010300	A13	2388	795	1013150–1016196	6.751	89.172
GhSMXL3-1-D	Gh_D04G021100	D04	2346	781	2840294–2843087	7.965	87.403
GhSMXL3-2-D	Gh_D09G145500	D09	2583	860	41594390–41597580	6.962	95.162
GhSMXL3-3-D	Gh_D13G010500	D13	2388	795	881251–884228	6.728	89.203
GhSMXL4-A	Gh_A13G235100	A13	2769	922	106078787–106082139	8.124	103.048
GhSMXL6-A	Gh_A10G247300	A10	3006	1001	114136161–114140859	8.334	110.726
GhSMXL6-D	Gh_D10G279400	D10	3018	1005	65267514–65272027	7.945	111.045
GhSMXL7-A	Gh_A05G297200	A05	3315	1116	43110739–43115841	6.784	122.859
GhSMXL7-D	Gh_D05G306200	D05	3351	1116	34015087–34020029	6.618	122.772
GhSMXL4-D	Gh_D13G238200	D13	2769	922	61213860–61217008	7.402	102.95
GrSMXL3-1	Grai_04G022220	Chr04	2583	860	3077802–3080358	6.962	95.293
GrSMXL6	Grai_05G000730	Chr05	3351	1116	113065942–113068498	6.618	122.716
GrSMXL7	Grai_05G033070	Chr05	3366	1121	114074546–114076917	6.507	123.658
GrSMAX1	Grai_08G030620	Chr08	3270	1089	2150749–2153436	7.65	111.174
GrSMXL3	Grai_09G016790	Chr09	2346	781	71185776–71188579	7.574	87.25
GrSMXL6	Grai_10G031250	Chr10	3270	1089	41926634–41929438	7.65	111.174
GrSMXL3-2	Grai_13G001050	Chr13	2388	795	1013713–1016286	6.665	89.293
GrSMXL4	Grai_13G027690	Chr13	2769	922	881673–884251	7.402	103.055

2.2. Phylogenetic Evolution and Gene Structure Analysis of SMXL Family

To investigate the evolutionary relationships of the SMXL gene family in cotton, this study selected SMXL protein sequences from the monocot model plant rice (*Oryza sativa*), the dicot model plant Arabidopsis (*Arabidopsis thaliana*), and four cotton species: the tetraploid cultivated species upland cotton (*Gossypium hirsutum*) and sea island cotton (*G. barbadense*), as well as the diploid wild species *G. raimondii* and *G. arboreum*. A maximum likelihood (ML) phylogenetic tree was constructed to analyze these sequences. Based on Bootstrap analysis with 1000 replicates (node support values $\geq$ 70%) and Bayesian posterior probabilities (BPP $\geq$ 0.95), all SMXL genes were classified into three subfamilies (Figure 1). Notably, the SMXL6/7/8 subfamily formed a monophyletic clade with OsD53, a core strigolactone (SL) signaling factor in rice (Bootstrap = 74%), indicating a functional conservation of these genes in the SL signaling pathway.

Further analysis revealed that the genetic distances among SMXL genes within the cotton genus (Gossypium) were significantly smaller than those between cotton and rice (average genetic distance: 0.15 within Gossypium vs. 0.55 between cotton and rice). Importantly, the clustering pattern of SMXL6/7/8 in cotton was consistent with homologous genes in maize (*GmSMXL8*) [22] and soybean (*MaSMXL6/7/8*) [23], supporting the universality of this subfamily classification across divergent plant lineages.

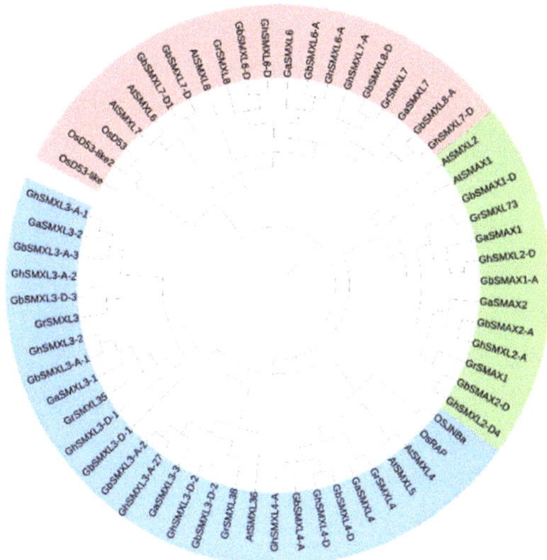

Figure 1. Phylogenetic tree of SMXL protein in *G. arboreum*, *G. barbadense*, *G. raymond*, *G. hirsutum*, *Arabidopsis thaliana*, and *Oryza sativa* L. Alignment of SMXL amino acid sequences was completed using ClustalW, and the phylogenetic tree was constructed by MEGA 7.0 using adjacent linkage method, with 1000 repeats. Gh, Gb, Ga, Gr, Os, and At represent *G. hirsutum*, *G. barbadense*, *G. raymond*, *G. raymond*, *Oryza sativa* L., and *Arabidopsis thaliana*.

2.3. Analysis of Protein Domains of SMXL Family in Cotton

To investigate the structural characteristics of SMXL family proteins in cotton, this study extracted four SMXL family protein sequences and analyzed their protein domains using the online program Pfam (http://pfam.xfam.org/) (accessed on 24 December 2024). The results were visualized using Tbtools software (accessed on 24 December 2024) (Figure 2). The analysis revealed that the SMXL gene family primarily contains two conserved domains: Clp_A and p-loop_NTase. The Clp_A domain is the N-terminal region of Clp proteases [24], while the p-loop_NTase domain is a critical region for protein kinase function, participating in various biological processes such as translation (initiation, elongation, and release factors), signal transduction, cell motility, and intracellular transport [25].

As shown in Figure 3, the domain distribution of SMXL family proteins exhibits notable features: Clp_A domain: highly conserved in all SMXL proteins, located at the N-terminus, and potentially involved in protein degradation or complex assembly. p-loop_NTase domain: Most SMXL proteins contain 1–2 copies of this domain, located downstream of the Clp_A domain. Its function may be related to ATP binding and hydrolysis, suggesting that SMXL proteins play a role in energy-dependent cellular processes [26]. Domain arrangement pattern: the sequential arrangement of Clp_A and p-loop_NTase domains in SMXL proteins is highly consistent, indicating potential functional synergy between these domains.

These structural features provide important clues for deciphering the functional mechanisms of SMXL proteins. For example, the p-loop_NTase domain may regulate ATP-dependent signal transduction processes, participating in the transmission and response of strigolactone (SL) signals.

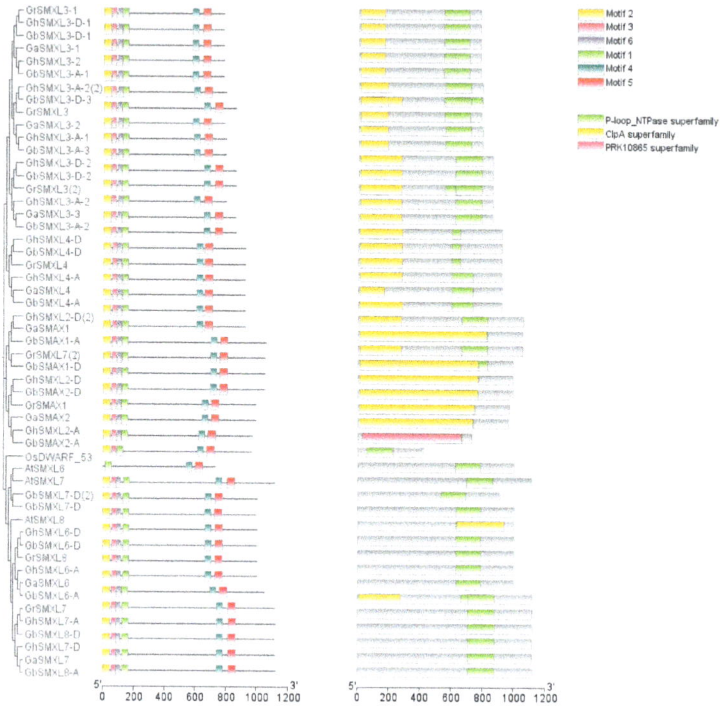

Figure 2. Genetic structure and conserved domains of SMXL protein in four cotton species.

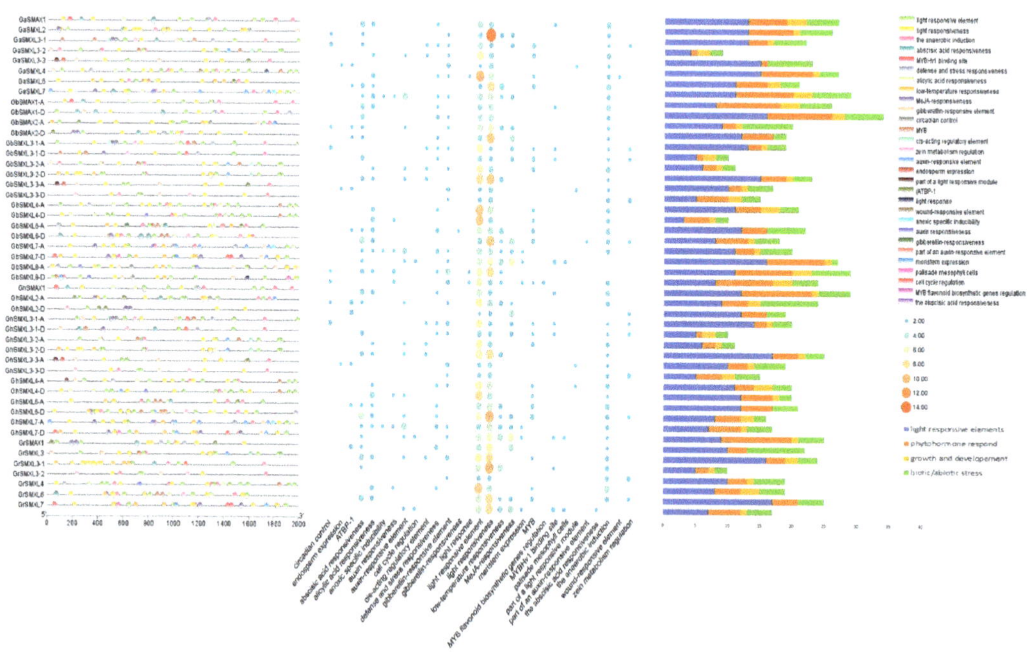

Figure 3. Analysis of cis-acting elements of SMXL in *G. hirsutum*, *G. barbadense*, *G. Raymond*, and *G. arboreum*.

2.4. Cis-Acting Elements Analysis

According to the results in Figure 3, promoter analysis revealed that the SMXL gene contains cis-acting elements associated with stress response and hormone response. Specifically, the ABRE element is linked to abscisic acid (ABA) signaling, while P-box, TATC-box, and GARE-motif are associated with gibberellin (GA) response. The TCA-element is related to salicylic acid (SA) signaling, and CGTCA-motif and TGACG-motif are involved in jasmonic acid (JA) response. Additionally, the ERE element is associated with ethylene signaling [27]. These hormone response elements underscore the critical role of the SMXL gene family in hormone signaling and transmission.

Furthermore, the presence of ARE and GC-Motif anaerobic-inducing elements suggests that SMXL genes may participate in information transduction in response to biological stress [27]. The promoter region of SMXL genes also contains MYB binding sites and the cell cycle regulatory element MSA-like, indicating potential roles in circadian rhythm regulation and cell cycle control [28]. Notably, the SMXL gene promoter harbors the highest number of photoregulatory responses and hormone-type acting elements, highlighting its central position in multiple response regulatory networks.

Collectively, these findings suggest that SMXL genes play a pivotal role in integrating hormonal, environmental, and developmental signals, making them key regulators in plant stress adaptation and growth regulation.

2.5. Analysis of SMXL8 Gene Expression Characteristics in Gossypium barbadense

The elite sea island cotton (*Gossypium barbadense*) cultivar A130768 is a variety that is renowned for its high-quality long fibers, and a comprehensive gene expression database has been established for its key fiber development stages (0–35 DPA). Therefore, it was selected as the core material for spatiotemporal expression analysis of *GbSMXL8* and in vitro ovule culture experiments to ensure the precision and reproducibility of research on fiber development regulatory mechanisms.

Transcriptome analysis revealed that the SMXL gene family exhibited significantly higher expression levels at 0 days post anthesis (DPA) and 35DPA, with *GbSMXL8* (Gbar_A05G03997800), showing marked differential expression across distinct *Gossypium barbadense* cultivars ($p < 0.01$). To further validate the spatiotemporal expression profile of *GbSMXL8*, qRT-PCR was performed to analyze its tissue-specific expression patterns. The results demonstrated that GbSMXL8 was predominantly expressed in roots, 20 DPA fibers, and 35 DPA fibers, with expression levels significantly higher than those in stems and leaves ($p < 0.05$; Figure 4). Notably, the peak expression of *GbSMXL8* coincided with critical stages of fiber development: rapid elongation (20 DPA) and secondary cell wall deposition (35 DPA). These findings suggest that *GbSMXL8* may regulate cotton fiber development through the strigolactone (SL) signaling pathway.

2.6. Effect of Auricolactone on Fiber Development

To investigate the role of SLs in cotton fiber development, this study utilized an in vitro ovule culture system of sea island cotton (*Gossypium barbadense*), combined with treatments of the exogenous SL analog GR24 (rac-GR24, 10 μM) and the SL biosynthesis inhibitor Tis108 (5 μM) (Figure 5A) [29,30]. The results showed that, compared to the untreated control group, GR24 significantly promoted fiber elongation ($p < 0.05$), while Tis108 treatment inhibited fiber elongation ($p < 0.05$) (Figure 5B). Further analysis revealed that SL content exhibited a dynamic increase during the rapid fiber elongation stage (20–30 DPA) and was significantly positively correlated with fiber length ($p < 0.001$) (Figure 5C). Notably, GR24 treatment significantly increased ovule weight ($p < 0.01$), whereas Tis108 treatment reduced ovule weight ($p < 0.05$), indicating that SLs promote fiber elongation by modulating material accumulation (Figure 5E,F).

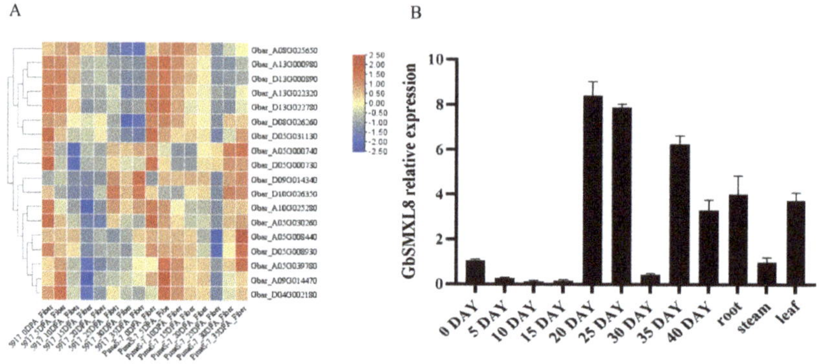

Figure 4. Expression analysis of SMXL during fiber development of sea island cotton. (**A**): Transcriptome expression heat map of different varieties of *G. barbadense*. (**B**): Relative expression level of *GbSMXL8* in different tissues of *G. barbadense*.

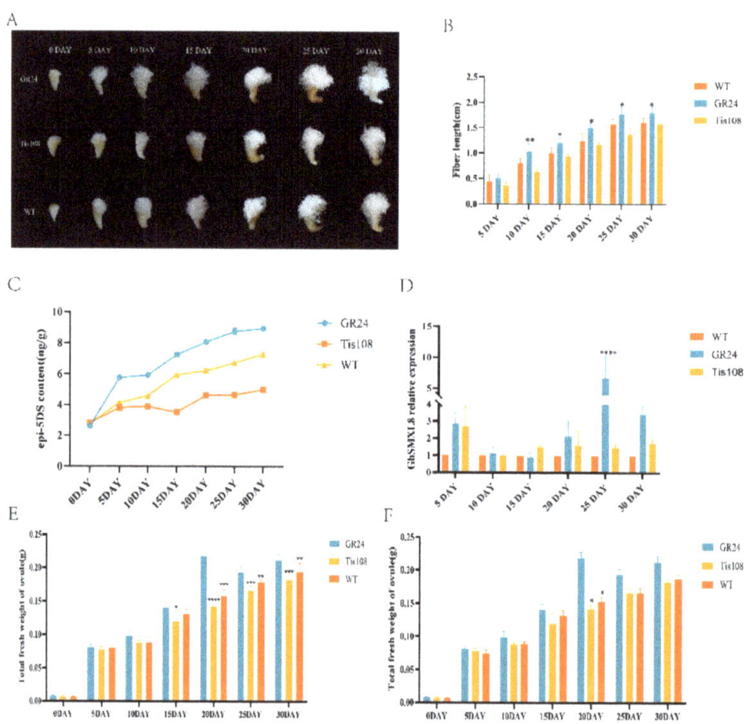

Figure 5. Positive regulation of polygalactone on the fiber development of *G. barbadense*. (**A**): Cotton fiber phenotype (collected at 1 DPA) cultured in vitro for 5, 10, 15, 20, 25, 30 days in a medium containing 15 μM SL synthetic analog GR24, 15 μM SL biosynthesis inhibitor Tis 108, and controls; (**B**): Average fiber length; (**C**): SL epi-5DS content in ovules at varied growth stages; (**D**): relative expression of GbSMXL8 in cotton fibers treated with 15 μM mGR24 or 15 μM Tis 108 for 5, 10, 15, 20, 25, and 30 days; (**E**): total fresh weight of ovule; (**F**): total ovule dry weight. * $p < 0.05$; ** $p < 0.01$; *** $p < 0.001$; **** $p < 0.0001$. Wild type (A130768).

To understand the role of *GbSMXL8* in SL signal transduction, RT-qPCR results showed that the relative expression level of *GbSMXL8* was much higher in GR24 treatment than that in Tis108 treatment at 25 and 30 days of ovule development (Figure 5D). It could be understood that upregulation of *GbSMXL8* gene expression would be triggered when the content of SLs accumulated to a certain extent in vivo. This upregulation could inhibit the tendency of the fibers to elongate. Therefore, we can infer that *GbSMXL8* can respond to the signal changes in SLs to mediate the process of fiber growth and development in cotton.

2.7. Impact of Editing GbSMXL8 on Fibers

Based on the known information, we can reasonably speculate that *GbSMXL8*, as a key inhibitor in the SLs signal transduction pathway, could regulate cotton fiber development. In order to further evaluate the specific role of *GbSMXL8* in cotton breeding, especially its potential to improve fiber quality, we planned to edit *GbSMXL8* in *G. hirsutum* varieties with relatively poor fiber quality and analyze its impact on cotton fiber quality. To validate the editing efficiency of CRISPR/Cas9-mediated *GbSMXL8* gene modification and its impact on cotton fiber quality, this study conducted target site mutation analysis on T_0-generation transgenic plants. Specific primers were designed to amplify the target site and flanking sequences of the *GbSMXL8* gene. PCR products were separated by 1.5% agarose gel electrophoresis and subsequently subjected to sequencing analysis (Figure 5E). Sequencing results revealed that 3 out of 7 positive transformants exhibited target site mutations (editing efficiency: 42.9%*), including single-base insertions (+1 bp) and deletions (−3 bp) (Figure 5B).

This study systematically compared the number and length of stem trichomes between *GbSMXL8* gene-edited transgenic plants and wild-type (WT) plants at the same developmental stage. As shown in Figure 6A, the trichome density of transgenic plants was significantly higher than that of WT ($p < 0.05$). To quantify phenotypic variation, we further measured trichome length for four genotypes (WT, smxl8-1, smxl8-2, and smxl8-3), with three biological replicates per genotype (n = 30 trichomes per sample). The results showed:WT: average length of 1.245 ± 0.102 mm (CV = 8.2%), indicating high phenotypic uniformity in WT plants. smxl8-1: average length of 1.744 ± 0.201 mm (CV = 11.5%), representing a 40.1% increase compared to WT ($p < 0.01$), but with a significantly higher coefficient of variation (CV) among samples. smxl8-2: average length of 1.639 ± 0.232 mm (CV = 14.1%), representing a 31.6% increase compared to WT ($p < 0.05$). smxl8-3: average length of 1.516 ± 0.131 mm (CV = 8.6%), representing a 21.8% increase compared to WT ($p < 0.05$) (Figure 7).

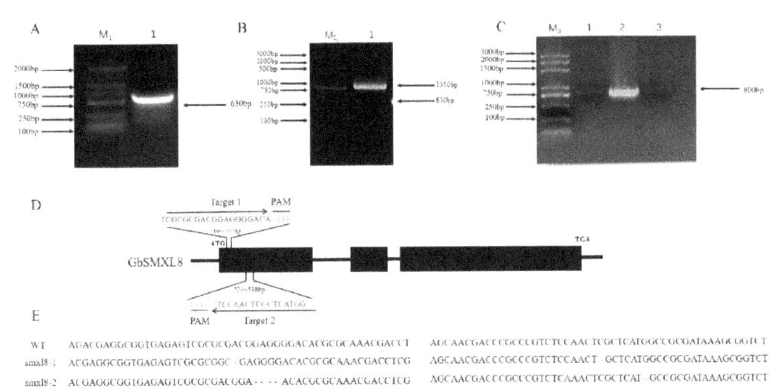

Figure 6. *Cont.*

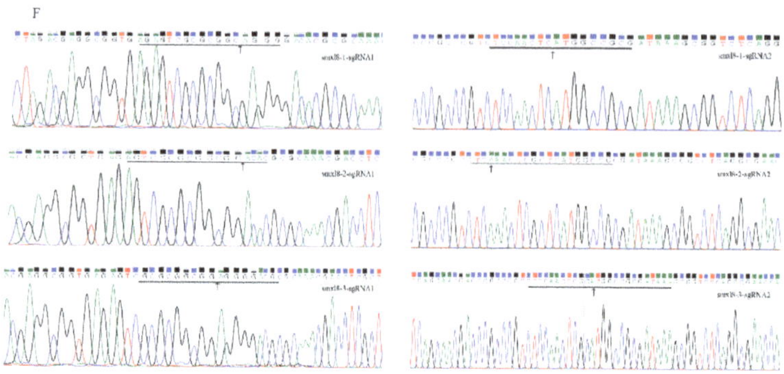

Figure 6. Gene editing SMXL8 target analysis and identification. (**A**): Cloning of double target sgRNA of *GbSMXL8* gene; (**B**): *GbSMXL8* gene; (**C**): PCR product of double target sgRNA Agrobacterium; (**D**): PCR product identification of transgenic plant; (**E**): Two sgRNA positions of *GbSMXL8* (**F**): Sequencing of SmXL8 edited plant.

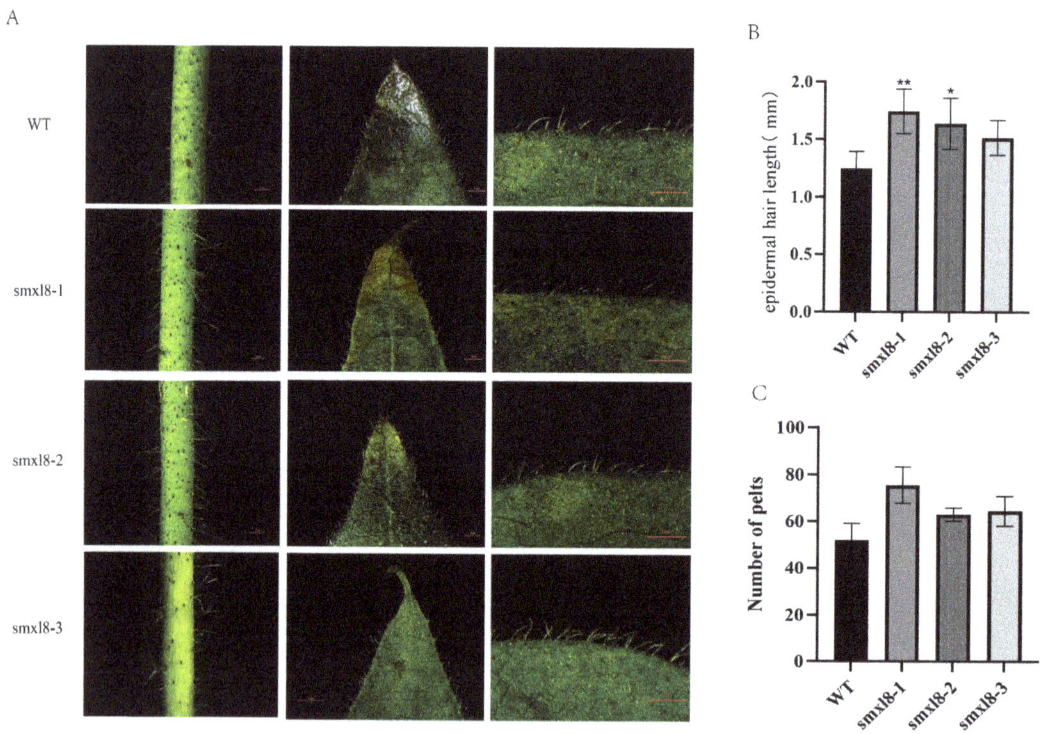

Figure 7. *GbSMXL8* inhibited stem epidermal hair growth (**A**): *GbSMXL8* edited cotton stem and leaf margin epidermal hair; (**B**): *GbSMXL8* edited cotton epidermal hair average length; (**C**): *GbSMXL8* edited cotton epidermal hair number. * $p < 0.05$; ** $p < 0.01$;. WT (ZM 49).

These results demonstrate that *GbSMXL8* gene editing not only significantly alters trichome length but also induces phenotypic heterogeneity, which may be related to CRISPR/Cas9 off-target effects or epigenetic regulation.

In addition to changes in length, the stem trichome density of smxl8 transgenic plants increased by 20.8% (smxl8-1), 7.01% (smxl8-2), and 5.26% (smxl8-3) compared to WT. Prediction of cis-acting elements revealed that they were mainly regulated by light.

Therefore, photosynthetic performance analysis of *GbSMXL8* transgenic plants showed that chlorophyll content and the Pn and Gs of transgenic plants were significantly increased compared with the WT ($p < 0.05$), indicating that knocking out *GbSMXL8* gene had positive effects on photosynthesis of plants. Meanwhile, the Tr of transgenic plants decreased significantly ($p < 0.05$), further highlighting the important role of *GbSMXL8* in regulating water use and growth of plants (Figure 8).

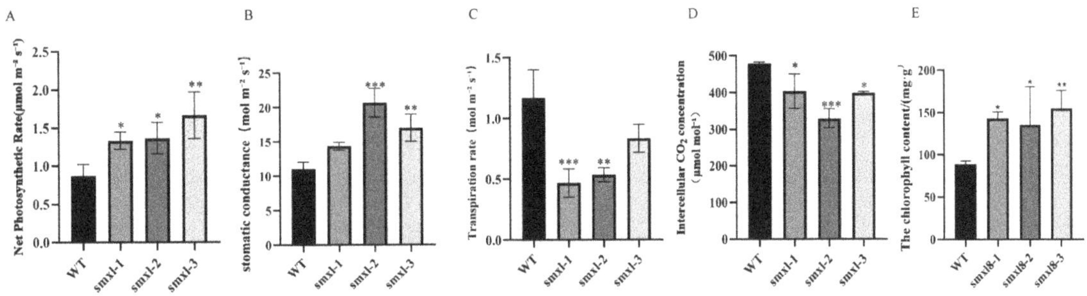

Figure 8. Phenotypic and physiological characterization of transgenic cotton plants. (**A**) Net photosynthetic rate (Pn) in leaves of wild-type (WT) and transgenic cotton lines. (**B**) Stomatal conductance (Gs) in leaves of WT and transgenic lines. (**C**) Transpiration rate (Tr) in leaves of WT and transgenic lines. (**D**) Intercellular CO_2 concentration (Ci) in leaves of WT and transgenic lines. (**E**) Chlorophyll content in leaves of WT and transgenic lines. Notes:WT: Wild-type control (cultivar ZM 49). Statistical significance: * $p < 0.05$; ** $p < 0.01$; *** $p < 0.001$ (Student's *t*-test). All data represent mean ± SD ($n = 3$ biological replicates).

There were three deletion mutations in the target sequence of *GbSMXL8* mutant cotton. The deletion fragment of target 1 and target 2 was 1–2 bp and 2 bp, respectively. The deletion of the target would lead to frameshift mutations in the amino acid sequence, resulting in phenotypic variants. The t0 plants numbered 1 had a deletion of 1 bp in the target region 1 fragment, a base mutation of 1 bp, and transgenic plants had a longer fiber phenotype (Figure 9A).

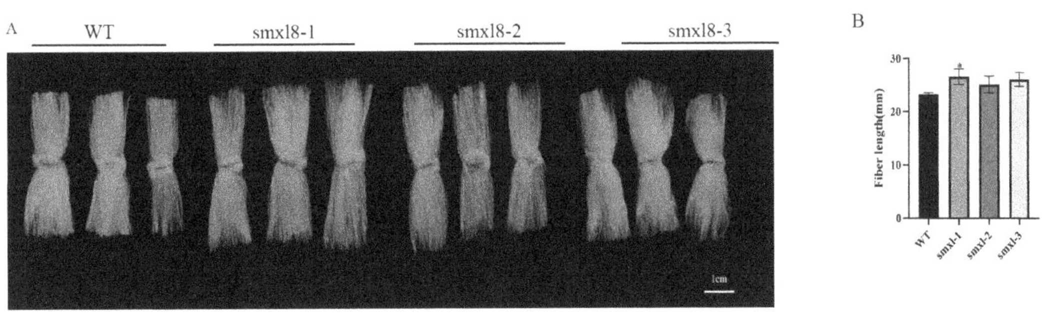

Figure 9. Effect of transgenic plants on fiber development. (**A**): WT, smxl8 transgenic plant fibers; (**B**): smxl8 transgenic cotton leaf mature fiber length. * $p < 0.05$; WT (ZM 49).

To further verify this, we randomly selected T1 cotton fibers with *GbSMXL8* edited, with three samples per group. The average fiber length of the control cotton fiber measured by carding was 23.32 mm, while that of the transgenic cotton increased significantly to 25.97 mm (Figure 9).

3. Discussion

G. hirsutum and *G. barbadense* are cultivated tetraploid species, domesticated from a common ancestor. Sea island cotton is known as the "king of cotton" with extensive use in high-end textiles because of its long fiber, high strength, and good luster. But the fiber quality of *G. hirsutum* is relatively low, although it has strong adaptability and high yield. Through transgenic technology, *G. hirsutum* can obtain the characteristics of the high-quality fiber of sea island cotton and a highly improved length, strength, and fineness of fiber [2]. There is a close interaction between unicornum lactone (SL) and various plant hormones, including auxins, cytokinins, abscisic acid, jasmonic acid, and GAs. Together with SL, these hormones can control plant growth by forming a complex and integrated regulatory network [7].

In this network, SLs play a key role in effectively increasing the size of the plant meristem and transition zone by inhibiting the transport of auxin so as to accurately regulate and control the elongation process of taproot. At the same time, the lactone of SLs can also inhibit lateral root primordia and its growth, depending on the cytokinin receptor *AHK3* [19]. In addition, the SLs not only has a regulatory effect by itself, but also promotes the synthesis of ethylene, and enhances the transport of auxin and the expression of its receptor *TIR1*, thus further regulating the elongation of root hair [20].

It is noteworthy that elongating cotton fiber is a polarized elongating cell sharing similar regulatory mechanism to root hair development in Arabidopsis [31]. Studies have shown that cotton fibers and stem epidermal hairs have high similarity in molecular regulation mechanism [32]. Several key genes (such as *GhMYB25* and *GhTTG1*) are involved in regulating the development of epidermal hair and cotton fiber at the same time. The functional deletion or abnormal expression of these genes will significantly affect the yield and quality of cotton fiber [33]. For example, deletion of the *GhMYB25* gene not only resulted in a decrease in stem epidermal hair density, but also significantly reduced fiber initiation and ultimately reduced lint percentage [34]. Based on the conservation of this developmental mechanism, the epidermal hair coat of stem and leaf is widely used as a model system for studying the genetic mechanism of cotton fiber development [35]. By comparing the transcriptome data of epidermal hair and fiber development, researchers found that both share core regulatory networks (such as gibberellin and single-leg auractone signaling pathways) in key biological processes such as cell elongation and secondary wall synthesis, which provides important clues for understanding the molecular mechanism of cotton fiber development [36]. In our experiment, the three single-cell models of cotton fiber, root hair, and epidermis exhibited partial regulation conservation in the development process, which could be explained by the regulatory role by similar genes [37]. As an important repressor of SL signaling, *SMXL8* degradation can induce the connection of key transcription factors between SL signaling perception and response. *SMXL8* may also interact with genes in other pathways, such as auxin, Gas, and other hormone signalings to regulate plant growth [36].

Following database-supported identification of 48 SMXL genes, we divided SMXL proteins into four species and into three groups based on similar functions shared by them. Proper functioning of SMXL proteins could be realized according to their specific domains or motifs. Members of the SMXL family have the Clp-N motif, a protein motif that is characteristic of the nucleoside triphosphate hydrolase superfamily and may be responsible for the signaling component of SL sensing. The promoter region of SMXL was analyzed for cis-acting elements. There were widespread cis-elements related to plant growth, especially photo-responsive elementspresent in every SMXL. Hormone-related elements and stress-related cis-elements (e.g., defense and stress-responsive, MYB-related and low-temperature-responsive) were detected in large amounts, providing clues for further exploring SMXL functions [26]. Hence, this experiment used *GbSMXL8* of subgroup III (with the closest evolutionary relationship), which was selected due to its closest evolutionary relationship to other known SMXL genes. The expression of *GbSMXL8* was analyzed in various tissues of *Gossypium barbadense* including fiber, root, stem, and leaf, to uncover its tissue-specific expression patterns. The results revealed that *GbSMXL8* was highly expressed during specific developmental stages of fiber growth, particularly at 20, 25, and 40 days post-anthesis (DPA). This suggests that *GbSMXL8* may play a significant role in fiber development, potentially regulating processes such as cell elongation, secondary cell wall biosynthesis, or other mechanisms critical for fiber maturation.

The fiber length of ovules significantly increased with the rise in strigolactone (SL) content, indicating that SLs play a promoting role in cotton fiber development. To further verify whether *GbSMXL8* is involved in the SL signaling pathway, we examined the expression level of *GbSMXL8* after treatment with the exogenous SL analog GR24. The results showed that the expression of *GbSMXL8* was significantly upregulated following GR24 treatment, confirming our hypothesis that *GbSMXL8* may act as a downstream responsive factor in the SL signaling pathway, participating in the regulation of fiber development. This finding not only supports the important role of SLs in fiber development but also enhances the accuracy of subsequent gene function screening. We constructed the *GbSMXL8* cotton editing vector and found that *GbSMXL8* modulated cotton growth. The phenotype observation of *GbSMXL8* edited plant with an increased number of epidermal hairs indicated that it could interfere with the development process of the cotton epidermal hair. The cis-acting element prediction determined the photosynthetic indexes of the *GbSMXL8* transgenic plants. Transgenic plants had higher chlorophyll content and photosynthetic rate, wherein the Pn represented the accumulation speed of the plants for organic matters.

A higher photosynthetic rate (Pn) indicates that plants can generate more organic matter by utilizing light energy, which promotes favorable growth, development, and stress resistance. Editing the *GbSMXL8* gene has been shown to improve cotton fiber quality. In conclusion, genetic regulation of the *GbSMXL8* gene can enhance cotton fiber quality without adversely affecting normal nutrient metabolism or reproductive growth. This experiment provides theoretical support for subsequent breeding efforts aimed at improving fiber quality in Gossypium hirsutum by utilizing screened superior genes.

The study primarily focused on the functional characterization of *GbSMXL8* in fiber development, but its broader roles in other physiological processes, such as stress responses or nutrient allocation, remain unexplored. Further investigations are needed to determine whether *GbSMXL8* has pleiotropic effects beyond fiber development.

The experiments were conducted under controlled conditions, which may not fully replicate field environments. Field trials are necessary to validate the practical applicability of *GbSMXL8* editing in improving fiber quality under natural growing conditions. The molecular mechanisms by which *GbSMXL8* regulates fiber development, particularly its interactions with other signaling pathways or downstream targets, are not yet fully understood. Additional mechanistic studies, such as transcriptomic or proteomic analyses, would provide deeper insights.

The findings suggest that *GbSMXL8* is a promising target for genetic improvement of cotton fiber quality. However, future studies should explore the potential of combining *GbSMXL8* editing with other fiber-related genes to achieve synergistic effects. Investigating the role of *GbSMXL8* in different cotton species, such as *Gossypium barbadense*, could reveal species-specific regulatory mechanisms and further enhance fiber quality breeding strategies. Long-term studies are needed to assess the stability of fiber quality improvements and any potential unintended effects of *GbSMXL8* editing on cotton agronomic traits. By addressing these limitations and perspectives, future research can build on the current findings to develop more robust and effective strategies for improving cotton fiber quality through genetic regulation.

4. Materials and Methods

In this experiment, the materials of the sea island cotton "A130768" of *G. barbadense* were preserved by the Key Laboratory of Agriculture and Biotechnology of Xinjiang Agricultural University, planted in cotton breeding base of Agricultural College of Xinjiang Agricultural University. Materials sampled in this study were petals, stamens, and pistils taken from the day of flowering, as well as young leaves of plants that had grown for about 2 months. All these samples were stored at −70 °C after freezing in liquid nitrogen for later use. *G. hirsutum* "ZM 49" and gene-edited cotton were planted in artificial climate chamber at (26 ± 2) °C with 16 h of illumination.

4.1. Structural Analysis of SMXL Gene Family in Cotton

The Cotton FGD database (https://cottonfgd.org/)(accessed on 21 December 2024) was visited to, respectively, download genome-wide translated sequences of *G. arboreum* (A2, CRIv1.0), *G. raymond* (D5, JGI_v2.0), *G. hirsutum* (AD1, ZJUv2.1), and *G. barbadense* (AD2, H7124_ZJUv1.1). Using SMXL protein sequences of Arabidopsis thaliana as a reference, SMXL protein was then searched by local blase tool (E value < 1.0×10^{-5}). Domain (PF02861) was identified using Hidden Markov Models (HMM) and local HMM searches were performed in the four databases using HMMER 3.0. Cross-screening candidate family genes were screened by these two methods to reduce false positives. Finally, domain analysis on all candidate sequences was completed in InterProScan and SMART after discarding sequences without this domain.

The amino acid residue number, relative molecular weight, hydrophilicity, and theoretical isoelectric point of cotton SMXL protein were analyzed by online ExPASy software (https://web.expasy.org/protparam/) (accessed on 23 December 2024).

4.2. Phylogenetic Analysis of Cotton SMXL Family

For this analysis, SMXL protein sequences of six species (Arabidopsis thaliana included) were aligned by ClustalW 2.1 software firstly, followed by the construction of rootless phylogenetic tree by MEGA MEGA 7.0.26. Neighborhood joining method was used to set the number of cycles to 1000.

4.3. Gene Structure and Conserved Motif Analysis

SMXL gene structure information was obtained from genome annotation files of *G. hirsutum*, *G. barbadense*, *G. arboreum*, and *G. raymond*, which was then entered into the online Gene Structure Display Server [38] to visually analyze exon/intron structure of SMXL gene. Protein Sequence Retrieval: The protein sequences of SMXL family members from the four cotton species were extracted from the genome annotation files. The protein sequences were submitted to the Pfam database, a widely used resource for protein family analysis. Pfam uses Hidden Markov Models (HMMs) to identify and annotate conserved protein domains. The SMXL protein sequences were analyzed to identify domains such as the Clp-N domain (characteristic of SMXL proteins) and other potential functional motifs. To further visualize the conserved domains, the protein sequences were analyzed using Tbtools (accessed on 28 December 2024) [39], a bioinformatics software that integrates multiple tools for genomic and transcriptomic data analysis. The domain architecture of each SMXL protein was graphically represented, showing the location and arrangement of conserved domains within the protein sequences. This visualization helped clarify the structural and functional conservation of SMXL proteins across the four cotton species.

4.4. Cotton SMXL Cis-Acting Element Prediction

The obtained promoter sequences were submitted to the PlantCARE database (http://bioinformations.psb.ugent.be/webtools/plantcare/html/) (accessed on 29 December 2024) for cis-acting element prediction using default parameters (core promoter element threshold Score > 8.0). The predicted results were screened manually to eliminate duplication and low confidence (Score < 5.0) elements, while functional elements related to hormone response, environmental stress, light regulation, and developmental regulation, were retained. The prediction results are visually analyzed by Tbtools software v0.09, and the types, quantities, and distribution positions of each component are counted.

4.5. Expression Characterization of the SMXL Family in Cotton

Transcriptome data of fiber development stage of 'pimas-7″5917' were retrieved from NCBI website and normalized by Log2 (FPKM + 1). Then, this study constructed the heatmaps using the pheatmap program in R(3.3.0), with the visualization of gene expression patterns simultaneously.

4.6. Ovule Culture and Observation

The basal medium was prepared using a specialized cotton ovule in vitro culture medium (PM1671, Coolaber, Beijing, China), consisting of MS basal salts (with vitamin B_5), 30 g/L glucose, and 0.5 g/L MES buffer. The pH of the medium was adjusted to 5.6 ± 0.1 using 0.1 mol/L NaOH, followed by autoclaving at 121 °C for 20 min (SuoTengYiLiao, STL-FE, Hangzhou, China). After sterilization, the following components were aseptically added under a laminar flow hood according to experimental design: Strigolactone analog: GR24 (Cat. 41012ES03, SESEN, Shanghai, China) at a final concentration of 15 µM. Strigolactone biosynthesis inhibitor: Tis108 (Cat. S25193, Yuanye, Shanghai, China) at a final concentration of 5 Mm [40].

Hormone concentrations were optimized based on pre-experimental trials and referenced to the cotton fiber in vitro culture system established by. The hormone-supplemented medium was aliquoted into sterile glass culture bottles (50 mL per bottle) and stored at 4 °C in the dark until use. Prior to inoculation, the medium was equilibrated to room temperature.

Ovules were collected from sea island cotton (*Gossypium barbadense* cv. 'A130768') at 2 days post anthesis (2 DPA). Surface sterilization was performed by immersing ovules in 10% (*w*/*v*) $HgCl_2$ solution for 15 min, followed by five rinses with sterile distilled water to remove residual disinfectant. Sterilized ovules were transferred to sterile Petri dishes containing filter paper using pre-chilled sterile forceps.

Sterilized ovules were inoculated into hormone-supplemented medium (20 ovules per bottle) and cultured in a growth chamber under dark conditions at 31 °C and 70% relative humidity for 35 days without agitation to prevent fiber damage. Fiber morphology was monitored at 5, 10, 15, 20, 25, and 30 days of culture using a Leica DM500 optical microscope (10× objective). Fiber length was quantified using ImageJ software (accessed on 2 January 2024) (n = 50 fibers per replicate). All experiments included three biological replicates.

4.7. Expression Pattern Analysis

A was extracted from cotton tissues (roots, stems, leaves, and fibers at key developmental stages: 5 DPA, 10 DPA, 20 DPA, 25 DPA, and 35 DPA) using the RNA Prep Pure Plant Kit (Cat. No. DP441; Tiangen, Beijing, China). RNA quality was assessed by 1.5% agarose gel electrophoresis, which revealed clear 28S and 18S rRNA bands, indicating high RNA integrity. Subsequently, 1 µg of total RNA was reverse-transcribed into cDNA using the FastKing-RT SuperMix Kit (Cat. No. KR118; Tiangen, Beijing, China) under the following conditions: 42 °C for 15 min and 95 °C for 3 min. The quality of cDNA was verified by PCR amplification of the reference gene *GhUBQ7* (primer sequences: F: 5′-GCTGCTGCTACTGCTACTGC-3′, R: 5′-CAGCTAGCTAGCTAGCTAGC-3′), and the amplification products were confirmed by 1.5% agarose gel electrophoresis.

Specific primers for GbSMXL8 (F: 5′-ATGGCTAGCTACGTAGCTAC-3′, R: 5′-TCAGCTAGCTAGCTAGCTAG-3′) were designed using Primer-BLAST (https://www.ncbi.nlm.nih.gov/tools/primer-blast/) (accessed on 12 December 2024), generating a 150 bp amplicon that spans an intron to avoid genomic DNA contamination. Quantitative real-time PCR (qRT-PCR) was performed in a 20 μL reaction system using SYBR Green I dye, with GhUBQ7 as the internal reference gene. Each experiment included three biological replicates (independent RNA extraction and reverse transcription) and three technical replicates (parallel reactions of the same cDNA template) per sample. The relative expression levels of the target gene were calculated using the ΔCt method, with the formula: ΔCt = Ct(target gene) − Ct(reference gene), and relative expression = 2^(−ΔCt). Data analysis was conducted using one-way analysis of variance (ANOVA) followed by Tukey's HSD multiple comparison test, with a significance level set at $p < 0.05$.

4.8. CRISPR/Cas9 System-Mediated Editing of Cotton

For the *GbSMXL8* gene, according to the sequence of the cotton *GbSMXL8* gene, the online CRISPR2.0 software was used to design editing targets in the first exon region (Figure 1), and the sequences of the two targets were aligned in the cotton genome database to eliminate non-specific target interference.

The plasmid of WMC016 vector was extracted and digested with Bsa I. The sgRNA fragment with target was amplified from the plasmid containing sgRNA vector and ligated with the vector WMC016 after single enzyme digestion. The plasmid was extracted and transformed into Agrobacterium EHA105. To ensure the reliability of transgenic cotton lines, a single bacterial colony harboring the target construct was selected for PCR amplification. Positive clones confirmed by sequencing were subsequently used for Agrobacterium-mediated cotton transformation. The genetic transformation was performed following the protocol established by Ge et al. [41], using the widely cultivated cotton cultivar 'ZM49' as the recipient. As a preferred transgenic receptor in China, 'ZM49' offers a mature genetic transformation system with an efficiency ≥10% and stable agronomic traits, including consistent fiber quality and yield performance across generations. This high-efficiency system ensures reproducible T_0 plant generation and minimizes somaclonal variation, making it ideal for functional studies of fiber development-related genes. We used a selectable marker gene (hygromycin resistance) for initial screening. Cotton plants were cultured on medium containing hygromycin (50 mg/L) to select transformants exhibiting antibiotic resistance. The editing efficiency of the target gene (*GbSMXL8*) was confirmed by PCR amplification of the target region, followed by Sanger sequencing. This allowed us to identify specific mutations, including insertions, deletions, or substitutions, in the transgenic lines. Edited lines were compared with non-transgenic controls to assess differences in trichome density, fiber length, and other agronomic traits. After obtaining transgenic plants, seedling DNA was extracted for PCR reaction, with the design of primers according to target sites. Mutations at target sites of positive transgenic plants were detected by sequencing.

4.9. Statistics of Fiber and Epidermal Hair Length of Transgenic Cotton

The fiber length of transgenic cotton was measured by combing method. The changes in the number and density of the pelts on the stem and petiole were observed microscopically. The number and length of the pelts on the stem were counted by ImageJ software V 1.54k.

4.10. Determination of Physiological and Biochemical Indexes

4.10.1. Measurement of Photosynthetic Rate

To assess the photosynthetic performance of the plants, we employed a portable photosynthesis system (e.g., LI-6800 or similar models), a highly accurate and widely used instrument for measuring photosynthetic parameters in real-time under field or controlled conditions. For each measurement, we selected three leaves of similar growth status to ensure consistency and reliability of the data. These leaves were sampled from closely positioned nodes on the plant to minimize variability due to environmental or developmental differences.

Each photosynthetic indicator was measured three times to ensure data accuracy and reproducibility, and the average value was calculated for further analysis. The key photosynthetic parameters measured included: net photosynthetic rate (Pn), transpiration rate (Tr), stomatal conductance (Gs), and intercellular carbon dioxide concentration (Ci).

4.10.2. Chlorophyll Content Determination

An amount of 0.1 g of the leaf sample was meticulously ground into a fine powder using liquid nitrogen to ensure homogeneous disruption of cellular structures. Subsequently, chlorophyll extraction was performed by adding precisely 1.5 milliliters of 95% ethanol (Sinopharm, 100092683, Beijing, China). The extract was then carefully transferred into a sterile 2-milliliter Eppendorf tube and centrifuged at 12,000 revolutions per minute (rpm) for a duration of 15 min at a controlled temperature of 4 °C to separate the supernatant from cellular debris.

Following centrifugation, the supernatant was meticulously collected, serving as the basis for subsequent chlorophyll content determination. The optical density (OD) values of the supernatant were measured at wavelengths of 665 nanometers (nm) and 649 nm using a spectrophotometer manufactured by UNICO (V1600; Shanghai, China). These specific wavelengths are known to correlate with the absorption peaks of chlorophyll a and b, respectively.

The concentrations of chlorophyll a (Ca) and chlorophyll b (Cb) were then calculated employing the well-established formulas: $Ca = 13.95 \times OD665 - 6.88 \times OD649$ and $Cb = 24.96 \times OD649 - 7.32 \times OD665$. These calculations were based on the OD values obtained from the spectrophotometric analysis, allowing for the quantitative assessment of chlorophyll content within the leaf samples with a high degree of precision and academic rigor.

4.11. Determination of Auricolactone Content

The sample was 0.5 g of the middle part of each leaf's plant line with uniform growth, and the frozen leaves in liquid nitrogen were tested for auricolactone contentby ELISA [42].

Author Contributions: L.C.: Writing—original draft, methodology, visualization; W.X.: conceptualization; L.Z.: data curation, formal analysis; Y.C.: validation; Q.C. (Qin Chen): supervision; Q.C. (Quanjia Chen): funding acquisition, K.Z.: writing—reviewing and editing. All authors have read and agreed to the published version of the manuscript.

Funding: Key R & D Project of Xinjiang Uygur Autonomous Region (2024B02002-1). Shenzhen Science and Technology Program (JSGG20220831104600002). "Tianshan Talent" Training Program (2023TSYCLJ0012). Cotton Industry Technical System Support of Xinjiang Uygur Autonomous Region (XJARS-03-07). Xinjiang Long-Staple Cotton Breeding Joint Research Project (2022MH-01).

Data Availability Statement: Data are contained within the article.

Conflicts of Interest: The authors declare no conflicts of interest.

References

1. Song, X.; Zhu, G.; Hou, S.; Ren, Y.; Amjid, M.W.; Li, W.; Guo, W. Genome-Wide Association Analysis Reveals Loci and Candidate Genes Involved in Fiber Quality Traits Under Multiple Field Environments in Cotton (*Gossypium hirsutum*). *Front. Plant Sci.* **2021**, *12*, 695503. [CrossRef] [PubMed]
2. Dimitrova, V.; Koleva, M.; Valkova, N. Heterosis and combining ability in intraspecific (*G. hirsutum* L.) and interspecific (*G. hirsutum* L. × *G. barbadense* L.) cotton hybrids. *Bulg. J. Crop Sci.* **2023**, *60*, 37–47. [CrossRef]
3. Malinga, L.N. The importance and economic status of cotton production. *Acad. Lett.* **2021**. [CrossRef]
4. Cao, J. C4H1 Gene Cloning and Expression Analysis between Sea-Island Cotton (*Gossypium barbadense* L.) Fiber and Upland Cotton (*Gossypium hirsutum* L.). *Fiber Xinjiang Agric. Sci.* **2015**, *52*, 1773–1781.
5. Chen, G.; Liu, Z.; Li, S.; Liu, L.; Lu, L.; Wang, Z.; Mendu, V.; Li, F.; Yang, Z. Characterization of chromatin accessibility and gene expression reveal the key genes involved in cotton fiber elongation. *Physiol. Plant.* **2023**, *175*, e13972. [CrossRef]
6. Xu, E.; Chai, L.; Zhang, S.; Yu, R.; Zhang, X.; Xu, C.; Hu, Y. Catabolism of strigolactones by a carboxylesterase. *Nat. Plants* **2021**, *7*, 1495–1504. [CrossRef] [PubMed]
7. Jiu, S.T.; Xu, Y.; Zhang, C.X.; Wang, L.; Ma, C.; Xu, W.P.; Wang, S.P. Research advancements on strigolactones and its regulatory effect on root growth and development in plants. *Mol. Plant Breed.* **2020**, *18*.
8. Zheng, J.; Hong, K.; Zeng, L.; Wang, L.; Kang, S.; Qu, M.; Dai, J.; Zou, L.; Zhu, L.; Tang, Z.; et al. Karrikin Signaling Acts Parallel to and Additively with Strigolactone Signaling to Regulate Rice Mesocotyl Elongation in Darkness. *Plant Cell* **2020**, *32*, 2780–2805. [CrossRef]
9. Xu, X.; Jibran, R.; Wang, Y.; Dong, L.; Flokova, K.; Esfandiari, A.; McLachlan, A.R.G.; Heiser, A.; Sutherland-Smith, A.J.; Brummell, D.A.; et al. Strigolactones regulate sepal senescence in Arabidopsis. *J. Exp. Bot.* **2021**, *72*, 5462–5477. [CrossRef]
10. Wu, F.; Gao, Y.; Yang, W.; Sui, N.; Zhu, J. Biological Functions of Strigolactones and Their Crosstalk with Other Phytohormones. *Front. Plant Sci.* **2022**, *13*, 821563. [CrossRef]
11. Zwanenburg, B.; Blanco-Ania, D. Strigolactones: New plant hormones in the spotlight. *J. Exp. Bot.* **2018**, *69*, 2205–2218. [CrossRef]
12. Qu, B.; Qin, Y.; Bai, Y. From signaling to function: How strigolactones regulate plant development. *Sci. China Life Sci.* **2020**, *63*, 1768–1770. [CrossRef] [PubMed]
13. Agusti, J.; Herold, S.; Schwarz, M.; Sanchez, P.; Ljung, K.; Dun, E.A.; Brewer, P.B.; Beveridge, C.A.; Sieberer, T.; Sehr, E.M.; et al. Strigolactone signaling is required for auxin-dependent stimulation of secondary growth in plants. *Proc. Natl. Acad. Sci. USA* **2011**, *108*, 20242–20247. [CrossRef] [PubMed]
14. Duan, J.; Yu, H.; Yuan, K.; Liao, Z.; Meng, X.; Jing, Y.; Liu, G.; Chu, J.; Li, J. Strigolactone promotes cytokinin degradation through transcriptional activation of *CYTOKININ OXIDASE/DEHYDROGENASE 9* in rice. *Proc. Natl. Acad. Sci. USA* **2019**, *116*, 14319–14324. [CrossRef]
15. Jiang, L.; Liu, X.; Xiong, G.; Liu, H.; Chen, F.; Wang, L.; Meng, X.; Liu, G.; Yu, H.; Yuan, Y.; et al. DWARF 53 acts as a repressor of strigolactone signalling in rice. *Nature* **2013**, *504*, 401–405. [CrossRef]
16. Kee, Y.J.; Ogawa, S.; Ichihashi, Y.; Shirasu, K.; Yoshida, S. Strigolactones in Rhizosphere Communication: Multiple Molecules with Diverse Functions. *Plant Cell Physiol.* **2023**, *64*, 955–966. [CrossRef] [PubMed]
17. Cao, Y.-Y.; Chen, Y.-C.; Guo, S.-H.; Gan, X.-Y.; Tian, L.; Huang, L.-Q.; Yuan, Y. Research progress in strigolactones and application prospect in medicinal plants. *Zhongguo Zhongyao Zazhi* **2023**, *48*, 3132–3139.
18. Hu, J.; Ji, Y.; Hu, X.; Sun, S.; Wang, X. BES1 functions as the co-regulator of D53-like SMXLs to inhibit BRC1 expression in strigolactone-regulated shoot branching in *Arabidopsis*. *Plant Commun.* **2020**, *1*, 100014. [CrossRef]
19. Zhou, F.; Lin, Q.; Zhu, L.; Ren, Y.; Zhou, K.; Shabek, N.; Wu, F.; Mao, H.; Dong, W.; Gan, L.; et al. D14-SCF(D3)-dependent degradation of D53 regulates strigolactone signalling. *Nature* **2013**, *504*, 406–410. [CrossRef]
20. Song, X.; Lu, Z.; Yu, H.; Shao, G.; Xiong, J.; Meng, X.; Jing, Y.; Liu, G.; Xiong, G.; Duan, J.; et al. IPA1 functions as a downstream transcription factor repressed by D53 in strigolactone signaling in rice. *Cell Res.* **2017**, *27*, 1128–1141. [CrossRef]
21. Ma, H.; Duan, J.; Ke, J.; He, Y.; Gu, X.; Xu, T.H.; Yu, H.; Wang, Y.; Brunzelle, J.S.; Jiang, Y.; et al. A D53 repression motif induces oligomerization of TOPLESS corepressors and promotes assembly of a corepressor-nucleosome complex. *Sci. Adv.* **2017**, *3*, e1601217. [CrossRef] [PubMed]
22. Zhang, H.; Wang, L.; Gao, Y.; Guo, Y.; Zheng, N.; Xu, X.; Xu, M.; Wang, W.; Liu, C.; Liu, W.; et al. Genome-Wide Identification of SMXL Gene Family in Soybean and Expression Analysis of GmSMXLs under Shade Stress. *Plants* **2022**, *11*, 2410. [CrossRef] [PubMed]
23. Li, R.; An, J.P.; You, C.X.; Wang, X.F.; Hao, Y.J. Genome-wide analysis and identification of the *SMXL* gene family in apple (*Malus* × *domestica*). *Tree Genet. Genomes* **2018**, *14*, 61.
24. Pan, Y.; Zhan, J.; Jiang, Y.; Xia, D.; Scheuring, S. A concerted ATPase cycle of the protein transporter AAA-ATPase Bcs1. *Nat. Commun.* **2023**, *14*, 6369. [CrossRef]

25. Matias, P.M.; Baek, S.H.; Bandeiras, T.M.; Dutta, A.; Houry, W.A.; Llorca, O.; Rosenbaum, J. The AAA+ proteins Pontin and Reptin enter adult age: From understanding their basic biology to the identification of selective inhibitors. *Front. Mol. Biosci.* **2015**, *2*, 17. [CrossRef]
26. Marand, A.P.; Eveland, A.L.; Kaufmann, K.; Springer, N.M. cis-Regulatory Elements in Plant Development, Adaptation, and Evolution. *Annu. Rev. Plant Biol.* **2023**, *74*, 111–137. [CrossRef]
27. Sakagami, J.I.; Iwata, Y.; Nurrahma, A.H.I.; Siaga, E.; Junaedi, A.; Yabuta, S. Plant adaptations to anaerobic stress caused by flooding. *IOP Conf. Ser. Earth Environ. Sci.* **2020**, *418*, 012080. [CrossRef]
28. Zhang, Z.; Liu, Z.; Wu, H.; Xu, Z.; Zhang, H.; Qian, W.; Gao, W.; She, H. Genome-Wide Identification and Characterization of MYB Gene Family and Analysis of Its Sex-Biased Expression Pattern in *Spinacia oleracea* L. *Int. J. Mol. Sci.* **2024**, *25*, 795. [CrossRef]
29. Krasylenko, Y.; Komis, G.; Hlynska, S.; Vavrdová, T.; Ovečka, M.; Pospíšil, T.; Šamaj, J. GR24, A Synthetic Strigolactone Analog, and Light Affect the Organization of Cortical Microtubules in Arabidopsis Hypocotyl Cells. *Front. Plant Sci.* **2021**, *12*, 675981. [CrossRef]
30. Ito, S.; Umehara, M.; Hanada, A.; Yamaguchi, S.; Asami, T. Effects of strigolactone-biosynthesis inhibitor TIS108 on *Arabidopsis*. *Plant Signal. Behav.* **2013**, *8*, e24193. [CrossRef]
31. Lee, J.A. Revision of the genetics of the hairiness-smoothness system of *Gossypium*. *J. Hered.* **1985**, *76*, 123–126. [CrossRef]
32. Serna, L.; Martin, C. Trichomes: Different regulatory networks lead to convergent structures. *Trends Plant Sci.* **2006**, *11*, 274–280. [CrossRef]
33. Hu, H.; He, X.; Tu, L.; Zhu, L.; Zhu, S.; Ge, Z.; Zhang, X. GhJAZ2 negatively regulates cotton fiber initiation by interacting with the R2R3-MYB transcription factor GhMYB25-like. *Plant J.* **2016**, *88*, 921–935. [CrossRef]
34. Xu, B.; Zhang, J.; Shi, Y.; Dai, F.; Jiang, T.; Xuan, L.; He, Y.; Zhang, Z.; Deng, J.; Zhang, T.; et al. *GoSTR*, a negative modulator of stem trichome formation in cotton. *Plant J.* **2023**, *116*, 389–403. [CrossRef]
35. Xiao, G.; Zhao, P.; Zhang, Y. A Pivotal Role of Hormones in Regulating Cotton Fiber Development. *Front. Plant Sci.* **2019**, *10*, 87. [CrossRef] [PubMed]
36. Chang, L.; Ramireddy, E.; Schmülling, T. Lateral root formation and growth of *Arabidopsis* is redundantly regulated by cytokinin metabolism and signalling genes. *J. Exp. Bot.* **2013**, *64*, 5021–5032. [CrossRef]
37. Mayzlish-Gati, E.; De-Cuyper, C.; Goormachtig, S.; Beeckman, T.; Vuylsteke, M.; Brewer, P.B.; Beveridge, C.A.; Yermiyahu, U.; Kaplan, Y.; Enzer, Y.; et al. Strigolactones Are Involved in Root Response to Low Phosphate Conditions in Arabidopsis. *Plant Physiol.* **2012**, *160*, 1329–1341. [CrossRef] [PubMed]
38. Hu, B.; Jin, J.; Guo, A.-Y.; Zhang, H.; Luo, J.; Gao, G. GSDS 2.0: An upgraded gene feature visualization server. *Bioinformatics* **2015**, *31*, 1296–1297. [CrossRef]
39. Chen, C.J.; Chen, H.; Zhang, Y.; Thomas, H.R.; Frank, M.H.; He, Y.H.; Xia, R. TBtools: An integrative toolkit developed for interactive analyses of big biological data. *Mol. Plant* **2020**, *13*, 1194–1202. [CrossRef]
40. T ian, Z.; Zhang, Y.; Zhu, L.; Jiang, B.; Wang, H.; Gao, R.; Friml, J.; Xiao, G. Strigolactones act downstream of gibberellins to regulate fiber cell elongation and cell wall thickness in cotton (*Gossypium hirsutum*). *Plant Cell* **2022**, *34*, 4816–4839.
41. Ge, X.; Xu, J.; Yang, Z.; Yang, X.; Wang, Y.; Chen, Y.; Wang, P.; Li, F. Efficient genotype-independent cotton genetic transformation and genome editing. *J. Integr. Plant Biol.* **2023**, *65*, 907–917. [CrossRef] [PubMed]
42. Konstantinou, G.N. Enzyme-Linked Immunosorbent Assay (ELISA). In *Methods in Molecular Biology*; Springer: Berlin/Heidelberg, Germany, 2022; pp. 115–134.

International Journal of
Molecular Sciences

MDPI

Article

Phylogeny and Functional Differentiation of the Terpene Synthase Gene Family in Angiosperms with Emphasis on *Rosa chinensis*

Qi Li [1,†], Yifang Peng [1,†], Tao Zhao [2], Qijing Dong [1], Qian Yang [1], Xiaoyu Liu [1] and Yu Han [1,*]

[1] Beijing Key Laboratory of Ornamental Plants Germplasm Innovation & Molecular Breeding, Beijing Laboratory of Urban and Rural Ecological Environment, Key Laboratory of Genetics and Breeding in Forest Trees and Ornamental Plants, Ministry of Education, State Key Laboratory of Efficient Production of Forest Resources, National Engineering Research Center for Floriculture, School of Landscape Architecture, Beijing Forestry University, Beijing 100083, China

[2] State Key Laboratory of Crop Stress Biology for Arid Areas, Shaanxi Key Laboratory of Apple, College of Horticulture, Northwest A&F University, Yangling 712100, China

* Correspondence: hanyubjfu@bjfu.edu.cn

† These authors contributed equally to this work.

Abstract: Terpenes are pivotal for plant growth, development, and adaptation to environmental stresses. With the advent of extensive genomic data and sophisticated bioinformatics tools, new insights into the evolutionary dynamics and functional diversification of terpene synthases (TPSs) have emerged. Despite genome-wide identifications of the TPS family in certain species, comprehensive cross-species analyses remain scarce. In this study, we conducted a genome-wide identification and subgroup classification of TPS families across 115 angiosperms with available genomic sequences. Our phylogenomic synteny network analysis elucidated the complex evolutionary history of TPS genes, revealing notable expansions and contractions among subgroups. Specifically, TPS-a showed significant expansion, while TPS-b was variably lost in some Poaceae, indicating adaptive responses. TPS-c maintained considerable conservation across species, whereas TPS-e/f diverged into distinct evolutionary trajectories despite functional overlap, with TPS-e further splitting into two angiosperm-specific clades. The TPS-g subgroup displayed lineage-restricted distribution, primarily in super-rosids and monocots. Notably, TPS-d and TPS-h subgroups were absent in angiosperms. Employing *Rosa chinensis* as a case study, we identified RcTPS23, a conserved bifunctional terpene synthase, highlighting the utility of cross-species synteny data in functional prediction. This comprehensive analysis elucidates the phylogenetic and functional landscape of TPS subgroups in angiosperms, providing a robust framework for predicting TPS function and guiding further functional investigations.

Keywords: genome wide identification; synteny analysis and networks; TPS; super rosids; bifunctional gene

1. Introduction

Terpenoids constitute a vast group of natural compounds, with over 50,000 distinct types identified across a broad spectrum of organisms including plants, bacteria, and invertebrates [1–3]. These compounds are primarily composed of concatenated C_5 isoprenoid units, with structural variations such as monoterpenes (C_{10}), sesquiterpenes (C_{15}), and diterpenes (C_{20}) defined by the number of these isoprenoid blocks [3,4]. The biosynthesis of terpenoids is facilitated through two principal pathways: the mevalonate (MVA) pathway and the methyl-erythritol phosphate (MEP) pathway, both of which produce the universal precursors, isopentenyl diphosphate (IPP) and dimethylallyl diphosphate (DMAPP) [5]. In plants, a large family of structurally related enzymes known as terpene synthases/cyclases (TPSs) orchestrates terpenoid biosynthesis. Typically, TPSs in plastids catalyze the cyclization of substrates such as geranyl/neryl diphosphate (GPP/NPP, C_{10}; neryl being the cis-isomer of geranyl) to yield monoterpenes, or convert geranylgeranyl/nerylneryl diphosphate (GGPP/NNPP, C_{20}) into diterpenes. Conversely, cytosolic TPSs primarily catalyze the transformation of trans/cis-farnesyl diphosphate (*E, E*-FPP/*Z, Z*-FPP, C_{15}) into sesquiterpenes [6–8]. However, the specific localization of terpene biosynthesis can vary, with instances of monoterpenes and diterpenes also being synthesized in the cytosol, and sesquiterpenes in the plastids [9–11]. Given the crucial role of terpenes in plant biology, research into TPSs has been extensive.

The protein sequence of TPS is characterized by two domains: the N-terminal domain (Pfam ID PF01397) includes the 'RRX$_8$W' (R, arginine; W, tryptophan; X, alternative amino acid) motif, and the C-terminal domain (Pfam ID PF03936) provides binding sites for Mg^{2+} and Mn^{2+} [4,12]. The C-terminal domain is further divided into α, β, and γ domains. The α domain is a highly α-helical "isoprenoid" or farnesyl diphosphate synthase (FPPS or FPPase) fold, with catalysis mediated via 'DDxxD' motifs using a Mg^{2+}-dependent "ionization-initiated" mechanism [13]. The β and γ domains are located between a pair of α-helical double-barrel domains and carry out a "protonation-initiated" mechanism utilizing a catalytic 'DxDD' motif [8]. The taxonomy, distribution, and products of TPSs in plants have been well studied (Table 1) [12–18].

Table 1. Taxonomy of plant TPS subgroups.

Subgroup	Classification	Domain Composition	Motif Features	Distribution	Subcellular Localization	Type of Products
TPS-a	Class I	βα	With 'RRX$_8$W' motif (the second 'R' is not conserved)	Angiosperms	M/P/C	Sesqui-TPSs
TPS-b	Class I	βα	With conserved 'RRX$_8$W' motif	Angiosperms	P/C	Mono-TPSs, Isoprene synthase
TPS-g	Class I	βα	Without 'RRX$_8$W' motif	Angiosperms	P/C	Acyclic mono-, Sesqui- and Di-TPSs
TPS-h	Class I/II	γβα	Contain both 'DxDD' and 'DDxxD' motifs	Mosses, Liverworts, Lycophytes and Ferns	unknown	Di-TPSs
TPS-e/f	Class I	γβα/βα	Without 'DxDD' motif	Vascular plants	M/P/C	Ent-kaurene synthase (KS); Mono-, Sesqui-, and Di-TPSs
TPS-c	Class II	γβα/βα	With conserved 'DxDD' motif	Land plants	M/P	Copalyl diphosphate synthase (CPS), CPS/KS; Mono-, Sesqui-, and Di-TPSs
TPS-d	Class I/II	βα/γβα	Contain both 'DxDD' and 'DDxxD' motifs	Gymnosperms	unknown	Mono-, Sesqui- and Di-TPSs

Abbreviation: Di-, diterpene; Mono-, monoterpene; Sesqui-, sesquiterpene; C, cytosolic; P, plastidic; M, mitochondrial.

The TPS family is mid-sized and highly diversified in plants. The ancestral TPS was likely a bifunctional Class I and II diterpene synthase producing the ent-kaurene required for phytohormone production [19]. The ancestral TPS genes underwent gene duplication at least twice, leading to three ancient TPS lineages: TPS-c, TPS-e/f, and the remaining TPS (h/d/a/b/g) subfamilies (also called subgroups) [8]. TPS-a is the largest subgroup and expanded after the split of the monocot and dicot lineages [20]. The TPS-c subfamily enzymes CPS (copalyl diphosphate synthase) and KS (ent-kaurene synthase) catalyze the committed steps in gibberellin biosynthesis, which governs stem elongation, seed germination, and flowering time across land plants [19]. The evolutionary characteristics of the other subgroups in complex angiosperms have not yet been fully elucidated.

TPS families have been extensively studied in diverse plant species, including *Arabidopsis thaliana* (L.) Heynh. [21], tomato (*Solanum lycopersicum* L.) [8,16], grapevine (*Vitis vinifera* L.) [22], *Eucalyptus grandis* W. Hill [23], apple (*Malus domestica* Borkh.) [24], *Mentha longifolia* (L.) Huds. [25], common wheat (*Triticum aestivum* L.) [26], and Orchidaceae [27], etc. Both transcriptome-based analysis and the isolation of individual genes have contributed to the identification of increasing numbers of plant *TPS* genes [28,29]. The increasing availability of high-quality genomic resources means that it should now be possible to comprehensively analyze the evolution and functional diversification of *TPSs*. We reasoned that building networks of the phylogenetic and syntenic relationships of *TPSs* across numerous species could provide insights into the evolution of these important genes, which exhibit variation and conservation in both gene sequence and biochemical function.

Monoterpene alcohols and other volatile terpene substances provide roses (*Rosa* spp.) with their scent [30–32], which is an important target for rose breeders. As a eudicot species within the super-rosids clade—a major evolutionary radiation of core dicots—rose exhibits genomic features characteristic of this advanced angiosperm group [33]. This phylogenetic context helps explain the observed expansion of specialized metabolic gene families like TPSs. In addition, highly volatile terpene compounds are widely used in essential oil production [34]. A unique monoterpene biosynthesis enzyme in rose, *Rosa hybrida* nudix hydrolase 1 (RhNUDX1), shows geranyl diphosphate diphosphohydrolase activity in vitro and is involved in geraniol biosynthesis [4]. However, little is known about rose TPSs. A high-quality genome sequence is now available for *R. chinensis* 'Old Blush': (1) Its status as a diploid progenitor (2n = 2x = 14) of modern tetraploid garden roses simplifies genomic analyses; (2) The cultivar's exceptional terpene volatile diversity provides a rich phenotypic trait for gene-function correlation; (3) As a founder genotype in rose breeding programs, functional insights gained here can directly inform fragrance enhancement strategies [33]. The availability of 'Old Blush' genome sequence provides an opportunity to explore the relationships between TPSs and thereby lay the foundation for analysis of terpene-related processes in rose, such as fragrance formation, plant–insect interaction, and phytohormone biosynthesis.

In this study, systematic characterization across diverse angiosperms were used effective methods: (1) Profile HMM searches (HMMER3.0) provide superior sensitivity for detecting divergent TPS homologs; (2) Dual-domain verification (PF01397/PF03936) ensures accurate family membership assignment; (3) Stringent *E*-value thresholds ($<10^{-3}$) balance discovery rates with false positives in large-scale analyses. We performed genome-wide identification and subgroup classification of the TPS families in 115 angiosperms with sequenced genomes (including super-asterids, super rosids, basal-eudicots, monocots, *Liriodendron chinense* (Hemsl.) Sarg. and *Amborella trichopoda* Baill.) and constructed a phylogenomic synteny network to investigate their evolutionary traits. We then used *R. chinensis* as an example species and performed a detailed comparative analysis of TPS subfamily members. Based on synteny analysis, we discovered and validated a bifunctional protein, RcTPS23, which has both linalool and nerolidol synthase activity. Our findings further expand our understanding of the phylogeny and functional differentiation of this important gene family in angiosperms and pave the way for analysis of functional diversity and conservation of TPSs from a new perspective.

2. Results

2.1. Genome-Wide Identification and Subgroup Classification of TPSs in 115 Angiosperms

To gain a comprehensive understanding of TPSs in angiosperms, we chose 115 angiosperms with available genomic sequences to construct a phylogenetic tree and used published genomes data to identify TPS sequences (Figure 1). Our genome-wide sequence homology search, focused on two conserved domains (PF01397 and PF03936), identified 3802 TPS genes (Table S1). The number of TPS genes varied significantly among the species studied, ranging from 91 in *Gossypium barbadense* L. to just two in *Zostera marina* L. Notably, five monocot species, including *Oropetium thomaeum* (L.f.) Trin. and *Elaeis guineensis* Jacq., harbored fewer than ten TPS genes each.

Based on conserved domain motif and phylogenetic analyses, we further divided the 3802 TPSs into five subgroups: TPS-a, TPS-b, TPS-c, TPS-g, and TPS-e/f; no TPS-d and TPS-h subgroup were identified (Table S2). We arranged the subgroups based on the phylogenetic order of the 115 angiosperms, as shown in Figure 1. TPS-a was the largest subgroup, with 1806 members, followed by TPS-b. For most (85 out of 115) angiosperms, the TPS-a subgroup accounts for the largest proportion of TPS families. The length of a branch in a phylogenetic tree has a certain relationship with the evolutionary times of the corresponding species [5,20,21]. We observed a significant trend where species positioned at the termini of longer evolutionary branches exhibit an expanded number of members in the TPS-a subgroup, such as the Poaceae branch in monocots (from *Triticum turgidum* L. to *Echinochloa crus-galli* (L.) P.Beauv.), the Solanaceae branch in super-asterids (from *Capsicum chinense* Jacq. to *Petunia axillaris* (Lam.) Britton, Sterns & Poggenb.), and the Rosaceae branch in super-rosids (from *Prunus mume* (Siebold) Siebold & Zucc. to *Fragaria vesca* L.) (orange bars in Figure 1). Other TPS subgroups did not exhibit such a pattern, especially TPS-b, which was the second largest subgroup. In the Poales branch, the size of TPS-b subgroup did not exceed four, and eight Poaceae plants lack TPS-b subgroup members (yellow bars in Figure 1). Our data also revealed the absence of some subgroups in various species. For example, the TPS-a, -b, and -f subgroups were not observed in the aquatic plants *Z. marina* and *Spirodela polyrhiza* (L.) Schleid. In addition, five basal-eudicots lack TPS-f subgroup members. The TPS-c and TPS-e/f subgroups contain TPSs involved in ent-kaurene production, which is required for primary metabolism [19]. We found that no angiosperm had lost both TPS-c and TPS-e/f subfamily members. In most angiosperms of our study, the expansion of TPS gene families was due to the expansion of the TPS-a and TPS-b subgroups, whereas the quantities of TPSs in TPS-c, TPS-e/f, and TPS-g subgroups had not fluctuated greatly.

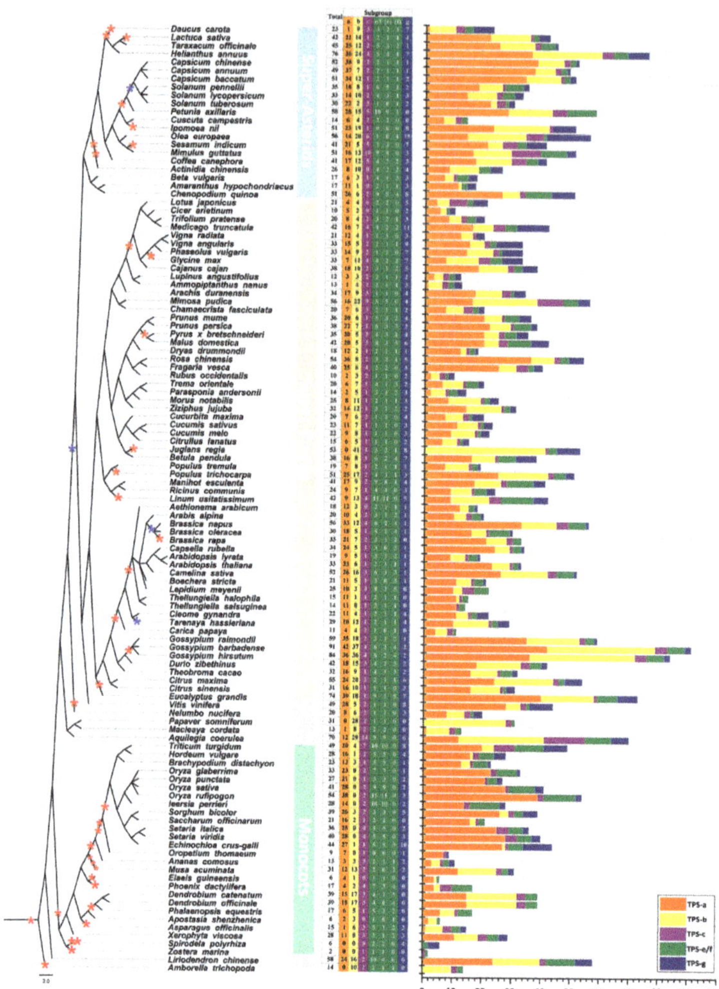

Figure 1. Phylogenetic distribution and genomic organization of TPS genes in angiosperms. (**Left**) Phylogenetic tree of 115 angiosperm species. Major clades are color-coded: super-asterids (light blue), super-rosids (light orange), basal eudicots (light purple), and monocots (light green). *Liriodendron chinense* (magnoliids) and *Amborella trichopoda* (basal angiosperm) are highlighted in light gray. Red asterisks mark whole-genome duplication (WGD) events, blue asterisks indicate whole-genome triplication (WGT) events based on published genomic data [35,36]. (**Center**) The number of TPS genes identified per species, with color gradients indicating subgroup composition: TPS-a (orange), TPS-b (yellow), TPS-c (purple), TPS-e/f (green), TPS-g (blue). Complete data in Table S1. (**Right**) Comparative distribution of TPS subgroups gene counts across species. With column lengths representing subgroup member counts and color scheme corresponds to TPS subgroups as in center panel.

2.2. *Synteny Analysis of TPS Gene Families Across 115 Plant Genomes*

To elucidate the evolutionary conservation and diversification of terpene synthase (TPS) genes, we constructed synteny networks for 115 plant genomes. Synteny networks are crucial for depicting homologous relationships between gene pairs and tracing the evolutionary paths of genes [37,38]. In these networks, nodes represent TPS genes with syntenic relationships, while edges denote the evolutionary lineages connecting these genes. Our analysis identified 22,035 edges and 1618 nodes, which we organized into 127 distinct communities or node clusters (Tables S3–S5). These clusters were subsequently mapped using phylogenetic profiling, which visually represents the presence or absence of specific syntenic TPS clusters across the 115 angiosperms (Figure S1, Table S6). Notably, Figure S1 reveals the presence of species-specific TPS clusters within particular families, such as Solanaceae, Brassicaceae, and Poaceae, suggesting these clusters may have originated from whole-genome duplication events. Conversely, the absence of specific TPS clusters in some clades may be attributed to the limited number of species examined, underscoring potential gaps in our current genomic coverage.

Syntenic relationships among species are invaluable for evolutionary analysis, providing insights into gene conservation and divergence over time [39]. To further uncover the potential evolutionary stories of TPSs in angiosperms, we constructed five subgroup-specific networks whose nodes represent subgroups TPS-a, -b, -c, -e/f, and -g (Figure 2A–E, Table S5). Each subgroup showed a specific synteny pattern, and the synteny patterns of certain subgroups were unique to specific plants [40]. The TPS-a network was the largest group, with 71 clusters and nodes that were widespread across angiosperms. A similar pattern emerged for the TPS-b network: some small clusters had the same outer circle color (species group), with interconnected nodes within the same species categories (Figure 2A,B). As shown in Figure 2C, the TPS-c network was highly interconnected, with nodes belonging to super-asterids, super-rosids, and basal eudicots and several nodes from monocots distributed in the periphery. Several nodes from super-asterids and monocots form four small clusters. Both the TPS-e/f and TPS-g networks were small and distributed in super-rosids and monocot species. The synteny pattern of TPS-c was different from the other subgroups. The tight syntenic junctions of TPS-c indicated that this subgroup was highly conserved within eudicots.

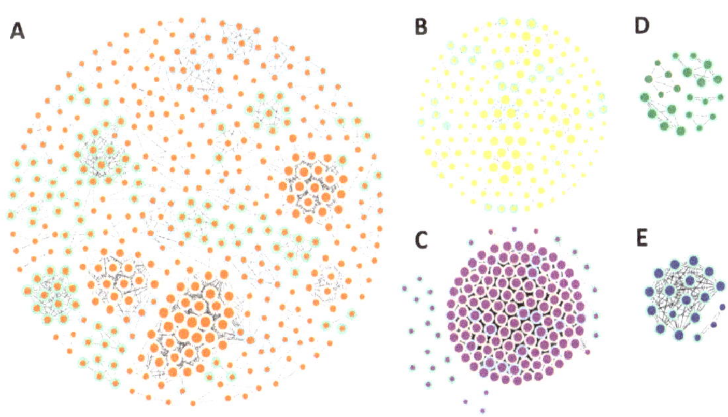

Figure 2. Syntenic network analysis of TPS gene family evolution. (**A–E**) Circos plots showing syntenic relationships within TPS subgroups: (**A**) TPS-a (orange), (**B**) TPS-b (yellow), (**C**) TPS-c (purple), (**D**) TPS-e/f (green), (**E**) TPS-g (blue). Each node represents a TPS gene, with inner ring color indicating subgroup and outer ring color denoting species group (super-asterids: light blue; super-rosids: light orange; basal eudicots: light purple; monocots: light green; *L. chinense* and *A. trichopoda*: light gray). Node sizes reflect the number of synteny connections per node, with detailed col-linear TPS gene pair information available in Table S5.

2.3. Phylogenetic Analysis of the TPS-c and TPS-e/f Subgroups

To elucidate the phylogenetic relationships and functional diversification within the TPS-c and TPS-e/f subgroups, conserved across the angiosperms yet demonstrating remarkable functional diversity, we employed a robust analytical framework. By conducting a comprehensive multiple sequence alignment of TPS genes from 115 angiosperm species, we constructed maximum-likelihood phylogenetic trees to dissect their evolutionary trajectories. The phylogenetic delineation, presented in Figure 3, Table S2, and Figure S2, clearly segmented these genes into three distinct clades: TPS-c, TPS-e, and TPS-f.

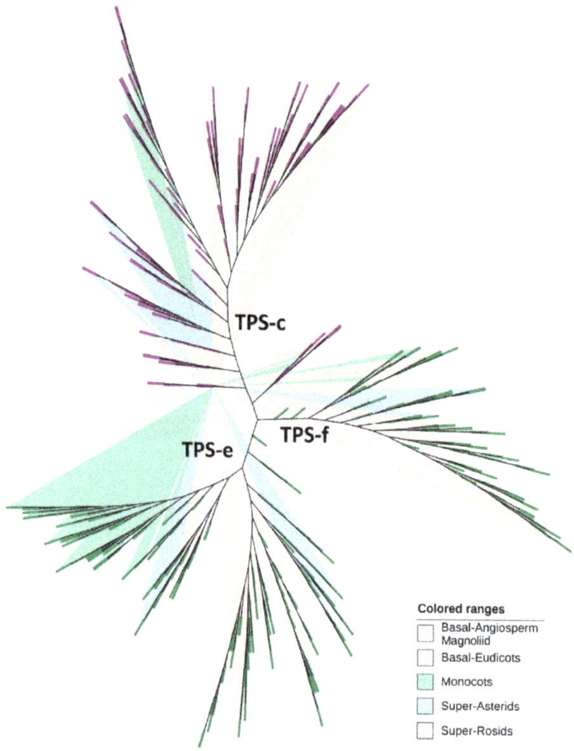

Figure 3. Phylogenetic relationships of TPS-c and TPS-e/f subgroups across 115 angiosperm species. The TPS-c clades are shown in purple, and the TPS-e/f clades are shown in green. The phylogenetic tree was constructed using the TPS-c and TPS-e/f members from 115 angiosperms (Table S1). The clades are highlighted in different colors: super-rosids (light orange), monocots (light green), super-asterids (light blue), basal-eudicots (light purple), basal-angiosperms and magnoliids (light gray).

Our results reveal a stable representation of TPS-c genes across diverse plant lineages, encompassing 21 super-asterids, 61 super-rosids, five basal-eudicots, 26 monocots, one basal-angiosperm, and one magnoliid. Conversely, the TPS-e and TPS-f subgroups, traditionally considered a single entity due to overlapping functional attributes and structural homologies, exhibited discernible phylogenetic bifurcation. Notably, TPS-e diverged into two subclades: one enriched with monocot genes, indicative of an expansion specific to this clade post the monocot-dicot evolutionary split, while the other was devoid of them. These findings accentuate the intricate evolutionary dynamics of the TPS gene family, reflecting both the retention of ancestral traits and the diversification essential for ecological adaptations.

2.4. Comparative Analysis of the TPS Family in R. chinensis

Among the 12 Rosales species examined, *R. chinensis* hosts the most extensive TPS gene family. We constructed a phylogenetic tree of 54 RcTPS genes, which grouped into three main clades as depicted in Figure 4A. The predominant clade comprises 36 genes from the TPS-a subgroup, significantly larger than the others, which include eight TPS-b and five TPS-g members in the second clade, and two TPS-c and three TPS-e/f members in the third clade. Detailed phylogenetic analysis revealed two sub-clades within the TPS-a subgroup, characterized by notably diverse branch lengths, indicating a significant expansion relative to other subgroups (Figure 4A). We further explored the chromosomal distribution of these 54 RcTPS genes, finding a pronounced uneven distribution across the seven chromosomes of *R. chinensis* (Figure 4B). Notably, chromosome 5 (Chr5) harbors 20 RcTPS genes, whereas chromosome 4 (Chr4) contains only one. All members of the TPS-g and TPS-c subfamilies are localized to Chr5 and Chr6, respectively. Except for Chr2, TPS-a gene clusters are present on six chromosomes. This genomic architecture, combined with our phylogenetic findings, underscores the expansion of the TPS-a genes. Additionally, gene structure analysis provides further evolutionary insights; for instance, most TPS-a, TPS-b, and TPS-g subgroup members feature seven exons, while TPS-c and TPS-e/f subgroup members typically have 13 to 15 exons (Figure S3). These structural patterns, similar to those observed in *A. thaliana*, *E. grandis*, and *M. domestica* [21,23,41], suggest a relative conservation of exon features across these species.

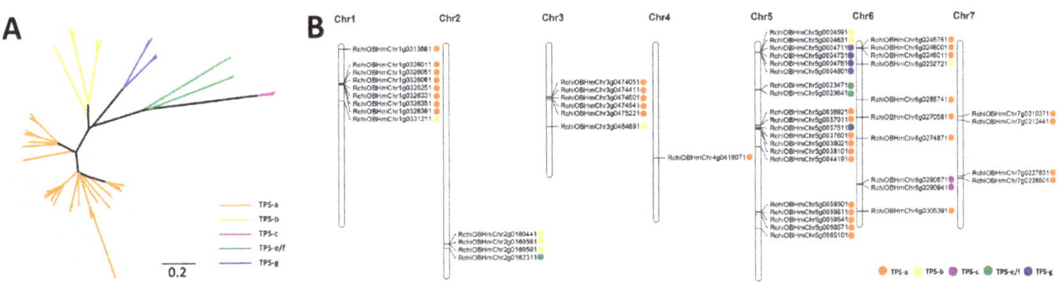

Figure 4. Phylogenetic and chromosomal distribution of RcTPSs in *R. chinensis*. (**A**) This panel shows a phylogenetic tree of TPS proteins from *R. chinensis*, constructed using the maximum-likelihood method with 1000 bootstrap replicates via RAxML. The tree segregates TPS proteins into six subgroups, each indicated by a distinct color. The composition and relationships of these subgroups are detailed in Table S7. (**B**) Chromosomal distribution of putative functional terpene synthase genes across the chromosomes (denoted as Chr) of *R. chinensis*. Each gene is represented by a small dot, color-coded by subfamily: TPS-a (orange), TPS-b (yellow), TPS-c (purple), TPS-e/f (green), and TPS-g (blue), providing a visual representation of gene location and subgroup prevalence.

We explored the syntenic relationships of 54 identified RcTPSs across 115 angiosperms, finding that 27 of these genes are syntenic homologs with TPSs from other species. The relationships involving RcTPSs were individually illustrated, with each RcTPS positioned centrally in the diagrams (Figure 5A). Notably, two RcTPSs (RchiOBHmChr5g0023471 and RchiOBHmChr2g0160561) demonstrated extensive syntenic connections. The majority of RcTPSs (23 out of 27) predominantly showed synteny with TPSs from species within the super-rosid and super-asterid groups. Moreover, two RcTPSs (RchiOBHmChr1g0326251 and RchiOBHmChr5g0004801) appeared as Rosaceae-specific syntelogs, indicating unique syntenic relationships confined to this family. Additionally, three distinct pairs of syntenic genes were identified between *R. chinensis* and *V. vinifera*, highlighting specific evolutionary links (items 9, 10, and 11 in Figure 5A).

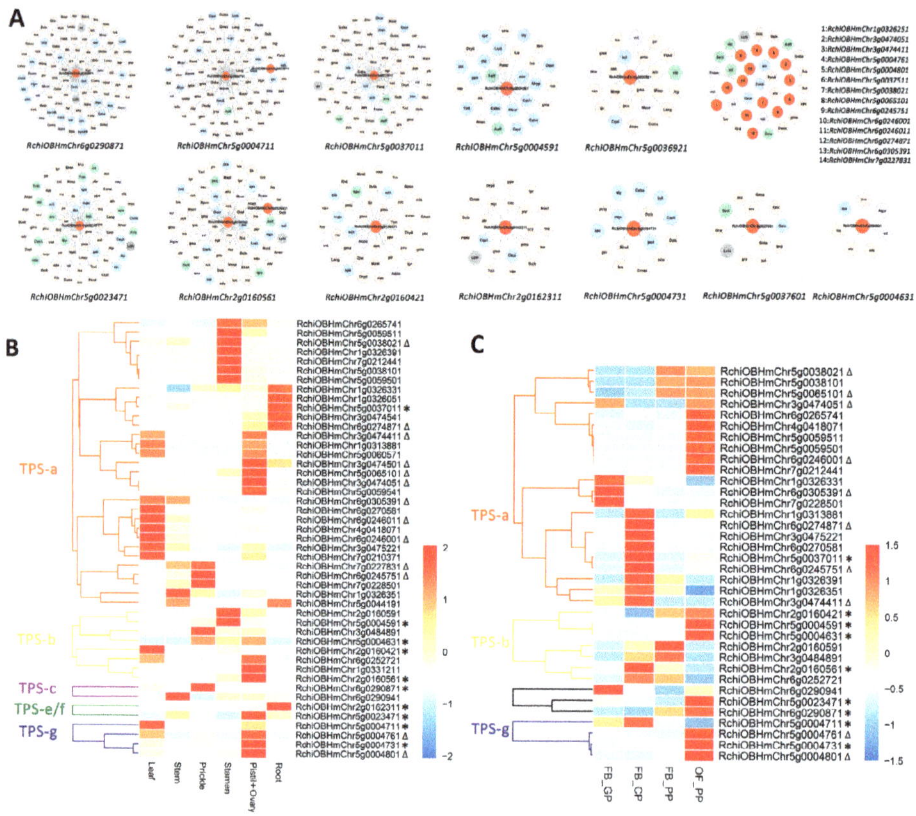

Figure 5. Synteny networks and spatiotemporal expression patterns of RcTPSs in *R. chinensis*. (**A**) Synteny clusters of TPSs across various species with *R. chinensis* highlighted. Each node represents one TPS, color-coded to indicate species classification: super-asterids (light blue), super-rosids (light orange), basal eudicots (light purple), monocots (light green), *L. chinense* and *A. trichopoda* (light gray). Nodes specific to *R. chinensis* are marked in red, with gene names labeled beneath each cluster. (**B**) Spatial expression profiles of *TPS* genes across various tissues of 'Old Blush' (root, leaf, stamen, pistil, ovary, stem, and prickle), using RNA-seq data sourced from NCBI (accession number PRJNA546486). Black triangles indicate RcTPSs with a single synteny connection; black asterisks denote genes with multiple synteny connections. (**C**) Temporal expression patterns of *TPS* genes during four developmental stages of rose petals in 'Old Blush': FB_GP (green petals in a flower bud), FB_CP (petals changing color in a flower bud), FB_PP (pink petals in a flower bud), and OF_PP (pink petals in an open flower), based on RNA-seq data from NCBI (accession number PRJNA351281). Symbols are used consistently as in (**B**) to indicate the connectivity of synteny edges among the RcTPSs.

We analyzed the gene expression patterns of *RcTPSs* using RNA-seq data from various rose tissues and across four critical developmental stages of petals. From two distinct RNA-seq datasets, 47 and 36 *RcTPSs* were identified, as detailed in Tables S8 and S9. The expression patterns and synteny relationships for each subgroup are illustrated in Figure 5B,C. Predominantly, *RcTPSs* exhibited heightened expression levels in specific tissues, with several genes presenting multiple synteny edges (marked by black asterisks), particularly within subgroups TPS-b, TPS-c, TPS-e/f, and TPS-g. A notable observation was that only one *RcTPS* (*RchiOBHmChr5g0037011*) from the TPS-a subgroup demonstrated syntenic relationships with more than one species and showed expression across various rose tissues. Additionally, eleven and eight RcTPSs from the TPS-a subgroup exhibited a single synteny edge with genes from other species, as shown in Figure 5B,C. Petals are the main tissues from which fragrance is released in rose, and many terpenes related to flower fragrance are mainly synthesized in petals. As the rose flower opens, the amount of fragrance released increases [42]. Each subgroup contains members that were highly expressed in OF_PP stage petals. These finding lay the foundation for further in-depth study of RcTPS function.

2.5. Identification of the Bifunctional Terpene Synthase RcTPS23

TPS functions are often related to their subcellular localization (plastid or cytosol), and some TPSs can react with both GPP and FPP when expressed in vitro. The most common of these potentially bifunctional TPSs, whose products are monoterpene linalool and sesquiterpene nerolidol, have been reported in *A. thaliana*, strawberry, tomato, cotton, and other species (Figure 6A) [16,43–46]. BLAST (version 2.13.0) analysis of the protein sequences in our data revealed that most of these TPSs belong to the TPS-g subgroup and synnet Cluster126. We constructed a syntenic network of AtTPS14, FaNES1, GhTPS12, and SlyTPS39 and found that most nodes were in subgroup TPS-g and belong to various species (Figure 6B). Two rose TPSs were present in this syntenic network. We cloned these genes and named them *RcTPS23* (*RchiOBHmChr5g0004711*) and *RcTPS10* (*RchiOBHmChr2g0160421*). The sequence of *RcTPS10* contained a frameshift. *RcTPS23* reached its highest expression levels in stage FB_CP petals but was expressed at very low levels in stamens, stems, roots, and OF_PP stage petals (Figure 6C). We investigated the subcellular localization of RcTPS23 by expressing RcTPS23-eGFP fusion protein in *A. thaliana* protoplasts (Figure 6D). Confocal laser scanning microscopy revealed that RcTPS23-eGFP localized to the cytosol, suggesting that *RcTPS23* might function in sesquiterpene biosynthesis.

A

Protein_Name	Node_ID	Cluster_ID	Subgroup	Subcellular_localization	Substrate	Main_product	References
AtTPS14	ath_AT1G61680	Cluster126	TPS-g	plastid	GPP, FPP	linalool, nerolidol	Chen *et al.*, 2003
FaNES1	fve_30669	Cluster126	TPS-g	cytosol, plastid	GPP, FPP	linalool, nerolidol	Aharoni *et al.*, 2004
GhTPS12	GhirGH_D10G0184	Cluster126	TPS-g	plastid (prediction)	GPP, FPP	linalool, nerolidol	Huang *et al.*, 2018
SlyTPS39	sly_10g0053902sly	Cluster126	TPS-g	cytosol	GPP, FPP	linalool, nerolidol	Falara *et al.*, 2011
SlyTPS37	sly_10g0054102sly	none	TPS-g	cytosol	GPP, FPP	linalool, nerolidol	Falara *et al.*, 2011
SlyTPS5	sly_01g1058902sly	none	TPS-b	plastid	GPP, FPP	linalool, nerolidol	Van Schie *et al.*, 2007

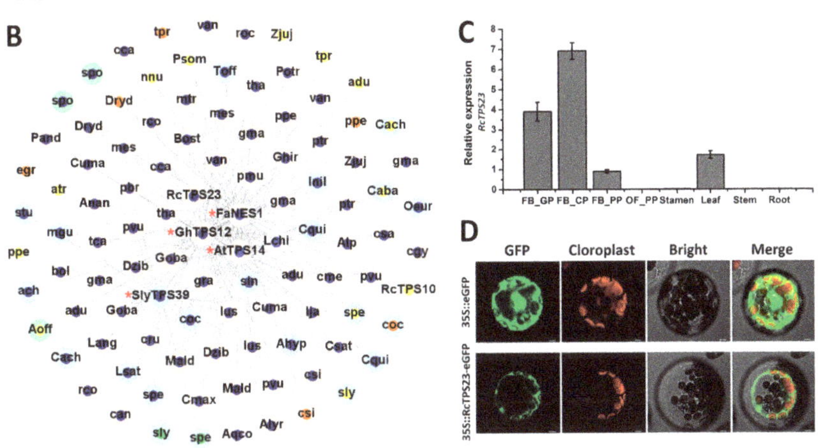

Figure 6. Syntenic localization, spatiotemporal expression, and subcellular localization of RcTPS23. (**A**) Catalog of TPS enzymes capable of catalyzing the substrates GPP and FPP to produce linalool or

nerolidol [16,43–46]. (**B**) Syntenic analysis of key TPSs, including AtTPS14, FaNES1, GhTPS12, and SlyTPS39 (highlighted with red asterisks). Each node is annotated with species abbreviations and color-coded: nodes in the outer loop represent different plant groups—super-asterids (light blue), super-rosids (light orange), basal eudicots (light purple), monocots (light green), *L. chinense* and *A. trichopoda* (light gray); nodes in the inner loop represent TPS subgroups, coded as TPS-a (orange), TPS-b (yellow), TPS-c (purple), TPS-e/f (green), and TPS-g (blue). (**C**) Expression analysis of *RcTPS23* across four developmental stages of rose petals and various tissues, conducted via qRT-PCR: FB_GP (green petals in a flower bud), FB_CP (petals changing color in a flower bud), FB_PP (pink petals in a flower bud), and OF_PP (pink petals in an open flower). Expression data are expressed as mean ± standard deviation from three biological replicates. (**D**) Subcellular localization of 35S::RcTPS23-eGFP fusion protein (green) in *A. thaliana* protoplasts. Using 35S::eGFP as the control. Red fluorescence corresponds to chloroplast autofluorescence detected under 488 nm excitation. Scale bars = 10 μm.

Finally, to perform functional analysis, we heterologous expressed *RcTPS23* in *Escherichia coli* strain BL21 (DE3) and analyzed the in vitro chemical products of the resulting recombinant proteins using different substrates by GC-MS. The product produced by RcTPS23 and the substrate FPP was (*E*)-nerolidol (Figure 7A,B). However, linalool was produced when GPP was used as the substrate (Figure 7C,D). In these assays, linalool and nerolidol were identified by comparing mass spectra. Each reaction had a blank control and three biological replications. In summary, we used the syntenic relationships of TPSs to identify the TPS-g subgroup member *RcTPS23* and analyzed its gene expression pattern in rose. We determined that RcTPS23 was a bifunctional terpene synthase that produces nerolidol and linalool in vitro and might possess nerolidol biosynthetic activity in vivo due to its localization in the cytoplasm.

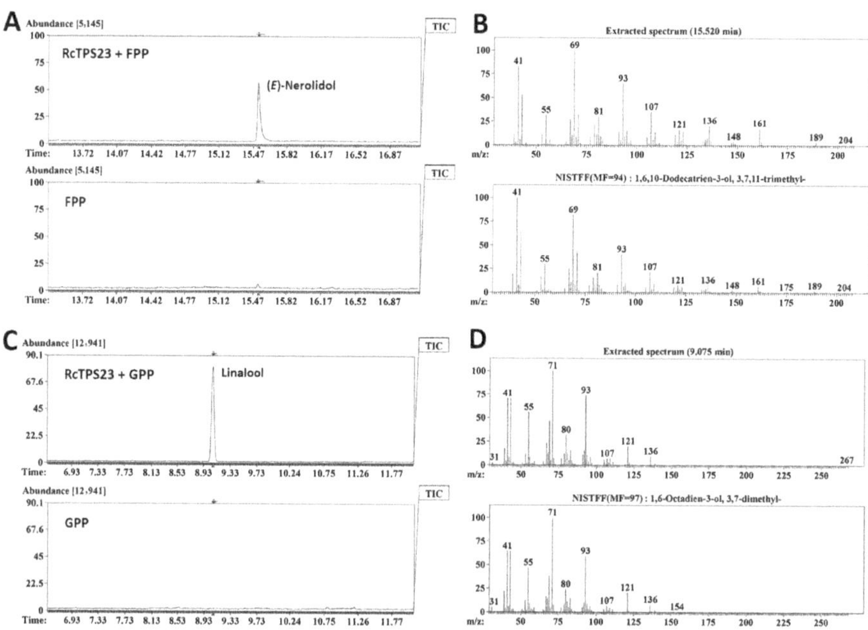

Figure 7. Dual catalytic activity of RcTPS23 in producing (*E*)-nerolidol and linalool in vitro. (**A**) Gas chromatography-mass spectrometry (GC-MS) chromatograms displaying sesquiterpene synthesis by recombinant RcTPS23 using farnesyl diphosphate (FPP) as a substrate (top panel). The blank control, containing only the substrate FPP, is shown for comparison. (**B**) Mass spectrum of (*E*)-nerolidol synthesized

by incubating FPP with purified RcTPS23 protein (top panel), compared against the mass spectrum of (*E*)-nerolidol from the NIST standard reference database (bottom panel). (**C**) GC-MS chromatograms illustrating monoterpene production by recombinant RcTPS23 with geranyl diphosphate (GPP) as the substrate (top panel). The corresponding blank control includes only the substrate GPP. (**D**) Mass spectrum of linalool produced following incubation of GPP with purified RcTPS23 protein (top panel). For comparative analysis, the bottom panel displays the mass spectrum of linalool from the NIST standard reference database.

3. Discussion

The TPS families in the plant kingdom are highly diversified [20]. In this study, we performed the first large-scale assessment of the number of TPS-subgroup members in 115 angiosperms (Tables S1 and S2). Despite the key roles that whole-genome duplication (WGD) and whole-genome triplication (WGT) events often have in increasing gene family size [47], we did not find evidence for such effects on TPS families (Figure 1). A previous study suggested that the large-scale expansion of the TPS family occurred after the divergence of dicot and monocot plants [48]. Our data support this conclusion. Moreover, we determined that the expansion of the TPS family was mainly due to the expansion of the TPS-a and TPS-b subgroups. The expansion of TPS-a suggests a correlation between the rate of evolutionary change—reflected by longer branch lengths—and the diversification of the TPS-a genes. These findings imply that species undergoing rapid evolutionary changes, as indicated by their placement on long branches of the phylogenetic tree, may have experienced selective pressures favoring the expansion of their TPS-a gene repertoire. Such expansions could be adaptive, potentially enhancing the species' ability to synthesize a diverse array of terpenoids, which are critical for various ecological functions including defense mechanisms and pollinator attraction. However, there were several exceptions. For instance, *L. chinense*, a magnoliid plant, contains 58 TPSs, including 24 TPS-a subgroup members. An expansion of the TPS-a subgroup is thought to have occurred after the split of monocots and dicots. The TPS-a subgroup mainly contains sesquiterpene synthases and is highly divergent in all seed plants [20]. We determined that the expansion of TPS-a occurred in monocots, super-rosids, and super-asterids after their split. The expansion of TPS-a in *L. chinense* appears to be unique compared to the other species, which is worth exploring. The significant expansion of the TPS-a subgroup likely reflects multiple biological mechanisms. First, ecological pressures such as co-evolution with specialized herbivores may drive selection for sesquiterpene diversity, as evidenced by TPS-a copy number variation across maize populations adapting to distinct pest pressures [49]. Second, genomic analyses in Arabidopsis reveal TPS-a clusters flanked by transposable elements, suggesting repeat-mediated duplication as a key expansion driver [50]. Finally, subfunctionalization through promoter divergence could enable tissue-specific expression partitioning of defense-related terpenoids [51]. These mechanisms collectively suggest that TPS-a expansions represent adaptive genomic responses to biotic challenges. The variable number of genes among taxa points to the independent loss or duplication of genes in different genomes. The number of TPSs in the TPS-b subgroup appears to be irregular and this subgroup has been lost in many Poaceae species (Figure 1). The lost functionalities due to the loss of TPS-b, such as mono-synthase activities, might be replaced by those of other subgroups. Notably, the absence of TPS-b raises questions about how Poaceae compensate for the loss of these volatile compounds. One possibility is that grasses have evolved alternative biosynthetic routes: TPS-g members might acquire monoterpene synthase activity through structural convergence, or non-TPS enzymes such as prenyltransferases could generate monoterpene precursors. Additionally, ecological shifts may reduce dependency on monoterpenes; the prevalence of wind pollination diminishes the need for floral volatiles, while silica-based physical defenses in grasses might offset the loss of monoterpene-mediated chemical protection.

Synteny could reflect important relationships between the genomic contexts of genes in terms of both function and regulation [52,53]. Therefore, analyzing the syntenic relationships of genes across a wide range of species provides important information about the evolution of gene families involved in plant growth and development. Phylogenomic synteny network analysis has been performed to examine the MADS-box, LEA (Late Embryogenesis Abundant), and MYB (v-myb avian myeloblastosis viral) gene families. Such analysis could be used to reveal genomic diversification, positional conservation, ancient tandem duplications, and lineage-specific transpositions [54–56]. However, this method did not appear to be so effective for analyzing the TPS gene family. First, among the 3802 TPSs of 115 angiosperms, only 1618s TPS were found to have syntenic relationships; this quantity is quite low. Second, in our syntenic cluster analysis, unlike the MADS-box and MYB gene families, many TPS edges could not be separated, and these edges were mixed and interwoven to form the huge synnet cluster126 (Tables S4 and S5). These features point to the complexity of TPS evolution. These TPSs without syntenic genes in other species represent "specialized gifts" prepared by angiosperms on their respective evolutionary journeys.

The ancestral TPS originated in land plants after their divergence from green algae and encoded a bifunctional copalyl/kaurene synthase (CPSKS) [8]. The TPS-c subgroup contains extant examples of bifunctional CPSKS, which are functionally analogous to the ancestral TPS required for the biosynthesis of gibberellins and related phytohormones [19,57]. As shown in Figure 2C, the syntenic relationships of TPS-c subfamily members are tightly connected. Moreover, phylogenetic analysis (Figure 3) indicated a high degree of conservation among TPS-c subfamily members in angiosperms. The TPS-e/f subgroup includes kaurene synthase genes, which are required for phytohormone biosynthesis. TPS-f is thought to be derived from TPS-e [22]. As shown in Figure 3, the TPS-e branch separated into two clades. These findings suggest that new functional domains may have appeared in the TPS-e subgroup of angiosperms.

Investigating TPSs is important for understanding the evolution of terpenoid biosynthesis in plants. Here, except for the important diterpenes (such as gibberellin and membrane sterols), we mainly focused on volatile monoterpenes and sesquiterpenes involved in floral scents. Rose is a great model in which to explore fragrance. We conducted a comprehensive analysis of the TPS gene family of *R. chinensis* (Figures 4 and 5). Our data indicate that the expansion of TPS-a in *R. chinensis* mainly occurred via tandem duplication, and only one TPS-a subfamily member (RchiOBHmChr5g0037011) shares syntenic relationships with members in other angiosperms. The functions of the expanded TPS-a subfamily in rose (the types and contents of sesquiterpenes might also have expanded) are being intensively studied. While transcriptomic data suggest diverse expression patterns across RcTPS paralogs, our focused analysis of RcTPS23 demonstrates how phylogenomic synteny can prioritize candidates for functional studies. Systematic validation of additional TPSs remains an important direction for future research.

If non-syntenic TPSs expand, they may confer new terpene synthesis activities in plants. Analyzing TPSs with syntenic relationships might also shed light on the conservation of TPS function. Most terpenes are classified as secondary metabolites that help plants better adapt in their local environments [58]. Recent studies further elucidate TPS-mediated defense mechanisms across diverse species: In maize, ZmTPS12-produced dolabralexins confer resistance against *Fusarium graminearum* through direct antifungal activity [59]. Dendrobium orchids upregulate DcTPS7 expression under herbivore attack, catalyzing germacrene D synthesis to repel insects [60]. Potato StTPS2 synthesizes bulnesol/elemol diterpenes that systemically prime jasmonate signaling upon wounding [61]. Notably, tea plant CsTPS1 and CsTPS2 produce defense-related sesquiterpenes (δ-cadinene and α-humulene) that synergistically enhance resistance to *Ectropis obliqua* [62]. These findings collectively highlight the evolutionary convergence of TPS diversification in biotic stress adaptation. The sesquiterpene alcohol nerolidol and its derivative can induce the accumulation of defense-related compounds with extensive natural anti-herbivore or anti-pathogen effects [63,64]. Linalool, an acyclic monoterpene alcohol, is extremely widespread in plants. This compound attracts natural enemies of herbivores, thereby participating in the complex interplay between pollinator attraction and plant defense [65–67]. Using SynNet analysis, we detected the TPS-g subgroup member RcTPS23, which has dual functions in both nerolidol and linalool biosynthesis. RcTPS23 localizes to the cytosol (Figures 6 and 7). A previous study of linalool nerolidol synthases in roses supports our results [68]. We speculate that this dual linalool nerolidol synthase activity is ubiquitous in angiosperms (Figure 6B). Although the subcellular localization of this bifunctional enzyme determines that it can only produce linalool or nerolidol in vivo (Figure 6A), both compounds can increase the ability of plants to resist biotic stress. Our findings lay the foundation for further exploring RcTPS23, and they increase our understanding of the phylogeny and functional differentiation of this important gene family in angiosperms. These insights directly inform rose breeding strategies. such as, overexpression of RcTPS23 in petal tissues could enhance floral fragrance (via linalool) while maintaining vegetative pest resistance (via nerolidol). The conserved TPS-g subgroup features identified here facilitate rapid ortholog discovery in hybrid rose cultivars for marker-assisted selection, etc.

4. Materials and Methods

4.1. Genomic Analysis of 115 Plant Species and the Genome-Wide Identification of TPSs

The publicly available genomes of 115 species were used for genome-wide identification of TPSs (Table S1). These 115 angiosperms include 21 super-asterids, 61 super-rosids, five basal-eudicots, 26 monocots, one magnoliids (*L. chinense*), and one basal-angiosperm (*A. trichopoda*). The download websites of whole-genome protein sequences this study used can be obtained in Table S1. The R package (version 0.6.5) plant list was used to check the species status [69]. The phylogenetic relationships were analyzed with Phylomatic (version 3), and the Figtree program (version 1.4.4) was used to visualize the phylogenetic tree. All-against-all comparisons between pairwise genomes and the detection of synteny blocks were performed using a GitHub script (https://github.com/zhaotao1987/SynNet-Pipeline) (accessed on 1 December 2021).

Candidate terpene synthase (TPS) genes were identified using the Hmmsearch program in the HMMER3.0 package [70] with default settings and were confirmed using the Pfam database. All TPSs were identified based on the presence of two specific domains: the Pfam N-terminal domain (PF01397) and the Pfam C-terminal domain (PF03936) [20]. Significant hits (E–value < 10^{-3}) were used to identify candidate TPSs that were encoded in the genomes of 115 angiosperms (nodes) (Table S2). The protein sequences of identified TPSs have been listed in the FASTA file "TPSs.fasta".

4.2. Subgroup Classification and Phylogenetic Analysis of TPSs

After domain verification, the 3802 TPS protein sequences were classified into six TPS subgroups. Based on the method of [20], TPS-e and TPS-f were combined into one subgroup: TPS-e/f. Multiple amino acid sequence alignment was performed using ClustalX (version 2.0) with standard parameters. The results were analyzed using the online program MEME (http://meme-suite.org/, (accessed on 30 January 2022)), and the alignments were manually adjusted to classify the conserved motifs of TPSs, such as 'RRX$_8$W', 'DDxxD', 'DxDD', and 'NSE/DTE'. The multiple sequence alignment results were saved in fasta format.

A phylogenetic tree of the chosen TPS protein sequences was constructed using RAxML [71] with the maximum-likelihood method based on the best substitution model Jones-Taylor-Thornton (JTT) [72], with 1000 bootstrap replicates. The phylogenetic tree was visualized using the Figtree program (version 1.4.4) and iTOL online software (version 5) [73].

4.3. Synteny Block Detection, Network Construction, and Network Clustering of TPSs

The SynNet pipeline described by Zhao [37] was used for syntenic block calculations, network construction, and analysis of cluster distribution (https://github.com/zhaotao1987/SynNet-Pipeline) (accessed on 1 December 2021). The synteny blocks in the 115 species were identified with MCScanX software (https://github.com/wyp1125/MCScanX) (accessed on 20 December 2021) [74] using default parameters (minimum gene number in a collinear block = 5, maximum gaps = 25). The output files of TPS genes syntenic blocks (edges, with headers Locus_1 and Locus_2) are provided in Tables S3 and S4. A synteny network (SynNet) of syntenic blocks was built containing all the TPS genes in the 115 species, and the SynNet was imported and visualized in Gephi (version 0.9.1) [75].

The clique percolation method in CFinder was used to locate all putative clique communities (k = 3) in the TPS gene synteny network (synnets) [76,77]. The synnets were used to identify communities (clusters of gene nodes) (Tables S4 and S5), and these synteny communities were further used for phylogenetic profiling. For each genome, the number of syntenic gene copies of each cluster was counted, as shown in Table S6. The Jaccard method of the vegan package [78] was carried out to calculate the dissimilarity index of all clusters, followed by separate hierarchical clustering and visualization of these clusters using "ward. D" and "pheatmap".

4.4. Chromosomal Localization and Intron-Exon Structure Analysis of RcTPSs

Information about chromosomal locations and gene structures was obtained from the gff3 file in the *R. chinensis* genome database (RchiOBHm-V2) [33] and is provided in Table S7. Mapchart [79] was used to visualize the chromosomal locations of all putative functional terpene synthases in *R. chinensis.* The online software GSDS (http://gsds.gao-lab.org/, (accessed on 15 January 2022)) was used to depict the introns and exons in the 54 *RcTPS* genes.

4.5. Gene Expression Analysis of RcTPSs

Two RNA-seq data sets were used to obtain rose TPS gene expression data by the TPM method (transcripts per kilobase million) [80]. One data set includes transcriptome data from rose petals at different stages of development (BioProject PRJNA351281, Table S8, Q30 $\geq$ 92%, mapping rate $\geq$ 85%, Error rate $\leq$ 0.02), and the other contains transcriptome data from different rose tissues (root, stem, leaf, prickle, stamen, pistil, and ovary tissue) (BioProject PRJNA546486, Table S9, Q30 $\geq$ 93%, mapping rate $\geq$ 90%, Error rate $\leq$ 0.03). Heat maps representing gene expression levels were drawn with the R package pheatmap (version 1.0.12).

4.6. Reagents and Materials

Commercial reagents included: Plant RNA Purification Reagent (Promega, Madison, WI, USA, Cat# A7280); PrimeScript RT Reagent Kit with gDNA Eraser and SYBR Premix Ex Taq (Takara Bio, Shiga, Otsu, Japan, Cat# RR047A/RR420A); KOD-Plus-Neo DNA Polymerase (TOYOBO, Osaka, Japan, Cat# KOD-401); KpnI/XhoI restriction enzymes (New England Biolabs, Ipswich, MA, USA, Cat# R3142/R0146); In-Fusion HD Cloning Kit (Takara Bio, Cat# 639649); Ni Sepharose 6 Fast Flow (GE Healthcare, Chicago, IL, USA, Cat# 17531801); isopropyl-β-D-thiogalactopyranoside (IPTG, Sigma-Aldrich, St. Louis, MO,

USA, Cat# 367-93-1); farnesyl diphosphate (FPP, Cat# 44270) and geranyl diphosphate (GPP, Cat# G6772) from Sigma-Aldrich. Instrumentation comprised: CFX96™ Real-Time PCR system (Bio-Rad Laboratories, Hercules, CA, USA); Leica TCS SP8 confocal microscope (Leica Microsystems, Wetzlar, Germany); Agilent 7890A/5975C GC-MS system with HB-5MS column (5% phenyl methyl silox, Agilent J&W Scientific, Folsom, CA, USA). Oligonucleotide primers were synthesized by Sunbiotech Co., Ltd. (Beijing, China). General chemical reagents were purchased from Sangon Biotech (Shanghai, China).

4.7. Validation of Gene Expression Data by qRT-PCR

Quantitative reverse-transcription PCR (qRT-PCR) was performed to validate the gene expression levels of *RcTPS23* in eight tissues of *R. chinensis* 'Old Blush' (petals at four stages: FB_GP, FB_CP, FB_PP, OF_PP, stamen, leaf, stem, and root tissues). A Plant RNA Isolation Kit was used to extract total RNA from each sample. A PrimeScript RT Reagent Kit with gDNA Eraser was used to synthesize first-strand cDNA from 1.0 μg total RNA. The reaction mixture contained 10 μL SYBR Premix Ex Taq, 0.4 μL 10 μM forward and reverse transcript-specific primers (forward primer: 5′-TTTAGAGAAGCAATACACCAGGTC-3′, reverse primer: 5′-CAGGGAAGCCTTGTTGTCTTA-3′), 2 μL cDNA, and 7.2 μL sterile distilled water. The qRT-PCR was performed using a CFX96™ Real-Time PCR system with the following program: 95 °C for 30 s; 40 cycles of 95 °C for 5 s and 60 °C for 30 s; and a final melting curve analysis of 95 °C for 15 s, 60 °C for 1 min, and 95 °C for 15 s. The relative gene expression levels were normalized against the expression level of the endogenous reference gene *RcActin* [81] and calculated using the $2^{-\Delta\Delta Ct}$ method [82]. Each sample was examined with three biological replicates. The histograms were generated using the Origin9 program (OriginLab, Northampton, MA, USA). All commercial reagents and kits used in this section are detailed in Section 4.6 (Reagents and Materials).

4.8. Subcellular Localization

The cDNA sequence of *RcTPS23* was amplified using the forward primer 5′-GGTACCATGACGTTCATTCTTCAAGC-3′ and the reverse primer 5′-CTCGAGGCTAGCC ATGAGTGATGC-3′ under the following conditions: 94 °C for 2 min, followed by 35 cycles of 94 °C for 10 s, 54 °C for 15 s, and 72 °C for 2 min, with a final extension of 72 °C for 5 min, using KOD-Plus-Neo DNA Polymerase. The PCR product was cloned into the pEZS-NL vector, which had been cleaved with KpnI/XhoI. *A. thaliana* protoplast isolation and transformation were carried out according to a standard protocol [83]. Using empty vector-transfected protoplasts as the control. Reagent sources are provided in Section 4.6. The transformed protoplasts were incubated at 23 °C for 16 h and observed under a laser confocal microscope (Leica TCS SP8, Wetzlar, Germany) to detect eGFP fluorescence; eGFP was excited at 488 nm.

4.9. Generation of Recombinant RcTPS23 and In Vitro Enzyme Activity Assays

The cDNA sequence of *RcTPS23* was cloned into the *pET-32a* vector by In-Fusion cloning using the forward primer 5′-GGCTGATATCGGATCCATGACGTTCATTCTTCAAGC-3′ and the reverse primer 5′-CAAGCTTGTCGACGGAGCTCGCTAGCCATGAGTGATGC-3′ and transformed into *E. coli* strain BL21 (DE3). Recombinant protein production was induced by adding 0.8 mM isopropyl-b-d-thiogalactopyranoside (IPTG) and incubating at 16 °C for 24 h. The *E. coli* cells were sonicated and centrifuged to obtain the supernatant, and proteins were purification on a Ni Sepharose column. The purified proteins were collected and concentrated.

In vitro enzyme activity assays of RcTPS23 recombinant protein were carried out using 50 μg purified protein in 200 μL of assay buffer [50 mM HEPES (pH 7.4), 25 mM $MgCl_2$, 5 mM DTT, 5% (v/v) glycerol, 50 μM FPP or 50 μM GPP]. The blank control contained the same buffer and substrates but no recombinant protein. The samples were incubated at 30 °C for 2 h, and the volatile products were absorbed on 100 μm PDMS fiber prior to GC-MS analysis. The PDMS fiber was transferred to the injection port (250 °C) of the GC-MS system and desorbed for 5 min. GC was performed using a GC system coupled with a mass spectrometer. For GC, an HB-5MS column (5% phenyl methyl silox: 30 m × 250 μm i.d., 0.25 μm) was used for all samples. The analytical conditions were as follows: The initial temperature was held at 50 °C for 1 min and raised to 210 °C for 2 min at 3 °C/min, maintained for 3 min, and increased to 230 °C at 15 °C/min. The analysis was conducted in split-less mode; helium was used as the carrier gas at a rate of 1.0 mL/min. The mass spectrometer was set as follows: 230 °C, electron-impact (EI) model with 70 eV electron power; 280 °C for the auxiliary temperature. The scan range was 80–500 m/z. Refer to Section 4.3 for reagent specifications. The volatile products were identified and characterized by comparing mass spectra using AMDIS software (Automated Mass Spectral Deconvolution and Identification System, http://www.amdis.net/, (accessed on 10 May 2022)) and the NIST Mass Spectral Library (https://www.nist.gov/, (accessed on 10 May 2022)). The assay conditions were established based on previously validated methods for terpene synthase characterization [84,85].

5. Conclusions

Our research across 115 angiosperms reveals the genomic diversity and evolutionary trajectories of the terpene synthase (TPS) gene family, identifying 3802 TPS genes and categorizing them into five distinct subgroups. This extensive classification highlights significant variability in gene numbers, with adaptive expansion observed in the TPS-a subgroup. Synteny analysis further delineated homologous relationships and evolutionary conservation, identifying 22,035 edges and 1618 nodes that map the genetic architecture and evolutionary pressures shaping these genes. Notably, lineage-specific expansions linked to whole-genome duplications suggest significant evolutionary roles. Our focused study on *R. chinensis* uncovered a large TPS gene family with a substantial expansion in the TPS-a subgroup. This study enhances our understanding of plant secondary metabolism and sets a foundation for future explorations into the ecological and adaptive significance of TPS genes. It should be acknowledged that our study has certain limitations: the angiosperm-centric sampling may underrepresent early plant lineages, functional validation was limited to RcTPS23, and ecological correlations remain unexplored. These findings hold biotechnological promise for fragrance compound biosynthesis (e.g., engineering TPS-a genes like RcTPS23 for industrial terpene production) and agricultural innovation (e.g., developing pest-resistant crops through volatile organic compound engineering). Future investigations should prioritize integrating multi-omics approaches to link genomic patterns with terpene metabolite profiles, and employing high-throughput CRISPR screening to decipher genotype–phenotype relationships across plant taxa.

Supplementary Materials: The following supporting information can be downloaded at: https://www.mdpi.com/article/10.3390/ijms26052113/s1.

Author Contributions: Conceptualization, Y.H.; Methodology, Q.L., Y.P., T.Z. and X.L.; Software, Q.L., T.Z. and X.L.; Formal analysis, Y.P. and Q.D.; Investigation, Q.Y.; Resources, Q.D. and Q.Y.; Data curation, Y.P., T.Z., X.L. and Y.H.; Writing—original draft, Q.L. and Y.P.; Writing—review & editing, T.Z. and Y.H.; Visualization, Q.L., Q.D. and Q.Y.; Supervision, Q.Y.; Funding acquisition, Y.H. All authors have read and agreed to the published version of the manuscript.

Funding: This work was supported by the National Key Research and Development Project of China (2023YFD1200105), China; National Natural Science Foundation of China (32471938, 31800594),

China; Fundamental Research Funds for the Central Universities (QNTD202306), China; and Beijing Natural Science Foundation (6174045), China.

Institutional Review Board Statement: Not applicable.

Informed Consent Statement: Not applicable.

Data Availability Statement: The data are contained within the article and Supplementary Materials.

Acknowledgments: Thanks to Xiang Gao of Northeast Normal University for his guidance on enzyme activity experimental techniques.

Conflicts of Interest: The authors declare no conflict of interest.

References

1. Gershenzon, J.; Dudareva, N. The function of terpene natural products in the natural world. *Nat. Chem. Biol.* **2007**, *3*, 408–414. [CrossRef] [PubMed]
2. Padovan, A.; Keszei, A.; Külheim, C.; Foley, W.J. The evolution of foliar terpene diversity in Myrtaceae. *Phytochem. Rev.* **2014**, *13*, 695–716. [CrossRef]
3. Yamada, Y.; Kuzuyama, T.; Komatsu, M.; Shin-ya, K.; Omura, S.; Cane, D.E.; Ikeda, H. Terpene synthases are widely distributed in bacteria. *Proc. Natl. Acad. Sci. USA* **2015**, *112*, 857–862. [CrossRef] [PubMed]
4. Degenhardt, J.; Köllner, T.G.; Gershenzon, J. Monoterpene and sesquiterpene synthases and the origin of terpene skeletal diversity in plants. *Phytochemistry* **2009**, *70*, 1621–1637. [CrossRef]
5. Bohlmann, J.; Meyer-Gauen, G.; Croteau, R. Plant terpenoid synthases: Molecular biology and phylogenetic analysis. *Proc. Natl. Acad. Sci. USA* **1998**, *95*, 4126–4133. [CrossRef]
6. Liao, P.; Hemmerlin, A.; Bach, T.J.; Chye, M.L. The potential of the mevalonate pathway for enhanced isoprenoid production. *Biotechnol. Adv.* **2016**, *34*, 697–713. [CrossRef]
7. Vranová, E.; Coman, D.; Gruissem, W. Network analysis of the MVA and MEP pathways for isoprenoid synthesis. *Annu. Rev. Plant Biol.* **2013**, *64*, 665–700. [CrossRef]
8. Zhou, F.; Pichersky, E. The complete functional characterisation of the terpene synthase family in tomato. *New Phytol.* **2020**, *226*, 1341–1360. [CrossRef]
9. Sallaud, C.; Rontein, D.; Onillon, S.; Jabès, F.; Duffé, P.; Giacalone, C.; Thoraval, S.; Escoffier, C.; Herbette, G.; Leonhardt, N.; et al. A novel pathway for sesquiterpene biosynthesis from *Z*,*Z*-farnesyl pyrophosphate in the wild tomato *Solanum habrochaites*. *Plant Cell* **2009**, *21*, 301–317. [CrossRef]
10. Dong, L.; Miettinen, K.; Goedbloed, M.; Verstappen, F.W.A.; Voster, A.; Jongsma, M.A.; Memelink, J.; van der Krol, S.; Bouwmeester, H.J. Characterization of two geraniol synthases from *Valeriana officinalis* and *Lippia dulcis*: Similar activity but difference in subcellular localization. *Metab. Eng.* **2013**, *20*, 198–211. [CrossRef]
11. Falara, V.; Alba, J.M.; Kant, M.R.; Schuurink, R.C.; Pichersky, E. Geranyllinalool synthases in Solanaceae and other angiosperms constitute an ancient branch of diterpene synthases involved in the synthesis of defensive compounds. *Plant Physiol.* **2014**, *166*, 428–441. [CrossRef] [PubMed]
12. Karunanithi, P.S.; Zerbe, P. Terpene synthases as metabolic gatekeepers in the evolution of plant terpenoid chemical diversity. *Front. Plant Sci.* **2019**, *10*, 1166. [CrossRef] [PubMed]
13. Cao, R.; Zhang, Y.; Mann, F.M.; Huang, C.C.; Mukkamala, D.; Hudock, M.P.; Mead, M.E.; Prisic, S.; Wang, K.; Lin, F.Y.; et al. Diterpene cyclases and the nature of the isoprene fold. *Proteins* **2010**, *78*, 2417–2432. [CrossRef] [PubMed]
14. Schnee, C.; Köllner, T.G.; Gershenzon, J.; Degenhardt, J. The maize gene terpene synthase 1 encodes a sesquiterpene synthase catalyzing the formation of (*E*)-β-Farnesene, (*E*)-Nerolidol, and (*E*, *E*)-Farnesol after herbivore damage. *Plant Physiol.* **2002**, *130*, 2049–2060. [CrossRef]
15. Thoma, R.; Schulz-Gasch, T.; D'Arcy, B.; Benz, J.; Aebi, J.; Dehmlow, H.; Hennig, M.; Stihle, M.; Ruf, A. Insight into steroid scaffold formation from the structure of human oxidosqualene cyclase. *Nature* **2004**, *432*, 118–122. [CrossRef]
16. Falara, V.; Akhtar, T.A.; Nguyen, T.T.H.; Spyropoulou, E.A.; Bleeker, P.M.; Schauvinhold, I.; Matsuba, Y.; Bonini, M.E.; Schilmiller, A.L.; Last, R.L.; et al. The tomato terpene synthase gene family. *Plant Physiol.* **2011**, *157*, 770–789. [CrossRef]
17. Gao, Y.; Honzatko, R.B.; Peters, R.J. Terpenoid synthase structures: A so far incomplete view of complex catalysis. *Nat. Prod. Rep.* **2012**, *29*, 1153. [CrossRef]
18. Dudareva, N.; Klempien, A.; Muhlemann, J.K.; Kaplan, I. Biosynthesis, function and metabolic engineering of plant volatile organic compounds. *New Phytol.* **2013**, *198*, 16–32. [CrossRef]
19. Jia, Q.; Brown, R.; Köllner, T.G.; Fu, J.; Chen, X.; Wong, G.K.S.; Gershenzon, J.; Peters, R.J.; Chen, F. Origin and early evolution of the plant terpene synthase family. *Proc. Natl. Acad. Sci USA* **2022**, *119*, e2100361119. [CrossRef]

20. Chen, F.; Tholl, D.; Bohlmann, J.; Pichersky, E. The family of terpene synthases in plants: A mid-size family of genes for specialized metabolism that is highly diversified throughout the kingdom. *Plant J.* **2011**, *66*, 212–229. [CrossRef]
21. Tholl, D.; Lee, S. Terpene Specialized Metabolism in *Arabidopsis thaliana*. *Arab. Book* **2011**, *9*, e0166. [CrossRef] [PubMed]
22. Bosman, R.N.; Lashbrooke, J.G. Grapevine Mono-and Sesquiterpenes: Genetics, Metabolism, and Ecophysiology. *Front. Plant Sci.* **2023**, *14*, 1111392. [CrossRef]
23. Myburg, A.A.; Smit, M.; Wang, J.P.; Mizrachi, E. Resolving the evolutionary dynamics of eucalypt terpene synthases through pan-genome analysis. *Plant Commun.* **2023**, *4*, 100528.
24. Du, L.S.; Qi, S.Y.; Ma, J.J.; Xing, L.B.; Fan, S.; Zhang, S.W.; Li, Y.M.; Shen, Y.W.; Zhang, D.; Han, M.Y. Identification of TPS family members in apple (*Malus × domestica* Borkh.) and the effect of sucrose sprays on TPS expression and floral induction. *Plant Physiol. Biochem.* **2017**, *120*, 10–23. [CrossRef]
25. Chen, Z.Q.; Vining, K.J.; Qi, X.W.; Yu, X.; Zheng, Y.; Liu, Z.Q.; Fang, H.L.; Li, L.; Bai, Y.; Liang, C.Y.; et al. Genome-wide analysis of terpene synthase gene family in *Mentha longifolia* and catalytic activity analysis of a single terpene synthase. *Genes* **2021**, *12*, 518. [CrossRef]
26. Liu, Y.; Li, D.; Liu, Y.; Wang, J.; Liu, C. Genome-wide identification and evolution-profiling analysis of TPS gene family in Triticum plants. *Int. J. Mol. Sci.* **2024**, *25*, 8546. [CrossRef]
27. Huang, L.M.; Huang, H.; Chuang, Y.C.; Chen, W.H.; Wang, C.N.; Chen, H.H. Evolution of terpene synthases in Orchidaceae. *Int. J. Mol. Sci.* **2021**, *22*, 6947. [CrossRef]
28. Tholl, D. Terpene synthases and the regulation, diversity and biological roles of terpene metabolism. *Curr. Opin. Plant Biol.* **2006**, *9*, 297–304. [CrossRef]
29. Clotault, J.; Geoffriau, E.; Lionneton, E.; Briard, M.; Peltier, D. Carotenoid biosynthesis genes provide evidence of geographical subdivision and extensive linkage disequilibrium in the carrot. *Theor. Appl. Genet.* **2010**, *121*, 659–672. [CrossRef]
30. Cairns, T.; American Rose Society. *Modern Roses XI: The World Encyclopedia of Roses: The Most Comprehensive Listing of Roses in the World, over 24,000 Roses Described, Includes a Comprehensive List of Roses of Historical and Botanical Importance, Old Garden Roses at Sangerhausen Included in Listing Together with an Article on This Unique Rose Repository, Official Modern Rose Registrations Filed with the International Registrar for Roses, the American Rose Society, Includes Unregistered Roses in Commerce, Report on Classifications of Roses, Report on Registration Process of Roses, Report on the History of the 'Peace' Rose*; Academic Press: Cambridge, MA, USA, 2000.
31. Joichi, A.; Yomogida, K.; Awano, K.; Ueda, Y. Volatile components of tea-scented modern roses and ancient Chinese roses. *Flavour Fragr. J.* **2005**, *20*, 152–157. [CrossRef]
32. Jin, Y.X.; Cui, H.X.; Yuan, X.Y.; Liu, L.; Liu, X.J.; Wang, Y.L.; Ding, J.Q.; Xiang, H.; Zhang, X.X.; Liu, J.F.; et al. Identification of the main aroma compounds in Chinese local chicken high-quality meat. *Food Chem.* **2021**, *359*, 129930. [CrossRef] [PubMed]
33. Raymond, O.; Gouzy, J.; Just, J.; Badouin, H.; Verdenaud, M.; Lemainque, A.; Vergne, P.; Moja, S.; Choisne, N.; Pont, C.; et al. The Rosa genome provides new insights into the domestication of modern roses. *Nat. Genet.* **2018**, *50*, 772–777. [CrossRef] [PubMed]
34. Gupta, P.; Taneja, S.C.; Shah, B.A.; Sethi, V.K.; Qazi, G.N. Lipase-catalyzed separation of geometrical isomers: Geraniol–nerol. *Chem. Lett.* **2007**, *36*, 1110–1111. [CrossRef]
35. Clark, J.W.; Donoghue, P.C. Whole-genome duplication and plant macroevolution. *Trends Plant Sci.* **2018**, *23*, 933–945. [CrossRef]
36. Qiao, X.; Li, Q.; Yin, H.; Qi, K.; Li, L.; Wang, R.; Zhang, S.; Paterson, A.H. Gene duplication and evolution in recurring polyploidization–diploidization cycles in plants. *Genome Biol.* **2019**, *20*, 38. [CrossRef]
37. Zhao, T.; Schranz, M.E. Network approaches for plant phylogenomic synteny analysis. *Curr. Opin. Plant Biol.* **2017**, *36*, 129–134. [CrossRef]
38. Jiao, Y.; Paterson, A.H. Polyploidy-associated genome modifications during land plant evolution. *Phil. Trans. R. Soc. B.* **2014**, *369*, 20130355. [CrossRef]
39. Goel, M.; Sun, H.; Jiao, W.B.; Schneeberger, K. SyRI: Finding genomic rearrangements and local sequence differences from whole-genome assemblies. *Genome Biol.* **2019**, *20*, 277. [CrossRef]
40. Tang, H.; Bowers, J.E.; Wang, X.; Ming, R.; Alam, M.; Paterson, A.H. Synteny and collinearity in plant genomes. *Science* **2008**, *320*, 486–488. [CrossRef]
41. Trapp, S.C.; Croteau, R.B. Genomic organization of plant terpene synthases and molecular evolutionary implications. *Genetics* **2001**, *158*, 811–832. [CrossRef]
42. Baudino, S.; Sun, P.; Caissard, J.-C.; Nairaud, B.; Moja, S.; Magnard, J.-L.; Bony, A.; Jullien, F.; Schuurink, R.C.; Vergne, P.; et al. Rose floral scent. *Acta Hortic.* **2019**, *1232*, 69–80. [CrossRef]
43. Chen, F.; Tholl, D.; D'Auria, J.C.; Farooq, A.; Pichersky, E.; Gershenzon, J. Biosynthesis and emission of terpenoid volatiles from Arabidopsis flowers. *Plant Cell* **2003**, *15*, 481–494. [CrossRef] [PubMed]
44. Aharoni, A.; Giri, A.P.; Verstappen, F.W.A.; Bertea, C.M.; Sevenier, R.; Sun, Z.; Jongsma, M.A.; Schwab, W.; Bouwmeester, H.J. Gain and loss of fruit flavor compounds produced by wild and cultivated strawberry species. *Plant Cell* **2004**, *16*, 3110–3131. [CrossRef]

45. van Schie, C.C.; Haring, M.A.; Schuurink, R.C. Tomato linalool synthase is induced in trichomes by jasmonic acid. *Plant Mol. Biol.* **2007**, *64*, 251–263. [CrossRef]
46. Huang, X.Z.; Xiao, Y.T.; Köllner, T.G.; Jing, W.X.; Kou, J.F.; Chen, J.Y.; Liu, D.F.; Gu, S.H.; Wu, J.X.; Zhang, Y.J.; et al. The terpene synthase gene family in *Gossypium hirsutum* harbors a linalool synthase GhTPS12 implicated in direct defence responses against herbivores. *Plant Cell Environ.* **2018**, *41*, 261–274. [CrossRef]
47. Cannon, S.B.; Mitra, A.; Baumgarten, A.; Young, N.D.; May, G. The roles of segmental and tandem gene duplication in the evolution of large gene families in *Arabidopsis thaliana*. *BMC Plant Biol.* **2004**, *4*, 10. [CrossRef]
48. Jiang, S.Y.; Jin, J.; Sarojam, R.; Ramachandran, S. A comprehensive survey on the terpene synthase gene family provides new insight into its evolutionary patterns. *Genome Biol. Evol.* **2019**, *11*, 2078–2098. [CrossRef]
49. Sun, Y.; Xiao, W.; Wang, Q.; Zhang, Y.; Liu, H.; Song, J.; Li, K.; Chen, J.; Li, X. Multiple variation patterns of terpene synthases in 26 maize genomes. *BMC Genomics* **2023**, *24*, 46. [CrossRef]
50. Tholl, D. Biosynthesis and biological functions of terpenoids in plants. In *Biotechnology of Isoprenoids*; Schrader, J., Bohlmann, J., Eds.; Advances in Biochemical Engineering/Biotechnology; Springer: Cham, Switzerland, 2015; Volume 148, pp. 63–106. [CrossRef]
51. Boutanaev, A.M.; Moses, T.; Zi, J.; Nelson, D.R.; Mugford, S.T.; Peters, R.J.; Osbourn, A. Investigation of terpene diversification across multiple sequenced plant genomes. *Proc. Natl. Acad. Sci. USA* **2015**, *112*, E81–E88. [CrossRef]
52. Dewey, C.N. Positional orthology: Putting genomic evolutionary relationships into context. *Brief. Bioinform.* **2011**, *12*, 401–412. [CrossRef]
53. Lv, J.; Havlak, P.; Putnam, N.H. Constraints on genes shape long-term conservation of macro-synteny in metazoan genomes. *BMC Bioinform.* **2011**, *12*, S9–S11. [CrossRef] [PubMed]
54. Zhao, T.; Holmer, R.; de Bruijn, S.; Angenent, G.C.; van den Burg, H.A.; Schranz, M.E. Phylogenomic synteny network analysis of MADS-Box transcription factor genes reveals lineage-specific transpositions, ancient tandem duplications, and deep positional conservation. *Plant Cell* **2017**, *29*, 1278–1292. [CrossRef] [PubMed]
55. Artur, M.A.S.; Zhao, T.; Ligterink, W.; Schranz, E.; Hilhorst, H.W.M. Dissecting the genomic diversification of Late Embryogenesis Abundant (LEA) protein gene families in plants. *Genome Biol. Evol.* **2019**, *11*, 459–471. [CrossRef] [PubMed]
56. Han, Y.; Yu, J.Y.; Zhao, T.; Cheng, T.R.; Wang, J.; Yang, W.R.; Pan, H.T.; Zhang, Q.X. Dissecting the genome-wide evolution and function of R2R3-MYB transcription factor family in *Rosa chinensis*. *Genes* **2019**, *10*, 823. [CrossRef]
57. Morrone, D.; Chambers, J.; Lowry, L.; Kim, G.; Anterola, A.; Bender, K.; Peters, R.J. Gibberellin biosynthesis in bacteria: Separate ent-copalyl diphosphate and ent-kaurene synthases in *Bradyrhizobium japonicum*. *FEBS Lett.* **2009**, *583*, 475–480. [CrossRef]
58. Fraenkel, G.S. The raison d'Être of secondary plant substances: These odd chemicals arose as a means of protecting plants from insects and now guide insects to food. *Science* **1959**, *129*, 1466–1470. [CrossRef]
59. Block, A.K.; Vaughan, M.M.; Schmelz, E.A.; Christensen, S.A. Biosynthesis and function of terpenoid defense compounds in maize (*Zea mays*). *Planta* **2018**, *247*, 1051–1066. [CrossRef]
60. Zhan, X.; Qian, Y.; Mao, B. Metabolic profiling of terpene diversity and the response of prenylsynthase-terpene synthase genes during biotic and abiotic stresses in *Dendrobium catenatum*. *Int. J. Mol. Sci.* **2022**, *23*, 6398. [CrossRef]
61. Dwivedi, V.; Rao, S.; Bomzan, D.P.; Kumar, S.R.; Shanmugam, P.V.; Olsson, S.B.; Nagegowda, D.A. Functional characterization of a defense-responsive bulnesol/elemol synthase from potato. *Physiol. Plant.* **2020**, *171*, 7–21. [CrossRef]
62. Liu, G.; Yang, M.; Fu, J. Identification and characterization of two sesquiterpene synthase genes involved in volatile-mediated defense in tea plant (*Camellia sinensis*). *Plant Physiol. Biochem.* **2020**, *155*, 650–658. [CrossRef]
63. Lee, S.; Badieyan, S.; Bevan, D.R.; Herde, M.; Gatz, C.; Tholl, D. Herbivore-induced and floral homoterpene volatiles are biosynthesized by a single P450 enzyme (CYP82G1) in *Arabidopsis*. *Proc. Natl. Acad. Sci. USA* **2010**, *107*, 21205–21210. [CrossRef] [PubMed]
64. Chen, S.; Zhang, L.; Cai, X.; Li, X.; Bian, L.; Luo, Z.; Li, Z.; Chen, Z.; Xin, Z. (*E*)-Nerolidol is a volatile signal that induces defenses against insects and pathogens in tea plants. *Hortic. Res.* **2020**, *7*, 52. [CrossRef] [PubMed]
65. Knudsen, J.T.; Eriksson, R.; Gershenzon, J.; Ståhl, B. Diversity and distribution of floral scent. *Bot. Rev.* **2006**, *72*, 1. [CrossRef]
66. Kamatou, G.P.P.; Viljoen, A.M. Linalool—A review of a biologically active compound of commercial importance. *Nat. Prod. Commun.* **2008**, *3*, 1183–1192. [CrossRef]
67. Ginglinger, J.-F.; Boachon, B.; Höfer, R.; Paetz, C.; Köllner, T.G.; Miesch, L.; Lugan, R.; Baltenweck, R.; Mutterer, J.; Ullmann, P.; et al. Gene coexpression analysis reveals complex metabolism of the monoterpene alcohol linalool in *Arabidopsis* flowers. *Plant Cell* **2013**, *25*, 4640–4657. [CrossRef] [PubMed]
68. Magnard, J.-L.; Bony, A.R.; Bettini, F.; Campanaro, A.; Blerot, B.; Baudino, S.; Jullien, F. Linalool and linalool nerolidol synthases in roses, several genes for little scent. *Plant Physiol. Biochem.* **2018**, *127*, 74–87. [CrossRef]
69. Zhang, J. Plantlist: Looking up the Status of Plant Scientific Names Based on the Plant List Database. R package version 0.6.5. 2020. Available online: https://github.com/helixcn/plantlist/ (accessed on 1 September 2020).

70. Finn, R.D.; Clements, J.; Eddy, S.R. HMMER web server: Interactive sequence similarity searching. *Nucleic Acids Res.* **2011**, *39*, W29–W37. [CrossRef]
71. Stamatakis, A. RAxML version 8: A tool for phylogenetic analysis and post-analysis of large phylogenies. *Bioinformatics* **2014**, *30*, 1312–1313. [CrossRef]
72. Guindon, S.; Gascuel, O. A simple, fast, and accurate algorithm to estimate large phylogenies by Maximum Likelihood. *Syst. Biol.* **2003**, *52*, 696–704. [CrossRef]
73. Letunic, I.; Bork, P. Interactive Tree Of Life (iTOL) v5: An online tool for phylogenetic tree display and annotation. *Nucleic Acids Res.* **2021**, *49*, W293–W296. [CrossRef]
74. Wang, Y.P.; Tang, H.B.; DeBarry, J.D.; Tan, X.; Li, J.P.; Wang, X.Y.; Lee, T.H.; Jin, H.Z.; Marler, B.; Guo, H.; et al. MCScanX: A toolkit for detection and evolutionary analysis of gene synteny and collinearity. *Nucleic Acids Res.* **2012**, *40*, e49. [CrossRef]
75. Bastian, M.; Heymann, S.; Jacomy, M. Gephi: An Open Source Software for Exploring and Manipulating Networks. *Proc. Int. AAAI Conf. Web Soc. Media* **2009**, *3*, 361–362. [CrossRef]
76. Palla, G.; Derényi, I.; Farkas, I.; Vicsek, T. Uncovering the overlapping community structure of complex networks in nature and society. *Nature* **2005**, *435*, 814–818. [CrossRef] [PubMed]
77. Fortunato, S. Community detection in graphs. *Phys. Rep.* **2010**, *486*, 75–174. [CrossRef]
78. Dixon, P. VEGAN, a package of R functions for community ecology. *J. Veg. Sci.* **2003**, *14*, 927–930. [CrossRef]
79. Voorrips, R.E. MapChart: Software for the graphical presentation of linkage maps and QTLs. *J. Hered.* **2002**, *93*, 77–78. [CrossRef]
80. Conesa, A.; Madrigal, P.; Tarazona, S.; Gomez-Cabrero, D.; Cervera, A.; McPherson, A.; Szcześniak, M.W.; Gaffney, D.J.; Elo, L.L.; Zhang, X.G. A survey of best practices for RNA-seq data analysis. *Genome Biol.* **2016**, *17*, 13. [CrossRef]
81. Han, Y.; Wan, H.H.; Cheng, T.R.; Wang, J.; Yang, W.R.; Pan, H.T.; Zhang, Q.X. Comparative RNA-seq analysis of transcriptome dynamics during petal development in *Rosa Chinensis*. *Sci. Rep.* **2017**, *7*, 43382. [CrossRef]
82. Livak, K.J.; Schmittgen, T.D. Analysis of relative gene expression data using real-time quantitative PCR and the $2^{-\Delta\Delta CT}$ method. *Methods* **2001**, *25*, 402–408. [CrossRef]
83. Abel, S.; Theologis, A. Transient transformation of Arabidopsis leaf protoplasts: A versatile experimental system to study gene expression. *Plant J.* **1994**, *5*, 421–427. [CrossRef]
84. Yang, S.; Wang, N.; Kimani, S.; Li, Y.; Bao, T.; Ning, G.; Li, L.; Liu, B.; Wang, L.; Gao, X. Characterization of Terpene synthase variation in flowers of wild Aquilegia species from Northeastern Asia. *Hortic. Res.* **2022**, *9*, uhab020. [CrossRef] [PubMed]
85. Lv, S.; Wu, Y.; Zhou, J.; Lian, M.; Li, C.; Xu, Y.; Meng, Q. The study of fingerprint characteristics of Dayi Pu-erh tea using a fully automatic HS-SPME/GC–MS and combined chemometrics method. *PLoS ONE* **2014**, *9*, e116428. [CrossRef]

International Journal of
Molecular Sciences

MDPI

Article

Transcriptome Analysis and Resistance Identification of *bar* and *BPH9* Co-Transformation Rice

Sanhe Li [1,†], Changyan Li [1,†], Jianyu Wang [1], Lei Zhou [1], Bian Wu [1], Zaihui Zhou [2], Xiaolei Fan [3], Aiqing You [1,*] and Kai Liu [1,*]

1 Hubei Key Laboratory of Food Crop Germplasm and Genetic Improvement, Laboratory of Crop Molecular Breeding, Food Crops Institute, Hubei Academy of Agricultural Sciences, Ministry of Agriculture and Rural Affairs, Wuhan 430064, China; lisanhe1@hbaas.com (S.L.); lichangyan@hbaas.com (C.L.); 202221107012051@stu.hubu.edu.cn (J.W.); zhoulei@hbaas.com (L.Z.); wubian@hbaas.com (B.W.)

2 College of Life Science and Technology, Huazhong Agricultural University, Wuhan 430070, China; zaihuizhou@mail.hzau.edu.cn

3 Jiangsu Key Laboratory of Crop Genomics and Molecular Breeding, College of Agriculture, Yangzhou University, Yangzhou 225009, China; xlfan@yzu.edu.cn

* Correspondence: aq_you@hbaas.com (A.Y.); liukai11153@hbaas.com (K.L.)

† These authors contributed equally to this work.

Abstract: Insect pests and weeds are the two major biotic factors affecting crop yield in the modern agricultural system. In this study, a brown planthopper (BPH) resistance gene (*BPH9*) and glufosinate tolerance gene (*bar*) were stacked into a single T-DNA cassette and transformed into an indica rice (*Oryza sativa* L.) line H23. The present study employed a gene stacking process that combines more than one gene/trait into an individual transgenic plant to meet the increasing cropping demands under complex conditions. The transgenic rice H23 (H23R) co-expressing *bar* and *BPH9* genes demonstrated both glufosinate tolerance and BPH resistance. We utilized transcriptome data to reveal the mechanism of *BPH9*-mediated brown planthopper resistance and to analyze the impact of exogenous transgenic fragments on upstream and downstream genes at insertion sites. The evaluation of insect resistance and glufosinate tolerance confirmed H23R as an excellent double-resistant transgenic rice. These findings indicate that H23R can satisfy insect management and weed control in the modern rice agricultural system. However, a deregulation study will help with prospective commercial planting.

Keywords: transgenic rice; *bar*; *BPH9*; resistance; transcriptome

1. Introduction

Rice is one of the most significant food crops for mankind, with the yield accounting for more than 40% of the total output of food crops. Two-thirds of the world's population depend on rice as their staple food. The safe and stable production of rice is the key to guaranteeing world food security. Various adversities (such as pathogens, pests, drought, and salinification) affect the production process, causing great economic losses; meanwhile, population growth also poses an enormous challenge to food security [1–3]. The demand brought about by the continuous growth of the population can only be met by constantly updating varieties and assuring stability and increase in the grain yield [4,5].

Weeds in the field compete with crops for water, fertilizer, light energy, and growing space, directly influencing crop yield and quality. In addition, some weeds are intermediate hosts of crop pathogens and pests, and are one of the significant biological limiting factors for enhancing crop yields, so the effective control of weeds in the field is one of the important measures to increase grain yields. As the rural population migrates to cities at an accelerated rate, the mass production and mechanization of agricultural planting is a foreseeable trend, which makes traditional manual weeding impractical [6,7]. The popularization and use of herbicides can dramatically decrease the labor required for rice field management as well as labor intensity [8]. Since the first commercialization of transgenic crops in 1996, herbicide tolerance has always been the most important trait of transgenic crops. Currently, the trait of herbicide tolerance is widely used in soybeans, maize, oilseed rape, cotton, sugar beet, and alfalfa, with the area accounting for more than 60% of the transgenic crop area in the globe. The major functional genes are *PAT/bar* and *EPSPS* [9–11].

Rice planthoppers are the major pests that harm rice growth [12]. In recent years, the outbreak frequency of brown planthoppers has enhanced significantly, and the damage has become increasingly serious, severely threatening the superior quality and high yield of rice [13,14]. Now chemical control is still a crucial method to control brown planthoppers. However, many countries in Asia have reported that brown planthoppers are resistant to conventional organophosphorus and carbamate insecticides [15–17]. Utilizing the pest resistance of insect-resistant rice varieties is one of the most economical and valid ways to control brown planthopper populations. The breeding and application of insect-resistant varieties possess many advantages, such as less input, simple operation, no harm to natural enemies, no pollution to the environment, and coordinated use with other technologies [18]. Due to the development of molecular cloning technology, some major genes for resistance to brown planthoppers have been successfully cloned, which facilitates the study of the insecticidal mechanism of resistance genes and the application of resistance genes to breed insect-resistant varieties [19–22].

The breeding of resistant rice through direct transgenic methods has greater advantages than traditional breeding. Appropriate regulatory elements can be used to control the expression level of resistance genes, thereby achieving faster and better pest resistance effects [23–26]. At present, around 30 *BPH* genes have been identified, and multiple *BPH* genes have been cloned [27]. The *BPH9* gene is a special *NBS-LRR* gene located on the long arm of chromosome 12 of rice. The protein it encodes is located in the endomembrane system, including the NB (nucleotide-binding) region and LRR (Leucine-rich repeat) region [28]). The *BPH9* gene of Pokkali, an indica rice variety, was cloned. In some other insect-susceptible varieties, the *BPH9* gene region is divided by two transposable elements, leading to abnormal expression of the gene. *BPH9* may produce avoidance and antibiotic effects on brown planthoppers by activating the salicylic acid and jasmonic acid signaling pathways [29].

In this study, two constitutive expression promoters were used to drive *BPH9*, the insect-resistant gene, and *bar*, the glufosinate tolerance gene, to construct the expression vector BU9-3301 and transfer into the improved rice two-line hybrid parent H23 through Agrobacterium-mediated method. Finally, the insect-resistant and herbicide-resistant rice transformant H23R was obtained. The herbicide-resistant gene and the insect-resistant gene were constructed into the same expression vector. During the transformation stage, the herbicide-resistant gene can be used to screen transformants. The obtained positive transformants have the composite trait of insect resistance and herbicide resistance and can be directly used as breeding parents, which avoids the trait aggregation process via multi-generation backcrossing and shortens the breeding cycle. Meanwhile, the *bar* gene has a history of application for many years, receiving wide evaluation and recognition in terms of safety; the *BPH9* gene is derived from Pokkali, a cultivated indica rice variety, and the safety risk has been markedly decreased. Transcriptome data were used to further analyze the *BPH9*-mediated broad-spectrum and long-lasting resistance mechanism to brown planthopper in H23R. Thus, the commercial planting of the insect-resistant and herbicide-tolerant rice transformant H23R created in this study, which has the composite trait of insect resistance and herbicide tolerance, can not only decrease the use of pesticides, reduce production costs, and protect the environment, but also retrieve yield losses caused by pests and weeds, enhance rice production efficiency, and maintain food security, which has tremendous economic, environmental, and social benefits.

2. Results

2.1. Identification of Resistance to Herbicides and Brown Planthoppers in H23R

The *bar* gene transferred in this study makes the transformant H23R resistant to the herbicide glufosinate-ammonium, which is a phenotypic trait widely used in transgenic crops. The active ingredient of glufosinate-ammonium, L-PPT (L-phosphinothricin), is an inhibitor of glutamine synthetase. Because glutamine synthetase is involved in the synthesis system of enzymes required for the ammonia metabolism pathway, the inhibition of glutamate synthase will decline the efficiency of ammonia metabolism. The rapid accumulation of ammonia will lead to the disintegration of plant cell membranes and block photosynthesis, eventually causing plant wilt or even death.

This experiment evaluated the tolerance of transformant H23R to target herbicides by manually spraying 1, 5, and 10 times the recommended field dosages. The results showed that the transgenic plant H23R also had good tolerance to 10 times the target herbicide glufosinate-ammonium, with a tolerance grade of "excellent" (Figure 1). In future commercial production, the transgenic rice H23R and its derived lines with the trait of glufosinate-ammonium tolerance can be treated with the herbicide glufosinate-ammonium for weed control and killing, which provides an available measure for weed control in the field after planting transgenic rice and further guarantees the environmental safety of transgenic rice.

Figure 1. Identification of glufosinate-ammonium resistance of transgenic materials in the field. (**A**). H23-CK one week after 1× dose of the Basta treatment. (**B**). H23R one week after 1× dose of the Basta treatment. (**C**). H23R one week after 5× dose of the Basta treatment. (**D**). H23R one week after 10× dose of the Basta treatment.

2.2. Resistance of H23R to Brown Planthopper During Seedling and Mature Stages

Brown planthoppers were inoculated at the seedling stage (2 weeks after sowing) and the mature stage (the peak tillering stage) for insect resistance identification. The results indicated that H23R showed excellent insect resistance compared to the control in both the seedling stage and the mature stage, demonstrating that H23R could be used in rice breeding for resistance to brown planthoppers (Figure 2).

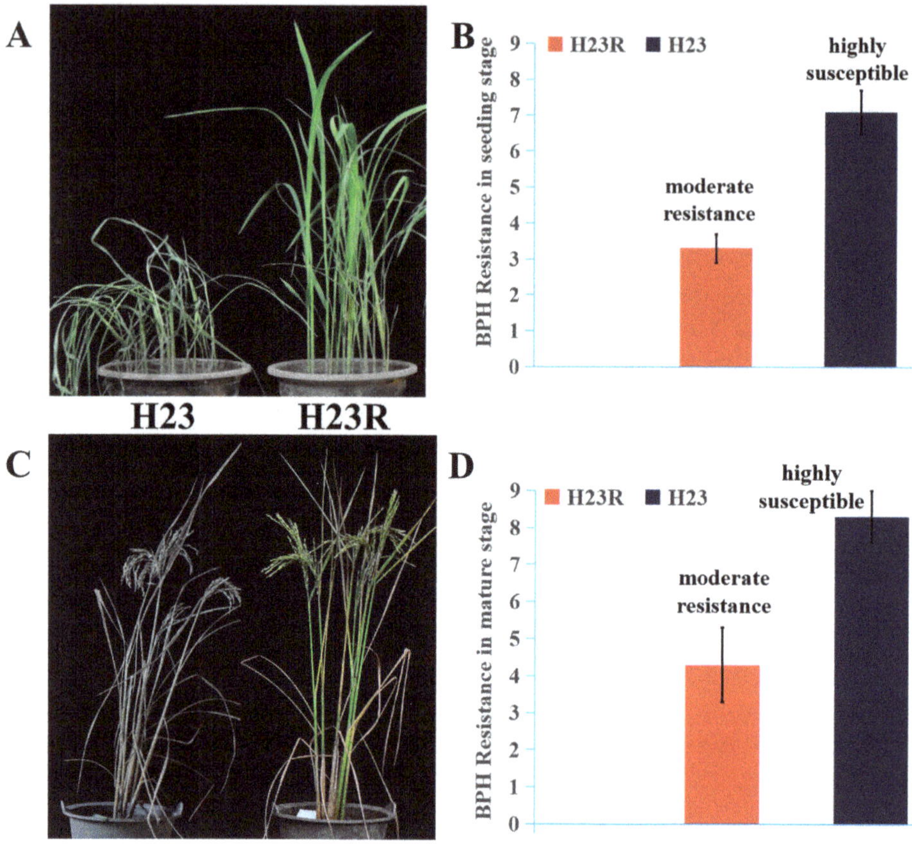

Figure 2. Resistance identification of brown planthopper. (**A**). Phenotypes of H23 and H23R resistant to brown planthopper at seeding stage. (**B**). The resistance levels of H23 and H23R to brown planthopper at the seedling stage. (**C**). Phenotypes of H23 and H23R resistant to brown planthopper at the mature stage. (**D**). The resistance levels of H23 and H23R to brown planthopper at mature stage. The methods and standards for identifying resistance of brown planthopper refer to Wang (2024) [25].

Currently, the breeding of resistant rice varieties is one of the most economical and effective ways to control brown planthopper populations. The breeding and application of insect-resistant varieties have many advantages, such as less input, simple operation, no harm to natural enemies, no pollution to the environment, and coordinated use with other technologies. The use of transgenic methods to rapidly and efficiently breed new rice germplasm with composite traits of insect resistance and herbicide resistance will significantly enhance rice production efficiency.

2.3. Identification of PAT/bar and BPH9 Protein Content in H23R

Indirect ELISA was used to detect BPH9 protein. The results showed that there was essentially no BPH9 protein detected in the H23 wild-type control (the protein content of BPH9 in the leaves, stems, and seeds of H23 was 0.57 μg/g, ND, and 0.70 μg/g, the small amount detected in leaves and seeds may be the interference of endogenous homologous proteins). The protein content of BPH9 in the leaves, stems, and seeds of H23R was 0.57 μg/g, 0.72 μg/g, and 0.70 μg/g, respectively, as shown in Figure 3A.

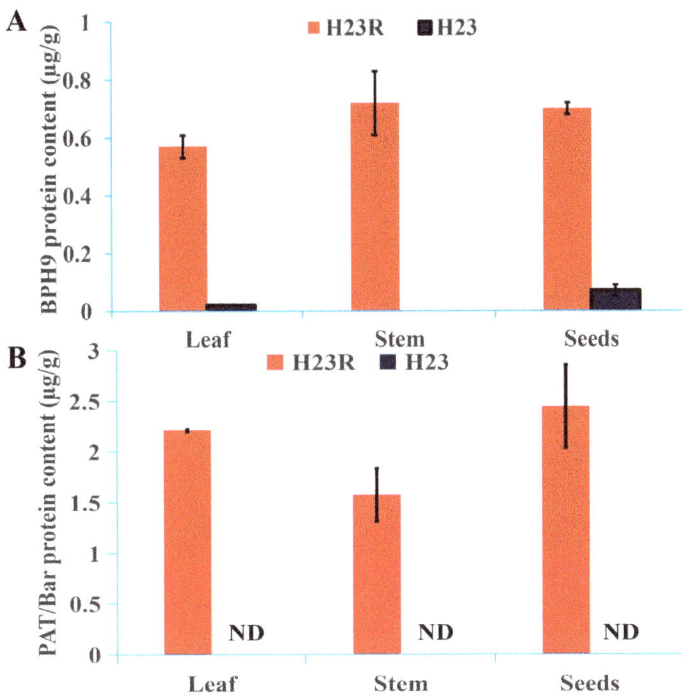

Figure 3. Target protein content in different rice tissues of T_2 generation transformant H23R. (**A**). BPH9 protein content in different tissues. (**B**). PAT/bar protein content in different tissues.

The experiment adopted the PAT/bar Enzyme-linked immunoassay quantitative detection kit (Shanghai Youlong Biotech AA1041) to detect the PAT protein expressed by the bar gene. The results showed that no PAT/bar protein was detected in the H23 wild-type control. The protein content of PAT in leaves, stems, and seeds of H23R was 2.21 μg/g, 1.57 μg/g, and 2.44 μg/g, respectively, as shown in Figure 3B.

2.4. Transcriptome Differential Gene Analysis

By comparing the transcriptome data of the transgenic and control groups, significant differences in the expression of a total of 1275 genes were observed, of which 605 genes were up-regulated and 670 genes were down-regulated (Figure 4A). The fold of difference was divided into three intervals: 1.5- to 5-fold, 5- to 10-fold, and 10-fold and above. The data showed that the expression changes in most genes were within the range of 1.5- to 5-fold, indicating that the effect of *BPH9* transgene on rice gene expression was mainly moderate up-regulation or down-regulation. Nevertheless, many genes showed more than 10-fold significant differential expression, indicating that these genes played key roles in the *BPH9* transgenic rice (Figure 4B). These highly variable genes were possibly related to specific biological functions or stress response mechanisms of the *BPH9* transgenic rice.

In the three biological replicates of H23R and H23, the differential gene expression clustering was clear and obvious, indicating good biological repeatability and reliable data results (Figure 4C,D). Eight differentially expressed genes related to resistance were selected for qRT PCR expression level validation (Figure 4E). All differential expression gene information in the transcriptome can be found in Supplementary S3.

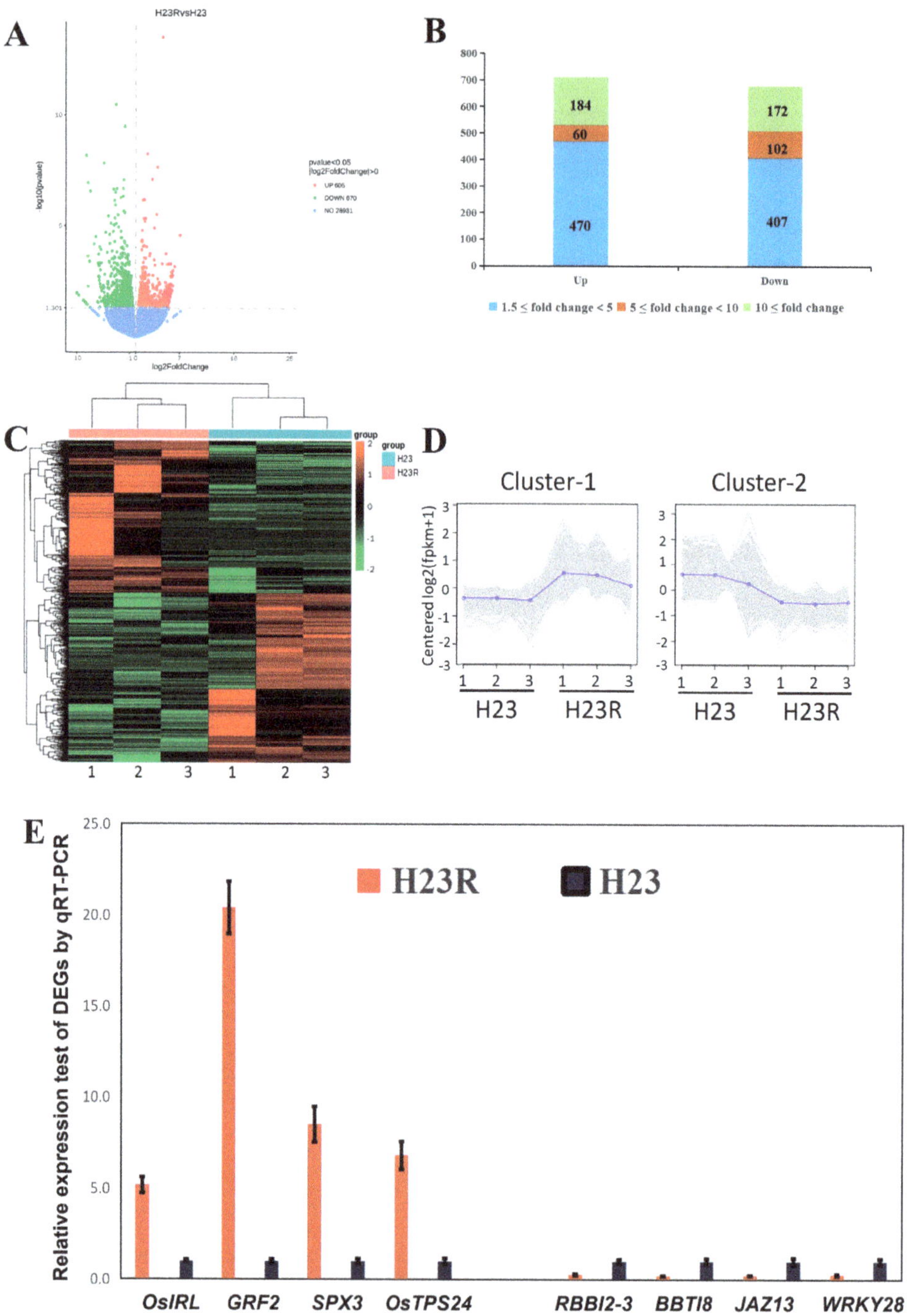

Figure 4. Differential expression gene analysis of H23R/H23 transcriptome. (**A**). Volcano plot of differentially expressed genes. (**B**). Differential fold distribution of differentially expressed genes. (**C**). Clustering of all the differentially expressed genes. (**D**). Expression pattern clustering of all the differentially expressed genes. (**E**). Q-RT PCR detection of partially differentially expressed genes.

2.5. GO Analysis of H23R/H23 DEGs

In the analysis of all the differential genes, multiple significantly enriched biological processes and molecular functions were observed. For instance, genes relevant to cell wall synthesis such as "cellulose synthase activity" and "cellulose biosynthetic process" were significantly enriched, indicating that the transfer of the *BPH9* gene may affect the synthesis and organization of rice cell walls. Moreover, the enrichment of "antioxidant activity" and "peroxidase activity" showed that the expression of the *BPH9* gene possibly enhanced the antioxidant capacity of rice, which played an important role in resisting brown planthopper infestation. In terms of molecular function, the enrichment of "flavin adenine dinucleotide binding" and "oxidoreductase activity" indicated that the *BPH9* gene probably influenced the activity of enzymes related to energy metabolism and redox reactions. These changes might be connected with the resistance mechanism of rice to brown planthoppers (Figure 5).

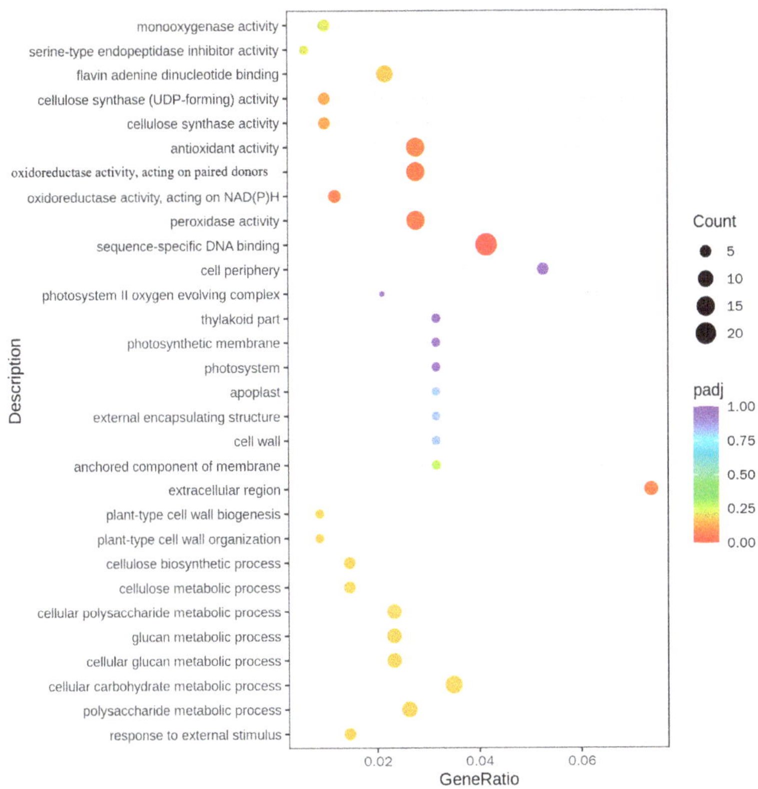

Figure 5. GO enrichment analysis of differentially expressed genes in H23R/H23.

In the analysis of up-regulated differential genes, the up-regulated expression of the genes related to energy metabolism and transport, such as "ATPase activity" and "proton-transporting ATP synthase complex", was observed, indicating that the transfer of the *BPH9* gene strengthened the energy metabolism capacity of rice, which was necessary to support the physiological activities of resistance to brown planthoppers. The enrichment of "unfolded protein binding" and "dephosphorylation" showed that the transfer of *BPH9* gene possibly affected the folding and phosphorylation state of the protein, which might be related to the signal transduction pathway that regulated rice against the infestation of brown planthoppers (Supplementary Figure S1A).

In the analysis of down-regulated differential genes, the down-regulated expression of the genes related to photosynthesis, such as "photosystem II oxygen evolving complex" and "photosynthetic membrane", was observed. This suggested that the transfer of the *BPH9* gene affected the photosynthetic efficiency of rice, which probably adjusted energy allocation to enhance resistance to brown planthoppers. In addition, the enrichment of "sequence-specific DNA binding" and "response to oxidative stress" showed that the transfer of the *BPH9* gene probably influenced the gene expression regulation and response to oxidative stress in rice. These changes might be connected with the adaptability of rice to the infestation of brown planthoppers (Supplementary Figure S1B).

2.6. KEGG Analysis of H23R/H23 DEGs

In the analysis of all the differential genes, significant enrichment of multiple metabolic pathways was observed. For instance, the enrichment of "Ubiquinone and other terpenoid-quinone biosynthesis" and "Oxidative phosphorylation" showed that the transfer of the *BPH9* gene may affect the energy metabolism and electron transport chain of rice. These changes may be connected with the enhancement of stress resistance and energy utilization efficiency of rice. The enrichment of "Carotenoid biosynthesis" and "Linoleic acid metabolism" may be related to the improvement of antioxidant capacity and membrane lipid fluidity of rice. The regulation of these metabolic pathways may help rice better adapt to environmental stress, including brown planthopper infestation. The enrichment of "Plant-pathogen interaction" showed that the transfer of *BPH9* gene may activate the defense mechanism of rice and strengthen the resistance to brown planthoppers, which may involve the activation of signal transduction pathways and the expression of defense-related genes (Figure 6 and Supplementary S4).

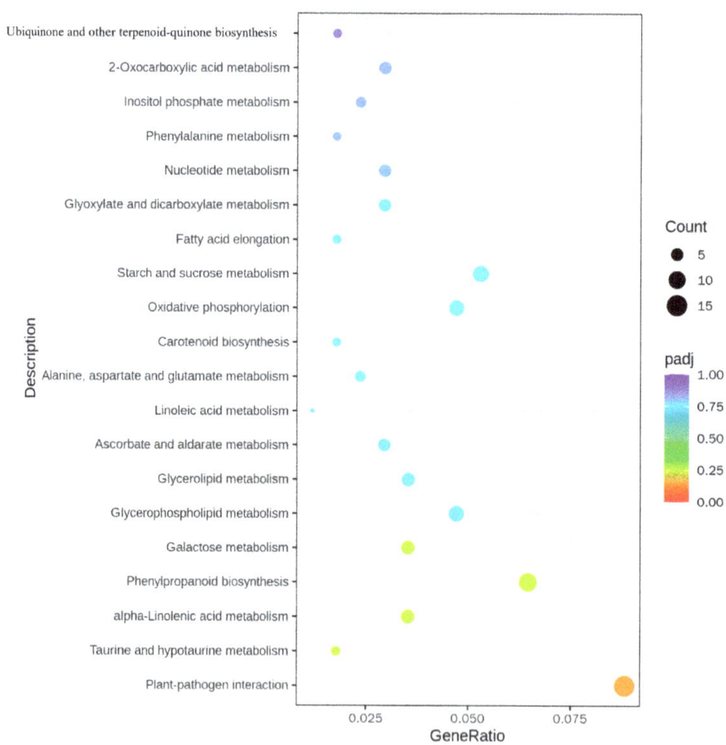

Figure 6. KEGG pathway enrichment analysis of differentially expressed genes in H23R/H23.

In the analysis of up-regulated differential genes, the enrichment of "Endocytosis" and "ABC transporters" may indicate that the transfer of the *BPH9* gene strengthens the membrane transport and material endocytosis capabilities of rice. This may help rice more availably deal with and eliminate harmful substances secreted by brown planthoppers. The enrichment of "Sesquiterpenoid and triterpenoid biosynthesis" may be related to the enhancement of secondary metabolism and synthesis of defense compounds in rice. These compounds possibly have anti-insect, anti-bacterial, or anti-fungal effects, thereby strengthening rice resistance. The enrichment of "Glycine, serine, and threonine metabolism" and "Glycerolipid metabolism" may be connected with the regulation of intracellular amino acid and lipid metabolism, which probably contribute to the maintenance of cellular function and structural integrity when rice is exposed to brown planthoppers infestation (Supplementary Figure S2A).

In the analysis of the down-regulated differential genes, the enrichment of "Photosynthesis-antenna proteins" and "Galactose metabolism" may indicate that the transfer of the *BPH9* gene affects photosynthesis and carbohydrate metabolism in rice. These changes may be connected with adjustments in energy allocation and metabolic demands to support resistance responses. The enrichment of "Sulfur relay system" and "Homologous recombination" may be connected with the regulation of the DNA repair and signal transduction pathways, which may help rice maintain genome stability and integrity in the face of brown planthopper infestation (Supplementary Figure S2B).

3. Discussion

Traditional weeding methods consume a mass of labor, while the cultivation of herbicide-resistant rice can sharply cut down the workload of manual weeding. This trait is particularly important today as labor costs are rising. By breeding herbicide-resistant rice varieties, herbicides can be used to control weeds during cultivation without damaging the growth of rice. The transgenic technology was used to make rice resistant to herbicides, enhance the competitiveness of crops with weeds in the field, and guarantee the growth stability and yield of crops. However, the long-term use of the same herbicide may cause weeds to develop resistance, so the scientific and rational use of herbicides is required. Breeding herbicide-resistant rice is of great significance to promoting the progress of agricultural science and technology, ensuring food security, and facilitating agricultural modernization. Simultaneously, it is necessary to comprehensively consider various factors, such as technology, environment, economy, and society, to guarantee healthy and orderly development. Glufosinate, a non-selective, broad-spectrum, highly efficient, and low-toxic organophosphorus herbicide, strongly inhibits the activity of glutamine synthetase (GS), an amino acid biosynthetic enzyme in bacteria and plants. GS plays a vital role in plant ammonia assimilation and metabolic regulation and is the only detoxification enzyme in plants, which can remove the toxicity of ammonia released by nitric acid reduction, amino acid degradation, and photorespiration. PAT protein is a phosphinothricin acetyltransferase (PAT, EC 2.3.1.183) which belongs to the acetyltransferase family. The substrate of PAT is phosphinothricin (PPT), which is the active ingredient of the herbicide glufosinate-ammonium. PAT can catalyze acetyl-CoA to acetylate the free amino group of PPT, the active ingredient of the herbicide glufosinate-ammonium, thereby detoxifying PPT, so that it cannot inhibit the activity of GS [11].

Since the Green Revolution, brown planthoppers have become the primary pest in global rice production. At present, most rice varieties are not insect-resistant, especially some major hybrid rice varieties even suitable for the growth and development of brown planthoppers, resulting in a widespread outbreak of brown planthoppers under the conditions of suitable climate. Brown planthoppers have developed resistance to dozens of the main insecticides (such as imidacloprid), which have special effects on pest control in the past decade, so the effectiveness of insecticide control has significantly declined. Despite the large-scale spraying of pesticides every year, serious yield losses of rice still occur in many places. Because the damage caused by brown planthoppers mostly occurs during the maturity and grain-filling stage of rice, the large-scale use of pesticides at this time also leads to severe pollution of the rice. Research has proven that using brown planthopper resistance genes to breed insect-resistant rice varieties is the most cost-effective method for comprehensive control of brown planthoppers. Combining transgenic technology with conventional breeding technology to breed brown planthoppers-resistant rice varieties and effectively control the occurrence and damage of brown planthoppers is an urgent task to guarantee green and efficient rice production and national food security.

Bph9 encodes a CC-NB-NB-LRR protein containing two NB domains located in the intracellular membrane system of rice cells. *Bph9* is principally expressed in parenchyma cells around the xylem and sieve tubes, and is not induced by brown planthoppers feeding. After brown planthoppers feeding, the contents of SA, JA, and JA-Ile in *Bph9*-resistant materials are significantly enhanced. Gene chips, quantitative PCR, and exogenous hormone treatment have verified the role of SA and JA hormones in the insect resistance of *Bph9*. *Bph9* is located on the brown planthoppers-resistant gene cluster on chromosome 12, and there are seven other brown planthoppers-resistant genes in this interval. Sequence analysis shows that these eight genes are multiple allelic forms of the same gene and can be divided into four allelic types: *Bph1/9-1*, *Bph1/9-2*, *Bph1/9-7*, and *Bph1/9-9*. The research results on *Bph9* allelic variation and different resistance spectra will provide a reference for the mining and utilization of resistance allelic types [28,29].

The transcriptome data show that *BPH9* may be upstream of the defense response to brown planthoppers. The GO analysis shows that the transfer of the *BPH9* gene has an impact on cell wall synthesis, antioxidant capacity, photosynthesis, energy metabolism, and the signal transduction of rice. These changes may work together to strengthen rice resistance to brown planthoppers. The strengthening of the cell wall may directly impede the invasion of brown planthoppers, while the enhanced antioxidant capacity and energy metabolism may support the physiological defense response of rice in the face of invasion. In addition, the response to oxidative stress and the regulation of signal transduction may involve more intricate molecular mechanisms that jointly coordinate the resistance of rice to brown planthoppers; according to KEGG analysis, the transfer of the *BPH9* gene poses a wide impact on the metabolic pathways of rice, involving energy metabolism, antioxidation, activation of defense mechanisms, adjustment of photosynthesis and carbohydrate metabolism, regulation of signal transduction and DNA repair, etc. In particular, the enhancement of energy metabolism pathways may provide rice with more energy to support its defense responses. The improvement of antioxidant capacity is conducive to protecting rice from oxidative stress caused by brown planthoppers. Meanwhile, activated defense mechanisms may involve the recognition of pathogens, activation of defense gene expression, and production of defense compounds. Adjustments in photosynthesis and carbohydrate metabolism may be to optimize energy allocation in response to infestation. The regulation of signal transduction and DNA repair is of great importance to the maintenance of genome stability and the response to external stresses. Furthermore, enhanced membrane transport and material endocytosis capabilities probably contribute to the elimination or disposal of harmful substances secreted by brown planthoppers in rice. Enhanced synthesis of secondary metabolites may provide additional anti-insect, anti-bacterial, or anti-fungal effects. These analysis data also show the complicacy of the resistance mechanism mediated by *BPH9*. There are also many challenges for brown planthoppers to overcome rice resistance. *BPH9* is a resistance gene worthy of widespread promotion and use in production.

4. Materials and Methods

4.1. Generation of H23R Transgenic Rice Plants, Molecular Identification, and Screening of Single-Copy Homozygous Families

The plant expression vector used was BU9-3301, with a full length of 14,143 bp, and the T-DNA region including the left and right border sequences was 7760 bp. Its skeleton was pcambia3301, a vector commonly used in plant genetic engineering, with a kanamycin resistance gene expressed in bacteria. Agrobacterium-mediated genetic transformation was used to transfer it into H23. Single-copy homozygous positive family H23R was screened out in the T_2 generation for further study via PCR positive detection and Southern hybridization copy number detection. The process of obtaining BPH and glufosinate-resistance rice is shown in Supplementary S1.

4.2. Identification of Resistance to Brown Planthoppers in H23R

1. Identification method of resistance to brown planthoppers at the seedling stage: The material to be identified was sown in the culture box. When the rice material grew to the two-leaf period, 7~8 brown planthopper nymphs with the age of 2~3 days old were added to each plant. When more than 90% of the control H23 died, each plant was scored and photographed according to the established grading criteria of 1 to 9. The final meaning value of each line was the resistance level of the material to be identified.
2. Identification method for brown planthoppers at the mature stage: The material to be identified was grown in the culture bucket until the peak tillering stage. Then, around 100 brown planthopper nymphs with an age of 2~3 days were added to each plant, and photos were taken two weeks later.

4.3. Identification of Herbicide Resistance

The non-transgenic control H23 and the transgenic family H23R were grown for about 3 weeks after sowing in the field. The H23R seedlings were sprayed with Basta herbicide at concentrations of 1× (720 g/ha glufosinate-ammonium), 5×, and 10×. The control H23 was sprayed with 1× glufosinate-ammonium. One week after applying glufosinate-ammonium, the growth status of the transgenic and control plants was evaluated and photographed.

4.4. Quantitative Protein Analysis of PAT and BPH9 in Different Parts of H23R and H23

The ELISA method was used to quantitatively measure the expression of different parts (seeds, stems, and leaves) of the transgenic rice.

The experiment used the PAT/bar enzyme-linked immunoassay quantitative detection kit (Shanghai Youlong Biotech AA1041, Shanghai, China) to detect the PAT protein expressed by the *bar* gene.

Indirect ELISA was adopted to detect BPH9 protein. The extracted sample solution and standards were coated on the microporous strip of the enzyme-labeled plate. When the BPH9 antibody was added, the samples/standards coated on the microporous strip captured it to form an antigen–antibody complex (Ag-Ab). The enzyme-labeled antibody (Ab-HRP) bound to the antigen–antibody complex to finally form an antigen–antibody-enzyme-labeled antibody complex (Ag-Ab-Ab-HRP), and then the color was developed by the enzyme-catalyzed TMB substrate reagent. After termination, the value was read by a microplate reader. The absorbance value of the sample is positively correlated with the BPH9 content, and the BPH9 content in the sample can be calculated by comparing it with the standard curve. The experimental method is described in Supplementary S2.

4.5. Transcriptome Analysis of H23R and Control H23

In this study, one leaf of H23R and H23 was taken as samples during the peak tillering stage (three biological replicates). Total RNA from each sample was extracted using TRIzol-reagent (Thermo Fisher Scientific, Waltham, MA, USA) according to the manufacturer's instructions and treated with RNase-free DNase I to remove genomic DNA contamination. After the determination of the quantity and quality using Agilent 2100 bioanalyzer (Santa Clara, CA, USA) and 1% denaturing gel electrophoresis, the RNA samples were sent to Smart Biotechnology Co., Ltd. (Tianjin, China) for RNA sequencing (RNA-seq).

High-quality reads were first aligned to the rice reference genome (https://rice.plantbiology.msu.edu/) using "HISAT2". The results were subjected to "featureCounts v2.0.0" to obtain the read counts of all the samples. Differentially expressed genes (DEGs) were identified using "DESeq2 1.2.4R" ($|\log2(FC)| > 1$ and Padj < 0.05). Gene Ontology (GO) (including biological process, cellular component, and molecular function) and KEGG pathway enrichment analyses of DEGs across the samples were performed through the clusterProfiler R package(4.14.4). Heatmap was illustrated using R/PHEATMAP and ComplexHeatmap, and the correlation coefficient matrices were generated and displayed using R/corrplot.

All differential expression gene information in the transcriptome can be found in Supplementary S3.

5. Conclusions

The insect-resistant and herbicide-tolerant rice H23R obtained in this study has the composite traits of insect resistance and herbicide tolerance. The planting of H23R can not only decrease the use of pesticides, reduce production costs, and protect the environment, but also reverse the yield loss caused by pests and weeds, improve rice production efficiency, and maintain food security, which has huge economic, environmental, and social benefits. *BPH9* is an allele of *BPH26*, while other BPH resistance genes in the cluster are alleles of *BPH9/26*. This gene locus exhibits extensive sequence diversity in rice germplasm resources. This discovery has significant implications for the co-evolution of host–pest interactions and the breeding of resistant varieties. Transcriptome analysis shows that corresponding to the phenotype of the high resistance to brown planthoppers of *BPH9*, its resistance mechanism involves multiple different signaling pathways and biological processes, which is beneficial to delaying the overcoming of brown planthoppers in production. *BPH9* is also a significant supplement to similar resistance gene resources in the existing production.

Supplementary Materials: The following supporting information can be downloaded at https://www.mdpi.com/article/10.3390/ijms26041762/s1.

Author Contributions: A.Y. and K.L. designed research; S.L., C.L., J.W. and L.Z. performed research; S.L., C.L., B.W., Z.Z. and X.F. analyzed data; and S.L., C.L., A.Y. and K.L. wrote the paper. All authors have read and agreed to the published version of the manuscript.

Funding: This research was supported by the STI 2030-Major Projects (2023ZD04074), the Science and Technology Major Program of Hubei Province (NO. 2022ABA001), the open funds of the Jiangsu Key Laboratory of Crop Genomics and Molecular Breeding (NO. PL202304), the Hubei Key Laboratory of Food Crop Germplasm and Genetic Improvement (NO. 2024lzjj13), and the National Natural Science Foundation of China (31701399).

Institutional Review Board Statement: Not applicable.

Informed Consent Statement: Not applicable.

Data Availability Statement: The original contributions presented in the study are included in the article and Supplementary Materials, further inquiries can be directed to the corresponding authors.

Conflicts of Interest: All the authors contributed to the article and approved the submitted version. The authors declare that the research was conducted in the absence of any commercial or financial relationships that could be construed as a potential conflict of interest.

References

1. Sinha, R.; Peláez-Vico, M.Á.; Shostak, B.; Nguyen, T.T.; Pascual, L.S.; Ogden, A.M.; Lyu, Z.; Zandalinas, S.I.; Joshi, T.; Fritschi, F.B.; et al. The effects of multifactorial stress combination on rice and maize. *Plant Physiol.* **2024**, *194*, 1358–1369. [CrossRef]
2. Zandalinas, S.I.; Peláez-Vico, M.Á.; Sinha, R.; Pascual, L.S.; Mittler, R. The impact of multifactorial stress combination on plants, crops, and ecosystems: How should we prepare for what comes next? *Plant J. Cell Mol. Biol.* **2024**, *117*, 1800–1814. [CrossRef]
3. Peláez-Vico, M.Á.; Sinha, R.; Induri, S.P.; Lyu, Z.; Venigalla, S.D.; Vasireddy, D.; Singh, P.; Immadi, M.S.; Pascual, L.S.; Shostak, B.; et al. The impact of multifactorial stress combination on reproductive tissues and grain yield of a crop plant. *Plant J. Cell Mol. Biol.* **2024**, *117*, 1728–1745. [CrossRef]
4. Zandalinas, S.I.; Sengupta, S.; Fritschi, F.B.; Azad, R.K.; Nechushtai, R.; Mittler, R. The impact of multifactorial stress combination on plant growth and survival. *New Phytol.* **2021**, *230*, 1034–1048. [CrossRef] [PubMed]
5. Shi, J.; An, G.; Weber AP, M.; Zhang, D. Prospects for rice in 2050. *Plant Cell Environ.* **2023**, *46*, 1037–1045. [CrossRef] [PubMed]
6. An, S. Insect pest control using biotechnology-preface. *Arch. Insect Biochem. Physiol.* **2019**, *102*, e21624. [CrossRef]
7. Radhakrishnan, R.; Alqarawi, A.A.; Abd Allah, E.F. Bioherbicides: Current knowledge on weed control mechanism. *Ecotoxicol. Environ. Saf.* **2018**, *158*, 131–138. [CrossRef] [PubMed]
8. Mukandiwa, L.; Naidoo, V.; Katerere, D.R. The use of Clausena anisata in insect pest control in Africa: A review. *J. Ethnopharmacol.* **2016**, *194*, 1103–1111. [CrossRef] [PubMed]
9. Sivamani, E.; Nalapalli, S.; Prairie, A.; Bradley, D.; Richbourg, L.; Strebe, T.; Liebler, T.; Wang, D.; Que, Q. A study on optimization of pat gene expression cassette for maize transformation. *Mol. Biol. Rep.* **2019**, *46*, 3009–3017. [CrossRef]
10. Nguyen, T.D.; La, V.H.; Nguyen, V.D.; Bui, T.T.; Nguyen, T.T.; Je, Y.H.; Chung, Y.S.; Ngo, X.B. Convergence of Bar and Cry1Ac Mutant Genes in Soybean Confers Synergistic Resistance to Herbicide and Lepidopteran Insects. *Front. Plant Sci.* **2021**, *12*, 698882. [CrossRef] [PubMed]
11. Li, Z.; Zhang, Z.; Liu, Y.; Ma, Y.; Lv, X.; Zhang, D.; Gu, Q.; Ke, H.; Wu, L.; Zhang, G.; et al. Identification and Expression Analysis of EPSPS and BAR Families in Cotton. *Plants* **2023**, *12*, 3366. [CrossRef] [PubMed]
12. Li, J.; Li, S.; Li, J.; Tan, X.; Zhao, Z.; Jiang, L.; Hoffmann, A.A.; Fang, J.; Ji, R. Egg-associated secretions from the brown planthopper (*Nilaparvata lugens*) activate rice immune responses. *Insect Sci.* **2024**, *31*, 1135–1149. [CrossRef] [PubMed]
13. Guo, J.; Wang, H.; Guan, W.; Guo, Q.; Wang, J.; Yang, J.; Peng, Y.; Shan, J.; Gao, M.; Shi, S.; et al. A tripartite rheostat controls self-regulated host plant resistance to insects. *Nature* **2023**, *618*, 799–807. [CrossRef]
14. Zheng, X.; Zhu, L.; He, G. Genetic and molecular understanding of host rice resistance and *Nilaparvata lugens* adaptation. *Curr. Opin. Insect Sci.* **2021**, *45*, 14–20. [CrossRef]
15. Zhang, X.; Liao, X.; Mao, K.; Zhang, K.; Wan, H.; Li, J. Insecticide resistance monitoring and correlation analysis of insecticides in field populations of the brown planthopper Nilaparvata lugens (stål) in China 2012–2014. *Pestic. Biochem. Physiol.* **2016**, *132*, 13–20. [CrossRef]
16. Wu, S.F.; Zeng, B.; Zheng, C.; Mu, X.C.; Zhang, Y.; Hu, J.; Zhang, S.; Gao, C.F.; Shen, J.L. The evolution of insecticide resistance in the brown planthopper (*Nilaparvata lugens* Stål) of China in the period 2012–2016. *Sci. Rep.* **2018**, *8*, 4586. [CrossRef] [PubMed]
17. Liao, X.; Mao, K.; Ali, E.; Zhang, X.; Wan, H.; Li, J. emporal variability and resistance correlation of sulfoxaflor susceptibility among Chinese populations of the brown planthopper *Nilaparvata lugens* (Stål). *Crop Prot.* **2017**, *102*, 141–146. [CrossRef]
18. Fan, F.; Li, N.; Chen, Y.; Liu, X.; Sun, H.; Wang, J.; He, G.; Zhu, Y.; Li, S. Development of Elite BPH-Resistant Wide-Spectrum Restorer Lines for Three and Two Line Hybrid Rice. *Front. Plant Sci.* **2017**, *8*, 986. [CrossRef] [PubMed]
19. Hu, L.; Wu, Y.; Wu, D.; Rao, W.; Guo, J.; Ma, Y.; Wang, Z.; Shangguan, X.; Wang, H.; Xu, C.; et al. The Coiled-Coil and Nucleotide Binding Domains of BROWN PLANTHOPPER RESISTANCE14 Function in Signaling and Resistance against Planthopper in Rice. *Plant Cell* **2017**, *29*, 3157–3185. [CrossRef] [PubMed]
20. Shi, S.; Wang, H.; Nie, L.; Tan, D.; Zhou, C.; Zhang, Q.; Li, Y.; Du, B.; Guo, J.; Huang, J.; et al. Bph30 confers resistance to brown planthopper by fortifying sclerenchyma in rice leaf sheaths. *Mol. Plant* **2021**, *14*, 1714–1732. [CrossRef] [PubMed]
21. Cheng, X.; Wu, Y.; Guo, J.; Du, B.; Chen, R.; Zhu, L.; He, G. A rice lectin receptor-like kinase that is involved in innate immune responses also contributes to seed germination. *Plant J. Cell Mol. Biol.* **2013**, *76*, 687–698. [CrossRef] [PubMed]
22. Liu, Y.; Wu, H.; Chen, H.; Liu, Y.; He, J.; Kang, H.; Sun, Z.; Pan, G.; Wang, Q.; Hu, J.; et al. A gene cluster encoding lectin receptor kinases confers broad-spectrum and durable insect resistance in rice. *Nat. Biotechnol.* **2015**, *33*, 301–305. [CrossRef] [PubMed]
23. Guo, J.; Xu, C.; Wu, D.; Zhao, Y.; Qiu, Y.; Wang, X.; Ouyang, Y.; Cai, B.; Liu, X.; Jing, S.; et al. Bph6 encodes an exocyst-localized protein and confers broad resistance to planthoppers in rice. *Nat. Genet.* **2018**, *50*, 297–306. [CrossRef]

24. Li, C.; Zhou, Z.; Xiong, X.; Li, C.; Li, C.; Shen, E.; Wang, J.; Zha, W.; Wu, B.; Chen, H.; et al. Development of a multi-resistance and high-yield rice variety using multigene transformation and gene editing. *Plant Biotechnol. J.* 2024; *advance online publication*. [CrossRef]
25. Wang, J.; Ma, Z.; Fu, D.; Wu, Y.; Zhou, Z.; Li, C.; Shen, J. Identification, Cloning, and Characterization of Two Acupuncture-Injury-Inducing Promoters in Rice. *Int. J. Mol. Sci.* **2024**, *25*, 10564. [CrossRef] [PubMed]
26. Li, C.; Wang, J.; Ling, F.; You, A. Application and Development of Bt Insect Resistance Genes in Rice Breeding. *Sustainability* **2023**, *15*, 9779. [CrossRef]
27. Du, B.; Chen, R.; Guo, J.; He, G. Current understanding of the genomic, genetic, and molecular control of insect resistance in rice. *Mol. Breed.* **2020**, *40*, 24. [CrossRef]
28. Wang, Y.; Jiang, W.; Liu, H.; Zeng, Y.; Du, B.; Zhu, L.; He, G.; Chen, R. Marker assisted pyramiding of Bph6 and Bph9 into elite restorer line 93-11 and development of functional marker for Bph9. *Rice* **2017**, *10*, 51. [CrossRef] [PubMed]
29. Zhao, Y.; Huang, J.; Wang, Z.; Jing, S.; Wang, Y.; Ouyang, Y.; Cai, B.; Xin, X.F.; Liu, X.; Zhang, C.; et al. Allelic diversity in an NLR gene BPH9 enables rice to combat planthopper variation. *Proc. Natl. Acad. Sci. USA* **2016**, *113*, 12850–12855. [CrossRef]

International Journal of
Molecular Sciences

MDPI

Article

The Mulberry *WRKY* Transcription Factor *MaWRKYIIc7* Participates in Regulating Plant Drought Stress Tolerance

Xueqiang Su †, Manli Zhao †, Rong Zhou, Cuimin Xu, Ran Zhang, Ruixue Li * and Taichu Wang *

Sericultural Research Institute, Anhui Academy of Agricultural Sciences, Hefei 230000, China;
suxueqiangaaas@163.com (X.S.); 13645657446@163.com (M.Z.); mushroomrz@163.com (R.Z.);
xu_cuimin2022@163.com (C.X.); 18326199114@163.com (R.Z.)

* Correspondence: li-ruixue@163.com (R.L.); wangtaichuaaas@163.com (T.W.)

† These authors contributed equally to this work.

Abstract: The sericulture industry is an important component of the agricultural industry. Drought stress can cause yellowing, premature ageing, and the shrinkage of mulberry (*Morus alba* L.) leaves, greatly damaging their quality and restricting the high-quality development of the sericulture industry. *WRKY* transcription factors play important roles in the plant drought stress response. In this study, we found that *MaWRKYIIc7* of the mulberry *WRKY* TFs, had significantly higher expression levels in leaves than in other tissues and was induced to be expressed under drought stress. The *MaWRKYIIc7* protein is located in the nucleus and plasma membrane, and its transcriptional activity depends mainly on the N-terminal sequence. The overexpression of *MaWRKYIIc7* in *Arabidopsis* resulted in better drought tolerance. An analysis of the transient overexpression of *MaWRKYIIc7* in mulberry seedlings under drought stress revealed that the transgenic seedlings presented decreased stomatal opening, decreased MDA content, increased ROS clearance ability, and increased the expression of ABA biosynthesis-related genes. The Y1H and Dual-luc results indicate that *MaWRKYIIc7* can bind W-boxes to positively regulate *MaNCED1* and *MaRD29A*, synergistically regulating the drought tolerance of mulberry. Overall, our research suggests that *MaWRKYIIc7* can increase plant drought tolerance by promoting ROS clearance, adjusting stomatal opening, and activating the ABA signalling pathway.

Keywords: mulberry; WRKY TFs; drought stress; transcriptional regulation

1. Introduction

Droughts are one of the greatest threats to crop production, and there is a large area of arid or semiarid farmland worldwide, accounting for approximately 50% of the total farmland area. Crops under drought stress experience an average yield reduction of 48% [1,2]. Drought stress can cause damage to normal physiological processes in plants, leading to adverse reactions such as yellowing, atrophy, and a decreased photosynthetic rate [3,4]. As the main source of food for silkworms and a potential health food for humans, mulberry plants mainly produce leaves, which are relatively sensitive to drought or high temperatures [5,6]. At present, silkworm breeding is developing towards industrialisation. Artificial diets for silkworms, made from fresh mulberry leaves, overcome the seasonal limitations of traditional breeding and provide new opportunities for the development of related industries [7]. However, drought stress can reduce the yield of mulberry leaves per unit area, leading to premature ageing, shrinkage, and insufficient nutrient supply, greatly damaging the quality of mulberry leaves and affecting their direct feeding and artificial feed production. This is very unfavourable for the development of the industry and the expansion of its economic benefits.

Analysing the regulatory genes related to plant drought resistance and selecting new plant varieties with high drought resistance through precision breeding are highly important for effectively reducing yield losses caused by drought and stabilising agricultural income [8,9]. Multiple transcription factors are involved in the plant drought stress response, among which *WRKY* transcription factors (*WRKY* TFs) play important roles [10]. The *WRKY* TFs are a class of functionally diverse plant protein superfamilies that are involved in various biological processes, such as plant growth, embryonic morphogenesis, biosynthesis, and stress response [11]. Typical *WRKY* TFs include the WRKY domain (WRKYGQK motif) located at the N-terminus and two types of zinc finger structures ($C\text{-}X_{4\text{-}5}\text{-}C\text{-}X_{22\text{-}23}\text{-}H\text{-}X\text{-}H$, C_2H_2 and $C\text{-}X_{5\text{-}8}\text{-}C\text{-}X_{25\text{-}28}\text{-}H\text{-}X_{1\text{-}2}\text{-}C$, C_2HC) [12]. Owing to their different genetic structures, *WRKY* TFs can be divided into three subfamilies: group I contains two WRKY domains and one C_2H_2; group II has one WRKY domain and one C_2H_2 (which can be further divided into five subtypes); and group III consists of one WRKY domain and one C_2HC [13]. The *WRKY* TFs can exert regulatory functions by binding to the W-box *cis*-acting elements (C/TTGACT/C) in the downstream gene promoters [14].

Since the first *WRKY* TFs were reported in sweet potato in 1994, research on the *WRKY* TFs has been ongoing [15]. At present, researchers have completed the whole-genome identification of *WRKY* TFs in multiple species, such as *Arabidopsis*, rice, and wheat [11]. Among them, the *WRKY* TFs of *Glycine max*, *Brassica napus* and *Panicum virgatum* are relatively large in scale, with the number of family members ranging from 275 to 296 [12,13]. Multiple studies have shown that *WRKY* TFs play important roles in the plant drought stress response and adaptation. The overexpression of *AtWRKY57* significantly improved the drought tolerance of both transgenic *Arabidopsis* and *Oryza sativa* [16]. The heterologous expression of banana *MaWRKY80* in *Arabidopsis* can lead to a greater survival rate, faster growth rate, and lower malondialdehyde and reactive oxygen species contents in transgenic plants under drought stress. *MaWRKY80* can bind to W-box and regulate ABA biosynthesis by activating the expression of *NCED*, thereby promoting leaf stoma motility and water retention [17]. Sorghum *SbWRKY30* can regulate the expression of the drought stress response-related gene *SbRD19*, thereby increasing plant drought tolerance [18]. Soybean *GmWRKY54* can reduce water loss by regulating the opening and closing of stomata. *GmWRKY54* activates the expression of genes such as *PYL8*, *SRK2A*, *CIPK11*, and *CPK3* by directly binding to the promoter region and exerting its drought stress response-related functions through the ABA and Ca^{2+} signalling pathways [19]. Cotton *GhWRKY59* regulates the expression of *GhDREB2* (dependent on W-box *cis*-acting element) and mediates the independent stress response of abscisic acid (ABA) [20]. In summary, *WRKY* TFs constitute an important class of plant drought stress response regulatory genes that can play crucial roles in the precision breeding of highly drought-tolerant crop species.

Baranwal identified 54 *WRKY* TFs in mulberry and investigated the expression patterns of these *WRKY* TFs [21]. However, the expression of mulberry *WRKY* TFs under drought stress and the biological functions of the family members involved in the drought stress response are still unclear. In this study, we further divided the duration of drought stress in mulberry plants into 0, 3, 5, 8, and 10 days and investigated the changes in the expression levels of *WRKY* TFs in mulberry plants under drought stress and in different tissues. These results suggest that *MaWRKYIIc7* is a potential regulatory factor for drought stress in mulberry. We identified the subcellular localization and transcriptional activation activity of *MaWRKYIIc7*. By investigating the transient overexpression of *MaWRKYIIc7* in mulberry seedlings, including changes in stomatal opening and closing, stress-related physiological and biochemical indicators, ROS clearance, and expression patterns of ABA biosynthesis-related genes, we revealed the biological function of *MaWRKYIIc7* in enhancing the plant response to drought stress. Yeast one-hybrid (Y1H), dual-luciferase (Dual-luc) assay results revealed that *MaWRKYIIc7* can rely on a W-box to bind to *MaNCED1* and *MaRD29A* and can participate in the stress response by activating the expression of these genes. In summary, our results enrich functional research on *WRKY* TFs and lay the foundation for elucidating the molecular mechanisms of the drought stress response in mulberry plants.

2. Results

2.1. WRKY TF Bioinformatics Analysis

We divided the *WRKY* TFs into three subfamilies according to their classification method: Groups I, II, and III. Group I contains nine members, whereas group II is subdivided into five categories: IIa (3), IIb (7), IIc (13), IId (5), and IIe (8). Group III has nine family members. The number of amino acids (AAs) of the mulberry *WRKY* TFs ranged from 180 to 769, with *MaWRKYIIc6* and *MaWRKYIIb6* having the lowest and highest AA numbers, respectively. The results of the calculation of physicochemical parameters revealed a significant difference in the kDa values of mulberry *WRKY* TFs, with members between 20 and 30 kDa accounting for 38.5% of the total (only *MaWRKYI2* and *MaWRKYIIb6* exceeded 80 kDa) (Table S1).

We identified 16 members with pI values greater than 7.5, which were mainly concentrated in the group II subfamily, with a maximum value of 9.68 (*MaWRKYIId4*). The pI values of the nine members of the Group III subfamily are less than 6.5, all of which are acidic amino acids. Only six *WRKY* TFs had a grand average hydropathicity (GRAVY) greater than -1, and the maximum GRAVY value (−0.508) was observed for *MaWRKYIId4*. The subcellular localization prediction results revealed that *MaWRKYIIc4* may be expressed in the chloroplast (chlo) or nucleus (nucl), whereas the remaining *WRKY* TFs were predicted to be expressed in the nucleus. Sequence analysis (Figure 1) revealed that all *WRKY* TFs have typical WRKY domains, and the conservation of the WRKYGQK sequence is high. However, a glutamine mutation at the Q site of *MaWRKYIIc3* and *MaWRKYIIc8* to lysine (K) is the only conserved sequence mutation, indicating the possible functional divergence of these two genes.

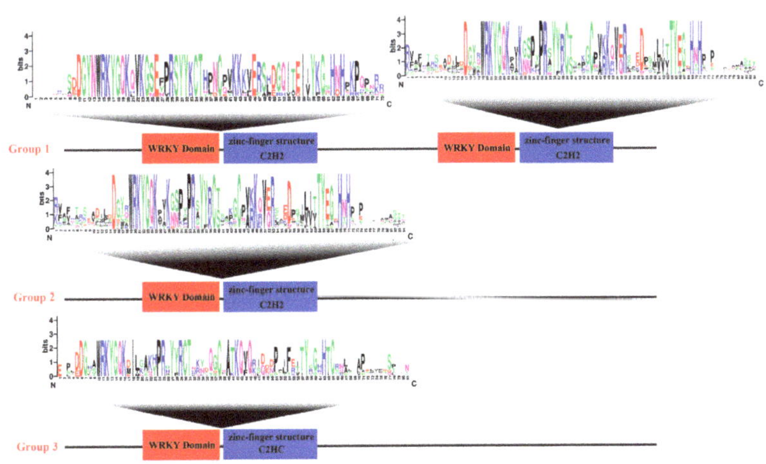

Figure 1. Conserved WRKY and zinc finger domain composition. Alignment analysis of *WRKY* TFs sequences using the ClustalW tool. These domain diagrams were plotted using the online WebLogo tool.

2.2. Environmental Selection Pressure Analysis of the WRKY TFs

To investigate the main driving forces and environmental selection pressures of *WRKY* TFs, we calculated the Ka, Ks, and Ka/Ks ratios for the 21 duplicated gene pairs in mulberry (Table 1). Ka/Ks > 1 indicates positive selection, whereas Ka/Ks < 1 indicates negative selection [22]. Among the 21 duplicated gene pairs, only *MaWRKYIIe3-MaWRKYIIe4* experienced positive selection (Ka/Ks = 1.204), while the Ka/Ks ratios of the remaining duplicated gene pairs did not exceed 0.761, indicating that these duplicated gene pairs experienced strong negative selection. We subsequently collected the *WRKY* TFs sequences of *Arabidopsis*, *Eucalyptus grandis*, *Brachypodium distachyon*, and *Medicago sativa* and compared them with those of mulberry for 4DTV value analysis (Figure S1). Among the five species, WRKY TFs were associated with large-scale duplication events, the peak of 4DTV occurred at values of approximately 0.5.

Table 1. Ka/Ks analysis for the duplicated *WRKY* TFs paralogues from mulberry.

Orthologous Gene	Ka	Ks	Ka/Ks
MaWRKYI3-MaWRKYI5	0.317	1.432	0.221
MaWRKYI7-MaWRKYI9	0.352	1.822	0.193
MaWRKYIIa1-MaWRKYIIa2	0.424	2.134	0.199
MaWRKYIIb1-MaWRKYIIb3	0.452	1.819	0.249
MaWRKYIIb1-MaWRKYIIb4	0.370	1.376	0.269
MaWRKYIIb2-MaWRKYIIb6	0.543	2.010	0.270
MaWRKYIIb3-MaWRKYIIb4	0.496	1.883	0.263
MaWRKYIIc2-MaWRKYIIc6	0.259	1.348	0.192
MaWRKYIIc3-MaWRKYIIc8	0.507	1.802	0.281
MaWRKYIIc10-MaWRKYIIc11	0.739	1.279	0.577
MaWRKYIIe3-MaWRKYIIe4	0.625	0.519	1.204
MaWRKYIIe3-MaWRKYIIe5	0.541	2.103	0.257
MaWRKYIIe4-MaWRKYIIe5	0.641	0.843	0.761
MaWRKYIII1-MaWRKYIII4	0.485	1.673	0.290
MaWRKYIII1-MaWRKYIII8	0.417	1.668	0.250
MaWRKYIII2-MaWRKYIII3	0.138	0.239	0.578
MaWRKYIII2-MaWRKYIII6	0.732	1.994	0.367
MaWRKYIII3-MaWRKYIII5	0.138	0.239	0.578
MaWRKYIII3-MaWRKYIII6	0.756	2.042	0.370
MaWRKYIII4-MaWRKYIII8	0.432	1.400	0.309
MaWRKYIII5-MaWRKYIII6	0.732	1.994	0.367

2.3. Analysis of WRKY TFs Expression Patterns

To comprehensively understand the expression specificity of *WRKY* TFs in mulberry, we investigated the changes in the expression patterns of *WRKY* TFs in leaves, fruits, roots, stems, staminate flowers, and pistillate flowers and leaves after drought stress treatment (Figures S2 and S3). The expression levels of *MaWRKYI4* and *MaWRKYIId3* in different tissue parts were relatively low, and *MaWRKYI6*, *IIc1*, *IIc8*, *IId4*, *IId5*, *IIe5*, and *IIe8* were highly expressed mainly in leaves. We identified potential regulatory factors that may be involved in fruit development in various subgroups. These *WRKY* TFs, such as *MaWRKYI5*, *IIa1*, *III1*, and *III4*, contain highly expressed levels in fruits. Most *WRKY* TFs did not have the strongest expression signals in roots or stems, with only *MaWRKYIIc2*, *IId1*, *IId2*, and *III6* (roots) and *MaWRKYIIe3* and *III4* (stems) having higher expression levels than those in other tissues. *MaWRKYI1*, *IIa2*, and *IIe1* presented high expression peaks in staminate or pistillate flowers, which may be involved in the sexual reproduction of mulberry plants (Figure S2).

Mulberry seedlings presented a severe stress phenotype after 10d of drought stress treatment. To fully explain this phenomenon, we conducted in-depth research on the changes in the expression patterns of *WRKY* TFs induced by drought stress (drought stress treatment times were divided into 0, 3, 5, 8, and 10 d) (Figure S3). The results revealed that the transcription of 24 family members (including *MaWRKYI2, IIa1, IIc4, IIc7, IIe2,* III8, etc.) was activated by drought stress; the difference was that the expression level of *MaWRKYIIc2* rapidly increased to its highest point at 3 d, the peaks of *MaWRKYIIc7, III8* occurred at 8 d, the other *WRKY* TFs in response to drought induction presented the most active transcription at 5 d. Our previous research revealed that *MaWRKYIII8* may play a role as an important regulatory factor in the mulberry drought stress response [23]. The expression pattern of *MaWRKYIIc7* is similar to that of *MaWRKYIII8*, and its expression level is significantly greater in leaves than in other tissues. Therefore, in this study, *MaWRKYIIc7* was used as a potential drought stress regulatory factor to further investigate its biological functions.

2.4. Subcellular Localization and Transcription Activity Analysis of MaWRKYIIc7

MaWRKYIIc7 is potentially involved in the response of mulberry to drought stress. We found that some core cis-*elements* in the promoter regions of *MaWRKYIIc7*, such as one MBS (MYB binding site involved in drought stress regulation), CGTCA motif and TGACG motif, are associated with the MeJA response. Notably, the promoter regions contain ABRE, indicating that *MaWRKYIIc7* may also be induced by ABA expression (Table S2). To further clarify the expression patterns of *MaWRKYIIc7*, we constructed a MaWRKYIIc7-eGFP recombinant protein, and green spontaneous fluorescence was detected from both eGFP and MaWRKYIIc7-eGFP via laser confocal microscopy. Transient MaWRKYIIc7-eGFP protein expression was localised in the nucleus and plasma membrane of tobacco leaf cells (Figure 2a).

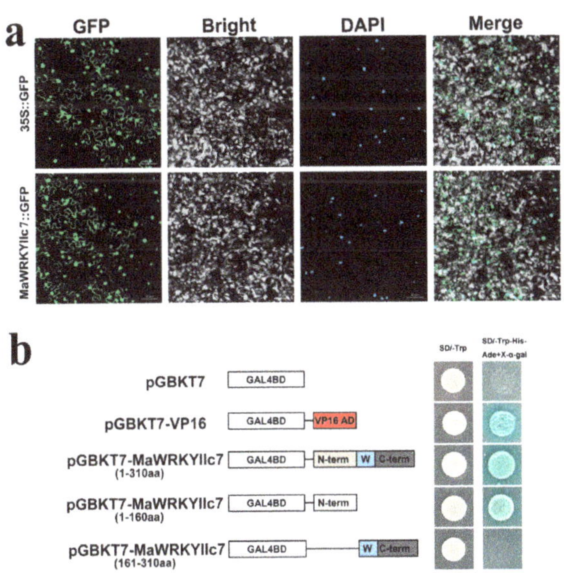

Figure 2. Subcellular localization and transcriptional activation assay of *MaWRKYIIc7*. (**a**) Subcellular localization of *MaWRKYIIc7* in the tobacco epidermis. Fluorescence detection of tobacco leaf epidermal cells, with 35S:GFP transformed tobacco epidermis as control; DAPI is the nuclear localization marker. (**b**) Schematic diagram of different fragments of *MaWRKYIIc7* connected to pGBKT7 vector. Yeast cells are cultured on tryptophan deficient medium (SD-T) and selective medium lacking tryptophan, histidine, and adenine (add X-α-D galactosidase, SD-T/H/A+X-α-gal).

We attempted to explain the potential transcriptional activity of *MaWRKYIIc7* in the absence of bait proteins. pGBKT7-*MaWRKYIIc7* (1–310 AA, 1–160 AA, and 161–310 AA) was transferred to yeast cells, which were subsequently cultured on SD/Trp-His-Ade+X-α-gal (SD-T/H/A+X-α-gal) plates, together with a pGBKT7 negative control and a pGBKT7-VP16 positive control. The growth of the yeast cells showed that pGBKT7 and pGBKT7-*MaWRKYIIc7* (161–310 AA) could not grow normally on the culture media, but the yeast cells transformed with pGBKT7-VP16 and pGBKT7-*MaWRKYIIc7* (1–310 AA, 1–160 AA) grew well (Figure 2b).

2.5. The Overexpression of MaWRKYIIc7 in Arabidopsis Enhances Drought Tolerance

Through screening with hygromycin plates and qRT-PCR, *MaWRKYIIc7* was identified in 3 lines (OEIIc7-1, -2, and -3); the expression levels of exogenous genes were highest in OEIIc7-2 (the control group was transferred to the pCAMBIA1305 empty carrier, OE1305) (Figures 3 and S4). When mannitol was used to treat transgenic *Arabidopsis* seedlings, the root growth of OEIIc7-1, -2, -3, and OE1305 plants was affected when the concentration of mannitol was 100 mM, but the root length of OEIIc7-1, -2, and -3 plants was greater than that of OE1305 plants. With increasing mannitol concentration (200 mM), the growth of all the transgenic *Arabidopsis* was further inhibited, OE1305 were still more severely affected than the other transgenic *Arabidopsis* (Figure 3).

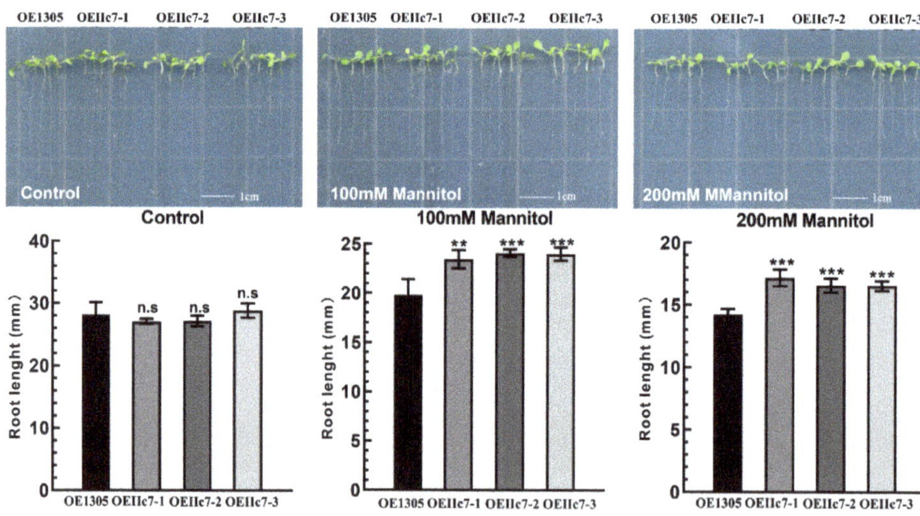

Figure 3. Overexpression of *MaWRKYIIc7* in *Arabidopsis* enhances drought tolerance. Effect of mannitol treatment on *MaWRKYIIc7* overexpression in *Arabidopsis*. Determination of main root length after **5 d** of mannitol treatment (0, 100, and 200 mM). Asterisks indicate the significant difference among OE1305, OEIIc7-1, OEIIc7-2, and OEIIc7-3 lines under the same treatment (** significant difference at $p < 0.01$, *** significant difference at $p < 0.001$, n.s indicates no significant difference).

To better understand the roles of *MaWRKYIIc7* in plants exposed to drought stress, we investigated the changes in the expression of osmotic stress marker genes (including *AtAFP2*, *AtDREB1C*, *AtDREB1D*, *AtABA1*, and *AtARCK1*) (Figure S5). The results revealed that the expression levels of most response genes were significantly upregulated and greater in OEIIc7-2 than in OE1305. Among them, *AtDREB1D* was increased by 2-fold compared with OE1305, *AtABA1* was increased by approximately 3-fold, and *AtAFP2* was also upregulated significantly. In addition, *AtARCK1* and *AtDREB1C* were not activated by *MaWRKYIIc7*.

2.6. Transient Overexpression of MaWRKYIIc7 in Mulberry Leaves Enhances Plant Drought Resistance

To clarify the biological functions of *MaWRKYIIc7*, we transformed the recombinant plasmids pCAMBIA1305 and pCAMBIA1305-*MaWRKYIIc7* into *Agrobacterium* cells (with transient overexpression in the leaves of mulberry seedlings via the injection method, TOE-*MaWRKYIIc7* and TOE-1305 were obtained). After overexpression, mulberry seedlings began to exhibit a significant atrophy phenotype 10 d after the last watering (Figure 4a). Under drought stress, the stress symptoms of TOE-*MaWRKYIIc7* were significantly lower than those of TOE-1305. For example, the development of mulberry seedling leaves in TOE-1305 was significantly inhibited, with severe shrinkage and curling, and some upright stems were damaged. The leaves of TOE-*MaWRKYIIc7* developed slightly better than those of TOE-1305 did, with some degree of wrinkling, and the middle leaves presented yellowing and curling phenomena. Upon inspection of the stomatal opening of the leaves, TOE-*MaWRKYIIc7* exhibited a significant decrease in the stomatal opening after being subjected to drought stress. Although the degree of stomatal opening in TOE-1305 also decreased, it was still greater than that in TOE-*MaWRKYIIc7* (Figure 4b).

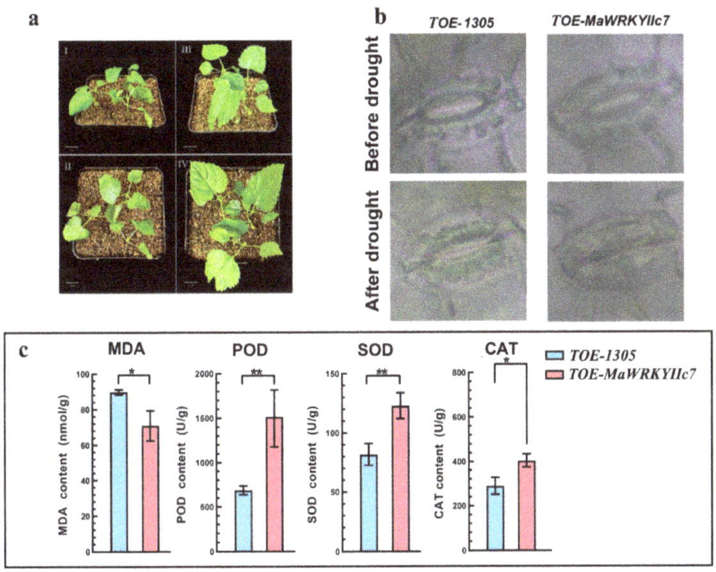

Figure 4. Phenotype of *MaWRKYIIc7* transient overexpression transformed mulberry seedlings after drought treatment. (**a**) Phenotypic analysis of *MaWRKYIIc7* transient overexpression mulberry seedlings after 10 d of drought treatment. I, II image depicts the plants subjected to control treatment (TOE-1305) and III, IV is a genetically modified plant (TOE-*MaWRKYIIc7*). (**b**) Observation of stomatal opening of TOE-1305 and TOE-*MaWRKYIIc7* before and after drought treatment. (**c**) Effects of drought treatment on the contents of MDA, POD, SOD, and CAT in TOE-1305 and TOE-*MaWRKYIIc7* (* significant difference at $p < 0.05$; ** significant difference at $p < 0.01$).

The MDA, POD, SOD, CAT activities (contents) were significantly different between transgenic plants and control group (Figure 4c). The trends of POD, SOD, and CAT activities were similar, with the TOE-*MaWRKYIIc7* content being greater than that of TOE-1305. The MDA content in TOE-1305 was the highest (89.77 nmol/g), whereas that in TOE-*MaWRKYIIc7* was only 70.96 nmol/g. These results indicate that the overexpression of *MaWRKYIIc7* enhances the defence of mulberry seedlings against drought stress, effectively reducing the adverse effects caused by stress.

2.7. MaWRKYIIc7 Activates the Expression of Drought Stress-Responsive Genes

The overexpression of transcription factors may activate (inhibit) multiple genes or pathways. We investigated the changes in the expression levels of several stress-responsive genes related to the drought stress response to better evaluate the roles of *MaWRKYIIc7* in the drought stress response (TOE-1305 as the standard reference) (Figure 5). The expression levels of genes related to proline accumulation (*MaArg, MaP5CS1, MaP5CS2,* and *MaproC*) and ROS clearance (*MaSOD2-1, MaPRX2,* and *MaCAT2*) were greater in TOE-*MaWRKYIIc7* than TOE-1305. The key genes in the ABA synthesis pathway also underwent changes. These results indicate that the overexpression of *MaWRKYIIc7* affects the expression of drought stress responsive genes in mulberry, increasing the drought tolerance of transgenic plants.

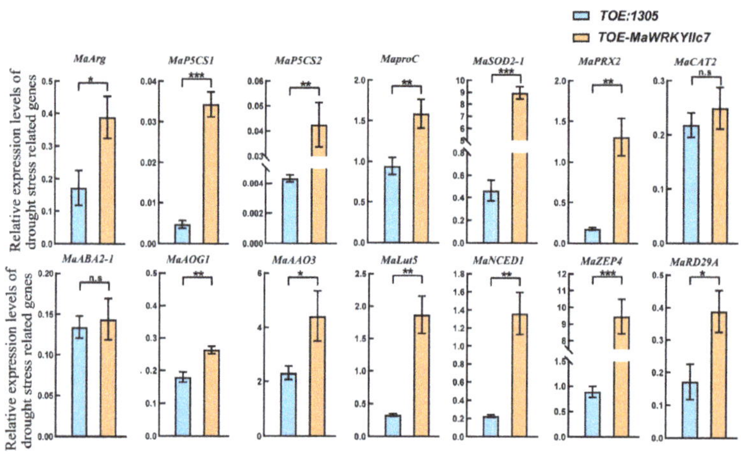

Figure 5. The expression pattern of drought stress related genes in *MaWRKYIIc7* overexpression mulberry (* significant difference at $p < 0.05$, ** significant difference at $p < 0.01$, *** significant difference at $p < 0.001$, n.s indicates no significant difference).

Researchers have demonstrated that *WRKY* TFs can regulate plant drought tolerance by modulating the expression of genes such as *RD29A*, *STZ*, and *DREB2* [24]. To investigate the molecular mechanism of *MaWRKYIIc7*, this study revealed that drought stress induced the expression of the mulberry *MaRD29A* (L484_015784) and *MaNCED1* (L484_021647), in which the promoter regions retained *cis*-acting elements related to ABA or stress (ABRE, MBS, and TC-rich repeats) and W-box elements (Figure S6) [23]. On the basis of the different positions of the promoter elements, this study divided the promoters of *MaRD29A* and *MaNCED1* into two segments (A1 and A2) and seven segments (N1-N7), and connected them to the Pabai vector (Table S3). They were subsequently cotransformed into yeast cells with the pGADT7-*MaWRKYIIc7* recombinant vector. The results showed that the yeast combination of Pabai-A2/Pabai-N7 and pGADT7-*MaWRKYIIc7* could grow normally on SD/-Leu/ABA (300 ng/mL or 500 ng/mL) media (Figure 6). Transient transcription assays were then performed to confirm the binding effect of *MaWRKYIIc7* on *MaRD29A*-A2 and *MaNCED1*-N7 via a Dual-luc reporter system (Figure 7a). The *MaRD29A*-A2 and *MaNCED1*-N7 sequences were inserted into the reporter vector to drive the expression of the firefly *luciferase* gene. The *MaRD29A*-A2-0800 and *MaNCED1*-N7-SK recombinant plasmids were transformed into *Agrobacterium*, and the bacterial cells were enriched and cultured to infect tobacco leaves (Figure 7b). The results revealed that the coexpression of *MaRD29A*-A2-Luc and *MaWRKYIIc7* resulted in significantly stronger fluorescence intensity than did the coexpression of *MaRD29A*-A2-Luc and the empty effect vector. Similarly, co-expressing *MaNCED1*-N7-Luc with *MaWRKYIIc7* also resulted in the highest luminescence intensity, indicating that *MaWRKYIIc7* could bind to the promoter regions to activate the transcription of *MaNCED1* (Figure 7c,d). The presence of a W-box in both the A2 and N7 sequences indicates that *MaWRKYIIc7* can bind to the promoters of *MaRD29A* and *MaNCED1* through a W-box.

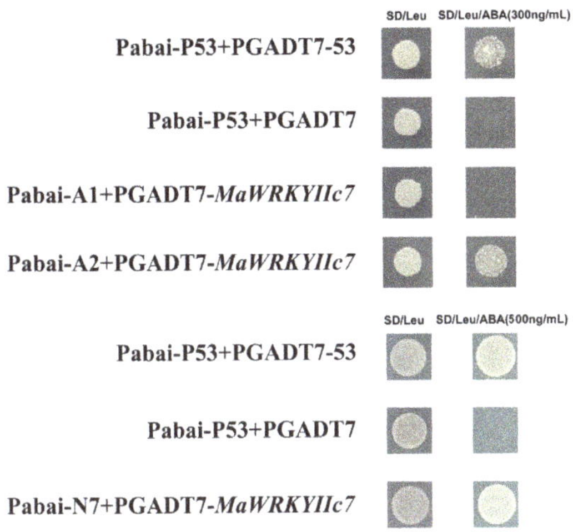

Figure 6. Y1H detection of binding activity between *MaWRKYIIc7* and interactive candidates. A1 and A2 represent fragments of *proMaRD29A*, while N7 belongs to *proMaNCED1*.

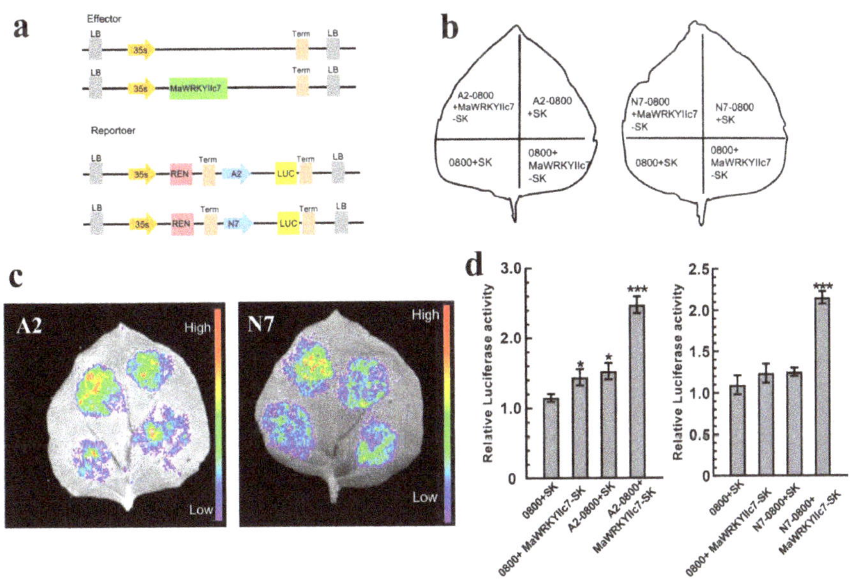

Figure 7. Dual-luc assays validate *MaWRKYIIc7* binding to *proMaRD29A* and *proMaNCED1* to activate their transcription activity. (**a**) Schematic diagrams of the effector and reporter vectors used in Dual-luc assays. (**b**) Layout pattern diagram of tobacco leaf infection. (**c**) Fluorescence intensity measurement of Dual-luc assays. (**d**) Relative luciferase activity detection. Asterisks indicate significant differences (* significant difference at $p < 0.05$, *** significant difference at $p < 0.001$).

3. Discussion

Drought is an unavoidable topic in agricultural production, which can lead to reduced crop yields [1]. Previous researchers have extensively studied drought resistance genes in plants and analysed the drought stress regulatory network centred on transcription factors such as *MYB*, *AP2/ERF*, and *NAC* [10,25]. *WRKY* TFs constitute one of the largest transcription factor families in plants and are involved in regulating various abiotic stress responses, especially the drought stress response [26,27]. In *Giardia lamblia*, *Dictyostelium discoideum*, *Chlamydomonas reinhardtii*, and other lower species, *WRKY* TFs were identified, indicating that the *WRKY* gene appeared earlier than plant differentiation did [28]. The small size of the *WRKY* gene family in mulberry may be due to large-scale chromosome breakage, duplication, or loss during two major gene replication events in plants [29], which results in some *WRKY* TFs not being retained in the genome.

In this study, we traced 21 duplicated gene pairs in mulberry, all these gene pairs related to duplication events come from the same group, such as *MaWRKYI3*/*MaWRKYI5* and *MaWRKYIIb1*/*MaWRKYIIb3* (Table 1). A study in *Pisum sativum* revealed that 16 *PsWRKYs* involved in 9 tandem duplication events all originated from the same subfamily, but 19 pairs of segmented duplication events involved some combinations of members of different subfamilies [30]. Similar conclusions can also be drawn in *Avena sativa*: 111 *WRKY* TFs duplicated gene pairs are composed of 108 fragment repeats and 3 tandem repeats. Fragment repeats are the main means of amplifying the *Avena sativa WRKY* family [31]. These results indicate that the expansion mechanism of the 111 *WRKY* TFs is different among species and that segmental duplication events may not be the main driving force for the expansion of mulberry *WRKY* TFs. This contrasts with the results of some species with larger transcription factor families, which rely on segmental duplication as a driving force for expansion [32–34]. In-depth analysis revealed that other 21 duplicated gene pairs in mulberry, except for *MaWRKYIIe3-MaWRKYIIe4* (Ka/Ks = 1.204), underwent strong purifying selection (Table 1). The Ka/Ks values of *MaWRKYIIe3-MaWRKYIIe4* are much greater than 1, indicating that under environmental pressure, *MaWRKYIIe3* and *MaWRKYIIe4* have derived new functions after gene duplication events, which is of great significance for the evolution of species [35]. An analysis of the *WRKY* TFs sequence of mulberry revealed that the length range of amino acid sequences was 180–769 amino acids (AA) (Table S1). There were significant differences in kDa values (minimum 20.9, maximum 88.3), which is consistent with the results in *Pisum sativum* (154–710 AA, 17.68–79.12 kDa) [30], indicating that environmental pressure during the evolution of mulberry *WRKY* TFs led to changes in gene structure. We found that approximately 70% of the family members had pI values less than 7.5 (all III group members had values less than 6.5), and most mulberry *WRKY* TFs tended to contain acidic amino acids.

WRKY TFs regulate growth and development and secondary metabolism and respond to external pressure in different plant organs. In our study, *WRKY* TFs presented different expression patterns in leaf, fruit, root, stem, staminate flower and pistillate flower, indicating their functional diversity in mulberry (Figure S3). For example, *MaWRKYI5*, *IIa1*, *III1*, *III4*, etc., are highly expressed in fruits and are potential regulatory factors related to fruit development; *MaWRKYI1*, *IIa2*, and *IIe1* are transcriptionally active in flowers and may be involved in regulating the development of flower organs; and *MaWRKYIIe3*, *III4* may play important roles in the morphogenesis of mulberry, as their expression levels in the stem are higher than those in other tissue parts. Unlike other gene families with close evolutionary relationships and similar expression patterns [36], mulberry *WRKY* TFs exhibit significant differences in expression patterns within each subfamily. The peak expression of nine members in group III was distributed in four tissue sites: leaves (4), fruits (1), roots (1), stems (1), and pistil flowers (2). These genes may have undergone functional divergence as a result. As important response factors for plants to cope with abiotic stress, *WRKY* TFs exhibit subfamily-specific involvement in regulation. Each group of the *WRYK* TFs has different biological functions, among which group II may be involved in plant abiotic stress response [37]. For example, *LchiWRKY18*, *EjWRKY17*, and *TaWRKY31* all belong to group II; these genes regulate drought tolerance through various pathways in response to drought stress induction [37–39]. We focused our attention on group II mulberry *WRKY* TFs, and found that **6** subfamily members were induced by drought stress, reaching peak expression levels 3–8 d after treatment, which was more than 10 times greater than those in the control group (*MaWRKYIIc2* and *IIc7*) (Figure S3). Notably, none of the six members of IIb were induced by drought stress, indicating that these genes may be involved in the response to other abiotic stresses. Among the *WRKY* TFs in which the expression was induced by drought, *MaWRKYIIc7* expression gradually increased with increasing treatment time, and the expression level in the leaves was significantly greater than that in the other tissues. According to our research as well as those in *Arabidopsis* and *Triricum aestivum*, *MaWRKYIIc7* is closely related to *TaWRKY31* and *AtWRKY57*, and these genes may increase plant drought tolerance [37].

This study further elucidated the biological functions of *MaWRKYIIc7*. We identified ABRE *cis*-acting elements in the *MaWRKYIIc7* promoter region, indicating that this gene may be induced by ABA expression (Table S2). TFs regulate gene transcription inside the nucleus, but extra-nuclear TFs have been found in many TF families, including plant-specific *NAC*, *ARF*, WRKY, etc. These TFs exist in a transcriptionally inactive state outside the nucleus and enter the nucleus through nuclear localisation signals or other pathways when the body is subjected to biotic or abiotic stress [40]. These extra-nuclear TFs can be found in the cytosol or associated with various membrane systems, including the endoplasmic reticulum and plasma membrane. Similarly to the results of soybean (*GmMYB183*) and eggplant (*SmNST1*) [41,42], the recombinant MaWRKYIIc7-eGFP we constructed produced fluorescence signals in both the nucleus and plasma membrane. This result suggests that *MaWRKYIIc7* may be an extranuclear TF that can remain outside the nucleus to control transcriptional activity. However, the mechanisms of *MaWRKYIIc7* nuclear import and export still need further research. (Figure 2a). The transcriptional activation results revealed that full-length *MaWRKYIIc7* and N-terminal segments ranging from 1 to 160 AA grew normally on SD-T/H/A media, whereas C-terminal segments containing conserved WRKY domains (161–310 AA) did not exhibit transcriptional activity (Figure 2b). Research in wheat also showed that conserved WRKY domains are not necessary for the transcriptional activation activity of *WRKY* TFs [25]. We overexpressed *MaWRKYIIc7* in *Arabidopsis* to preliminarily evaluate its biological functions. When mannitol was used to simulate drought stress, as the concentration of mannitol increased, the growth of all the transgenic plants significantly decreased, OEIIc7 seedlings were resistant to stress, the length of the roots of the OEIIc7 seedlings was significantly greater than that of the OE1305 seedlings (Figure 3). Some stress genes related to osmotic stress in *Arabidopsis*, such as *AtDREB1D*, *AtABA1*, and *AtAFP2*, were significantly upregulated in the transgenic lines overexpressing *MaWRKYIIc7* (compared with OE1305) (Figure S5). The overexpression of *MaWRKYIIc7* can increase the drought resistance of overexpressing *Arabidopsis*, and this improved ability to cope with stress may be due to the widespread activation of some related genes. This phenomenon can also be observed in other species, such as *HaWRKY76*, *TaWRKY19*, and *GmWRKY16*, which can confer stronger drought resistance in transgenic *Arabidopsis* via the upregulation of stress and ABA-responsive genes [43–45].

We constructed *MaWRKYIIc7* transiently overexpressing mulberry (TOE-*MaWRKYIIc7*) and subjected them to drought treatment under natural conditions (control group transferred to pCAMBIA1305 empty load, TOE-1305) (Figure 4a). After 10 d of natural drought,

TOE-*MaWRKYIIc7* presented better drought tolerance than TOE-1305, and the stress symptoms of leaf development damage, atrophy, and curling were also relatively mild. Transcription factors, including *WRKY* TFs, regulate the plant drought stress response, which often relies on the ABA signalling pathway. The ABA signalling pathway regulates the opening and closing of leaf stomata, which can effectively prevent rapid water loss and is an important strategy for plants to cope with drought stress [46]. *Arabidopsis AtWRKY1* can regulate the expression of *MYB2, ABCG40, DREB1A*, and other ABA signalling pathways, affecting the perception of ABA signals by leaf guard cells to control stomatal opening and closing [47]. *MfWRKY17* has been shown to regulate ABA biosynthesis and stress-related gene expression, controlling stomatal opening and closing to maintain the relative water content [48]. We observed stress phenotypes similar to those associated with the genes mentioned above. Under drought stress, the overexpression of *MaWRKYIIc7* induced stomatal closure in mulberry leaves, resulting in less stomatal opening and water dispersion loss than in TOE-1305 plants (Figure 4b). Moreover, the expression of genes related to the ABA signalling pathway (*Lut5, ZEP4*, and *NCED1*) was significantly upregulated in TOE-*MaWRKYIIc7* (Figure 5). These results showed that *MaWRKYIIc7* can activate the ABA signalling pathway in mulberry, regulate leaf stomatal opening and closing, and play a positive role in the plant's response to drought stress. When plants are subjected to abiotic stresses, excessive amounts of ROS accumulate, leading to the oxidation of lipids, proteins, nucleic acids, and even cell death [49]. The MDA content can effectively reflect the degree of lipid peroxidation and is an important indicator for monitoring plant stress capacity [50]. We found that TOE-*MaWRKYIIc7* has greater antioxidant activity than TOE-1305, with higher levels of POD, SOD, and CAT, and a small accumulation of MDA (Figure 4c). The qRT-PCR results confirmed this result, as the expression of ROS scavenging-related genes such as *MaSOD2-1, MaPRX2*, and *MaCAT2* was significantly upregulated in TOE-*MaWRKYIIc7* (Figure 5). This finding is consistent with the strategy of activating genes such as *HbWRKY82, ZmWRKY40*, and *IbWRKY2* to clear ROS and increase the response of plants to drought stress [51–53]. The results of phenotype and physiological indicators measurement showed that the leaves of *MaWRKYIIc7* overexpressing plants had stronger water retention ability and ROS scavenging ability, while the degree of lipid peroxidation in cells was weakened. In summary, *MaWRKYIIc7* can activate ROS scavenging-related genes and regulate stomatal opening and closing through the ABA signalling pathway to increase plant drought tolerance.

Numerous studies have shown that *WRKY* TFs can directly regulate the expression of genes such as *RD29A, NCED*, and *ABA3* and can participate in the regulation of the plant drought stress response. *AtWRKY57* and *MaWRKYIIc7* are homologous genes; *AtWRKY57* relies on W-box activation to express *RD26A* and *NCED3*, thereby increasing drought tolerance in *Arabidopsis* [54]. An improvement in wheat drought tolerance can be achieved by combining *TaWRKY19* with the W-box of *DREB2A* and *Cor6.6*, thereby activating *DREB2A, RD29A, DR29B*, and *Cor6.6* [44]. We identified W-boxes from both *proMaRD29A* and *proMaNCED1*. Y1H and Dual-luc confirm the molecular mechanism of *MaWRKYIIc7*-dependent W-box regulation by *MaRD29A* and *MaNCED1* (Figures 6 and 7). The results of qRP-PCR further confirmed that the upregulation of *MaRD29A* and *MaNCED1* expression was induced by drought and *MaWRKYIIc7* (Figure 5).

On the basis of the above research results, we believe that *MaWRKYIIc7* may interact with *MaRD29A* and *MaNCED1* synergistically and positively regulate the drought tolerance of mulberry through the ABA pathway. Our research enriches the results of plant stress regulation studies, demonstrating the potential of *MaWRKYIIc7* in crop resistance breeding and laying a theoretical foundation for the rapid development of the sericulture industry.

4. Materials and Methods

4.1. Plant Materials

The mulberry materials used in this study were sourced from the Mulberry Germplasm Resources Nursery (Hefei City, China), and the same water and fertiliser system and management plan were used throughout the entire process. Mulberry seedlings were obtained through seed germination. Drought stress treatment was carried out by controlling irrigation. After the last watering, sampling began when the soil moisture decreased to 25%. The first collection of mulberry seedling materials was recorded at 0 days (d), followed by a continuous collection of drought-treated materials for 3, 5, 8, and 10 d.

4.2. WRKY TFs Sequences Bioinformatics Analysis

DNATOOLS1.2.3 software is used to create a local database containing information on mulberry amino acids and nucleic acids. We used the WRKY domain sequence as a search probe and queried the mulberry *WRKY* TFs sequence via the default E value in the TBlastN search. SMART online software (http://smart.embl.de/, 30 January 2025) was used to filter out sequences without the WRKY domain or zinc finger structure, and redundant and repetitive sequences were also removed. The ExPASy website was used to investigate information such as the molecular weights and pI values of *WRKY* TFs (https://www.expasy.org/, 30 January 2025) [55]. The subcellular localisation results were predicted via the WoLF PSORT website tool (https://wolfpsort.hgc.jp/, 30 January 2025) [56]. The *cis*-elements of the *MaWRKYIIc7* promoter were analysed using the Plantcare online website (https://bioinformatics.psb.ugent.be/webtools/plantcare/html/, 30 January 2025).

4.3. Ka/Ks Ratio and 4DTV Value Analysis

The appropriate version of OrthoFinder2.0 software was downloaded via Anaconda (https://github.com/davidemms/OrthoFinder/, 30 January 2025), which was built into the backend of the server to identify the orthologous genes of mulberry *WRKY* TFs [57]. The website process can execute the following command: orthofinder-f/input-S diamond-M msa-T fasttree-t 20. The results were analysed via the online tool OrthoVenn3 (https://orthovenn3.bioinfotoolkits.net/, 30 January 2025), and the ClusterVenn module was used to statistically analyse gene clusters and perform sequence alignment of homologous gene pairs via the FFT-NS-i programme of MAFFT7.526 software [58]. The Ka/Ks ratio calculation was performed via the KaKs_Calculator2.0 tool for ParaAT2.0 (parallel alignment and back translation). Simultaneously, the batch_4DTV_calculation script was used to batch calculate the value of 4DTV on the result file processed by ParaAt2.0 [59].

4.4. Real-Time Quantitative RT-PCR

Total RNA from different tissues was extracted via TRIzol reagent (Magen Biotechnology, Guangzhou, China). The first strand of cDNA from each sample was obtained via reverse transcriptase (Vazyme, Nanjing, China). The corresponding primers used for RT-PCR are listed in Table S4. Each 20 μL qRT-PCR system consisted of 10 μL of SYBR Premix ExTaq^{TM} II (2×) (Takara, Kusatsu, Japan), 2 μL of cDNA, 6.4 μL of water, and 0.8 μL of upstream and downstream primers. *MaActin* and *MaGAPDH* were used as reference genes, and the relative expression values were calculated ($10^{-(\Delta Ct/3)}$). Three biological replicates were performed for RT-PCR detection, and the quantitative results are shown as the means ± SDs [60].

4.5. Determination of Physiological and Biochemical Indicators Related to Drought Stress

The malondialdehyde (MDA) content and activity of peroxidase (POD), superoxide dismutase (SOD), and catalase (CAT) were determined in 1 g of each leaf sample [23]. MDA determination was carried out using the thiobarbituric acid method. Quartz sand and 0.05 mol/L phosphate buffer were added to a precooled mortar and ground into a homogenate, after which 5 mL of 0.5% thiobarbituric acid solution was added. After shaking well, the mixture was boiled in a boiling water bath for 10 min, cooled, and centrifuged, and the resulting supernatant was taken as the test solution. A 0.5% thiobarbituric acid solution was used as a blank control, and the absorbance of the solution was measured at 532, 600, and 450 nm via a spectrophotometer. The MDA content in the sample was calculated via the formula $C = 6.45 (A_{532}\text{-}A_{600}) - 0.56A_{450}$ and $Y = CV/W$, where C = MDA concentration (μmol/L), V = extraction volume (mL), W = fresh tissue weight (g), Y = MDA content (nmol/g).

The activity of POD was determined using the guaiacol method. Quartz sand, calcium carbonate, and distilled water were added to the mortar, and the leaf samples were ground thoroughly. The homogenate was transferred to a 50 mL volumetric flask, adjusted to a constant volume, shaken well and centrifuged. The supernatant, which is the POD test solution, was extracted. An amount of 1 mL of the test solution was mixed with 1 mL of 0.1% guaiacol, 6.9 mL of distilled water, and 1 mL of 0.18% hydrogen peroxide. The mixture was shaken well and reacted at 25 °C for 10 min, ensuring the accuracy of the reaction time. To terminate the reaction, 0.2 mL of 5% phosphoric acid was added, the absorbance of the solution was measured at 470 nm, the value on the standard curve was read, and the POD activity in the sample was calculated.

The activity of SOD was determined using the nitro-blue tetrazolium method. Vinylpyrrolidone phosphate buffer was added to the sample in a precooled mortar, which was ground into a homogenate and transferred to a 10 mL volumetric flask and adjusted to a constant volume. The homogenised material was centrifuged, and the supernatant was collected to obtain the crude extract of SOD. A colour reagent was added for the colour reaction, the absorbance of the solution was measured at 560 nm, and the SOD activity in the sample was calculated.

When conducting CAT activity testing, the sample was ground into a homogenate in a phosphate buffer solution and diluted to 10 mL. The mixture was incubated at 4 °C for 10 min and then centrifuged to collect the supernatant, which was the CAT activity test solution. Then, 1.5 mL of phosphate buffer, 1 mL of distilled water, and 0.2 mL of crude enzyme solution were added (the crude enzyme was heated and boiled as a control). The test samples were incubated at 25 °C, and 0.3 mL of hydrogen peroxide (0.1 mol/L) was added. The absorbance was immediately measured at 240 nm, and it was recorded every minute for a total of 5 measurements.

4.6. Transformation of *MaWRKYIIc7* in *Arabidopsis* and Mulberry

The pCAMBIA1305-*MaWRKYIIc7* recombinant plasmid was transferred into *Agrobacterium* and enrichment culture (bacterial cells were collected by centrifugation). The infection buffer (0.02% Silwet L-77, 1/2 MS, 5% sucrose) was prepared, and the collected bacterial cells were suspended. A UV spectrophotometer was used to limit the OD_{600} value of the infection buffer to between 0.6 and 0.8 for subsequent infection experiments. When infecting *Arabidopsis*, first, the already grown pods were removed, the flower buds were in full bloom, and fresh flower buds were chosen for infection. During infection, the flower buds that were in good growth condition were soaked for 45 s in the invasion medium, and any remaining infection solution on the flower buds was gently removed after completion. The infected *Arabidopsis* plants were transferred to a dark environment for 1 d of cultivation, ensuring that sufficient water was provided during this period. After being cultivated in the dark, the *Arabidopsis* plants were transferred to a normal environment for cultivation and then infected once every week. During the peak flowering period, a total of three infections were performed. The *Arabidopsis* plants were continuously cultivated, and seeds were collected after they matured.

After the seeds were collected, they were disinfected with 75% alcohol for 1 min and 10% sodium hypochlorite for 13 min. After disinfection, the seeds were rinsed with sterile water four times and then evenly sown on MS solid media supplemented with hygromycin. After alternating between light and dark cultivation for 15 d, the normally growing seedlings were transferred to nutrient soil for further cultivation, and the other seedlings were discarded. The qRT-PCR detection *MaWRKYIIc7* expression levels in positive transgenic plants. The instantaneous transformation of mulberry seedlings is achieved via seedlings that have grown for 15 d after sowing and germination, and the same infection solution as described above was used for infection. A syringe was used to aspirate the infection solution, which was thoroughly injected into the lower epidermis of the mulberry leaves, and the leaves were then incubated in the dark for 36 h.

4.7. Subcellular Localization and Transcriptional Activation Analysis

Agrobacterium was enriched and cultured with the pCAMBIA1305-*MaWRKYIIc7* recombinant plasmid and the pCAMBIA1305 empty vector, and the bacterial cells were collected via centrifugation. A tobacco leaf infection solution was prepared (10 mM MES, 10 mM $MgCl_2$, and 0.1 mM AS) and this infection solution was used to resuspend the Agrobacterium cells. A UV spectrophotometer was used to set the OD_{600} value of the infection solution between 0.6 and 0.8. Flat tobacco leaves in good condition that had grown for 30 d were selected for infection. A syringe was used to administer the injection solution, which was injected thoroughly into the lower epidermis of the tobacco leaves. The mixture was incubated in the dark for 36 h, ensuring sufficient moisture during this period. After the dark cultivation was complete, the leaf tissue near the injection hole was removed and placed on a glass slide, pure water was added dropwise, the sample was covered with a cover glass, and absorbent paper was used to absorb excess water. After production was complete, the cover glass was placed downwards under a confocal laser microscope to observe the spontaneous fluorescence of eGFP.

The full-length and truncated *MaWRKYIIc7* gene was inserted into the pGBKT7 vector. All of the specific primers used are listed in supplementary Table S2. The various expression vectors, pGBKT7-*MaWRKYIIc7* (1–310 AA), pGBKT7-*MaWRKYIIc7* (1–160 AA), pGBKT7-*MaWRKYIIc7* (161–310 AA), the empty vector pGBKT7 (as a negative vector), and pGBKT7-VP16 (as a positive vector), were transferred into the yeast strain AH109, which was inoculated onto SD/-Trp and SD/-Trp/-His/Ade X-a-Gal plates and cultured for 3 d.

4.8. Yeast One-Hybrid Assay

The promoter sequence containing W-box elements was ligated into the pAbAi vector and transformed into yeast Y1H receptor cells after digestion. The minimum AbA concentration was determined, and the recombinant plasmid pGADT7-*MaWRKYIIc7* was translated into Y1H receptor cells containing the abovementioned plasmid [61]. The empty AD plasmid was used as a negative control, and yeast growth was observed on SD/−Leu plates with or without AbA.

4.9. Dual-Luciferase Imaging Assays

For the firefly luciferase complementary imaging experiment, the promoter sequence containing the W-box element was inserted into the pGreenII 0800-Luc vector containing the luciferase reporter gene, and the full-length encoding sequence of *MaWRKYIIc7* was inserted into the pGreenII 62-SK vector. The combination of different vectors was infiltrated into tobacco leaves via syringes and cultured in the dark for 2–3 d. After a luciferin (100 μM) spray, the leaves were kept in darkness for 5 min before fluorescence observation.

The luciferase and Renilla activities of the infiltrated leaves were measured on a 96-well microplate luminometer (wavelength is 560 nm). The final activity was expressed as the luciferase/Renilla activity ratio [62].

5. Conclusions

This study involved evolutionary and expression pattern analysis of the *WRKY* TFs family in mulberry and clarified the molecular mechanism of *MaWRKYIIc7*'s involvement in drought stress response. We believe that the small size of the *WRKY* TFs family in mulberry may be due to a lack of segmental duplication. *MaWRKYIIc7* is induced to express under drought stress and reached its peak expression at 8 d of stress treatment. We created *MaWRKYIIc7* transiently overexpressing lines in mulberry seedlings, and by analysing the changes in stress phenotype, stomatal opening and closing degree, ROS clearance ability, and expression patterns of stress-related genes, we clarified the biological function of *MaWRKYIIc7* in enhancing the plant response to drought stress via the ABA signalling pathway. The Y1H and Dual-luc results indicated that *MaWRKYIIc7* can bind to the W-box, positively regulating *MaNCED1* and *MaRD29A* and synergistically regulating the drought tolerance of mulberry. This study supplements the theoretical research on plant stress regulation and provides a theoretical basis and technical support for the molecular breeding of mulberry. We will use precision breeding methods in the future to create new mulberry varieties with better agronomic traits and apply our research results to production practice.

Supplementary Materials: The following supporting information can be downloaded at https://www.mdpi.com/article/10.3390/ijms26041714/s1.

Author Contributions: T.W. and R.L. designed all the details of this work. X.S., R.Z. (Rong Zhou) and M.Z. completed experiments and data analysis. X.S. and R.Z. (Ran Zhang) drafted the manuscript. C.X. provided assistance on language issues in the manuscript. All authors have read and agreed to the published version of the manuscript.

Funding: We would like to express our gratitude to all members of the research institute for their valuable feedback and inspiration. This work was financially supported by the Anhui Provincial Natural Science Foundation General Project (2308085MC97), the Youth Talent Project of Anhui Academy of Agricultural Sciences (QNYC-202214), the Open Project of Key Laboratory of Silkworm and Mulberry Genetic Improvement and the Ministry of Agriculture and Rural Affairs (KL202211), the Technology System of Characteristic Agricultural Industry of Anhui Province (ahtsnycytx-12), Supported by China Agriculture Research System of MOF and MARA, and the Open Project of Anhui Province Key Laboratory of Resource Insect Biology and Innovative Utilization (FKLRIB202406 and FKLRIB202408).

Institutional Review Board Statement: This article does not contain any studies with human participants or animals performed by any of the authors.

Data Availability Statement: The following information was supplied regarding data availability: the mulberry genome can be found in the Mulberry Genome Database (https://morus.biodb.org/links, accessed on 30 January 2025). The following information is available at https://phytozome-next.jgi.doe.gov/, 30 January 2025: *Arabidopsis*, *Eucalyptus grandis*, *Brachypodium distachyon*, and *Medicago sativa*.

Acknowledgments: The authors are deeply grateful to RSR, Bing Li, and Yongjing Deng, who provided the sample used in the study and very effective direction. The authors also thank Feng Chen and Xin Sun for providing valuable suggestions and comments.

Conflicts of Interest: The authors declare there are no competing interests.

References

1. Cohen, I.; Zandalinas, S.I.; Huck, C.; Fritschi, F.B.; Mittler, R. Meta-analysis of drought and heat stress combination impact on crop yield and yield components. *Physiol. Plant.* **2020**, *171*, 66–76. [CrossRef]
2. Liu, B.S.; Hu, Y.H.; Li, M.; Xue, H.H.; Wang, Y.; Li, Z.H. Effects of drought stress on the functional traits and rhizosphere microbial community structure of *Cyperus esculentus*. *Grassl. Sci.* **2024**, *70*, 109–120. [CrossRef]
3. Gupta, A.; Rico-Medina, A.; Caño-Delgado, A.I. The physiology of plant responses to drought. *Science* **2020**, *368*, 266–269. [CrossRef] [PubMed]
4. Martignago, D.; Rico-Medina, A.; Blasco-Escáez, D.; Fontanet-Manzaneque, J.B.; Cano-Delgado, A.I. Drought resistance by engineering plant tissue-specific responses. *Front. Plant Sci.* **2020**, *10*, 1676. [CrossRef] [PubMed]
5. Abdelrahman, M.; El-Sayed, M.; Jogaiah, S.; Burritt, D.J.; Tran, L.S.P. The "STAY-GREEN" trait and phytohormone signaling networks in plants under heat stress. *Plant Cell Rep.* **2017**, *36*, 1009–1025. [CrossRef]
6. Lei, P.; Yu, F.; Liu, X.Y. Recent advances in cellular degradation and nuclear control of leaf senescence. *J. Exp. Bot.* **2023**, *74*, 5472–5486. [CrossRef] [PubMed]
7. Wang, G.; Xu, X.X.; Ma, L.; Yang, J.M.; Yang, B.C.; Peng, R.J.; Pan, Z.H.; Sima, Y.H.; Xu, S.Q. Nonsteaming method improves the nutritional value and utilization efficiency of silkworm artificial diets. *J. Econ. Entomol.* **2024**, *117*, 388–400. [CrossRef]
8. Zareen, S.; Ali, A.; Yun, D.J. Significance of ABA biosynthesis in plant adaptation to drought stress. *J. Plant Biol.* **2024**, *67*, 175–184. [CrossRef]
9. Prasad, V.B.R.; Govindaraj, M.; Djanaguiraman, M.; Djalovic, I.; Shailani, A.; Rawat, N.; Singla-Pareek, S.L.; Pareek, A.; Prasad, P.V.V. Drought and high temperature stress in sorghum: Physiological, genetic, and molecular insights and breeding approaches. *Int. J. Mol. Sci.* **2021**, *22*, 9826. [CrossRef]
10. Manna, M.; Thakur, T.; Chirom, O.; Mandlik, R.; Deshmukh, R.; Salvi, P. Transcription factors as key molecular target to strengthen the drought stress tolerance in plants. *Physiol. Plant.* **2020**, *172*, 847–868. [CrossRef] [PubMed]
11. Khoso, M.A.; Hussain, A.; Ritonga, F.N.; Ali, Q.; Channa, M.M.; Alshegaihi, R.M.; Meng, Q.L.; Ali, M.; Zaman, W.; Brohi, R.D.; et al. *WRKY* transcription factors (TFs): Molecular switches to regulate drought, temperature, and salinity stresses in plants. *Front. Plant Sci.* **2022**, *13*, 1039329. [CrossRef] [PubMed]
12. Eulgem, T.; Rushton, P.J.; Robatzek, S.; Somssich, I.E. The *WRKY* superfamily of plant transcription factors. *Trends Plant Sci.* **2000**, *5*, 199–206. [CrossRef] [PubMed]
13. Chen, L.G.; Song, Y.; Li, S.J.; Zhang, L.P.; Zou, C.S.; Yu, D.Q. The role of *WRKY* transcription factors in plant abiotic stresses. *Biochim. Biophys. Acta (BBA)-Gene Regul. Mech.* **2012**, *1819*, 120–128. [CrossRef] [PubMed]
14. Ciolkowski, I.; Wanke, D.; Birkenbihl, R.P.; Somssich, I.E. Studies on DNA-binding selectivity of *WRKY* transcription factors lend structural clues into WRKY-domain function. *Plant Mol. Biol.* **2008**, *68*, 81–92. [CrossRef]
15. Ishiguro, S.; Nakamura, K. Characterization of a cDNA encoding a novel DNA-binding protein, *SPF1*, that recognizes SP8 sequences in the 5′ upstream regions of genes coding for sporamin and beta-amylase from sweet potato. *Mol. Gen. Genet.* **1994**, *244*, 563–571. [CrossRef]
16. Jiang, Y.J.; Qiu, Y.P.; Hu, Y.R.; Yu, D.Q. Heterologous expression of *AtWRKY57* confers drought tolerance in *Oryza sativa*. *Front. Plant Sci.* **2016**, *7*, 145. [CrossRef] [PubMed]
17. Liu, G.Y.; Li, B.; Li, X.; Wei, Y.X.; He, C.Z.; Shi, H.T. *MaWRKY80* positively regulates plant drought stress resistance through modulation of abscisic acid and redox metabolism. *Plant Physiol. Biochem.* **2020**, *156*, 155–166. [CrossRef] [PubMed]
18. Yang, Z.; Chi, X.Y.; Guo, F.F.; Jin, X.Y.; Luo, H.L.; Hawar, A.; Chen, Y.X.; Feng, K.K.; Wang, B.; Qi, J.L.; et al. *SbWRKY30* enhances the drought tolerance of plants and regulates a drought stress-responsive gene, *SbRD19*, in sorghum. *J. Plant Physiol.* **2020**, *246*, 153142. [CrossRef]
19. Wei, W.; Liang, D.W.; Bian, X.H.; Shen, M.; Xiao, J.H.; Zhang, W.K.; Ma, B.; Lin, Q.; Chen, X.; Chen, S.Y.; et al. *GmWRKY54* improves drought tolerance through activating genes in abscisic acid and Ca^{2+} signaling pathways in transgenic soybean. *Plant J.* **2019**, *100*, 384–398. [CrossRef] [PubMed]
20. Li, F.J.; Li, M.Y.; Wang, P.; Cox, K.L.; Duan, L.S.; Dever, J.K.; Shan, L.B.; Li, Z.H.; He, P. Regulation of cotton (*Gossypium hirsutum*) drought responses by mitogen-activated protein (MAP) kinase cascade-mediated phosphorylation of *GhWRKY59*. *New Phytol.* **2017**, *215*, 1462–1475. [CrossRef] [PubMed]
21. Baranwal, V.K.; Negi, N.; Khurana, P. Genome-wide Identification and Structural, Functional and evolutionary analysis of *WRKY* components of mulberry. *Sci. Rep.* **2016**, *6*, 30794. [CrossRef] [PubMed]
22. Bitocchi, E.; Rau, D.; Benazzo, A.; Bellucci, E.; Goretti, D.; Biagetti, E.; Panziera, A.; Laidò, G.; Rodriguez, M.; Gioia, T.; et al. High level of nonsynonymous changes in common bean suggests that selection under domestication increased functional diversity at target traits. *Front. Plant Sci.* **2017**, *7*, 2005. [CrossRef] [PubMed]
23. Li, R.X.; Su, X.Q.; Zhou, R.; Zhang, Y.P.; Wang, T.C. Molecular mechanism of mulberry response to drought stress revealed by complementary transcriptomic and iTRAQ analyses. *BMC Plant Biol.* **2022**, *22*, 36. [CrossRef]

24. Zhang, Y.L.; Du, P.; Xiong, F.Q.; Zhang, X.J.; Song, H. *WRKY* genes improve drought tolerance in *Arachis duranensis*. *Front. Plant Sci.* **2022**, *13*, 910480. [CrossRef]
25. Shinozaki, K.; Yamaguchi, S.K. Gene networks involved in drought stress response and tolerance. *J. Exp. Bot.* **2007**, *58*, 221–227. [CrossRef]
26. Fan, F.; Wang, Q.; Li, H.; Ding, G.; Wen, X. Transcriptome-wide identification and expression profiles of masson pine *WRKY* transcription factors in response to low phosphorus stress. *Plant Mol. Biol. Rep.* **2021**, *39*, 1–9. [CrossRef]
27. Yu, Y.; Song, T.Q.; Wang, Y.K.; Zhang, M.F.; Li, N.; Yu, M.; Zhang, S.X.; Zhou, H.W.; Guo, S.H.; Bu, Y.N.; et al. The wheat *WRKY* transcription factor *TaWRKY1-2D* confers drought resistance in transgenic *Arabidopsis* and wheat (*Triticum aestivum* L.). *Int. J. Biol. Macromol.* **2023**, *226*, 1203–1217. [CrossRef] [PubMed]
28. Zhang, Y.J.; Wang, L.J. The *WRKY* transcription factor superfamily: Its origin in eukaryotes and expansion in plants. *BMC Evol. Biol.* **2005**, *5*, 1. [CrossRef]
29. Flagel, L.E.; Wendel, J.F. Gene duplication and evolutionary novelty in plants. *New Phytol.* **2009**, *183*, 557–564. [CrossRef]
30. Xiong, R.Q.; Peng, Z.H.; Zhou, H.; Xue, G.X.; He, A.L.; Yao, X.; Weng, W.F.; Wu, W.J.; Ma, C.; Bai, Q.; et al. Genome-wide identification, structural characterization and gene expression analysis of the *WRKY* transcription factor family in pea (*Pisum sativum* L.). *BMC Plant Biol.* **2024**, *24*, 141.
31. Liu, K.Q.; Ju, Z.L.; Jia, Z.F.; Liang, G.L.; Ma, X.; Liu, W.H. Genome-wide identification and characterization of the oat (*Avena sativa* L.) *WRKY* transcription factor family. *Genes* **2022**, *13*, 1918. [CrossRef]
32. Du, D.L.; Hao, R.J.; Cheng, T.R.; Pan, H.T.; Yang, W.R.; Wang, J.; Zhang, Q.X. Genome-wide analysis of the *AP2/ERF* gene family in *Prunus mume*. *Plant Mol. Biol. Rep.* **2013**, *31*, 741–750. [CrossRef]
33. He, S.T.; Hao, X.M.; He, S.L.; Hao, X.G.; Zhang, P.; Chen, X.N. Genome-wide identification, phylogeny and expression analysis of *AP2/ERF* transcription factors family in sweet potato. *BMC Genom.* **2022**, *23*, 139. [CrossRef]
34. Nakano, T.; Suzuki, K.; Fujimura, T.; Shinshi, H. Genome-wide analysis of the *ERF* gene family in *Arabidopsis* and rice. *Plant Physiol.* **2006**, *140*, 411–432. [CrossRef]
35. Hurst, L.D. The Ka/Ks ratio: Diagnosing the form of sequence evolution. *Trends Genet.* **2002**, *18*, 486–487. [CrossRef] [PubMed]
36. Wu, W.H.; Yang, J.C.; Yu, N.; Li, R.S.; Yuan, Z.X.; Shi, J.S.; Chen, J.H. Evolution of the *WRKY* family in angiosperms and functional diversity under environmental stress. *Int. J. Mol. Sci.* **2024**, *25*, 3551. [CrossRef]
37. Ge, M.M.; Tang, Y.; Guan, Y.J.; Lv, M.C.; Zhou, C.; Ma, H.L.; Lv, J.Y. *TaWRKY31*, a novel *WRKY* transcription factor in wheat, participates in regulation of plant drought stress tolerance. *BMC Plant Biol.* **2024**, *24*, 27. [CrossRef]
38. Wang, D.; Chen, Q.Y.; Chen, W.W.; Liu, X.Y.; Xia, Y.; Guo, Q.G.; Jing, D.L.; Liang, G.L. A *WRKY* transcription factor, *EjWRKY17*, from *Eriobotrya japonica* enhances drought tolerance in transgenic *Arabidopsis*. *Int. J. Mol. Sci.* **2021**, *22*, 5593. [CrossRef] [PubMed]
39. Wu, W.H.; Zhu, S.; Xu, L.; Zhu, L.M.; Wang, D.D.; Liu, Y.; Liu, S.Q.; Hao, Z.D.; Lu, Y.; Yang, L.M.; et al. Genome-wide identification of the *Liriodendron chinense WRKY* gene family and its diverse roles in response to multiple abiotic stress. *BMC Plant Biol.* **2022**, *22*, 25. [CrossRef] [PubMed]
40. Marathe, S.; Grotewold, E.; Otegui, M.S. Should I stay or should I go? Trafficking of plant extra-nuclear transcription factors. *Plant Cell* **2024**, *36*, 1524–1539. [CrossRef] [PubMed]
41. Wei, Y.M.; Han, R.R.; Yu, Y.X. *GmMYB183*, a R2R3-MYB Transcription factor in tamba black soybean (*Glycine max*. cv. Tamba), conferred aluminum tolerance in *Arabidopsis* and soybean. *Biomolecules* **2024**, *14*, 724. [CrossRef]
42. Wang, J.L.; Zhang, X.X.; Yang, H.Q.; Li, S.R.; Hu, Y.; Wei, D.Y.; Tang, Q.L.; Yang, Y.; Tian, S.B.; Wang, Z.M. Eggplant NAC domain transcription factor *SmNST1* as an activator promotes secondary cell wall thickening. *Plant Cell Environ.* **2024**, *47*, 4293–4304. [CrossRef]
43. Ma, Q.B.; Xia, Z.L.; Cai, Z.D.; Li, L.; Cheng, Y.B.; Liu, J.; Nian, H. *GmWRKY16* enhances drought and salt tolerance through an ABA-mediated pathway in *Arabidopsis thaliana*. *Front. Plant Sci.* **2019**, *9*, 1979. [CrossRef] [PubMed]
44. Niu, C.F.; Wei, W.; Zhou, Q.Y.; Tian, A.G.; Hao, Y.J.; Zhang, W.K.; Ma, B.A.; Lin, Q.; Zhang, Z.B.; Zhang, J.S.; et al. Wheat *WRKY* genes *TaWRKY2* and *TaWRKY19* regulate abiotic stress tolerance in transgenic *Arabidopsis* plants. *Plant Cell Environ.* **2012**, *35*, 1156–1170. [CrossRef] [PubMed]
45. Raineri, J.; Ribichich, K.F.; Chan, R.L. The sunflower transcription factor *HaWRKY76* confers drought and flood tolerance to *Arabidopsis thaliana* plants without yield penalty. *Plant Cell Rep.* **2015**, *34*, 2065–2080. [CrossRef]
46. Huang, G.T.; Ma, S.L.; Bai, L.P.; Zhang, L.; Ma, H.; Jia, P.; Liu, J.; Zhong, M.; Guo, Z.F. Signal transduction during cold, salt, and drought stresses in plants. *Mol. Biol. Rep.* **2012**, *39*, 969–987. [CrossRef] [PubMed]
47. Qiao, Z.; Li, C.L.; Zhang, W. *WRKY1* regulates stomatal movement in drought-stressed *Arabidopsis thaliana*. *Plant Mol. Biol.* **2016**, *91*, 53–65. [CrossRef] [PubMed]
48. Huang, Z.; Guo, H.D.; Liu, L.; Jin, S.H.; Zhu, P.L.; Zhang, Y.P.; Jiang, C.Z. Heterologous expression of dehydration-inducible *MfWRKY17* of myrothamnus flabellifolia confers drought and salt tolerance in *Arabidopsis*. *Int. J. Mol. Sci.* **2020**, *21*, 4603. [CrossRef] [PubMed]

49. Liu, C.X.; Mi, H.C.; Liu, H.W.; Xie, S.Z.; Wu, Y.J.; Ning, P.; Liang, T.S.; Lv, J.Y. Response to water deficit in glume of wheat: Expression profiling by microarray analysis. *Euphytica* **2017**, *213*, 26. [CrossRef]
50. Mittler, R.; Vanderauwera, S.; Gollery, M.; Van Breusegem, F. Reactive oxygen gene network of plants. *Trends Plant Sci.* **2004**, *9*, 490–498. [CrossRef]
51. Kang, G.J.; Yan, D.; Chen, X.L.; Yang, L.F.; Zeng, R.Z. *HbWRKY82*, a novel IIc WRKY transcription factor from *Hevea brasiliensis* associated with abiotic stress tolerance and leaf senescence in *Arabidopsis*. *Physiol. Plant.* **2021**, *171*, 151–160. [CrossRef] [PubMed]
52. Wang, C.T.; Ru, J.N.; Liu, Y.W.; Yang, J.F.; Li, M.; Xu, Z.S.; Fu, J.D. The maize *WRKY* transcription factor *ZmWRKY40* confers drought resistance in transgenic *Arabidopsis*. *Int. J. Mol. Sci.* **2018**, *19*, 2580. [CrossRef] [PubMed]
53. Zhu, H.; Zhou, Y.Y.; Zhai, H.; He, S.Z.; Zhao, N.; Liu, Q.C. A novel sweetpotato *WRKY* transcription factor, *IbWRKY2*, positively regulates drought and salt tolerance in transgenic *Arabidopsis*. *Biomolecules* **2020**, *10*, 506. [CrossRef] [PubMed]
54. Jiang, Y.J.; Liang, G.; Yu, D.Q. Activated expression of *WRKY57* confers drought tolerance in *Arabidopsis*. *Mol. Plant* **2012**, *5*, 1375–1388. [CrossRef]
55. Artimo, P.; Jonnalagedda, M.; Arnold, K.; Baratin, D.; Csardi, G.; Castro, E.; Duvaud, E.; Flegel, V.; Fortier, A.; Gasteiger, E.; et al. ExPASy: SIB bioinformatics resource portal. *Nucleic Acids Res.* **2012**, *40*, W597–W603. [CrossRef]
56. Horton, P.; Park, K.J.; Obayashi, T.; Fujita, N.; Harada, H.; Adams-Collier, C.J.; Nakai, K. WoLF PSORT: Protein localization predictor. *Nucleic Acids Res.* **2007**, *35*, W585–W587. [CrossRef] [PubMed]
57. Emms, D.M.; Kelly, S. OrthoFinder: Phylogenetic orthology inference for comparative genomics. *Genome Biol.* **2019**, *20*, 238. [CrossRef] [PubMed]
58. Sun, J.H.; Lu, F.; Luo, Y.J.; Bie, L.Z.; Xu, L.; Wang, Y. OrthoVenn3: An integrated platform for exploring and visualizing orthologous data across genomes. *Nucleic Acids Res.* **2023**, *51*, W397–W403. [CrossRef] [PubMed]
59. Zhang, Z.; Xiao, J.F.; Wu, J.Y.; Zhang, H.Y.; Liu, G.M.; Wang, X.M.; Dai, L. ParaAT: A parallel tool for constructing multiple protein-coding DNA alignments. *Biochem. Bioph. Res. Commun.* **2012**, *419*, 779–781. [CrossRef]
60. Zhou, F.W.; Xu, L.; Shi, C.G.; Wu, F.Y.; Yang, S.Z. Identification of the optimal quantitative RT-PCR reference gene for paper mulberry (*Broussonetia papyrifera*). *Curr. Issues Mol. Biol.* **2024**, *46*, 10779–10794. [CrossRef] [PubMed]
61. Gaudinier, A.; Tang, M.; Bagman, A.M.; Brady, S.M. Identification of protein-DNA interactions using enhanced yeast one-hybrid assays and a semiautomated approach. *Plant Genom. Methods Protoc.* **2017**, *1610*, 187–215.
62. Nakajima, Y.; Ohmiya, Y. Bioluminescence assays: Multicolor luciferase assay, secreted luciferase assay and imaging luciferase assay. *Expert Opin. Drug Dis.* **2010**, *5*, 835–849. [CrossRef] [PubMed]

International Journal of
Molecular Sciences

MDPI

Article

Genome-Wide Identification and Expression Analysis of the Mediator Complex Subunit Gene Family in Cassava

Lingling Zhou [1,†], Shuhui Sun [1,†], Linlong Zhu [1], Xian Chen [1], Ran Xu [1], Lian Wu [1,2,*] and Shuang Gu [1,2,*]

1 School of Breeding and Multiplication, Sanya Institute of Breeding and Multiplication, Hainan University, Sanya 572025, China; m15237352151@163.com (L.Z.); yx13515281250@163.com (S.S.); zhull22@126.com (L.Z.); datoua1116@163.com (X.C.); xuran@hainanu.edu.cn (R.X.)
2 School of Tropical Agriculture and Forestry, Hainan University, Haikou 570228, China
* Correspondence: wulian@hainanu.edu.cn (L.W.); gushuang@hainanu.edu.cn (S.G.)
† These authors contributed equally to this work.

Abstract: The Mediator complex (MED) functions as a co-activator in plants, transmitting transcriptional signals to regulate gene expression, including responses to environmental stresses. While the MED gene family has been identified in several species, it has not yet been reported in cassava. In this study, we identified 32 members of the *MeMED* gene family in cassava (*Manihot esculenta* Crantz) distributed across 13 chromosomes. These genes were categorized into distinct Mediator subunits based on their similarity to *Arabidopsis* modules. Promoter analysis revealed the presence of various cis-regulatory elements, which likely play key roles in regulating plant growth, development, and stress responses. RNA-seq data showed tissue-specific expression patterns for the *MeMED* genes, with significant expression observed in leaves, roots, petioles, stems, friable embryogenic callus, and shoot apical meristems. Further RT-qPCR analysis under various abiotic stress conditions—including drought, exogenous hydrogen peroxide, cold, heat, and salt—demonstrated that 10 selected *MeMED* genes exhibited significant differential expression, indicating their potential functional involvement in stress adaptation. These findings offer insights into the biological roles of the *MeMED* gene family in cassava, with implications for improving stress tolerance in future breeding programs.

Keywords: Mediator complex subunit; cassava; abiotic stress; expression profile; gene family

1. Introduction

The Mediator complex is an evolutionarily conserved multi-protein complex found across eukaryotes, playing a crucial role in regulating plant growth, development, defense, and hormone signaling [1,2]. It facilitates transcription by recruiting RNA polymerase II (Pol II) to specific gene promoters, linking transcription factors (TFs) bound at activators and repressors to the pre-initiation complex (PIC). Upon receiving regulatory signals, the Mediator complex undergoes conformational changes, creating a flexible surface that helps assemble the PIC. Acting as a molecular bridge, Mediator interacts with both the PIC and TFs to promote transcriptional activation [3,4].

Structurally, the core Mediator complex is divided into three main modules, namely the head, middle, and tail modules [5]. Each module comprises distinct subunits, contributing to the unique functional roles in transcription regulation [6]. While the number of subunits varies across species, plants typically have around 34 Mediator subunits [7]. The head module mainly interacts with Pol II to facilitate transcription, while the tail module is integral for binding gene-specific TFs. The middle module is responsible for transferring transcriptional signals between the tail and head modules and may also interact with Pol II directly [8]. Additionally, the kinase module, known as the CDK8 module, consists of CDK8, C-type cyclin (CycC), MED12, and MED13.

In recent years, multiple studies have highlighted the essential role of the Mediator complex in activating gene transcription in plants, particularly under environmental stresses such as salt, drought, cold, and heat [9–11]. Among its subunits, MED25 plays a positive role in salt stress responses by activating salt stress-responsive genes through interactions with key regulatory proteins, including DREB2A, ZINC FINGER HOMEODOMAIN 1 (ZFHD1), and MYB-like. The function of *MED25* in salt tolerance appears to be conserved across land plants. For instance, knocking out *PpMED25A* in Physcomitrella patens (a moss) increases the plant's sensitivity to salt stress [12]. Similarly, the *med18* mutant in Arabidopsis exhibits salt sensitivity, and MED18 interacts with NUCLEOPORIN85 (NUP85) to regulate stress-responsive genes [13]. A recent study identified four Mediator subunits—MED9, MED16, MED18, and CDK8—each representing a distinct module, as essential regulators of transcriptional responses to salt and heat stress in Arabidopsis, as revealed by RNA-seq analysis [14]. The *med19a* mutant exhibits reduced drought resistance, characterized by faster water loss from its leaves and a lower survival rate under drought conditions [15]. Additionally, *MED16*, *MED14*, and *MED2* are crucial for activating the ABRE motif in response to ABA signaling. Mutants of these subunits are insensitive to ABA-induced growth inhibition and lose water more rapidly than wild-type plants, highlighting their specific role in drought adaptation [16]. Recent studies have also linked *CDK8* to drought stress regulation. Mutations in *CDK8* lead to increased stomatal density, impaired stomatal aperture control, and reduced drought tolerance. Conversely, the overexpression of *CDK8* enhances drought resistance [17].

Among the Mediator subunits, *MED16* is one of the earliest identified components involved in abiotic stress responses. It plays a critical role in cold stress tolerance by initiating physiological and metabolic adjustments that sustain cellular homeostasis. Before being classified as part of the Mediator complex, *MED16* was referred to as *SENSITIVE TO FREEZING6* (*SFR6*) and was recognized for its contribution to cold acclimation and enhanced freezing tolerance [18]. Additionally, *VuMED16* genes in asparagus bean have been found to influence cell wall modifications under cold stress, suggesting that the function of *MED16* in cold stress responses may be evolutionarily conserved across different plant species [19]. Further investigations into the rao1 mutant of Arabidopsis, which carries a mutation in *CDK8*, have revealed that *CDK8* is also involved in mitochondrial retrograde signaling under H_2O_2-induced oxidative stress and cold stress conditions [20].

More recently, the Mediator complex has been implicated in heat stress responses. In Arabidopsis, *MED14* and *MED17* mutants exhibit lower survival rates after heat treatment, reduced heat tolerance, and impaired acquired thermotolerance. Notably, approximately 25% of heat stress-induced genes are significantly downregulated in these mutants, further emphasizing the role of the Mediator complex in heat stress adaptation [21].

Overall, these studies collectively highlight the crucial role of the Mediator complex and its associated subunits in plant responses to abiotic stresses, including salt, drought, cold, and heat. Further research is needed to elucidate the precise molecular mechanisms through which these Mediator subunits coordinate stress adaptation in plants.

Cassava (*Manihot esculenta* Crantz), a member of the *Euphorbiaceae* family, is the sixth-largest food crop globally [22] and the third-largest source of caloric energy [23]. It provides essential dietary energy to nearly one billion people in tropical regions, including Africa and Latin America, playing a vital role in smallholder economies and food security. Known for its tolerance to drought and poor soils, cassava has become increasingly important as a climate-resilient crop in response to growing population pressures. However, improving cassava yield and quality remains challenging due to both biotic and abiotic stresses, as well as limited breeding technologies.

Members of the *MED* gene family have been relatively well characterized in species such as *Arabidopsis*, rice, and asparagus bean through genomic analysis [7,19]. However, the *MeMED* gene family in cassava has not been thoroughly studied. In this study, we conducted a comprehensive analysis of the *MeMED* gene family in cassava using assembled genome data. We examined the phylogenetic relationships, gene structures, chromosomal locations, synteny, and cis-elements to better understand their evolutionary patterns and potential functions. Additionally, we explored the expression profiles of *MeMED* genes across various cassava tissues using publicly available transcriptome data. We also assessed the responses of *MeMED* genes to different abiotic stresses through quantitative real-time RT-PCR (qRT-PCR). Our findings provide valuable insights that will guide the future functional characterization of *MeMED* genes in cassava.

2. Results

2.1. Identification of the MeMED Gene Family in Cassava

Through BlastP sequence alignment and querying the Conserved Domains Database (CDD), we identified a total of 32 members of the *MeMED* gene family in the cassava genome (Table 1). These 32 *MeMED* genes are distributed across 13 chromosomes (LG1, LG3, LG4, LG5, LG6, LG8, LG9, LG11, LG13, LG14, LG15, LG16, and LG17), and were named *MeMED33_3* to *MeCdk8_1* based on their *Arabidopsis* homologs (Table 1). The physicochemical characteristics of the 32 *MeMED* genes are summarized in Table 1.

Table 1. Information on the *MeMED* genes in cassava.

	Gene Name	Chr	CDS (bp)	Protein Length (A.A)	Protein Molecular Weight (KDa)	pI	Gravy	Mediator Module
Manes.01G210100.1	*MeMED33_3*	1	3987	1329	144.55	6.71	0.169	Tail
Manes.01G220201.1	*MeMED28_2*	1	318	106	12.41	6.13	−0.185	Head
Manes.03G077300.1	*MeMED22_1*	3	477	159	17.00	6.12	−0.226	Head
Manes.03G185600.9	*MeMED33_1*	3	3969	1323	144.45	6.68	0.227	Tail
Manes.03G199800.1	*MeMED20_1*	3	666	222	25.36	6.08	−0.266	Head
Manes.03G210700.11	*MeMED21_1*	3	405	135	14.69	4.4	−0.37	Middle
Manes.04G136400.1	*MeMED6_1*	4	741	247	27.51	5.18	−0.378	Head
Manes.05G034300.10	*MeMED12_1*	5	6798	2266	250.60	8.92	−0.27	Kinase
Manes.05G061600.1	*MeMED28_1*	5	438	146	17.12	5.78	−0.886	Head
Manes.05G073300.1	*MeMED33_4*	5	3993	1331	143.11	7.22	0.185	Tail
Manes.06G049700.1	*MeMED17_1*	6	1998	666	74.30	5.65	−0.3	Head
Manes.06G129600.1	*MeMED31_1*	6	621	207	23.70	9.28	−0.637	Middle
Manes.08G002200.1	*MeMED4_1*	8	1239	413	44.99	4.97	−0.542	Middle
Manes.08G033200.3	*MeMED27_1*	8	1263	421	46.41	8.76	−0.292	Tail
Manes.08G039900.1	*MeMED7_1*	8	507	169	19.64	9.16	−0.675	Middle
Manes.08G061400.1	*MeMED26_1*	8	1431	477	53.66	5.56	−0.948	Middle
Manes.08G118700.1	*MeMED37_1*	8	1983	661	73.14	5.28	−0.463	Unknow
Manes.09G000400.6	*MeMED21_2*	9	405	135	14.74	4.42	−0.407	Middle
Manes.09G002800.2	*MeMED26_2*	9	1443	481	54.11	5.94	−0.844	Middle
Manes.09G101000.3	*MeMED10_1*	9	564	188	20.41	5.23	−0.384	Middle
Manes.09G152800.6	*MeMED14_1*	9	5478	1826	197.87	7.96	−0.202	Middle
Manes.11G030300.3	*MeMED6_2*	11	744	248	27.74	5.2	−0.468	Head
Manes.13G095300.2	*MeMED18_1*	13	690	230	24.59	6.71	0.137	Head
Manes.14G098600.1	*MeMED25_1*	14	2511	837	89.26	8.79	−0.343	Tail
Manes.14G121800.7	*MeMED13_1*	14	5925	1975	213.55	5.64	−0.208	Kinase
Manes.15G022100.1	*MeMED33_2*	15	3972	1324	143.98	6.29	0.185	Tail
Manes.16G001300.1	*MeMED11_1*	16	351	117	13.26	5.78	−0.521	Head
Manes.17G043400.1	*MeMED16_1*	17	3756	1252	135.29	6.1	−0.232	Tail

Table 1. *Cont.*

	Gene Name	Chr	CDS (bp)	Protein Length (A.A)	Protein Molecular Weight (KDa)	pI	Gravy	Mediator Module
Manes.17G082000.1	*MeMED23_1*	17	4839	1613	180.90	6.64	−0.08	Tail
Manes.17G099700.2	*MeMED32_1*	17	441	147	15.40	4.69	−0.098	Tail
Manes.13G050900.1	*MeCycC_1*	13	762	254	29.76	6.54	−0.098	Kinase
Manes.09G052700.5	*MeCdk8_1*	9	1434	478	53.00	9.24	−0.491	Kinase

The CDS lengths of these genes ranged from 318 bp (*MeMED28_2*) to 6798 bp (*MeMED12_1*), with an average CDS length of 2004 bp. Corresponding amino acid lengths varied from 106 (MeMED28_2) to 2266 (MeMED12_1), with an average length of 668.22 amino acids. The molecular weights of the proteins ranged from 12.41 kD (MeMED28_2) to 250.6 kD (MeMED12_1), with an average weight of 73.33 kD. The theoretical isoelectric points (pI) ranged from 4.4 (MeMED21_1) to 9.28 (MeMED31_1), with 24 proteins (75%) having a pI lower than seven, indicating they are weakly acidic, and eight proteins (25%) having a pI greater than seven, indicating they are alkaline. The hydrophilicity of the proteins encoded by the *MeMED* genes ranged from −0.948 to 0.227.

The Mediator subunits in cassava were categorized according to the Arabidopsis Mediator modules. Among them, nine subunits were assigned to the head module, nine to the middle module, nine to the tail module, four to the kinase module, and one subunit could not be classified (Table 1). Chromosomes 8 and 9 each contained a relatively large number of genes, with five genes on each chromosome. Chromosomes 4, 11, 15, and 16 each harbored one *MeMED* gene. Additionally, chromosomes 1, 5, and 9 each contained a gene cluster, with the following gene pairs: *MeMED33_3* and *MeMED28_2*, *MeMED28_1* and *MeMED33_4*, and *MeMED21_2* and *MeMED26_2* (Figure 1).

2.2. Phylogenetic Analysis of Cassava MeMED Proteins

We constructed a phylogenetic tree for 125 MeMED proteins (42 from *Arabidopsis*, 32 from *Manihot esculenta* Crantz, and 51 from *Oryza sativa*) to study the evolutionary relationships of MeMED proteins across these species (Figure 2). Based on their evolutionary divergence, the MeMED proteins in cassava can be classified into three subclades. The first subclade includes nine proteins, namely MeMED11_1, MeMED7_1, MeMED6_1, MeMED6_2, MeMED16_1, MeMED25_1, MeMED21_2, MeMED21_1, and MeMED4_1. Among them, three pairs of orthologous MED members are shared between cassava and rice as follows: MeMED11_1 and OsMED11_1, MeMED25_1 and OsMED25_1, and MeMED4_1 and OsMED4_1. The second subclade includes 10 proteins, namely MeMED22_1, MeMED27_1, MeMED32_1, MeMED12_2, MeMED10_1, MeMED13_1, MeMED14_2, MeMED20_1, MeMED31_1, and MeCdk8_1. Among these, one pair of orthologous MED members is shared between cassava and rice, which is MeMED13_1 and OsMED13_1. The third subclade includes 13 proteins, with two pairs of orthologous MED members shared between cassava and rice as follows: MeMED23_1 and OsMED23_1, and MeMED17_1 and OsMED17_1. Based on the Ka/Ks ratio, the selective pressure on genes can generally be inferred: a Ka/Ks ratio greater than one suggests positive selection, a ratio less than one indicates purifying selection, and a ratio of one is associated with neutral selection. The Ka/Ks ratios for the *MeMED* gene pairs range from 0.08 to 0.59, reflecting that all the *MeMED* gene pairs are under negative or purifying selection (Ka/Ks < 1) (Table S1). These results indicate that purifying selection has played a role in preserving the conservation of the *MeMED* gene structure during domestication or evolutionary processes.

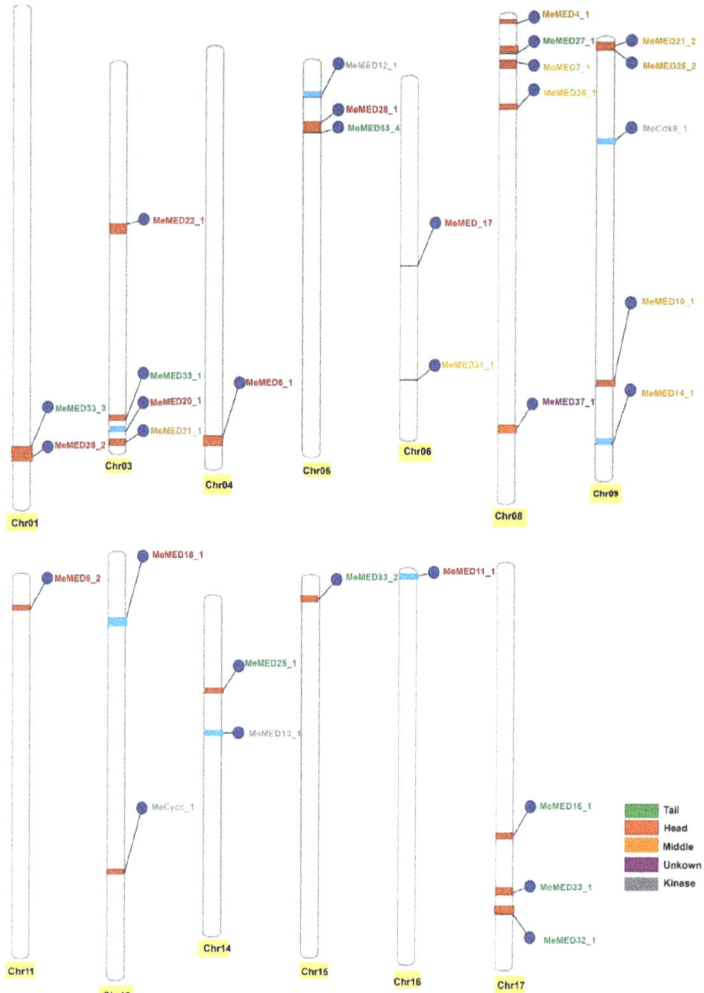

Figure 1. Chromosomal localization and distribution of *MeMED* Gene in Cassava. The yellow color was used solely to enhance the visibility of the chromosomes, providing a clear contrast. It does not convey any additional specific meaning or indication.

Gene synteny analysis across species revealed that 12 *MeMED* members in cassava exhibit synteny with *AtMED* members in *Arabidopsis*. These cassava genes are distributed on chromosomes 1, 4, 5, 8, 9, 11, 13, 14, 15, and 17, while the corresponding Arabidopsis genes are located on five chromosomes (Figure 3). Additionally, five *MeMED* members in cassava show synteny with *OsMED* members in *Oryza sativa*. These cassava genes are located on chromosomes 6, 8, 13, and 17, while the corresponding rice genes are distributed on chromosomes 2, 5, 9, and 10 (Figure 3).

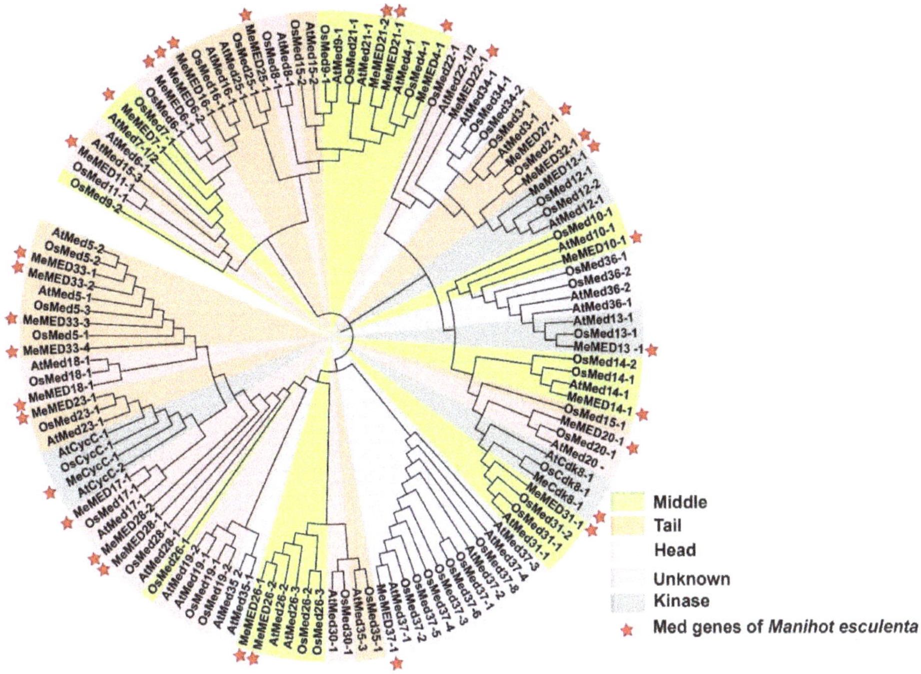

Figure 2. Phylogenetic analysis of MeMED proteins from three plant species, namely *Manihot esculenta*, *Arabidopsis thaliana*, and *Oryza sativa*. The phylogenetic tree was constructed by using mega 7 with the maximum likelihood method and a bootstrap value of 1000 replicates. The different colors in the tree represent different MED modules. The red-colored star represents the MED genes in linebreak emphManihot esculenta.

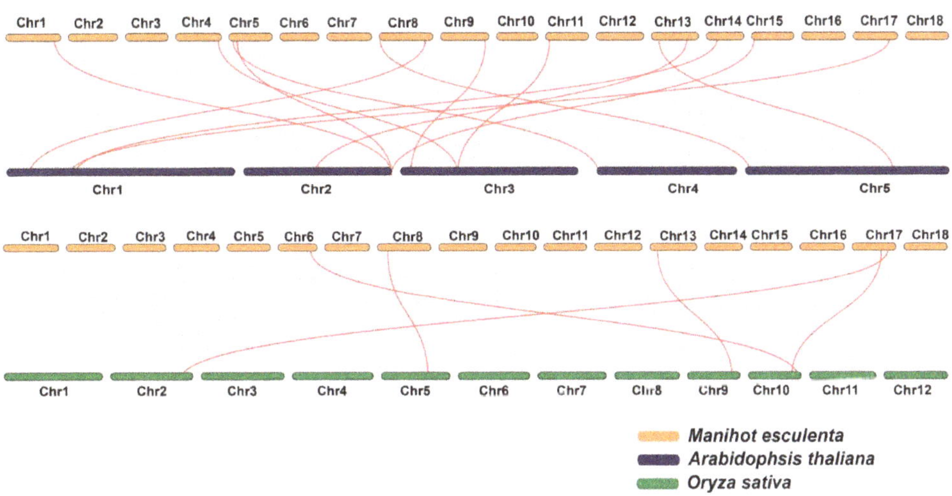

Figure 3. Collinear analysis between the cassava *MeMED* genes and the genomes of *Arabidopsis* and *Oryza sativa*.

2.3. Analysis of Gene Structure and Cis-Acting Elements of the Cassava MeMED Genes

Based on gene sequence similarity, we classified the 32 *MeMED* genes into three distinct groups. The largest group contains 19 *MeMED* genes, while the other two groups consist of nine and four *MeMED* genes, respectively (Figure 4). The number of exons in *MeMED* genes ranges from 1 to 23. For example, *MeMED4_1*, *MeCdk8_1*, and *MeMED32_1* each contain a single exon, while *MeMED13_1* contains 23 exons (Figure 4). These results indicate significant functional divergence among the *MeMED* genes during their evolutionary process.

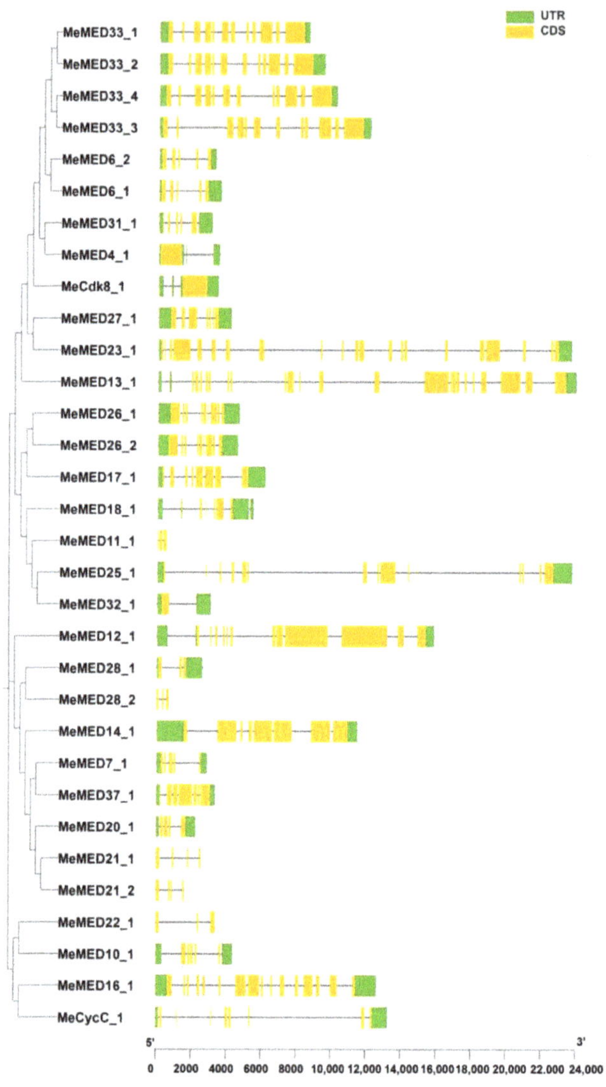

Figure 4. Gene structure analysis of the cassava *MeMED* gene family.

Cis-regulatory elements play a crucial role in regulating gene expression. To identify the types of regulatory elements, we analyzed the 2 kb promoter sequences of all *MeMED* genes using the PlantCARE database. The promoter regions of *MeMED* genes primarily contain four categories of *cis*-regulatory elements, including light-responsive, hormone-responsive, stress-inducible, and core elements/binding sites (Figure 5). Most *MeMED* genes are associated with light-responsive elements, such as G-Box, GA-motif, GT1-motif, GATA-motif, TCT-motif, ACE, I-box, Gap-box, TCCC-motif, AE-box, 3-AF1 binding site, chs-CMA1a, Sp1, LAMP-element, and ATCT-motif. Hormone-responsive elements include abscisic acid-responsive elements (ABRE), methyl jasmonate-responsive elements (TGACG-motif, CGTCA-motif), salicylic acid-responsive elements (TCA-element), auxin-responsive elements (AuxRE, TGA-element), and gibberellin-responsive elements (GARE-motif, P-box, TATC-box). Stress-inducible elements include low-temperature-responsive elements (LTR) and defense/stress-related elements (TC-rich repeats, WUN-motif). Core elements and binding sites include O_2-site, CCAAT-box, and MBS (Table S2). These findings suggest that *MeMED* genes may play significant roles in plant growth, development, and stress resistance.

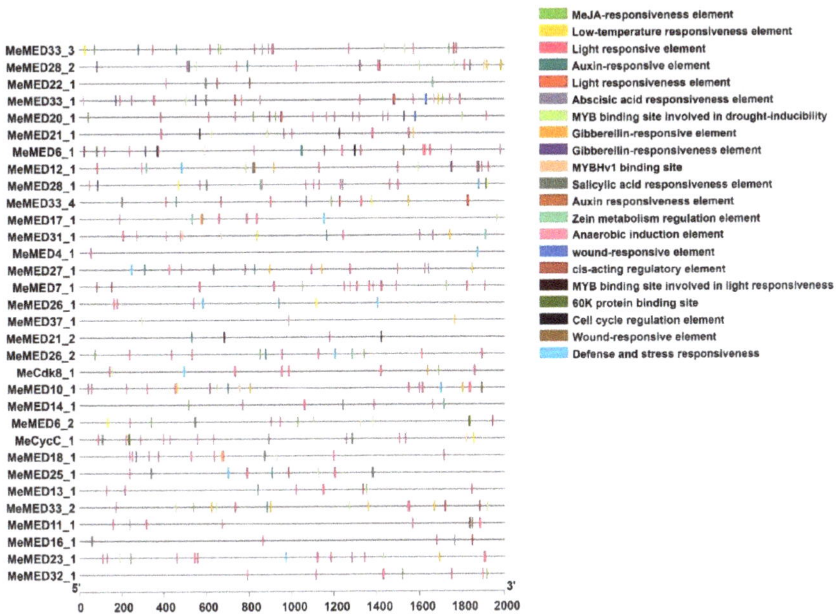

Figure 5. Evaluation of cis-regulatory elements in the *MeMED* gene promoters' regions. The colored boxes denote different cis-regulatory elements, each linked to specific biological responses or functions.

2.4. Expression Profiles of MeMED Genes in Different Tissues of Cassava Based on RNA-Seq

To explore the potential function of the *MeMED* gene in cassava, we analyzed RNA-seq data from the GEO database to examine the expression patterns of *MeMED* gene family members across seven different tissues. Transcriptome analysis revealed that 32 *MeMED* genes exhibited detectable transcriptional abundance across seven different organs, with significant expression differences observed among genes in various tissues (Figure 6). Specifically, *MeMED22_1*, *MeMED10_1*, *MeMED7_1*, and *MeMED11_1* were highly expressed in leaves; *MeMED22_1*, *MeMED10_1*, and *MeMED11_1* showed high expression levels in roots; *MeMED22_1*, *MeMED10_1*, *MeMED32_1*, *MeCycC_1*, *MeMED7_1*, and *MeMED11_1* were highly expressed in petioles; *MeMED22_1*, *MeMED10_1*, and *MeMED21_1* were highly expressed in stems; and *MeMED23_1*, *MeMED22_1*, and *MeMED10_1* were highly expressed in friable embryogenic callus. Notably, *MeMED22_1* and *MeMED10_1* were highly expressed in shoot apical meristems. Additionally, *MeMED28_2* and *MeMED37_1*, except in friable embryogenic callus, showed relatively low expression across the seven organs, indicating that they may be non-functional (Figure 6). These findings suggest that members of the *MeMED* gene family may exhibit functional differentiation in cassava.

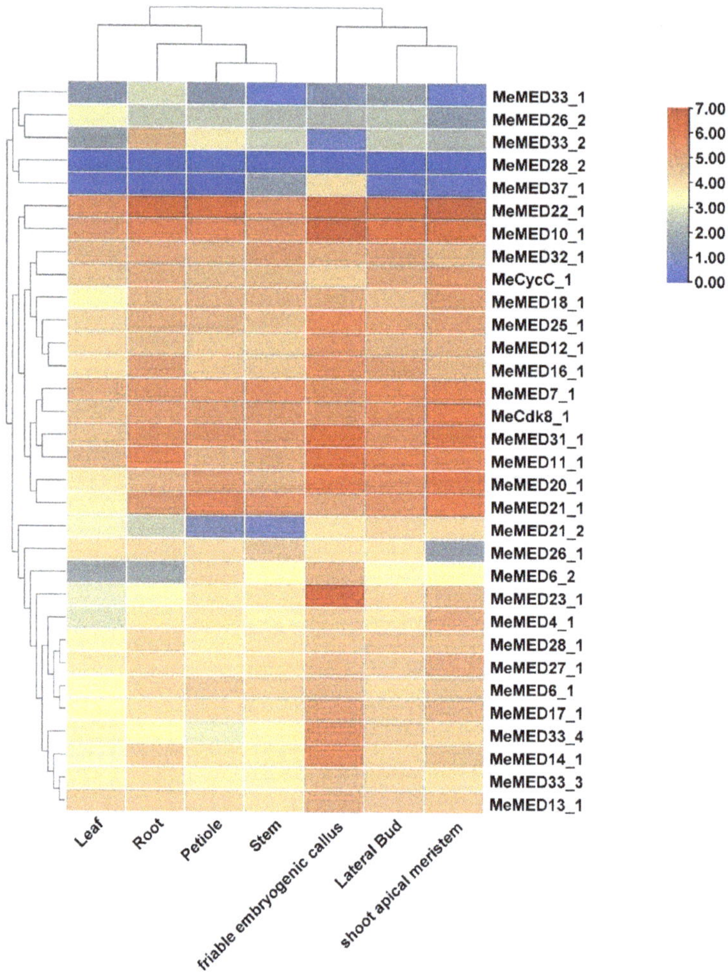

Figure 6. Tissue-specific expression analysis of the *MeMED* gene family in cassava. The red color represents higher expression, the yellow color indicated low expression, whereases the blue color indicated no expression in the RNA-seq expression data.

2.5. Expression Analysis of MeMED Genes in Response to Abiotic Stresses

Based on the RNA-seq expression profile, we selected 10 *MeMED* genes that are highly expressed across various tissues to analyze their expression in response to abiotic stresses. To investigate the expression of these genes under abiotic stresses (drought, exogenous hydrogen peroxide, cold, heat, and salt), tissues were analyzed at 4 h, 8 h, 12 h, and 24 h after treatment using RT-qPCR. The results indicated that the 10 *MeMED* genes were relatively sensitive to drought, exogenous hydrogen peroxide, cold, heat, and salt treatments (Figure 7). For example, the expression of *MeCdk8_1* significantly increased from 4 h to 24 h hours under drought and heat treatments, whereas it significantly decreased under exogenous hydrogen peroxide treatment. The expression of *MeMED7_1* was significantly upregulated after 4 h of salt treatment, reaching its highest level at 8 h. The expression of *MeMED16_1* was significantly upregulated at 4 h post-cold treatment and then decreased to its lowest level at 24 h. *MeMED11_1* showed a significant upregulation at 12 h after heat treatment. Interestingly, *MeMED20_1*, *MeMED22_1*, *MeMED23_1*, and

MeMED31_1 exhibited similar expression patterns after drought treatment, with significant upregulation, while their expression levels decreased under exogenous hydrogen peroxide, cold, heat, and salt treatments. Furthermore, the expression of *MeMED21_1* displayed a unique pattern, significantly increasing after 8 h of drought and cold treatments. These findings suggest that the *MeMED* genes may exhibit functional differentiation in response to various stress conditions.

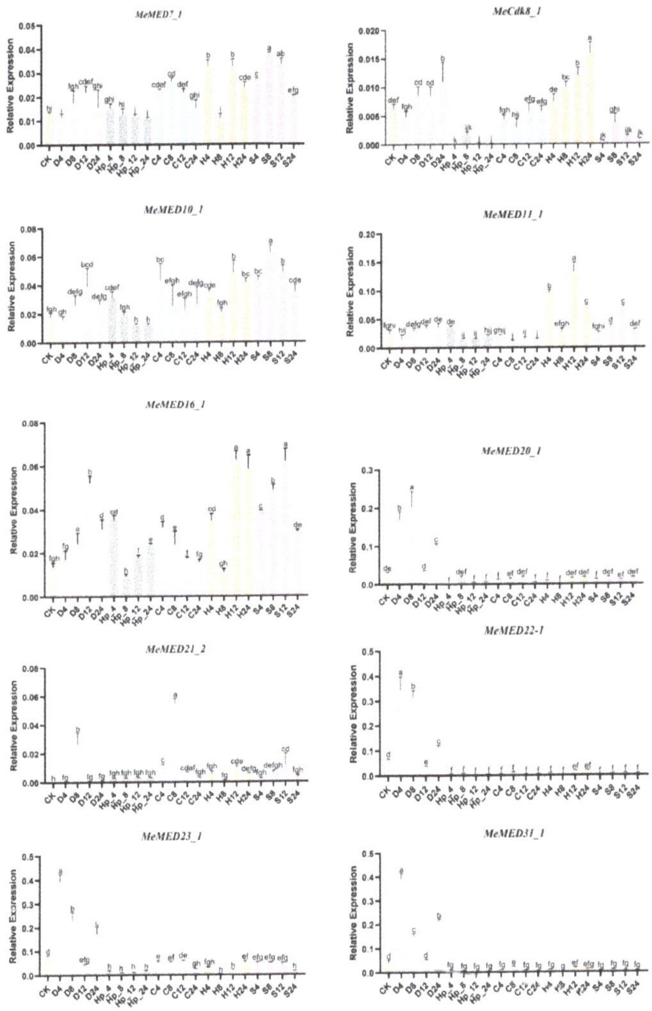

Figure 7. Expression patterns of selected *MeMED* genes under different abiotic stresses. Data are presented as means ± standard errors. Statistically significant differences are indicated by asterisks ($p \leq 0.05$). CK, control; D, drought; HP, exogenous hydrogen peroxide; C, cold; H, heat; S, salt. Time intervals of 4, 8, 12, and 24 h are denoted by the numbers 4, 8, 12, and 24, respectively. Samples collected at 0 h served as the control (CK). Statistical significance was determined using a one-way ANOVA followed by Tukey's HSD post hoc test. Different letters indicate significant differences ($p < 0.05$) within each group.

3. Discussion

Given the crucial role of *MeMED* genes in various physiological processes, including responses to biotic and abiotic stresses [16], it is important to investigate their potential functions in cassava. Currently, *MeMED* genes have been identified and phylogenetically analyzed in only a few plants [7,19]. However, *MeMED* genes have not yet been identified in cassava. In this study, we identified 32 *MeMED* genes in the cassava genome (Table 1 and Figure 1). This number is lower than that found in *Arabidopsis*, rice, and asparagus bean. The variation in the number of *MeMED* genes across species not only reflects differences in their evolutionary history but may also be linked to factors such as genome characteristics, ecological adaptation, and stress response mechanisms.

3.1. Phylogenetic Relationships and Evolution of the MeMED Gene Family

Analyzing the phylogenetic relationships and evolutionary patterns of gene families across different species provides valuable insights into gene conservation, functional diversification, and the evolutionary history of these genes. In our phylogenetic analysis of the 32 *MeMED* genes in cassava, we compared them with homologous genes from *Arabidopsis* and rice. The results revealed that the cassava *MeMED* gene family is closely related to the *MED* gene families of other plant species, suggesting a conserved evolutionary history. For example, three pairs of orthologous *MED* genes were identified between cassava and rice, namely *MeMED11_1* and *OsMED11_1*, *MeMED25_1* and *OsMED25_1*, and *MeMED4_1* and *OsMED4_1* (Figure 2).

Furthermore, the presence of orthologous gene pairs between cassava and rice strengthens the hypothesis of a shared ancestral origin for these genes. Despite the divergence of cassava (Euphorbiaceae) and rice (Poaceae) into separate plant families, the observed synteny between the two species suggests that the *MeMED* gene family in both plants likely originated from a common ancestor. This highlights the conserved nature of the *MeMED* family and provides insights into its evolutionary conservation across distantly related species.

3.2. Regulation of MeMED Genes by Cis-Acting Elements

Analyzing the cis-regulatory elements in the promoter regions of plant genes is essential for understanding the mechanisms that regulate gene expression [24]. The analysis of cis-regulatory elements in the promoter regions of *MeMED* genes revealed that many of these genes are associated with light-responsive, hormone-responsive, and stress-inducible elements (Figure 5). These findings suggest that *MeMED* genes play crucial roles in regulating plant growth, development, and responses to various environmental stresses. The observation that most *MeMED* genes contain light-responsive elements further supports the possibility that they are involved in regulating processes that respond to changing environmental light conditions, such as photomorphogenesis or circadian rhythm regulation.

The presence of stress-related elements, such as low-temperature-responsive (LTR) motifs and defense-/stress-related sequences, underscores the potential importance of these genes in cassava's adaptation to abiotic stresses, particularly low temperature. This is especially relevant considering the increasing frequency and intensity of climate-induced stresses affecting crop production globally. In other plants, some *MeMED* genes associated with low temperature have been well studied. For example, *AtMED16* in *Arabidopsis* is involved in the CBF (C-repeat binding factor) pathway, which is a central mechanism in cold stress response [25]. Furthermore, *AtMED16*, along with *AtMED14* and *AtMED2*, is required for the expression of certain cold-responsive genes that are induced by low temperatures. However, it is important to note that not all cold-responsive genes are regulated by these factors, indicating a complex and specific regulation of gene expression under cold stress [16]. This emphasizes the multifaceted role of *MeMED* genes, which may function similarly in cassava, regulating a subset of cold-responsive genes while leaving others unaffected.

3.3. Expression Characteristics of MeMED Genes in Response to Stress

The Mediator complex plays a pivotal role as a co-activator in transcriptional regulation in plants, participating in nearly all gene transcription processes. Numerous studies have emphasized its involvement in abiotic stress responses, highlighting the complex's role in modulating stress-related genes [7,26]. In this study, we investigated the expression changes in ten cassava *MeMED* genes under various stress treatments (Figure 7). Most of these genes exhibited significant alterations in expression throughout the stress treatments, underscoring the crucial role of *MeMED* genes in cassava's response to environmental stresses. Previous studies have confirmed the crucial role of CDK8 in responses to drought, salt, and cold stresses [14,17,20]. Our study yielded similar results, demonstrating that *MeCdk8_1* exhibited a significant increase in expression under both drought and heat stress, suggesting its potential involvement in regulating the cell cycle and stress responses. Notably, its expression was markedly lower under hydrogen peroxide treatment, indicating a possible role in oxidative stress regulation. These findings highlight the versatility of *MeCdk8_1* in responding to various stress signals and suggest that it may play a similarly critical role in multiple stress responses in cassava. Further research is needed to elucidate its precise function and underlying mechanisms in stress tolerance.

Interestingly, *MeMED16_1* was significantly upregulated at 4 h post-cold treatment before its expression declined to its lowest level at 24 h, suggesting its involvement in early cold stress signaling. These findings align with previous studies, which demonstrated that *At-MED16* detects nuclear signals, participates in cold response pathways, and enhances nuclear replication and cell growth in *Arabidopsis* [25,27]. Moreover, our results are consistent with those of *VuMED16* in asparagus bean, which has been shown to modulate cell wall responses to cold stress [19].

Multiple studies in Arabidopsis have also reported the role of *AtMED* genes in salt stress, including *AtMED25* [12], *AtMED18* [13], and *Atcdk8* genes [14]. In *Arabidopsis, AtMED16, AtMED14*, and *AtMED2*, as essential Mediator subunits, participate in the activation of ABRE motif-mediated regulation. Mutant plants lacking these genes are insensitive to ABA-induced growth inhibition but exhibit a significantly higher water loss rate in detached leaves compared to the wild-type, underscoring the specific role of these genes in drought response [16]. In contrast, *MeMED11_1* showed a marked upregulation after heat stress, further suggesting that it may be implicated in the heat stress response. The consistent upregulation of *MeMED20_1, MeMED22_1, MeMED23_1*, and *MeMED31_1* under drought conditions, followed by a decline in expression under other stresses, indicates that these genes may play more specific roles in drought tolerance and possibly share common regulatory pathways related to water deficit.

Of particular interest is *MeMED21_1*, which exhibited a unique expression pattern: it significantly increased after 8 h of drought and cold treatments, suggesting that it may play a dual role in mediating responses to both drought and cold stresses. This dual regulation could enhance cassava's adaptability to multiple environmental stresses, potentially contributing to its stress tolerance. These findings are consistent with previous research on *Arabidopsis*, where *AtMED25* has been shown to regulate responses to both drought and salt stresses [12]. Further research on *MeMED21_1* could elucidate its precise molecular mechanisms and provide valuable insights for improving stress resilience in cassava cultivation.

4. Material and Methods

4.1. Mining of MED Family in Manihot esculenta

In our study, we searched the TAIR database to obtain the MED protein sequences for *Arabidopsis* (https://www.arabidopsis.org/, accessed on 2 July 2024) [28]. These sequences were used as query sequences and were blasted against the cassava genome (*Manihot esculenta* v8.1) available online at phytozome (https://phytozome-next.jgi.doe.gov/info/Mesculenta_v8_1, accessed on 1 July 2024) using the BlastP tool (https://phytozome-next.jgi.doe.gov/blast-search, accessed on 2 July 2024). After removing duplicates, the non-redundant protein sequences were analyzed using SMART (http://smart.embl-heidelberg.de/, accessed on 2 July 2024) and NCBI conserved domain searches (https://www.ncbi.nlm.nih.gov/Structure/cdd/wrpsb.cgi, accessed on 4 July 2024) [29]. The final MED proteins were renamed according to their physical positions in the cassava genome (*Manihot esculenta* v8.1). The MED modules (head, tail, middle, unknown, and kinase) were named according to the *A. thaliana* MED genes module [7]. Additionally, the physicochemical properties of the 32 MeMED proteins, including protein length (aa), theoretical isoelectric point (pI), molecular weight (kDa), and grand average of hydropathicity (GRAVY), were computed using ProtParam (https://web.expasy.org/protparam/, accessed on 8 July 2024) [30].

4.2. Phylogeny of Cassava MED Proteins

To construct the phylogenetic tree of MeMED proteins, the protein sequences of cassava, *A. thaliana*, and *Oryza sativa* were first aligned using MEGA software (Version 7) [31]. The maximum likelihood tree was then generated using IQ-TREE software, version 2.0 (http://www.iqtree.org, accessed on 11 July 2024) [32] with bootstrap values set to 1000 replicates. The resulting phylogeny was further refined using ITOL (https://itol.embl.de/, accessed on 14 July 2024) [33].

4.3. Gene Structural Analysis of Cassava MED Proteins

The intron and exon organization of the *MeMED* genes was predicted using TBtools (version V2.056) [34]. To identify the cis-acting elements within the promoters of the MeMED genes, a 2 kb sequence upstream of the start codon was submitted to the Plant-CARE website [35]. The graphical representation of the cis-regulatory elements in *MeMED* genes was generated using TBtools.

4.4. Evolutionary Analysis of MeMED Genes

The synteny relationships among *Cassava*, *Oryza sativa*, and *Arabidopsis* were assessed using MCScanX, MCScanX software, version 1.0 and visualization was performed with TBtools (v2.012) [36]. Additionally, the nonsynonymous (Ka) and synonymous (Ks) substitution rates between gene pairs were calculated. The divergence time (T, in millions of years ago; MYA) was estimated using the following formula: $T = Ks/2R$, where $R = 15 \times 10^{-8}$ represents the rate of synonymous substitutions per site per year [37].

4.5. Expression Analyses of MeMED Genes in Different Tissues

The Bart Lab Cassava Atlas database (https://shiny.danforthcenter.org/cassava_atlas/, accessed on 28 July 2024) was searched to obtain the expression data for the *MeMED* genes. Temporal and spatial expression patterns of the MeMED proteins were evaluated using RNA-seq data from various cassava tissues, including the root, stem, leaf, and petiole, at different developmental stages. The heatmap was generated using TBtools. The FPKM values were calculated as described in previous studies [38].

4.6. RNA Extraction, cDNA Synthesis, and RT-qPCR Analysis

Total RNA was extracted using the RNAprep Pure Plant Kit (Lot no. Q5510, Tiangen, Beijing, China), following the manufacturer's instructions. The first complementary DNA strand (cDNA) was synthesized using the HiScript 1st Strand cDNA Synthesis Kit V221 (Vazyme Biotech Co., Ltd., Sanya, China). RT-qPCR was performed using ChamQ™ SYBR RT-qPCR Master Mix (Vazyme Biotech Co., Ltd., Sanya, China). Each reaction had a total volume of 20 μL, comprising 0.4 μL forward primer, 0.4 μL reverse primer, 2 μL cDNA, 10 μL SYBR mix, and 7.2 μL ddH_2O. The cycling program included an initial denaturation step at 95 °C for 30 s, followed by 40 cycles of denaturation at 95 °C for 10 s and annealing/extension at 60 °C for 30 s. *Me*-actin was used as an internal control. The data were analyzed using the $2^{-\Delta\Delta Ct}$ method [39]. The primers for the selected *MeMED* genes are listed in Table S3. The heatmap was generated using GraphPad Prism 8 software.

4.7. Plant Material and Stress Treatment Conditions

In this study, the typical cultivated cassava variety SC9 (South China 9) was used for stress treatments. The SC9 seeds were obtained from the National Cassava Germplasm Repository (NCGR) at the Tropical Crops Genetic Resources Institute, Chinese Academy of Tropical Agricultural Sciences, HaikouHainan Province, China. Vigorous seeds were selected and sterilized with a 10% hypochlorous acid solution for 5 min. The seeds were then grown on a water-saturated filter paper in a growth chamber (25 °C day/night with a 16 h/8 h light/dark cycle), after which the plants were transferred to pots containing soil. For stress treatments, the plants were subjected to 300 mM NaCl for salt stress, 20% PEG6000 for drought stress, 4 °C for cold stress, 42 °C for heat stress, and 10% hydrogen peroxide (H_2O_2) for oxidative stress. Samples were collected at 0 (control), 4, 8, 12, and 24 h post-treatment. Three biological replicates were performed for each treatment. All samples were rapidly frozen in liquid nitrogen and stored at −80 °C for subsequent expression analysis.

5. Conclusions

This study provides the first genome-wide analysis of the *MeMED* gene family in cassava. We identified 32 *MeMED* genes and examined their basic features, gene structure, evolutionary relationships, and expression patterns. RT-qPCR analysis showed significant changes in the expression of the selected 10 *MED* genes, including *MeMED7-1, MeCdk8-1, MeMED10-1, MeMED11-1, MeMED16-1, MeMED20-1, MeMED21-1, MeMED22-1, MeMED23-1, and MeMED31-1* (under various abiotic stresses, suggesting their potential role in enhancing cassava's stress tolerance). The *MeMED21_1*, which exhibited a unique expression pattern towards abiotic stress, could offer new strategies to improve cassava resilience. These findings have a significant implication for climate change and sustainable agriculture. The identified genes in our study may contribute to new cassava varieties that may show more tolerance towards the harsh environmental conditions, which are becoming more prevalent due to climate change. This study can help to ensure food production and contribute to the sustainable production of cassava. Overall, our research provides valuable insights into the *MeMED* gene family and lays the groundwork for future studies on important crops like cassava.

Supplementary Materials: The following supporting information can be downloaded at: https://www.mdpi.com/article/10.3390/ijms26041666/s1.

Author Contributions: S.G., R.X. and L.W. conceived and designed the experiments. L.Z. (Lingling Zhou) performed all the bioinformatics analyses. S.S., L.Z. (Linlong Zhu) and X.C. conducted the experiments. S.G. and L.W. wrote the manuscript. R.X. supervised the project and provided funding. All authors have read and agreed to the published version of the manuscript.

Funding: This study was supported by the Project of Sanya Yazhou Bay Science and Technology City (SCKJ-JYRC-2024-28), the Natural Science Foundation of Hainan Province (322RC592), and the National Natural Science Foundation of China (32360083 and 31960159).

Data Availability Statement: The authors confirm that the data supporting the findings of this study are available within the article and its Supplementary Materials.

Acknowledgments: The authors thank Stephen Cong and Li Rui for their continuous support and manuscript invitations, as well as all the editors and reviewers for their valuable and insightful comments on this manuscript.

Conflicts of Interest: The authors declare no conflicts of interest.

References

1. Yang, Y.; Li, L.; Qu, L.J. Plant Mediator complex and its critical functions in transcription regulation. *J. Integr. Plant Biol.* **2016**, *58*, 106–118. [CrossRef] [PubMed]
2. Zhai, Q.Z.; Li, C.Y. The plant Mediator complex and its role in jasmonate signaling. *J. Exp. Bot.* **2019**, *70*, 3415–3424. [CrossRef] [PubMed]
3. Schilbach, S.; Hantsche, M.; Tegunov, D.; Dienemann, C.; Wigge, C.; Urlaub, H.; Cramer, P. Structures of transcription pre-initiation complex with TFIIH and Mediator. *Nature* **2017**, *551*, 204. [CrossRef] [PubMed]
4. Kornberg, R.D. Mediator and the mechanism of transcriptional activation. *Trends Biochem. Sci.* **2005**, *30*, 235–239. [CrossRef]
5. Plaschka, C.; Larivière, L.; Wenzeck, L.; Seizl, M.; Hemann, M.; Tegunov, D.; Petrotchenko, E.V.; Borchers, C.H.; Baumeister, W.; Herzog, F.; et al. Architecture of the RNA polymerase II-Mediator core initiation complex. *Nature* **2015**, *518*, 376–380. [CrossRef]
6. Robinson, P.J.; Trnka, M.J.; Bushnell, D.A.; Davis, R.E.; Mattei, P.J.; Burlingame, A.L.; Kornberg, R.D. Structure of a Complete Mediator-RNA Polymerase II Pre-Initiation Complex. *Cell* **2016**, *166*, 1411–1422. [CrossRef] [PubMed]
7. Mathur, S.; Vyas, S.; Kapoor, S.; Tyagi, A.K. The Mediator Complex in Plants: Structure, Phylogeny, and Expression Profiling of Representative Genes in a Dicot (Arabidopsis) and a Monocot (Rice) during Reproduction and Abiotic Stress. *Plant Physiol.* **2011**, *157*, 1609–1627. [CrossRef] [PubMed]
8. Chadick, J.Z.; Asturias, F.J. Structure of eukaryotic Mediator complexes. *Trends Biochem. Sci.* **2005**, *30*, 264–271. [CrossRef] [PubMed]
9. Chen, J.; Yang, S.; Fan, B.; Zhu, C.; Chen, Z. The Mediator Complex: A Central Coordinator of Plant Adaptive Responses to Environmental Stresses. *Int. J. Mol. Sci.* **2022**, *23*, 6170. [CrossRef] [PubMed]
10. Chong, L.; Guo, P.; Zhu, Y. Mediator Complex: A Pivotal Regulator of ABA Signaling Pathway and Abiotic Stress Response in Plants. *Int. J. Mol. Sci.* **2020**, *21*, 7755. [CrossRef]
11. Kidd, B.N.; Cahill, D.M.; Manners, J.M.; Schenk, P.M.; Kazan, K. Diverse roles of the Mediator complex in plants. *Semin. Cell Dev. Biol.* **2011**, *22*, 741–748. [CrossRef] [PubMed]
12. Elfving, N.; Davoine, C.; Benlloch, R.; Blomberg, J.; Brännström, K.; Müller, D.; Nilsson, A.; Ulfstedt, M.; Ronne, H.; Wingsle, G.; et al. The Arabidopsis thaliana Med25 mediator subunit integrates environmental cues to control plant development. *Proc. Natl. Acad. Sci. USA* **2011**, *108*, 8245–8250. [CrossRef]
13. Pei, Z.-M.; Zhu, Y.; Wang, B.; Tang, K.; Hsu, C.-C.; Xie, S.; Du, H.; Yang, Y.; Tao, W.A.; Zhu, J.-K. An Arabidopsis Nucleoporin NUP85 modulates plant responses to ABA and salt stress. *PLoS Genet.* **2017**, *13*, e1007124.
14. Crawford, T.; Karamat, F.; Lehotai, N.; Rentoft, M.; Blomberg, J.; Strand, Å.; Björklund, S. Specific functions for Mediator complex subunits from different modules in the transcriptional response of Arabidopsis thaliana to abiotic stress. *Sci. Rep.* **2020**, *10*, 5073. [CrossRef] [PubMed]
15. Li, X.H.; Yang, R.; Gong, Y.F.; Chen, H.M. The Arabidopsis Mediator Complex Subunit MED19a is Involved in ABI5-mediated ABA Responses. *J. Plant Biol.* **2018**, *61*, 97–110. [CrossRef]
16. Lee, M.; Dominguez-Ferreras, A.; Kaliyadasa, E.; Huang, W.J.; Antony, E.; Stevenson, T.; Lehmann, S.; Schäfer, P.; Knight, M.R.; Ntoukakis, V.; et al. Mediator Subunits MED16, MED14, and MED2 Are Required for Activation of ABRE-Dependent Transcription in Arabidopsis. *Front. Plant Sci.* **2021**, *12*, 649720. [CrossRef]
17. Zhu, Y.F.; Huang, P.C.; Guo, P.C.; Chong, L.; Yu, G.B.; Sun, X.L.; Hu, T.; Li, Y.; Hsu, C.C.; Tang, K.; et al. CDK8 is associated with RAP2.6 and SnRK2.6 and positively modulates abscisic acid signaling and drought response in Arabidopsis. *New Phytol.* **2020**, *228*, 1573–1590. [CrossRef] [PubMed]
18. Knight, H.; Veale, E.L.; Warren, G.J.; Knight, M.R. The sfr6 mutation in Arabidopsis suppresses low-temperature induction of genes dependent on the CRT/DRE sequence motif. *Plant Cell* **1999**, *11*, 875–886. [CrossRef]
19. Liang, L.; Wang, D.; Xu, D.; Xiao, J.; Tang, W.; Song, X.; Yu, G.; Liang, Z.; Xie, M.; Xu, Z.; et al. Comparative phylogenetic analysis of the mediator complex subunit in asparagus bean (*Vigna unguiculata* ssp. *sesquipedialis*) and its expression profile under cold stress. *BMC Genom.* **2024**, *25*, 149. [CrossRef] [PubMed]
20. Ng, S.; Giraud, E.; Duncan, O.; Law, S.R.; Wang, Y.; Xu, L.; Narsai, R.; Carrie, C.; Walker, H.; Day, D.A.; et al. Cyclin-dependent Kinase E1 (CDKE1) Provides a Cellular Switch in Plants between Growth and Stress Responses. *J. Biol. Chem.* **2013**, *288*, 3449–3459. [CrossRef] [PubMed]
21. Ohama, N.; Moo, T.L.; Chua, N.H. Differential requirement of MED14/17 recruitment for activation of heat inducible genes. *New Phytol.* **2021**, *229*, 3360–3376. [CrossRef] [PubMed]

22. Zhao, P.J.; Liu, P.; Shao, J.F.; Li, C.Q.; Wang, B.; Guo, X.; Yan, B.; Xia, Y.J.; Peng, M. Analysis of different strategies adapted by two cassava cultivars in response to drought stress: Ensuring survival or continuing growth. *J. Exp. Bot.* **2015**, *66*, 1477–1488. [CrossRef] [PubMed]
23. Li, S.B.; Cui, Y.Y.; Zhou, Y.; Luo, Z.T.; Liu, J.D.; Zhao, M.M. The industrial applications of cassava: Current status, opportunities and prospects. *J. Sci. Food Agric.* **2017**, *97*, 2282–2290. [CrossRef] [PubMed]
24. Marand, A.P.; Eveland, A.L.; Kaufmann, K.; Springer, N.M. cis-Regulatory Elements in Plant Development, Adaptation, and Evolution. *Annu. Rev. Plant Biol.* **2023**, *74*, 111–137. [CrossRef] [PubMed]
25. Knight, H.; Mugford, S.G.; Ülker, B.; Gao, D.H.; Thorlby, G.; Knight, M.R. Identification of SFR6, a key component in cold acclimation acting post-translationally on CBF function. *Plant J.* **2009**, *58*, 97–108. [CrossRef] [PubMed]
26. Chen, K.; Li, G.J.; Bressan, R.A.; Song, C.P.; Zhu, J.K.; Zhao, Y. Abscisic acid dynamics, signaling, and functions in plants. *J. Integr. Plant Biol.* **2020**, *62*, 25–54. [CrossRef] [PubMed]
27. Liu, Z.P.; Chen, G.; Gao, F.; Xu, R.; Li, N.; Zhang, Y.Y.; Li, Y.H. Transcriptional Repression of the APC/C Activator Genes CCS52A1/A2 by the Mediator Complex Subunit MED16 Controls Endoreduplication and Cell Growth in Arabidopsis. *Plant Cell* **2019**, *31*, 1899–1912. [CrossRef] [PubMed]
28. Rhee, S.Y.; Beavis, W.; Berardini, T.Z.; Chen, G.; Dixon, D.; Doyle, A.; Garcia-Hernandez, M.; Huala, E.; Lander, G.; Montoya, M.; et al. The Arabidopsis Information Resource (TAIR): A model organism database providing a centralized, curated gateway to Arabidopsis biology, research materials and community. *Nucleic Acids Res.* **2003**, *31*, 224–228. [CrossRef]
29. Marchler-Bauer, A.; Derbyshire, M.K.; Gonzales, N.R.; Lu, S.N.; Chitsaz, F.; Geer, L.Y.; Geer, R.C.; He, J.; Gwadz, M.; Hurwitz, D.I.; et al. CDD: NCBI's conserved domain database. *Nucleic Acids Res.* **2015**, *43*, D222–D226. [CrossRef] [PubMed]
30. Garg, V.K.; Avashthi, H.; Tiwari, A.; Jain, P.A.; Ramkete, P.W.; Kayastha, A.M.; Singh, V.K. MFPPI—Multi FASTA ProtParam Interface. *Bioinformation* **2016**, *12*, 74–77. [CrossRef]
31. Tamura, K.; Stecher, G.; Kumar, S. MEGA11 Molecular Evolutionary Genetics Analysis Version 11. *Mol. Biol. Evol.* **2021**, *38*, 3022–3027. [CrossRef] [PubMed]
32. Minh, B.Q.; Schmidt, H.A.; Chernomor, O.; Schrempf, D.; Woodhams, M.D.; von Haeseler, A.; Lanfear, R. IQ-TREE 2: New Models and Efficient Methods for Phylogenetic Inference in the Genomic Era. *Mol. Biol. Evol.* **2020**, *37*, 1530–1534. [CrossRef] [PubMed]
33. Letunic, I.; Bork, P. Interactive Tree Of Life (iTOL) v4: Recent updates and new developments. *Nucleic Acids Res.* **2019**, *47*, W256–W259. [CrossRef] [PubMed]
34. Chen, C.J.; Chen, H.; Zhang, Y.; Thomas, H.R.; Frank, M.H.; He, Y.H.; Xia, R. TBtools: An Integrative Toolkit Developed for Interactive Analyses of Big Biological Data. *Mol. Plant* **2020**, *13*, 1194–1202. [CrossRef] [PubMed]
35. Lescot, M.; Dehais, P.; Thijs, G.; Marchal, K.; Moreau, Y.; Van de Peer, Y.; Rouze, P.; Rombauts, S. PlantCARE, a database of plant cis-acting regulatory elements and a portal to tools for in silico analysis of promoter sequences. *Nucleic Acids Res.* **2002**, *30*, 325–327. [CrossRef] [PubMed]
36. Goodstein, D.M.; Shu, S.Q.; Howson, R.; Neupane, R.; Hayes, R.D.; Fazo, J.; Mitros, T.; Dirks, W.; Hellsten, U.; Putnam, N.; et al. Phytozome: A comparative platform for green plant genomics. *Nucleic Acids Res.* **2012**, *40*, D1178–D1186. [CrossRef]
37. Koch, M.A.; Haubold, B.; Mitchell-Olds, T. Comparative evolutionary analysis of chalcone synthase and alcohol dehydrogenase loci in Arabidopsis, Arabis, and related genera (Brassicaceae). *Mol. Biol. Evol.* **2000**, *17*, 1483–1498. [CrossRef] [PubMed]
38. Trapnell, C.; Roberts, A.; Goff, L.; Pertea, G.; Kim, D.; Kelley, D.R.; Pimentel, H.; Salzberg, S.L.; Rinn, J.L.; Pachter, L. Differential gene and transcript expression analysis of RNA-seq experiments with TopHat and Cufflinks. *Nat. Protoc.* **2012**, *7*, 562–578. [CrossRef] [PubMed]
39. Livak, K.J.; Schmittgen, T.D. Analysis of relative gene expression data using real-time quantitative PCR and the 2(-Delta Delta C(T)) Method. *Methods* **2001**, *25*, 402–408. [CrossRef]

International Journal of *Molecular Sciences*

MDPI

Article

Genome-Wide Identification and Expression Profile of *Farnesyl Pyrophosphate Synthase* (*FPS*) Gene Family in *Euphorbia Hirta* L.

Xinyi Bian [1,2,†], Tingkai Wu [2,3,*,†], Runrun Qiang [2,3], Zhi Deng [2,3], Fazal Rehman [2,3], Qiyu Han [2,3], Dong Xu [2,3], Yuan Yuan [2,3], Xiaobo Wang [2,3], Zewei An [2,3], Wenguan Wu [2,3], Huasen Wang [1] and Han Cheng [2,3,*]

1 College of Horticulture Science, Zhejiang Agriculture and Forestry University, Hangzhou 311300, China; bxyabcd2024@163.com (X.B.); whsych66@163.com (H.W.)

2 State Key Laboratory of Tropical Crop Breeding, Sanya Institute, Rubber Research Institute, Chinese Academy of Tropical Agricultural Sciences, Sanya 572025, China; qrrsicau@163.com (R.Q.); zizip@163.com (Z.D.); fazlpbg@gmail.com (F.R.); hanqiyu163@163.com (Q.H.); xudongzhuanyong@163.com (D.X.); yuanyuan_2023@126.com (Y.Y.); xiaobo_wang15@126.com (X.W.); anzewei@163.com (Z.A.); wenguanwu88@163.com (W.W.)

3 Rubber Research Institute, Chinese Academy of Tropical Agricultural Sciences, Haikou 571100, China

* Correspondence: wtksicau@163.com or wutk@catas.cn (T.W.); forcheng@gmail.com (H.C.)

† These authors contributed equally.

Abstract: The biosynthesis of isopentenyl diphosphate (IPP) and dimethylallyl diphosphate (DMAPP), which are essential for sesquiterpenes and triterpenes, respectively, is primarily governed by the mevalonate pathway, wherein *farnesyl pyrophosphate synthase* (*FPS*) plays a pivotal role. This study identified eight members of the FPS gene family in *Euphorbia hirta*, designated *EhFPS1–EhFPS8*, through bioinformatics analysis, revealing their distribution across several chromosomes and a notable tandem gene cluster. The genes exhibited strong hydrophilic properties and key functional motifs crucial for enzyme activity. An in-depth analysis of the *EhFPS* genes highlighted their significant involvement in isoprenoid metabolism and lipid biosynthesis, with expression patterns influenced by hormones such as jasmonic acid and salicylic acid. Tissue-specific analysis demonstrated that certain *FPS* genes, particularly *EhFPS1*, *EhFPS2*, and *EhFPS7*, showed elevated expression levels in latex, suggesting their critical roles in terpenoid biosynthesis. Furthermore, subcellular localization studies have indicated that these proteins are primarily found in the cytoplasm, reinforcing their function in metabolic processes. These findings provide a foundational understanding of the *FPS* genes in *E. hirta*, including their gene structures, conserved domains, and evolutionary relationships. This study elucidates the potential roles of these genes in response to environmental factors, hormone signaling, and stress adaptation, thereby paving the way for future functional analyses aimed at exploring the regulation of terpenoid biosynthesis and enhancing stress tolerance in this species.

Keywords: Farnesyl diphosphate synthase; *Euphorbia hirta* L.; bioinformatics; subcellular localization; expression analysis

1. Introduction

Euphorbia hirta L. is a common herbaceous plant, also known as a hairy spurge, asthma weed, snakeweed, or ara tanah, belonging to the genus *Euphorbia* in the family Euphorbiaceae. It is native to Central America, was later introduced to Southeast Asia, and is widely distributed in tropical and subtropical regions worldwide [1]. In China, it is mainly distributed in the south and southwest, including Hainan, Fujian, Guangdong,

Guangxi, Guizhou, Yunnan, and Taiwan. It has a unique morphology, with an erect branched stem covered with brown or yellowish-brown coarse hairs; opposite leaves, oblong or lanceolate, with serrated lobes on the leaf margins; numerous inflorescences densely packed in the axils of the leaves; and triangular capsules covered with short, soft hairs [2]. *E. hirta* has a strong reproductive capacity and grows on grassy and shrubby hillsides, mostly in sandy soils [3].

As a traditional Chinese herbal medicine, the whole plant is rich in various active substances, including polyphenolic flavonoids such as luteolin, quercetin, 3′-methylmyricetin, quercetin, and gallic acid [4]; tannins such as protocatechuic acid, tannic acid, protochebulic acid, chlorogenic acid, and 3,4-di-o-galloylquinic acid; and terpenoids and triterpenoids such astaraxacin,23–25-methoxycyclophane-3-ol,tinyatoxin-12-deoxy-4β-Hydroxyphorbol-13-dedocanoate-20-aceate, and 12-deoxy-4β-hydroxyphorbol [5,6]. These compounds provide the herbs with unique medicinal properties. Modern pharmacological studies have shown that E. hirta has antibacterial, anticancer, anti-inflammatory, antiasthmatic, and diuretic effects [7]. It is mainly used to treat lung abscess, furuncle, dysentery, diarrhea, stranguria, hematuria, eczema, tinea pedis, skin itching, postpartum low milk production, and other diseases [8]. Terpenoids are compounds derived from mevalonate with isoprene units as the basic structural unit of the molecular skeleton and their extensions. This process involves multiple enzyme catalyses and complex regulatory networks.

Farnesyl pyrophosphate synthase (FPS) is a key enzyme that regulates the synthesis of farnesyl diphosphate (FPP) from isopentenyl pyrophosphate (IPP) and dimethylallyl diphosphate (DMAPP) via the mevalonate pathway (MVA). Its catalytic synthesis products are precursors of various terpenoid derivatives such as steroids, saponins, sesquiterpenes, and triterpenes in plants [9]. Therefore, because of their link to an important rate-limiting enzyme in the terpene synthesis pathway, cloning and functional analysis of *FPS* genes has always been a research hotspot. Researchers have cloned and identified farnesyl pyrophosphate synthase in many plants, including rubber trees, *Eucommia ulmoides*, *Panax notoginseng*, *Arabidopsis thaliana*, and tobacco [10,11]. The functional loss of *AtFPS1* and *AtFPS2* in *A. thaliana* can cause plant death [10].

It was also found that the content of carotenoids and terpene aroma substances in the leaves of transgenic FPS tobacco lines were significantly increased compared with the wild type [11]. Another study revealed heterologous expression of the *FPS* gene in tobacco, with significantly enhanced resistance to brown star and black neck diseases [12]. It was also identified that *EuWRKY30* can directly bind to the W-box element in the EuFPS1 promoter and activate its expression. In addition, overexpression of *EuWRKY30* significantly upregulates the expression of *EuFPS1*, further increasing the density of rubber particles and the content of Eucommia [13]. Using the yeast single hybrid screening library method, the upstream transcriptional regulatory factors GISNF2 and GIMHR of the *FPS* gene in the triterpene synthesis pathway of *Ganoderma lucidum* were identified, providing a basis for further investigation into the expression and regulation mechanism of the *FPS* gene [14]. However, to date, no genes in the *E. hirta FPS* gene family have been identified or systematically studied.

In this study, full-length sequences of three *FPS* genes from rubber trees were used as references. Eight EhFPS proteins were identified in the *E. hirta* genome by BLASTP. Bioinformatic and expression pattern analyses were performed to lay a theoretical foundation for further exploring the function of *EhFPS* and analyzing the synthesis pathways of hemiterpenes and triterpenes in *E. hirta*.

2. Results

2.1. Identification and Physicochemical Properties of EhFPS

The FPS protein domains in the rubber tree were identified using the conservative structure databases InterPro and CDD, and BLASTP comparison was performed in the constructed protein sequence library of *E. hirta*. Resultantly, eight *FPS* candidate genes, namely *EhFPS1, EhFPS2, EhFPS3, EhFPS4, EhFPS5, EhFPS6, EhFPS7*, and *EhFPS8*, were found as per position sequence in the *E. hirta* genome (Table 1). Analysis of the physicochemical properties of EhFPS proteins showed that the encoded amino acid lengths ranged from 323 to 389 aa, and the molecular weight of the protein was between 37.44 and 45.24 kDa. The smallest member was EhFPS5, consisting of 323 amino acids, with a molecular weight of 37.44 kDa, whereas the largest member was EhFPS3, consisting of 389 amino acids, with a molecular weight of 45.24 kDa. The isoelectric points were between 5.11 and 6.52, with mostly acidic proteins, and the average hydrophobicity index (AHI) was less than 0, ranging from −0.353 to −0.255, indicating that EhFPS proteins were all hydrophilic. In the secondary structure of the EhFPS protein, the highest proportion was the α-helix, and 75% of the proteins had an α-helix index greater than 60%; EhFPS5 had the highest index of 65.94%, and EhFPS4 had the lowest index of 56.72%. β-turn had the lowest percentage, with EhFPS8 accounting for the highest percentage (3.76%). Subcellular localization analysis showed that except for EhFPS3, EhFPS4, and EhFPS8, which were localized in chloroplasts, the others were localized in the cytoplasm.

Table 1. The basic information of *EhFPS* genes.

Gene Name	No. of Amino Acids	Molecular Weight (Kda)	Isoelectric Point (PL)	Total Average Hydrophobicity	Secondary Structure (%)				Subcellular Localization
					Alpha Helix (%)	Beta Turn (%)	Extendstrand (%)	Randomcoli (%)	
EhFPS1	342	39.52	5.5	−0.264	63.16	2.92	7.02	26.90	Cytoplasm
EhFPS2	350	40.50	5.11	−0.255	65.43	3.14	5.71	25.71	Cytoplasm
EhFPS3	389	45.24	5.91	−0.305	59.90	3.34	6.17	30.59	Chloroplast
EhFPS4	335	38.64	5.96	−0.348	56.72	2.99	8.96	31.34	Chloroplast
EhFPS5	323	37.44	5.14	−0.293	65.94	2.48	7.74	23.84	Cytoplasm
EhFPS6	342	39.73	5.49	−0.353	65.50	2.92	6.43	25.15	Cytoplasm
EhFPS7	336	38.97	6.52	−0.317	63.99	2.98	7.14	25.89	Cytoplasm
EhFPS8	346	39.95	6.52	−0.323	61.27	3.76	5.49	29.48	Chloroplast

2.2. Chromosomal Location and Gene Duplication Analysis of EhFPS

To reveal the distribution of the *FPS* genes, their sequences were aligned to the genomic data of *E. hirta*, and chromosome mapping was performed using gene location information, as shown in Figure 1. These genes were unevenly distributed on the four chromosomes, with one member each on Ch1, Ch3, and Ch8. The number of *EhFPS* genes in Ch6 was the highest, with five members. A tandem gene cluster comprising *EhFPS4* and *EhFPS7* was identified on chromosome 6 within a 1 Mb region, along with fragment replication involving *EhFPS1* (Figure 2A). Evolutionary analysis of these tandem repeat genes revealed that their Ka/Ks ratio was significantly less than 1 (Figure 2B). A low Ka/Ks ratio suggests that nonsynonymous mutations are largely excluded, likely because of their detrimental effects on protein function. This phenomenon is typically observed in functionally important and conserved genes such as those encoding metabolically critical enzymes or structural proteins. Furthermore, analysis of gene duplication events indicated that the *FPS* genes of both *E. hirta* and rubber trees contained a pair of duplicate genes. Specifically, *EhFPS4* on Ch6 in *E. hirta*, *EhFPS1* on Ch3, and *EhFPS2* on Ch5 in *Hevea brasiliensis* were identified as fragment repeat genes (Figure 2C). These findings suggest that homologous gene pairs emerged through gene replication during evolution.

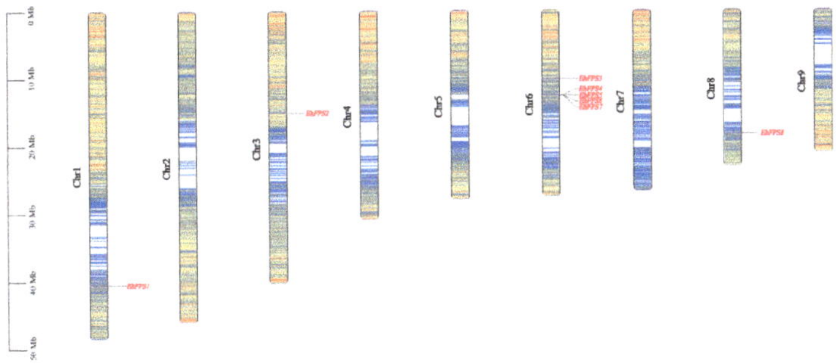

Figure 1. Genetic map of *E. hirta* and chromosome mapping of *FPS* genes on *E. hirta* genome. The scale to the left is in megabase units (Mb). The chromosome number is indicated to the left of each chromosome. Red-colored letters indicate gene positions on each chromosome. Blue lines on chromosomes indicate gene density. The greater the density of genes, the denser the blue line area.

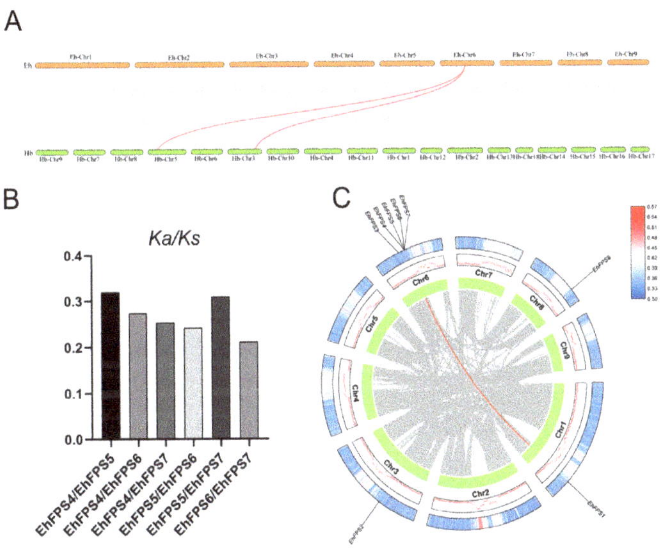

Figure 2. Series replication and selection pressure analysis of *FPS* genes. (**A**) Collinearity analysis of *FPS* genes between *E. hirta* and *H. brasiliensis*. Gray lines in the background highlight collinear blocks within the *E. hirta* and other plant genomes, while collinear *FPS* gene pairs are linked with red lines. (**B**) Selection pressure analysis of *FPS* gene cluster. (**C**) Analysis of tandem replication of *FPS* genes in the genome of *E. hirta*. Chromosomes 1~9 are represented by rectangles. Lines and histograms along the rectangle represent the gene density on the chromosome. The gray lines represent the isolinear blocks in the genome of *E. hirta*, while the red lines between chromosomes represent the gene pairs with fragment repeats.

2.3. Conserved Domains and Gene Structural Analysis of EhFPS

Cluster analysis showed that *EhFPS* genes could be divided into three categories, of which *EhFPS2* and *EhFPS4* were clustered into one category, *EhFPS3* alone was in one category, and the remaining five members were in one category (Figure 3A). The conserved domains of each *EhFPS* gene were predicted using NCBI-CDD, and the results showed that each EhFPS protein contained a single polyprenyl synthase domain of a similar length (Figure 3B). The structures of *EhFPS* genes were similar and contained more exons; among them, *EhFPS2* and *EhFPS5* had 12 exons, and the remaining members had 11 exons. However, only *EhFPS6*, *EhFPS1*, *EhFPS3*, and *EhFPS2* had UTR ends, whereas the other four genes did not have UTR ends (Figure 3C). Comparing the conserved domains and motifs of all genes, except EhFPS8, the other seven members all contained the important functional domain motif II DDXXD, which binds Mg2+ to regulate enzyme activity, whereas Motif2 (DDYXD) was shared by all members (Figure 4). In addition, it contained GKLNR (Motif1), which binds to the substrate, and KT (Motif4), which is the active site of the enzyme. Through protein tertiary structure modelling, the sequence identity of the EhFPS protein and Eucommia farnesyl diphosphate synthase 1 chimera (template number: 7bux.1.A) from the PDB database (https://www.rcsb.org/, accessed on 7 May 2024) was 78.36%, the sequence similarity was 0.81, and the coverage was 0.86. Modelling based on this template showed that EhFPS is composed of homodimeric oligonucleotides formed by α-helices and that the spatial structure contains a "pocket" for binding to the substrate. At the same time, comparisons revealed that *EhFPS1*, *EhFPS2*, *EhFPS5*, and *EhFPS6* had seven α-helical chains with similar overall structures, *EhFPS3* and *EhFPS7* had similar side chain structures, and *EhFPS4* and *EhFPS8* had similar tertiary structures (Figure 5), indicating that changes in the protein structure of members may lead to protein functional diversity.

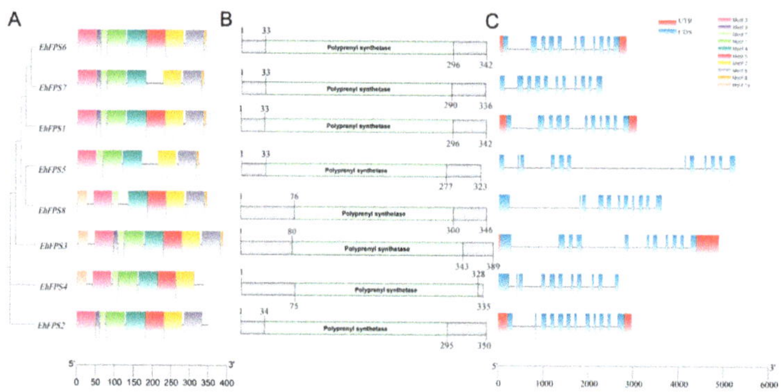

Figure 3. The sequence motif, conserved domains, and gene structural analysis of EhFPS. (**A**) The motif prediction of gene structures. In motif pattern, different colored boxes indicate different motifs; motif1–motif10 were represented by different numbers in the boxes (1–10). (**B**) Conserved domain prediction. The conserved structure of *FPS* genes with only one polyprene synthetase domain was predicted by NCBI-CDD, and the gray color represents the conserved structure of polyprene synthetase domain. (**C**) Gene structure map. The blue bar indicates the coding sequence (CDS), the black line indicates the intron, and the red bar indicates the untranslated region (UTR).

2.4. Sequence and Phylogenetic Analysis of Proteins Encoded by the EhFPS

To clarify the phylogenetic relationship between the *EhFPS* genes, the amino acid sequences of the *FPS* genes of *E. hirta*, rubber tree, cassava, *Eucommia ulmoides*, *Taraxacum koksaghyz* Rodin, *Taraxacum mongolicum*, and *A. thaliana* were used to construct a phylogenetic tree using the neighbor-joining method (Figure 6). *EhFPS* genes were divided into two subtribes, of which the branch of subtribe I contained all EhFPS proteins other than *EhFPS2*; *EhFPS1* was independently separated in this branch. The other subtribe in which *EhFPS2* was located also contained multiple proteins, such as *Taraxacum koksaghyz* Rodin, *Taraxacum mongolicum*, and *Eucommia ulmoides*, among which *EhFPS2* was closely related to *HbFPS3* in the rubber tree and *MeFPS3* in cassava. *MeFPS1* and *HbFPS1*, as well as *MeFPS2* and *HbFPS2*, were independently evolved subclades. In summary, there were homologous proteins in *Rhizoma Dioscoreae* that were closely related to the rubber tree, *Eucommia ulmoides*, and cassava, but most differentiated into independent branches. Therefore, it is speculated that it exercises a different functional mechanism from that of other plants such as rubber and cassava.

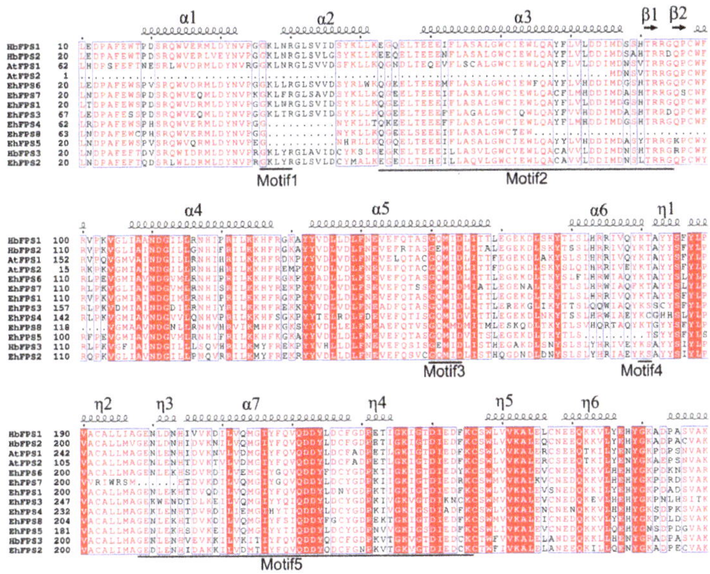

Figure 4. Multiple sequence alignment of a conserved region from *E. hirta*, *H. brasiliensis*, and *A. thaliana*. Motif1 (GKXXR), motif2 (EXXXXXXLXXDDXXDXXXXRRG), motif3 (GQXXD), motif4 (KT), and motif5 (GXXFQXXDDXXDXXXXXXXXGKXXXDXXXXK). Among them, the lysine residue of motif1 can promote the binding of FPPS with isopentenyl pyrophosphate (IPP). Motif2 contains the first aspartic-acid-rich region, DDXXD, which determines the length of the product chain. The amino acid residues in this region participate in the formation of the C-C bond between IPP and the allylic substrate and determine the length of the product chain. In motif3, among the three fixed conserved amino acids, G (1st position), Q (2nd position), and D (5th position) are all conserved in all plant FPPS sequences. Motif4 is a conserved amino acid in all plant FPPS sequences. Motif5 contains the second aspartic-acid-rich region, DDXXD. The marked helices represent the α-helices in the FPS protein, which play a role in forming the catalytic pocket or stabilizing the protein structure.

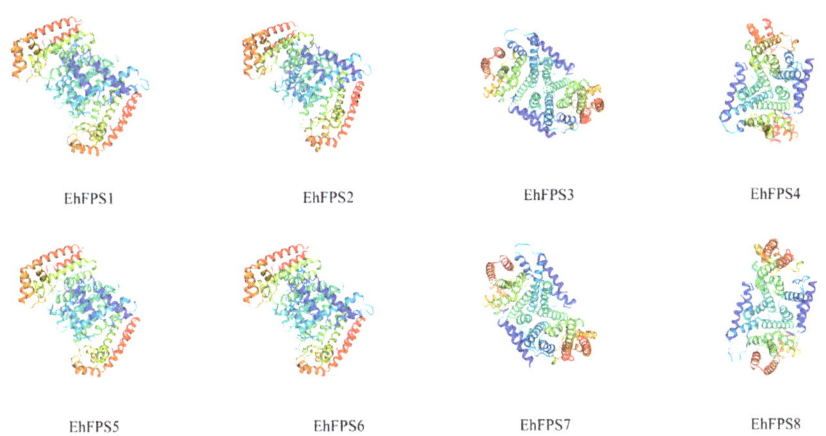

Figure 5. Modeling of eight EhFPS proteins. The structure of eight EhFPS proteins with >90% confidence level are shown. Using the SWISS-MODEL, the EhFPS protein sequence was used for modeling. The structures of eight EhFPS proteins with >90% confidence level are shown. The finest place in the figure is the loop region, which is divided into three channels: red, green, and blue. The range of each color channel is 0–255; 0 indicates that there is no color channel, and 255 indicates the maximum intensity of the color channel. The higher the channel value, the stronger the three colors.

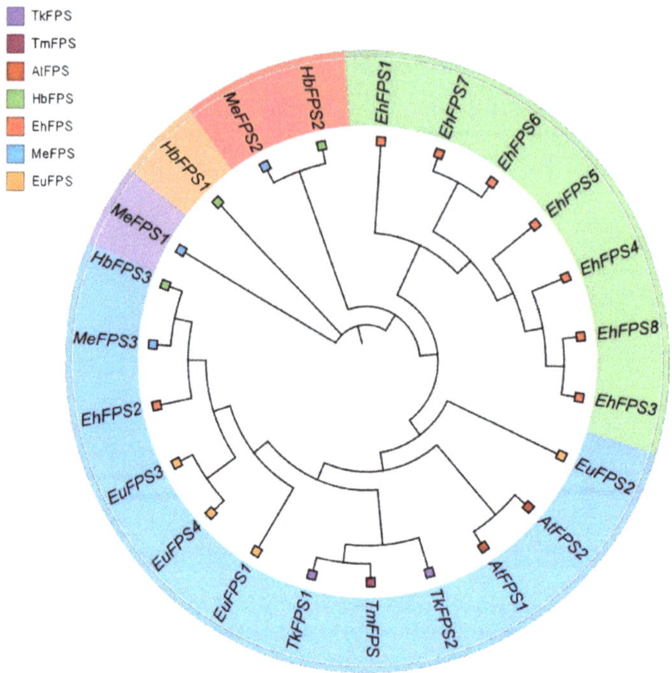

Figure 6. Phylogenetic tree of gene members of the FPS family in seven species. Squares of different colors represent *FPS* genes of different species. The five different subfamilies were distinguished by yellow and green bands.

2.5. Gene Ontology and Interaction Network Analysis of EhFPS

Studies of *FPS* genes in various latex-producing plants have hypothesized that these genes are involved in a wide range of biological processes. GO enrichment analysis was conducted to investigate the functional roles of *FPS* genes in *E. hirta*; a gene ontology (GO) enrichment analysis was conducted (Figure 7A). The results revealed that eight genes were enriched in 18 GO terms (Table S1), all of which were categorized under biological processes. The top five significantly enriched terms were isoprenoid metabolic process (GO:0006720), isoprenoid biosynthetic process (GO:0008299), cellular lipid metabolic process (GO:0044255), lipid biosynthetic process (GO:0008610), and lipid metabolic process (GO:0006629). These findings suggest that *EhFPS* genes are likely to be involved in the synthesis and metabolism of isoprenoid compounds, as well as in lipid biosynthesis and metabolism. Overall, these results highlight the critical roles of *FPS* genes in the key physiological functions of *E. hirta*, including growth, development, stress adaptation, and maintenance of the cell membrane structure.

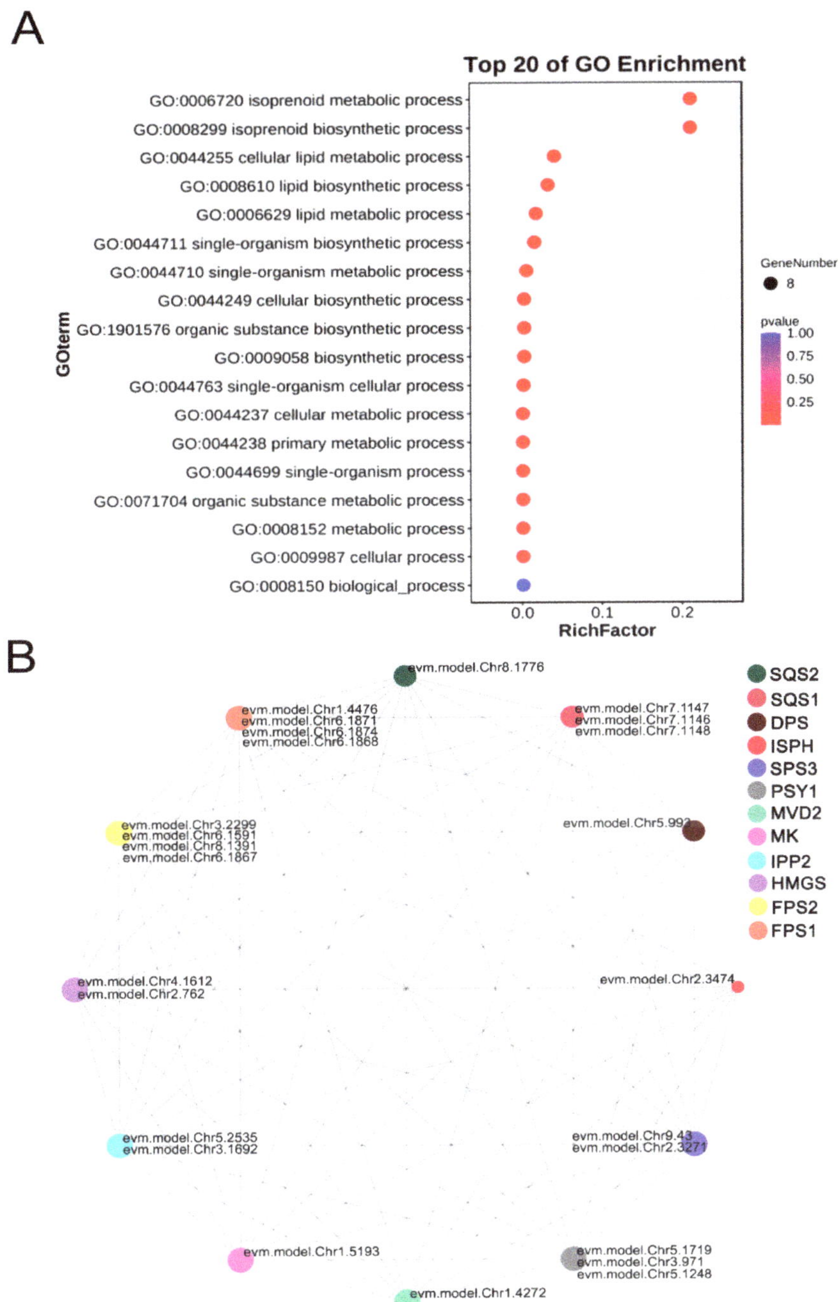

Figure 7. The gene ontology enrichment (GO) analysis and protein–protein interactions (PPI) of *EhFPS* genes. (**A**) GO analysis. The 20 enrichments with the highest scores were selected for display. The horizontal coordinate represents the gene ratio, and the circle represents the number of genes. (**B**) Protein interaction network of *E. hirta* genes mapped to *Arabidopsis* genes. The circle size indicates the number of interaction partners, with larger circles representing more extensive interaction networks.

The protein-regulatory network of *A. thaliana* has been systematically studied as a model plant. FPS proteins were located in homologues of *A. thaliana*, and the interaction networks and their interaction networks with other proteins were deduced (Table S2). Farnesyl pyrophosphate synthases (FPS1 and FPS2) are key enzymes in the isoprene biosynthesis pathway in *A. thaliana*. They catalyze the condensation of isoprene pyrophosphate (FPP) from isoprene IPP and dimethylallyl pyrophosphate (DMAPP), which are direct precursors of isoprene compounds, such as carotenoids, sterols, and sesquiterpenes. In the protein–protein interaction (PPI) network maps constructed with FPS1 and FPS2, the protein showed extensive connections with multiple proteins, including upstream and downstream key enzymes, indicating that the FPS protein is not only the main producer of FPP, an intermediate product of the isoprene pathway, but the regulatory hub of the metabolic network. Among the proteins involved, HMGS, IPP2, and ISPH serve as upstream regulators of FPS, which utilizes IPP and DAMPP as substrates to directly influence metabolic flow in a manner dependent on HMGS and IPP2. Conversely, SQS1, SQS2, DPS, and PSY1 were identified as direct downstream proteins of FPS. The close association between FPS, SQS1, and SQS2 highlights their collaborative role in the sterol biosynthesis pathway. Specifically, SQS1 catalyzes the conversion of farnesyl pyrophosphate (FPP) to squalene, an intermediate product in the sterol biosynthesis pathway, and a precursor for the synthesis of gas triterpenes. Farnesyl pyrophosphate synthase (FPS) and lycopene synthase (PSY1) are critical components of carotenoid biosynthesis. FPS facilitates the transformation of isopentene pyrophosphate (IPP) and dimethylallyl pyrophosphate (DMAPP) into farnesyl pyrophosphate (FPP), whereas PSY1 employs FPP as a substrate to catalyze the synthesis of lycopene.

In summary, the FPS family proteins serve as the central hub of isoprene metabolism. Their functions are heavily reliant on the availability of upstream precursors, including HMGS, MK, and IPP2, which influence the flow direction of downstream metabolic pathways such as those leading to sterols, diterpenes, and carotenoids. Additionally, by synergizing with enzymes, such as SQS, DPS, and PSY1, FPS proteins play a crucial role in plant growth, development, and metabolic resistance (Figure 7B).

2.6. Analysis of Cis-Acting Elements in the Promoter of EhFPS

To understand the expression regulation mode of the *FPS* genes in *E. hirta*, the cis-acting elements of the 2000 bp sequence upstream of the *EhFPS* gene promoter were analyzed (Figure 8). The promoters of the eight *EhFPS* genes contained a large number of cis-acting elements related to the light response, plant hormone induction, and stress response. Statistics for all cis-elements revealed that the *EhFPS* promoter sequence contains 1–27 light-responsive and regulatory elements, which mainly include Box4, G-Box, and GT1-motif elements. Among them, 87.5% of the members had Box4 and G-box elements, and in the hormone response module, they mainly included the jasmonate response elements CGTCA-motif and TGACG-motif and the abscisic acid response element (ABRE). Among them, 30 jasmonic acid (JA) response elements were enriched. Except for the *EhFPS4* promoter, which did not detect the response element of this hormone, the other promoters contained it to varying degrees. Antioxidant response elements (AREs) were significantly enriched in the stress response module, and this element was not detected only in the promoter of *EhFPS2* among the members. In summary, *EhFPS* plays an important role in biological processes such as light induction, hormone induction, and adversity stress response in *E. hirta*.

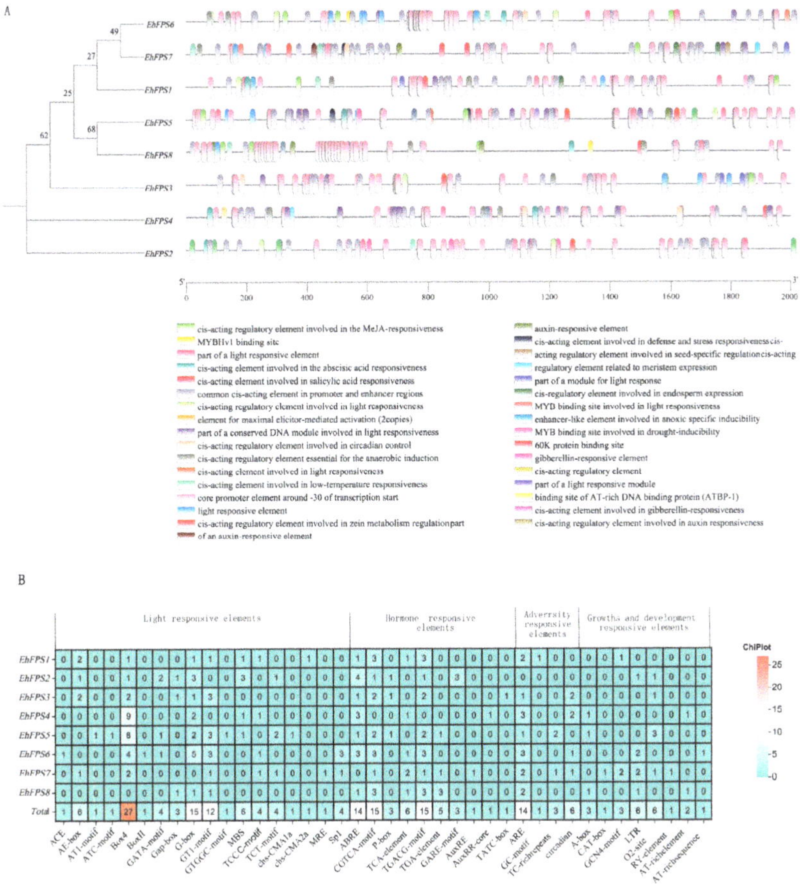

Figure 8. The cis-element in the promoters of *EhFPS* genes. (**A**) Prediction of cis-acting elements in the *FPS* gene promoter region of *E. hirta*. The distribution of 33 cis-acting elements with key functions. (**B**) Statistics of the numbers of cis-elements on the promoters. The various numbers represent the numbers of cis-acting elements.

2.7. EhFPS Genes Responds to Exogenous SA and JA Signals

Thirty cis-elements that responded to JA signals and six that responded to salicylic acid (SA) signals were detected in the promoters of the eight *EhFPS* genes (Figure 8). Therefore, we tested the expression levels of eight *EhFPS* genes treated with 1.0 mM SA and JA. The results indicated that all *EhFPS* genes, except *EhFPS5*, responded significantly to SA signals with markedly upregulated expression levels ($FC \geq 3$, Q value ≤ 0.01) (Figure 9). Based on the expression trends, *EhFPS* genes were divided into three categories under SA treatment (Figure 9A). Group I included *EhFPS1*, *EhFPS2*, and *EhFPS7*, all of which exhibited a "rise-then-fall" expression pattern. Group II, comprising EhFPS3, *EhFPS4*, and *EhFPS8*, exhibited a gradual increase in expression over time. Group III contained only *EhFPS6* and showed no significant changes, whereas the other family members exhibited an upregulation trend. Under JA treatment, *EhFPS1*, *EhFPS2*, *EhFPS7*, and *EhFPS8* demonstrated a similar "rise-then-fall" expression pattern, with *EhFPS2* peaking at 4 h and the other members peaking at 8 h (Figure 9B). *EhFPS5* did not respond to signals associated with salicylic acid (SA) processing, whereas *EhFPS3* and *EhFPS4* did not respond to signals related to jasmonic acid (JA) processing. In summary, *EhFPS1*, *EhFPS2*, *EhFPS7*, and *EhFPS8* responded rapidly and strongly to SA and JA signals, suggesting that they are critically important for terpenoid biosynthesis.

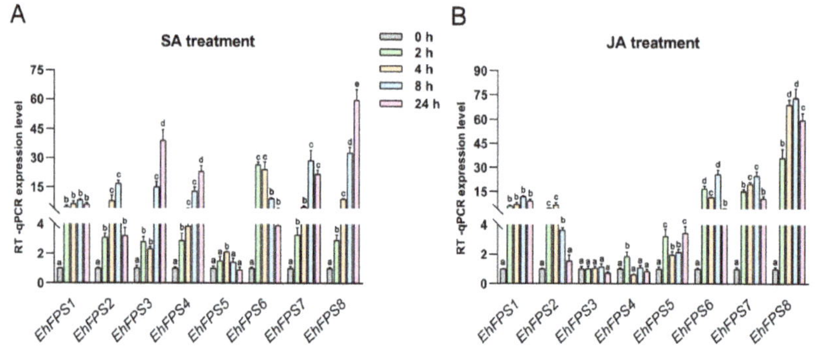

Figure 9. Expression analysis of *EhFPS* genes under SA and JA treatment. (**A**) The gene expression level after SA treatment. (**B**) The expression level of genes after JA treatment. One-way ANOVA and Tukey's multiple comparison test were used to evaluate the statistical significance. The relative qRT-PCR expression level is shown on the left y-axis. The difference between different letters in the same group was significant at $p < 0.05$. The error bar represents the standard deviation of three biological replicates. Differently colored tiles represent time scale.

2.8. Expression Analysis of EhFPS

Gene expression is a critical determinant of gene functions. To study the function of FPS in the sesquiterpenes and triterpenes of *E. hirta*, we selected five *E. hirta* tissue samples for transcriptomic analysis. Subsequently, we calculated the average of the expression results and constructed a heatmap to show the expression levels (Figure 10A). RT-qPCR analysis showed *EhFPS* gene-specific expression patterns. *EhFPS1, EhFPS2, EhFPS3,* and *EhFPS5* were highly expressed in the roots, stems, leaves, and flowers, whereas *EhFPS1, EhFPS2, EhFPS3,* and *EhFPS7* were highly expressed in the latex (Figure 10B). The transcriptome was found to be roughly consistent with the quantitative experimental results, and genes with high expression in latex were further screened, namely *EhFPS1, EhFPS2,* and *EhFPS7*. In summary, *EhFPS1, EhFPS2,* and *EhFPS7* from the eight genes were screened for subsequent functional analyses.

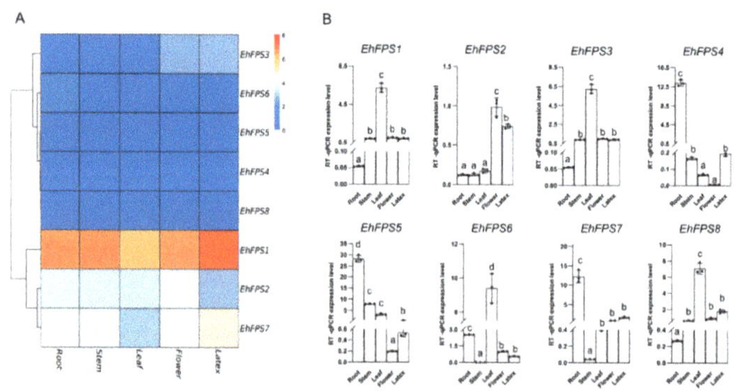

Figure 10. Expression patterns of *EhFPS* in different tissues of *E. hirta*. (**A**) Heat map of transcriptome based on FPKM values. (**B**) Expression profile of eight genes based on average of five tissues. a > 0.05, 0.05 > b > 0.01, 0.01 > c > 0.001, 0.001 > d > 0.0001.

2.9. Subcellular Localization of EhFPS1, EhFPS2, and EhFPS7

The leaves of *N. benthamiana* were infected with the empty pCAMBIA2300-35S-eGFP(C) vector as the control and the experimental group fusion expression vectors pCAMBIA2300-*EhFPS1*-eGFP, pCAMBIA2300-*EhFPS2*-eGFP, and pCAMBIA2300-*EhFPS7*-eGFP. The transformed *N. benthamiana* were placed in a 26 °C incubator under weak light for 2 d, and the green fluorescence and cytoplasmic markers were observed under the 488 nm and 600 nm channels of a laser confocal scanning microscope. The results indicated that the empty pCAMBIA2300-35S-eGFP construct (C) exhibited green fluorescence signals in the nucleus, cytoplasm, cell membrane, and cell wall. In contrast, the GFP fluorescence of the fusion proteins pCAMBIA2300-*EhFPS1*-eGFP, pCAMBIA2300-*EhFPS2*-eGFP, and pCAMBIA2300-*EhFPS7*-eGFP overlapped with the fluorescence of the cytoplasmic RFP protein, which differed from the continuously distributed green fluorescence observed on the cell membrane and cell wall as well as the uniform distribution in the cytoplasm (Figure 11).

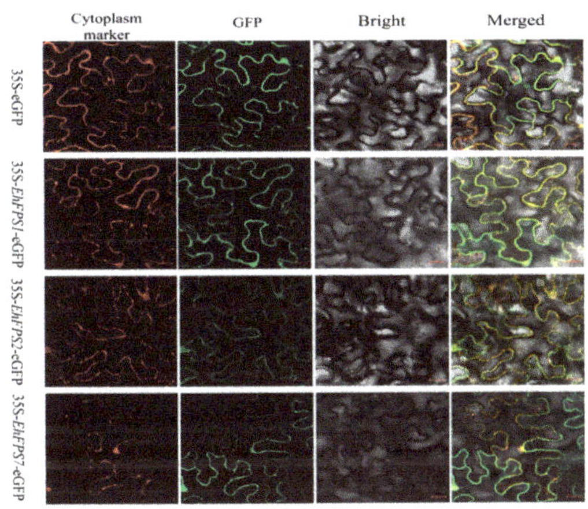

Figure 11. The subcellular localization of EhFPS-GFP fusion protein was transiently expressed in tobacco leaves. Images were captured using a confocal microscope. Control group: 35S-eGFP:35S-pCAMBIA2300-eGFP; cytoplasm marker: 35S-Os1sSrke-RFP, 35S-*EhFPS1*-eGFP, 35-*EhFPS2*-eGFP, and 35S-*EhFPS7*-eGFP were constructed into the 35S-pCAMBIA2300-eGFP vector by homologous recombination of *EhFPS1*, *EhFPS2*, and *EhFPS7*, respectively. The images include the cytoplasmic marker as a red fluorescence channel (first panels) and GFP as a green fluorescence channel (second panels). The bright-field (third panel) and merged (fourth panel) images are shown on the right. Bar = 20 μm.

3. Discussion

Triterpenoids are a class of natural products formed by condensation of isoprene as the basic unit. They are widely found in medicinal plants such as licorice, ginseng, and eleuthero and have significant anti-inflammatory, antibacterial, and antitumor properties. To date, researchers have identified six triterpenoids in *E. hirta*: 3β-taraxacum, taraxerone, tirucallol, cycloart-23-ene-3β,25-diol, 3β-9,19-cyclolanost-23-ene-3,25-diol, and vomifoliol [15]. The *FPS* gene family serves as a crucial rate-limiting enzyme in the terpene synthesis pathway in plants and significantly contributes to the production of hemiterpenes and triterpenoids [16]. Its members have been identified in various medicinal plants and have been shown to play important roles in the synthesis of isoprene and triterpene saponins. The mRNA level of dammarane synthase in *Centella asiatica* was significantly elevated in a similar study that examined the heterologous expression of the *PgFPS* gene cloned from ginseng roots in *C. asiatica* hairy roots. This increased production of saponins and triterpenes in *C. asiatica* results in the rapid accumulation of madecassoside [17]. *FPS* plays a crucial role in the synthesis of bioactive compounds essential for growth and stress responses in plant species such as *Matricaria recutita* [18] and *Blumea balsamifera* [19]. Significant conservation of the *FPS* sequence across plant families, particularly within Asteraceae, was indicated by the discovery of cDNA for *FPS* in various species through sequencing studies. Consistently with our findings, another study reported that the full-length cDNA of the *FPS* gene (*CnFPPS*) in *Chamaemelum nobile* was 1239 bp long and encoded for a 342 amino acid with a molecular weight of approximately 39.38 kDa [20].

An investigation of the sesquiterpene lactone drug artemisinin revealed that overexpression of the endogenous *FPS* gene in *Artemisia annua* can increase endogenous artemisinin content by 2 to 25 times [21]. The simultaneous transfer of two copies of the *FPS* gene reduces artemisinin production in transgenic plants [22]. However, cotransformation of *A. annua* with the *FPS* and *DBR2* genes can significantly increase artemisinin content by 3.36 times [23]. Compared with the wild type, the contents of four ginsenosides, Rh1, Rg1, Re, and Rd, detected in the *FPS* gene overexpression lines of *Panax notoginseng* were significantly increased by 2.66, 1.76, 4.35, and 2.90 times, respectively [24]. However, the identification and expression analysis of *FPS* genes in *E. hirta* have not yet been reported.

In this study, using comparative genomic methods, eight *EhFPS* genes were successfully identified and were found to be located on chromosomes 1, 3, 6, and 8, with chromosome 6 containing five of these genes. Notably, a tandem gene cluster comprising *EhFPS4* and *EhFPS7* was identified. Furthermore, the results of the gene duplication analysis suggest that the evolution of these genes may have resulted in homologous gene pairs through tandem duplication and fragment duplication events, further confirming the evolutionary mechanism of *FPS* genes in plant genomes. In addition, gene duplication analysis results suggest that the evolution of these genes may have been formed by tandem duplication and fragment duplication events. Through comparative analysis, we found that the functional domains of EhFPS protein family members were highly conserved and contained DDXXD, GQXXD, and GKLNR domains that regulate enzyme activities. Previous studies have shown that plant FPPS have five conserved domains like those of other organisms, namely domains I (GKXXR), II (EXXXXXXLXXDDXXDXXXXRRG), III (GQXXD), IV (KT), and VI (GXXFQXXDDXXDXXXXXXXXGKXXXDXXXXK) [25].

Furthermore, it has been found that FPS contains two aspartate-rich regions, namely the DDXXD motifs in domains II and V. This motif plays a crucial role in the catalytic activity of FPS, which is responsible for binding magnesium ions (Mg^{2+}), thereby stabilizing pyrophosphate groups and promoting substrate binding and reaction processes. The amino acid residues in this region participate in the formation of C-C bonds between IPP and allylic substrates and determine the chain length of the product. The aspartate and arginine residues in domain II are essential for FPPS function, and their substitution can lead to a decrease in the Vmax of IPP and geranyl pyrophosphate (GPP) while leaving Km largely unchanged. Conversely, mutation of the first aspartate residue in domain V significantly reduced FPS activity and increased the Km value for IPP, confirming the critical role of aspartate in enzymatic activity [26].

In our study, EhFPS8 lacked the DDXXD motif in the conserved domain II, suggesting a weakening of its core catalytic function. However, further experimental validation is required. Through physical and chemical property analyses, EhFPS proteins were found to be hydrophilic and have a high proportion of α-helices. This is similar to the FPS proteins reported in rubber trees and other plants, the α-helical structures of which are often closely related to their catalytic activity and substrate binding ability [27,28]. Previous studies have shown that prerequisite C5 units (DMAPP and IPP) are synthesized through both the plastidial methyl-D-erythritol phosphate (MEP) and cytosolic MVA pathway.

However, the subsequent biosynthetic pathways for different terpenoid classes vary [29]. Specifically, sesquiterpenes and triterpenes are synthesized via the cytosolic MVA pathway, whereas monoterpenes, diterpenes, and tetraterpenes are synthesized through the plastidial MEP pathway [30]. For example, cytoplasmic localized *AtFPS1S* can participate in the synthesis of sesquiterpenes, such as (+)-δ-cadinene, and in inducible defense against aphid infestation in *Arabidopsis* [31]. In ginseng, *PcFPS* localized in the cytosol is capable of regulating triterpenoids such as pachymic acid and eburicoic acid [32]. *AtFPS2* encodes an FPS enzyme that localizes to the chloroplast. This enzyme is involved in the synthesis of monoterpenes and diterpenes, which are essential components of chlorophylls, carotenoids, and gibberellins.

In this study, previous predictions using subcellular localization showed that *EhFPS1, EhFPS2, EhFPS5, EhFPS6,* and *EhFPS7* were localized in the cytoplasm, whereas *EhFPS3, EhFPS4,* and *EhFPS8* were localized in the chloroplasts. In addition, the subcellular localizations of *EhFPS1, EhFPS2,* and *EhFPS7* proteins in *N. benthamiana* were confirmed (Figure 11). Subcellular localization prediction analysis indicated that, except for *EhFPS3, EhFPS4,* and *EhFPS8,* the remaining EhFPS proteins were primarily localized in the cytoplasm. This finding is consistent with their role in terpenoid biosynthesis, since most terpenoid synthases are active in the cytoplasm. In contrast, those localized in the chloroplast, such as *EhFPS3, EhFPS4,* and *EhFPS8,* may play roles in the metabolic pathways associated with photosynthesis in plants. In this study, GO enrichment analysis revealed that *EhFPS* genes were significantly enriched in biological processes related to isoprenoid metabolism and lipid biosynthesis, indicating that these genes play crucial roles in the secondary metabolism of *E. hirta.* This finding aligns well with our hypothesis regarding the function of the FPS protein family, suggesting their important role in the synthesis of isoprenoids and lipid compounds. Previous studies have demonstrated that *FPS* genes play a pivotal role in isoprenoid metabolism in plants, particularly in the biosynthesis of terpenoids and carotenoids [18–20,32]. Moreover, comparative analysis showed that the functional domains of *EhFPS* genes are highly conserved, including the enzyme activity-regulating motifs DDXXD and GKLNR, further supporting their hypothesized functions as terpene synthases. The type and number of cis-acting elements in the promoter region play important regulatory roles in gene expression [33].

The promoter of the *FPS* gene of *E. hirta* contains a large number of cis-acting elements related to light response, plant hormone induction, and stress response, indicating that members of this family are involved in a variety of biological processes. Studies have reported that spectral quality mainly changes the distribution of terpenoids, whereas the intensity and period affect their abundance [34]. For example, illumination duration and temperature have an important impact on the growth of *Ganoderma lucidum* (a Chinese medicine) and synthesis of triterpenes. The combination of 36 °C and 16 h of illumination significantly increased the yield of *Ganoderma lucidum* triterpenes, which was 98.32% higher than that of the control. Light treatment can also induce an increase in the mycelial biomass of *Ganoderma lucidum* and promote the synthesis of triterpenoids by upregulating the expression of key enzyme genes such as HMGR, SQS, and LAS [35]. Another study on *Eleutherococcus senticosus* revealed that saponin content significantly increased under varying light conditions. Four *EsbZIP* genes associated with saponin biosynthesis were identified through correlation analysis: *EsbZIP1, EsbZIP2, EsbZIP4,* and *EsbZIP5.* The promoter region of *EsbZIP* contains numerous photoresponsive elements, and its activity varies with light intensity [36].

We observed that the promoters of the eight *EhFPS* genes contained different numbers of light-responsive cis-elements (Box4, G-box, and GT1-motif). Therefore, it can be speculated that the transcription level of *EhFPS* may be regulated by transcription factors involved in terpene synthesis, such as AP2/ERF, BHLH, MYB, NAC, WRKY, and bZIP, which ultimately affect the synthesis and accumulation of triterpenoids [37]. Plant hormones have been reported to affect terpenoid synthesis through various mechanisms. For example, plant hormones such as JA, MeJA, and SA can regulate the differential expression of key enzyme activities or genes (HMRG, DXR, and FPS) in the terpenoid biosynthesis pathway by acting as signaling molecules or inducers, thereby affecting the accumulation of the final product [38]. Moreover, plant hormones, such as GA, BR, and ABA, are also involved in regulating the synthesis of terpenoids [39,40]. Seven genes linked to ginsenoside Rg3 synthesis (FPS, UGT, CYP339, etc.) were screened using trait correlation analysis. Exogenous MeJA treatment markedly increased the expression levels. Six hours after induction, the expression level of the *FPS* gene peaked and was 7.1 times higher than that in the control group [41]. It has been discovered that treating birch trees with exogenous MeJA at a concentration of 0.5 mmol/L and SA at a concentration of 50 mmol/L can greatly raise the *BpFPS* gene's expression level and encourage the buildup of total triterpenoids in the leaves and outer bark of the stem [42].

In addition, promoter analysis and RT-qPCR quantitative experiments revealed that the promoters of birch *BpMYB21* and *BpMYB61* contain plant hormone response elements, including ABRE, CGTCA-motif, TCA-element, and GARE motifs. These elements can be induced by ABA, JA, and SA and subsequently control the production of triterpenoids by regulating the expression of important enzyme genes (SS and SE) [43]. A MeJA response element was found in the promoter sequence of the *CbWRKY24* gene of *Ursula bile*, which responds to MeJA induction. In overexpressed plants, it was found that the expression levels of genes such as *HMGR*, *FPS*, *SE*, and *β-AS* were significantly upregulated, thereby improving the total saponin content of bear bile grass [44]. Treatment of *Panax notoginseng* with three plant elicitors (JA, ABA, and SA) can upregulate the expression levels of the key genes *PgFPS* and *PgIPPI* for terpenoid synthesis, promote the synthesis of triterpenoid compounds, and promote the accumulation of triterpenoid saponins [45]. Gene expression analysis of *M. recutita* revealed that *MrFPS* was significantly upregulated in response to methyl jasmonate. The expression patterns varied considerably among different plant tissues, with the highest expression levels observed in flowers and stems. This tissue-specific distribution implies that different plant sections have distinct roles for the FPS enzyme in the production of secondary metabolites [18]. *CnFPPS* expression was consistently observed across various tissues in *C. nobile*, with the highest concentrations found in the root system. This distribution pattern indicated that *CnFPPS* may play a key role in regulating terpenoid production [20]. Functional analyses of *AvFPS* revealed that *Aconitum vilmorinianum* was significantly upregulated in response to methyl jasmonate. Moreover, genetically modified expression of *AvFPS* increased the plant's resistance to environmental stressors, which increases its ability to be cultivated under less ideal circumstances [46].

In summary, plant hormones affect the biosynthesis and accumulation of triterpenoid compounds by affecting key enzyme activities, differential expression of key enzyme genes, regulation of transcription factors and promoter activities, and synergistic interactions between plant hormones [32]. In this study, a large number of cis-elements (ABRE, TCA-element, CGTCA-motif, and TGACG-motif) that responded to the plant hormones ABA, SA, and MeJA were also found in the promoters of *E. hirta FPS* genes. Through exogenous JA and SA treatments, we further investigated the expression patterns of *EhFPS* genes under SA and JA signaling. It was found that, except for *EhFPS5*, all other genes responded to SA treatment, exhibiting significant upregulation of expression. The expression trends of

EhFPS genes could be divided into three subgroups, with *EhFPS1*, *EhFPS2*, and *EhFPS7* showing similar expression patterns, whereas *EhFPS3*, *EhFPS4*, and *EhFPS8* displayed a consistent trend. This is consistent with the known relationship between terpenoid biosynthesis and hormone signaling in plants. Under exogenous JA treatment, genes such as *EhFPS1*, *EhFPS2*, *EhFPS6*, *EhFPS7*, and *EhFPS8* rapidly responded to JA signals, showing a significant increase in expression levels.

We also observed that some genes, such as *EhFPS1*, *EhFPS2*, and *EhFPS7*, responded quickly to both SA and JA signals, demonstrating expression patterns similar to those of terpenoid synthases under various stress conditions, as reported in the literature [39]. This result is also consistent with our findings from cis-acting element analysis of the promoters. These experiments further support the hypothesis that these genes play a key role in terpenoid biosynthesis. Therefore, it is speculated that terpenoid compounds are mainly synthesized in the latex and leaves. Terpenoids are widely present in nature and constitute the main components of essences, resins, and pigments in some plants [47]. Tissue-specific expression analysis revealed that *EhFPS1*, *EhFPS2*, and *EhFPS7* exhibited significantly higher expression levels in the latex of *E. hirta* than the other genes, suggesting that these genes may be involved in the regulatory pathway of latex synthesis. This finding was consistent with our hypothesis regarding terpenoid biosynthesis. Natural rubber is a natural polymer composed primarily of cis-1,4-polyisoprene, which belongs to the terpenoid class and is a major component of resins. Specific expression of these genes in latex is crucial for terpenoid biosynthesis. Similar tissue-specific expression patterns have been reported in rubber trees and other plants, further supporting our understanding of the functional roles of *EhFPS* genes [48]. Importantly, the FPS gene family is essential for metabolic activities that produce different terpenoids, which serve as defenses against infections and organisms that consume plants. For example, the *FPS* found in *M. recutita* contributes to the synthesis of α-bisabolol, a pharmacologically active substance essential for its medicinal effects [18]. Investigation of the *FPS* gene family in several species demonstrated its essential function in plant metabolism, specifically in the biosynthesis of terpenoids, which are essential for development, growth, and stress reactions [18–20]. The need for further research to fully understand the biological importance of *FPS* genes is underscored by their high degree of conservation, diverse expression patterns, and functional implications for secondary metabolism.

4. Materials and Methods

4.1. Experimental Materials

Wild-type *E. hirta* used in this study was provided by the Chinese Academy of Tropical Agricultural Sciences, Haikou, Hainan Province, China (110°33′ E,19°98′ N). We selected 20 samples of *E. hirta* at the same growth stage, cleaned the plant surfaces with 0.1 M PBS buffer, and subsequently separated the root, stem, leaves, flowers, latex, and other tissue samples while maintaining low temperatures. Equal amounts of these mixed samples were stored on dry ice and sent to Shanghai Ouyi Biotechnology Co, Ltd. (Shanghai, China) for transcriptome sequencing. The wild-type *E. hirta* plants utilized in this study exhibited uniform growth 45 days after sowing. Indoor culture conditions were maintained at 25 °C with a light intensity of 400 μmol m^{-2} s^{-1} for 14 h, followed by 8 h of darkness. We weighed 0.021 g of jasmonic acid (JA) and 0.0138 g of salicylic acid (SA) as dry powders, which were then dissolved in anhydrous ethanol. The solution was titrated with double-distilled water (ddH_2O) to achieve a final volume of 1 L, with the pH adjusted to 6.8. This procedure resulted in a working solution of JA and SA at concentration of 0.1 mM. To prevent degradation, it is crucial to avoid exposure to light and to store the solution at 4 °C. In a low-light environment, the solution was sprayed onto both surfaces of the plant leaves at a distance of 20 cm. The treatments were applied at 0, 2, 4, 8, and 24 h, with three biological replicates established at each time point. *N. benthamiana* used in the subcellular localization experiment was provided by the Molecular Breeding Laboratory of the Rubber Research Institute, Chinese Academy of Tropical Agricultural Sciences, Haikou, Hainan Province, China.

4.2. Identification and Physicochemical Analysis of EhFPS Genes

The domains contained in the rubber tree *FPS* gene were analyzed using CDS-Search (https://www.ncbi.nlm.nih.gov/Structure/cdd/cdd.shtml, accessed on 7 May 2024) and InterPro (https://www.ebi.ac.uk/interpro/ accessed on 7 May 2024). FPS family members were identified using BLASTP of the genome and transcriptome data of *E. hirta* (unpublished) and named according to their physical distance. The chromosomal location of *EhFPS* genes and identification of tandemly repeated genes were determined using a previously described method [49]. The *E. hirta* GFF annotation file was input into the Gene Location Visualization from the GTF/GFF program of TBtools-II v. 2.136 software, which was used to obtain a rough distribution map of *EhFPS* genes on chromosomes. Referring to the criteria for determining gene clusters, two adjacent genes were no more than 250 kb apart on the chromosome, the tandemly arranged *FPS* genes on each chromosome were identified, and their Ka/Ks were calculated. The online tool ExPASy ProtParam (https://web.expasy.org/protparam/ accessed on 17 July 2024) was used to predict and analyze the physicochemical indices of *EhFPS* genes, including amino acid molecular weight, protein length, theoretical isoelectric point, α-helix, and β-turn. The online tool SWISS-MODEL (https://swissmodel.expasy.org/interactive accessed on 18 June 2024) was used to predict the secondary structure of the *EhFPS* genes. The WOLF PSORT website (https://wolfpsort.hgc.jp/ accessed on 19 July 2024) was used to predict the subcellular localization of the *EhFPS* genes.

4.3. Phylogenetic Analysis

The amino acid sequences of *EhFPS* genes were added to the NCBI database (https://www.ncbi.nlm.nih.gov/ accessed on 21 July 2024), and the highly homologous FPS protein sequences in the rubber tree, cassava, *Eucommia*, dandelion, and *A. thaliana* were retrieved and downloaded through BLASTP comparison ("1×10^{-5}", "1×10^{-10}"). The biological software Clustalw v. 1.83 was used to perform a complete multiple sequence alignment of the screened amino acid sequences, and a phylogenetic tree was constructed using the neighbor-joining method (bootstrap value was set to 1000) in MEGA v. 7.0. Optimization was performed using the online software Evolview v. 4 (https://www.evolgenius.info/evolview accessed on 21 July 2024).

4.4. Gene Structural Analysis of EhFPS

The MEME online tool (https://meme-suite.org/meme/tools/meme accessed on 22 June 2024), NCBI (https://www.ncbi.nlm.nih.gov/Structure/cdd/wrpsb.cgi accessed on 22 June 2024), and the InterPro database (https://www.ebi.ac.uk/interpro/ accessed on 22 June 2024) were used to analyze the conserved motifs of proteins, with the base number set to 10 and other parameters as default. The exon–intron structure of *EhFPS* was analyzed using TBtools-II v. 2.119 [50], and the phylogenetic tree, conserved domains, and gene structures were visualized.

4.5. Collinearity Analysis and Visualization of FPS Genes Between H. brasiliensis and E. hirta

Referring to the experimental method of Xue et al. [51], a script was used to extract the information files required by TBtools such as chromosome length, chromosome gene information, and link information between chromosomes based on collinearity and GFF annotation files. The files were then input into the Advanced Circus plug-in TBtools to complete the collinearity analysis between *E. hirta* and *H. brasiliensis*, and to draw a collinearity map between chromosomes.

4.6. Cis-Acting Elements Analysis in the Promoter of EhFPS

The plug-in Fasta extract in TBtools was used to extract the 2000 bp sequence upstream of the gene transcription start site as the promoter analysis region. The cis-acting elements were predicted using the promoter analysis online website PlantCare (https://bioinformatics.psb.ugent.be/webtools/plantcare/html/ accessed on 21 June 2024), and the prediction results were visualized using the plug-in Basic Biosequence View.

4.7. Expression Pattern Analysis

Transcriptome data from our laboratory were used to analyze the expression levels of the *FPS* gene family members in different tissues of *E. hirta*. The transcriptome data obtained by sequencing were uploaded to the local server (SSH-2.0-OpenSSH_7.4:10.38.0.1) the RSEM software v. 2.1 package was used for expression analysis to obtain the FPKM value of each gene, and the HeatMap plugin in TBtools was used to draw a heat map.

4.8. Gene Ontology Enrichment Analysis

Gene ontology (GO) enrichment analysis was performed to understand the biological roles of the identified genes. Functional annotations were derived from molecular breeding laboratory databases, providing a foundation for categorizing genes based on their biological processes, molecular functions, and cellular components. The analysis was conducted using the Cluster Profiler R package v. 4.14.0, which facilitated systematic identification of overrepresented GO terms among the target gene sets. Statistical significance was determined using thresholds of $Q < 0.05$ and $p < 0.05$. Visualization of enriched GO terms was performed using the same package.

4.9. Interaction Network of E. Hirta Homologues in A. Thaliana

Homologous genes of *E. hirta* were identified in *A. thaliana* using OrthoVeen v. 3.0 (https://orthovenn3.bioinfotoolkits.net/ accessed on 21 November 2024). Further analysis of the *EhFPS* homologous gene set using STRING (https://cn.string-db.org/ accessed on 21 November 2024) with a minimum interaction score of 0.6 was required to limit the results to no more than 30 interactors. Genes with homologues in *E. hirta* were compared, filtered from the interaction results, and visualized using bioinformatics analysis performed using OECloud tools.

4.10. RNA Extraction and RT-qPCR Analysis

Reverse transcription quantitative PCR validation and expression analysis of *FPS* genes were performed using transcriptome data from *E. hirta*. Total RNA from five tissues (root, stem, leaf, flower, and latex) of wild-type *E. hirta* was extracted using a SteadyPure Plant RNA Extraction Kit (AG21019; Aikerui, Changsha, China). The concentration and integrity of the RNA samples were determined using a NanoDrop 2000 spectrophotometer and gel electrophoresis. cDNA was synthesized by reverse transcription using the PrimeScript™ RT Reagent Kit and the gDNA Eraser Kit (Takara, Dalian, China). RT-qPCR primers for *EhFPS* genes were designed using the online Integrated DNA Technologies website (https://www.idtdna.com/scitools/Applications/RealTimePCR/default.aspx accessed on 21 November 2024) (Table S3). RT-qPCR was performed using *ACT1.2373* as the internal reference gene and the reverse-transcribed cDNA obtained above as the template, according to the instructions of the 2 × Q3 SYBR qPCR Master Mix (Universal) kit (2205,

TOLOBIO, Shanghai, China). The relative expression level of *EhFPS* was calculated using the $2^{-\Delta\Delta Ct}$ method, and the results were plotted using the data statistics software Prism v. 10.

4.11. Cloning of CDS Sequences of EhFPS1, EhFPS2, and EhFPS7 and Construction of Fusion Expression Vector

Total RNA was extracted from wild-type *E. hirta* leaves, and its concentration and integrity were determined using a NanoDrop 2000 spectrophotometer and gel electrophoresis. The extracted total RNA sample was analyzed by electrophoresis, which clearly showed two bands, 28S and 18S, indicating that the extracted total RNA was of good quality and met the requirements for reverse transcription. Agarose gel electrophoresis (0.8%) was used to detect the linearized vector and the CDS sequence amplification products of *EhFPS1*, *EhFPS3*, and *EhFPS7* (Figure S1A). The empty pCAMBIA2300-35S-eGFP(C) plasmid was a single band with a size of approximately 10,233 bp, while the electrophoretic migration rate of the pCAMBIA2300-35S-eGFP(C) double enzyme digestion product was significantly lower than that of the circular plasmid. Specific primers designed according to the CDS region were used for amplification, and the product sizes of *FPS1*, *FPS3*, and *FPS7* were target bands of 1029, 1053, and 1011 bp, respectively (Figure S1B). Single clones of the recombinant plasmids were selected from *E. coli* for PCR identification. The results showed that the PCR products of the two selected single clones contained single bands of 1029, 1053, and 1011 bp, confirming that they were both positive (Figure S1C), and were subsequently sent to Shanghai Sangon Biotechnology Co., Ltd. (Shanghai, China) for Sanger sequencing.

4.12. Transient Expression of EhFPS1, EhFPS2, and EhFPS

According to the CDS sequences of *EhFPS1*, *EhFPS3*, and *EhFPS7*, CE Design v. 1.04 software was used to design specific amplification primers for the homology arms of the vector using BamHI and SalI restriction sites (Table S4). The CDS sequences of *EhFPS1*, *EhFPS2*, and *EhFPS7* were amplified using reverse-transcribed cDNA as a template. The system was as follows: Primer-F 0.5 μL, Primer-R 0.5 μL, cDNA 1 μL (200 ng/μL); TakaRa 2 × PrimeSTAR® Mix 25 μL, ddH_2O 23 μL. Amplification program: 98 °C/3 min, 98 °C/10 s, 65 °C/10 s, 72 °C/1 min (35 cycles), 72 °C/1 min, stored at 4 °C. The plasmid was then recombined and transformed into DH5α competent cells. Following successful sequencing, the recombinant plasmid was transformed into GV3101-p19 competent cells in accordance with Zhao's method [52]. The transformed cells were then used to infect *Nicotiana benthamiana*, which were maintained in a dark culture room at 26 °C for 48 h. Observations were made and recorded using a laser confocal microscope (ZEISS LSM800, Göttingen, Germany) under a 488 nm channel.

5. Conclusions

This study employed the genome and transcriptome data of *E. hirta* and systematically identified eight *FPS* genes distributed on chromosomes 1, 3, 6, and 8. Through the identification of conserved domains and phylogenetic analysis, it was found that seven family members contained DDIMD motifs, all of which contained GQMXD and DDYXD motifs. The gene structure was highly conserved and the relationship was close. In the promoter region of *EhFPS* genes, we identified large numbers of light-responsive elements (Box4, G-box, and GT1-motif) and hormone-responsive elements (ABRE, CGTCA-motif, TGACG-motif, and TCA-element) and verified whether their expression levels were induced by relevant hormones. In addition, expression pattern analysis revealed that *EhFPS1*, *EhFPS2*, and *EhFPS7* were expressed specifically in various tissues and their encoded proteins were localized in the cytoplasm. These results provide a theoretical basis for in-depth research into the regulatory mechanisms and biological functions of *EhFPS* genes in the synthesis of terpenoids in *E. hirta*.

Supplementary Materials: The following supporting information can be downloaded at: https://www.mdpi.com/article/10.3390/ijms26020798/s1, Figure S1: Gene amplification and vector construction. (A) RNA detection; (B) 35::2300::eGFP vector linearization and amplification of EhFPS1, EhFPS2 and EhFPS7 genes; (C) Colony PCR verification, 1–5: recombinant EhFPS1, 6–10: recombinant EhFPS2, 11–15: recombinant EhFPS3. Bands are shown at specific length above 10,000 bp and 1000 bp; Table S1: Identified genes under gene ontology enrichment analysis; Table S2: FPS proteins homologues and their interaction network with other proteins; Table S3: List of primers used for qRT-PCR; Table S4: List of fusion expression vector primers.

Author Contributions: H.C. and T.W. conceived and designed the study. T.W. and X.B. performed data analysis and drafted the manuscript. H.C., T.W. and F.R. reviewed and revised the manuscript and formal analysis. Z.D., R.Q., Q.H., Y.Y., D.X., X.W., W.W., Z.A. and H.W. contributed to sample collection. All authors have read and agreed to the published version of the manuscript.

Funding: This research was financially supported by the Nanhai New Star Technology Innovation Talent Platform Project of Hainan Province (No. NHXXRCXM202330), supported by the Hainan Provincial Natural Science Foundation of China (No. 324QN304), the Chinese Academy of Tropical Agricultural Sciences for Science and Technology Innovation Team of the National Tropical Agricultural Science Center (No. CATASCXTD202401) and the Project of the National Key Laboratory for Tropical Crop Breeding (No. NKLTCBCXTD01).

Institutional Review Board Statement: Not applicable.

Informed Consent Statement: Not applicable.

Data Availability Statement: All analyzed data can be found in the article or in the Supplementary Material.

Conflicts of Interest: The authors declare no conflicts of interest.

References

1. Zhang, X. Screening of the Anti-Diarrhea Active Ingredients from *Euphorbia hirta* L. and Preliminary Study on Their Mechanism. Master's Thesis, Northeast Agricultural University, Heilongjiang, China, 2022. [CrossRef]
2. Flora of China Editorial Committee, Chinese Academy of Sciences. *Flora of China*; Science Press: Beijing, China, 1993; p. 18. Available online: https://www.hanspub.org/reference/reference?referenceid=10899 (accessed on 7 November 2024).
3. *State Administration of Traditional Chinese Medicine Editorial Committee of Chinese Materia Medica*; Chinese Materia Medica Shanghai Science and Technology Press: Shanghai, China, 1999; Volume 14, pp. 490–502. Available online: https://www.hanspub.org/reference/referencepapers?referenceid=55477&&paperid=13621 (accessed on 7 November 2024).
4. Mallavadhani, U.V.; Narasimhan, K. Two novel butanol rhamnosides from an Indian traditional herb, *Euphorbia hirta* L. *Nat. Prod. Res.* **2009**, *23*, 644–651. [CrossRef] [PubMed]
5. Wang, L.; Li, Y.; Yang, M.; Tang, R.; Zang, Z.; Zhao, Y. Study on the chemical components of *Herba lycopodii* (II). *Chin. Tradit. Patent Med.* **2014**, *36*, 1687–1692. [CrossRef]
6. Wu, Y.; Qu, W.; Geng, D.; Liang, J.; Luo, Y. Phenols and flavonoids from the aerial part of *Euphorbia hirta*. *Chin. J. Nat. Med.* **2012**, *10*, 40–42. [CrossRef]
7. Zhao, X.; Li, B.; Wang, J.; Zhu, L.; Zhang, G. Advances in chemical components, pharmacological activities and clinical applications of *Euphorbia hirta* L. and *Euphorbia thymifolia* L. *J. Shenyang Pharm. Univ.* **2023**, *40*, 381–393. [CrossRef]
8. Song, L.; Xu, H.; Yang, L.; Wang, Z. Overview of research on chemical components and pharmacological activities of *Euphorbia hirta* L. *J. Chin. Med. Mat.* **2012**, *35*, 1003–1009. [CrossRef]
9. Zhang, G.; Wu, Y.; Chen, H.; Yu, J.; Yang, D.; Zhang, J.; Yang, Y. Cloning and Bioinformatics Analysis of Farnesyl Diphosphate Synthase (FPS) Gene in *Pogostemon cablin*. *Mol. Plant Breed.* **2019**, *17*, 2163–2170. [CrossRef]
10. Closa, M.; Vranová, E.; Bortolotti, C.; Bigler, L.; Arró, M.; Ferrer, A.; Gruissem, M. The *Arabidopsis thaliana* FPP synthase isozymes have overlap and specific functions in isoprenoid biosynthesis, and complete loss of FPP synthase activity causes early developmental arrest. *Plant J.* **2010**, *63*, 512–525. [CrossRef]

11. Li, X.; Cui, H.; Liu, H.; Wang, Y. Studies on cartenoids and aromatic components of the transgenic flue-cured tobacco expressing fps gene. *Chin. Tobacco Sci.* **2006**, *27*, 25–27. [CrossRef]
12. Cui, H.; Liu, H.; Li, X. Study on resistance to brown spot disease of transgenic tobacco with exogenous farnesyl pyrophosphate synthase gene. *J. Crops* **2006**, *32*, 817–820. [CrossRef]
13. Zhang, S.; Chen, H.; Wang, S.; Du, K.; Song, L.; Xu, T.; Xia, Y.; Guo, R.; Kang, X.; Li, Y. Positive regulation of the Eucommia rubber biosynthesis-related gene EuFPS1 by EuWRKY30 in *Eucommia ulmoides*. *Inter. J. Biol. Macromol.* **2024**, *2*, 131751. [CrossRef] [PubMed]
14. Xu, X.; Lai, R.; Chen, T.; Shi, L.; Chen, S. Construction of yeast one-hybrid library of FPS gene from *Ganoderma lucidum* and screening of its upstream transcription factors. *Chin. Herbal. Med.* **2020**, *51*, 3770–3776. [CrossRef]
15. Wang, Y.; Jiang, J.; Chen, Y.; Zhao, Y. Study on chemical constituents of *Euphorbia hirta* L. *Anhui Agric. Sci.* **2012**, *40*, 4060–4062. [CrossRef]
16. Gao, W.; Meng, Q.; Wang, X.; Chen, F.; He, M.; Zhou, Y. Overexpression of GPS and FPS from *Chrysanthemum indicum* var. aromaticum resulted in modified trichome formation and terpenoid biosynthesis in Tobacco. *Plant Growth Reg.* **2023**, *99*, 553–566. [CrossRef]
17. Kim, O.T.; Kim, S.H.; Ohyama, K.; Muranaka, T.; Chop, Y.E.; Lee, H.Y.; Kim, M.Y.; Hwang, B. Upregulation of phytosterol and triterpene biosynthesis in *Centella asiatica* hairy roots overexpressed ginseng farnesyl diphosphate synthase. *Plant Cell Rep.* **2010**, *29*, 403–411. [CrossRef] [PubMed]
18. Su, S.; Zhang, H.; Liu, X.; Pan, G.; Ling, S.; Zhang, X.; Yang, X.; Tai, Y.; Yuan, Y. Cloning and characterization of a farnesyl pyrophosphate synthase from *Matricaria recutita* L. and its upregulation by methyl jasmonate. *Genet. Mol. Res.* **2015**, *14*, 349–361. [CrossRef]
19. Pang, Y.; Guan, L.; Wu, L.; Chen, Z.; Wang, K.; Xie, X.; Yu, F.; Chen, X.; Zhang, Y.; Jiang, Q. Cloning and sequence analysis of the Blumea balsamifera DC farnesyl diphosphate synthase gene. *Genet. Mol. Res.* **2014**, *13*, 9874–9882. [CrossRef] [PubMed]
20. Liu, X.; Tao, T.; Meng, X.; Zhang, W.; Chang, J.; Xu, F. Cloning and Expression Analysis of a Farnesyl Diphosphate Synthase (FPPS) Gene from *Chamaemelum nobile*. *Not. Bot. Horti Agrobot. Cluj-Napoca* **2017**, *45*, 358–364. [CrossRef]
21. Chen, D.; Ye, H.; Li, G. Expression of a chimeric farnesyl diphosphate synthase gene in *Artemisia annua* L. transgenic plants via *Agrobacterium tumefaciens*-mediated transformation. *Plant Sci.* **2000**, *155*, 179–185. [CrossRef] [PubMed]
22. Han, J.; Liu, B.; Ye, H.; Wang, H.; Li, Z.; Li, G. Effects of overexpression of the endogenous farnesyl diphosphate synthase on the artemisinin content in *Artemisia annua* L. *J. Integr. Plant Biol.* **2006**, *48*, 482–487. [CrossRef]
23. Tang, K.; Shi, P.; Fu, X.; Shen, Q.; Jiang, W.; Ma, Y.; Hao, X.; Sun, X. The Method of Co-Transforming FPS and DBR2 Genes to Improve the Content of Artemisinin and the Preparation of *Artemisia annua*. CN201611012726.2, 31 May 2017.
24. Zhang, X.; Ge, F.; Deng, B.; Shah, T.; Huang, Z.; Liu, D.; Chen, C. Molecular cloning and characterization of PnbHLH1 transcription factor in *Panax notoginseng*. *Molecules* **2017**, *22*, 1268. [CrossRef]
25. Wang, P.; Liao, Z.; Guo, L.; Li, W.; Tang, K. Cloning and functional analysis of acDNA encoding *Ginkgo biloba* farnesyl diphosphatesynthase. *Mol. Cells* **2004**, *18*, 150–156. [CrossRef] [PubMed]
26. Anna, S.; Danuta, P. Farnesyl diphosphate synthase; regulation of product specificity. *Acta Biochim. Pol.* **2005**, *52*, 45–55. [CrossRef]
27. Cheng, H.; Li, L.; Cheng, S.; Cao, F.; Wang, Y.; Yuan, H. Genome-wide identification and expression analysis of the farnesyl pyrophosphate synthase (FPS) gene family in *Hevea brasiliensis*. *Ind. Crops Prod.* **2020**, *145*, 112102.
28. Wang, G.; Li, T.; Aziz, N.; Broun, P.; Dai, X.; He, J.; King, A.; Zhao, P.; Dixon, Z. Terpene biosynthesis in glandular trichomes of hop. *Plant Physiol.* **2008**, *158*, 499–511. [CrossRef] [PubMed]
29. Wang, J.; Meng, H.; Wang, Y. Advances in synthetic biology of terpenoids. *Chin. Sci. Life Sci.* **2015**, *45*, 1040–1050. [CrossRef]
30. Wang, L.; Fang, X.; Yang, C.; Li, J.; Chen, X. Plant terpenoid secondary metabolism and its regulation. *Chin. Sci. Life Sci.* **2013**, *43*, 1030–1046. [CrossRef]
31. Closa, M.; Gómez-Benáitez, J. Molecular characterization of two isoforms of a farnesyl pyrophosphate synthase in *Arabidopsis thaliana*: Involvement in sesquiterpene biosynthesis and inducible defense against aphid infestation. *New Phyt.* **2010**, *188*, 835–846.
32. Zhou, X.; Zhang, X.; Liu, L.; Fu, C. Cloning and characterization of farnesyl diphosphate synthase gene from *Poria cocos*. *Int. J. Mol. Sci.* **2014**, *15*, 22188–22197.
33. Yao, X.; Zhao, J.; Hu, S.; Li, X.; Pang, Q.; Yu, Z. Identification and expression pattern analysis of SAPK gene family in *Dendrobium officinale*. *Chin. J. Biotech.* **2024**, *40*, 1195–1210. [CrossRef]
34. Contreras-Avilés, W.; Heuvelink, E.; Marcelis, L.; Kappers, I. Ménage à trois: Light, terpenoids, and quality of plants. *Trends Plant Sci.* **2024**, *29*, 572–588. [CrossRef]
35. Li, J. Effects of Temperature and Light Conditions on the Content of Triterpenoids in *Ganoderma lucidum* During Liquid Fermentation. Master's Thesis, Huazhong Agricultural University, Wuhan, China, 2018. [CrossRef]
36. Li, C.; Jiao, M.; Zhao, X.; Ma, J.; Cui, Y.; Kou, X.; Long, Y.; Xing, Z. bZIP transcription factor responds to changes in light quality and affects saponins synthesis in *Eleutherococcus senticosus*. *Int. J. Biol. Macromol.* **2024**, *279*, 135273. [CrossRef] [PubMed]

37. Dong, Y.; Zhang, W.; Ling, Z.; Li, J.; Bai, H.; Li, H.; Shi, L. Research progress on the regulation of plant terpenoid biosynthesis by transcription factors. *Acta Bot. Sin.* **2020**, *55*, 340–350. [CrossRef]
38. Su, W.; Jiang, Y.; Bai, Y.; Gan, X.; Liu, Y.; Lin, S. Research Progress on Transcription Factors Regulating Plant Terpenoids Biosynthesis. *J. Agric. Biotechnol.* **2019**, *27*, 919–926.
39. Dai, H.; Bai, X.; Li, X.; Zhang, X.; Luo, L.; Zhang, X.; Cao, G.; He, S. Research progress on the role and regulatory mechanism of plant hormones in the biosynthesis of triterpenoids. *Chin. Med. Herbol.* **2021**, *52*, 6391–6402. [CrossRef]
40. Ma, Y.; Chen, Q.; Wang, Y.; Zhang, F.; Wang, G. Heteromerization of short-chain trans-prenyltransferase controls precursor allocation within a plastidial terpenoid network. *J. Integr. Plant Biol.* **2023**, *65*, 1170–1182. [CrossRef] [PubMed]
41. Liu, Y.; Luo, S.; Hua, J.; Li, D.; Ling, Y.; Lou, Q. Characterization of defensive cadinenes and a novel sesquiterpene synthase responsible for their biosynthesis from the invasive *Eupatorium adenophorum*. *New Phytol.* **2021**, *229*, 1740–1754. [CrossRef] [PubMed]
42. Wang, K.; Zhang, Z.; Li, S.; Hu, J.; Liu, T.; Jiang, Y.; Wu, J.; Lu, M.; Zhao, M.; Li, L.; et al. Transcriptome-Wide Analysis for Ginsenoside Rb3 Synthesis-Related Genes and Study on the Expression of Methyl Jasmonate Treatment in *Panax Ginseng*. *Life* **2021**, *11*, 387. [CrossRef]
43. Li, X. Regulation of MeJA and SA on Triterpene Synthesis and FPS Gene Cloning in *Betula platyphylla* Saplings. Master's Thesis, Northeast Forestry University, Harbin, China, 2012.
44. Yin, J.; Sun, L.; Li, Y.; Xiao, J.; Wang, S.; Yang, J.; Qu, Z.; Zhan, Y. Functional identification of BpMYB21 and BpMYB61 transcription factors responding to MeJA and SA in birch triterpenoid synthesis. *BMC Plant Biol.* **2020**, *20*, 374. [CrossRef] [PubMed]
45. Sun, W.; Zhan, J.; Zheng, T.; Sun, R.; Wang, T.; Tang, Z.; Bu, T.; Li, C.; Wu, Q.; Chen, H. The jasmonate-responsive transcription factor CbWRKY24 regulates terpenoid biosynthetic genes to promote saponin biosynthesis in *Conyza blinii* H. Lév. *J. Genet.* **2018**, *97*, 1379–1388. [CrossRef] [PubMed]
46. Wang, X.; Li, G.; Wang, B.; Xu, F.; Ma, X. Cloning and functional verification of farnesyl pyrophosphate synthase gene (AvFPS) from *Aconitum vilmorinianum*. *Chin. J. Trad. Chin. Med.* **2024**, *49*, 4387–4395. [CrossRef]
47. Liang, Z.; Fang, Y.; Yang, D. Advances in chemical constituents, pharmacological activities and clinical application of *Euphorbiae Herba*. *J. Shenyang Pharm. Univ.* **2017**, *37*, 255–264.
48. Liu, H.; Yang, X.; Chen, Z.; Zhang, L.; Wang, Y. Tissue-specific expression of terpenoid biosynthetic genes in rubber-producing plants. *J. Plant Physiol.* **2017**, *213*, 62–71. [CrossRef]
49. Ma, J.; Gu, X.; Wang, S.; Chen, Z.; Zhang, B.; Li, L. Identification and spatio-temporal expression analysis of KCS gene family in *Acer truncatum*. *Res. Econ. For.* **2022**, *40*, 201–214. [CrossRef]
50. Chen, C.; Wu, Y.; Li, J.; Wang, X.; Zeng, Z.; Xu, J.; Liu, Y.; Feng, J.; Chen, H.; He, Y.; et al. TBtools-II: A "one for all, all for one" bioinformatics platform for biological big-data mining. *Mol. Plant* **2023**, *16*, 1733–1742. [CrossRef] [PubMed]
51. Xue, J.; Zhao, T.; Liu, Y.; Zhang, Y.; Shao, Z. Genome-Wide Analysis of the Nucleotide Binding Site Leucine-Rich Repeat Genes of Four Orchids Revealed Extremely Low Numbers of Disease Resistance Genes. *Front. Genet.* **2020**, *10*, 1286. [CrossRef] [PubMed]
52. Zhao, T.; Huang, C.; Li, N.; Ge, Y.; Wang, L.; Tang, Y.; Wang, Y.; Li, Y.; Zhang, C. Ubiquitin ligase VvPUB26 in grapevine promotes proanthocyanidin synthesis and resistance to powdery mildew. *Plant Physiol.* **2024**, *195*, 2891–2910. [CrossRef] [PubMed]

International Journal of
Molecular Sciences

MDPI

Article

Genome-Wide Identification, Phylogenetic Evolution, and Abiotic Stress Response Analyses of the Late Embryogenesis Abundant Gene Family in the Alpine Cold-Tolerant Medicinal *Notopterygium* Species

Xuanye Wu †, Xiaojing He †, Xiaoling Wang, Puyuan Liu, Shaoheng Ai, Xiumeng Liu, Zhonghu Li * and Xiaojuan Wang *

Key Laboratory of Resource Biology and Biotechnology in Western China, Ministry of Education, College of Life Sciences, Northwest University, Xi'an 710069, China; 202221372@stumail.nwu.edu.cn (X.W.); heexj13@163.com (X.H.); wxiaoling98@163.com (X.W.); liupuyuan111@stumail.nwu.edu.cn (P.L.); 17329448694@163.com (S.A.); xiumengliu@163.com (X.L.)

* Correspondence: lizhonghu@nwu.edu.cn (Z.L.); xiaojwang@nwu.edu.cn (X.W.); Tel./Fax: +86-29-88302411 (Z.L.)

† These authors contributed equally to this work.

Abstract: Late embryogenesis abundant (LEA) proteins are a class of proteins associated with osmotic regulation and plant tolerance to abiotic stress. However, studies on the *LEA* gene family in the alpine cold-tolerant herb are still limited, and the phylogenetic evolution and biological functions of its family members remain unclear. In this study, we conducted genome-wide identification, phylogenetic evolution, and abiotic stress response analyses of *LEA* family genes in *Notopterygium* species, alpine cold-tolerant medicinal herbs in the Qinghai–Tibet Plateau and adjacent regions. The gene family identification analysis showed that 23, 20, and 20 *LEA* genes were identified in three *Notopterygium* species, *N. franchetii*, *N. incisum*, and *N. forrestii*, respectively. All of these genes can be classified into six LEA subfamilies: LEA_1, LEA_2, LEA_5, LEA_6, DHN (Dehydrin), and SMP (seed maturation protein). The LEA proteins in the three *Notopterygium* species exhibited significant variations in the number of amino acids, physical and chemical properties, subcellular localization, and secondary structure characteristics, primarily demonstrating high hydrophilicity, different stability, and specific subcellular distribution patterns. Meanwhile, we found that the members of the same LEA subfamily shared similar exon–intron structures and conserved motifs. Interestingly, the chromosome distributions of *LEA* genes in *Notopterygium* species were scattered. The results of the collinearity analysis indicate that the expansion of the *LEA* gene family is primarily driven by gene duplication. A *Ka/Ks* analysis showed that paralogous gene pairs were under negative selection in *Notopterygium* species. A promoter *cis*-acting element analysis showed that most *LEA* genes possessed multiple *cis*-elements connected to plant growth and development, stress response, and plant hormone signal transduction. An expression pattern analysis demonstrated the species-specific and tissue-specific expression of *NinLEAs*. Experiments on abiotic stress responses indicated that the *NinLEAs* play a crucial role in the response to high-temperature and drought stresses in *N. franchetii* leaves and roots. These results provide novel insights for further understanding the functions of the *LEA* gene family in the alpine cold-tolerant *Notopterygium* species and also offer a scientific basis for in-depth research on the abiotic stress response mechanisms and stress-resistant breeding.

Keywords: *Notopterygium*; *LEA* gene family; genome-wide analysis; molecular evolution; expression analysis

1. Introduction

Plants are subjected to an array of extraneous environments that exert influences on their ontogeny, development, and maturation. In order to accommodate complex and volatile environment conditions, plants have evolved numerous gene families possessing analogous architectures and functionalities. The late embryogenesis abundant protein (LEA) gene family is responsible for encoding LEA proteins [1]. LEA proteins constitute a diverse family of small, hydrophilic polypeptides (10–30 kD); their initial discovery was made during the terminal phases of embryonic development in cotton (*Gossypium hirsutum* L.) seeds [2]. LEA proteins are highly hydrophilic and thermally stable, with their higher-order structures comprising non-periodic linear and α-helix conformations [3], and present in different tissues of plants, including seeds, roots, stems, and buds [4]. Under abiotic stress conditions such as drought, salinity, and extreme temperatures, many plants accumulate highly hydrophilic LEA proteins [5,6] to stabilize vastly different target biomolecules, such as enzymes, membranes, and DNA/RNA, protecting plant tissues from various abiotic stresses [7,8].

Based on sequence characteristics and structural domains, the *LEA* gene family is generally categorized into eight subfamilies: LEA_1, LEA_2, LEA_3, LEA_4, LEA_5, LEA_6, Dehydrin (DHN), and seed maturation protein (SMP) [9]. *LEA* genes have been identified in many plant species, for instance, 51 *AtLEA* were identified in the model plant *Arabidopsis thaliana* (L.) Heynh., and most of them had abscisic acid response elements (ABREs) and temperature response elements (LTREs) in their promoters; many genes containing the respective promoter elements were induced by abscisic acid, temperature, or drought [5]. Meanwhile, an evolutionary analysis showed that the genome of *Oryza sativa* L. and *Triticum aestivum* L. contained 34 and 281 *LEA* family genes [10,11]. A total of 123 *NtLEA* genes were also identified in tobacco (*Nicotiana tabacum* L.), which were significantly induced under NaCl stress [12]. Additionally, some studies have shown that there were significant differences in *LEA* gene expression among species of the same genus with different ploidy levels. Some other studies have conclusively shown that LEA proteins played a pivotal role in plants' adaptive responses to abiotic stressors, particularly those associated with temperature and water. For example, a study showed that *OsEm1* (LEA protein family) in rice could enhance the tolerance of species to drought stress [13]. In potatoes, ABA induction leads to high expressions of *StLEA1-3* and *StDHN-1*, while *StLEA3-1*, *StDHN-1*, *StDHN-3*, and *StASR-4* are significantly induced by salt stress [12]. The transformation of tobacco with the key gene *PgLEA2-50* enhanced osmoregulation and antioxidant levels in transgenic lines, significantly improving their resistance to abiotic stress [14]. However, the mechanisms of abiotic stress responses of LEA family genes in alpine cold-tolerant herb plants remain unclear.

Notopterygium, a genus endemic to China, which belongs to the Apiaceae and is primarily distributed in high-altitude mountainous regions of the Qinghai–Tibet Plateau and its adjacent provinces, such as Shaanxi, Qinghai, Gansu, and Sichuan. This genus encompasses four species clades exhibiting distinct morphological variations, *Notopterygium franchetii*, *Notopterygium incisum*, *Notopterygium oviforme*, and *Notopterygium forrestii* [15], among which the first two are particularly recognized as essential perennial medicinal herbs in China. According to the Chinese Pharmacopoeia, the rhizomes and roots of these herbs are primarily utilized in the treatment of ailments such as wind-cold syndromes with exterior manifestations, headache accompanied by stiffness of the neck, and rheumatic paralysis. However, due to human exploitation and destruction, coupled with the impacts of global climate change on their unique natural habitats, wild resources are scarce, and there is an urgent need to develop resistant varieties. The identification of resistant genes and/or family genes is very important to develop novel resources or varieties. In a previous study, Jia et al. (2017) conducted transcriptome sequencing on *N. incisum* and *N. franchetii* to explore environmental adaptation genes, and their research revealed the presence of various environmental stress genes regulating substance and energy metabolism in both species [16]. In a subsequent study by Jia et al. (2019), transcriptome sequencing was performed on the flowers, leaves, and stems of *N. incisum*, and 21 genes significantly associated with cold tolerance were identified [17]. Currently, there remains a scarcity of research on the molecular mechanisms of abiotic stress responses in *Notopterygium* species. It is strongly necessary to conduct a genome-wide identification of resistant family genes in *Notopterygium* species and to further investigate the evolutionary history of this gene family and its relationship with the diversity in the sequence and function of LEA proteins. In this study, we conducted the first genome-wide analysis of *LEA* genes in *Notopterygium* species, aiming to characterize their sequences, evolutionary relationship, putative function, and expression profiles in response to drought and temperature stresses. Our research provides invaluable insights that will pave the way for the future exploration of *LEA* genes in the alpine cold-tolerant *Notopterygium* species.

2. Results

2.1. The Identification of the LEA Genes in Notopterygium

A total of 20, 20, and 23 *LEA* gens were identified in *N. franchetii* (*NfrLEA1-NfrLEA23*), *N. forrestii* (*NfoLEA1-NfoLEA23*), and *N. incisum* (*NinLEA1-NinLEA23*), which were classified into six subfamilies, LEA_1, LEA_2, LEA_5, LEA_6, DHN, and SMP, based on protein conserved domains and a phylogenetic analysis. SMP and LEA_5 are the top two subfamilies with the largest numbers of members (Figures 1 and S1).

The protein lengths of coding sequences range from 84 (*NfrLEA4* and *NfoLEA16*) to 455 amino acids (*NfrLEA11* and *NfrLEA15*) with an average sequence length of 218 amino acids. The theoretical molecular weight of LEA proteins ranges from 9.05 to 50.33 kiloDaltons (kDs), and the theoretical isoelectric point (pI) values of 15 out of 23 LEA proteins in *N. franchetii*, 13 out of 20 LEA proteins in *N. forrestii*, and 13 out of 20 LEA proteins in *N. incisum* are less than 7.0, indicating that most of the LEA proteins in *Notopterygium* species are acid proteins. The instability indices range from 8.15 to 63.36, and LEA proteins in the LEA_5 subfamily are unstable proteins. The aliphatic index ranges from 24.3 to 110.68, with 94% of LEA proteins having an aliphatic index below 100 and a grand average of hydropathicity (GRAVY) value less than 0, indicating high hydrophilicity (ranging from −1.385 to −0.062). Subcellular localization prediction indicated that LEA proteins are mainly distributed in the nucleus, particularly those belonging to the DHN subfamily and the LEA_6 subfamily, while the remaining LEA proteins are distributed in the nucleus, cytoplasm, mitochondrion, and plasma membrane (Table S2). The results of secondary structure predictions for the amino acid sequences encoded by *LEA* genes are shown in Table S3. Among the amino acid sequences, the secondary structure is predominantly a random coil structure with a water content of more than 20%, ranging from 24.66% to 79.76%, while β-turn has the lowest proportion, ranging from 0.00% to 17.7%. The secondary structure of LEA proteins in the LEA_2 subfamily is relatively regular, and it has the highest proportion of extended strands among the six subfamilies.

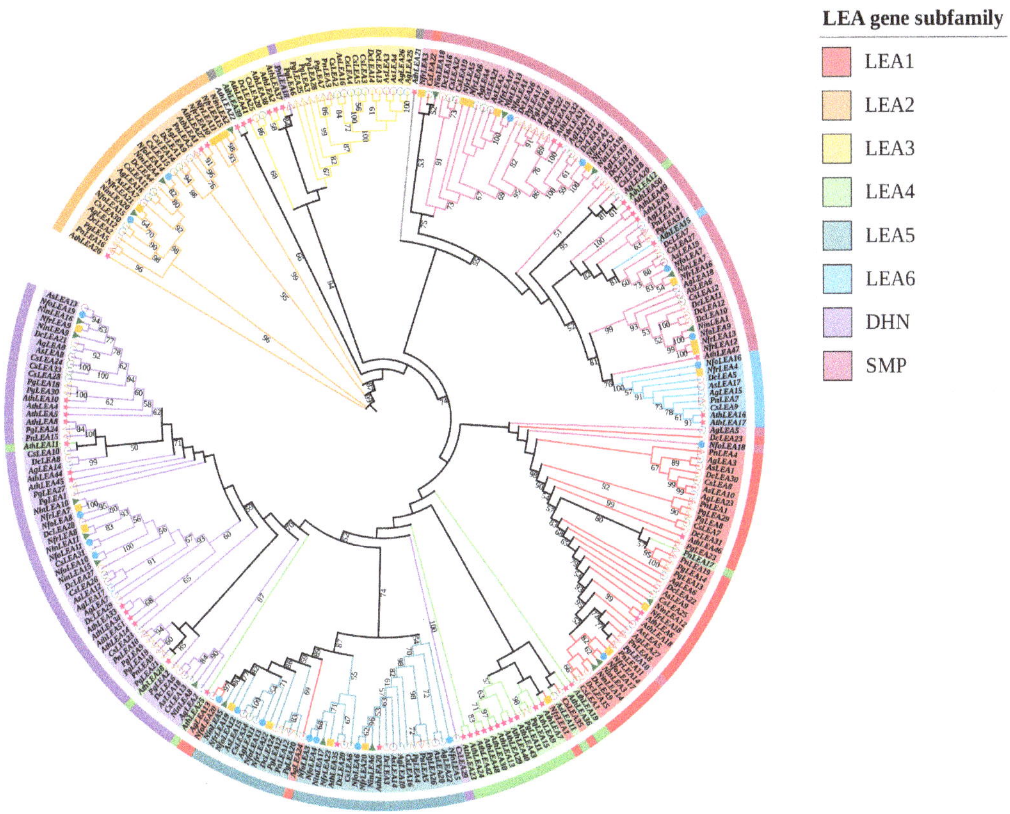

Figure 1. A phylogenetic evolutionary tree of *LEA* gene families in *Notopterygium* species and seven other species. The differently shaped symbols preceding the gene IDs represent different species, with solid symbols denoting the three species of *Notopterygium* and *A. thaliana* (represented by a star), the differently colored hollow circles representing the four species of Apiaceae, and the differently colored hollow triangles representing the two species of Araliaceae.

2.2. Phylogenetic Analysis and Classification of LEA Gene Family

To determine the classification and evolutionary relationships of the identified *LEA* gene family, a phylogenetic tree was constructed based on the multiple sequence alignment of 227 LEA protein sequences, including 23 in *N. franchetii*, 20 in *N. forrestii*, 20 in *N. incisum*, 51 in *A. thaliana*, 35 in *C. sativum*, 27 in *A. graveolens*, 31 in *D. carota*, 20 in *A. sinensis*, 31 in *P. ginseng*, and 17 in *Panax notoginseng* (Figure 1). According to the constructed phylogenetic relationship, the 227 LEA proteins were classified into eight major phylogenetic groups: LEA_1, LEA_2, LEA_3, LEA_4, LEA_5, LEA_6, DHN, and SMP. Of these, DHN and SMP are the largest groups with 60 and 58 *LEA* gene family members, respectively, and LEA_6 is the least abundant member of the subfamily with 10 members. The distribution ratio of *LEA* gene family members among various subfamilies follows this pattern consistently across every species. In the LEA_4 subfamily, only four species, *A. sinensis*, *P. ginseng*, *P. notoginseng*, and *A. thaliana*, contain members of the *LEA* gene family, while no other species do. This suggests that gene loss events have occurred during the evolution of LEA_4 subfamily members. Furthermore, *AthLEA21* and *AthLEA22* do not belong to any subfamily, indicating that these two genes exhibit significant sequence divergence and are phylogenetically distant from *LEA* gene family members in other species. In the earliest classification of the *A. thaliana LEA* gene family, these two genes were assigned to the AtM subfamily [5]. No AtM members were found in the *Notopterygium* species and the other six species examined, suggesting that these two genes may have undergone gene loss events during evolution. Additionally, it was observed that individual members of other subfamilies appear in the branches of each subfamily. This could be due to the duplication and loss of a small portion of *LEA* genes during evolution, accompanied by gene neofunctionalization and subfunctionalization. Paralogous genes of various species show considerable divergence, ultimately leading to low sequence similarity among some *LEA* gene family members across different species.

2.3. Gene Structure, Protein Conserved Motif, and Cis-Acting Element Analysis

The exon–intron structures of all the identified *LEA* genes were investigated to better understand the structural diversity of these genes in *Notopterygium* species. The results reveal (Figure 2D) that the majority of members possess 1–2 introns, with the SMP subfamily exhibiting the highest number of introns among *LEA* genes, where each member contains at least two introns. There are some differences in the number of exons and the length of introns among members of the *LEA* gene family. For example, two members of the LEA_6 subfamily lack introns, while members of the LEA_1, LEA_2, LEA_5, and DHN subfamilies predominantly contain a single intron. Furthermore, it was observed that, except for *NinLEA18*, which has a unique structure characterized by a notably long intron segment, the intron–exon structural features of *LEA* gene family members belonging to the same subfamily are highly similar. These findings suggest that the phylogenetic relationships and subfamily classifications in the *LEA* gene family are closely associated with their gene structures.

To further reveal the structural and functional characteristics of LEA proteins, the online website MEME was employed to analyze the conserved structural motifs, and detailed motif information is provided in Table S4. A total of 15 conserved motifs were identified and named motif 1 to motif 15. As shown in Figure 2B, no member has a complete set of 15 conserved motifs, and the motif numbers of the LEA family members range from one to seven. Sequences grouped in the same subfamily share an identical number of conserved motifs and types of motifs. Notably, the LEA_6 subfamily lacks any conserved motifs and exhibits a short sequence length, suggesting a potential loss of conserved motifs during evolutionary processes for members of this subfamily. In contrast, the LEA_1 subfamily displays a lower number and variety of conserved motifs in their protein sequences, with some members possessing only one conserved motif. All members of this subfamily contain motif 14, and three members also harbor motif 4, so they were grouped into a distinct category. LEA members of the DHN subfamily consistently contain motif 7 and motif 11, while some members additionally possess motif 12, and others have motifs 14 and 5. These findings suggest that *LEA* genes in the same subfamily exhibit higher homology and closer evolutionary relationships.

In addition, 12 putative conserved domains were identified and inhibited, as shown in Figure 2C. In general, the 63 conserved domains of LEA proteins in *Notopterygium* species lack high similarity, but unique conserved segments exist in each *LEA* gene subfamily. Specifically, the SMP subfamily contains the SMP domain, and some members also include the ACD_sHsps_p23-like domain. The DHN subfamily is characterized by the DHN domain. The domains of LEA_2 subfamily members are divided into two categories: those containing only the LEA_2 superfamily and those containing both the LEA_2 and Why domains. Members of the LEA_5 subfamily are classified into those that contain either the LEA_5 or LEA_5 superfamily domain. Members of the LEA_1 and LEA_6 subfamilies each contain only a single domain, the LEA_1 or LEA_6 domain, respectively. These results reveal that members in the *LEA* gene family that are phylogenetically close have similar compositions of conserved domain and similar functions. The conserved domains and motifs of LEA proteins in the same subfamily are highly consistent, while there are significant differences between different subfamilies.

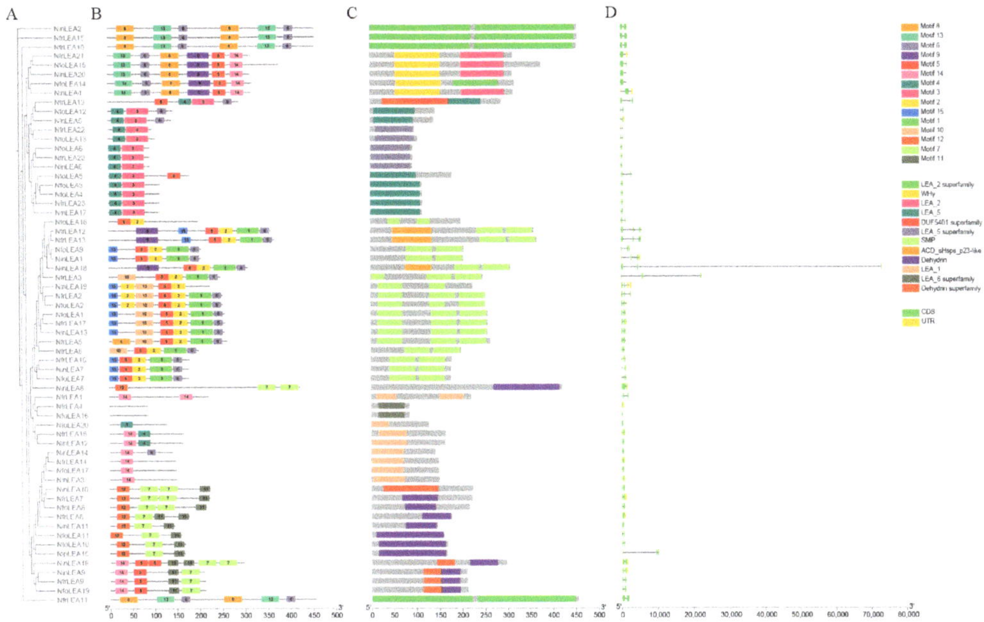

Figure 2. The phylogenetic relationships, distribution of conserved motifs, domain prediction, and gene structure analysis of *Notopterygium LEA* genes. (**A**) A phylogenetic tree of 23 *NfrLEA*, 20 *NfoLEA*, and 20 *NinLEA* genes. (**B**) The distribution of conserved motifs in LEA proteins. Fifteen conservative motifs in LEA are displayed in different color boxes. (**C**) The domains of LEA proteins. These denote the matching types that represent various confidence levels (specific matching and non-specific matching) and domain model ranges (superfamily and multi-domain). (**D**) *LEA* gene structures. The description of exons and introns is obtained by adopting image tools. The yellow and green boxes and the green lines represent the non-coding region (UTR) and exons and introns, respectively.

An analysis of the 2000 bp sequence upstream of the transcription start site of *LEAs* in *Notopterygium* identified 1718 putative *cis*-elements related to stress response (Figures 3 and S2). The promoter region of the LEA gene family mainly contains responsive elements such as light-regulated elements, abscisic acid (ABA), light, methyl jasmonate (MeJA), and anaerobic induction. Hormone-responsive elements were found in the promoter regions of most *LEA* genes. Among them, 84% of the *LEA* genes contain ABA-responsive elements, participating in the regulation of the ABA signaling pathway involved in abiotic stress responses. Additionally, 67.6% of *LEA* gene promoters harbor MeJA-responsive elements. Furthermore, auxin-, gibberellin-, and salicylic acid-responsive elements were also detected, all of which are associated with plant stress resistance processes. Simultaneously, a large number of *cis*-elements directly related to drought, anaerobiosis, low temperature, defense, and salt stresses were identified. In summary, *LEA* genes may participate in *Notopterygium* species growth and development through multiple pathways and respond to various abiotic stresses.

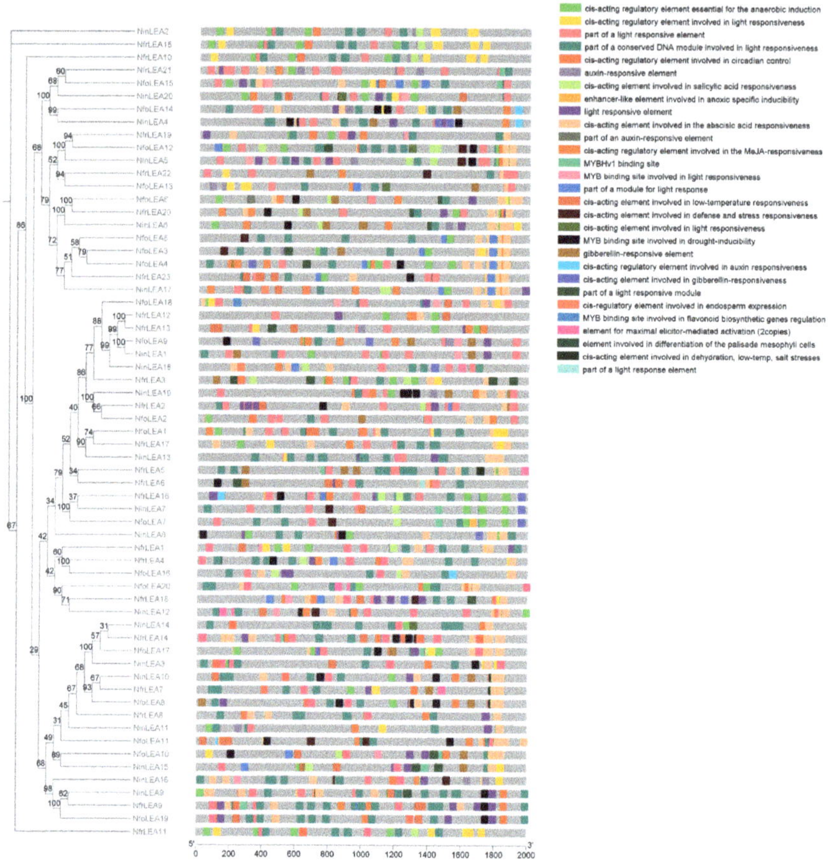

Figure 3. Prediction of *cis*-acting elements in promoter region of *LEA* genes in *Notopterygium* species.

2.4. PPI Networks and GO Annotation

The PPI network of LEA proteins is shown in Figure 4A; 11 LEA proteins, namely NinLEA2, NinLEA6, NinLEA7, NinLEA12, NinLEA17, NinLEA18, NinLEA19, NinLEA20, NfoLEA16, NfoLEA18, and NfrLEA19, were mapped onto the PPI network, and 6 LEA proteins were significantly interconnected among them. Specifically, NinLEA6 and NinLEA17 function as molecular chaperones, stabilizing protein structures and protecting proteins and membranes from aggregation, which is crucial for stress tolerance, particularly against dehydration and low-temperature stress. NinLEA12 is involved in dehydration tolerance, while NinLEA19 enhances plant tolerance to osmotic stress during germination and seedling stages. Additionally, strong interactions were observed between NinLEA6 and NinLEA12, between NinLEA6 and NinLEA19, as well as between NinLEA19 and NinLEA20, indicating that these proteins share similar functions and participate in biological processes related to plant responses to external stress through their interactions. The remaining five LEA proteins were relatively isolated, and their protein functions remain unclear.

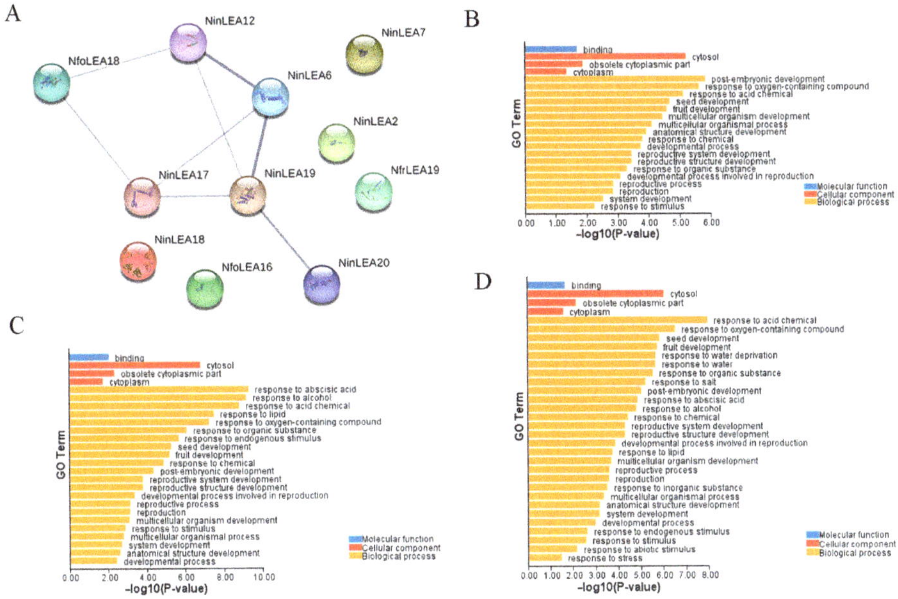

Figure 4. Protein–protein interaction analysis of LEA proteins and result of GO annotation of LEA genes in *Notopterygium* species. (**A**) PPI network in *Notopterygium* species. (**B**) GO functional annotation of *NfrLEAs*. (**C**) GO functional annotation of *NfoLEAs*. (**D**) GO functional annotation of *NinLEAs*.

The GO annotation results reveal that *LEA* genes are broadly involved in annotations related to biological processes, molecular functions, and cellular components (Figure 4B–D). In biological processes, they are all implicated in key processes such as responses to chemicals, growth and development, and seed and fruit development. Additionally, binding functions in molecular functions and cellular components such as cytosol and cytoplasm are also common annotation characteristics shared by these three species. These commonalities indicate a certain degree of conservation in *LEA* genes in terms of basic life activities and cellular constitution. A total of 11 *NfrLEA* genes were mapped to different GO functional nodes, with a relatively large proportion of annotations related to metabolic processes, cellular processes, and growth and development, suggesting their specific functions in these aspects (Figure 4B). Fourteen mapped *NfoLEA* genes exhibited prominent responses to chemicals such as ABA oxygenated compounds, which may be relevant to their hormone regulation and antioxidant functions (Figure 4C). Thirteen *NinLEA* genes were mapped and are associated with osmotic regulation, demonstrating their unique functions in resisting abiotic stress (Figure 4D). The differences in the GO annotation results reflect the specific physiological strategies developed by the *Notopterygium* species when adapting to their environment.

2.5. Chromosomal Distribution, Collinearity, and Selective Pressure Analysis

Twenty-one *NfrLEA* genes were precisely mapped to eight chromosomes of *N. franchetii*. Chromosome 1 (Chr1) harbored the largest number of genes, with six *NfrLEA* genes concentrated at the chromosomal ends (Figure 5). Five *NfrLEA* genes were distributed on Chr3, and most of them on chromosomes were located at their ends. All 20 *NfoLEA* genes were also specifically localized to chromosomes, with Chr10 having the highest number of genes at six, clustered together. All 20 *NinLEA* genes were precisely mapped to ten chromosomes, with 3 *NinLEA* genes each on Chr1, Chr2, Chr4, and Chr10. Overall, *NinLEA* genes were more dispersed across the chromosomes. Notably, no *LEA* genes from three species were found on Chr7.

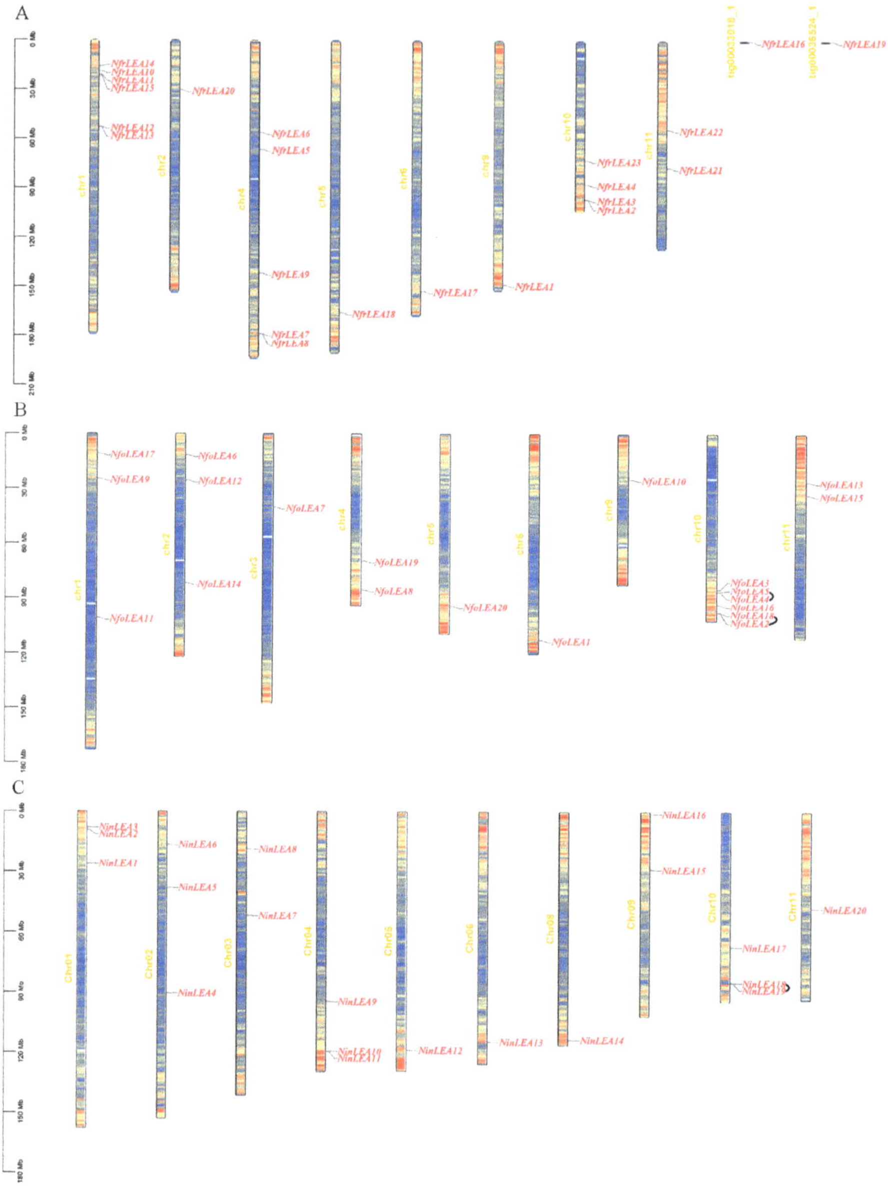

Figure 5. A chromosomal localization analysis of *Notopterygium LEA* genes. (**A**) The localization of *NfrLEAs*. (**B**) The localization of *NfoLEAs*. (**C**) The localization of *NinLEAs*.

Eighteen and twenty-four pairs of paralogous genes encompassed all members of *NfrLEAs and NfoLEAs* (Figure 6A,B). In *N. incisum*, the *LEA* gene family comprised approximately 50 pairs of homologous *NinLEAs* (Figure 6C), double the number found in both *N. franchetii* and *N. forrestii*. To explore potential evolutionary clues of the *LEA* gene family, an interspecific collinearity graph of *Notopterygium* associated with model plant *A. thaliana* was constructed (Figure 6D). The results indicate that there are 39 and 35 orthologous pairs between *N. franchetii* and *N. incisum* and between *N. incisum* and *N. forrestii*, respectively, and collinearity was mainly distributed on Chr1, Chr2, Chr4, Chr9, and Chr 10. Only 11 pairs of LEA genes showed collinearity between *A. thaliana* and *N. franchetii*. These findings suggest that *LEA* gene families in *Notopterygium* species are more closely related and may share a common ancestor and that *LEA* genes may not originate from *A. thaliana*.

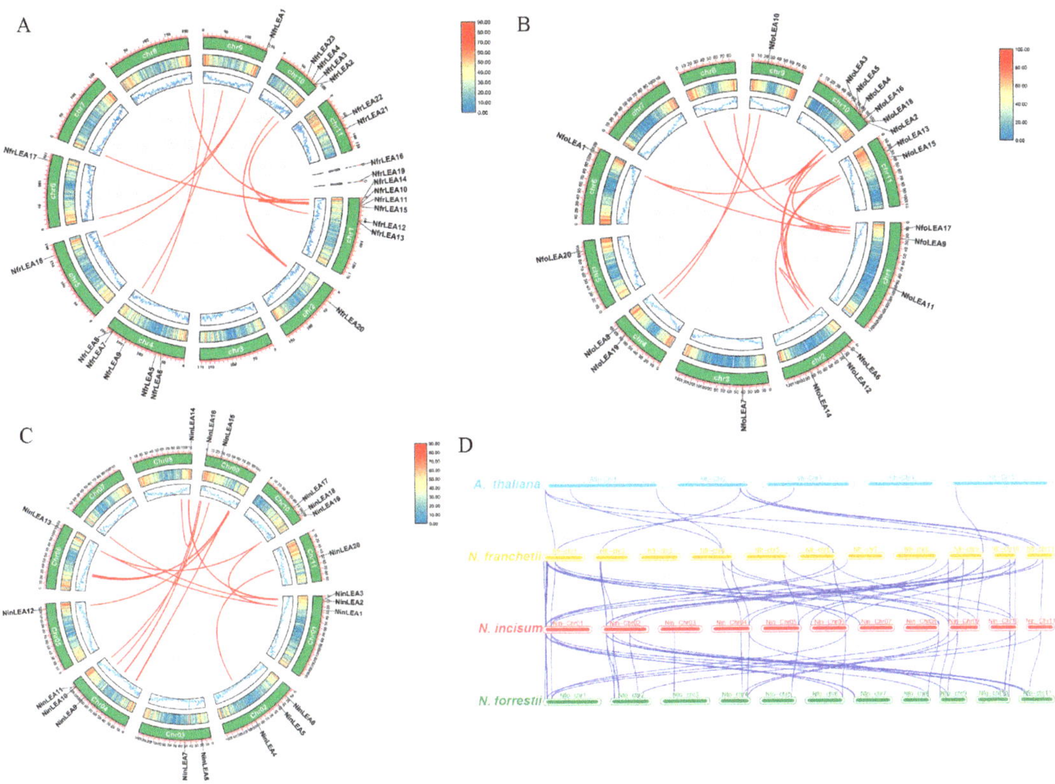

Figure 6. Interchromosomal relationships of *Notopterygium LEA* genes. (**A**) Intra-species collinearity analysis of *NfrLEAs*. (**B**) Intra-species collinearity analysis of *NfoLEAs*. (**C**) Intra-species collinearity analysis of *NinLEAs*. (**D**) Collinearity of *LEA* gene families in *Notopterygium* species. Gray lines indicate all collinearity blocks in three *Notopterygium* genomes, and red and blue lines indicate duplicated *LEA* gene pairs.

The *Ka/Ks* values were calculated to investigate potential selective pressure during the evolution of the *LEA* gene family in *Notopterygium* species. The results show that the *Ka/Ks* values of all gene pairs exhibit *Ka/Ks* < 1 (Tables S5–S7), suggesting that the repetitive *LEA* genes are primarily constrained by intense purification selection pressure in *Notopterygium* species.

2.6. Expression Pattern Analyses of NinLEA in Notopterygium Species

To reveal the tissue-specific and species-specific expression of *LEA* genes, this study conducted a transcriptome analysis on different tissues (roots, stems, leaves, flowers, and fruits) of four *Notopterygium* species (Figure 7). The results indicate that *NinLEA10*, *NinLEA11*, and *NinLEA13* were not expressed in any of the four species. Among the

NinLEAs with detectable expression levels, most were expressed at different levels in different tissues and species, and the expression of some genes was highly tissue-specific and species-specific. In *N. franchetii*, *NinLEA2*, *NinLEA3*, and *NinLEA6* were significantly more expressed in root tissues than in other tissues, while *NinLEA1*, *NinLEA5*, *NinLEA7*, *NinLEA9*, *NinLEA18*, and *NinLEA20* had the highest expression levels in flowers, and *NinLEA4*, *NinLEA15*, *NinLEA16*, and *NinLEA19* had the highest expression levels in leaves. In *N. incisum* and *N. oviforme*, *NinLEAs* were mainly highly expressed in the roots and flowers, with lower expression being found in the leaves (Figure 7B,D). Different *NinLEAs* were highly expressed in various tissues of *N. forrestii* (Figure 7C). These results indicate that the tissue-specific expression of *NinLEAs* exists and differs among *Notopterygium* species. Additionally, *NinLEA14* and *NinLEA17* were only expressed in *N. forrestii*, but it was also the only species in which *NinLEA20* was not expressed (Figure 7C), while *NinLEA8* was only not expressed in *N.franchetii* (Figure 7A). Notably, *NinLEA2* and *NinLEA3* had relatively high expression levels in the roots of all species, suggesting that they may be essential for root development in *Notopterygium* species.

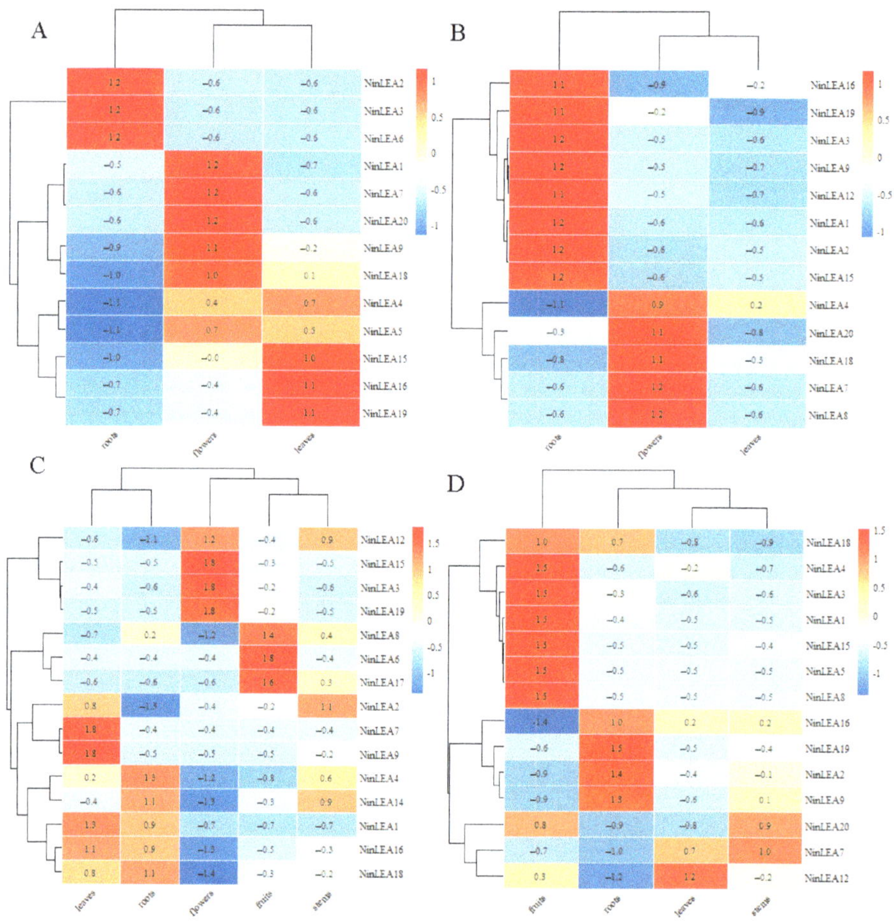

Figure 7. Gene expression heat map of *NinLEAs* in different tissues of four *Notopterygium* species. (**A**) Expression patterns of *NinLEAs* in different tissues of *N. franchetii*. (**B**) Expression patterns of *NinLEAs* in different tissues of *N. incisum*. (**C**) Expression patterns of *NinLEAs* in different tissues of *N. forrestii*. (**D**) Expression patterns of *NinLEAs* in different tissues of *N. oviforme*.

2.7. Expression Analysis of NinLEAs in N. franchetii Under Different Abiotic Stress

Based on a transcriptome expression pattern clustering analysis, six genes (*NinLEA1*, *NinLEA2*, *NinLEA3*, *NinLEA9*, *NinLEA19*, and *NinLEA20*) that exhibit stable expression across different tissues and species yet demonstrate high specificity in *N. franchetii* were selected. The responses of these genes to high-temperature, low-temperature, and drought stresses were analyzed using qRT-PCR. The results indicate that the relative expression levels of the *NinLEAs* underwent significant changes under three stress treatments, suggesting their involvement in the response of *N. franchetii* to high-temperature, low-temperature, and drought stresses (Figure 8). Furthermore, variations in the relative expression levels of different *NinLEAs* were observed. For instance, under high-temperature stress, the relative expression of *NinLEA19* in leaves continuously increased over a period of 0–12 h, while the expression trends of the other genes first rose, then declined, and subsequently rose again within the same time frame. In roots, the relative expression levels of *NinLEA1*, *NinLEA2*, *NinLEA3*, and *NinLEA19* first decreased and then increased, whereas those of *NinLEA9* and *NinLEA20* exhibited the opposite trend (Figure 8A). Under continuous low-temperature conditions, the relative expression levels of *NinLEA1*, *NinLEA3*, and *NinLEA20* in the leaves exhibited a trend of initially increasing and then decreasing (Figure 8C). We also found that *NinLEA3* exhibits a similar expression pattern under both low- and high-temperature stresses. Additionally, the relative expression levels of the same gene in different tissues varied under the same stress. For example, under drought stress, the relative expression of *NinLEA3* in the leaves first increased and then decreased, while in the roots, it first decreased and then increased, showing the complete opposite trend. Under high-temperature stress, the changes in all *NinLEAs* were more pronounced in the leaves compared to the roots (Figure 8B).

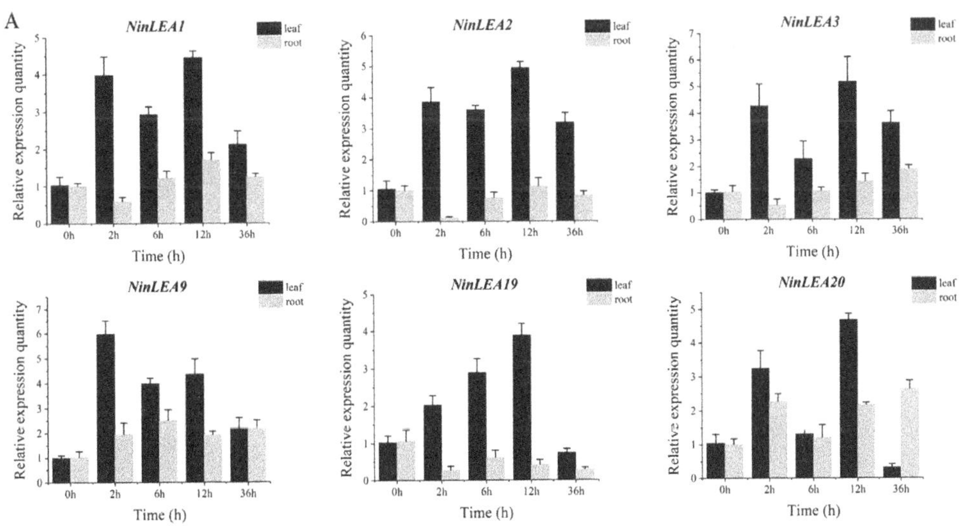

Figure 8. *Cont.*

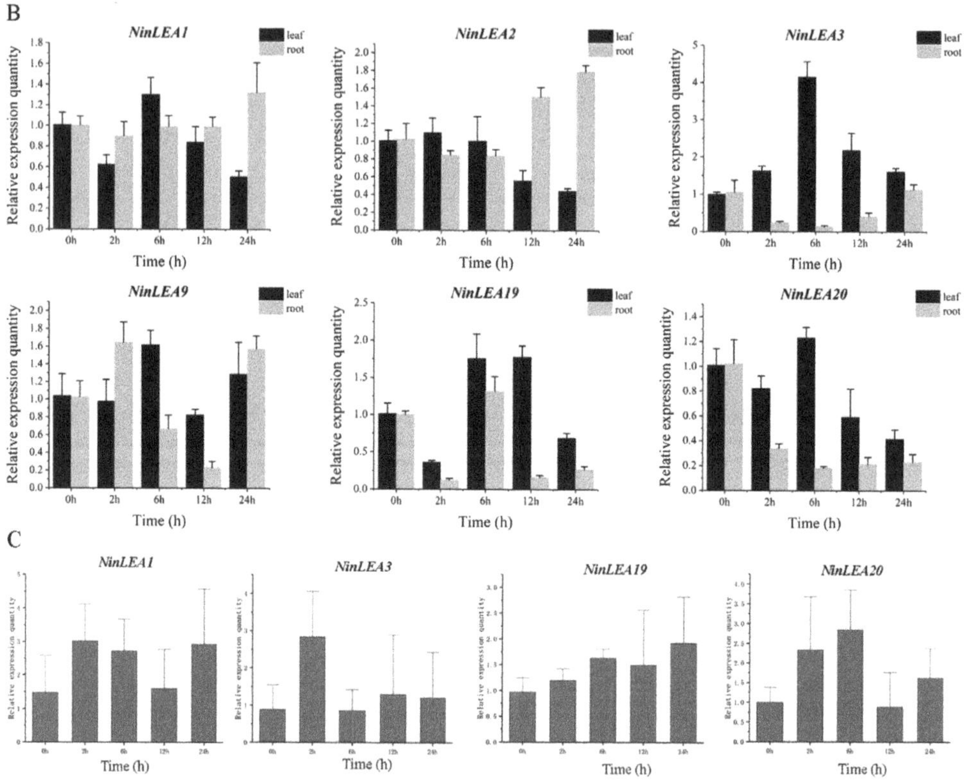

Figure 8. Expression analysis of *NinLEAs* in *N. franchetii* under abiotic stress. (**A**) Relative expression levels of *NinLEAs* in leaves and roots under high-temperature stress. (**B**) Relative expression levels of *NinLEAs* in leaves and roots under drought stress. (**C**) Relative expression levels of *NinLEAs* in leaves under low-temperature stress.

3. Discussion

The *LEA* gene family occupies a central stage in both the intricate process of plant embryonic development and the vital mechanisms underlying plant resilience to abiotic stressors such as drought and cold [18]. The proteins encoded by *LEA* genes, characterized as multifaceted functional proteins in plants, exhibit remarkable capabilities in scavenging intracellular ROS and safeguarding macromolecular structures, thereby effectively mitigating the detrimental impacts of abiotic stresses on plant organisms [19]. *Notopterygium* species are currently confined to the ecologically unique Qinghai–Tibet Plateau and its adjacent high-altitude mountainous habitats [20], where they thrive under exacting environmental conditions and are particularly susceptible to a variety of influences from abiotic stresses. Notably, *N. franchetii* and *N. incisum* are esteemed medicinal herbs of considerable significance, and the meticulous identification of genes associated with their remarkable stress resistance holds profound scientific and practical value.

In this study, we identified a total of 63 *LEA* genes belonging to six LEA subfamilies from the genomes of three *Notopterygium* species (23 *NfrLEAs*, 20 *NinLEAs*, and 20 *NfoLEAs*), with a member count comparable to that observed in tomato, rice, tea plant, and other plants such as *A. graveolens*, *A. sinensis*, and *P. notoginseng* (Figure 1). However, in species such as tobacco, cotton, and ginseng, the number of LEA gene family members exceeds 100 [21–23]. This may be attributed to the different selective pressures and genetic variations in these species during evolution, leading to the expansion of the *LEA* gene family and the accumulation of more *LEA* genes to enhance stress resistance. Among the six LEA subfamilies, the SMP subfamily dominates with 18 members. The LEA_3 and LEA_4 subfamilies were not observed (Figure 1). Similarly, members of the LEA_4 subfamily were also absent in the *LEA* gene families of Apiaceae species, such as *C. sativum*, *A. graveolens*, and *D. carota* in this study, and the number of LEA_3 subfamily members was extremely low (Figure 1). This suggests that a significant gene loss may have occurred for LEA_3 and LEA_4 subfamily members during the evolution of Apiaceae plants. Variations in the number and types of *LEA* genes exist among different species. For example, the LEA4 subfamily dominates in *Brassica campestris* [5], while in *Prunus mume* and wheat, the DHN subfamily has the largest number of members [24,25]. In tobacco and ginseng, the LEA_2 subfamily has the highest number of members [12,14], indicating the complexity of LEA gene evolution.

Subcellular localization prediction indicates that LEA proteins are primarily distributed in the nucleus, cytoplasm, and mitochondria, with a smaller proportion located in the plasma membrane. These findings align with previous studies on *A. thaliana* and *Pterocarya stenoptera* [5,26], reflecting the diversity of intracellular LEA protein functions and mechanisms in *Notopterygium* species (Table S2). An analysis of physical and chemical properties revealed that, except for the LEA_5 subfamily, the proteins encoded by *LEA* genes from the other subfamilies are stable. Furthermore, 94% of LEA proteins exhibit high hydrophilicity. LEA proteins with high hydrophilicity and thermostability can redirect water molecules, bind salt ions, and subsequently eliminate reactive oxygen species (ROS) that accumulate in cells due to external stress stimulation [27]. These properties contribute to the tolerance to dehydration and other stresses in plants [9].

In this study, members of the same subfamily are mostly clustered in the same branch, indicating a high degree of evolutionary conservation across species (Figure 1). The intron–exon structural characteristics of *LEA* gene family members belonging to the same subfamily are highly similar. A total of 89% of LEA gene family members have 0–2 introns. Members of the LEA_6 subfamily lack introns, while members of the LEA_1, LEA_2, LEA_5, and DHN subfamilies predominantly contain only one intron (Figure 2D). Previous studies have found that genes related to stress responses generally have a lower number of introns, and genes with fewer introns are beneficial for reducing transcription costs and enabling faster transcription and expression processes, thereby allowing cells to rapidly respond to abiotic stresses [28]. For example, most *LEA* genes in *A. thaliana* contain only one intron [5], and more than half of the *LEA* genes in wheat lack introns [10].

During the course of prolonged evolution, plants have developed a set of complex strategies to adapt and survive under harsh environmental conditions, with transcriptional changes being a primary adaptive mechanism. *Cis*-acting elements on gene promoters regulate the mRNA abundance of numerous downstream target genes through interactions with transcription factors, playing a crucial role in plant responses to abiotic stresses [29]. For instance, the MBS element is associated with drought stress responses [30], while TC-rich repeats are related with salt stress responses [31]. Therefore, analyzing the *cis*-acting elements in gene promoter regions aids in identifying specific functions of genes. In this study, 53 *LEA* genes contained ABA-responsive elements, 67.6% of LEA gene promoters harbored MeJA-responsive elements, and 91.6% of LEA genes contained MYB and light-responsive elements. Studies on wheat, tea plants, and ginseng have found similar regulatory elements in the upstream regulatory regions of *LEA* genes [14,23,25]. Additionally, a large number of elements are involved in responses to other compounds and non-ABA-dependent stress responses. These elements may directly or indirectly participate in stress responses, regulate growth and development, and maintain cellular homeostasis, enabling *LEA* genes to respond to abiotic stresses through multiple pathways. For *Notopterygium* species, the same gene may play a role in different tissues, for example, a gene is expressed in the root, flower, and leaf of *N. franchetii* (Figure S3); three genes are expressed in the root, flower, and leaf of *N. incisum* (Figure S4); one gene is expressed in various tissues of *N. forbesii* (Figure S5); three genes are expressed in various tissues of *N. oviforme* (Figure S6). Meanwhile, members of the same gene family may also be expressed in the same tissues across different species (Figure S7).

LEA proteins can prevent the collapse of cellular structures by binding to cell membranes [32], and they can act as molecular chaperones by binding to misfolded proteins and repairing misassembled proteins to restore their biological activity [33]. The PPI network analysis conducted in this study revealed that NinLEA6 and NinLEA17 proteins function as molecular chaperones to stabilize protein structures (Figure 4A). Both *NinLEA6* and *NinLEA17* genes belong to the LEA_5 subfamily, and research on *A. thaliana* has shown that LEA_5 regulates organellar translation to enhance respiration relative to photosynthesis in response to stress [34]. The GO enrichment analysis results indicate that there are partial differences in the functions of *LEA* genes among the three *Notopterygium* species. The functions of *LEA* genes in *N. franchetii* were primarily relevant to metabolic processes and growth and development, while those in *N. forrestii* were mainly involved in hormone regulation and antioxidant activities. The *LEA* genes in *N. incisum* were primarily associated with osmoregulation (Figure 4B–D). However, these biological processes were all related to the function of combating abiotic stresses, confirming the results of the *cis*-acting element analysis in the promoter regions, which indicate that the pathways through which *LEA* genes in *Notopterygium* species respond to abiotic stresses are not singular.

In this study, the results of chromosomal distribution show that most members of the *LEA* gene family in each species are scattered at different locations on different chromosomes in the three *Notopterygium* species (Figure 5). The formation of this phenomenon is significantly related to chromosome segment duplication and tandem duplication events, which have led to changes in the chromosomal locations of gene family members [35,36]. The dispersed distribution of genes may indicate that these genes have undergone different selective pressures during evolution, resulting in functional differentiation. Alternatively, it may increase the complexity of gene expression regulation, where different gene family members may be controlled by distinct regulatory mechanisms to achieve specific temporal and spatial expression patterns [37]. This regulatory complexity aids organisms in achieving precise gene expression regulation during growth and development. The *Ka/Ks* values for paralogous gene pairs were less than 1 (Tables S5–S7), indicating that the LEA gene family members in the three *Notopterygium* species underwent rigorous purifying selection during their evolution and expansion. Similarly, strong purifying selection effects have been observed in the evolution of *LEA* gene families in other species, such as *Populus trichocarpa, Triticum aestivum*, and maize [25,38,39]. The *NfrLEAs* exhibited higher homology with *NfoLEAs* and *NinLEAs*, with similar numbers of orthologous gene pairs, but showed the lowest homology with *AtLEAs*. This suggests that there is a closer genetic relationship among the *LEA* gene families in *Notopterygium* species.

Through the analysis of RNA-Seq data, we found that not all *NinLEAs* are expressed in the tissues of the three *Notopterygium* species, and there are differences in expression abundance and tissue specificity among the co-expressed *NinLEAs* (Figure 7). In most species, *LEA* genes exhibit distinct tissue-specific expression patterns. For example, in *N. forresii, NinLEA2, NinLEA3*, and *NinLEA6* are highly expressed in root tissues, while *NinLEA1, NinLEA5*, and others are predominantly expressed in flowers, and *NinLEA4* is abundant in the leaves (Figure 7A). In *N. incisum*, most of the expressed *LEA* genes have the highest expression levels in the roots, whereas *NinLEA4* and *NinLEA7* are highly expressed in the flowers (Figure 7B). Similar tissue-specific expression patterns were observed in *N. forrestii* and *N. oviforme*, although specific genes and tissues with high expression differed. Across species, there were variations in the expression patterns of *LEA* genes. Genes such as *NinLEA5, NinLEA10*, and *NinLEA11* were not expressed in *N. forrestii* and *N. oviforme* but showed different expression profiles in *N. franchetii* and *N. incisum*. The same gene may also have different tissues with high expression in different species. For instance, *NinLEA3* was highly expressed in flowers in both *N. franchetii* and *N. forrestii* but was abundant in the roots in *N. incisum* (Figure 7). Additionally, *NinLEA8* and *NinLEA10* were not expressed or were expressed at extremely low levels in multiple tissues across multiple species. The *LEA* genes in the four *Notopterygium* species exhibited complex expression patterns and characteristics in different tissues, which may reflect their different functions and roles in plant growth and development and in the response to environmental stress. In contrast, most *NnLEA* genes in lotus are primarily expressed during the later stages of cotyledon and plumule development, indicating their crucial roles in seed maturation [40]. The *LEA* genes in *Arachis hypogaea* usually exhibit relatively consistent expression but at generally low levels, while genes in the LEA2, LEA3, and DHN subfamilies are strongly expressed in the roots, leaves, and flowers [41], which is similar to the expression pattern of *NinLEA* in *N. franchetii*. Most *LEA* genes in *Juglans regia* are expressed in leaf and flower tissues [42], resembling the expression pattern of *NinLEAs* in *N. forrestii*. We further analyzed the responses of *NinLEA* genes to high-temperature, low-temperature, and drought stresses through RNA-Seq data and qRT-PCR validation (Figure 8). The relative expression levels of *NinLEA* genes underwent significant changes after three stress treatments. Compared with high-temperature stress, the changes in the relative expression levels of *NinLEA* genes in leaf tissues under drought stress were relatively insignificant, and the variation ranges of expression levels in both leaf and root tissues were generally similar. Additionally, we found that *NinLEA3* exhibited a similar expression pattern in the leaves under both high- and low-temperature conditions, and the relative expression levels of *LEA* genes showed a significant increase over a period of 0–2 h under low-temperature conditions. These results suggest that *LEA* genes play a significant role in helping *Notopterygium* species cope with extreme environments. *Notopterygium* species grow in high-altitude regions and often face challenges posed by extreme environmental conditions, such as low temperatures and water scarcity. Therefore, identifying the members of the LEA gene family in *Notopterygium* species and understanding their expression patterns under extreme conditions such as low temperatures, high temperatures, and drought will help us understand the response mechanisms of *Notopterygium* species in extreme environments. This study not only provides valuable stress-resistant gene resources in *Notopterygium* species but also provides profound insights into how these species respond to extreme climate events.

4. Materials and Methods

4.1. Identification and Characterization of LEA Genes in Notopterygium

The whole genome sequencing data of the three species of *Notopterygium* were obtained from the research group. Based on two methods, *LEA* gene family members were identified. Firstly, the hidden Markov models (HMMs) corresponding to the LEA domain (PF03760, PF03168, PF03242, PF02987, PF00477, PF00257, PF04927, and PF10714) were downloaded from the Pfam database v34.0 (http://pfam.sanger.ac.uk/, accessed on 22 March 2024). The LEA protein sequences of *Notopterygium* species were aligned using the HMM model in HMMER3.0 (http://plants.ensembl.org/hmmer/index.html, accessed on 24 March 2024), using an Expect value (E value) of 10^{-5} [43]. Then, the protein sequences of *A. thaliana* were obtained from the TAIR database (https://www.arabidopsis.org/, accessed on 25 March 2024) and were used to perform a BLASTP search in the *Notopterygium* species protein database for proteins exhibiting E values less than 10^{-5}. Afterwards, the above two results were intersected, incomplete sequences and redundant sequences were removed, and the NCBI-CDD database (https://www.ncbi.nlm.nih.gov/, accessed on 4 April 2024) was used for verification.

ProtParam (https://web.expasy.org/protparam/, accessed on 7 April 2024) was used to predict the physical and chemical properties. The WoLF PSORT web tool (https://web.expasy.org/protparam/, accessed on 8 April 2024) [44] and the SOPMA (https://npsa-prabi.ibcp.fr/cgi-bin/npsa_automat.pl?page=/NPSA/npsa_sopma.html, accessed on 21 April 2024) [45] were used to predict the subcellular localization and the secondary structure composition of LEA proteins, respectively.

4.2. Sequence Alignment and Phylogenetic Analysis

To gain further understanding of the evolutionary relationships among the *LEA* genes, we compared the genes in *Notopterygium*, *A. thalian*, *Coriandrum sativum* L., *Apium graveolens* L., *Daucus carota* L., *Angelica sinensis* (Oliv.) Diels, *ginseng* (*Panax ginseng* C. A. Mey.), and *Panax notoginseng* (Burkill) F. H. Chen ex C. H. Chow. Full-length alignment of the *LEA* gene sequences was performed using MUSCLE [46], and Maximum Likelihood (ML) trees were constructed using the MEGA X software (version 11.0.13) with a bootstrap value of 1000 [47].

4.3. Analyses of Gene Structure, Protein Conserved Motifs, and Promoter Cis-Acting Elements

Conserved motifs were predicted for each LEA protein sequence using the MEME web server v5.33 (https://meme-suite.org/meme/tools/meme, accessed on 24 April 2024), with the maximum number of motifs set to 15 and all other parameters set to default values [48].

The promoter sequences of all *LEA* genes, approximately 2000 bp upstream of the transcription start site (TSS), were extracted from the genome data of *Notopterygium* species. The *cis*-acting elements in the promoter region were predicted using the PlantCARE database (https://bioinformatics.psb.ugent.be/webtools/plantcare/html/, accessed on 28 April 2024) [49]. All of the above results were visualized using TBtools software (version 2.056) [50].

4.4. Prediction of Protein–Protein Interaction (PPI) Networks and Gene Ontology (GO) Annotation

STRING (https://string-db.org/, accessed on 25 April 2024) program was used to predict the functional interacting network models of LEA proteins; the required score was set at a threshold of 0.40. The eggNOG-mapper (http://eggnog-mapper.embl.de/, accessed on 25 April 2024) was used to obtain functional annotation information for *LEA* genes [51]. Finally, enrichment analysis results were generated using TBtools (version 2.056) [50].

4.5. Chromosomal Distribution, Collinearity, and Selective Pressure Analysis

Genome data for *Arabidopsis* were downloaded from the Ensembl Plants database (http://plants.ensembl.org/info/data/ftp/index.html, accessed on 3 May 2024). The TBtools software (version 2.056) was used to analyze and visualize the chromosomal distribution. The MCScanX software (version 11.0.13) was used to search the gene level collinearity [52]. The *Ka/Ks* Calculator v3.0 tool was used to calculate the ratio of non-synonymous to synonymous substitution (*Ka/Ks*) of duplicated gene pairs [53].

4.6. Expression Pattern Analysis

To gain insights into the expression patterns of *LEA* genes, transcriptome datasets of *Notopterygium* were analyzed. The transcriptomic data of *LEA* genes in five tissues (roots, stems, leaves, fruits, and flowers) were obtained by RNA-seq to detect the expression profile. Raw data were filtered using fastp with the *N. incisum* genome serving as the reference. Alignment was performed using Hisat2 (version 2.2.0) [54], and unqualified alignments were removed with SAMtools (version 1.17) [55]. StringTie (version 2.2.3) was used for single-sample assembly and quantification [56]. Differential expression analysis was conducted using edgeR (version 3.32.1) with Trimmed Mean of M-values (TMM) normalization and a Biological Coefficient of Variation (BCV) set to 0.1. Venn diagrams, set plots, and cluster analysis heatmaps were generated using R language.

4.7. Plant Materials, Growth Condition, and Abiotic Stress in N. franchetii

Three-year-old *N. franchetii* seedlings, with vigorous growth, similar morphology, and intact root systems, were selected as plant material and placed in an intelligent light-controlled incubator for the following three treatments: (1) high-temperature stress treatment, where cultivation conditions were 40 °C with a 12 h light/dark cycle, and leaf and root samples were collected at 0 h, 2 h, 6 h, 12 h, and 36 h; (2) low-temperature stress treatment, where cultivation conditions were 0 °C with a 12 h light/dark cycle, and leaf samples were collected at 0 h, 2 h, 6 h, 12 h, and 24 h; (3) and drought stress treatment, where cultivation conditions were 15 ± 2 °C with a 12 h light/dark cycle. The plants were treated with 40% PEG solution, and leaf and root samples were collected at 0 h, 2 h, 6 h, 12 h, and 24 h.

4.8. RNA Extraction and Quantitative Real-Time PCR

Total RNA was extracted from the collected *N. franchetii* samples using a widely recognized TRIzol kit (Accurate Biology, Changsha, China) following the manufacturer's instructions. RNA quality was assessed using an ultra-micro spectrophotometer, NanoDrop2000 (Kaiao Technology Development Co., Ltd., Beijing, China), which can detect the integrity and concentration, and the results were confirmed using RNase agarose-free gel electrophoresis. After RNA isolation, cDNA synthesis was performed using Hifair®III 1st Strand cDNA Synthesis SuperMix for qRT-PCR (gDNA digester plus) (Yeasen, Shanghai, China). For quantitative real-time PCR (Bioer 9600 FQD-96C, Hangzhou, China), primers were designed using Primer Premier 6.0 software (Premier Biosoft Inc., San Francisco, CA, USA (https://premierbiosoft.com/, accessed 26 April 2024)), and the primer design sequences are shown in Table S1. The qRT-PCR reaction mixture consisted of 3.6 μL ddH_2O, 1 μL cDNA, 0.4 μL specific primers, and 5 μL ChamQ SYBR qPCR Master Mix (Vazyme Biotech Co., Ltd., Nanjing, China). The amplicon length of the target gene ranges from 453 bp to 1368 bp. The PCR procedure was as follows: pre-incubation at 95 °C for 5 min, followed by 40 cycles of denaturation at 95 °C for 10 s, and annealing at 60 °C for 30 s, with a melting curve analysis to confirm primer and gene specificity. Each procedure included three biological replications and two technical replications. The relative expression levels of *NinLEA* genes were calculated using the $2^{-\Delta\Delta CT}$ method [57]. Actin8 from *N. franchetii* was used as internal reference gene. We calculated the stability of *NinLEA* genes and reference gene using Bestkeeper [58].

5. Conclusions

In this study, a comprehensive and systematic characterization of the *LEA* gene family in three *Notopterygium* species was conducted. Totals of 23, 20, and 20 *LEA* genes were identified in the genomes of *N. franchetti*, *N. incisum*, and *N. forrestii*, respectively. A phylogenetic analysis revealed that the identified *LEA* genes in *Notopterygium* could be classified into six major groups, with no apparent pattern of distribution on the chromosomes. Members in the same group exhibited similar gene structures and motif compositions. A molecular evolutionary analysis suggested that the *LEA* gene families in *Notopterygium* species evolved under the influence of purifying selection, which eliminated deleterious mutation sites during evolution. An analysis of promoter *cis*-elements, expression patterns, and expression characteristics under abiotic stress conditions indicated that some *NinLEAs* may be involved in the growth and development of *Notopterygium* species as well as their response to high-temperature, low-temperature, and drought stresses. These findings provide theoretical support for further investigation into the functions and mechanisms of *LEA* genes in response to abiotic stress in the alpine cold-tolerant *Notopterygium* species, laying the foundation for molecular breeding in these species.

Supplementary Materials: The following supporting information can be downloaded at https://www.mdpi.com/article/10.3390/ijms26020519/s1.

Author Contributions: Writing—review, editing, supervision, and funding acquisition, Z.L. and X.W. (Xiaojuan Wang); methodology, X.W. (Xuanye Wu), X.H., and X.W. (Xiaoling Wang); software, X.W. (Xuanye Wu), X.H., and X.W. (Xiaoling Wang); validation, P.L. and S.A.; formal analysis, X.W. (Xuanye Wu), X.H., and X.W. (Xiaoling Wang); investigation, X.L. and P.L.; data curation, S.A. and X.L.; writing—original draft preparation, X.W. (Xuanye Wu) and X.H. All authors have read and agreed to the published version of the manuscript.

Funding: This research was funded by the National Natural Science Foundation of China (32470392, 31970359, and 32371694), the Key Program of Research and Development of Shaanxi Province (2022ZDLSF06–02) and the Graduate Innovation Project of Northwest University (CX2023058).

Institutional Review Board Statement: Not applicable.

Informed Consent Statement: Not applicable.

Data Availability Statement: All data are contained within the article and the Supplementary Materials.

Conflicts of Interest: The authors declare no conflicts of interest.

References

1. Ramanjulu, S.; Bartels, D. Drought- and Desiccation-Induced Modulation of Gene Expression in Plants. *Plant Cell Environ.* **2002**, *25*, 141–151. [CrossRef] [PubMed]
2. Dure, L.; Greenway, S.C.; Galau, G.A. Developmental Biochemistry of Cottonseed Embryogenesis and Germination: Changing Messenger Ribonucleic Acid Populations As Shown by in Vitro and in Vivo Protein Synthesis. *Biochemistry* **1981**, *20*, 4162–4168. [CrossRef] [PubMed]
3. Sun, X.; Rikkerink, E.H.A.; Jones, W.T.; Uversky, V.N. Multifarious Roles of Intrinsic Disorder in Proteins Illustrate Its Broad Impact on Plant Biology. *Plant Cell* **2013**, *25*, 38–55. [CrossRef] [PubMed]
4. Shao, H.B.; Liang, Z.S.; Shao, M.A. LEA Proteins in Higher Plants: Structure, Function, Gene Expression and Regulation. *Colloids Surf. B Biointerfaces* **2005**, *45*, 131–135.
5. Hundertmark, M.; Hincha, D.K. LEA (Late Embryogenesis Abundant) Proteins and Their Encoding Genes in *Arabidopsis thaliana*. *BMC Genom.* **2008**, *9*, 118. [CrossRef]
6. Bies-Ethève, N.; Gaubier-Comella, P.; Debures, A.; Lasserre, E.; Jobet, E.; Raynal, M.; Cooke, R.; Delseny, M. Inventory, Evolution and Expression Profiling Diversity of the LEA (Late Embryogenesis Abundant) Protein Gene Family in *Arabidopsis thaliana*. *Plant Mol. Biol.* **2008**, *67*, 107–124. [CrossRef]
7. Tunnacliffe, A.; Wise, M.J. The Continuing Conundrum of the LEA Proteins. *Naturwissenschaften* **2007**, *94*, 791–812. [CrossRef] [PubMed]
8. Hernandez-Sanchez, I.E.; Maruri Lopez, I.; Martinez-Martinez, C.; Janis, B.; Jimenez-Bremont, J.F.; Covarrubias, A.A.; Menze, M.A.; Graether, S.P.; Thalhammer, A. LEAfing through Literature: Late Embryogenesis Abundant Proteins Coming of Age—Achievements and Perspectives. *J. Exp. Bot.* **2022**, *73*, 6525–6546. [CrossRef] [PubMed]

9. Hunault, G.; Jaspard, E. LEAPdb: A Database for the Late Embryogenesis Abundant Proteins. *BMC Genom.* **2010**, *11*, 221. [CrossRef]
10. Zan, T.; Li, L.Q.; Li, J.T.; Zhang, L.; Li, X.J. Genome-Wide Identification and Characterization of Late Embryogenesis Abundant Protein-Encoding Gene Family in Wheat: Evolution and Expression Profiles during Development and Stress. *Gene* **2020**, *736*, 144422. [CrossRef]
11. Kentaro, S. Relative Predicativity and Dependent Recursion in Second-Order Set Theory and Higher-Order Theories. *J. Symb. Log.* **2014**, *79*, 712–732. [CrossRef]
12. Geng, W.B.; Wang, Y.N.; Zhang, J.; Liu, Z.H.; Chen, X.Y.; Qin, L.T.; Yang, L.; Tang, H. Genome-Wide Identification and Expression Analyses of Late Embryogenesis Abundant (LEA) Gene Family in Tobacco (*Nicotiana tabacum* L.) Reveal Their Function in Abiotic Stress Responses. *Gene* **2022**, *836*, 146665. [CrossRef] [PubMed]
13. Yu, J.; Lai, Y.M.; Wu, X.; Wu, G.; Guo, C.K. Overexpression of *OsEm1* Encoding a Group I LEA Protein Confers Enhanced Drought Tolerance in Rice. *Biochem. Biophys. Res. Commun.* **2016**, *478*, 703–709. [CrossRef] [PubMed]
14. Wang, Q.; Lei, X.J.; Wang, Y.H.; Di, P.; Meng, X.R.; Peng, W.Y.; Rong, J.B.; Wang, Y.P. Genome-Wide Identification of the *LEA* Gene Family in *Panax ginseng*: Evidence for the Role of *PgLEA2-50* in Plant Abiotic Stress Response. *Plant Physiol. Biochem.* **2024**, *212*, 108742. [CrossRef] [PubMed]
15. Yang, J.; Feng, L.; Yue, M.; He, Y.L.; Zhao, G.F.; Li, Z.H. Species Delimitation and Interspecific Relationships of the Endangered Herb Genus *Notopterygium* Inferred from Multilocus Variations. *Mol. Phylogenet. Evol.* **2019**, *133*, 142–151. [CrossRef]
16. Jia, Y.; Liu, M.L.; Yue, M.; Zhao, Z.; Zhao, G.F.; Li, Z.H. Comparative Transcriptome Analysis Reveals Adaptive Evolution of *Notopterygium incisum* and *Notopterygium franchetii*, Two High-Alpine Herbal Species Endemic to China. *Molecules* **2017**, *22*, 1158. [CrossRef]
17. Jia, Y.; Bai, J.Q.; Liu, M.L.; Jiang, Z.F.; Wu, Y.; Fang, M.F.; Li, Z.H. Transcriptome Analysis of the Endangered *Notopterygium incisum*: Cold-Tolerance Gene Discovery and Identification of EST-SSR and SNP Markers. *Plant Divers.* **2019**, *41*, 1–6. [CrossRef] [PubMed]
18. Hincha, D.K.; Thalhammer, A. LEA Proteins: IDPs with Versatile Functions in Cellular Dehydration Tolerance. *Biochem. Soc. Trans.* **2012**, *40*, 1000–1003. [CrossRef]
19. Yu, X.M.; Yue, W.R.; Yang, Q.; Zhang, Y.N.; Han, X.M.; Yang, F.Y.; Wang, R.G.; Li, G.J. Identification of the LEA Family Members from *Caragana korshinskii* (Fabaceae) and Functional Characterization of *CkLEA2-3* in Response to Abiotic Stress in *Arabidopsis*. *Braz. J. Bot.* **2019**, *42*, 227–238. [CrossRef]
20. Jia, Y.; Liu, M.L.; López-Pujol, J.; Jia, R.W.; Kou, Y.X.; Yue, M.; Guan, T.X.; Li, Z.H. The Hybridization Origin of the Chinese Endemic Herb Genus *Notopterygium* (Apiaceae): Evidence from Population Genomics and Ecological Niche Analysis. *Mol. Phylogenet. Evol.* **2023**, *182*, 107736. [CrossRef] [PubMed]
21. Cao, J.; Li, X. Identification and Phylogenetic Analysis of Late Embryogenesis Abundant Proteins Family in Tomato (*Solanum lycopersicum*). *Planta* **2015**, *241*, 757–772. [CrossRef] [PubMed]
22. Wang, X.S.; Zhu, H.B.; Jin, G.L.; Liu, H.L.; Wu, W.R.; Zhu, J. Genome-Scale Identification and Analysis of LEA Genes in Rice (*Oryza sativa* L.). *Plant Sci.* **2007**, *172*, 414–420. [CrossRef]
23. Wang, W.D.; Gao, T.; Chen, J.F.; Yang, J.K.; Huang, H.Y.; Yu, Y.B. The Late Embryogenesis Abundant Gene Family in Tea Plant (*Camellia sinensis*): Genome-Wide Characterization and Expression Analysis in Response to Cold and Dehydration Stress. *Plant Physiol. Biochem.* **2019**, *135*, 277–286. [CrossRef]
24. Du, D.L.; Zhang, Q.X.; Cheng, T.R.; Pan, H.T.; Yang, W.R.; Sun, L.D. Genome-Wide Identification and Analysis of Late Embryogenesis Abundant (LEA) Genes in *Prunus mume*. *Mol. Biol. Rep.* **2013**, *40*, 1937–1946. [CrossRef]
25. Liu, H.; Xing, M.Y.; Yang, W.B.; Mu, X.Q.; Wang, X.; Lu, F.; Wang, Y.; Zhang, L.S. Genome-Wide Identification of and Functional Insights into the Late Embryogenesis Abundant (*LEA*) Gene Family in Bread Wheat (*Triticum aestivum*). *Sci. Rep.* **2019**, *9*, 13375. [CrossRef] [PubMed]
26. Li, Y.; Qi, S.Z.; Chen, S.Y.; Li, H.X.; Zhang, T.; Bao, F.; Zhan, D.J.; Pang, Z.W.; Zhang, J.F.; Zhao, J. Genome-Wide Identification and Expression Analysis of Late Embryogenesis Abundant (*LEA*) Genes Reveal Their Potential Roles in Somatic Embryogenesis in Hybrid Sweetgum (*Liquidambar styraciflua* × *Liquidambar formosana*). *For. Res.* **2023**, *3*, 12. [CrossRef]
27. Tompa, P.; Bánki, P.; Bokor, M.; Kamasa, P.; Kovács, D.; Lasanda, G.; Tompa, K. Protein-Water and Protein-Buffer Interactions in the Aqueous Solution of an Intrinsically Unstructured Plant Dehydrin: NMR Intensity and DSC Aspects. *Biophys. J.* **2006**, *91*, 2243–2249. [CrossRef] [PubMed]
28. Jeffares, D.C.; Penkett, C.J.; Bähler, J. Rapidly Regulated Genes Are Intron Poor. *Trends Genet.* **2008**, *24*, 375–378. [CrossRef] [PubMed]
29. Liu, J.H.; Peng, T.; Dai, W.S. Critical *cis*-Acting Elements and Interacting Transcription Factors: Key Players Associated with Abiotic Stress Responses in Plants. *Plant Mol. Biol. Rep.* **2014**, *32*, 303–317. [CrossRef]
30. Tao, Y.; Wang, F.T.; Jia, D.M.; Li, J.T.; Zhang, Y.M.; Jia, C.G.; Wang, D.P.; Pan, H.Y. Cloning and Functional Analysis of the Promoter of a Stress-Inducible Gene (ZmRXO1) in Maize. *Plant Mol. Biol. Rep.* **2015**, *33*, 200–208. [CrossRef]

31. Sun, Q.H.; Gao, F.; Zhao, L.; Li, K.P.; Zhang, J.R. Identification of a New 130 Bp *cis*-Acting Element in the TsVP1 Promoter Involved in the Salt Stress Response from *Thellungiella halophila*. *BMC Plant Biol.* **2010**, *10*, 90. [CrossRef]
32. Candat, A.; Paszkiewicz, G.; Neveu, M.; Gautier, R.; Logan, D.C.; Avelange-Macherel, M.H.; Macherel, D. The Ubiquitous Distribution of Late Embryogenesis Abundant Proteins across Cell Compartments in *Arabidopsis* Offers Tailored Protection against Abiotic Stress. *Plant Cell* **2014**, *26*, 3148–3166. [CrossRef]
33. Chakrabortee, S.; Tripathi, R.; Watson, M.; Kaminski Schierle, G.S.; Kurniawan, D.P.; Kaminski, C.F.; Wise, M.J.; Tunnacliffe, A. Intrinsically Disordered Proteins as Molecular Shields. *Mol. Biosyst.* **2012**, *8*, 210–219. [CrossRef]
34. Karpinska, B.; Razak, N.; Shaw, D.S.; Plumb, W.; Van De Slijke, E.; Stephens, J.; De Jaeger, G.; Murcha, M.W.; Foyer, C.H. Late Embryogenesis Abundant (LEA)5 Regulates Translation in Mitochondria and Chloroplasts to Enhance Growth and Stress Tolerance. *Front. Plant Sci.* **2022**, *13*, 875799. [CrossRef] [PubMed]
35. Freeling, M. Bias in Plant Gene Content Following Different Sorts of Duplication: Tandem, Whole-Genome, Segmental, or by Transposition. *Annu. Rev. Plant Biol.* **2009**, *60*, 433–453. [CrossRef]
36. Magadum, S.; Banerjee, U.; Murugan, P.; Gangapur, D.; Ravikesavan, R. Gene Duplication as a Major Force in Evolution. *J. Genet.* **2013**, *92*, 155–161. [CrossRef] [PubMed]
37. Xie, T.; Chen, C.J.; Li, C.H.; Liu, J.R.; Liu, C.Y.; He, Y.H. Genome-Wide Investigation of WRKY Gene Family in Pineapple: Evolution and Expression Profiles during Development and Stress. *BMC Genom.* **2018**, *19*, 490. [CrossRef] [PubMed]
38. Cheng, Z.H.; Zhang, X.M.; Yao, W.J.; Zhao, K.; Liu, L.; Fan, G.F.; Zhou, B.R.; Jiang, T.B. Genome-Wide Search and Structural and Functional Analyses for Late Embryogenesis-Abundant (LEA) Gene Family in Poplar. *BMC Plant Biol.* **2021**, *21*, 110. [CrossRef]
39. Zhang, Y.P.; Zhang, X.J.; Zhu, L.J.; Wang, L.X.; Zhang, H.; Zhang, X.H.; Xu, S.T.; Xue, J.Q. Identification of the Maize LEA Gene Family and Its Relationship with Kernel Dehydration. *Plants* **2023**, *12*, 3674. [CrossRef]
40. Chen, L.; Xin, J.; Song, H.Y.; Xu, F.; Yang, H.; Sun, H.; Yang, M. Genome-Wide Study and Functional Characterization Elucidates the Potential Association of Late Embryogenesis Abundant (LEA) Genes with Lotus Seed Development. *Int. J. Biol. Macromol.* **2023**, *226*, 1–13. [CrossRef]
41. Huang, R.L.; Xiao, D.; Wang, X.; Zhan, J.; Wang, A.Q.; He, L.F. Genome-Wide Identification, Evolutionary and Expression Analyses of LEA Gene Family in Peanut (*Arachis hypogaea* L.). *BMC Plant Biol.* **2022**, *22*, 155. [CrossRef]
42. Ma, J.Y.; Zuo, D.J.; Ye, H.; Yan, Y.J.; Li, M.D.; Zhao, P. Genome-Wide Identification, Characterization, and Expression Pattern of the Late Embryogenesis Abundant (LEA) Gene Family in *Juglans regia* and Its Wild Relatives *J. mandshurica*. *BMC Plant Biol.* **2023**, *23*, 80. [CrossRef] [PubMed]
43. Finn, R.D.; Clements, J.; Eddy, S.R. HMMER Web Server: Interactive Sequence Similarity Searching. *Nucleic Acids Res.* **2011**, *39*, W29–W37. [CrossRef] [PubMed]
44. Horton, P.; Park, K.J.; Obayashi, T.; Fujita, N.; Harada, H.; Adams-Collier, C.J.; Nakai, K. WoLF PSORT: Protein Localization Predictor. *Nucleic Acids Res* **2007**, *35*, W585–W587. [CrossRef] [PubMed]
45. Geourjon, C.; Deleage, G. SOPMA: Significant Improvements in Protein Secondary Structure Prediction by Consensus Prediction from Multiple Alignments. *Bioinformatics* **1995**, *11*, 681–684. [CrossRef] [PubMed]
46. Edgar, R.C. MUSCLE: Multiple Sequence Alignment with High Accuracy and High Throughput. *Nucleic Acids Res.* **2004**, *32*, 1792–1797. [CrossRef]
47. Kumar, S.; Stecher, G.; Li, M.; Knyaz, C.; Tamura, K. MEGA X: Molecular Evolutionary Genetics Analysis across Computing Platforms. *Mol. Biol. Evol.* **2018**, *35*, 1547–1549. [CrossRef] [PubMed]
48. Bailey, T.L.; Williams, N.; Misleh, C.; Li, W.W. MEME: Discovering and Analyzing DNA and Protein Sequence Motifs. *Nucleic Acids Res.* **2006**, *34*, W369–W373. [CrossRef]
49. Lescot, M.; Déhais, P.; Thijs, G.; Marchal, K.; Moreau, Y.; Van De Peer, Y.; Rouzé, P.; Rombauts, S. PlantCARE, a Database of Plant *cis*-Acting Regulatory Elements and a Portal to Tools for *in Silico* Analysis of Promoter Sequences. *Nucleic Acids Res.* **2002**, *30*, 325–327. [CrossRef] [PubMed]
50. Chen, C.J.; Chen, H.; Zhang, Y.; Thomas, H.R.; Frank, M.H.; He, Y.H.; Xia, R. TBtools: An Integrative Toolkit Developed for Interactive Analyses of Big Biological Data. *Mol. Plant* **2020**, *13* (Suppl. 2), 1194–1202. [CrossRef]
51. Huerta-Cepas, J.; Forslund, K.; Coelho, L.P.; Szklarczyk, D.; Jensen, L.J.; Von Mering, C.; Bork, P. Fast Genome-Wide Functional Annotation through Orthology Assignment by eggNOG-Mapper. *Mol. Biol. Evol.* **2017**, *34*, 2115–2122. [CrossRef]
52. Wang, Y.P.; Tang, H.B.; Debarry, J.D.; Tan, X.; Li, J.P.; Wang, X.Y.; Lee, T.H.; Jin, H.Z.; Marler, B.; Guo, H.; et al. *MCScanX*: A Toolkit for Detection and Evolutionary Analysis of Gene Synteny and Collinearity. *Nucleic Acids Res.* **2012**, *40*, e49. [CrossRef]
53. Zhang, Z.; Li, J.; Zhao, X.Q.; Wang, J.; Wong, G.K.S.; Yu, J. KaKs_Calculator: Calculating Ka and Ks Through Model Selection and Model Averaging. *Genom. Proteom. Bioinform.* **2006**, *4*, 259–263. [CrossRef]
54. Kim, D.; Paggi, J.M.; Park, C.; Bennett, C.; Salzberg, S.L. Graph-Based Genome Alignment and Genotyping with HISAT2 and HISAT-Genotype. *Nat. Biotechnol.* **2019**, *37*, 907–915. [CrossRef]
55. Li, H.; Handsaker, B.; Wysoker, A.; Fennell, T.; Ruan, J.; Homer, N.; Marth, G.; Abecasis, G.; Durbin, R. The Sequence Alignment/Map Format and SAMtools. *Bioinformatics* **2009**, *25*, 2078–2079. [CrossRef]

56. Pertea, M.; Pertea, G.M.; Antonescu, C.M.; Chang, T.C.; Mendell, J.T.; Salzberg, S.L. StringTie Enables Improved Reconstruction of a Transcriptome from RNA-Seq Reads. *Nat. Biotechnol.* **2015**, *33*, 290–295. [CrossRef]
57. Livak, K.J.; Schmittgen, T.D. Analysis of Relative Gene Expression Data Using Real-Time Quantitative PCR and the 2−ΔΔCT Method. *Methods* **2001**, *2*, 402–408. [CrossRef]
58. Pfaffl, M.W.; Tichopad, A.; Prgomet, C.; Neuvians, T.P. Determination of Stable Housekeeping Genes, Differentially Regulated Target Genes and Sample Integrity: BestKeeper–Excel-Based Tool Using Pair-Wise Correlations. *Biotechnol. Lett.* **2004**, *26*, 509–515. [CrossRef]

International Journal of *Molecular Sciences*

MDPI

Article

Novel Allelic Mutations in *Dw3* Gene That Affect the Height of Sorghum Plants

Ping Wang [1], Bingbing Liang [1], Zhengjun Li [2], Chunyu Wang [2], Lixia Zhang [2,*] and Xiaochun Lu [2,*]

[1] Institute of Plant Protection, Liaoning Academy of Agricultural Sciences, Shenyang 110161, China; pingw-556@163.com (P.W.); liangbb2121@163.com (B.L.)

[2] Institute of Sorghum, Liaoning Academy of Agricultural Sciences, Shenyang 110161, China; 18608900201@163.com (Z.L.); wangchunyu88899@126.com (C.W.)

* Correspondence: zhanglx993@sohu.com (L.Z.); luxiaocun2000@126.com (X.L.)

Abstract: Breeding for dwarfing traits in sorghum is crucial. However, only three genes (*Dw1–Dw3*) that control plant height have been mapped. In this study, 634 sorghum cultivars were collected to investigate plant height and genotypes. Four were genotyped *Dw1DW2Dw3* (wild type) but with different plant heights, and they were selected to construct two populations and map new dwarf genes. Bulked segregant analysis with whole-genome resequencing of the two populations identified the candidate gene in one same genomic region—on chromosome 7. Then, it was narrowed down to a region containing nine genes. Amino acid and DNA sequence analysis of the parent and offspring plants revealed that two novel allelic mutations in the *Dw3* gene play a role in reducing the plant height—8R262 or 8R417, including 1 bp substitution and 2 bp deletions. Furthermore, we sequenced 19 cultivars that primarily exhibited a "one-dwarf" hybrid or wild-type and presumed another allelic mutation via the amino acid alignment of 8R019, 8R100, and 8R402, which was another one-base substitution. These results indicate that multiple types of allelic mutations in the *Dw3* gene should be considered when identified or applied.

Keywords: BSA-seq; *Dw3* gene; novel mutations; sorghum

1. Introduction

A variety of biomass energy sources can produce biofuels, including crops, energy crops, aquatic plants, and forestry waste [1,2]. Among them, Sorghum [*Sorghum bicolor* (L.) Moench.] is a staple food crop for half a billion people and an emerging biofuel crop [3]. Furthermore, sorghum is an important crop species for farmers in semiarid and arid regions. Because rising temperatures and reduced precipitation due to climate change make some areas unsuitable for maize and rice production, the importance of drought-tolerant crops like sorghum is likely to increase [4].

Breeding for dwarfing traits in sorghum is crucial because it enables mechanical harvesting, improves lodging resistance, and increases fertilizer use efficiency. Four major dwarfing genes, identified as *Dw1–Dw4*, control height by modifying the internode length [5]. In 2016, the gene corresponding to *Dw1* underwent map-based cloning using an F2 population and HIFs derived from Hegari and 80M. It was shown that the Dw1 protein could affect cell proliferation activity in the internodes [6,7]. As a result of mutations, *dw2*, located on chromosome 6, greatly reduces internode lengths [8,9]. Similarly, *Dw3*, located on chromosome 7, was finally shown to encode a transmembrane protein of the adenosine triphosphate-binding cassette (ABC) transporter superfamily involved in auxin efflux, which was cloned decades ago. The *Dw3* mutant allele contained an 882 bp tandem duplication in exon 5 that abolished the enzyme's function. This allele is known to be unstable, with the progeny of a homozygous *dw3* plant reverting to a tall *Dw3* wild type at a rate of around 1/600 [10]. In 2024, three InDel mutations in the *Dw3* gene, including an 82 bp deletion, a 6 bp duplication, and a 15 bp deletion, were discovered [11].

Although *Dw4* has not been cloned, it is inferred to be located on chromosome 4 or 6 [12,13]. In the early stages of hybrid utilization in China, sorghum primarily exhibited a "one-dwarf", reaching heights of 2.5–3 m, leading to serious lodging. Recognizing the need for dwarf hybrids (approximately 2 m in height) for efficient production, Liang et al. advocated for sorghum dwarf breeding [14].

Bulked Sergeant Analysis (BSA) is a fast, convenient, and efficient method of determining the linkage between markers and traits by constructing DNA pools of extreme traits. It has been widely adopted in gene-mapping studies in recent years [15–19]. Compared with traditional linkage mapping, where progeny are individually genotyped and phenotyped, BSA requires the preparation of fewer samples for genotyping and allows for more efficient phenotyping. This is because only individuals that represent the extreme phenotypes of a population need to be identified [15].

China is the world's eighth-largest producer of sorghum, which is mainly used as food, feed biofuels, and alcoholic beverages. It was introduced through the spreading of Indian cultivars during the silkworm trade or directly from the eastern coast of Africa by returning Chinese seamen [19,20]. Between the 1930s and 1950s, sorghum was grown as a traditional cereal on 10 million hectares, but the production area drastically declined to about 2 million hectares by 1990 and then further to 0.6 million hectares by 2008. Instead, sorghum has become increasingly popular as animal feed, and in recent years, the Chinese government has invested in research into improved varieties for this purpose [3]. In addition, sorghum is also a major ingredient in liquor production in China's food industry [21]. Almost all famous Chinese brands of distilled liquors, such as Moutaijiu (Moutai aroma liquor), Luzhoulaojiao, Wuliangye (strong aroma liquor), and Fenjiu (light aroma liquor), are brewed with sorghum grain as a key ingredient [22]. Sorghum has been planted in China for a long time and is rich in germplasm resources. Therefore, finding new genes by typing the genes affecting plant height in common sorghum cultivars in China is helpful. The corresponding research is of great benefit to sorghum breeding.

In this study, the genotypes were systematically investigated in 634 sorghum cultivars, which contain a number of Chinese sorghum landraces, and 4 cultivars were identified, *Dw1Dw2Dw3*, with significant differences in plant height. Subsequently, two groups of plants were generated for BSA-seq analysis to map the new genes affecting plant height. By using a recombinant-derived progeny testing strategy [23,24], the candidate gene was narrowed to a region that includes nine genes between 59.75 and 59.86 Mbp. Amino acid sequence analysis of the parent and offspring plants revealed that the novel mutated *Dw3* genes, rather than the novel genes, play a role in reducing the plant height of 8R262 or 8R417. The genotypes of 19 sorghum cultivars primarily exhibited a "one-dwarf" hybrid or wild-type (*Dw1Dw2Dw3*), and a plant height of less than 2 m was identified. Meanwhile, two novel allelic mutations were identified in Chinese sorghum landraces, and another one was inferred through the sequencing results of the Indian cultivars. Consequently, this discovery pointed to the potential existence of different dwarf mutations in the *dw3* gene, some of which could be additional stable sources of the trait. Overall, our study provides new insights into the *Dw3* gene and novel theoretical support for its applications.

2. Results

2.1. Phenotyping and Genetic Analysis of Dwarf Sorghum Cultivars

In order to seek the fourth or more genes that affect sorghum height, 634 sorghum cultivars were planted in Shenyang city (42° N, 123° E) and numbered to investigate plant height and genotypes. We genotyped *Dw1* and *Dw2* via Sanger sequencing and *Dw3* using the PCR amplification method, which has been commonly employed because the *dw3* mutant allele contains an insertion of a fragment that can be clearly identified [10,25].

As shown in Table 1 and Table S1, the plant heights of three Chinese sorghum landraces (8R262, 8R277, and 8R417) and one sorghum cultivar (8R037) from Morocco were investigated. The plant heights of 8R262 and 8R417 were 107 $\pm$ 10 cm and 133 $\pm$ 6 cm, while those of 8R277 and 8R037 were 170 $\pm$ 10 cm and 175 $\pm$ 7 cm, respectively. Although they differed significantly in plant height, all four cultivars were identified as *Dw1Dw2Dw3* and selected for subsequent hybridization manipulations because they had relatively close heading times.

Table 1. Genotyping of genes affecting plant height and statistics of plant height and heading date.

Cultivar Number	Plant Height (cm)	Heading Date (Day)
8R277	170 ± 10 a	68
8R262	107 ± 10 c	65
8R037	175 ± 7 a	72
8R417	133 ± 6 b	70

Data marked with different letters (a–c) indicate significant differences, as analyzed by SPSS 23.0 software (Duncan's multiple range test, $\alpha = 0.05$).

The tall cultivars, 8R277 and 8R037, were used as the female parents to cross the dwarf cultivars 8R262 and 8R417, respectively, in an inheritance study and during the construction of the mapping population (Table 2, Figure 1). All F1 plants showed a tall plant height, and the F2 plants displayed different heights corresponding to the parents. Therefore, the dwarf-to-tall was a completely recessive trait in two groups of populations. According to the law of segregation, the ratio of dominant-trait individuals to recessive homozygous-trait individuals was 3 to 1 in F2 populations. Among the 708 F2 individuals in the 8R262 × 8R277 population, 536 exhibited a tall plant height, and 172 showed a dwarf plant height, corresponding to a segregation ratio of 3:1 determined using the Chi-square test ($\chi^2 = 0.19 < 3.84$, p-value = 0.67 > 0.05). $\chi^2 \leq 3.84$ indicates that the theoretical segregation ratio significantly correlates with the actual ratio. Similarly, in the 8R037 × 8R417 population, F2 individuals also showed a 3:1 segregation ratio of tall and dwarf ($\chi^2 = 0.99 < 3.84$, p-value = 0.32 > 0.05) (Table 2). These results demonstrate that a single recessive gene should control the dwarf trait.

Table 2. The phenotype analysis results of tall/dwarf plant height in the two sorghum populations.

Population	Generation	Total Plants	Tall	Dwarf	[a] Expected Ratio	[b] χ^2	[c] p-Value
8R262 × 8R277	F_2	708	536	172	3:1	0.19	0.67
8R037 × 8R417	F_2	594	456	138	3:1	0.99	0.32

[a] Expected ratio: dominant-trait individual: recessive-trait individual = 3:1. [b] χ^2: $\chi^2 < 3.84$ is considered significantly correlated. [c] p-value: p-value > 0.05 indicates no statistically significant difference.

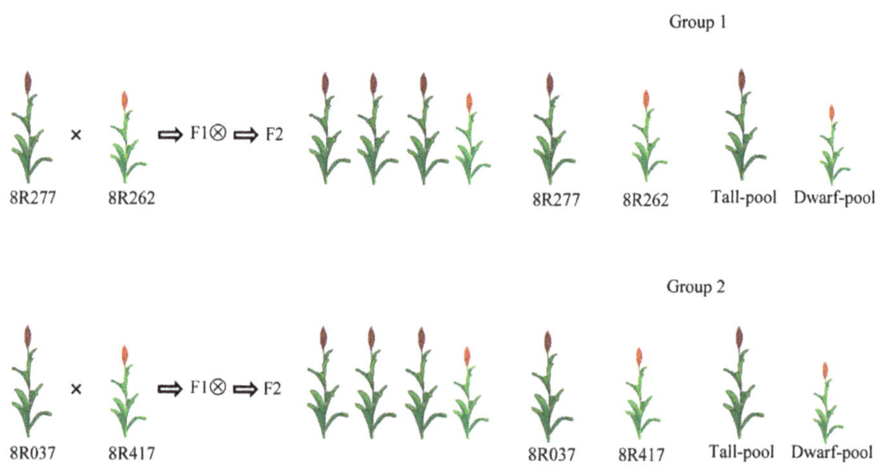

Figure 1. Crossing and extreme pool design.

2.2. BSA-Seq Analysis and Mapping of Candidate Gene

To map the new genes affecting plant height, we generated two groups of plants for BSA-seq analysis, as shown in Figure 1. Fifty plants with an extreme plant height were selected as the highly tall pool or extreme dwarf pool in both groups, respectively. On the Illumina HiSeq sequencing platform, a total of 725,474,232 raw reads for group 1 and 771,050,148 raw reads for group 2 were generated after sequencing the two groups of parent plants and extreme mixing pools (Table 3). By using quality-control measures, such as removing joints and low-quality reads, 108,821,134,800 and 115,657,522,200 clean bases for groups 1 and 2 were obtained, respectively. The quality of the sequencing data was evaluated, and the sequencing error rate and statistical Q20, Q30, and GC contents were calculated, as shown in Table 2. The quality evaluation showed that the GC contents of these reads were 43.64–43.78% (group 1) and 44.62~45.55% (group 2). The two groups of sequencing data had high quality (Q20 $\geq$ 93%, Q30 $\geq$ 97) and could be adopted for further analysis.

Table 3. Sequencing statistics for BSA-seq samples.

	Sample	Clean Reads	Clean Bases	Mapped (%)	Properly Mapped (%)	Average Depth	Q20 (%)	Q30 (%)	GC Content (%)
Group 1	B8R262	106,444,214	15,966,632,100	98.40	90.03	21	97.71	93.38	43.64
	B8R277	110,755,784	16,613,367,600	98.33	89.73	22	97.80	93.62	43.78
	Dwarf pool	256,010,212	38,401,531,800	96.04	89.16	49	98.18	94.43	44.63
	Tall pool	252,264,022	37,839,603,300	97.81	90.06	49	97.88	93.66	43.71
Group 2	B8R037	59,041,644	8,856,246,600	96.15	88.82	11	98.09	94.15	45.21
	B8R417	62,894,480	9,434,172,000	96.94	90.73	12	97.71	93.37	45.55
	Dwarf pool	338,573,164	50,785,974,600	94.91	87.23	62	98.29	94.77	44.81
	Tall pool	310,540,860	46,581,129,000	96.38	89.46	57	98.31	94.86	44.62

The GATK software toolkit (Version 4.1.8.1) was used for SNP detection and filtration. A total of 862,470 (group 1) and 1,285,629 (group 2) SNPs were identified, of which 45,340 were located in the exon regions, including 29,271 non-synonymous mutations, 545 stop gain mutations, and 75 stop loss mutations (group 1). The data from group 2 suggest that 67,802 were located in the exon regions, containing 47,584 non-synonymous mutations, 847 stop gain mutations, and 133 stop loss mutations. To determine the genomic region linked to plant height, we utilized the ΔSNP-index method to assess the allele segregation of the SNPs and InDels between the two DNA pools representing the extreme groups. According to the fitted △SNP-index computer simulation experiment, taking the threshold of 99% confidence (red line in the figure), the association region was located in the range of 10.5 Mb on chromosome 7 of 8R262 plants. Meanwhile, it was located within an 8.2 Mb region on chromosome 7 of 8R417 plants, as indicated by the arrows in Figure 2.

After several screenings, we successively obtained 123 recombinants and narrowed the candidate gene to an interval flanked by the markers InDel-6 and InDel-15. Coincidentally, the interval almost coincided with the region of the most credible prediction in BSA-seq analysis. To further locate the candidate gene, another 10 InDel markers were applied to seek the limited individuals' recombinants. Due to the finding of the last four recombinants, the candidate gene was narrowed to a region that includes nine genes between 59.75 and 59.86 Mbp (Figure 3). *Dw3* was among them, as annotated and listed in Table 4. Moreover, a similar experiment was carried out in group 2, and consistent results were obtained.

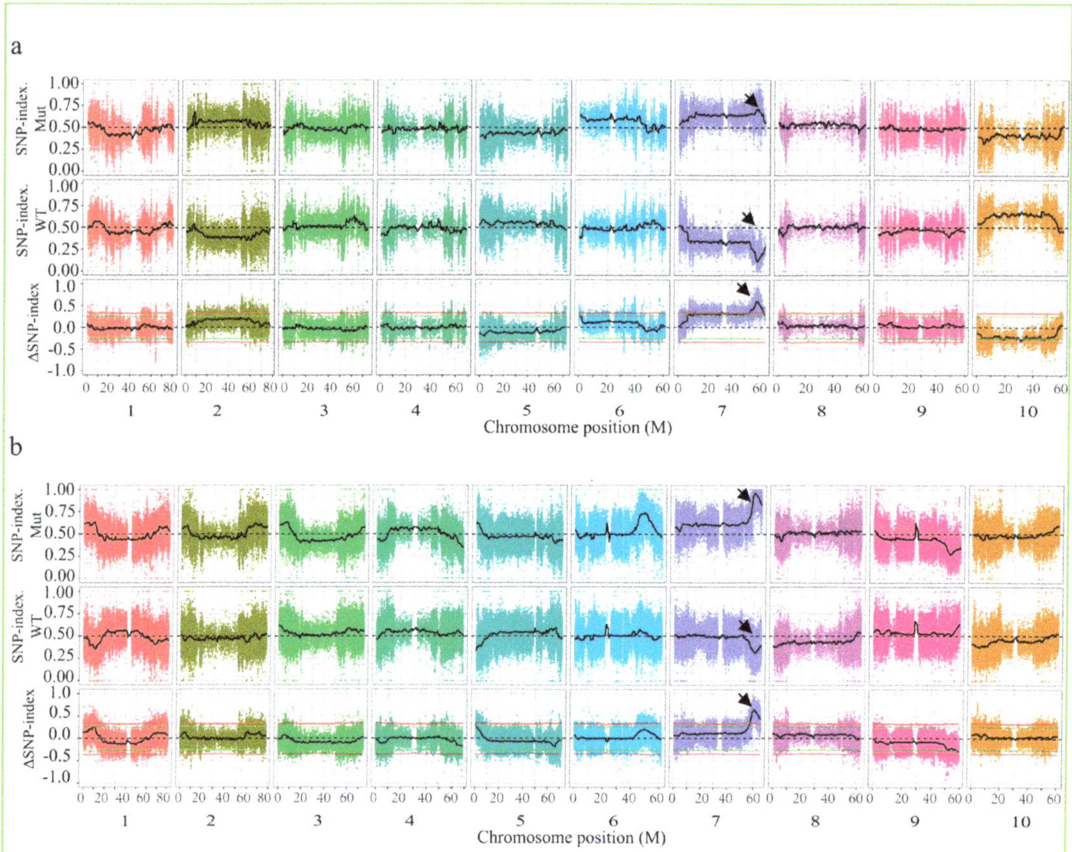

Figure 2. Genome-wide association mapping of plant height. The candidate SNPs and InDels of group 1 (**a**) and group 2 (**b**) using the ΔSNP-index algorithm with a cutoff of ΔSNP-index > 0.5. The arrow indicates the region where the predicted candidate gene was located.

Table 4. Gene descriptions from the NCBI of 9 genes in the predictive region.

Gene Number	Annotation
SORBI_3007G163300	NA
SORBI_3007G143400	Tetraspanin-8, *Arabidopsis thaliana* (Mouse-ear cress).
SORBI_3007G163433	NA
SORBI_3007G163466	NA
SORBI_3007G143500	Uncharacterized protein, *Arabidopsis thaliana* (Mouse-ear cress).
SORBI_3007G153600	Protein HVA22, *Hordeum vulgare* (Barley).
SORBI_3007G163700	NA
SORBI_3007G163800	ABC transporter B family member 1 (*Dw3*, Sorghum)
SORBI_3007G163901	NA

Genes were blasted on the NCBI website (https://www.ncbi.nlm.nih.gov/ (accessed on 4 December 2023)). "NA" indicates no annotations.

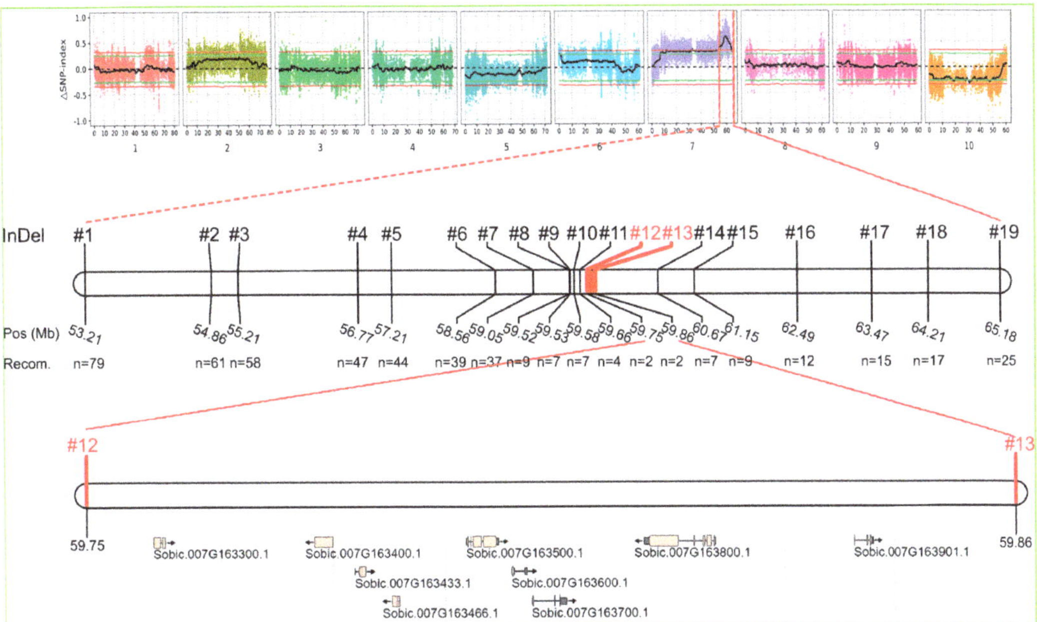

Figure 3. Fine mapping of the candidate gene. Narrowing the candidate region through recombinant-derived progeny testing strategy produced the genetic linkage map of a candidate region of Chr.7. The genes and diagnostic markers in the 0.11 Mb interval.

2.3. Identification of the Candidate Gene Related to Plant Height in Sorghum Plants

We performed Sanger sequencing for all nine genes in the male and female parents to identify the candidate gene. Although the DNA sequences of the nine genes were not identical, only Dw3 had a non-synonymous mutation in the amino acid sequence of 8R262 or 8R417 plants, as shown in Figure 4. To illustrate the variations, CDS sequences (Figure 4a,c) and amino acid sequences (Figure 4b,d) of Dw3 were arranged for further analysis. In contrast to 8R277 and other cultivars, the amino acid sequence of 8R262 had a substitution of proline for glutamine at position 898 (Figure 4b), which was caused by the conversion of "A" to "C" at position 2693 in the CDS sequence (Figure 4a). The CDS sequence 8R417 also had a two-base deletion at position 3254 (Figure 4c), resulting in the frameshift mutation starting at position 1085 (Figure 4d).

Furthermore, we identified the *Dw3* gene of 8R262 and 8R417 varieties again using the traditional PCR method, and the results showed that the *Dw3* gene of these two varieties did not have fragment insertion compared with the BTx623 sorghum reference genome (Figure S1). The integrated results revealed that the novel mutated *Dw3* genes, rather than a novel gene, play a role in determining the plant height of 8R262 and 8R417. Hence, the genotypes of some common sorghum cultivars primarily exhibiting a "one-dwarf" hybrid or wild-type (*Dw1Dw2Dw3*) and plant heights of less than 2 m are re-summarized and listed in Table 5 since two novel mutant forms of *Dw3* were identified. Furthermore, the *Dw3* genes of these cultivars were also determined using Sanger sequencing. As shown in Table S2, both the CDS sequences of 8R258 and 8R428 also had a two-base deletion at position 3254, consistent with 8R417, and a significantly lower plant height. Therefore, it can be inferred that this type of mutation is not an isolated phenomenon. In addition, in contrast to other cultivars, the amino acid sequences of 8R019, 8R100, and 8R402 had a substitution of arginine for lysine at position 985, which was caused by the conversion of "A" to "G" at position 2954 in the CDS sequence (Table S2). At the same time, the plant height of the cultivars decreased significantly, which may be a new mutation type and needs further determination.

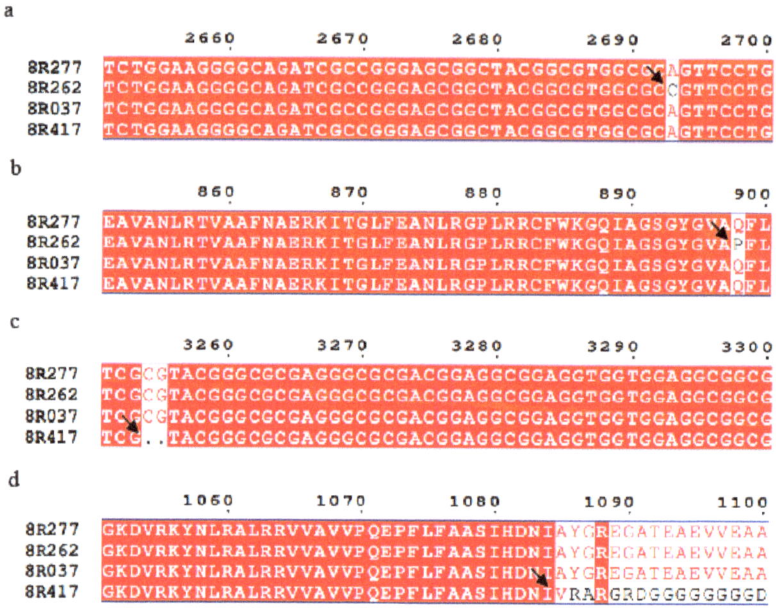

Figure 4. CDS and amino acid sequence analysis of *Dw3* genes from different cultivars. The CDS (**a**,**c**) and amino acid (**b**,**d**) sequences of Dw3 were arranged to show the variations, respectively. Multiple sequence alignment was performed using the espript 3.0 online tool (https://espript.ibcp.fr/ESPript/cgi-bin/ESPript.cgi (accessed on 1 September 2024)). Arrows point to locations where there are variations in the sequences.

Table 5. The genotyping of genes affecting plant height.

Cultivar Number	Origin	Genotype	Plant Height (cm)
8R019	India	*Dw1dw2dw3*	111 ± 16 ef
8R037	Morocco	*Dw1Dw2Dw3*	170 ± 10 abc
8R082	South Africa	*Dw1Dw2Dw3*	192 ± 11 ab
8R100	India	*Dw1dw2bdw3*	122 ± 12 ef
8R105	Sultan	*dw1Dw2Dw3*	121 ± 21 ef
8R151	India	*Dw1Dw2Dw3*	199 ± 48 a
8R205	Zimbabwe	*Dw1dw2Dw3*	142 ± 11 cde
8R206	Zimbabwe	*Dw1dw2Dw3*	144 ± 8 cdef
8R219	South Korea	*Dw1dw2Dw3*	165 ± 7 abcd
8R260	China	*Dw1Dw2Dw3*	184 ± 17 ab
8R258	China	*Dw1dw2Dw3*	129 ± 6 def
8R262	China	*Dw1Dw2dw3*	107 ± 10 f
8R277	China	*Dw1Dw2Dw3*	169 ± 10 abc
8R281	China	*Dw1Dw2Dw3*	160 ± 24 bcd
8R315	India	*Dw1dw2Dw3*	157 ± 22 bcd
8R390	The United States	*Dw1dw2Dw3*	159 ± 11 bcd
8R402	China	*Dw1Dw2dw3*	122 ± 11 ef
8R417	China	*Dw1Dw2dw3*	120 ± 29 ef
8R428	China	*Dw1Dw2Dw3*	170 ± 4 abc

The genotypes of some common sorghum cultivars primarily exhibiting a "one-dwarf" hybrid or wild-type (*Dw1Dw2Dw3*) and a plant height of less than 2 m are listed in the table. Data marked with different letters (a–f) indicate significant differences, as analyzed using SPSS 23.0 software (Duncan's multiple range test, α = 0.05).

3. Discussion

Over the past few years, the application of BSA-seq technology in sorghum gene mapping has gradually attracted the attention of researchers. In 2022, Mendu et al. identified and characterized three allelic thick leaf mutants (*thl1*, *thl2*, and *thl3*). BSA-seq analysis showed that the causal mutation for the *thl* phenotype is in the endo-1,4-β-glucanase gene (*SbKOR1*) [26]. De Riseis et al. identified two distinct mutations in the Sorghum bicolor *Id1* (*SbId1*) homolog through BSA-seq analysis, which both delayed the flowering time [27]. Moreover, Demeke B. Mewa reported mapping via BSA-seq of ANTHRACNOSE RESISTANCE GENE 2 (ARG2) encoding a nucleotide-binding leucine-rich repeat (NLR) protein that confers race-specific resistance to *Colletotrichum sublineola* strains [28]. These successful examples indicate the feasibility of using BSA-seq in this study to map genes affecting the height of sorghum plants.

A great many factors can affect plant height. In sorghum, plant height is controlled by a few major dwarfing genes. Quinby and Karper identified four loci (*Dw1*–*Dw4*) that control the internode length by measuring the height of the stem from the ground to the flag leaf. At each Dw locus, the dominant allele increased the internode length [5,29]. Only three of the four genes (*Dw1*–*3*) were mapped and cloned, and they were located on chromosomes 9, 6, and 7, respectively [6,8,10].

The fourth classical dwarfing locus in sorghum, *Dw4*, is known to be unlinked to the other dwarfing loci and is believed to be on chromosome 4 or 6, but associated genes remain uncloned [12,13]. In our study, 19 markers were developed after BSA-seq analysis. After several rounds of screening, the gene affecting plant height was mapped on chromosome 7 (Figure 2), which was quite different from previous inferences for *Dw4* [12].

Since the nonfunctional *dw3* allele is the result of an 882 bp tandem duplication in exon 5 that disrupts protein function and the plant's ability to establish an auxin gradient in the intermediate internodes, the PCR amplification method was adopted to identify mutations [10]. As illustrated in Table 1, although no fragment insertion was observed in the *Dw3* gene sequences, plant height was lower in several cultivars. Subsequently, in the BSA-seq analysis, genes affecting plant height were mapped to an interval containing nine open reading frames (ORFs). At the same time, the Sanger sequencing results showed that only the *Dw3* gene had a non-synonymous mutation in the amino acid sequence (Figure 4). The mutant form of the *Dw3* gene differed from previous studies [10,25]. This allele is known to be unstable, with the progeny of a homozygous *dw3* plant reverting to a tall *Dw3* wild type at a rate of around 1/600 [10].

All the results above suggest that mutations in the *Dw3* gene reduce plant height, and two novel mutation forms were identified.

In China, sorghum has been a critical ingredient for brewing liquors since the Yuan Dynasty (1270-1368 CE) [30,31]. Sorghum is mainly cultivated in China's southwest, north, and northeast regions, and more than 80% of sorghum grain is used for liquor-making [18,32]. Therefore, there are abundant sorghum germplasm resources in China, which provide more convenience for the selection and breeding of new cultivars. This study systematically identified the main genotypes affecting plant height genes of 19 sorghum cultivars (Table 5), which contain a number of Chinese sorghum landraces. In our study, two novel allelic mutations identified using BSA-seq were only found in Chinese sorghum landraces. In addition, we also inferred another allelic mutation by studying the sequencing results of the Indian cultivars (8R019 and 8R100) or 8R402 (a Chinese sorghum landrace). Our article pointed to the potential existence of different dwarf mutations in the *dw3* gene, some of which could be additional stable sources of the trait.

4. Materials and Methods

4.1. Plant Materials, Phenotyping, DNA Extraction, and Sanger Sequencing

To investigate plant height and genotypes, 397 sorghum cultivars were obtained from the China National Center for Sorghum Improvement. Similarly, 237 sorghum cultivars were obtained from the Sorghum Mini Core Collection (MCC) to identify additional genes influencing sorghum height [33]. The field trials to evaluate the phenotypic performance of sorghums were conducted in Shenyang city (42° N, 123° E) in the 2014 summer season.

In order to improve pollination success and find suitable crosses, the heading date (HD) was recorded when 50% of the plants in the plot showed 50% flowering. Three plots of sorghum plants of each cultivar were planted. Three individual plants from the middle of each plot were selected for phenotype evaluation. Plant height was evaluated at maturity, as described by Zou et al. [34].

Genomic DNA (5–6 plants per accession) was extracted from 30-day-old seedlings by using the CTAB protocol [35]. DNA concentration and quality were estimated with an ND-1000 spectrophotometer (NanoDrop, Wilmington, DE, USA) and by electrophoresis on 1.0% agarose gels with a lambda DNA standard.

Fragments of *Dw1*, *Dw2*, and *Dw3* were amplified. PCR was performed in a 25 μL reaction containing 2× EXtaqMix (Takara, Kusatsu-shi, Japan), 0.4 μM of each primer, 0.125 μg of DNA, and double-distilled water. A PCR protocol was used to amplify the fragments, including an initial denaturation cycle at 95 °C for 4 min, followed by 30 cycles of denaturation at 95 °C for 30 s, annealing at 57 °C for 30 s, and extension at 72 °C for 2.5 min. Primers for Sanger sequencing and PCR-amplification were listed in Table S3, and PCR products were sequenced by Biomarker Technologies Corporation (Beijing, China).

4.2. SNP Library Construction and High-Throughput Sequencing

The specific length amplified fragment sequencing (SLAF-seq) employed in this study was performed following Sun et al. with minor modifications [36]. The SLAF library was constructed as follows. First, a SLAF pre-design experiment was performed to establish conditions to optimize SLAF yield, avoid repetitive SLAFs, and obtain an even distribution of SLAFs for maximum SLAF-seq efficiency. These considerations improved the efficiency of SLAF-seq. Next, the SLAF library was constructed in accordance with the pre-designed scheme. Genomic DNA was incubated at 37 °C with *Mse* I (New England Biolabs, NEB.; Ipswich, MA, USA), T4 DNA ligase (NEB), ATP (NEB), and *Mse* I adapter. The reaction was heat-inactivated at 65 °C and then digested for the additional restriction enzyme *Alu* I at 37 °C. Then, polymerase chain reactions (PCRs) were carried out using diluted restriction ligation samples, dNTP, Taq DNA polymerase (NEB), and *Mse* I primer containing barcode1. Using E.Z.N.A.H Cycle Pure Kit (Omega, Biel/Bienne, Switzerland), the PCR productions were purified and pooled. The pooled samples were incubated at 37 °C with *Mse* I, T4 DNA ligase, ATP, and Solexa adapter, purified using a Quick Spin column (Qiagen, Hilden, Germany), and then run out on a 2% agarose gel. Using a Gel Extraction Kit (Qiagen, Hilden, Germany), fragments of 450–500 bp (with indexes and adaptors) in size were isolated. These fragment productions were then subjected to PCR amplification with Phusion Master Mix (NEB) and Solexa Amplification primer mix (Illumina, Inc., San Diego, CA, USA) to add barcode 2 according to the Illumina sample preparation guide.

Samples were gel-purified by excising DNA of 450–500 bp and diluted for paired-end sequencing on an Illumina HiSeq 2000 sequencing platform (Illumina, Inc., San Diego, CA, USA) in Beijing (http://www.biomarker.com.cn (accessed on 7 November 2024)). SNP genotyping and evaluation were then performed.

All SLAF paired-end reads with clear index information were clustered according to sequence similarity, which was detected using one-to-one alignment by BLAT (−tileSize = 10 − stepSize = 5) [37]. Sequences with over 90% identity were grouped in one SLAF locus. Using the minor allele frequency (MAF) evaluation, alleles were defined in each SLAF.

In mapping populations of sorghum, one locus can contain at most four genotypes, so the groups containing more than four tags were filtered out as repetitive SLAFs. SLAFs with a sequence depth of less than 107 were filtered out of the following analysis. SLAFs with 2–4 tags were identified as polymorphic SLAFs, which were considered potential markers. Tags with at least 10× total coverage were retained. Sequences were mapped to the BTx623 sorghum reference genome using BWA [38,39], and SNPs were called with the

TASSEL 3.0 GBS pipeline (www.maizegenetics.net/tassel/ (accessed on9 January 2024)). Missing data were imputed with NPUTE [40].

5. Conclusions

Three novel allelic mutations were discovered by BSA-seq analysis and amino acid sequence alignment. Two of them were only identified from Chinese sorghum landraces. Multiple types of allelic mutations in the *Dw3* gene should be considered when identified or applied.

Supplementary Materials: The following supporting information can be downloaded at: https://www.mdpi.com/article/10.3390/ijms252212000/s1.

Author Contributions: P.W. and B.L.: Formal analysis, Writing—original draft. X.L.: Conceptualization. Z.L. and C.W.: Methodology, Formal analysis, Investigation. L.Z.: Methodology, Formal analysis, Data curation. P.W. and X.L.: Writing—review and editing, Supervision, Project administration. All authors have read and agreed to the published version of the manuscript.

Funding: This research was supported by the Liaoning Province Special Program for Germplasm Innovation and Store Grain in Technology (2023JH1/10200009) and the China Agricultural Research System (CARS-06-14.5-A3).

Institutional Review Board Statement: Not applicable.

Informed Consent Statement: Not applicable.

Data Availability Statement: The data presented in this study are available upon request from P.W.

Acknowledgments: We thank Yibin Zhang (Cotton Research Institute, Chinese Academy of Agricultural Sciences) for useful discussions.

Conflicts of Interest: The authors declare no competing interests.

References

1. Demirbas, A. Biofuels sources, biofuel policy, biofuel economy and global biofuel Projections. *Energy. Convers. Manag.* **2008**, *49*, 2106–2116. [CrossRef]
2. Stamenković, O.S.; Siliveru, K.; Veljković, V.B.; Banković-Ilić, I.B.; Tasić, M.B.; Ciampitti, I.A.; Đalović, I.G.; Mitrović, P.M.; Sikora, V.Š.; Vara, P.V. Prasad Production of biofuels from sorghum. *Renew. Sust. Energ. Rev.* **2020**, *124*, 109769. [CrossRef]
3. Mundia, C.W.; Secchi, S.; Akamani, K.; Wang, G. A regional comparison of factors affecting global Sorghum production: The case of north America, Asia and Africa's sahel. *Sustainability* **2019**, *11*, 2135. [CrossRef]
4. Lobell, D.B.; Burke, M.B.; Tebaldi, C.; Mastrandrea, M.D.; Falcon, W.P.; Naylor, R.L. Prioritizing climate change adaptation needs for food security in 2030. *Science* **2008**, *319*, 607–610. [CrossRef] [PubMed]
5. Karper, E.R.; Quinby, R.J. Sorghum-Its production, utilization and breeding. *Econ. Bot.* **1947**, *1*, 355–371. [CrossRef]
6. Hilley, J.; Truong, S.; Olson, S.; Morishige, D.; Mullet, J. Identification of *Dw1*, a regulator of sorghum stem internode length. *PLoS ONE* **2016**, *11*, e0151271. [CrossRef]
7. Yamaguchi, M.; Fujimoto, H.; Hirano, K.; Araki-Nakamura, S.; Ohmae-Shinohara, K.; Fujii, A.; Tsunashima, M.; Song, X.J.; Ito, Y.; Nagae, R.; et al. Sorghum *Dw1*, an agronomically important gene for lodging resistance, encodes a novel protein involved in cell proliferation. *Sci Rep.* **2016**, *6*, 28366. [CrossRef]
8. Klein, R.R.; Mullet, J.E.; Jordan, D.R.; Miller, F.R.; Rooney, W.L.; Menz, M.A.; Franks, C.D.; Klein, P.E. The effect of tropical sorghum conversion and inbred development on genome diversity as revealed by high-resolution genotyping. *Crop. Sci* **2008**, *48*, S12. [CrossRef]
9. Morishige, D.T.; Klein, P.E.; Hilley, J.L.; Sahraeian, S.M.E.; Sharma, A.; Mullet, J.E. Digital genotyping of sorghum—A diverse plant species with a large repeat-rich genome. *BMC. Genom.* **2013**, *14*, 1–9. [CrossRef]
10. Multani, D.S.; Briggs, S.P.; Chamberlin, M.A.; Blakeslee, J.J.; Murphy, A.S.; Johal, G.S. Loss of an MDR transporter in compact stalks of maize *br2* and sorghum *dw3* mutants. *Science* **2003**, *302*, 81–84. [CrossRef]
11. Diatta-Holgate, E.; Bergsma, B.; Tuinstra, M.R. Mutations in the *dwarf3* gene confer height stability in sorghum. *Plant. Genome* **2024**, *17*, e20466. [CrossRef] [PubMed]
12. Morris, G.P.; Ramu, P.; Deshpande, S.P.; Hash, C.T.; Shah, T.; Upadhyaya, H.D.; Riera-Lizarazu, O.; Brown, P.J.; Acharya, C.B.; Mitchell, S.E.; et al. Population genomic and genome-wide association studies of agroclimatic traits in sorghum. *Proc. Nat. Acad. Sci. USA* **2013**, *110*, 453–458. [CrossRef] [PubMed]

13. Xin, L.; Xian, R.L.; Eyal, F.; Jian, M.Y. Dissecting repulsion linkage in the dwarfing gene *Dw3* region for sorghum plant height provides insights into heterosis. *Proc. Natl. Acad. Sci. USA* **2015**, *112*, 11823–11828.
14. Liang, T.; Bi, N. Scientific research road in the sorghum field-Chinese famous sorghum breeder cattle paradise. *Seed. Technol.* **2014**, *15*, 1–5.
15. Michelmore, R.W.; Paran, I.; Kesseli, R. Identification of markers linked to disease-resistance genes by bulked segregant analysis: A rapid method to detect markers in specific genomic regions by using segregating populations. *Proc. Nat. Acad. Sci. USA* **1991**, *88*, 9828–9832. [CrossRef]
16. Zhang, K.; Yuan, M.; Xia, H.; He, L.Q.; Ma, J.; Wang, M.X.; Zhao, H.L.; Hou, L.; Zhao, S.Z.; Li, P.C.; et al. BSA-seq and genetic mapping reveals *AhRt2* as a candidate gene responsible for red testa of peanut. *Theor. Appl. Genet.* **2022**, *135*, 1529–1540. [CrossRef]
17. Klein, H.; Xiao, Y.G.; Conklin, P.A.; Govindarajulu, R.; Kelly, J.A.; Scanlon, M.J.; Whipple, C.J.; Bartlett, M. Bulked-segregant analysis coupled to whole genome sequencing (BSA-Seq) for rapid gene cloning in maize. *G3-Genes. Genome Genet.* **2018**, *8*, 3583–3592. [CrossRef]
18. Vogel, G.; Laplant, K.E.; Mazourek, M.; Gore, M.A.; Smart, C.D. A combined BSA-Seq and linkage map approach identifies genomic regions associated with Phytophthora root and crown rot resistance in squash. *Theor. Appl. Genet.* **2021**, *134*, 1015–1031. [CrossRef]
19. USDA. *World Agricultural Production*; Circular Series WAP 3–19; USDA: Washington, DC, USA, 2019.
20. Iqbal, M.A.; Iqbal, A. Overview on Sorghum for food, feed, forage and fodder: Opportunities and problems in Pakistan's perspectives. *Am. Eurasian. J. Agric. Environ. Sci.* **2015**, *15*, 1818–1826.
21. Chen, B.R.; Wang, C.Y.; Wang, G.P.; Zhu, Z.X.; Xu, N.; Shi, G.S.; Yu, M.; Wang, N.; Li, J.H.; Hou, J.M. Genome-wide association study for starch content and constitution in sorghum (*Sorghum bicolor* (L.) Moench). *J. Integr. Agric.* **2019**, *18*, 2446–2456. [CrossRef]
22. Zhang, L.Y.; Xu, X.J.; Ding, Y.Q.; Cao, N.; Gao, X.; Feng, Z.; Li, K.Y.; Cheng, B.; Zhou, L.B.; Ren, M.J.; et al. GWAS of grain color and tannin content in Chinese sorghum based on whole-genome sequencing. *Theor. Appl. Genet.* **2023**, *136*, 77. [CrossRef] [PubMed]
23. Yang, Q.; Zhang, D.F.; Xu, M.L. A sequential quantitative trait locus fine-mapping strategy using recombinant-derived progeny. *J. Integr. Plant. Biol.* **2012**, *54*, 228–237. [CrossRef] [PubMed]
24. Zuo, W.L.; Chao, Q.; Zhang, N.; Ye, J.R.; Tan, G.Q.; Li, B.L.; Xing, Y.X.; Zhang, B.Q.; Liu, H.J.; Fengler, K.A.; et al. A maize wall-associated kinase confers quantitative resistance to head smut. *Nat. Genet.* **2015**, *47*, 151–157. [CrossRef] [PubMed]
25. Barrero Farfan, I.D.; Bergsma, B.R.; Johal, G.; Tuinstra, M.R. A stable *dw3* allele in sorghum and a molecular marker to facilitate selection. *Crop. Sci.* **2012**, *52*, 2063–2069. [CrossRef]
26. Mendu, L.; Jalathge, G.; Dhillon, K.K.; Singh, N.P.; Balasubramanian, V.K.; Fewou, R.; Gitz, D.C.; Chen, J.; Xin, Z.; Mendu, V. Mutation in the endo-β-1,4-glucanase (KORRIGAN) is responsible for thick leaf phenotype in sorghum. *Plants* **2022**, *11*, 3531. [CrossRef]
27. De Riseis, S.; Chen, J.; Xin, Z.; Harmon, F.G. Sorghum bicolor INDETERMINATE1 is a conserved primary regulator of flowering. *Front. Plant. Sci.* **2023**, *14*, 1304822. [CrossRef]
28. Mewa, D.B.; Lee, S.; Liao, C.J.; Adeyanju, A.; Helm, M.; Lisch, D.; Mengiste, T. Anthracnose Resistance GENE2 confers fungal resistance in sorghum. *Plant. J.* **2023**, *113*, 308–326. [CrossRef]
29. Quinby, J.R. *Sorghum Improvement and the Genetics of Growth*; Texas A&M University Press: College Station, TX, USA, 1974.
30. Shinoda. *Baijiu—The Introduction of Sorghum*; Taiwan Elite Publishing Co., Ltd.: Taipei, Taiwan, 1958.
31. Yang, L.; Wang, Q.; Guo, X.K.; Guo, R.; Shao, Q.; Liu, Q.S. Effect of tannin content of sorghum on bacterial community in fermented grains of Fen-flavor Daqu Baijiu. *China Brew.* **2020**, *39*, 6.
32. Cui, Z.H.; Luo, J.H.; Qi, C.H.Y.; Ruan, Y.Y.; Li, J.; Zhang, A.; Yang, X.H.; He, Y. Genome-wide association study (GWAS) reveals the genetic architecture of four husk traits in maize. *BMC Genom.* **2016**, *17*, 946–960. [CrossRef]
33. Upadhyay, H.D.; Pundir, R.P.S.; Dwivedi, S.L.; Gowda, C.L.L.; Reddy, V.G.; Singh, S. Developing a mini core collection of sorghum for diversified utilization of germplasm. *Crop. Sci.* **2009**, *49*, 1769–1780. [CrossRef]
34. Zou, G.H.; Yan, S.; Zhai, G.W.; Zhang, Z.P.; Zou, J.Q.; Tao, Y.Z. Genetic variability and correlation of stalk yield related traits and sugar concentration of stalk juice in a sweet sorghum (*Sorghum bicolor* L. Moench) population. *Australian. J. Crop. Sci.* **2011**, *5*, 1232–1238.
35. Mace, E.S.; Buhariwalla, K.K.; Buhariwalla, H.K.; Crouch, J.H. A high-throughput DNA extraction protocol for tropical molecular breeding programs. *Plant. Mol. Biol. Rep.* **2003**, *21*, 459–460. [CrossRef]
36. Sun, X.W.; Liu, D.Y.; Zhang, X.F.; Li, W.B.; Liu, H.; Hong, W.G.; Jiang, C.B.; Guan, N.; Ma, C.X.; Zeng, H.P.; et al. SLAF-seq: An efficient method of large-scale De novo SNP discovery and genotyping using high-throughput sequencing. *PLoS ONE* **2013**, *8*, e58700. [CrossRef] [PubMed]
37. Kent, W.J. BLAT—The BLAST-like alignment tool. *Genome. Res.* **2002**, *12*, 656–664. [PubMed]
38. Paterson, A.H.; Bowers, J.E.; Bruggmann, R.; Dubchak, I.; Grimwood, J.; Gundlach, H.; Haberer, G.; Hellsten, U.; Mitros, T.; Poliakov, A.; et al. The Sorghum bicolor genome and the diversification of grasses. *Nature* **2009**, *457*, 551–556. [CrossRef]

39. Li, H.; Handsaker, B.; Wysoker, A.; Fennell, T.; Ruan, J.; Homer, N.; Marth, G.; Abecasis, G.; Durbin, R. The sequence alignment/map format and SAMtools. *Bioinformatics* **2009**, *25*, 2078–2079. [CrossRef]
40. Roberts, A.; McMillan, L.; Wang, W.; Parker, J.; Rusyn, I.; Threadgill, D. Inferring missing genotypes in large SNP panels using fast nearest-neighbor searches over sliding windows. *Bioinformatics* **2007**, *23*, 401–407. [CrossRef]

International Journal of
Molecular Sciences

MDPI

Article

Transcriptomic and Metabolomic Profiling Reveals the Variations in Carbohydrate Metabolism between Two Blueberry Cultivars

Haiyan Yang [1], Zhiwen Wei [2], Yaqiong Wu [1], Chunhong Zhang [1], Lianfei Lyu [1], Wenlong Wu [1,*] and Weilin Li [2,*]

1 Jiangsu Key Laboratory for the Research and Utilization of Plant Resources, Institute of Botany, Jiangsu Province and Chinese Academy of Sciences (Nanjing Botanical Garden Mem. Sun Yat-Sen), Nanjing 210014, China; haiyanyang@cnbg.net (H.Y.); ya_qiong@126.com (Y.W.); chzhang@cnbg.net (C.Z.); njbglq@163.com (L.L.)

2 Co-Innovation Center for Sustainable Forestry in Southern China, College of Forestry, Nanjing Forestry University, Nanjing 210037, China; wzw0709wzw@163.com

* Correspondence: wuwenlong@jib.ac.cn (W.W.); wlli@njfu.edu.cn (W.L.); Tel.: +86-25-8434-7063 (W.W.); +86-25-8542-8531 (W.L.)

Abstract: Blueberry is a high-quality fruit tree with significant nutritional and economic value, but the intricate mechanism of sugar accumulation in its fruit remains unclear. In this study, the ripe fruits of blueberry cultivars 'Anna' and 'Misty' were utilized as experimental materials, and physiological and multi-omics methodologies were applied to analyze the regulatory mechanisms of the difference in sugar content between them. The results demonstrated that the 'Anna' fruit was smaller and had less hardness than the 'Misty' fruit, as well as higher sugar content, antioxidant capability, and lower active substance content. A total of 7067 differentially expressed genes (DEGs) (3674 up-regulated and 3393 down-regulated) and 140 differentially abundant metabolites (DAMs) (82 up-regulated and 58 down-regulated) were identified between the fruits of the two cultivars. According to KEGG analysis, DEGs were primarily abundant in phenylpropanoid synthesis and hormone signal transduction pathways, whereas DAMs were primarily enriched in ascorbate and aldarate metabolism, phenylpropanoid biosynthesis, and the pentose phosphate pathway. A combined multi-omics study showed that 116 DEGs and 3 DAMs in starch and sucrose metabolism (48 DEGs and 1 DAM), glycolysis and gluconeogenesis (54 DEGs and 1 DAM), and the pentose phosphate pathway (14 DEGs and 1 DAM) were significantly enriched. These findings suggest that blueberries predominantly increase sugar accumulation by activating carbon metabolism network pathways. Moreover, we identified critical transcription factors linked to the sugar response. This study presents new understandings regarding the molecular mechanisms underlying blueberry sugar accumulation and will be helpful in improving blueberry fruit quality through breeding.

Keywords: blueberry; fruit quality; sugar accumulation; starch and sucrose metabolism; glycolysis/gluconeogenesis; pentose phosphate pathway

1. Introduction

Blueberry, a plant of the genus *Vaccinium* in the Ericaceae, is an economic fruit tree with significant nutritional and health care value [1,2]. Blueberries are usually divided into four types, namely, lowbush blueberry, highbush blueberry, semi-highbush blueberry, and rabbiteye blueberry [3]. Blueberry fruits are oblate or spherical in shape and are dark blue or scarlet when ripe. Blueberry fruit is soft and juicier. It has an excellent level of antioxidant active ingredients, including polyphenols, flavonoids, anthocyanins, and ellagic acid, which have unique effects such as preventing vascular aging, protecting eyesight, anti-inflammation, anti-oxidation, and anti-cancer [1,2]. Blueberry fruit can be consumed fresh or processed into a number of items, including sauce, dried fruit, fruit juice, and fruit wine [4].

The manual cultivation of blueberries originated in North America [5]. In the past twenty years, blueberry cultivation and consumption have spread fast from North America to the rest of the world. Blueberry has been booming in China in recent years, and it is now readily available throughout the country. The farmed area is extensive in the Liaoning, Shandong, Jiangsu, and Yunnan provinces, resulting in significant economic benefits [6]. People in China are becoming more aware of the nutritional value of blueberries as the economy improves and consumer demand increases [7]. As a result, in order to better adapt to the needs of consumers, cultivating excellent blueberry varieties and improving the quality of blueberry fruits have become the focus of blueberry development and research in China [6,7].

Excellent fruit quality is a key factor in ensuring fruit commerciality and attracting consumers [8,9]. Fruit quality is defined primarily by the external appearance and internal quality of the fruit. The external quality is mostly determined by its size, hardness, and color, while the interior quality is determined by sugar, organic acids, aroma, and antioxidant active components [10,11]. Sugar is the main component of fruit quality and flavor, as well as the primary raw material for vitamin, aromatic compound, and pigment synthesis [11,12]. Recent studies have demonstrated that sugars can be employed as signaling molecules to participate in the metabolic processes of hormones, anthocyanins, nitrogen, and other substances [13–15]. As a result, in-depth research into the carbohydrate accumulation mechanism in the fruit is the foundation of the quality formation theory, and it is critical to increasing the quality of blueberry fruit.

The sugars in the mature fruits are mostly sucrose, fructose, and glucose [16,17]. Various investigations have discovered that starch accumulation and sugar metabolism in fruits are closely related to enzyme activity, but the mechanisms of sugar accumulation and the regulation of important enzymes vary with plant species [17]. Carbohydrate accumulation and metabolism have been studied in a variety of fruits, such as strawberry, citrus, pear, apricot, and pitaya [18–22]. Starch is the most important storage polysaccharide in plants [17]. The accumulation of starch maintains adequate carbon source reserves, while the decomposition products produced by amylase are utilized for sucrose synthesis [23]. Sucrose is the most commonly used form of assimilation product transit from "source" to "sink" in most plants [24]. Sucrose is a fruit quality component as well as a metabolic regulator and a signaling factor that either activates or inhibits the transcription of specific genes [24]. In plants, the enzymes closely related to sucrose metabolism include sucrose phosphate synthase (SPS), sucrose synthase (SUS), and invertase [25]. SPS is mainly responsible for sucrose biosynthesis; SUS catalyzes the reversible conversion of sucrose and fructose; and invertase is in charge of sucrose hydrolysis [20,22,26]. Wang et al. [26] discovered that increased SPS and SUS gene expression resulted in a considerable accumulation of sucrose in squash. There was a positive correlation between plum sucrose content and ChSPS1 activity [20]. Zhang et al. [22] reported that acid invertase was engaged in the regulation of fructose formation in pitaya through a transcriptome investigation.

Currently, research on carbohydrate metabolism in blueberry fruits mainly focuses on the variations in sugar content and related enzyme activity during fruit development and rarely involves the molecular mechanism [16,27]. Therefore, the understanding of the mechanism of carbohydrate accumulation in different cultivars of blueberry fruits is still in its infancy. Comprehensive analysis of transcriptomics and metabolomics is vital for understanding the complex regulatory network of carbohydrate metabolism in fruits. In recent years, multi-omics integration has been used to better understand the signaling pathways and mechanisms of sugar accumulation in fruits. Yang et al. [28] investigated the main causes of variations in sweet cherry quality at different developmental stages, including sugar, organic acids, and flavonoids, through the combination of transcriptome and metabolomics. Gao et al. [29] comprehensively analyzed the molecular mechanism of the variation in sugar accumulation at different developmental stages of the two pineapple cultivars through transcriptome and metabolomics. However, the mechanisms underlying sugar accumulation in blueberry fruits have not been well studied. Based on a comprehensive study of fruit quality, an association analysis of transcriptomics and metabolomics was performed to investigate the gene function and gene regulatory network associated with carbohydrate metabolism in different blueberry cultivars. In the present work, we explored the response mechanisms of different genes and metabolites in carbohydrate metabolism in different blueberry cultivars and created a carbohydrate metabolism regulation network. The findings will help to deepen the understanding of the molecular regulatory mechanisms of blueberry fruit sugar metabolism and facilitate the molecular breeding of high-sugar blueberry cultivars.

2. Results

2.1. Fruit Appearance, Antioxidant System, Bioactive Compounds and Flavor Indicators

Sugars are the primary components that influence the sweetness and taste of blueberry fruit. Blueberries vary in their sugar composition and quantity. In order to understand the accumulation mechanism of sugar substances in blueberry fruits, in this experiment, two blueberry cultivars, 'Anna' and 'Misty' (Figure 1), were selected, with obvious differences in sugar content, and the differences in appearance and quality of these two cultivars were systematically studied. In this study, the single fruit weight of 'Misty' was much greater than that of 'Anna', which was 71.2% heavier than that of 'Anna' (Figure 2A). Furthermore, the transverse diameter was 14.87% larger than 'Anna', while the vertical diameter was 10.96% larger (Figure 2B,C).

Excessive reactive oxygen species (ROS) production during ripening and senescence can cause oxidative damage to fruits. To maintain sufficient ROS homeostasis, plants produce a variety of free radical scavengers, primarily composed of antioxidant compounds or antioxidant enzymes. In the present study, the generation rate of $O_2^{\cdot-}$ and H_2O_2 content in the fruits of the two cultivars did not differ dramatically. As compared to 'Misty' fruit, 'Anna' fruit had higher SOD and POD activities, as well as total antioxidant capacity (T-AOC), which could be attributed to its higher MDA level (Figure 2D–I).

Firmness is the main indicator used to assess fruit softening, and polyphenols, anthocyanins, flavonoids, and ellagic acid are the main active components of blueberry fruit. The study found that the hardness and the main active components concentrations of 'Anna' fruits were obviously lower than those of 'Misty' fruits, but the soluble solids content and solid-to-acid ratio of 'Anna' fruit were considerably greater, by 49.3% and 72.8%, respectively (Figure 2J–Q). In addition, 'Anna' fruit had a lower total acid content than 'Misty' fruit. The concentrations of sucrose, glucose, and fructose in 'Anna' fruit were considerably greater than those in 'Misty' fruit, which were 49.65%, 62.69%, and 33.16% higher, respectively (Figure 2R–T). These results indicated that the 'Anna' fruit has a sweet taste and good flavor, but the content of the functional components is slightly lower.

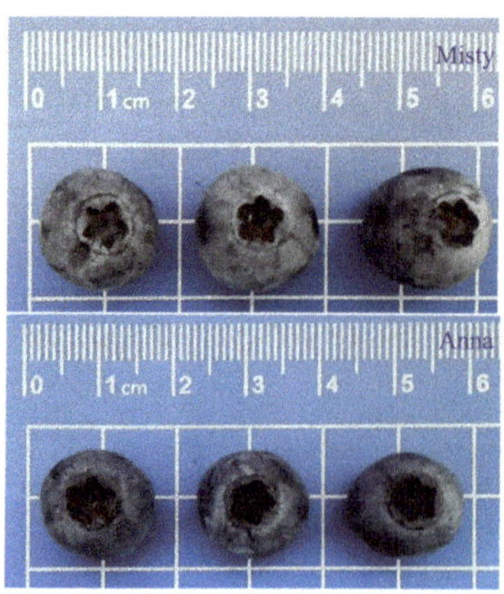

Figure 1. Fruit appearance of 'Anna' and 'Misty' blueberry cultivars.

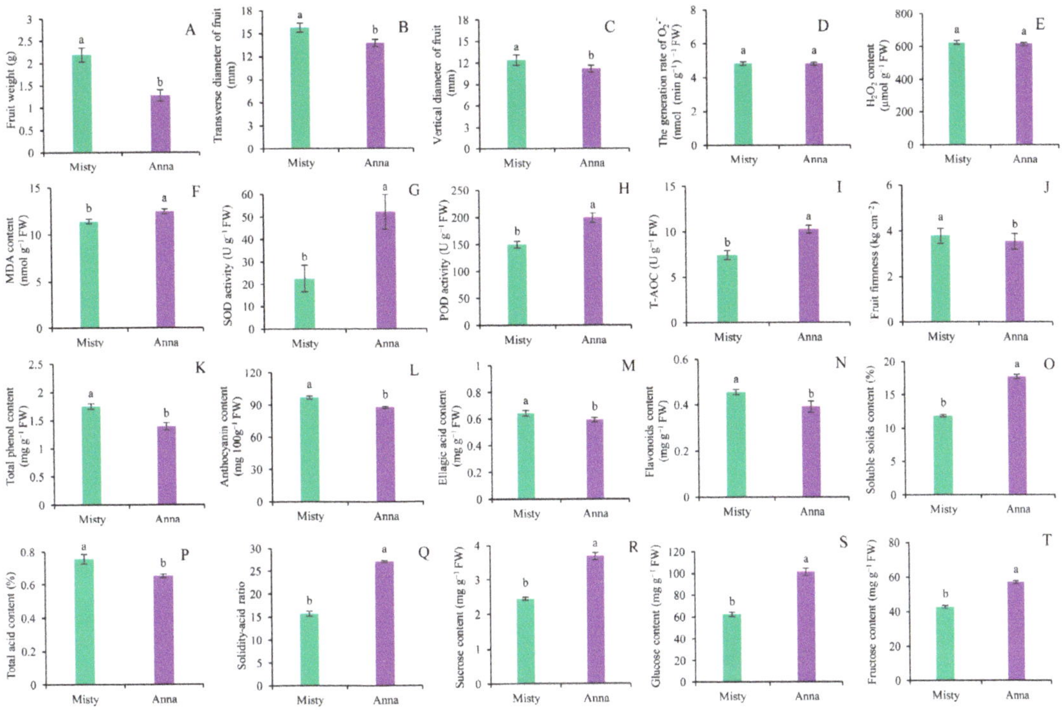

Figure 2. Antioxidant system, bioactive substances, and flavor indexes of 'Anna' and 'Misty' blueberry cultivars. (**A**) fruit weight, (**B**) transverse diameter, (**C**) vertical diameter, (**D**) the generation rate of $O_2^{\cdot-}$, (**E**) H_2O_2 content, (**F**) MDA content, (**G**) SOD activity, (**H**) POD activity, (**I**) T-AOC, (**J**) fruit firmness, (**K**) total phenol content, (**L**) anthocyanin content, (**M**) ellagic acid content, (**N**) flavonoids content, (**O**) soluble solids content, (**P**) total acid content, (**Q**) solidity-acid ratio, (**R**) sucrose content, (**S**) glucose content, and (**T**) fructose content were measured in blueberry fruits. Columns with different letters indicate significant differences between 'Anna' and 'Misty' ($p < 0.05$).

2.2. Transcriptomic Analysis

Transcriptome sequencing was used to investigate changes in gene expression among blueberry fruits of different cultivars. The sequencing data quality was high, with 42.9–46.1 million clean reads and 6.7–6.72 G clean bases acquired; Q30 was between 92.47 and 93.63%, and GC content was between 46.27 and 46.77%. The total mapped reads were greater than 81%, indicating that the sequencing data were sufficiently aligned for further analysis (Table S1). According to Figure 3A, the two samples co-expressed a total of 16,772 genes, with 2402 and 2294 genes expressed specifically in the 'Anna' and 'Misty' groups of samples, respectively.

With 'Misty' fruit as the control group and $\log_2$ (fold change) ≥ 1 and padj ≤ 0.05 genetic variations for the filter screen, 7067 differentially expressed genes (DEGs) were identified, of which 3674 DEGs were up-regulated and 3393 DEGs were down-regulated. They made up 51.99% and 48.01% of the total (Figure 3B,C). A hierarchical clustering analysis of DEGs indicated considerable variations between the two cultivars in gene expression patterns (Figure 3D). With strong repeatability, the three replicates of each group clustered into one class (Figure 3D).

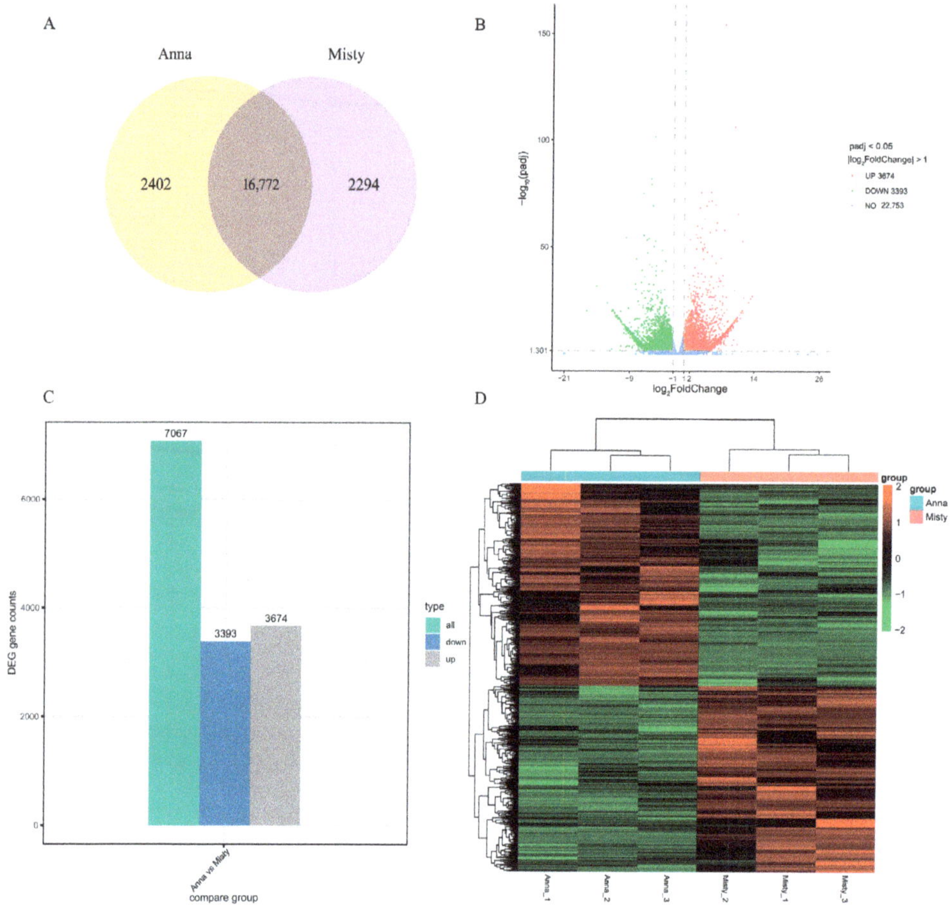

Figure 3. Transcriptomic analysis of the 'Anna' and 'Misty' blueberry cultivars. (**A**) Venn diagram of all identified genes. (**B**) Volcano plot of all identified genes; red represents up-regulated genes, green represents down-regulated genes, and blue represents genes that did not change significantly. (**C**) The number of DEGs. (**D**) Heatmap of all identified genes.

2.3. GO and KEGG Enrichment Analysis

As shown in Figure 4A, the top five significantly enriched GO terms among the up-regulated DEGs were 'ADP binding', 'UDP-glycosyltransferase activity', 'pyridine-containing compound metabolic process', 'carbohydrate catabolic process', and 'nicotinamide nucleotide biosynthetic process'. Among the down-regulated DEGs, 'ATPase activity', 'active transmembrane transporter activity', 'ATPase activity, coupled to transmembrane movement of substances', 'ATPase activity, coupled to movement of substances', and 'ATPase activity, coupled' were the top five highly enriched GO terms (Figure 4B).

According to KEGG enrichment analysis, the top five significantly enriched pathways among the up-regulated DEGs were 'glycolysis/gluconeogenesis', 'plant hormone signal transduction', 'zeatin biosynthesis', 'MAPK signaling pathway-plant', and 'fatty acid degradation' (Figure 4C). Among the down-regulated DEGs, the top five significantly enriched pathways were 'fatty acid elongation', 'ABC transporters', 'phenylpropanoid biosynthesis', 'starch and sucrose metabolism', and 'the phosphatidylinositol signaling system' (Figure 4D).

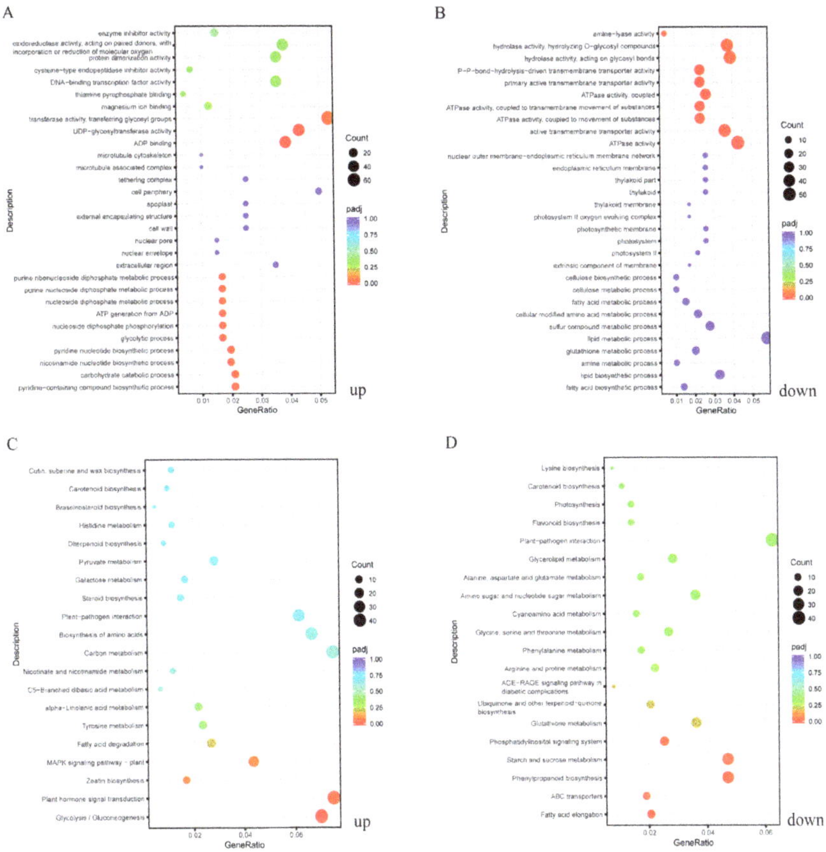

Figure 4. GO enrichment analysis of the up-regulated DEGs (**A**) and down-regulated DEGs (**B**). KEGG enrichment analysis of the up-regulated DEGs (**C**) and down-regulated DEGs (**D**).

2.4. DEGs in Phenylpropanoid Biosynthesis and Plant Hormone Signal Transduction Pathways

The phenylpropanoid biosynthesis pathway is a major source of phenolic compound accumulation in plants. Based on the KEGG enrichment analysis, this pathway was considerably enriched, and 45 genes were differentially expressed, with 15 genes significantly elevated and 30 genes greatly down-regulated (Figure S1). Among them, two 4-coumarate-CoA ligase (4CL) genes, four beta-glucosidase (BG1) genes, one caffeate O-methyltransferase (COMT) gene, one shikimate o-hydroxycinnamoyltransferase (HCT) gene, one cytochrome P450 monooxygenase 98 (CYP98) gene, four peroxidase (POD) genes, and two cinnamyl-alcohol dehydrogenase (CAD) genes were highly up-regulated. Hormone signaling is closely linked to sugar metabolism. In the plant hormone signal transduction pathway, 70 DEGs were identified (45 up-regulated and 25 down-regulated). The auxin and brassinosteroid hormone signal transduction pathways were highly enriched in 21 (14 up-regulated and 7 down-regulated) and 17 (11 up-regulated and 6 down-regulated) DEGs, respectively (Figure S2). In the auxin signal transduction pathway, one auxin influx carrier (AUX1) gene, four auxin-responsive protein IAA (AUX/IAA) genes, one auxin response factor (ARF) gene, two auxin-responsive GH3 gene family (GH3) genes, and six SAUR family protein (SAUR) genes were highly up-regulated. In the brassinosteroid signal transduction pathway, four brassinosteroid insensitive 1-associated receptor kinase 1 (BAK1) genes, one protein brassinosteroid insensitive 1 (BRI) gene, one protein brassinosteroid insensitive 2 (BIN1) gene, four xyloglucosyl transferase (TCH4) genes, and one cyclin D3 (CYCD3) gene were considerably up-regulated. These up-regulated genes may play a major role in sugar metabolism.

2.5. Transcription Factors (TFs)

Transcription factors (TFs) can either promote or suppress target gene transcription. To further understand the differential expression of TFs between the fruits of these two cultivars, TF families with a DEG count in the top 20 were identified (Table S2). As shown in Figure 5, the MYB family had the most DEGs (34 genes), with 18 of them being up-regulated and 16 being down-regulated. The number of DEGs that were up-regulated outnumbered those that were down-regulated in 13 transcription factor families. There were more than 15 DEGs in numerous transcription factor families, including MYB, AP2/ERF, bHLH, WRKY, B3, and MDAS-box. Within these six TF families, there were 18, 15, 10, 10, 13, and 7 significantly up-regulated genes, respectively. Among these up-regulated genes, a total of ten genes had $\log_2$ (fold change) values greater than six, including two MYB genes (gene-Vadar_022042 and gene-Vadar_028064), one AP2/ERF gene (gene-Vadar_012138), three bHLH genes (gene-Vadar_024786, gene-Vadar_021192, and gene-Vadar_000396), one WRKY gene (gene-Vadar_014305), two B3 genes (novel.480 and gene-Vadar_002518), and one MADS-box gene (gene-Vadar_003970) (Table S3).

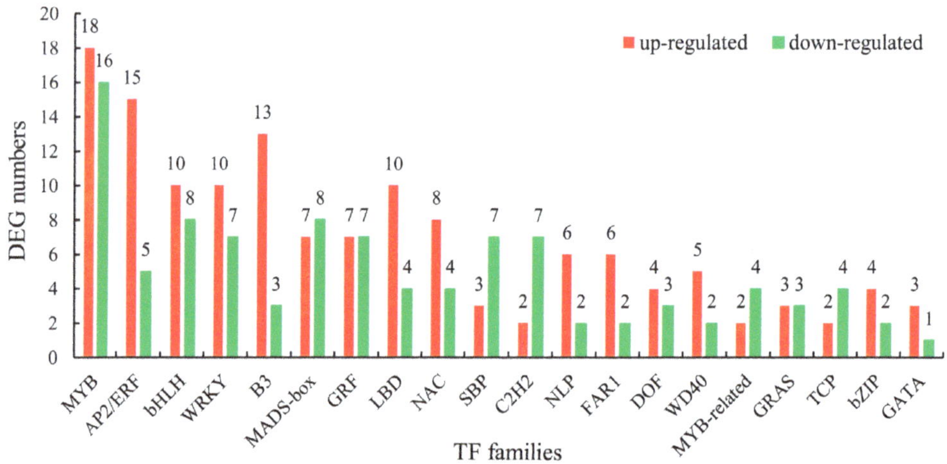

Figure 5. Number of DEGs involved in the top 20 TF families.

2.6. Metabolic Profiling

In all, 845 metabolites were found in blueberry fruits, with 536 detected in the positive ion mode and 309 found in the negative ion mode (Table S4). These metabolites were further annotated and functionally categorized employing the HMDB, KEGG, and LIPID MAPS databases (Figure 6). According to the HMDB annotation, the metabolites were mostly lipids and lipid-like molecules, organic acids and derivatives, phenylpropanoids and polyketides, organoheterocyclic compounds, organic oxygen compounds, and benzenoids. The KEGG pathway annotation found that in the positive ion mode, the metabolites were primarily abundant in amino acid metabolism, biosynthesis of other secondary metabolites, and lipid metabolism pathways. In the negative ion mode, the metabolites were mainly enriched in the biosynthesis of other secondary metabolites, carbohydrate metabolism, and amino acid metabolism. The annotation of LIPID MAPS indicated that the metabolites were primarily flavonoids, isoprenoids, and fatty acids and conjugates.

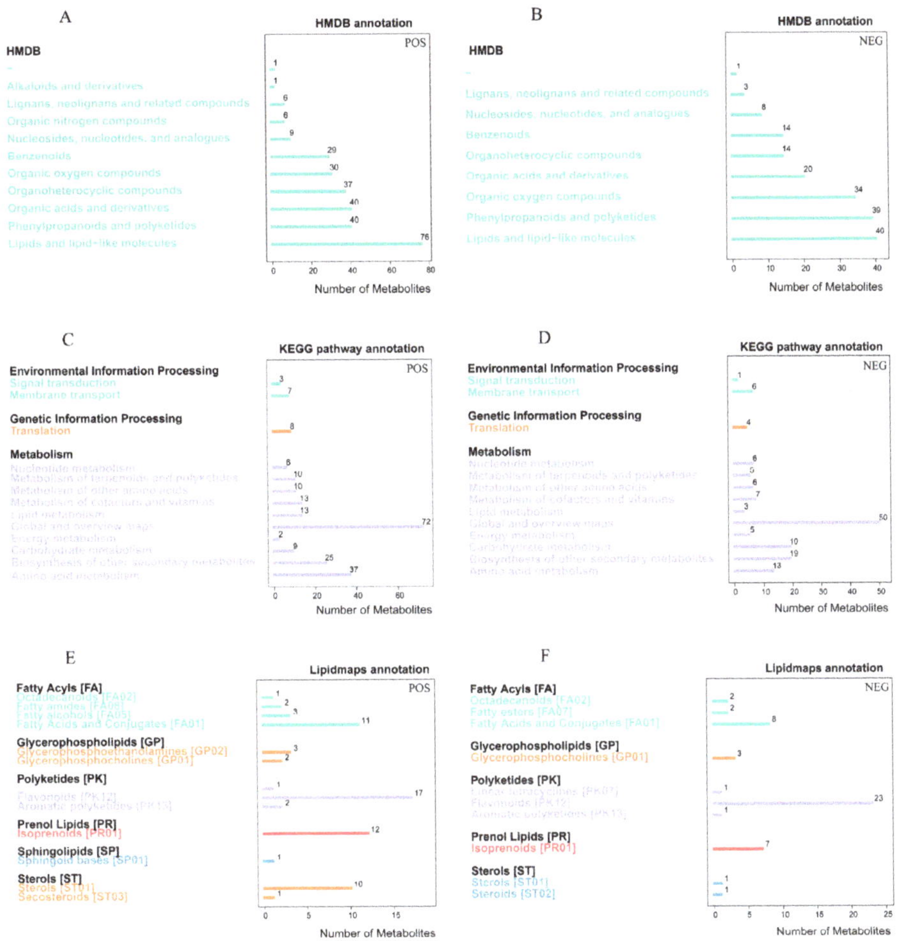

Figure 6. Metabolite profiling and classification using the HMDB, KEGG, and LIPID MAPS databases. The HMDB annotation of the metabolites in the positive (**A**) and negative (**B**) ion modes. The KEGG pathway annotation of the metabolites in the positive (**C**) and negative (**D**) ion modes. The LIPID MAPS annotation of the metabolites in the positive (**E**) and negative (**F**) ion modes.

2.7. Metabolite Multivariate Statistical Analysis

Principal component analysis (PCA) was used on the two sample groups to differentiate between changes in metabolite accumulation in the fruits. As shown in Figure S3A,B, the PC1 can explain 35.62% and 38.16% of the total variation in the positive and negative ion modes, respectively, and the PC2 can explain 26.89% and 27.83% of the total variance, respectively. There is a clear differentiation between the samples (Figure S3). PLS-DA score plot analysis showed that the two samples were clearly separated and clustered inside the group (Figure S3C,D). In the positive and negative ion modes, the R2Y (cum) values of the PLS-DA model were 0.99 and 1, respectively, demonstrating that the model could effectively display metabolite differences between the sample groups. Furthermore, 200 response permutation tests were used to validate the accuracy of the PLS-DA model, and the findings revealed that the constructed model was stable and dependable, and that it could be used for subsequent differential metabolite screening (Figure S3E,F).

2.8. DAM Screening and Analysis

In this study, VIP > 1.0, FC > 1.5, or FC < 0.667 and p-value < 0.05 were used as screening criteria to screen DAMs. In the positive ion mode, 70 DAMs were screened, 44 of which were up-regulated and 26 of which were down-regulated. A total of 70 DAMs were screened in the negative ion mode, with 38 DAMs up-regulated and 32 DAMs down-regulated (Figure S4A,B). The results of the hierarchical clustering analysis showed that there was a substantial difference between the two sample groups and that the DAMs within the same group were clustered together. The similarity within groups was high, and these DAMs could be used for group difference analysis. Correlation analysis of the top 20 DAMs showed that, with the exception of di-C,C-pentosyl-apigenin, betulin, 4-oxododecanedioic acid, Bruceine D, stigmasterol, and combretastatin A4, most DAMs were strongly positively associated with each other in the positive ion mode. N6-isopentenyladenosine was found to be positively associated to 4-chloro-5-morpholino-2-quinoxalin-2-ylpyridazin-3(2H)-one and N-[(4-hydroxy-3-methoxyphenyl)methyl]-8-methylnonanamide. In the negative ion mode, ixoside, geniposidic acid, benzyl 6-O-beta-D-glucopyranosyl-beta-D-glucopyranoside, and 8-epiloganic acid were positively correlated with each other. Stevioside was only positively correlated with hydroxysafflor yellow A (Figure S4C,D).

2.9. KEGG Enrichment Analysis of DAMs

Pathways hosting DAMs in blueberry fruits were analyzed by the KEGG database. The study found that DAMs were primarily enriched in 17 metabolic pathways in the positive ion mode, including ascorbate and aldarate metabolism, phenylpropanoid biosynthesis, the pentose phosphate pathway, galactose metabolism, and starch and sucrose metabolism, which were the top five significantly enriched metabolic pathways (Figure 7A). The DAMs were primarily enriched in 10 metabolic pathways in the negative ion mode, including valine, leucine, and isoleucine biosynthesis, terpenoid backbone biosynthesis, monoterpenoid biosynthesis, glycolysis/gluconeogenesis, and pyruvate metabolism, which were the top five significantly enriched metabolic pathways (Figure 7B).

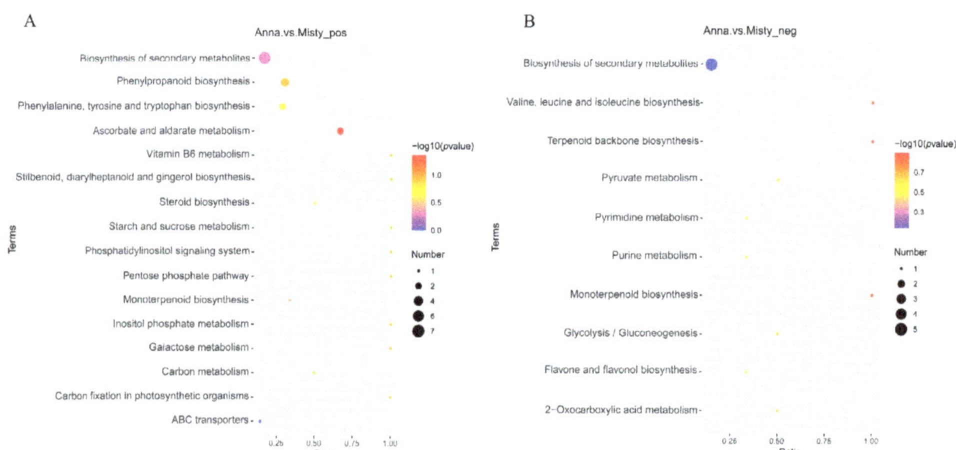

Figure 7. KEGG enrichment analysis of DAMs in the positive (**A**) and negative (**B**) ion modes.

2.10. Transcriptomics and Metabolomics Correlation Analysis

In order to gain a deeper comprehension of the differences in carbohydrate accumulation between the two blueberry cultivars, we performed a combined transcriptome and metabolomic analysis on three pathways (starch and sucrose metabolism, glycolysis/gluconeogenesis, and pentose phosphate pathway) in the carbon metabolism network (Figure 8).

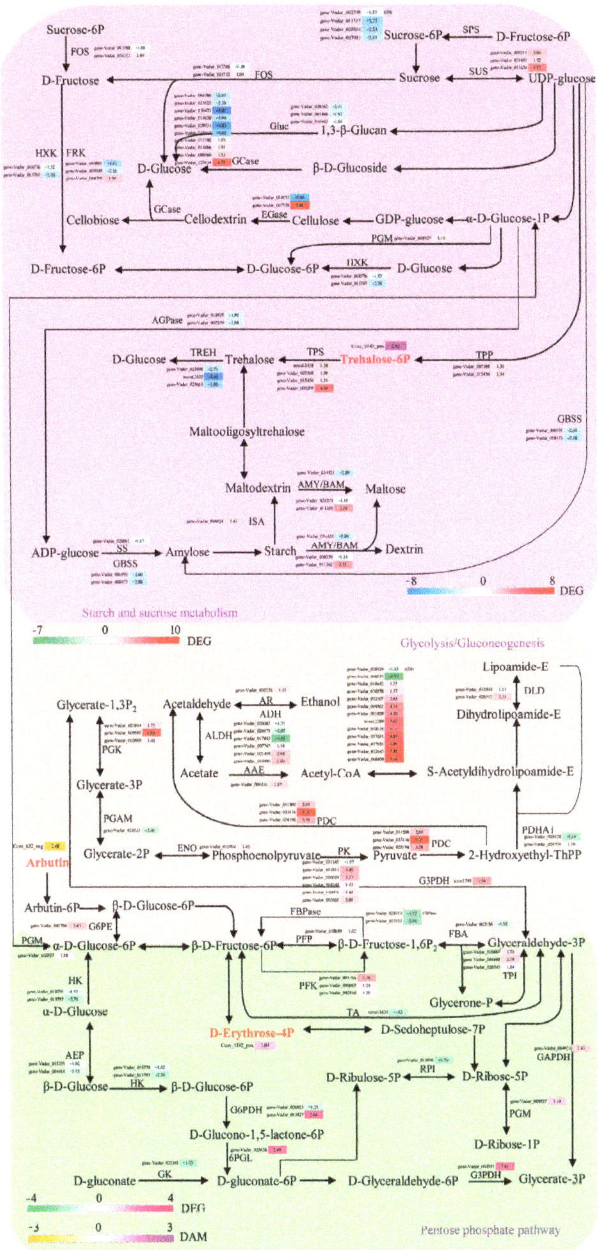

Figure 8. DAMs and DEGs association studies in starch and sucrose metabolism, glycolysis/gluconeogenesis, and the pentose phosphate pathway. The scales were generated using $\log_2$ (fold change) values. The expression levels of DEGs and DAMs are shown from green to red and yellow to purple (low to high) in the comparison of 'Anna' vs. 'Misty', respectively. HXK, hexokinase; FRK, fructokinase; FOS, beta-fructofuranosidase; Gluc, glucan endo-1,3-beta-glucosidase; GCase, beta-glucosidase; EGase, endoglucanase; SPS, sucrose-phosphate synthase; SUS, sucrose synthase; PGM, phosphoglucomutase; AGPase, ADP-glucose pyrophosphorylase; TPS, trehalose 6-phosphate synthase;

TPP, trehalose 6-phosphate phosphatase; TREH, alpha,alpha-trehalase; GBSS, granule-bound starch synthase; AMY, alpha-amylase; BAM, beta-amylase; ISA, isoamylase; SS, starch synthase; HK, hexokinase; AEP, aldose 1-epimerase; G6PE, glucose-6-phosphate 1-epimerase; FBPase, fructose-1,6-bisphosphatase; PFK, ATP-dependent phosphofructokinase; PFP, pyrophosphate-dependent fructose-6-phosphate phosphotransferase; FBA, fructose-bisphosphate aldolase; TPI, triosephosphate isomerase; G3PDH, glyceraldehyde 3-phosphate dehydrogenase; GAPDH, glyceraldehyde-3-phosphate dehydrogenase; PGK, phosphoglycerate kinase; PGAM, 2,3-bisphosphoglycerate-dependent phosphoglycerate mutase; ENO, enolase; PK, pyruvate kinase; PDHA1, pyruvate dehydrogenase E1 component subunit alpha; PDC, pyruvate decarboxylase; DLD, dihydrolipoyl dehydrogenase; AAE, butyrate-CoA ligase; ALDH, aldehyde dehydrogenase; ADH, alcohol dehydrogenase; AR, aldose reductase; GK, gluconokinase; G6PDH, glucose-6-phosphate 1-dehydrogenase; 6PGL, 6-phosphogluconolactonase; TA, transaldolase; RPI, ribose 5-phosphate isomerase A.

The study found that, in the starch and sucrose metabolism pathway, there were 48 DEGs, i.e., two hexokinase (HXK) genes, three fructokinase (FRK) genes, two beta-fructofuranosidase (FOS) genes, three glucan endo-1,3-beta-glucosidase (Gluc) genes, ten beta-glucosidase (GCase) genes, two endoglucanase (EGase) genes, four sucrose-phosphate synthase (SPS) genes, three sucrose synthase (SUS) genes, one phosphoglucomutase (PGM) gene, two ADP-glucose pyrophosphorylase (AGPase) genes, two trehalose 6-phosphate phosphatase (TPP) genes, four trehalose 6-phosphate synthase (TPS) genes, three alpha-trehalase (TREH) genes, two granule-bound starch synthase (GBSS) genes, one alpha-amylase (AMY) gene, two beta-amylase (BAM) genes, one isoamylase (ISA) gene, and one starch synthase (SS) gene and one DAM (Trehalose 6-phosphate (Com_2145_pos)) were significantly enriched. In this pathway, 21 DEGs were substantially up-regulated.

In the glycolysis/gluconeogenesis pathway, 54 DEGs, i.e., one PGM gene, two hexokinase (HK) genes, two aldose 1-epimerase (AEP) genes, one glucose-6-phosphate 1-epimerase (G6PE) gene, two fructose-1,6-bisphosphatase (FBPase) genes, three 6-phosphofructokinase (PFK) genes, one diphosphate-dependent phosphofructokinase (PFP) gene, one fructose-bisphosphate aldolase (FBA) gene, three triosephosphate isomerase (TPI) genes, one glyceraldehyde 3-phosphate dehydrogenase (G3PDH) gene, one glyceraldehyde-3-phosphate dehydrogenase (GAPDH) gene, one phosphoglycerate kinase (PGK) gene, one 2,3-bisphosphoglycerate-dependent phosphoglycerate mutase (PGAM) gene, one enolase (ENO) gene, six pyruvate kinase (PK) genes, two pyruvate dehydrogenase E1 component subunit alpha (PDHA1) genes, three pyruvate decarboxylase (PDC) genes, two dihydrolipoyl dehydrogenase (DLD) genes, one acetate/butyrate-CoA ligase (AAE) gene, six aldehyde dehydrogenase (ALDH) genes, twelve alcohol dehydrogenase (ADH) genes, and one aldose reductase (AR) gene, and one DAM (arbutin (Com_632_neg)) showed synergistic changes. The majority of the DEGs in this pathway, particularly the ADH genes, were considerably up-regulated.

In the pentose phosphate pathway, 14 DEGs, i.e., one gluconokinase (GK) gene, two glucose-6-phosphate 1-dehydrogenase (G6PDH) genes, one 6-phosphogluconolactonase (6PGL) gene, one GAPDH gene, two FBPase genes, one PFP gene, three PFK genes, one transaldolase (TA) gene, one PGM gene, and one ribose 5-phosphate isomerase A (RPI) gene and one DAM (D-erythrose 4-phosphate (Com_1192_pos)) demonstrated synergistic alterations.

Among the DEGs in these three pathways, six had $\log_2$ (fold change) values larger than six, including one EGase gene (gene-Vadar_007558), one PGK gene (gene-Vadar_009565), one PDC gene (gene-Vadar_033116), and three ADH genes (gene-Vadar_017055, gene-Vadar_022642, and gene-Vadar_006809). These genes may be crucial for fruit sugar metabolism (Table S3). Furthermore, to corroborate the RNA-seq data, 12 DEGs from the starch and sucrose metabolism pathway and 12 DEGs from the glycolysis/gluconeogenesis pathway were selected for qRT-PCR analysis. The results of qRT-PCR were consistent with those of transcriptomics (Figure S5), indicating the high degree of accuracy of the transcriptome sequencing results.

3. Discussion

Blueberry quality has become increasingly important as people's living levels have improved in recent years [11]. Improving blueberry fruit quality has emerged as a key research avenue for adapting to market development and an essential objective of modern agricultural blueberry production [9,10]. The purpose of this study was to gain insight into the changes in genes and metabolites in ripe fruits of the high-sugar blueberry cultivar 'Anna' and the common blueberry cultivar 'Misty', as well as their link to fruit quality. This research might be used to choose molecular genetic targets for enhancing carbohydrate accumulation in blueberries.

Blueberry fruit quality refers to the total of several positive features that suit people's needs [10]. In this study, it was found that the size and firmness of the 'Anna' fruit were significantly lower than those of the 'Misty' fruit. Active compounds, such as polyphenols, anthocyanins, ellagic acid, and flavonoids, were substantially lower in 'Anna' fruit than in 'Misty' fruit. The appearance of fruits is an essential aspect of attracting customers to buy them [30]. Fruits with beautiful shapes have a higher market value, and most consumers prefer large fruit varieties. The hardness of fruits influences not only their storage properties but also their taste [31]. Antioxidant active substances with various health care functions are also important indicators of blueberry quality [1,2]. According to this study, the commercial performance of 'Anna' fruit was much lower than that of 'Misty' fruit, which limited the marketing and cultivation of this cultivar in China. The analysis of the expression levels of the genes involved in the phenylpropanoid biosynthesis pathway revealed that 45 genes were variously expressed, and the expression levels of the majority of genes were drastically reduced. The study concluded that the ability of 'Anna' fruit to synthesize secondary metabolites was significantly lower than that of 'Misty' fruit, which was the main reason for the lower content of active substances in this cultivar. In addition, the MDA content, SOD activity, POD activity, and T-AOC value of 'Anna' fruit were much higher than those of 'Misty' fruit. The higher antioxidant capability may be used to repair membrane lipid peroxidation damage and maintain the equilibrium of reactive oxygen species (ROS) metabolism inside the fruit [11]. This could be related to the lower levels of polyphenols, anthocyanins, ellagic acid, and flavonoids in 'Anna' fruit, as these active components can help plants resist free radical stress.

The sugar content and composition in fruits are primary indicators of fruit flavor [16]. Sugars found in ripe fruits are primarily sucrose, fructose, and glucose [22,26]. According to recent research, ripening 'O'Neal' blueberry fruits primarily accumulate fructose and glucose, followed by sucrose [27]. Similarly, soluble solids, sugar-acid ratio, fructose, glucose, and sucrose contents of 'Anna' fruits were found to be much higher than those of 'Misty' fruits, but titratable acid content was significantly lower. Sugar accumulation is caused not only by the "sink/source" relationship between photosynthetic organs and fruits [24], but also by the respiratory metabolism process, which promotes the improvement of fruit quality, the transformation of flavor substances, and the presentation of ripening quality [32–35].

To explore the variations in sugar accumulation in blueberry fruits, this study used combined transcriptome and metabolome analysis to identify three key metabolic pathways related to carbon metabolism: starch and sucrose metabolism, glycolysis/gluconeogenesis, and the pentose phosphate pathway. The metabolism of starch and sucrose is a complicated biochemical process that relies on the synergistic action of multiple enzymes [17,23,24]. In the starch and sucrose metabolism pathway, 48 DEGs and one DAM (trehalose 6-phosphate) were discovered. SPS is a crucial enzyme in sucrose production, whereas SUS is involved in reversible sucrose metabolism and primarily performs a catabolic role [20,26]. As compared to 'Misty' fruit, the expression of four SPS genes was considerably decreased, and three SUS genes were greatly promoted in 'Anna' fruit, which could promote sucrose breakdown and increase fructose and glucose levels. In plants, SS and GBSS are responsible for starch synthesis, while AMY and BAM are the main players in starch degradation [23]. Significant down-regulation of SS (gene-Vadar_020861) and GBSS (gene-Vadar_000393, gene-Vadar_008473) genes inhibited starch synthesis. The up-regulation of the BAM (gene-Vadar_011365) gene further promoted starch degradation. The down-regulation of AGPase (gene-Vadar_018937, gene-Vadar_002579) genes reduced the production of ADP-glucose, a substrate for starch synthesis [17]. In addition, the trehalose 6-phosphate (T6P) content of the 'Anna' fruit was substantially higher than that of the 'Misty' fruit. The considerable overexpression of the TPP genes (gene-Vadar_007308, gene-Vadar_015456) should be related to this. T6P functions as a signal substance for the sufficiency of carbon sources [36], as well as delaying leaf senescence and influencing fruit setting rate [37,38]. Therefore, its function in 'Anna' fruit deserves further investigation.

Sugar metabolism and respiratory metabolism in fruits are tightly connected. Fructose 6-phosphate (F6p), produced by starch and sucrose metabolism, enters the glycolysis/gluconeogenesis and the pentose phosphate pathway to produce energy and serves as an intermediary in other metabolic processes to ensure fruit development [39,40]. The study identified 54 and 14 DEGs and two DAMs (arbutin and D-erythrose 4-phosphate) from glycolysis/gluconeogenesis and the pentose phosphate pathway, respectively. The glycolysis/gluconeogenesis pathway is a set of processes that convert glucose to pyruvate while producing ATP [39]. Phosphofructokinase and pyruvate kinase (PK) are two key enzymes in this pathway. Phosphofructokinase catalyzes the interconversion of F6P to fructose-1,6-diphosphate (FDP). There are two kinds of phosphofructokinases in plants: diphosphate-dependent phosphofructokinase (PFP) and ATP-dependent phosphofructokinase (PFK) [40]. Transient *PbPFP1* gene transformation in pear fruit resulted in a considerable rise in fructose and sorbitol [35]. MdMYB108 in apples could bind to the MdPFK promoter and induce soluble sugar accumulation in the pulp [41]. PK is in charge of converting the high-energy phosphate group of phosphoenolpyruvate (PEP) to ADP in order to create ATP and pyruvate [42]. By modulating cellular metabolic flux and ATP synthesis, PK regulates several aspects of plant growth and development [34,43]. Qin et al. [43] found that the expression level of *EjPK* in high-sugar loquat varieties was significantly higher than that in low-sugar loquat varieties, suggesting its potential role in the regulation of sugar content in loquat fruits. Guan et al. [34] demonstrated that PK was involved in the natural deastringence process of persimmon at the late stage of ripening. In this study, one PFP, three PFK, and five PK genes were considerably up-regulated, indicating that 'Anna' fruit has a high glycolysis rate. The increased sugar content of 'Anna' fruit may be due to the differential expression of PFP, PFK, and PK genes.

In addition, one PDC gene (gene-Vadar_033116) and three ADH genes (gene-Vadar_017055, gene-Vadar_022642, and gene-Vadar_006809) were found to be significantly up-regulated between the two blueberry cultivars. PDC catalyzes pyruvate decarboxylation to acetaldehyde, whereas ADH mainly promotes acetaldehyde and ethanol interconversion [44]. 'Anna' fruits produce tiny amounts of ATP via the ethanol fermentation pathway, which converts NAD(P)H to NAD(P) and maintains the metabolic balance of the glycolysis/gluconeogenesis pathway [45,46]. Alcohols are one of the limiting factors for ester production, and previous research revealed that ADH is also involved in the ripening of plant fruits. ADH activity increased in the late stages of tomato and melon fruit development, regulating alcohol production and providing fruits a more mature flavor [32,33]. The upregulation of ADH genes may also contribute to the flavor development of 'Anna' fruit.

The primary physiological functions of the pentose phosphate pathway are to produce NADPH for reducing biosynthesis, pentose phosphate for nucleic acid metabolism, and some intermediates involved in amino acid and fatty acid synthesis [47–49]. Glucose-6-phosphate dehydrogenase (G6PDH) is a key rate-limiting enzyme that catalyzes the dehydrogenation of glucose-6-phosphate to gluconic acid 6-phosphate and NADPH [47]. The considerable upregulation of G6PDH (gene Varadar_013427) in our study may further increase NADPH synthesis in 'Anna' fruit. NADPH also functions as a cofactor of glutathione reductase, ensuring enough reduced glutathione content [49]. Because of the low quantity of active compounds, such as polyphenols and anthocyanins, in 'Anna' fruit, NADPH may have a key role in modulating oxidative stress in the fruit. Furthermore, the significantly enhanced D-erythrose 4-phosphate may stimulate fructose-6P synthesis, accelerating glycolysis and increasing sugar accumulation in 'Anna' fruits.

Sugar is not only an energy source and an important component of plant structural substances; it also acts as a signal molecule that regulates various metabolic activities in plants in conjunction with signals such as hormones and nitrogen, and it participates in the expression of related genes via signal transduction [13,15,50]. According to studies, auxin and BR are hormones with obvious links to sugar sensing and signaling. Glucose regulates the expression of genes responsible for auxin synthesis, auxin transport, and auxin signal transduction [51,52]. Sugar modulates *Arabidopsis* hypocotyl elongation along the BR signaling pathway [53]. Hao et al. [53] found that the *LecRK-VIII.2* gene regulated hypocotyl elongation by integrating the signaling cross-talk among BR, sucrose, glucose, and fructose in *Arabidopsis*. This gene played a pivotal role in mediating BZR1 target genes, which were co-regulated by BR and sugar. BR has a significant effect on sugar accumulation in fruits and increasing endogenous BR levels greatly increased sugar accumulation in grape and citrus fruits [54]. In this study, sugars in 'Anna' fruit were hypothesized to be critical signaling molecules that regulate auxin and brassinosteroid signaling. In tomato, *SlARF10* regulates fruit chlorophyll and sugar accumulation via regulating the transcriptional upregulation of *AGPase, SlGLK1, POR, CBP1*, and *CBP2* [55]. In Arabidopsis, BRI1 and BAK1 function genetically with G-protein subunits to regulate sugar-responsive growth and development [56]. In this study, there were 21 and 17 DEGs that were significantly enriched in the auxin and the brassinosteroid signal transduction pathways, respectively. One ARF gene (gene-Vadar_000924) was highly up-regulated in the auxin signal transduction pathway. Four BAK1 genes (gene-Vadar_023654, gene-Vadar_031910, gene-Vadar_012080, and gene-Vadar_028865) and one BRI gene (gene-Vadar_021721) were considerably up-regulated in the brassinosteroid signal transduction pathway. These genes were considered to be able to promote sugar accumulation in 'Anna' fruit.

Sugar synthesis, transport, and distribution occur throughout the plant growth and development process. The complicated sugar metabolism and transduction regulatory network integrates various internal and external factors to maintain the growth and development process [17,57–61]. A variety of transcription factors have been implicated in sugar response process regulation. *MdMYB305* has been shown to increase the sugar content of red-fleshed apple fruit by modulating the expression of the sugar-related genes *MdCWI1, MdVGT3*, and *MdTMT2* [57]. *CitERF16* positively regulates the expression of *CitSWEET11d*, thereby facilitating sucrose accumulation in citrus fruit [58]. *MdbHLH3* promotes soluble sugar accumulation in apples by activating *MdPFP* expression [59]. *HpWRKY3* is linked to sugar accumulation in pitaya fruit by promoting the expression of *HpINV2* and *HpSuS1* [60]. Co-transformation of *MuMADS1* and *MaOFP1* in tomato ovate mutants results in a considerable rise in sugar content [61]. In this study, the transcription factor families with DEG counts in the top 20 were identified. Among the transcription factor families MYB, AP2/ERF, bHLH, WRKY, and MDAS-box, there were 18, 15, 10, 10, and 7 significantly up-regulated genes, respectively. Among them, two MYB genes (gene-Vadar_022042 and gene-Vadar_028064), one AP2/ERF gene (gene-Vadar_012138), three bHLH genes (gene-Vadar_024786, gene-Vadar_021192, and gene-Vadar_000396), one WRKY gene (gene-Vadar_014305), and one MADS-box gene (gene-Vadar_003970), which had $\log_2$ (fold change) values greater than six, may be closely related to the sugar response of blueberry fruits. These TFs may interact with genes involved in sugar metabolism, regulating the accumulation of sugars in blueberry fruits.

4. Materials and Methods

4.1. Plant Materials

The fruits of 4-year-old blueberry 'Anna' and 'Misty' cultivars were used as experimental materials, which were collected from Lishui Baima Industrial Park (119°11′ E, 31°36′ N), Nanjing, Jiangsu Province, China. The cultivation and management conditions of the chosen plants were consistent. After the fruits had fully matured, three trees from each cultivar were chosen, and fruits without any signs of disease were collected randomly. The fruits were snap-frozen in liquid nitrogen and transported to the laboratory, where they were maintained in a refrigerator at −80 °C and used for physiological and biochemical assessments, as well as transcriptome and metabolomic analysis.

4.2. Fruit Appearance Indicators and Firmness

The fruit weight was assayed with an electronic balance. A digital Vernier caliper and a fruit firmness meter (Catalog No. 9300 (KM-5), Kyoto, Japan) were used to determine the fruit's transverse and longitudinal diameters and firmness, respectively.

4.3. Antioxidant-Related Parameters

The hydroxylamine chloride protocol was applied to determine the generation rate of $O_2^{\cdot -}$ [62]. The concentration of malondialdehyde (MDA) in fruit was determined by monitoring thiobarbituric acid reactive compounds [63]. Hydrogen peroxide (H_2O_2) and total antioxidant capacity (T-AOC) testing kits were used to determine the H_2O_2 level and T-AOC (Nanjing Jiancheng Bioengineering Institute, Nanjing, China). Fresh fruit (1 g) was homogenized in 10 mL of filtered water and centrifuged at 8000 rpm for 10 min at 4 °C. The supernatant was used to determine the H_2O_2 concentration. First, 0.1 mL of supernatant was added to 1 mL of titanium sulfate solution. After reaction at 37 °C for 1 min, 1 mL of the ammonia was added to the mixture. The absorbance was measured at 405 nm. Fresh fruit (1 g) was ground on ice with 10 mL of 0.01 M phosphate buffer (pH 7.4) and then centrifuged at 8000 rpm for 10 min at 4 °C. The T-AOC was determined using the supernatant. A quantity of 0.1 mL of the extract was incubated in the water at 37 °C for 30 min with 1 mL of 0.01 M phosphate buffer (pH 7.4), 2 mL of chromogenic agent, and 0.5 mL of ferric salt solution. Then 0.2 mL of stop solution and 0.2 mL of clarificant were added to the mixture. After standing at room temperature for 10 min, the absorption value of the reaction solution was measured at 520 nm. The activity of superoxide dismutase (SOD) was analyzed using the nitro blue tetrazolium (NBT) technique [64]. The activity of peroxidase (POD) was assessed by determining the guaiacol oxidation rate at 470 nm [65].

4.4. Fruit-Quality-Related Parameters

The content of total phenol and anthocyanin was measured using the Folin-Ciocalteu [66] and the pH differential methods [67], respectively. A digital refractometer (Atago, WYT-A, Tokyo, Japan) was used to determine the soluble solid content. The content of total acid was assessed using potentiometric titration in accordance with the state standard GB/T12456-2008, with citric acid serving as a standard. Then the ratio of the soluble solid to the total acid was determined.

The state standard GB/T 20574-2006 was used to measure the total flavonoid content. First, the fresh fruit sample (3 g) was extracted for 45 min at 65 °C in 30 mL of 95% ethanol. Then, the mixture was centrifuged for 5 min at 5000 rpm at 4 °C. The reaction mixture contained 3 mL of the supernatant, 1 mL of 100 g L^{-1} $Al(NO)_3$, 10 mL of 95% ethanol, and 1 mL of 9.8 g L^{-1} CH_3COOK, and was diluted to 50 mL with deionized water. The absorbance was determined at 415 nm after 1 h of incubation at room temperature.

The ellagic acid content was measured using the method presented by Maas et al. [68]. The samples were extracted in 40% ethanol by ultrasonic shaking in a water bath for 20 min at 80 °C. The resultant solution was then centrifuged at room temperature for 10 min at 7000 rpm. The reaction mixture contained 1 mL of the supernatant and 4 mL of 0.1 M NaOH. The absorbance was obtained at 357 nm.

4.5. Transcriptomic Analysis

Total RNA was extracted from blueberry fruits with the Tengen polysaccharide polyphenol kit (QIAGEN, Frankfurt, Germany). To check whether the RNA samples extracted in this test met the requirements, a Nano Drop1000 spectrophotometer and an Agilent 2100 Bioanalyzer were used for analysis. After qualified detection, the library was constructed using a NEBNext® Ultra™ Directional RNA Library Prep Kit for Illumina®. The main steps were performed by referring to the specifications.

Sequencing was performed on the platform of an illumina NovaSeq 6000 (illumina, San Diego, CA, USA). The original sequences after high-throughput sequencing were filtered and screened to obtain clean reads that could be used for subsequent analysis. Sequence comparison was conducted between clean reads and the blueberry reference genome (*Vaccinium darrowii*), using HISAT2 v2.0.5 [69]. Differentially expressed genes (DEGs) were screened under the conditions of $|\log_2$ (fold change)$| \geq 1$ and padj ≤ 0.05 [70]. Functional annotation and classification of DEGs were carried out with reference to the GO (Gene Ontology) database, and the metabolic pathways involved in DEGs were further analyzed with reference to the KEGG database. The sequencing data are available at the NCBI (https://www.ncbi.nlm.nih.gov/bioproject/PRJNA895491/) under the accession number PRJNA895491 on 29 October 2022.

4.6. Metabolite Extraction and Isolation

With liquid nitrogen, the blueberry fruit sample was ground into powder, and 100 mg of it was dissolved in 500 L of 80% methanol extract. It was placed in a cold bath for 5 min after complete disintegration. The supernatant was taken and passed through a 0.22 μm microporous filter membrane after centrifugation at 4 °C (at a speed of 15,000× *g* for 20 min). Metabolite identification was accomplished using ultra-high-performance liquid chromatography (UPLC) and tandem mass spectrometry (MS/MS) analyzers. Chromatographic separation was accomplished using a Hypesil Gold column (C18). The column temperature was 40 °C, and the flow rate was 0.2 mL min^{-1}. In positive ionization mode, the mobile phase A and the mobile phase B were 0.1% formic acid and methanol, respectively, while in negative ionization mode, the mobile phase A was changed to 5 mM ammonium acetate (pH 9.0). The gradient elution program was set up with reference to the protocol settings previously published. ESI was employed as the ion source for small compounds eluted from the column, and positive and negative ion scanning modes were used for sample mass spectrometry signal acquisition, respectively. The mass spectrum parameters were referred to using the previously reported method [11].

4.7. Metabolomic Data Processing and Analysis

Mass spectrometry data were integrated and corrected by Compound Discoverer 3.1 (CD3.1, Thermo Fisher, Waltham, MA, USA). Functional and taxonomic annotation of identified metabolites, primarily utilizing KEGG, HMDB, and LIPID MAPS databases. The processed data were first analyzed by principal component analysis (PCA) and partial least squares discrimination analysis (PLS-DA). Then, the differential abundant metabolites (DAMs) were selected based on their contribution value (VIP), fold change (FC), and significance (P) of the difference between groups. The screening criteria were VIP > 1.0, FC > 1.5 or FC < 0.667, and $p < 0.05$ [71,72]. The KEGG database was used for functional annotation of DAMs, and the annotation results were applied to perform KEGG pathway enrichment analysis.

4.8. Association Analysis of Transcriptomic and Metabolomic Data

The correlation analysis of DEGs and DAMs was based on the Pearson statistical method and was primarily carried out by calculating the correlation coefficient (corr) and *p* value of DEGs and DAMs. A positive correlation was defined as a corr value greater than 0, a negative correlation as a corr value less than 0, the closer the absolute value of corr was to 1, the stronger the correlation, and a significant correlation was defined as a *p* value less than 0.05.

4.9. qRT-PCR Analysis

For qRT-PCR amplification, a GS AntiQ qPCR SYBR Fast Mix (Genesand Biotech Co., Ltd., Beijing, China) and a QuantStudioTM 3 System real-time system were employed.

The following were the reaction conditions: 95 °C for 10 s, 60 °C for 30 s, for a total of 45 cycles; 95 °C for 15 s; 60 °C for 1 min; and 95 °C for 15 s. The GAPDH gene was chosen as the reference gene [73], and the $2^{-\Delta\Delta Ct}$ value method was used to calculate the relative expression of the related genes. The relevant primers used in this study are shown in Table S5.

4.10. Statistical Analysis

All experiments included at least three replicates. All data were analyzed by ANOVA using SPSS 25.0 statistical software (IBM SPSS, Inc., Chicago, IL, USA). A Tukey test or *t*-test was employed, and differences were considered statistically significant at $p < 0.05$.

5. Conclusions

In summary, we performed physiological, transcriptomic, and metabolomic investigations on different blueberry fruits to uncover the mechanisms driving the variability in sugar concentration between blueberry cultivars. DEGs and DAMs associated with carbohydrate metabolism were identified in blueberries. The findings revealed that: (1) the active substance level in 'Anna' fruit was low, but the sugar content was high, and the antioxidant system was active, which could be employed to maintain the balance of ROS metabolism in the fruit; (2) 'Anna' fruit increased sugar content by activating three critical metabolic processes (starch and sucrose metabolism, glycolysis/gluconeogenesis, and the pentose phosphate pathway); (3) sugars in 'Anna' fruit play important roles in regulating auxin and brassinosteroid signaling; (4) eight genes from MYB, AP2/ERF, bHLH, WRKY, and MDAS-box family transcription factors were significantly enriched in blueberry fruit, which were closely related to sugar response. However, the functions of the newly identified DEGs and the regulatory roles of transcription factors need to be further studied. This study provides new insights into understanding the synthesis of carbohydrates affecting blueberry fruits and will be useful for improving fruit quality in future breeding or cultivation efforts.

Supplementary Materials: The following supporting information can be downloaded at: https://www.mdpi.com/article/10.3390/ijms25010293/s1.

Author Contributions: Conceptualization, H.Y.; software, Y.W. and H.Y.; formal analysis, H.Y., Z.W., Y.W., C.Z. and L.L.; investigation, Z.W. and H.Y.; resources, W.W.; writing—original draft preparation, H.Y.; writing—review and editing, W.W. and W.L. All authors have read and agreed to the published version of the manuscript.

Funding: This work was supported by the "JBGS" Project of Seed Industry Revitalization in Jiangsu Province (JBGS[2021]021) and the Chinese Central Financial Project for Extension and Demonstration of Forestry Technology (SU[2021]TG08).

Institutional Review Board Statement: Not applicable.

Informed Consent Statement: Not applicable.

Data Availability Statement: The data and materials supporting the conclusions are contained within the article and the Supplementary Materials. The raw sequence data are available at the NCBI (https://www.ncbi.nlm.nih.gov/bioproject/PRJNA895491/) under the accession number PRJNA895491 on 29 October 2022.

Conflicts of Interest: The authors declare no conflicts of interest.

References

1. Michalska, A.; Łysiak, G. Bioactive compounds of blueberries: Post-harvest factors influencing the nutritional value of products. *Int. J. Mol. Sci.* **2015**, *16*, 18642–18663. [CrossRef] [PubMed]
2. Olas, B. Berry phenolic antioxidants—Implications for human health? *Front. Pharmacol.* **2018**, *9*, 78. [CrossRef] [PubMed]
3. Forney, C.F.; Qiu, S.; Jordan, M.A.; McCarthy, D.; Fillmore, S. Comparison of volatile compounds contributing to flavor of wild lowbush (*Vaccinium augustifolium*) and cultivated highbush (*Vaccinium corymbosum*) blueberry fruit using gas chromatography-olfactometry. *Foods* **2022**, *11*, 2516. [CrossRef] [PubMed]

4. Su, M.-S.; Silva, J.L. Antioxidant activity, anthocyanins, and phenolics of rabbiteye blueberry (*Vaccinium ashei*) by-products as affected by fermentation. *Food Chem.* **2006**, *97*, 447–451. [CrossRef]
5. Montecchiarini, M.L.; Silva-Sanzana, C.; Valderramo, L.; Alemano, S.; Gollán, A.; Rivadeneira, M.F.; Bello, F.; Vázquez, D.; Blanco-Herrera, F.; Podestá, F.E.; et al. Biochemical differences in the skin of two blueberries (*Vaccinium corymbosum*) varieties with contrasting firmness: Implication of ions, metabolites and cell wall related proteins in two developmental stages. *Plant Physiol. Bioch.* **2021**, *162*, 483–495. [CrossRef]
6. Yang, H.Y.; Wu, Y.Q.; Duan, Y.K.; Zhang, C.H.; Huang, Z.J.; Wu, W.L.; Lyu, L.F.; Li, W.L. Metabolomics combined with physiological and transcriptomic analyses reveal regulatory features associated with blueberry growth in different soilless substrates. *Sci. Hortic.* **2022**, *302*, 111145. [CrossRef]
7. Yang, Q.; Liu, E.; Fu, Y.; Yuan, F.; Zhang, T.; Peng, S. High temperatures during flowering reduce fruit set in rabbiteye blueberry. *J. Am. Soc. Hortic. Sci.* **2019**, *144*, 339–351. [CrossRef]
8. Musacchi, S.; Serra, S. Apple fruit quality: Overview on pre-harvest factors. *Sci. Hortic.* **2018**, *234*, 409–430. [CrossRef]
9. Sater, H.; Ferrão, L.F.V.; Olmstead, J.; Munoz, P.R.; Bai, J.; Hopf, A.; Plotto, A. Exploring environmental and storage factors affecting sensory, physical and chemical attributes of six southern highbush blueberry cultivars. *Sci. Hortic.* **2021**, *289*, 110468. [CrossRef]
10. Liu, B.H.; Wang, K.F.; Shu, X.G.; Liang, J.; Fan, X.L.; Sun, L. Changes in fruit firmness, quality traits and cell wall constituents of two highbush blueberries (*Vaccinium corymbosum* L.) during postharvest cold storage. *Sci. Hortic.* **2019**, *246*, 557–562. [CrossRef]
11. Yang, H.; Duan, Y.; Wei, Z.; Wu, Y.; Zhang, C.; Wu, W.; Lyu, L.; Li, W. Integrated physiological and metabolomic analyses reveal the differences in the fruit quality of the blueberry cultivated in three soilless substrates. *Foods* **2022**, *11*, 3965. [CrossRef] [PubMed]
12. Montero, T.M.; MollÁ, E.M.; Esteban, R.M.; López-Andréu, F.J. Quality attributes of strawberry during ripening. *Sci. Hortic.* **1996**, *65*, 239–250. [CrossRef]
13. Qin, G.; Zhu, Z.; Wang, W.; Cai, J.; Chen, Y.; Li, L.; Tian, S. A tomato vacuolar invertase inhibitor mediates sucrose metabolism and influences fruit ripening. *Plant Physiol.* **2016**, *172*, 1596–1611. [CrossRef] [PubMed]
14. Ai, T.N.; Naing, A.H.; Arun, M.; Lim, S.-H.; Kil Kim, C. Sucrose-induced anthocyanin accumulation in vegetative tissue of Petunia plants requires anthocyanin regulatory transcription factors. *Plant Sci.* **2016**, *252*, 144–150. [CrossRef] [PubMed]
15. Yang, H.Y.; Wei, Z.W.; Duan, Y.K.; Wu, Y.Q.; Zhang, C.H.; Wu, W.L.; Lyu, L.F.; Li, W.L. Transcriptomic and metabolomic investigation of the adaptation mechanisms of blueberries to nitrogen deficiency stress. *Sci. Hortic.* **2023**, *321*, 112376. [CrossRef]
16. Zhang, J.; Nie, J.-Y.; Li, J.; Zhang, H.; Li, Y.; Farooq, S.; Bacha, S.A.S.; Wang, J. Evaluation of sugar and organic acid composition and their levels in highbush blueberries from two regions of China. *J. Integr. Agric.* **2020**, *19*, 2352–2361. [CrossRef]
17. Wang, C.; Jiang, H.C.; Gao, G.; Yang, F.; Guan, J.Y.; Qi, H.Y. CmMYB44 might interact with CmAPS2-2 to regulate starch metabolism in oriental melon fruit. *Plant Physiol. Biochem.* **2023**, *196*, 361–369. [CrossRef]
18. Basson, C.E.; Groenewald, J.-H.; Kossmann, J.; Cronjé, C.; Bauer, R. Sugar and acid-related quality attributes and enzyme activities in strawberry fruits: Invertase is the main sucrose hydrolysing enzyme. *Food Chem.* **2010**, *121*, 1156–1162. [CrossRef]
19. Lin, Q.; Wang, C.Y.; Dong, W.C.; Jiang, Q.; Wang, D.L.; Li, S.J.; Chen, M.; Liu, C.R.; Sun, C.D.; Chen, K.S. Transcriptome and metabolome analyses of sugar and organic acid metabolism in Ponkan (*Citrus reticulata*) fruit during fruit maturation. *Gene* **2015**, *554*, 64–74. [CrossRef]
20. Wang, J.; Du, J.J.; Mu, X.P.; Wang, P.F. Cloning and characterization of the Cerasus Humilis sucrose phosphate synthase gene (ChSPS1). *PLoS ONE* **2017**, *12*, e0186650. [CrossRef]
21. Aslam, M.M.; Deng, L.; Wang, X.B.; Wang, Y.; Pan, L.; Liu, H.; Niu, L.; Lu, Z.H.; Cui, G.C.; Zeng, W.F.; et al. Expression patterns of genes involved in sugar metabolism and accumulation during peach fruit development and ripening. *Sci. Hortic.* **2019**, *257*, 108633. [CrossRef]
22. Zhang, Z.K.; Xing, Y.M.; Ramakrishnan, M.; Chen, C.B.; Xie, F.F.; Hua, Q.Z.; Chen, J.N.; Zhang, R.; Zhao, J.T.; Hu, G.B.; et al. Transcriptomics-based identification and characterization of genes related to sugar metabolism in 'Hongshuijing' Pitaya. *Hortic. Plant J.* **2022**, *8*, 450–460. [CrossRef]
23. Kötting, O.; Kossmann, J.; Zeeman, S.C.; Lloyd, J.R. Regulation of starch metabolism: The age of enlightenment. *Curr. Opin. Plant Biol.* **2010**, *13*, 320–328. [CrossRef] [PubMed]
24. Chai, L.; Wang, H.; Yu, H.; Pang, E.; Lu, T.; Li, Y.; Jiang, W.; Li, Q. Girdling promotes tomato fruit enlargement by enhancing fruit sink strength and triggering cytokinin accumulation. *Front. Plant Sci.* **2023**, *14*, 1174403. [CrossRef] [PubMed]
25. Fan, Y.; Li, C.; Li, Y.; Huang, R.; Guo, M.; Liu, J.; Sun, T.; Ge, Y. Postharvest melatonin dipping maintains quality of apples by mediating sucrose metabolism. *Plant Physiol. Biochem.* **2022**, *174*, 43–50. [CrossRef]
26. Wang, C.J.; Wang, Y.L.; Wang, M.M.; Han, H.Y.; Luo, Y.S.; Ding, W.Q.; Xu, W.L.; Zhong, Y.J.; Huang, H.X.; Qu, S.P. Soluble sugars accumulation and related gene expression during fruit development in Cucurbita maxima Duchesne. *Sci. Hortic.* **2020**, *272*, 109520. [CrossRef]
27. Li, X.B.; Li, C.N.; Sun, J.; Jackson, A. Dynamic changes of enzymes involved in sugar and organic acid level modification during blueberry fruit maturation. *Food Chem.* **2020**, *309*, 125617. [CrossRef]
28. Yang, H.; Tian, C.; Ji, S.; Ni, F.; Fan, X.; Yang, Y.; Sun, C.; Gong, H.; Zhang, A. Integrative analyses of metabolome and transcriptome reveals metabolomic variations and candidate genes involved in sweet cherry (*Prunus avium* L.) fruit quality during development and ripening. *PLoS ONE* **2021**, *16*, e0260004. [CrossRef]

29. Gao, Y.; Yao, Y.; Chen, X.; Wu, J.; Wu, Q.; Liu, S.; Guo, A.; Zhang, X. Metabolomic and transcriptomic analyses reveal the mechanism of sweet-acidic taste formation during pineapple fruit development. *Front. Plant Sci.* **2022**, *13*, 971506. [CrossRef]
30. Saftner, R.; Polashock, J.; Ehlenfeldt, M.; Vinyard, B. Instrumental and sensory quality characteristics of blueberry fruit from twelve cultivars. *Postharvest Biol. Technol.* **2008**, *49*, 19–26. [CrossRef]
31. Gallardo, R.K.; Zhang, Q.; Dossett, M.; Polashock, J.J.; Rodriguez-Saona, C.; Vorsa, N.; Edger, P.P.; Ashrafi, H.; Babiker, E.; Finn, C.E.; et al. Breeding trait priorities of the blueberry industry in the United States and Canada. *HortScience* **2018**, *53*, 1021–1028. [CrossRef]
32. Tesnière, C.; Verriès, C. Molecular cloning and expression of cDNAs encoding alcohol dehydrogenases from *Vitis vinifera* L. during berry development. *Plant Sci.* **2000**, *157*, 77–88. [CrossRef] [PubMed]
33. Chen, H.; Cao, S.; Jin, Y.; Tang, Y.; Qi, H. The relationship between CmADHs and the diversity of volatile organic compounds of three aroma types of melon (*Cucumis melo*). *Front. Physiol.* **2016**, *7*, 254. [CrossRef] [PubMed]
34. Guan, C.F.; Du, X.Y.; Zhang, Q.L.; Ma, F.W.; Luo, Z.R.; Yang, Y. DkPK genes promote natural deastringency in C-PCNA persimmon by up-regulating DkPDC and DkADH expression. *Front. Plant Sci.* **2017**, *8*, 149. [CrossRef] [PubMed]
35. Lü, H.M.; Li, J.M.; Huang, Y.H.; Zhang, M.Y.; Zhang, S.L.; Wu, J. Genome-wide identification, expression and functional analysis of the phosphofructokinase gene family in Chinese white pear (*Pyrus bretschneideri*). *Gene* **2019**, *702*, 133–142. [CrossRef] [PubMed]
36. Yadav, U.P.; Ivakov, A.; Feil, R.; Duan, G.Y.; Walther, D.; Giavalisco, P.; Piques, M.; Carillo, P.; Hubberten, H.-M.; Stitt, M.; et al. The sucros-trehalose 6-phosphate (Tre6P) nexus: Specificity and mechanisms of sucrose signalling by Tre6P. *J. Exp. Bot.* **2014**, *65*, 1051–1068. [CrossRef]
37. Wingler, A.; Delatte, T.L.; O'Hara, L.E.; Primavesi, L.F.; Jhurreea, D.; Paul, M.J.; Schluepmann, H. Trehalose 6-phosphate is required for the onset of leaf senescence associated with high carbon availability. *Plant Physiol.* **2012**, *158*, 1241–1251. [CrossRef]
38. Botton, A.; Eccher, G.; Forcato, C.; Ferrarini, A.; Begheldo, M.; Zermiani, M.; Moscatello, S.; Battistelli, A.; Velasco, R.; Ruperti, B.; et al. Signaling pathways mediating the induction of apple fruitlet abscission. *Plant Physiol.* **2011**, *155*, 185–208. [CrossRef]
39. Plaxton, W.C. The organization and regulation of plant glycolysis. *Annu. Rev. Plant Physiol. Plant Mol. Biol.* **1996**, *47*, 185–214. [CrossRef]
40. Basson, C.E.; Groenewald, J.-H.; Kossmann, J.; Cronjé, C.; Bauer, R. Upregulation of pyrophosphate: Fructose 6-phosphate 1-phosphotransferase (PFP) activity in strawberry. *Transgenic Res.* **2011**, *20*, 925–931. [CrossRef]
41. Mei, Z.X.; Li, Z.Q.; Lu, X.; Zhang, S.H.; Liu, W.J.; Zou, Q.; Yu, L.; Fang, H.C.; Zhang, Z.Y.; Mao, Z.Q.; et al. Supplementation of natural light duration promotes accumulation of sugar and anthocyanins in apple (*Malus domestica* Borkh.) fruit. *Environ. Exp. Bot.* **2023**, *205*, 105133. [CrossRef]
42. Ambasht, P.K.; Kayastha, A.M. Plant pyruvate kinase. *Biol. Plant.* **2002**, *45*, 1–10. [CrossRef]
43. Qin, Q.P.; Kaas, Q.; Zhang, L.L.; Xu, K.; Li, N.Y.; Zheng, W.W.; Lai, Q.X. Isolation and characterization of a cytosolic pyruvate kinase cDNA from loquat (*Eriobotrya japonica* Lindl.). *Plant Mol. Biol. Rep.* **2012**, *31*, 109–119. [CrossRef]
44. Kennedy, R.A.; Rumpho, M.E.; Fox, T.C. Anaerobic metabolism in plants. *Plant Physiol.* **1992**, *100*, 1–6. [CrossRef] [PubMed]
45. Drew, M.C. Oxygen deficiency and root metabolism: Injury and acclimation under hypoxia and anoxia. *Annu. Rev. Plant Physiol. Plant Mol. Biol.* **1997**, *48*, 223–250. [CrossRef] [PubMed]
46. Strommer, J. The plant ADH gene family. *Plant J.* **2011**, *66*, 128–142. [CrossRef]
47. Eicks, M.; Maurino, V.; Knappe, S.; Flügge, U.-I.; Fischer, K. The plastidic pentose phosphate translocator represents a link between the cytosolic and the plastidic pentose phosphate pathways in plants. *Plant Physiol.* **2002**, *128*, 512–522. [CrossRef]
48. Chen, Y.; Sun, J.; Lin, H.; Lin, M.; Lin, Y.; Wang, H.; Hung, Y.-C. Salicylic acid reduces the incidence of Phomopsis longanae Chi infection in harvested longan fruit by affecting the energy status and respiratory metabolism. *Postharvest Biol. Technol.* **2020**, *160*, 111035. [CrossRef]
49. Tao, S.K.; Zhu, Y.; Pan, Y.G.; Zhang, Z.K.; Huang, L.J. Enhancement of respiratory metabolism of the pentose phosphate pathway (PPP) strengthens the chilling tolerance of postharvest papaya fruit stored at 1 °C. *Postharvest Biol. Technol.* **2022**, *191*, 111988. [CrossRef]
50. Kunz, S.; Pesquet, E.; Kleczkowski, L.A. Functional dissection of sugar signals affecting gene expression in *Arabidopsis thaliana*. *PLoS ONE* **2014**, *9*, e100312. [CrossRef]
51. Mishra, B.S.; Singh, M.; Aggrawal, P.; Laxmi, A. Glucose and auxin signaling interaction in controlling *Arabidopsis thaliana* seedlings root growth and development. *PLoS ONE* **2009**, *4*, e4502. [CrossRef] [PubMed]
52. Mishra, B.S.; Sharma, M.; Laxmi, A. Role of sugar and auxin crosstalk in plant growth and development. *Physiol. Plant.* **2022**, *174*, e13546. [CrossRef] [PubMed]
53. Hao, N.J.; Zou, X.X.; Lin, X.X.; Cai, R.Q.; Xiao, W.J.; Tong, T.; Yin, H.P.; Sun, A.; Guo, X.H. LecRK-VIII.2 mediates the cross-talk between sugar and brassinosteroid during hypocotyl elongation in *Arabidopsis*. *Reprod. Breed.* **2021**, *1*, 55–63. [CrossRef]
54. Li, J.; Quan, Y.; Wang, L.; Wang, S. Brassinosteroid promotes grape berry quality-focus on physicochemical qualities and their coordination with enzymatic and molecular processes: A Review. *Int. J. Mol. Sci.* **2023**, *24*, 445. [CrossRef] [PubMed]
55. Yuan, Y.; Mei, L.; Wu, M.; Wei, W.; Shan, W.; Gong, Z.; Zhang, Q.; Yang, F.; Yan, F.; Zhang, Q.; et al. SlARF10, an auxin response factor, is involved in chlorophyll and sugar accumulation during tomato fruit development. *J. Exp. Bot.* **2018**, *69*, 5507–5518. [CrossRef] [PubMed]
56. Peng, Y.; Chen, L.; Li, S.; Zhang, Y.; Xu, R.; Liu, Z.; Liu, W.; Kong, J.; Huang, X.; Wang, Y.; et al. BRI1 and BAK1 interact with G proteins and regulate sugar-responsive growth and development in *Arabidopsis*. *Nat. Commun.* **2018**, *9*, 1522. [CrossRef]

57. Zhang, S.; Wang, H.; Wang, T.; Liu, W.; Zhang, J.; Fang, H.; Zhang, Z.; Peng, F.; Chen, X.; Wang, N. MdMYB305-MdbHLH33-MdMYB10 regulates sugar and anthocyanin balance in red-fleshed apple fruits. *Plant J.* **2023**, *113*, 1062–1079. [CrossRef] [PubMed]
58. Hu, X.; Li, S.; Lin, X.; Fang, H.; Shi, Y.; Grierson, D.; Chen, K. Transcription factor CitERF16 is involved in citrus fruit sucrose accumulation by activating CitSWEET11d. *Front. Plant Sci.* **2021**, *12*, 809619. [CrossRef]
59. Yu, J.Q.; Gu, K.D.; Zhang, L.L.; Sun, C.H.; Zhang, Q.Y.; Wang, J.H.; Wang, C.K.; Wang, W.Y.; Du, M.C.; Hu, D.G. MdbHLH3 modulates apple soluble sugar content by activating phosphofructokinase gene expression. *J. Integr. Plant Biol.* **2022**, *64*, 884–900. [CrossRef]
60. Wei, X.; Liu, F.; Chen, C.; Ma, F.; Li, M. The Malus domestica sugar transporter gene family: Identifications based on genome and expression profiling related to the accumulation of fruit sugars. *Front. Plant Sci.* **2014**, *5*, 569. [CrossRef]
61. Liu, J.; Zhang, J.; Wang, J.; Zhang, J.; Miao, H.; Jia, C.; Wang, Z.; Xu, B.; Jin, Z. MuMADS1 and MaOFP1 regulate fruit quality in a tomato ovate mutant. *Plant Biotechnol. J.* **2018**, *16*, 989–1001. [CrossRef] [PubMed]
62. Wang, A.G.; Luo, G.H. Quantitative relation between the reaction of hydroxylamine and superoxide anion radicals in plants. *Plant Physiol. Commun.* **1990**, *6*, 55–57.
63. Hodges, D.M.; Delong, J.M.; Forney, C.F.; Prange, R.K. Improving the thiobarbituric acid-reactive-substances assay for estimating lipid peroxidation in plant tissues containing anthocyanin and other interfering compounds. *Planta* **1999**, *207*, 604–611. [CrossRef]
64. Stewart, R.R.C.; Bewley, J.D. Lipid peroxidation associated with accelerated aging of soybean axes. *Plant Physiol.* **1980**, *65*, 245–248. [CrossRef] [PubMed]
65. Maehly, A.C. Plant peroxidase. *Methods Enzymol.* **1995**, *2*, 801–813. [CrossRef]
66. Ainsworth, E.A.; Gillespie, K.M. Estimation of total phenolic content and other oxidation substrates in plant tissues using Folin-Ciocalteu reagent. *Nat. Protoc.* **2007**, *2*, 875–877. [CrossRef] [PubMed]
67. Cheng, G.W.; Breen, P.J. Activity of phenylalanine ammonialyase (PAL) and concentrations of anthocyanins and phenolics in developing strawberry fruit. *J. Am. Soc. Hortic. Sci.* **1991**, *116*, 865–869. [CrossRef]
68. Maas, J.L.; Wang, S.Y.; Galletta, G.J. Evaluation of strawberry cultivars for ellagic acid content. *HortScience* **1991**, *26*, 66–68. [CrossRef]
69. Kim, D.; Langmead, B.; Salzberg, S.L. HISAT: A fast spliced aligner with low memory requirements. *Nat. Methods* **2015**, *12*, 357–360. [CrossRef]
70. Anders, S.; Huber, W. Differential expression analysis for sequence count data. *Genome Biol.* **2010**, *11*, R106. [CrossRef]
71. Heischmann, S.; Quinn, K.; Cruickshank-Quinn, C.; Liang, L.-P.; Reisdorph, R.; Reisdorph, N.; Patel, M. Exploratory metabolomics profiling in the kainic acid rat model reveals depletion of 25-hydroxyvitamin D3 during epileptogenesis. *Sci. Rep.* **2016**, *6*, 31424. [CrossRef] [PubMed]
72. Rao, G.; Sui, J.; Zhang, J. Metabolomics reveals significant variations in metabolites and correlations regarding the maturation of walnuts (*Juglans regia* L.). *Biol. Open* **2016**, *5*, 829–836. [CrossRef] [PubMed]
73. Yang, H.; Wu, Y.; Zhang, C.; Wu, W.; Lyu, L.; Li, W. Comprehensive resistance evaluation of 15 blueberry cultivars under high soil pH stress based on growth phenotype and physiological traits. *Front. Plant Sci.* **2022**, *13*, 1072621. [CrossRef] [PubMed]

International Journal of
Molecular Sciences

MDPI

Article

Hub Genes in Stable QTLs Orchestrate the Accumulation of Cottonseed Oil in Upland Cotton via Catalyzing Key Steps of Lipid-Related Pathways

Beena Alam [1,†], Ruixian Liu [1,†], Juwu Gong [1,2], Junwen Li [1,2], Haoliang Yan [1,2], Qun Ge [1,2], Xianghui Xiao [1], Jingtao Pan [1], Haihong Shang [1,2], Yuzhen Shi [1], Youlu Yuan [1,2,*] and Wankui Gong [1,*]

[1] National Key Laboratory of Cotton Bio-Breeding and Integrated Utilization, Institute of Cotton Research, Chinese Academy of Agricultural Sciences, Anyang 455000, China; shiyuzhen@caas.cn (Y.S.)

[2] Zhengzhou Research Base, National Key Laboratory of Cotton Bio-Breeding and Integrated Utilization, Zhengzhou University, Zhengzhou 450001, China

* Correspondence: yuanyoulu@caas.cn (Y.Y.); gongwankui@caas.cn (W.G.)

† These authors contribute equally to this work.

Abstract: Upland cotton is the fifth-largest oil crop in the world, with an average supply of nearly 20% of vegetable oil production. Cottonseed oil is also an ideal alternative raw material to be efficiently converted into biodiesel. However, the improvement in kernel oil content (KOC) of cottonseed has not received sufficient attention from researchers for a long time, due to the fact that the main product of cotton planting is fiber. Previous studies have tagged QTLs and identified individual candidate genes that regulate KOC of cottonseed. The regulatory mechanism of oil metabolism and accumulation of cottonseed are still elusive. In the current study, two high-density genetic maps (HDGMs), which were constructed based on a recombinant inbred line (RIL) population consisting of 231 individuals, were used to identify KOC QTLs. A total of forty-three stable QTLs were detected via these two HDGM strategies. Bioinformatic analysis of all the genes harbored in the marker intervals of the stable QTLs revealed that a total of fifty-one genes were involved in the pathways related to lipid biosynthesis. Functional analysis via coexpression network and RNA-seq revealed that the hub genes in the co-expression network that also catalyze the key steps of fatty acid synthesis, lipid metabolism and oil body formation pathways (*ACX4, LACS4, KCR1,* and *SQD1*) could jointly orchestrate oil accumulation in cottonseed. This study will strengthen our understanding of oil metabolism and accumulation in cottonseed and contribute to KOC improvement in cottonseed in the future, enhancing the security and stability of worldwide food supply.

Keywords: quantitative trait locus (QTL); kernel oil content (KOC); cottonseed; hub genes; *Gossypium hirsutum*

1. Introduction

Edible vegetable oils can not only be used as foods or feeds to alleviate global hunger and poverty but also as eco-friendly biofuels as alternatives to the increasingly scarce fossil fuels. *Gossypium* is the largest genus of the Malvaceae family [1,2]. Among the four domesticated species, upland cotton (*G. hirsutum*) is widely planted due to its wide adaptability and high yield, accounting for over 90% of the total cotton planting area. After cotton fiber, the first main product of cotton, which is ginned from the cottonseed, the second major product is cottonseed, which mainly consists of fuzz, hull and kernel. A cotton plant normally produces about 1.6 kg of seeds for every kilogram of lint produced. The cottonseed kernel accumulates 28.24–44.05% vegetable oil [1,3], which provides about 20% of worldwide oil product [2,4] and is recognized as the fifth-largest oil crop in the world, following soybean, palm, canola and sunflower [1]. The fatty acid (FA) composition of cottonseed oil makes it a vital source of vegetable oil, livestock feed and industrial bio-products, including lubricant oil and biodiesel [3,5–7]. The various unsaturation levels and carbon chain lengths of FAs can be used to define its nutritional and industrial values. Due to its flavor stability, it is also applied in a wide range of food processing [5]. Cottonseed oil is mainly accumulated in kernels of seeds and as the main product of cotton production is lint, the genetic regulatory mechanism of kernel oil content (KOC) accumulation and strategies for its improvement have been paid little attention. Some studies revealed that KOC is regulated in a complex quantitative genetic mechanism and significantly influenced by environmental parameters [7]. Molecular markers are crucial tools for genomic exploration, molecular description, and identification of phenotypic/genetic deviations. With the release of genome sequences of cotton species [8], plenty of reference genomes have been constructed [9,10]. These data have provided useful platforms for genetic marker development via various genotyping by sequencing (GBS) strategies, including specific locus amplified fragment sequencing (SLAF-seq), restriction site-associated DNA sequencing (RAD-seq) and microarray sequencing based on molecular hybridization [11]. Several high-density genetic maps (HDGMs) have been constructed using these markers and QTLs of KOC have thus been identified [5,7]. QTL mapping and candidate gene identification have been proven to be the basis of molecular marker-assisted selection (MAS), which has been confirmed to be a robust and operative breeding tool to improve the QTL-controlling traits [11]. However, in breeding practices, MAS is rarely used, mainly due to the lack of molecular markers closely linked to the target traits [12]. As far as the improvement in cottonseed KOC via MAS is concerned, it was estimated that the conversion of acetyl-CoA into seed oil directly requires at least 13 enzyme-dependent reactions [8]. This complex enzymatic reaction network makes it challenging to improve cottonseed KOC via conventional breeding strategies.

Recently, a natural mutant of FA desaturase-2 gene (*FAD2-1D*) was identified in accession GB-713 of *G. barbadense* and a functional link between the mutation and the increased oleic acid cottonseed oil was demonstrated [13]. In *G. hirsutum*, there are at least four different genes encoding *FAD2*, with *ghFAD2-1* playing a major role in the production of linoleic acid of cottonseed oil. The expression of *ghFAD2-1* is seed-specific and reaches its highest level at the middle maturity stage, between 25–35 days post-anthesis (dpa) [14,15]. Overexpression of this enzyme in cottonseed could increase the content of oleic acid [16]. In addition, double-bond introducing genes that turn oleic acid into linoleic acid play a key role in improving the oil quality via altering its FA composition from a perspective of health considerations. A class of desaturases, stearoyl-acyl carrier protein (ACP) desaturase (SAD), regulates the conversion of saturated FAs into unsaturated ones [17]. A previous study revealed that two *SADs* genes *GhSAD2* and *GhSAD4* in upland cotton are preferentially expressed in developing ovules [18]. Lysophosphatidic acid acyltransferase (LPAAT) was also reported to be involved in oil biosynthesis [7], catalyzing the rate-limiting step in the Kennedy pathway. Enhancing expression of Gh13LPAAT5 in Arabidopsis significantly increase the total oil accumulation in Arabidopsis seed [19]. *GhWRI1*, a transcription factor gene, was demonstrated to regulate oil accumulation in cottonseed. It may also participate in the responses of cotton to abiotic stresses [20]. These genes involved in cottonseed oil metabolism are mainly identified via linkage analysis [1,5,7], association analysis [21], or RNA expression analysis [22,23]. However, these studies still do not elucidate the mechanism by which the candidate genes regulate oil accumulation in cottonseed, especially how these individual genes form an interactive or co-expression network to perform their function.

In the current study, an RIL population consisting of 231 lines from a cross between a commercial cultivar with high fiber yield and low KOC of cottonseed, Lumianyan 28 (LMY28), and a cultivar with high fiber quality and high KOC, Xinluzao 24 (XLZ24), was investigated to answer the above question. Based on two HDGMs constructed via SLAF-seq genotyping and the Intl Cotton SNP Consortium_80k CHIP (CHIP) [24] in which we used both SLAF-seq genotyping technology and CHIP SNP marker set to genotype the RIL population. QTLs of KOC were identified and key candidate genes harbored in the QTLs were analyzed via RNA-seq data. Eventually, a co-expression network was identified, in which eight hub genes had dynamic expression during ovule development and cottonseed maturation. The hub genes *ACX4, LACS4, KCR1*, and *SQD1* located at the critical steps of both lipid biosynthesis and co-expression network suggest their functions in regulating oil accumulation in cottonseed. These results will help to better understand the genetic and molecular architecture of KOC of cottonseed, and will also provide useful sources for future research on KOC improvement projects.

2. Results

2.1. KOC Phenotypic Characteristics of the Parents and RIL Population

Descriptive statistics of KOC phenotypes of upland cottonseed of parents (LMY28 and XLZ24) and RIL population across six environments are presented in Table 1. Seeds of XLZ24 accumulated more KOC than LMY28 across all the environments and their differences reached $p < 0.01$ significance (Table 1, Figure 1a). The phenotypic distributions of KOC of the RILs in the Cotton Regions of the Yellow River Valley (YeRV) (namely 14AY, 15AY and 16AY) and of the Northwest Inland (NWI) (namely 14KEL and 15ALE) ranged from 24.21 to 37.13%. The mean values of the RILs in 14AY, 15AY, 16AY, 14KEL, 15ALE, and in BLUE were 28.60%, 26.86%, 27.7%, 29.20%, 27.31%, and 28.00% respectively. Likewise, the coefficients of variation (CV%) for KOC were 5.96%, 7.65%, 7.33%, 6.59%, 6.38%, and 6.19%, respectively (Table 1).

Table 1. Statistical analysis of KOC of parents and RILs.

Environment	Parents		RILs									
	LMY 28	XLZ 24	Mean ± SD	SE Mean	Range	Variance	CV(%)	Min	Max	Skewness	Kurtosis	H^2
14AY	27.59	27.95 **	28.60 ± 1.71	0.11	25.77–35.55	2.91	5.96	25.77	35.55	1.53	4.09	80.07
15AY	30.08	31.03 **	26.86 ± 2.06	0.14	23.47–36.00	4.22	7.65	23.47	36.00	1.91	5.46	83.81
16AY	28.75	29.59 **	27.71 ± 2.03	0.13	24.11–35.27	4.12	7.33	24.11	35.27	1.70	4.17	94.29
14KEL	26.06	27.88 **	29.20 ± 1.93	0.13	25.55–37.13	3.70	6.59	25.55	37.13	1.74	4.11	87.59
15ALE	25.37	27.78 **	27.32 ± 1.74	0.12	24.22–35.70	3.04	6.38	24.22	35.70	1.43	3.01	83.71
BLUE	27.17	28.59 **	28.01 ± 1.73	0.11	25.46–35.36	3.01	6.19	25.46	35.36	2.09	5.75	91.50

Note: LMY28, Lumianyan 28; XLZ24, Xinluzao 24; **, indicates the difference between kernel oil contents of LMY28 and XLZ24 reaches a significant level of $p < 0.01$; SD, standard deviation; SE, standard error; CV, coefficient of variation; Min, minimum value; Max, maximum value; H^2, broad-sense heritability.

Wide-ranging variations in the KOC phenotypes of cottonseeds of the 231 RIL accessions were observed across six environments from 2014 to 2016. ANOVA revealed that the variances of KOC phenotypes of the population sourced from genotype reached great significance, implying reliability of further analysis in the current study (Table S2). In each environment, both positive and negative transgressive segregations (the observed phenotypic KOC values higher than that of XLZ24 or lower than that of LMY28, respectively) in the RILs were also observed (Figure 1b). The absolute values of skewness and kurtosis of KOC were less than one in most of the environments, indicating that KOCs of individuals in the population did not deviate and that they significantly fitted the normal distribution of typical quantitative traits (Figures 1c and S1, Table S3). It was observed that the phenotypic performances of KOC across different environments were significantly correlated. The maximum correlation coefficient was identified between 16AY and the estimated BLUE (r = 0.93) (Figure 1c). The results also indicated that the KOC in 14KEL had comparatively lower correlation with those in other environments, with correlation coefficients ranging from 0.72 to 0.73 (Figure 1c). Strong broad-sense heritability was observed, which suggests a high contribution of genetic diversity to the phenotypic variation of upland cottonseed KOC under these environments (Figure 1d). All the statistical analysis proposed that cottonseed KOC traits in this research panel provided a good basis for further genetic analysis.

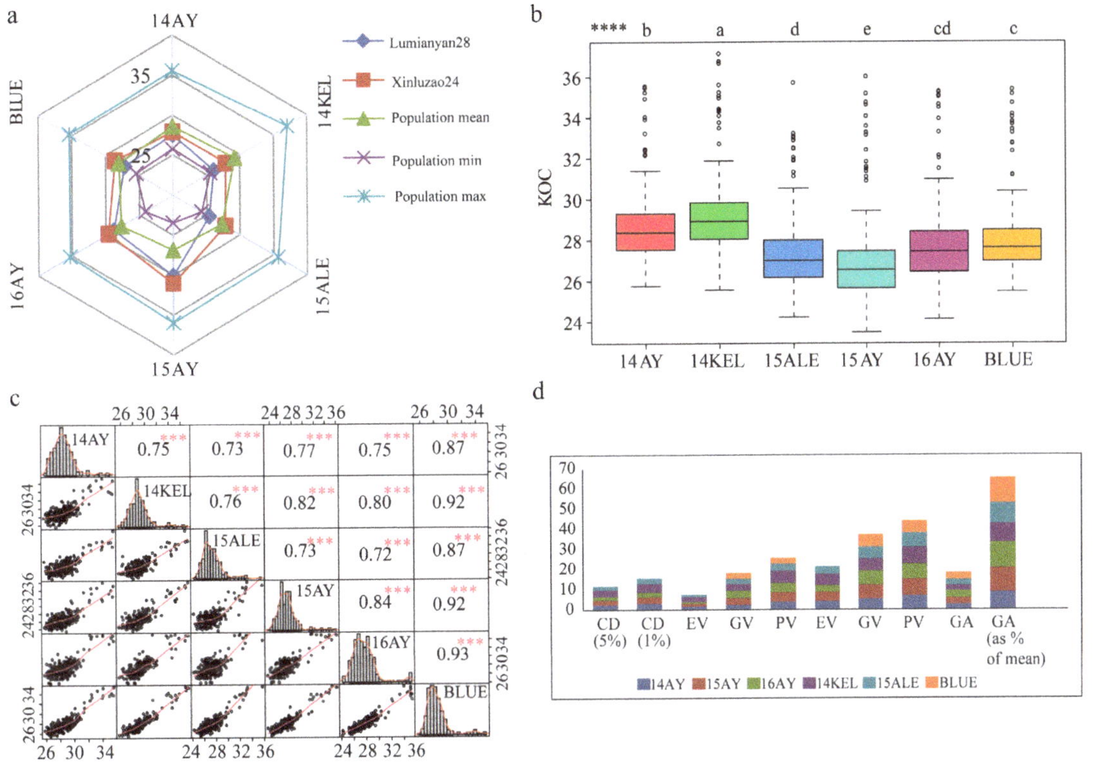

Figure 1. Phenotypic evaluation of kernel oil content (KOC) of cottonseed of the RILs and their parental lines across multiple environments. (**a**) Phenotypic performance of KOC of RILs, including highest, lowest and average values, and their parental lines. (**b**) Variance analysis of KOC of the RILs across six environments. (**c**) Scatterplots, distribution maps and correlation coefficients of KOC of RILs across different environments. (**d**) Genetic variability assessment of KOC of RILs across different environments. **Note:** population mean, mean value of KOC of RILs; population min, minimum value of KOC of RILs; population max, maximum value of KOC of RILs; 14AY, 2014 Anyang; 14KEL, 2014 Kuerle; 15ALE, 2015 Alaer; 15AY, 2015 Anyang; 16AY, 2016 Anyang; BLUE, the best linear unbiased estimate; CD, critical distance; EV, environmental variance; GV, genotypic variance; PV, phenotypic variance; ECV, environmental coefficient of variation; GCV, genotypic coefficient of variation; PCV, phenotypic coefficient of variation; GA, genetic advance. **** and a, b, c, d, e, indicate the significance of KOC phenotype differences between different environment at the $p < 0.05$ level in multiple comparisons; ***, indicates the correlation of KOC phenotypes between different environments reach a significant level of $p < 0.001$.

2.2. Hierarchical Cluster Analysis (HCA) and Discrimination of KOC Trait of the RILs

The individuals of the RILs underwent HCA based on their KOC performance across all environments, including their BLUE values. The results indicated that when the number of clusters k = 3, individuals of the RILs can be best discriminated based on their KOC level and phenotypic stability across the environments (Figures 2a and S2). Cluster 1 and cluster 2, which were categorized as medium- and low-KOC clusters and into which the two parental lines XLZ24 and LMY28 fell, consisted of 94 and 123 individuals, respectively. Cluster 3 consisted of 19 high-KOC individuals, which had accumulated 29.99–35.36%

of KOC in mature cottonseed (Figure 2b, Tables S3 and S4). The principal component (PC) analysis revealed that two PCs explained a total of 89.6% phenotypic variances, with PC1 84.5% and PC2 5.1%, respectively. The environmental vectors revealed that KOC phenotypes were positively correlated across different environments (Figure 2c). Phenotypic stability assessment of cluster 3 accessions via GGE biplot analysis revealed that these lines not only accumulated high KOC in mature cottonseeds but also that their accumulations had a high stability across different environments in the current study (Figure 2d).

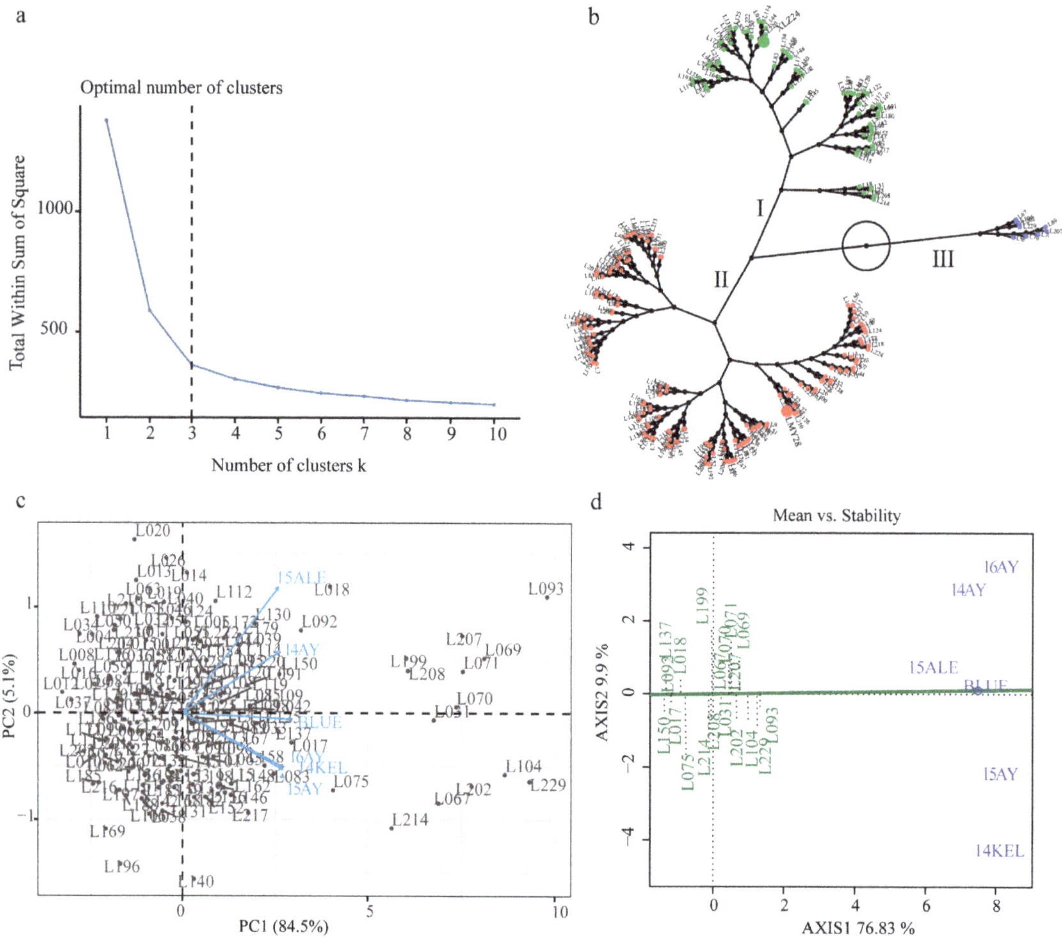

Figure 2. Discriminating analysis of KOC phenotypic performance of the RILs across multiple environments. (**a**) Analysis and determination of the optimal cluster number (k) of KOC of the RILs. (**b**) Clustering analysis of the RILs. (**c**) GGE Biplot analysis of KOC phenotypes of the RILs. (**d**) KOC stability analysis of high-KOC RILs (cluster 3) across multiple environments.

2.3. QTL Mapping of KOC in RIL Population

Based on the two HDGMs, which were constructed via SLAF-seq genotyping and CHIP genotyping strategies [24], respectively, a total of 105 KOC QTLs were identified, including 48 QTLs via CHIP (Figure 3a) and SLAF-seq 57 (Figure 3b) strategies. In sum, 43 QTLs were detected in at least two environments or by both strategies, which were regarded as stable ones (Figures 3c,d, S3 and S4, Tables S5 and S6).

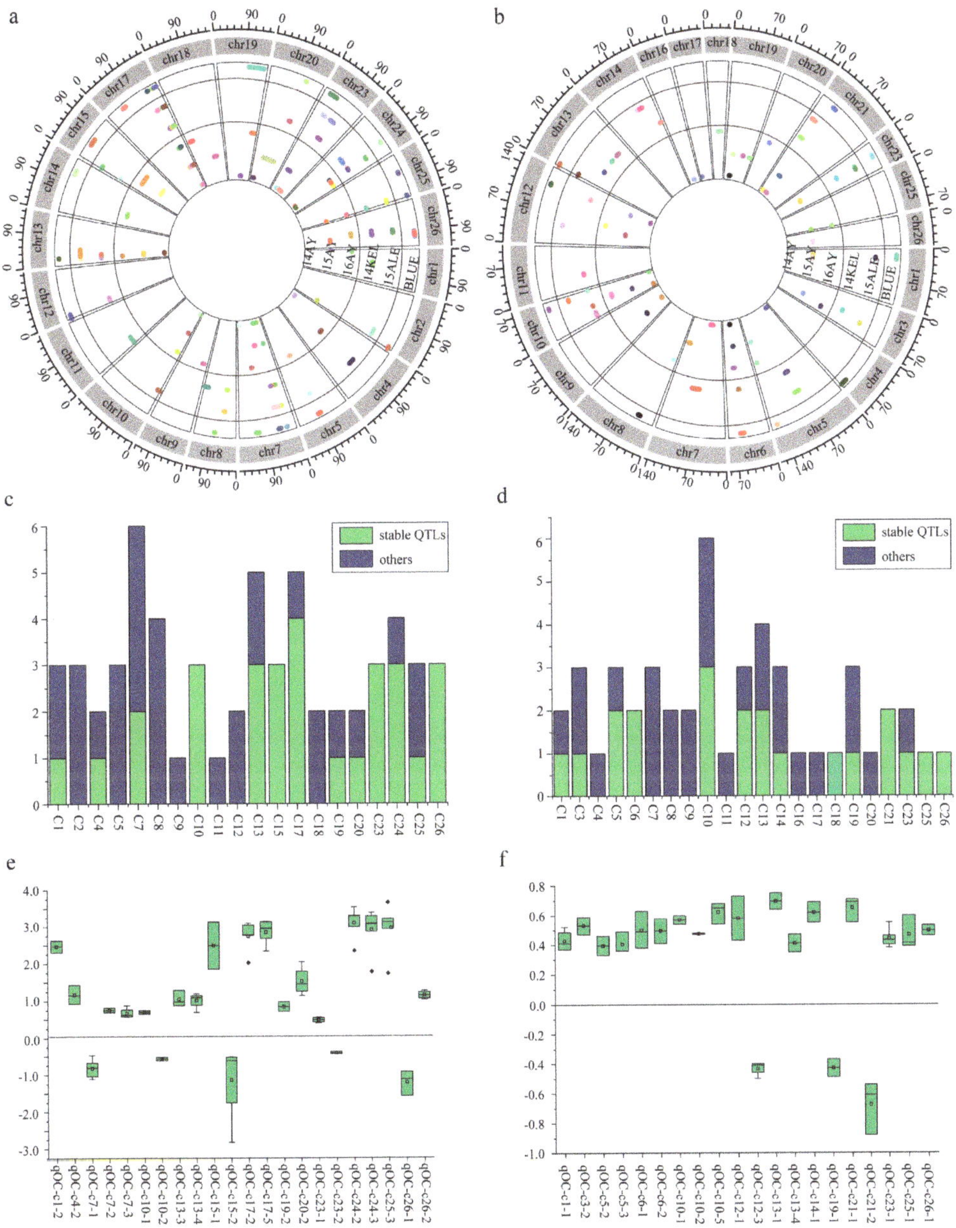

Figure 3. QTL identification via two high-density genetic maps (HDGMs) and their property analysis. (**a**) QTLs identified by HDGM of SLAF-seq strategy and their distribution across the genome. (**b**) QTLs identified by HDGM of CHIP strategy, and their distribution across the genome. (**c**) Stable QTLs identified by SLAF-seq strategy. (**d**) Stable QTLs identified by CHIP strategy. (**e**) Additive effect analysis of the stable QTLs identified by SLAF-seq strategy. (**f**) Additive effect analysis of the stable QTLs identified by CHIP strategy. **Note:** Colors represent different QTLs in each environment; The blue bars show the number of stable QTLs expressed in at least two environments and red bars indicate the number of other QTLs expressed in only one environment.

2.3.1. QTLs Identified via SLAF-Seq Strategy

A total of 57 QTLs of KOC were identified by the composite interval mapping (CIM) algorithm on 21 chromosomes, except c3, c6, c16, c21 and c22 across the cotton genome (Figure 3a, Table S5). In these QTLs, 23 were stable ones, which explained 3.20–59.79% of the total phenotypic variance, with LOD values ranging from 1.74 to 20.62 (Table S6). Four stable QTLs, qOC^{slaf}-chr7-1, qOC^{slaf}-chr17-3, qOC^{slaf}-chr24-3 and qOC^{slaf}-chr25-3 were detected in all six environments, explaining 3.57% to 59.01% of the total observed phenotypic variance, with LOD scores of 2.06–20.62. Four stable QTLs, qOC^{slaf}-chr17-2, qOC^{slaf}-chr17-4, qOC^{slaf}-chr17-5 and qOC^{slaf}-chr24-5 were detected in five environments, explaining 41.42 to 60.50% of the observed phenotypic variance, with LOD values of 3.08–19.95. Six stable QTLs, namely, qOC^{slaf}-chr17-3, qOC^{slaf}-chr13-4, qOC^{slaf}-chr15-2, qOC^{slaf}-chr20-2, qOC^{slaf}-chr23-1 and qOC^{slaf}-chr26-2, were detected in four environments (Figure 3c, Table S5). Eighteen QTLs had positive additive effects, indicating that LMY28 contributed alleles in these loci that favor KOC accumulation in cottonseed, while the remaining five QTLs had negative additive effects, indicating that XLZ24 contributed favorable alleles for KOC accumulation. Ten stable QTLs explained >10% phenotypic variation, the favorable alleles of which are all from LMY28 (Figure 3e).

2.3.2. QTLs Identified via CHIP Strategy

A total of 48 QTLs of KOC were identified by the CIM algorithm on 22 chromosomes, except c2, c15, c22 and c24 across the cotton genome (Figures 3b and S3, Table S6). In these QTLs, 20 were stable ones, which explained 2.99–11.10% of the total phenotypic variance, with LOD values ranging from 1.69 to 5.47 (Table S6). In these stable QTLs, qOC^{chip}-chr23-1 was detected in all six environments, explaining 3.36–6.97% of the total observed phenotypic variance (R^2), with LOD scores of 2.01–4.00. OC^{chip}-chr12-3 and qOC^{chip}-chr1-1 were detected in five and four environments, respectively, explaining 3.61–5.45% and 4.01–4.89% of the observed phenotypic variance with LOD values of 2.04–3.05 and 2.06–2.59, respectively. Six stable QTLs, namely qOC^{chip}-chr5-3, qOC^{chip}-chr6-1, qOC^{chip}-chr10-5, qOC^{chip}-chr21-1, qOC^{chip}-chr21-2 and qOC^{chip}-chr25-1 were detected in three environments (Figure 3d, Table S6). Seventeen QTLs had positive additive effects, indicating that LMY28 contributed alleles in these loci that favor KOC accumulation in cottonseed, while the remaining three QTLs had negative additive effects, indicating that XLZ24 contributed favorable alleles for KOC accumulation. Three QTLs, namely, qOC-chr6-1, qOC^{chip}-chr21-1 and qOC^{chip}-chr21-2, explained >10% of the phenotypic variation; the favorable alleles of the former two were from LMY28 and that of the latter one from XLZ24 (Figure 3f).

2.4. *Candidate Gene Identification and Functional Annotation*

Based on the physical position of the markers [9], a total of 5229 genes were obtained within the confidence marker interval of stable QTLs identified on the two HDGMs, of which 4159 genes were obtained on the HDGM constructed via SLAF-seq strategy and 1191 genes on the one via CHIP strategy. Of these 5229 gene, 121 genes were obtained simultaneously from both of the two maps (Figure 4a). When these gene were aligned to the Arabidopsis Acyl-Lipid Metabolism database (http://aralip.plantbiology.msu.edu/pathways/pathways, accessed on 23 December 2022) [25], a total of 162 genes including 134 coding genes and 1 TFs were identified to have homologues in the database (Table S7). These 162 genes might participate in lipid metabolism and were used for further functional annotation. GO term enrichment revealed that the genes were mainly enriched in the categories of biological process, cellular component and molecular function (Figure 4b,c). The first 20 significantly enriched GO terms were mainly accumulated in those of biological process and molecular function (Figure 4c). Kyoto Encyclopedia of Genes and Genomes (KEGG) analysis of these genes revealed that their functional annotations were mainly enriched in the pathways of metabolism, genetic information processing, and cellular processes (Figure 4d,e). A total of 51 genes were identified to be involved in the pathways related to lipid biosynthesis.

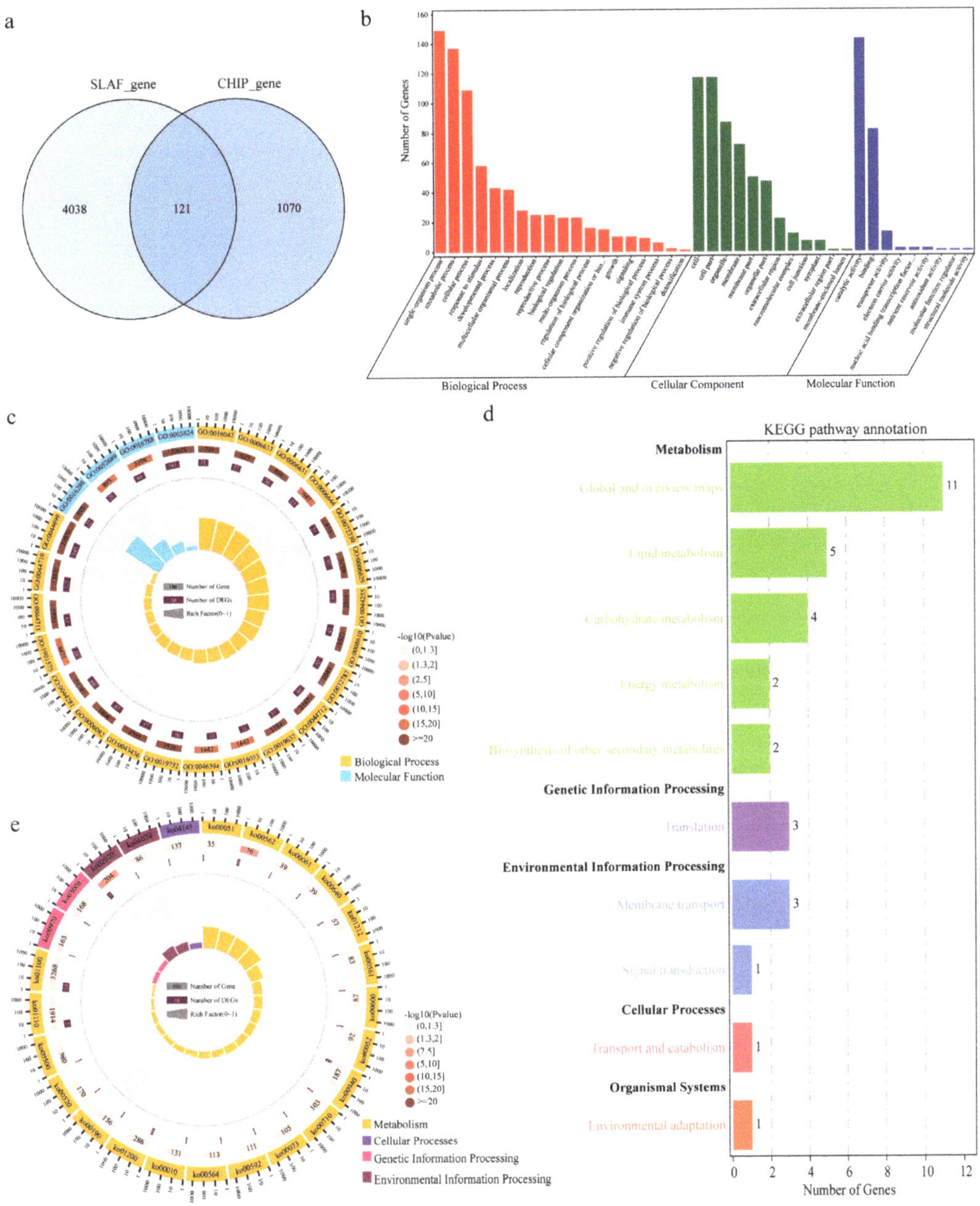

Figure 4. Bioinformatic analysis of the genes harbored in the stable QTLs identified by both high-density genetic maps (HDGMs). (**a**) Total number of the genes by the two HDGMs. (**b**) Gene Ontology enrichment analysis of the genes. (**c**) The first twenty terms significantly enriched in GO analysis. (**d**) Kyoto Encyclopedia of Genes and Genomes (KEGG) pathway annotations of the genes. (**e**) The first twenty pathways significantly enriched in KEGG pathway annotations.

To further capture how the possible candidate genes orchestrate KOC accumulation, the co-expression network among the proteins the 162 genes was explored via the Metascape database. The network identified four core MCODE components, confining a total of 63 hub genes, as well as complex interactions among genes within and between the components (Figure 5a). MCODE 1 consisted of 23 hub genes which were mainly involved in the metabolism of monocarboxylic, and FAs—carbon metabolism. MCODE 2 consisted of 21 hub genes that were mainly involved in monocarboxylic acid, organic phosphorus, and ADP metabolism. MCODE 3 consisted of 16 hub genes that were mainly involved in lipid biosynthesis, lipid metabolism, and signal transduction. MCODE 4 consisted of three hub genes that were mainly involved in glycolysis (Figure 5a).

2.5. Candidate Gene Expression Verification during Ovule Development

Further expression verification of these hub genes via RNA-seq data of different tissues/organs and different developmental stages of ovules of *Gossypium* germplasms TM-1 and Hai7124 revealed that eight hub genes have dynamic relative expression profiling during ovule development in both TM-1 and Hai7124 (Table 2, Figures 5b,c and S5). *GH_A05G3915* had a medium to high expression in developing ovules of TM-1 from 3 dpa to 20 dpa, while in those of Hai7124 only from 10 to 20 dpa (Figure 5b,c). *GH_D03G0050* and *GH_D05G0425* had a high expression in developing ovules at stages of 3 and 5 dpa of Hai7124 than those of TM-1. *GH_D03G0050* and *GH_A05G3915* had a higher expression in developing ovules of TM-1 than of H7124 at 20 dpa. *GH_D09G1772* had a higher expression in developing ovules of TM-1 than of Hai7124 at 10 dpa. *GH_D03G1283* had a higher expression in 20 dpa developing ovules and 25 dpa developing fibers of both TM-1 and Hai7124. Both *GH_D03G0050* and *GH_D05G0425* had a high expression in ovules of both germplasms TM-1 and Hai7124 from 0 to 1 dpa, however, the expression of *GH_D03G0050* was more highly induced in TM-1 ovules than in H7124 ovules, and in the latter, its expression may last longer. *GH_D03G1424, GH_D03G1431* and *GH_D03G1142* had high expression profiles in developing ovules from 3 to 10 dpa; however, their expression in TM-1 was higher in Hai7124 (Figure 5b,c). These dynamically expressed hub genes that catalyze the key steps in the MCODE networks may take responsible roles in KOC formation (Figure 5d).

Table 2. Brief description of 8 key genes that differentially expressed in ovule stages.

Gene ID	Gene Name	Functional Description	Pathways	Localization	MCODE	References
GH_A05G3915	*BCCP2*	Biotin carboxyl carrier protein of heteromeric ACCase	Plastid FA synthesis	Plastid	1	[23,26]
GH_D03G1283	*HSD1*	Steroleosin	Oil storage	Oil body	2	
GH_D03G0050	*PLDdelta*	Phospholipase D delta	Lipid signaling		2	
GH_D03G0938	*LACS4*	lipid biosynthesis and cellular lipid metabolic process	FA elongation	Plastid	3	[7]
GH_D03G1424/ GH_D03G1431	*KCR1*	Ketoacyl CoA reductase	FA elongation	Endomembrane	1	*GH_D03G1424/ GH_D03G1431*
GH_D03G1142	*ENO1*	Phosphoenyl pyruvate enolase	Glycolysis	Plastid	2	*GH_D03G1142*
GH_D05G0425	*SQD1*	UDP-Sulfoquinovose synthase	Plastidal glycerolipid, galactolipid and sulfolipid synthesis		3	*GH_D05G0425*
GH_D09G1772	*ACX4*	Acyl CoA oxidase	beta-oxidation		1	*GH_D09G1772*

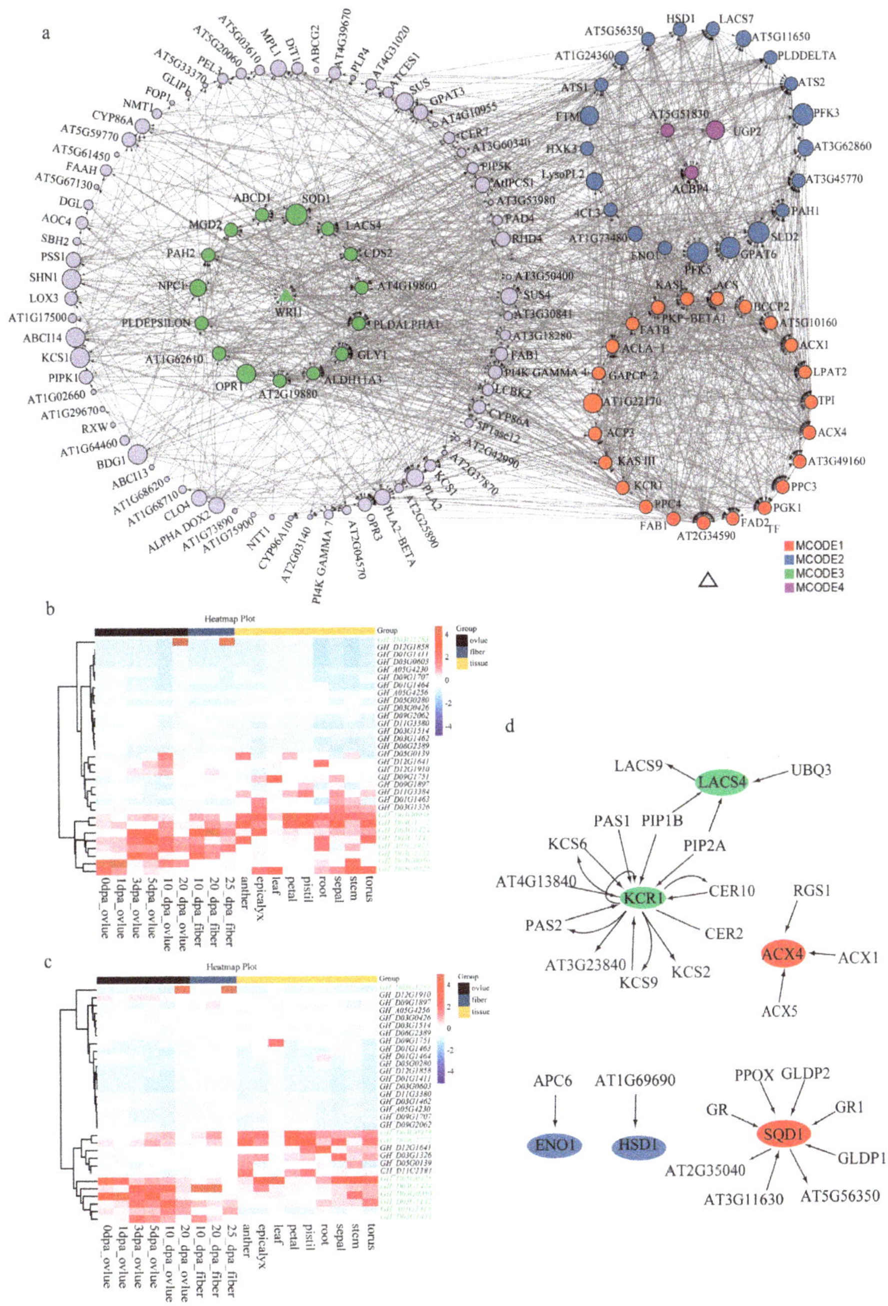

Figure 5. Co-expression network construction and hub gene expression verification during ovule development. (**a**) The network consists of four MCODEs, which are involved in pathways relating carbon metabolism, FAs and lipid biosynthesis, glycolysis. (**b**,**c**) Heatmaps of dynamic expression profiles of the hub genes in various organs and developing stages of upland cotton cultivars TM-1 and Hai7124, respectively. (**d**) Interactions of the hub genes that catalyze the key steps of lipid metabolic pathways.

3. Discussion

3.1. Plant Materials and Their Cottonseed KOC Performance across Various Environments

The RIL population of this study, together with their two parental lines, LMY28 and XLZ24, were planted in six different geographical and chronological environments. KOCs of LMY28 and XLZ24 were different across the different environments, especially in 15AY and 15ALE (Table 1). The diverse variations of climate factors in these experimental locations and across different years must have contributed to the KOC phenotypic variations of the materials used for QTL analysis. Therefore, planting in multiple environments expanded the phenotypic variations of these materials. Strong broad-sense heritability of the experimental materials under multiple environments suggests that the cottonseed KOC has a highly diverse genetic composition. The stability and correlation of KOC across different environments provides a good basis for further genetic analysis. Statistical analysis showed that the KOC of cottonseed in this research panel did not deviate and showed a roughly normal distribution, which confirmed that KOC was controlled by multiple genes and could be analyzed through linkage analysis [9]. Hierarchical cluster analysis (HCA) is essentially helpful to classify the relationship of the experimental materials according to the similarity or dissimilarity and the relationship of the characteristics of the variables [27]. HCA distinguished groups of individuals which maintained similar behavior patterns and further validated the PCA results [28]. The current study finally observed optimally three clustering groups (k = 3) of KOC phenotypes of the RIL population (Figure 2a,b). In phenotypic evaluations, the KOC of XLZ24 was higher than that of LMY28 (Figure 1a, Table 1), which is consistent with the fact that XLZ24 contributed favorable alleles of the majority of stable QTLs for oil accumulation in cottonseeds. However, the inferior parent LMY28 also contributed favorable alleles of some stable QTLs to enhance the trait value (Figure 3e,f). The recombination of the favorable alleles of these QTLs contributed to the transgressive segregation in progeny RILs and the relative consistency of the high KOC lines across different environments (cluster 3, Figure 2b). In previous QTL mapping reports, it was also detected that the beneficial alleles of QTLs of a target trait might come from different parental lines, that is to say, the beneficial alleles of the two parental lines co-determine the formation of the target trait in the population [1,29]. These consistent observations show that it is effective to improve cottonseed KOC through genetic improvement, and these high-KOC lines can play an effective role in future breeding projects to improve KOC of cottonseed. Due to the unique nature of cotton, its fiber quality plays a decisive role in cotton breeding research. In the current study, the materials used are derived from the varieties in practical cotton production [24]. Therefore, these high-KOC lines will have practical application value and can be applied to the simultaneous improvement of both fiber quality and KOC of cottonseed.

3.2. HDGM and QTL Mapping of Cottonseed KOC

Increasing the number of markers could improve the accuracy of QTL analysis and thereby identifying the potential candidate genes in these functional genome loci. The application of AFLP, RFLP, and SSR markers in QTL identification of oil content has previously been limited, resulting in inaccurate QTL region identification [9,21]. However, some studies have tackled the candidate genes regulating TAG accumulation in cottonseed via QTL analysis or GWAS [7,23,30]. In the current study, two HDGMs were applied to the KOC QTL analysis of the RIL population, a total of 43 stable QTLs (23 from SLAF-sequence and 20 from CHIP strategies) were thus detected. To confirm the constancy and reliability of the QTLs detected from the two HDGMs, we compared the KOC-QTLs with those published previously from linkage analysis of segregating populations with different genetic backgrounds or GWAS of natural accession panels. Five KOC-QTLs qOC^{slaf}-c17-3, qOC^{slaf}-c23-1, qOC^{chip}-c5-3, qOC^{chip}-c6-1 and qOC^{chip}-c23-1 were found to be overlapped with confidence intervals of published QTLs identified by other populations [7,21,31,32]. These QTLs are highly consistent and can be used for further marker assisted selection analysis. More interestingly, the current HDGMs based on SNPs had better resolution than the other maps used in previous reported studies. These results further provide a reliable roadmap for tracking the casual genes underlying KOC.

3.3. Candidate Genes Identification and Network Analysis

In plants, triacylglycerols (TAGs, esters of glycerol and FAs) are stored in oil bodies [33], which are dynamic spherical cytoplasmic organelles with 0.5–2.0 μm-diameter structures that contain the TAG matrix and its surrounding monolayer of phospholipids and oleosins [33,34]. They are integral to cell metabolism, with an effective lipids-turnover equilibrium of biogenesis and consumption [35,36]. During seed development, oil bodies are generated through the budding process of the smooth endoplasmic reticulum (ER), while TAGs and oleosins are produced and assembled [36]. Oleosins are 15–24 kDa alkaline proteins present in a variety of plant species. The structure of oleosins features in a C-terminal amphipathic α-helical domain, a central hydrophobic anti-parallel β-strand region, and an N-terminal amphipathic domain. It is reported that central hydrophobic domain anchors oleosin to oil bodies, while their N- and C-terminal amphipathic domains provide steric hindrance and electronegative repulsion to stabilize these lipid-storage organelles. The oxidation of FAs releases twice as much energy as the oxidation of carbohydrates or proteins, resulting in a significant reduction in carbon demand for energy storage in the form of FAs under the same storage conditions. Thus, TAGs represent a highly efficient form of energy storage. In seeds, the accumulation and stores of TAGs allow for enhanced germination and seedling growth until seedling photosynthesis is established [37]. Previous studies have tackled genes regulating TAG biosynthesis and oil body assembly [7]; however, they focused only on individual gene hunting through their specific strategies [38–41]. In the current study, we reported a co-expression network of candidate genes in the stable QTLs (Figure 5a). Expression analysis revealed that some hub genes in the network exhibited dynamic expressions during ovule development (Table 2), which strongly inferred that the interactions among these hub gene members in the network may jointly regulate ovule development and thereby determining the final accumulation of TAG in cottonseed. The hub genes, *ENO1* (*phosphoenolpyruvate enolase 1, GH_D03G1142*) [42], *ACX4* (*acyl-CoA oxidase-4, GH_D09G1772*) [43,44], *BCCP2* (*biotin carboxyl carrier protein 2, GH_A05G3915*) [23,41], *LACS4* (*long-chain acyl-CoA synthetase 4, GH_D03G0938*) [5,7,45,46] and *KCR1* (*ketoacyl CoA reductase 1, GH_D03G1424/GH_D03G1431*) [43] are involved in the steps of the pathways of FA biosynthesis and elongation, and the hub genes, *PLDδ* (*Phospholipase Dδ, GH_D03G0050*) [38], *SQD1* (*UDP-sulfoquinovose synthase 1, GH_D05G0425*) [47] and *HSD1* (*hydroxysteroid dehydrogenase 1, GH_D03G1283*) [48] are involved in TAG biosynthesis and oil body assembly (Figure 6). The hub genes *ACX4, LACS4, KCR1* and *SQD1* located at the key steps in the network (Figure 5d). *ACX4* encodes a peroxisomal acyl-CoA oxidase catalyzing b-oxidation of short-chain FAs, the conversion of Acyl-CoA to 2trans-Enoyl-CoA [43]. *ACX4* has been demonstrated to be the major enzyme that catalyzes short-chain substrate reactions [49]. The current study detected a dynamic increase in *ACX4* expression during ovule development, which is also consistent with previous observation on increased FA metabolism during embryonic development [44]. Very-long-chain FAs (VLCFAs), is demonstrated to play crucial roles in plant structure, physiology, and signaling. In upland cotton, modulating VLCFA biosynthesis during fiber development was demonstrated to regulate cotton fiber elongation [50]. *LACS4* was shown to provide CoA-activated VLCFAs (VLCFA-CoAs) to the pathways of FA elongation. *KCR1* functionally act on a membrane-bound FA elongation system, which multiple proteins might be involved [51]. A recent study demonstrated that KCR1 interacts with ketoacyl-CoA synthase (KCS) to form a FA elongase complex to catalyze biosynthesis of VLCFAs [52]. Our results suggest that *ACX4, LACS4* and *KCR1* might jointly catalyze a metabolism cascade of FA elongation and VLCFA synthesis (Figure 6), which further modulates oil accumulation in plant seeds [53]. *SQD1* was initially confirmed to mediate the biosynthesis of the sulfoquinovosyl headgroup of plant sulfolipids, catalyzing the transfer of SO_3^- to UDP-glucose [54]. In higher plant, sulfoquinovosyl diacylglycerol, a sulfolipid, is one of the four characteristic lipids present in chloroplast membranes [47], which is required for full maintenance of activity and stability of photosystem II [55]. Our results imply that the expression of *SQD1* is highly correlated to oil accumulation in cottonseed. How exactly SQD1 regulates the accumulation of oil in seeds still remains open to discussion; however, previous findings provided some valuable hints for this issue. A more recent study revealed that a gene, *LPPϵ1*, the expression of which was localized at the chloroplast outer envelope, could regulate ER lipid metabolism via interaction with the gene, *LPPα2*, which were expressed at ER system [56]. The correlation between dynamic homeostasis of lipid in oil body and plant responses to abiotic stress has been well established [57]. Biosynthesis of sulfoquinovosyl diacylglycerol (SQDG) coded by *SQD1* as a surrogate lipid for phosphatidylglycerol (PG) in chloroplast under phosphate deficiency [17] plays a role in abiotic stresses. Studies have revealed that genes with dynamic expression during plant embryo development and maturation are also active in plant responses to abiotic stresses [7,9,58]. In the abiotic stresses, including salt, drought, and freezing, a common phenomenon is the deficient availability of plant water status caused by external environmental factors [59]. The development and maturation of embryos may also undergo a process from sufficient to limited water supply caused by intrinsic physiological status of plant. These findings infer that the action mechanism of *SQD1* in seed oil accumulation might share a similarity to its role in regulating lipid dynamic homeostasis in plant responses to abiotic stresses. Interestingly, the oil body protein gene, *HSD1*, identified in the current study, is not involved in FA biosynthesis or TAG assembly (Figure 6). Studies have demonstrated that HSDs are minor components of oil body in oilseed crops [60,61] and that it is involved in lipid homeostasis [62] during embryo development, seed maturation, and plant responses to abiotic stresses [63].

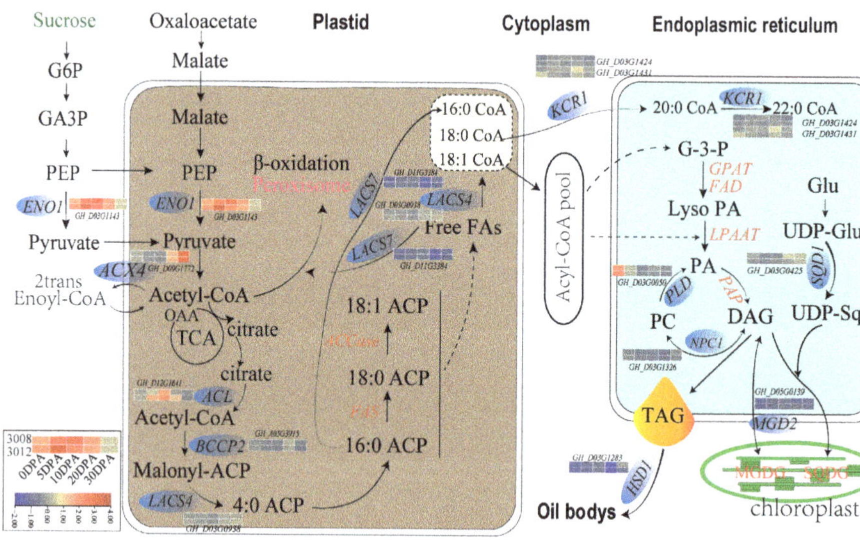

Figure 6. Hub genes in the expression network that have dynamic expressions in developing ovules and the catalytic steps of these hub genes in the pathways related to oil biosynthesis and oil body formation.

However, oil biosynthesis during embryo development is a complex and hierarchical biological process that is regulated by some complex gene co-expression networks [64]. In *Arabidopsis thaliana*, metabolic pathways associated with the biosynthesis and degradation of acyl-lipids have been summarized and most of the genes have been described [25]. Single gene mutations may affect the expression of genes involved in the common metabolic network process. Even though, any modification in oil biosynthesis controlling enzymes induced the alteration in TAG synthesis and even in oil yield [65]. Therefore, further identification of the network is necessary for the rapid capture of causal candidate genes. Here, the screening of potential candidate genes based on KOC-QTLs provides more information to eliminate the interference from a perspective of competition mechanism and metabolism [66]. The current study provided useful information for us to understand the intricate regulatory mechanism of oil accumulation process and provided a foundation for further identification of the causative loci using linkage and transcriptome analyses.

4. Materials and Methods

4.1. Plant Materials and Field Evaluation

Two upland cotton (*G. hirsutum*) field cultivars, LMY28 and XLZ24, and the RIL population consisting of 231 lines derived from a cross of LMY28×XLZ24 were used as experimental plant materials in this study. The procedure of the development of RIL population and the characteristics of two parental cultivars were detailed previously [24]. Briefly the cross LMY28×XLZ24 was made in 2008 summer season in the farm of the Institute of Cotton Research of Chinese Academy of Agricultural Sciences, Anyang, Henan province, China. The RIL population was reached till $F_{6:8}$ generation. The experimental materials were evaluated in field conditions across five different environments, including in Anyang, Henan province in 2014, 2015, and 2016 (designated as 14AY, 15AY and 16AY), in Kuerle in 2014 and in Alaer in 2015 in Xinjiang (designated as 14KEL and 15ALE respectively) [24]. The phenotypes of the target trait (KOC) of the experimental materials were evaluated after cottonseeds were ginned and husked. The best linear unbiased estimate (BLUE) of each line across the five environments, a theoretical value estimated by QTL ICIMapping 4.2 software [7], was regarded as the phenotype of another environment, which made a total of six different environments in the study. The field experiments were conducted in complete randomized block design with two biological replications in each environmental condition from 2014 to 2016.

4.2. Phenotypic Trait Evaluation

In harvest season, seed cotton of 50 typical bolls from each line was harvested. After dried, the seed cotton was ginned and the cottonseeds were delinted with dilute sulfuric acid solution. The delinted cottonseeds were dried and husked. The cottonseed kernels were stored at room temperature till KOC was evaluated. KOC was evaluated using the Niumag Imaging and Analyzing System (NMI20-Analyst, Shanghai, China), following a previously described protocol [5] with minor modifications.

4.3. Data Analysis

Descriptive statistics including the mean value, standard deviation, skewness and kurtosis, ANOVA of KOC in the experimental materials were performed. The recorded data of KOC traits were also subjected to evaluate descriptive statistics, ANOVA, correlation, principal component analysis (PCA), hierarchical cluster analysis (HCA), heat mapping and heritability (broad sense), critical distance in 5% and in 1%, phenotypic variance (PV), genotypic variance (GV), environmental variance (EV) and their coefficient of variation such as PCV, GCV and ECV, genetic advance (GA) and genetic advance as percentage of mean, in order to determine the significant genetic diversity among accessions. All the analyses were computed with Microsoft Excel 2016, IBM SPSS Statistics v. 25 (https://www.ibm.com/analytics/spss-statistics-software, accessed on 16 August 2022) and R-Studio v. 1.2.5001 (https://rstudio.com, accessed on 16 August 2022).

4.4. Linkage Mapping and QTL Identification

Two HDGMs were applied to the KOC QTL mapping of the RIL population. One HDGM was based on CHIP strategy, which contains 4851 molecular markers, including 122 SSR and 2719 SNP ones, with a total coverage of 2477.99 cM of *G. hirsutum* genome and an average marker interval of 0.51 cM between adjacent markers [24]. The other was based on the SLAF-seq strategy, which contains 4366 SNP markers with a total coverage of

3826.92 cM of *G. hirsutum* genome and an average marker interval of 0.73 cM (Liu et al., unpublished data). The brief depiction of the HDGMs of the RIL population is detailed in Table S1.

The CIM algorithms of Windows QTL Cartographer 2.5 software [67] with a mapping step of 1.0 cM and five control markers [68] was employed to detect the significance of a QTL existence in each individual environment. In the parameter setting, the threshold value of the logarithm of odds (LOD) was calculated with 1000 permutations (n = 1000) at the 0.05 significance level ($p < 0.05$). LOD score between 2.0 and permutation test LOD threshold were used to declare a suggestive QTL. QTLs identified in different environments that had fully or partially overlapping confidence intervals were regarded as the same QTL. The QTL detected in at least two environments was regarded as a potential stable one. Positive additive effect of a QTL means that its favorable allele was contributed by XLZ24 while negative additive effect means that its favorable allele contributed by LMY28. Stable QTLs were compared with the Cotton QTL database (https://www.cottonqtldb.org, accessed on 10 September 2022) [7] to determine whether they were previously identified or regarded as a new one. If the markers in the current study shared the same or overlapped physical position with those in the Cotton QTL database, then QTLs were regarded as previously published; otherwise, they were considered as new ones.

4.5. QTL Nomenclature

Nomenclature of QTL was designated following Sun's description [69]. The rule of QTL nomenclature starts with the letter "q" followed by the abbreviation of the trait, the number of chromosome where the QTL was identified, and the QTL serial number on that chromosome, and these elements are separated by the symbol "-" (for example, qKOC-c06-2 represents the second QTL of KOC on chromosome 6).

Candidate Gene Identification and Functional Annotation

The markers flanking the confidence intervals of the stable QTLs were selected to identify the candidate genes. These markers were aligned back to the physical position of upland cotton genome database (www.cottonfgd.org, accessed on 25 October 2022). Based on the position of these flanking markers, all the genes within the confidence interval were selected to perform candidate gene identification. For some of the QTLs with a large confidence interval, if the position of one marker flanking the confidence interval was too far from that of the nearest marker harbored in that confidence interval, the region between these two markers was excluded from the candidate gene identification. All the selected genes were categorized through the gene ontology (GO) analysis using agriGO 2.0 software, and the pathways correlated to the candidate genes were discovered by the Kyoto Encyclopedia of Genes and Genomes (KEGG) analysis using KOBAS 3.0 [7]. The pathways which related to oil and FAs biosynthesis were considered as the enriched terms (for GO) and enriched functional pathways (for KEGG). Then, identified genes were compared with the transcriptomic database available online at functional cotton genome database (www.cottonfgd.org, accessed on 25 October 2022). Finally, the *G. hirsutum* gene expression database (ZJU Assembly) was used to find the expression of the genes of candidate regions in relevant tissues.

4.6. Differential Expression Profiling and Construction of Co-Expression Network of Candidate Genes

All genes in the confidence interval of stable QTLs and their corresponding protein sequences were blasted against the Arabidopsis database (https://www.arabidopsis.org/, accessed on 25 October 2022). Cotton expression profile data were downloaded from the database (https://cotton.zju.edu.cn/, accessed on 25 October 2022). The database contains FPKM values of developing ovules at 0, 1, 3, 5, 10, and 20 dpa of "TM-1", a standard upland cotton line with a low KOC of 30.82%, and "Hai7124", a cultivar with a high KOC of 35.31%. The TM-1 database was used to analyze the expression profiles of candidate genes in KOC

QTLs, and the Hai7124 database was used to compare the differential expression of these genes during cotton ovule development between high and low KOC materials.

The Metascape database of Cytoscape 3.9.0 was used to conduct and visualize the regulatory co-expression network among the candidate genes encoding lipid biosynthesis and KOC related proteins. If the network contained 3–500 proteins, then the molecular complexity detection (MCODE) algorithm was used to identify the network components with density of node–node connections. For each MCODE component, pathway and process enrichment analyses were performed independently. The purpose of making this network was to elaborate the correlation among the genes encoding similar function performing proteins, especially the ones related to KOC of upland cotton [7].

5. Conclusions

The current study has identified 43 stable QTLs via two HDGMs constructed by SLAF-seq and CHIP strategies. Bioinformatics and co-expression network analysis revealed that a total of 51 genes were involved in the pathways related to lipid biosynthesis. Further expression analysis confirmed that eight hub genes in the network exhibited dynamic expression at different stages of cotton ovule development and seed maturation via RNA-seq verification. They are all involved in FA biosynthesis, TAG and oil body assembly. These findings not only deepen our understanding of the intricate regulatory processes that control oil accumulation but also provide a foundation for further identification of the causative loci using transcriptome and linkage analyses. These results also provide fundamental genetic resources for future breeding and genetic practices through MAS.

Supplementary Materials: The following supporting information can be downloaded at https://www.mdpi.com/article/10.3390/ijms242316595/s1.

Author Contributions: B.A.: investigation, data analysis, software, manuscript drafting. R.L.: data analysis, software, field investigation, data analysis, software. J.L.: investigation, field investigation, data analysis. H.Y.: investigation, formal analysis. Q.G.: data collection and analysis. X.X.: investigation, methodology, visualization. J.P.: data collection and analysis. H.S.: data collection and analysis. Y.S.: methodology, data collection. J.G.: field investigation, data collection, formal analysis, visualization. Y.Y.: supervision, funding acquisition, resources. W.G.: project conception, supervision, manuscript—drafting and editing. All authors have read and agreed to the published version of the manuscript.

Funding: This work was funded by grants from the Natural Science Foundation of Xinjiang Uygur Autonomous Region (2021D01B114), Special Fund for Basal Research of Central Public-interest Scientific Institutions (CN) (NO. 1610162023013, 1610162023002), Project of Kashgar Regional Science and Technology Plan (KS2023003), Agricultural Science and Technology Innovation Program of Chinese Academy of Agricultural Sciences China Agriculture Research System of MOF and MARA, China Agriculture Research System—Cotton (CARS-15-02)

Institutional Review Board Statement: Not applicable.

Informed Consent Statement: Not applicable.

Data Availability Statement: The datasets generated during and/or analyzed during the current study are available from the corresponding author upon reasonable request.

Acknowledgments: The authors are grateful for all the suggestions and support from laboratory members.

Conflicts of Interest: The authors declare no conflict of interest.

References

1. Wang, W.; Sun, Y.; Yang, P.; Cai, X.; Yang, L.; Ma, J.; Ou, Y.; Liu, T.; Ali, I.; Liu, D.; et al. A high density SLAF-seq SNP genetic map and QTL for seed size, oil and protein content in upland cotton. *BMC Genom.* **2019**, *20*, 599. [CrossRef]
2. Wendel, J.F.; Cronn, R.C. Polyploidy and the evolutionary history of cotton. *Adv. Agron.* **2003**, *78*, 78004–78008.
3. Shang, L.; Abduweli, A.; Wang, Y.; Hua, J. Genetic analysis and QTL mapping of oil content and seed index using two recombinant inbred lines and two backcross populations in Upland cotton. *Plant Breed.* **2016**, *135*, 224–231. [CrossRef]

4. Xu, Z.; Li, J.; Guo, X.; Jin, S.; Zhang, X. Metabolic engineering of cottonseed oil biosynthesis pathway via RNA interference. *Sci. Rep.* **2016**, *6*, 33342. [CrossRef]
5. Zhang, Z.; Gong, J.; Zhang, Z.; Gong, W.; Li, J.; Shi, Y.; Liu, A.; Ge, Q.; Pan, J.; Fan, S.; et al. Identification and analysis of oil candidate genes reveals the molecular basis of cottonseed oil accumulation in *Gossypium hirsutum* L. *Theor. Appl. Genet.* **2022**, *135*, 449–460. [CrossRef]
6. Dhamodaran, G.; Krishnan, R.; Pochareddy, Y.K.; Pyarelal, H.M.; Sivasubramanian, H.; Ganeshram, A.K. A comparative study of combustion, emission, and performance characteristics of rice-bran-, neem-, and cottonseed-oil biodiesels with varying degree of unsaturation. *Fuel* **2017**, *187*, 296–305. [CrossRef]
7. Gong, J.; Peng, Y.; Yu, J.; Pei, W.; Zhang, Z.; Fan, D.; Liu, L.; Xiao, X.; Liu, R.; Lu, Q.; et al. Linkage and association analyses reveal that hub genes in energy-flow and lipid biosynthesis pathways form a cluster in upland cotton. *Comput. Struct. Biotechnol. J.* **2022**, *20*, 1841–1859. [CrossRef]
8. Paterson, A.H.; Wendel, J.F.; Gundlach, H.; Guo, H.; Jenkins, J.; Jin, D.; Llewellyn, D.; Showmaker, K.C.; Shu, S.; Udall, J.; et al. Repeated polyploidization of *Gossypium* genomes and the evolution of spinnable cotton fibres. *Nature* **2012**, *492*, 423–427. [CrossRef]
9. Hu, Y.; Chen, J.; Fang, L.; Zhang, Z.; Ma, W.; Niu, Y.; Ju, L.; Deng, J.; Zhao, T.; Lian, J.; et al. *Gossypium barbadense* and *Gossypium hirsutum* genomes provide insights into the origin and evolution of allotetraploid cotton. *Nat. Genet.* **2019**, *51*, 739–748. [CrossRef]
10. Chen, Z.J.; Sreedasyam, A.; Ando, A.; Song, Q.; De Santiago, L.M.; Hulse-Kemp, A.M.; Ding, M.; Ye, W.; Kirkbride, R.C.; Jenkins, J.; et al. Genomic diversifications of five *Gossypium* allopolyploid species and their impact on cotton improvement. *Nat. Genet.* **2020**, *52*, 525–533. [CrossRef]
11. Ali, I.; Teng, Z.; Bai, Y.; Yang, Q.; Hao, Y.; Hou, J.; Jia, Y.; Tian, L.; Liu, X.; Tan, Z.; et al. A high density SLAF-SNP genetic map and QTL detection for fibre quality traits in *Gossypium hirsutum*. *BMC Genom.* **2018**, *19*, 879. [CrossRef]
12. Yang, H.; Li, C.; Lam, H.-M.; Clements, J.; Yan, G.; Zhao, S. Sequencing consolidates molecular markers with plant breeding practice. *Theor. Appl. Genet.* **2015**, *128*, 779–795. [CrossRef]
13. Shockey, J.; Dowd, M.; Mack, B.; Gilbert, M.; Scheffler, B.; Ballard, L.; Frelichowski, J.; Mason, C. Naturally occurring high oleic acid cottonseed oil: Identification and functional analysis of a mutant allele of *Gossypium barbadense* fatty acid desaturase-2. *Planta* **2017**, *245*, 611–622. [CrossRef]
14. Liu, Q.; Singh, S.P.; Brubaker, C.L.; Sharp, P.J.; Green, A.G.; Marshall, D. Molecular cloning and expression of a cDNA encoding a microsomal w-6 fatty acid desaturase from cotton (*Gossypium hirsutum*). *Funct. Plant Biol.* **1999**, *26*, 101–106. [CrossRef]
15. Sharif, I.; Farooq, J.; Chohan, S.M.; Saleem, S.; Kainth, R.A.; Mahmood, A.; Sarwar, G. Strategies to enhance cottonseed oil contents and reshape fatty acid profile employing different breeding and genetic engineering approaches. *J. Integr. Agr.* **2019**, *18*, 2205–2218. [CrossRef]
16. Chapman, K.D.; Austin-Brown, S.; Sparace, S.A.; Kinney, A.J.; Ripp, K.G.; Pirtle, I.L.; Pirtle, R.M. Transgenic cotton plants with increased seed oleic acid content. *J. Am. Oil Chem. Soc.* **2001**, *78*, 941–947. [CrossRef]
17. Hölzl, G.; Dörmann, P. Chloroplast lipids and their biosynthesis. *Annu. Rev. Plant Biol.* **2019**, *70*, 6.1–6.31. [CrossRef]
18. Shang, X.; Cheng, C.; Ding, J.; Guo, W. Identification of candidate genes from the SAD gene family in cotton for determination of cottonseed oil composition. *Mol. Genet. Genom.* **2017**, *292*, 173–186. [CrossRef]
19. Wang, N.; Ma, J.; Pei, W.; Wu, M.; Li, H.; Li, X.; Yu, S.; Zhang, J.; Yu, J. A genome-wide analysis of the lysophosphatidate acyltransferase (LPAAT) gene family in cotton: Organization, expression, sequence variation, and association with seed oil content and fiber quality. *BMC Genom.* **2017**, *18*, 218. [CrossRef]
20. Zhao, Y.; Liu, Z.; Wang, X.; Wang, Y.; Hua, J. Molecular Characterization and Expression Analysis of *GhWRI1* in Upland Cotton. *J. Plant Biol.* **2018**, *61*, 186–197. [CrossRef]
21. Du, X.; Liu, S.; Sun, J.; Zhang, G.; Jia, Y.; Pan, Z.; Xiang, H.; He, S.; Xia, Q.; Xiao, S.; et al. Dissection of complicate genetic architecture and breeding perspective of cottonseed traits by genome-wide association study. *BMC Genom.* **2018**, *19*, 451. [CrossRef]
22. Zhao, Y.; Huang, Y.; Wang, Y.; Cui, Y.; Liu, Z.; Hua, J. RNA interference of *GhPEPC2* enhanced seed oil accumulation and salt tolerance in Upland cotton. *Plant Sci.* **2018**, *271*, 52–61. [CrossRef]
23. Zhu, D.; Le, Y.; Zhang, R.; Li, X.; Lin, Z. A global survey of the gene network and key genes for oil accumulation in cultivated tetraploid cottons. *Plant Biotechnol. J.* **2020**, *19*, 1170–1182. [CrossRef]
24. Liu, R.; Gong, J.; Xiao, X.; Zhang, Z.; Li, J.; Liu, A.; Lu, Q.; Shang, H.; Shi, Y.; Ge, Q.; et al. GWAS Analysis and QTL Identification of Fiber Quality Traits and Yield Components in Upland Cotton Using Enriched High-Density SNP Markers. *Front. Plant Sci.* **2018**, *9*, 1067. [CrossRef]
25. Li-Beisson, Y.; Shorrosh, B.; Beisson, F.; Andersson, M.X.; Arondel, V.; Bates, P.D.; Baud, S.; Bird, D.; Debono, A.; Durrett, T.P.; et al. Acyl-lipid metabolism. *Arab. Book* **2013**, *11*, e0161. [CrossRef]
26. Turquetti-Moraes, D.K.; Moharana, K.C.; Almeida-Silva, F.; Pedrosa-Silva, F.; Venancio, T.M. Integrating omics approaches to discover and prioritize candidate genes involved in oil biosynthesis in soybean. *Gene.* **2022**, *808*, 145976. [CrossRef]
27. Banda, T.; Kumarasamy, M. Application of Multivariate Statistical Analysis in the Development of a Surrogate Water Quality Index (WQI) for South African Watersheds. *Water* **2020**, *12*, 1584. [CrossRef]

28. Rehman, F.; Saeed, A.; Yaseen, M.; Shakeel, A.; Ziaf, K.; Munir, H.; Tariq, S.A.; Raza, M.A.; Riaz, A. Genetic evaluation and characterization using cluster heat map to assess NaCl tolerance in tomato germplasm at the seedling stage. *Chil. J. Agr. Res.* **2019**, *79*, 56–65. [CrossRef]
29. Liu, X.; Teng, Z.; Wang, J.; Wu, T.; Zhang, Z.; Deng, X.; Fang, X.; Tan, Z.; Ali, I.; Liu, D.; et al. Enriching an intraspecific genetic map and identifying QTL for fiber quality and yield component traits across multiple environments in Upland cotton (*Gossypium hirsutum* L.). *Mol. Genet. Genom.* **2018**, *292*, 1281–1306. [CrossRef]
30. Zhao, Y.; Wang, Y.; Huang, Y.; Cui, Y.; Hua, J. Gene network of oil accumulation reveals expression profiles in developing embryos and fatty acid composition in Upland cotton. *J. Plant Physiol.* **2018**, *228*, 101–112. [CrossRef]
31. Zhao, W.; Kong, X.; Yang, Y.; Nie, X.; Lin, Z. Association mapping seed kernel oil content in upland cotton using genome-wide SSRs and SNPs. *Mol. Breed.* **2019**, *39*, 105. [CrossRef]
32. Zhu, D.; Li, X.; Wang, Z.; You, C.; Nie, X.; Sun, J.; Zhang, X.; Zhang, D.; Lin, Z. Genetic dissection of an allotetraploid interspecific CSSLs guides interspecific genetics and breeding in cotton. *BMC Genom.* **2020**, *21*, 431. [CrossRef]
33. Huang, A.H.C. Structure of plant seed oil bodies. *Curr. Opin. Struct. Biol.* **1994**, *4*, 493–498. [CrossRef]
34. Kim, H.U.; Hsieh, K.; Ratnayake, C.; Huang, A.H. A novel group of oleosins is present inside the pollen of *Arabidopsis*. *J. Biol. Chem.* **2002**, *277*, 22677–22684. [CrossRef]
35. Huang, A.H.C. Plant Lipid Droplets and Their Associated Proteins: Potential for Rapid Advances. *Plant Physiol.* **2018**, *176*, 1894–1918. [CrossRef]
36. Walther, T.C.; Chung, J.; Farese, R.V. Lipid Droplet Biogenesis. *Ann. Rev. Cell Dev. Biol.* **2017**, *33*, 491–510. [CrossRef]
37. Baud, S.; Lepiniec, L. Regulation of de novo fatty acid synthesis in maturing oilseeds of *Arabidopsis*. *Plant Physiol. Biochem.* **2009**, *47*, 448–455. [CrossRef]
38. Cai, G.; Fan, C.; Liu, S.; Yang, Q.; Liu, D.; Wu, J.; Li, J.; Zhou, Y.; Guo, L.; Wang, X. Nonspecific phospholipase C6 increases seed oil production in oilseed *Brassicaceae* plants. *N. Phytol.* **2020**, *226*, 1055–1073. [CrossRef]
39. Fatland, B.L.; Ke, J.; Anderson, M.D.; Mentzen, W.I.; Cui, L.W.; Allred, C.C.; Johnston, J.L.; Nikolau, B.J.; Wurtele, E.S. Molecular characterization of a heteromeric ATP-citrate lyase that generates cytosolic acetyl-coenzyme A in *Arabidopsis*. *Plant Physiol.* **2002**, *130*, 740–756. [CrossRef]
40. Fatland, B.L.; Nikolau, B.J.; Wurtele, E.S. Reverse genetic characterization of cytosolic acetyl-CoA generation by ATP-citrate lyase in *Arabidopsis*. *Plant Cell* **2005**, *17*, 182–203. [CrossRef]
41. Yin, Y.; Guo, Z.; Chen, K.; Tian, T.; Tan, J.; Chen, X.; Chen, J.; Yang, B.; Tang, S.; Peng, K.; et al. Ultra-High α-Linolenic Acid Accumulating Developmental Defective Embryo was Rescued by Lysophosphatidic Acid Acyltransferase 2. *Plant J.* **2020**, *103*, 2151–2167. [CrossRef]
42. Prabhakar, V.; Lottgert, T.; Gigolashvili, T.; Bell, K.; Flugge, U.I.; Hausler, R.E. Molecular and functional characterization of the plastid-localized Phosphoenolpyruvate enolase (ENO1) from *Arabidopsis thaliana*. *FEBS Lett.* **2009**, *583*, 983–991. [CrossRef]
43. Bao, B.; Chao, H.; Wang, H.; Zhao, W.; Zhang, L.; Raboanatahiry, N.; Wang, X.; Wang, B.; Jia, H.; Li, M. Stable, Environmental Specific and Novel QTL Identification as Well as Genetic Dissection of Fatty Acid Metabolism in *Brassica napus*. *Front. Plant Sci.* **2018**, *9*, 1018. [CrossRef]
44. Contento, A.L.; Bassham, D.C. Increase in catalase-3 activity as a response to use of alternative catabolic substrates during sucrose starvation. *Plant Physiol. Biochem.* **2010**, *48*, 232–238. [CrossRef]
45. Zhong, Y.; Wang, Y.; Li, P.; Gong, W.; Wang, X.; Yan, H.; Ge, Q.; Liu, A.; Shi, Y.; Shang, H.; et al. Genome-Wide Analysis and Functional Characterization of LACS Gene Family Associated with Lipid Synthesis in Cotton (*Gossypium* spp.). *Int. J. Mol. Sci.* **2023**, *24*, 8530. [CrossRef]
46. Jessen, D.; Olbrich, A.; Knüfer, J.; Krüger, A.; Hoppert, M.; Polle, A.; Fulda, M. Combined activity of LACS1 and LACS4 is required for proper pollen coat formation in *Arabidopsis*. *Plant J.* **2011**, *68*, 715–726. [CrossRef]
47. Shimojima, M. Biosynthesis and functions of the plant sulfolipid. *Prog. Lipid Res.* **2011**, *50*, 234–239. [CrossRef]
48. Baud, S.; Dichow, N.R.; Kelemen, Z.; d'Andréa, S.; To, A.; Berger, N.; Canonge, M.; Kronenberger, J.; Viterbo, D.; Dubreucq, B.; et al. Regulation of HSD1 in seeds of *Arabidopsis thaliana*. *Plant Cell Physiol.* **2009**, *50*, 1463–1478. [CrossRef]
49. Khan, B.R.; Adham, A.R.; Zolman, B.K. Peroxisomal Acyl-CoA oxidase 4 activity differs between *Arabidopsis* accessions. *Plant Mol. Biol.* **2012**, *78*, 45–58. [CrossRef]
50. Yang, Z.; Liu, Z.; Ge, X.; Lu, L.; Qin, W.; Qanmber, G.; Liu, L.; Wang, Z.; Li, F. Brassinosteroids regulate cotton fiber elongation by modulating very-long-chain fatty acid biosynthesis. *Plant Cell* **2023**, *35*, 2114–2131. [CrossRef]
51. Qu, C.; Jia, L.; Fu, F.; Zhao, H.; Lu, K.; Wei, L.; Xu, X.; Liang, Y.; Li, S.; Wang, R.; et al. Genome-wide association mapping and Identification of candidate genes for fatty acid composition in *Brassica napus* L. using SNP markers. *BMC Genom.* **2017**, *18*, 232.
52. Kogure, K.; Watanabe, A.; Ito, Y. Interaction of ONION2 ketoacyl CoA synthase with ketoacyl CoA reductase of rice. *Mol. Biol. Rep.* **2022**, *49*, 1643–1647. [CrossRef]
53. Gu, J.; Chao, H.; Wang, H.; Li, Y.; Li, D.; Xiang, J.; Gan, J.; Lu, G.; Zhang, X.; Long, Y.; et al. Identification of the relationship between oil body morphology and oil content by microstructure comparison combining with QTL analysis in *Brassica napus*. *Front. Plant Sci.* **2017**, *7*, 1989. [CrossRef]
54. Mulichak, A.M.; Theisen, M.J.; Essigmann, B.; Benning, C.; Garavito, R.M. Crystal structure of SQD1, an enzyme involved in the biosynthesis of the plant sulfolipid headgroup donor UDP-sulfoquinovose. *Proc. Natl. Acad. Sci. USA* **1999**, *96*, 13097–13102. [CrossRef]

55. Nakajima, Y.; Umena, Y.; Nagao, R.; Endo, K.; Kobayashi, K.; Akita, F.; Suga, M.; Wada, H.; Noguchi, T.; Shen, J.R. Thylakoid membrane lipid sulfoquinovosyl-diacylglycerol (SQDG) is required for full functioning of photosystem II in *Thermosynechococcus elongatus*. *J. Biol. Chem.* **2018**, *293*, 14786–14797. [CrossRef]
56. Nguyen, V.C.; Nakamura, Y. Distinctly localized lipid phosphate phosphatases mediate endoplasmic reticulum glycerolipid metabolism in *Arabidopsis*. *Plant Cell* **2023**, *35*, 1548–1571. [CrossRef]
57. Degenkolbe, T.; Giavalisco, P.; Zuther, E.; Seiwert, B.; Hincha, D.K.; Willmitzer, L. Differential remodeling of the lipidome during cold acclimation in natural accessions of *Arabidopsis thaliana*. *Plant J.* **2012**, *72*, 972–982. [CrossRef]
58. Hafeez, A.; Ge, Q.; Zhang, Q.; Li, J.; Gong, J.; Liu, R.; Shi, Y.; Shang, H.; Liu, A.; Iqbal, M.S.; et al. Multi-responses of O-methyltransferase genes to salt stress and fiber development of *Gossypium* species. *BMC Plant Biol.* **2021**, *21*, 37. [CrossRef]
59. Verslues, P.E.; Agarwal, M.; Katiyar-Agarwal, S.; Zhu, J.; Zhu, J.-K. Methods and concepts in quantifying resistance to drought, salt and freezing, abiotic stresses that affect plant water status. *Plant J.* **2006**, *45*, 523–539. [CrossRef]
60. Krawczyk, H.E.; Rotsch, A.H.; Herrfurth, C.; Scholz, P.; Shomroni, O.; Salinas-Riester, G.; Feussner, I.; Ischebeck, T. Heat stress leads to rapid lipid remodeling and transcriptional adaptations in *Nicotiana tabacum* pollen tubes. *Plant Physiol.* **2022**, *189*, 490–515. [CrossRef]
61. Li, F.; Asami, T.; Wu, X.; Tsang, E.W.T.; Cutler, A.J. A Putative Hydroxysteroid Dehydrogenase Involved in Regulating Plant Growth and Development. *Plant Physiol.* **2007**, *145*, 87–97. [CrossRef]
62. Zhang, Z.; Cheng, Z.-J.; Gan, L.; Zhang, H.; Wu, F.-Q.; Lin, Q.-B.; Wang, J.-L.; Wang, J.; Guo, X.-P.; Zhang, X.; et al. OsHSD1, a hydroxysteroid dehydrogenase, is involved in cuticle formation and lipid homeostasis in rice. *Plant Sci.* **2016**, *249*, 35–45. [CrossRef]
63. Shao, Q.; Liu, X.; Su, T.; Ma, C.; Wang, P. New Insights into the Role of Seed Oil Body Proteins in Metabolism and Plant Development. *Front. Plant Sci.* **2019**, *10*, 1568. [CrossRef]
64. Kunst, L.; Samuels, L. Plant cuticles shine: Advances in wax biosynthesis and export. *Curr. Opin. Plant Biol.* **2009**, *12*, 721–727. [CrossRef]
65. Voelker, T.; Kinney, A.J. Variation in the Biosynthesis of Seed-Storage Lipids. *Annu. Rev. Plant Physiol. Plant Mol. Biol.* **2001**, *52*, 335–361. [CrossRef]
66. Chao, H.; Wang, H.; Wang, X.; Guo, L.; Gu, J.; Zhao, W.; Li, B.; Chen, D.; Raboanatahiry, N.; Li, M. Genetic dissection of seed oil and protein content and identification of networks associated with oil content in *Brassica napus*. *Sci. Rep.* **2017**, *7*, 46295. [CrossRef]
67. Wang, S.; Basten, C.J.; Zeng, Z. *Windows QTL Cartographer*; v2.5.; North Carolina State University: Raleigh, NC, USA, 2012; Available online: http://statgen.ncsu.edu/qtlcart/WQTLCart.htm (accessed on 28 August 2022).
68. Zeng, Z.-B. Precision mapping of quantitative trait loci. *Genetics* **1994**, *136*, 1457–1468. [CrossRef]
69. Sun, F.-D.; Zhang, J.-H.; Wang, S.-F.; Gong, W.-K.; Shi, Y.-Z.; Liu, A.-Y.; Li, J.-W.; Gong, J.-W.; Shang, H.-H.; Yuan, Y.-L. QTL mapping for fiber quality traits across multiple generations and environments in upland cotton. *Mol. Breed.* **2012**, *30*, 569–582. [CrossRef]

International Journal of
Molecular Sciences

MDPI

Article

Transcriptome and Physiological Analysis of Rapeseed Tolerance to Post-Flowering Temperature Increase

Javier Canales [1,2,*,†], José F. Verdejo [3,4,†] and Daniel F. Calderini [4,*]

1 Institute of Biochemistry and Microbiology, Faculty of Sciences, Universidad Austral de Chile, Valdivia 5110566, Chile
2 ANID-Millennium Science Initiative Program-Millennium Institute for Integrative Biology (iBio), Santiago 8331150, Chile
3 Graduate School, Faculty of Agricultural Sciences, Universidad Austral de Chile, Valdivia 5110566, Chile; jose.verdejo@uach.cl
4 Plant Production and Plant Protection Institute, Faculty of Agricultural Sciences, Universidad Austral de Chile, Valdivia 5110566, Chile
* Correspondence: javier.canales@uach.cl (J.C.); danielcalderini@uach.cl (D.F.C.)
† These authors contributed equally to this work.

Abstract: Climate-change-induced temperature fluctuations pose a significant threat to crop production, particularly in the Southern Hemisphere. This study investigates the transcriptome and physiological responses of rapeseed to post-flowering temperature increases, providing valuable insights into the molecular mechanisms underlying rapeseed tolerance to heat stress. Two rapeseed genotypes, Lumen and Solar, were assessed under control and heat stress conditions in field experiments conducted in Valdivia, Chile. Results showed that seed yield and seed number were negatively affected by heat stress, with genotype-specific responses. Lumen exhibited an average of 9.3% seed yield reduction, whereas Solar showed a 28.7% reduction. RNA-seq analysis of siliques and seeds revealed tissue-specific responses to heat stress, with siliques being more sensitive to temperature stress. Hierarchical clustering analysis identified distinct gene clusters reflecting different aspects of heat stress adaptation in siliques, with a role for protein folding in maintaining silique development and seed quality under high-temperature conditions. In seeds, three distinct patterns of heat-responsive gene expression were observed, with genes involved in protein folding and response to heat showing genotype-specific expression. Gene coexpression network analysis revealed major modules for rapeseed yield and quality, as well as the trade-off between seed number and seed weight. Overall, this study contributes to understanding the molecular mechanisms underlying rapeseed tolerance to heat stress and can inform crop improvement strategies targeting yield optimization under changing environmental conditions.

Keywords: post-flowering temperature increase; *Brassica napus*; heat stress; transcriptome analysis; seed yield; seed number; seed weight; gene coexpression network analysis

1. Introduction

Climate-change-induced temperature fluctuations pose a significant threat to crop production, particularly in the Southern Hemisphere, where strong warming is predicted across various global warming scenarios [1–4]. Recently, Rivelli et al. (2021) reported a temperature increase in the Southern Cone of South America over the past 30–50 years, highlighting the urgency of understanding crop responses to these changes [5].

Southern Chile is a high-yield potential region for temperate crops, including rapeseed [6–9]. The increasing temperature due to climate change is a critical environmental factor affecting rapeseed, particularly during the grain filling period [10,11]. Although numerous studies have evaluated rapeseed responses to heat stress, most have been conducted under controlled conditions, limiting the applicability of their findings to field conditions [12].

Heat stress during the critical reproductive stages, such as flowering and seed filling, can result in significant losses in rapeseed seed yield [13]. High temperatures during flowering reduce pollen viability and fertilization success, leading to fewer seeds per pod [14]. Additionally, heat stress during seed filling shortens the duration of this stage, impairing seed development and lowering seed weights [15]. Both daytime and nighttime heat negatively impact reproductive development, with nighttime heat at flowering and seed filling substantially reducing the number of seeds per pod and overall seed yield [16].

Advancements in high-throughput RNA sequencing (RNA-seq) have illuminated the transcriptomic adaptations of rapeseed in response to heat stress [13]. Comparative analyses between rapeseed cultivars subjected to control and heat stress conditions have identified a conserved heat stress response mechanism, primarily mediated through the upregulation of heat shock proteins (HSPs) and downregulation of metabolic pathways [17]. Moreover, heat-tolerant and heat-sensitive cultivars display distinct stress-responsive transcriptional profiles, offering valuable insights into potential adaptive mechanisms [17].

Extending these insights, studies focusing on mature *B. napus* seeds have identified differentially expressed genes (DEGs) involved in protein processing, secondary metabolite pathways, and specific transcription factor families such as NAC, ERF, and bHLH [18,19]. These findings collectively elucidate the complex transcriptional landscape that contributes to compromised reproductive success under heat stress conditions.

The trade-off between seed number and seed weight is a critical factor in plant reproduction and crop yield improvement [20]. In rapeseed, seed weight is controlled by maternal genotype, which regulates seed size mainly through cell number [21]. The pod wall serves as a source of nutrients for seed development, with pod wall photosynthesis contributing to seed filling and final seed weight [21]. Transcriptome analysis of rapeseed has revealed genes involved in the interaction between seed weight (SW) and seed number (SN) [20], and genome-wide association studies have identified genetic loci controlling these traits [22].

Previous research on sensitivity of the seed yield to source-sink ratio reduction in southern Chile showed that rapeseed is more susceptible to source reduction between the start of flowering and 15 days after [8], which overlaps with the critical period of rapeseed [23]. Interestingly, seed yield showed partial or full resilience during the beginning of flowering from the start to 15 days after flowering (DAF) [8,23,24]. In a recent heat stress study conducted in Southern Chile [25], where air temperature increased after the beginning of flowering (from 10 to 20 DAF), seed weight of rapeseed was negatively affected, but not seed weight window. This background raises the question about the molecular mechanisms involved in the rapeseed response to both the increased temperature and source reduction during flowering and the beginning of grain filling.

The objective of this paper is to investigate the transcriptome and physiological responses of rapeseed to post-flowering temperature increases, providing valuable insights into the molecular mechanisms underlying rapeseed tolerance to heat stress. This knowledge can inform crop improvement strategies targeting yield optimization under changing environmental conditions.

2. Results

2.1. Effects of Heat Stress on Seed Yield and Quality Traits of Two Rapeseed Genotypes

The effects of heat stress on seed yield and quality traits of two rapeseed genotypes (Lumen and Solar) were evaluated in two experiments conducted under similar climate conditions during two consecutive growing seasons.

The heat stress treatment increased average temperature between 4.3 °C and 4.8 °C from the beginning of flowering to 15 days after flowering. The seed filling period was affected by heat stress across genotypes and seasons. For instance, seed filling period was shortened by three and six days in Lumen in Season 1 and 2, respectively, whereas Solar was delayed by one day in Season 1 and shortened by two days in Season 2.

Seed yield and seed number showed a significant interaction between genotype and heat stress ($p = 0.03$ and $p = 0.04$, respectively), indicating that both traits were negatively affected by heat stress but to different extents depending on the genotype (Figure 1). Lumen showed an averaged seed yield reduction of 9.3% and no significant change was found in seed number under heat stress, whereas Solar showed an average seed yield reduction of 28.7%, completely explained by the seed number reduction. Moreover, seed yield was negatively correlated with average temperature during the period of heat stress in Solar (R2 = 0.67; $p = 0.05$), but not in Lumen ($p > 0.05$).

Seed weight was affected by genotype ($p = 0.03$), with Lumen having higher seed weight than Solar. On the contrary, this trait did not show any significant response to heat stress (Figure 1). Seed oil concentration was also influenced by genotype ($p < 0.001$), with Lumen reaching higher values (50.5%) than Solar (48.7%). In addition, seed protein concentration was sensitive to both genotype ($p < 0.001$) and heat stress ($p = 0.038$). Contrary to oil, seed protein concentration was higher in Solar (17.0%) than in Lumen (15.4%) in control plants (Figure 1), supporting a trade-off, which has been often reported between these quality traits in rapeseed. Under heat stress, protein seed concentration increased in Solar by 13.1% only during the second growing season.

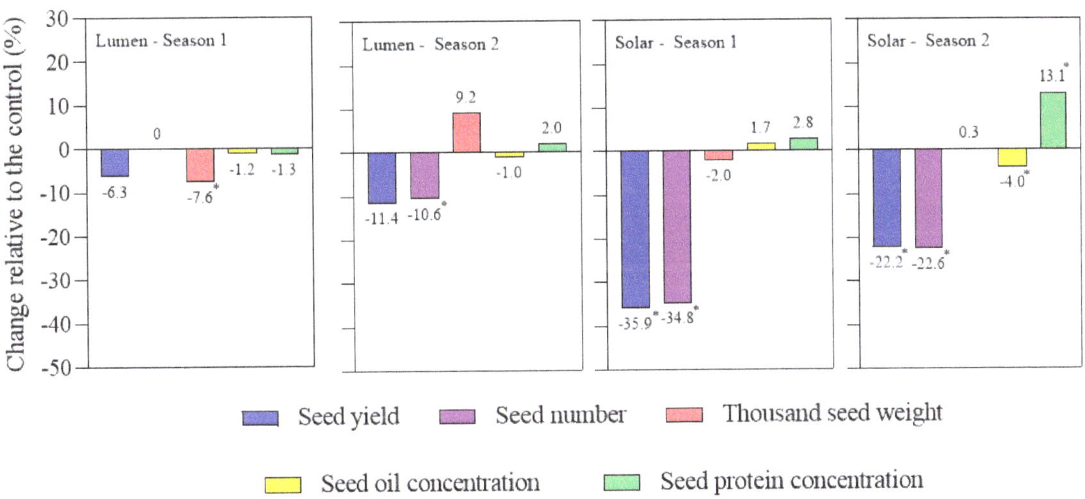

Figure 1. Response of seed yield, seed number, thousand seed weight, and quality traits (seed oil and protein concentrations) across seasons due to heat stress relative to the control. Asterisks indicate significant effect of heat stress treatment on the control by using Fisher's LSD test ($p < 0.05$).

2.2. Multivariate Analysis of RNA-Seq Data Reveals Differential Effect of Heat Stress on Source and Sink Tissues

To gain further insights into the molecular factors underlying heat stress response in rapeseed plants under field conditions, we conducted a multivariate RNA-seq analysis using pod and seed samples to distinguish the effects of stress on source and sink tissues. Our dataset incorporated samples from two independent seasons, with three biological replicates per condition during the first season and four biological replicates for the second season. In total, 112 samples were sequenced, generating 2500 million paired-end reads and averaging 22.3 million raw reads per sample, with a mean pseudoalignment of 16.4 million reads per sample (Table S1).

The multifactorial analysis revealed minimal to non-existent interaction between 2, 3, or 4 factors in both tissues, with most transcriptomic changes being attributable to the primary factors (Figure 2A,B). Notably, genotype was the predominant factor, affecting 28,012 and 29,091 genes in siliques and seeds, respectively (Figures 2A and 2B). Interestingly, the number of genes affected by other factors was also of similar magnitude in both tissues, except for heat treatment. We found 8.5 times more heat-regulated genes in siliques than in seeds (1698 vs. 202 genes, q-value < 0.01), suggesting that source tissues are considerably more sensitive to temperature stress. Moreover, the overlap between these two sets of heat-regulated genes is only 35 genes (Figure S1), indicating a tissue-specific response to this factor.

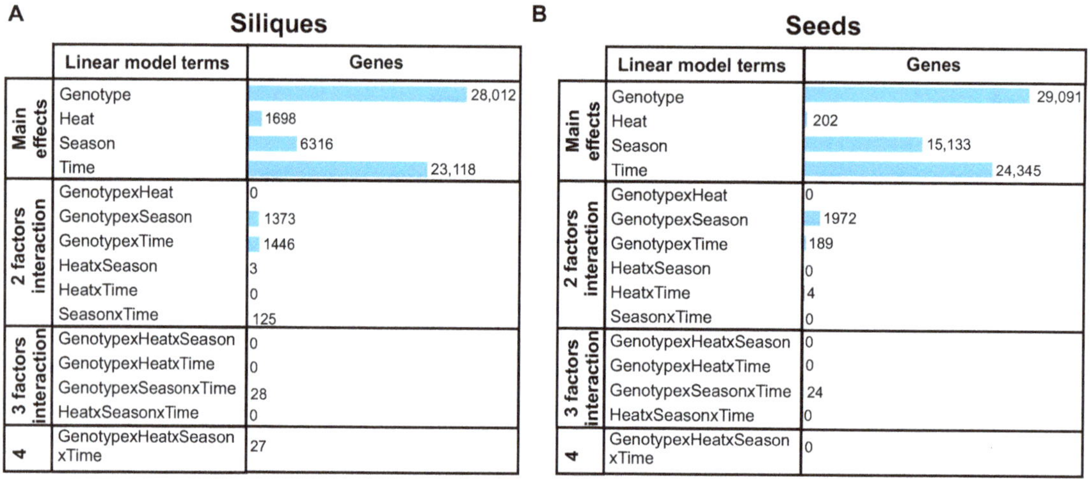

Figure 2. Multivariate analysis of RNA-seq data in siliques and seeds under heat stress. The bar plots display the number of differentially expressed genes (DEGs) attributed to various factors and their interactions in (**A**) siliques and (**B**) seeds based on sleuth analysis (q < 0.01). Genotype is the predominant factor affecting gene expression, whereas heat treatment has a more pronounced effect on siliques than seeds, indicating tissue-specific responses to temperature stress. Data were obtained from field experiments conducted in Valdivia, Chile, using two rapeseed genotypes (Lumen and Solar) subjected to control and heat stress conditions.

2.3. *Hierarchical Clustering Unveils Patterns of Heat-Induced Gene Expression in Siliques*

To investigate how heat-responsive genes are expressed in siliques, we performed a hierarchical clustering analysis on the 1698 genes that were significantly impacted by temperature (q-value < 0.01) in these maternal tissues. We used the "Dynamic Tree Cut" package in R to determine the optimal number of clusters, which was eight (Figure 3A). These clusters showed different patterns of gene expression in response to heat treatment and between genotypes.

Clusters 1, 2, and 5 comprised genes that were downregulated by heat stress, representing around 60% of the total heat-responsive genes (947 out of 1698 genes). Most of these genes had higher expression levels in Lumen than in Solar (Figure 3A). Cluster 1 contained genes that were more highly expressed at 7 DAF than at 14 DAF, and were enriched for biological processes related to photosynthesis and starch metabolism. Cluster 2 included genes that were associated with pigment metabolism and response to light. Cluster 5 had no significantly enriched biological process (Figure 3B).

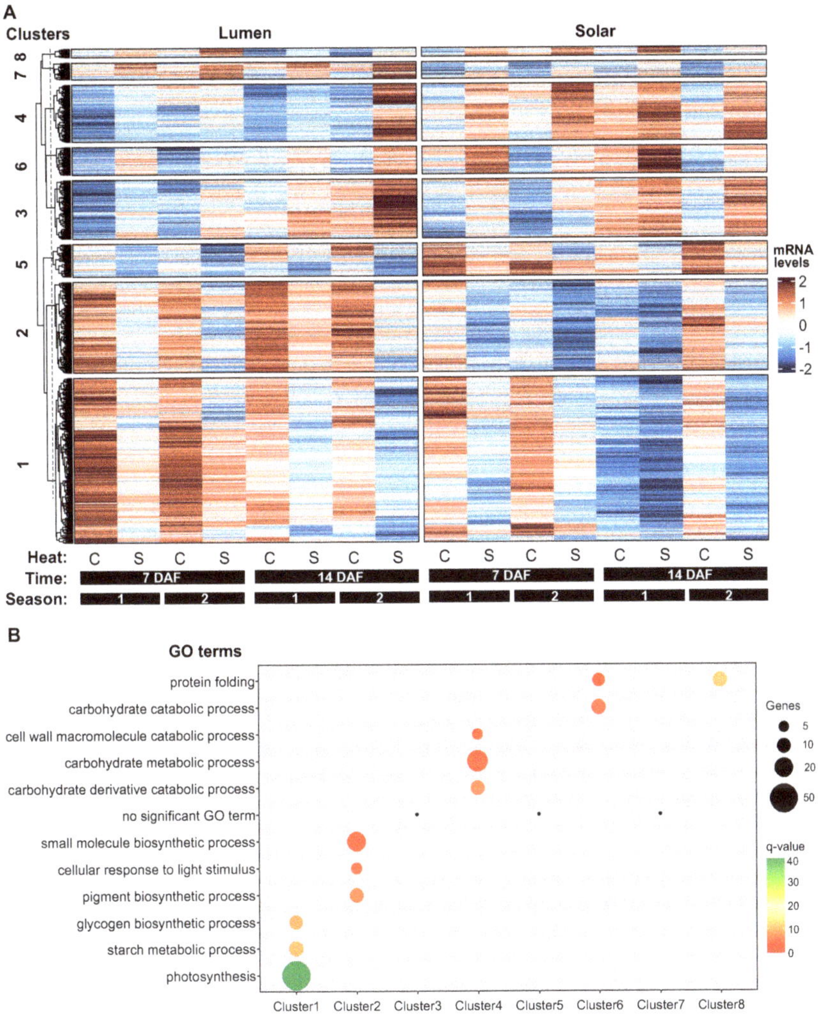

Figure 3. Hierarchical clustering analysis reveals distinct patterns of heat-responsive gene expression in siliques. (**A**) Heatmap showing hierarchical clustering of 1698 heat-responsive genes in siliques into eight distinct clusters based on their expression profiles across samples using ComplexHeatmap [26]. Rows represent genes and columns represent samples from two genotypes (Lumen and Solar), two heat treatments (control and heat stress), two time points (7 and 14 DAF), and two seasons. (**B**) Enrichment analysis of biological processes for each gene cluster. The number of genes and FDR-adjusted *p*-values are shown for the top enriched terms in each cluster. The black dots positioned above Clusters 3, 5, and 7 denote an absence of statistically significant GO terms. Clusters 1, 2, and 5 contain genes downregulated by heat stress and involved in photosynthesis, pigment metabolism, and carbohydrate metabolism. Clusters 3, 4, 6, 7, and 8 show heat-induced genes associated with various processes including protein folding and cell wall metabolism. Cluster 8 displays a genotype-specific temporal response, with decreased expression of protein folding genes in Lumen but not Solar at 14 DAF under heat stress.

Clusters 3, 4, 6, 7, and 8 consisted of genes that were upregulated by heat stress, but they were smaller (between 80 and 200 genes) and less functionally enriched than the downregulated clusters. Clusters 3 and 7 had no significantly enriched biological process (Figure 3B). Clusters 4 and 6 showed consistent differences in gene expression between genotypes, with higher levels in Solar than in Lumen. These clusters were enriched for biological processes involved in cell wall metabolism, carbohydrate metabolism, and protein folding (Figure 3B).

Cluster 8 was the most interesting cluster, as it was highly enriched for genes involved in protein folding and showed a genotype-specific temporal response to heat stress. At 7 DAF, the expression levels and heat response of these genes were similar between genotypes. However, at 14 DAF, these genes exhibited a strong decrease in expression in Lumen but not in Solar (Figure 3B).

Taking together, these findings suggest that heat treatment primarily has a negative impact on the expression of genes associated with photosynthesis in siliques. The results also highlight distinct gene clusters that reflect different aspects of heat stress adaptation in this tissue. In particular, the biological functions associated with cluster 8 genes suggest a role for protein folding in maintaining silique development and seed quality under high temperature conditions.

2.4. Hierarchical Clustering Reveals Three Distinct Patterns of Heat-Responsive Gene Expression in Seeds

In the case of seeds, the hierarchical clustering analysis on 202 heat-responsive genes revealed three clusters with distinct expression patterns (Figure 4A). Cluster 1 (81 genes) was negatively regulated by temperature, cluster 2 (63 genes) was positively regulated by temperature, and cluster 3 (58 genes) showed differential expression between the genotypes and positive response to heat treatment (Figure 4A).

Cluster 1 and 2 also exhibited a temporal effect on their expression levels. In cluster 1, most genes had lower expression at 14 DAF than at 7 DAF or 21 DAF, regardless of the temperature. In contrast, in cluster 2, most genes had higher expression at 14 DAF than at other time points, regardless of the temperature (Figure 4A). Gene ontology analysis indicated that cluster 1 genes were involved in "glycolipid metabolism" and "cellular response to phosphate starvation", whereas cluster 2 genes were involved in "carboxylic acid metabolic process" and "sucrose biosynthesis" (Figure 4B).

Cluster 3 was the only cluster that showed a clear genotypic difference in gene expression. However, this difference was not consistent across time or seasons (Figure 4A). For instance, at 7 DAF in the first season, the genes in this cluster had lower expression in Lumen than in Solar, but this trend was reversed at the same time point in the second season. Moreover, gene ontology analysis revealed that this cluster was enriched in genes related to "protein folding" and "response to heat" (Figure 4B).

2.5. Genotype-Specific Effects of Heat Stress on Photosynthesis-Related Genes in Siliques

Our physiological analysis revealed that the Lumen genotype exhibits increased resistance to heat stress compared to Solar in terms of seed yield. To investigate the genetic basis of this differential tolerance, we used the GeneOverlap R package to identify genes that were regulated by both genotype and heat stress in each group of tissues.

We found a significant intersection of 1015 genes in siliques (p-value = 5×10^{-36}, Figure 5A), but not in seeds, suggesting that siliques are more sensitive to genotype-dependent heat responses. We then mapped these 1015 genes to the clusters of siliques shown in Figure 3 and found that most of them belonged to clusters 1 and 2 (Figure 5B), which were enriched for photosynthesis and light response genes and showed a negative response to heat stress (Figure 3A).

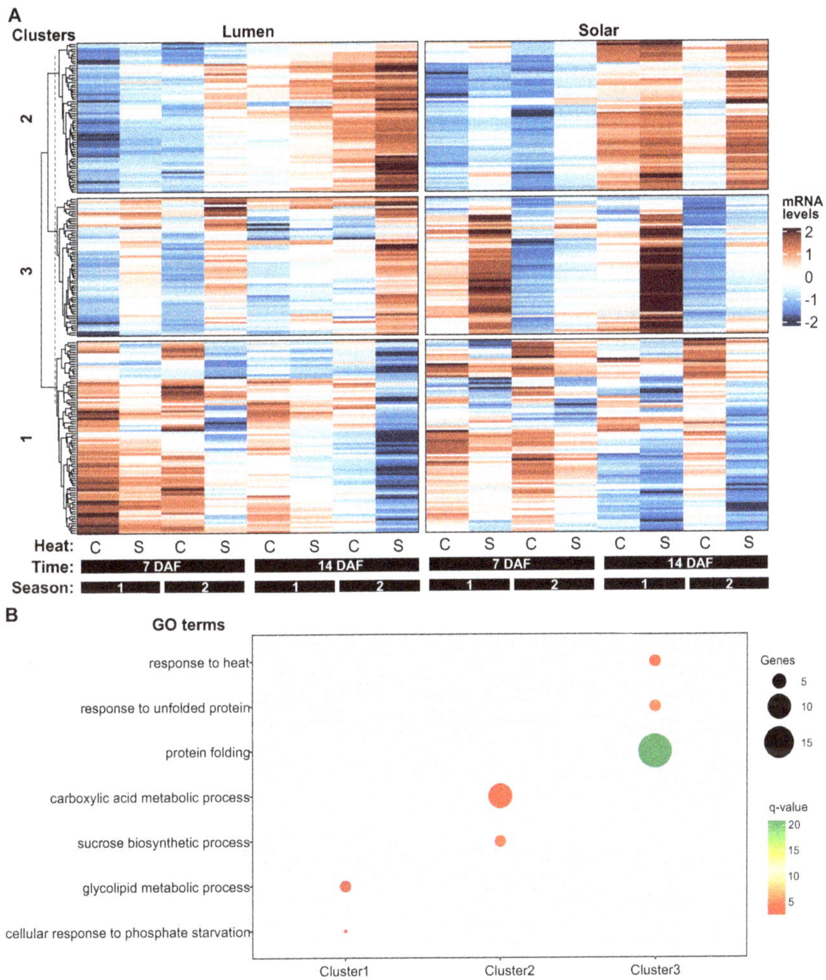

Figure 4. Hierarchical clustering reveals three distinct patterns of heat-responsive gene expression in rapeseed seeds. (**A**) Heatmap visualization of 202 differentially expressed genes (q-value < 0.01) in response to heat treatment, clustered into three groups with distinct expression profiles. Rows represent genes and columns represent different genotype (Lumen or Solar), timepoint (7 or 14 days after flowering), season (1 or 2), and temperature treatment (control or heat stress) conditions. Normalized expression values are color-coded based on row z-score. Cluster 1 (81 genes) shows downregulation by heat stress. Cluster 2 (63 genes) exhibits upregulation by heat. Cluster 3 (58 genes) displays genotype-specific responses to heat. (**B**) Gene ontology enrichment analysis indicates Cluster 1 is involved in glycolipid metabolism and phosphate starvation response, Cluster 2 in carboxylic acid metabolism and sucrose biosynthesis, and Cluster 3 in protein folding and response to heat. The small dot positioned above Cluster 1 denote an absence of statistically significant GO terms.

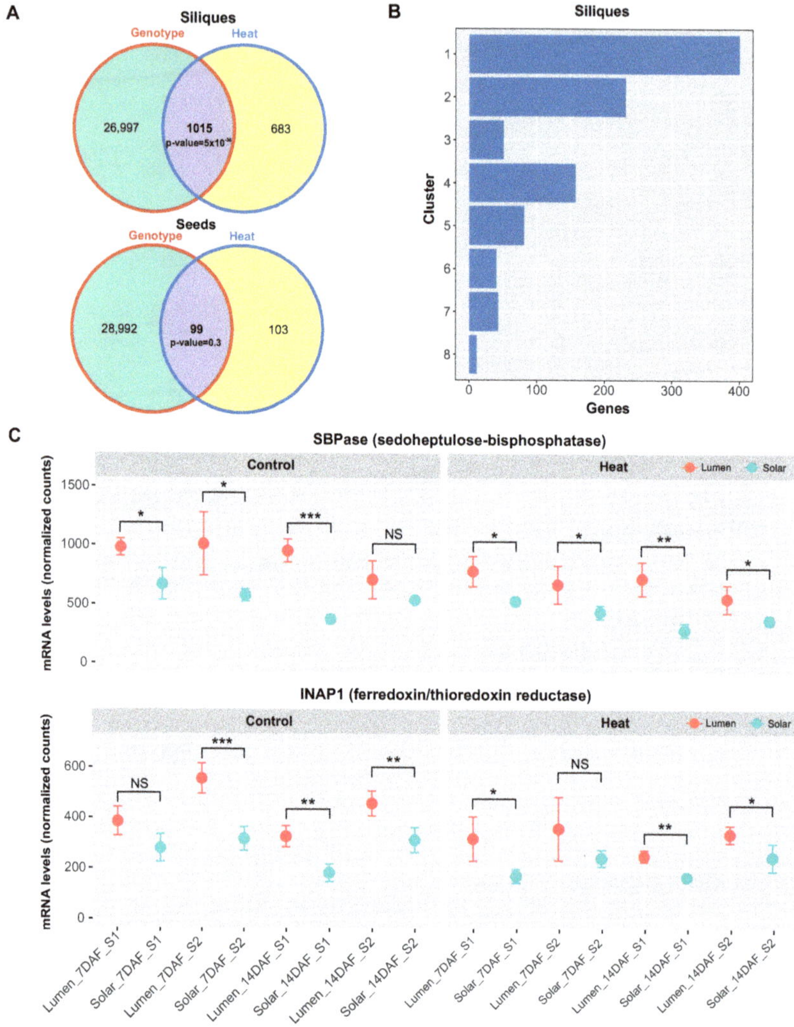

Figure 5. Genotype-dependent effects of heat stress on photosynthesis-related genes in siliques. (**A**) Venn diagrams showing the number of genes significantly regulated by heat, genotype, or both factors in siliques and seeds. Only siliques showed a significant overlap between the two factors ($p < 0.05$, GeneOverlap test). (**B**) Histogram showing the distribution of the 121 genes that were regulated by heat and genotype in siliques across clusters 1 and 2, which contain photosynthesis and light response genes. (**C**) Expression profiles of the photosynthesis-related genes *SBPase* (cluster 1) and *INAP1* (cluster 2) in silique samples under control and heat conditions. *SBPase* and *INAP1* mRNA levels were consistently higher in the Lumen genotype compared to Solar under heat stress based on RNA-seq data, suggesting improved maintenance of photosynthetic gene expression in the heat tolerant Lumen variety. Bars represent means ± SE (n = 3). Asterisks indicate significant differences between genotypes (NS, not significant, * $p < 0.05$, ** $p < 0.01$, *** $p < 0.001$).

Moreover, these clusters consistently exhibited higher expression levels of genes in Lumen than in Solar, indicating that Lumen maintains photosynthetic activity better than Solar under heat stress. To illustrate these differences, we selected two representative genes from each cluster: *SBPase* from cluster 1, which encodes the chloroplast enzyme sedoheptulose-1,7-bisphosphatase involved in the carbon reduction of the Calvin cycle [27], and *INAP1* from cluster 2, which encodes the catalytic subunit of the chloroplastic ferredoxin/thioredoxin reductase [28]. In both cases, we observed significant differences between genotypes in most of the samples analyzed, showing lower mRNA levels in Solar than in Lumen (Figure 5C,D).

2.6. Co-Expression Network Analysis Reveals Major Modules for Rapeseed Yield and Quality

To identify candidate genes related to rapeseed yield and quality, we performed a gene co-expression network analysis using the transcriptomes of siliques and seeds by the WGCNA method [29] as indicated in the materials and methods section. We constructed four co-expression networks corresponding to two different times after flowering (7 and 14 days) and two group of tissues (seeds and siliques) and compared them to select the network that showed the highest number of genes correlated with the agronomic parameters of yield and seed quality.

We found that the seed network at 7 DAF had the highest number of genes significantly correlated with the main agronomic traits of yield and quality (Figure S2). The seed network at 7 DAF showed more than five times as many genes correlated with seed number and yield than the silique networks at both 7 and 14 DAF (Figure S2). These results indicate that the co-expression network of seeds at 7 DAF is the most suitable for finding genes that explain the parameters of yield and seed quality.

The co-expression network of seeds at 7 DAF consisted of 35,432 genes distributed in 33 modules (Figure 6A). These traits included growth and development (plant height, branch per plant, siliques per plant, seeds per silique, weight per silique, biomass, and harvest index), seed quality (oil and protein content) and yield (seed number, thousand seed weight, yield). We identified eight co-expression modules with high correlation with some of these agronomic parameters (at least one absolute correlation of 0.7 or more and a p-value less than 0.0001, Figure 6B). Interestingly, five of these eight modules were correlated with quality and yield parameters and were also the largest ones in the network.

For seed quality, we found two modules positively correlated with protein content (the lightblue3 and firebrick modules) and one module positively correlated with oil content (lightskyblue2, Figure 6B). The overrepresentation analysis of biological functions showed that the lightblue3 and firebrick modules were enriched in genes related to protein synthesis and transport as well as photosynthesis and small molecule metabolism (Table S2). The lightskyblue2 module was enriched in genes related to glycolysis, purine metabolism, ribonucleoprotein complex biogenesis, protein folding, and mRNA splicing (Table S2).

2.7. Identification of Regulatory Factors Associated with the Compensation of SW and SN Flowering

In the case of yield, we found two highly correlated co-expression modules, "indianred2" and "palevioletred" with 3584 and 216 genes, respectively. Interestingly, both modules showed an inverse correlation between seed number and weight, suggesting that they could be related to the trade-off between these two yield-determining factors. Given the relevance of this topic for crop performance and improvement, we analyzed these modules in more detail.

As shown in Figure 7A, the genes in the indianred2 module showed a positive correlation with seed number and a negative correlation with weight. A biological function enrichment analysis revealed that it is especially enriched in genes associated with translation, as well as some functions related to transcription, protein ubiquitination and protein targeting (Figure 7B). To identify the regulators that control the expression of this module, we performed a regulatory network analysis using the R package Bionero and the transcription factor annotation from PlantTFDB 4.0 [30].

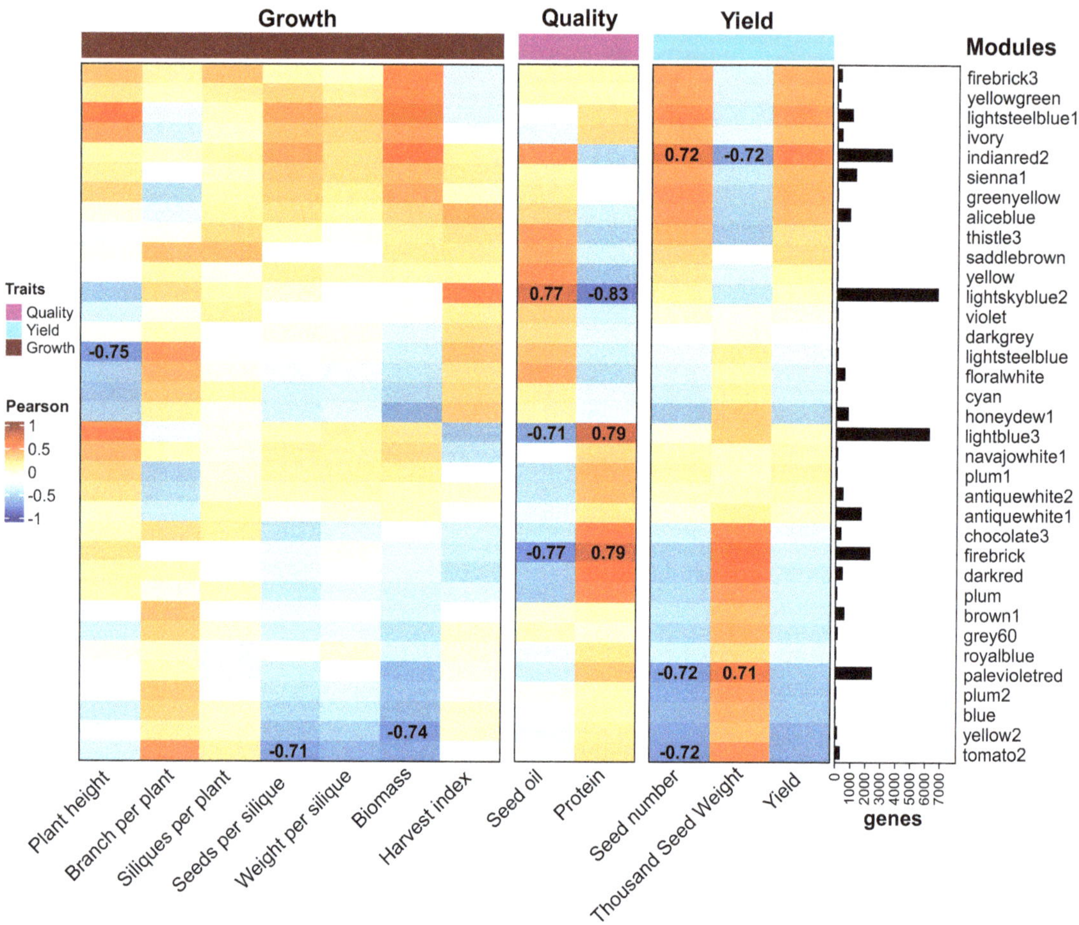

Figure 6. Correlation heatmap of gene coexpression modules with yield and quality traits in rapeseed. Heatmap showing Pearson correlation coefficients between module eigengenes and phenotypic traits related to plant growth, seed yield, and quality. The modules were generated by weighted gene coexpression network analysis of transcriptomic data from 28 samples of seeds at 7 days after flowering. Modules are grouped into categories and ordered based on strongest correlations. Red indicates a positive correlation, blue indicates a negative correlation. The largest modules positively correlated with protein content (lightblue3, firebrick), oil content (lightskyblue2), seed number (indianred2), and seed weight (palevioletred) are highlighted. This systems-level analysis reveals major coexpression modules associated with key agronomic traits that determine yield and quality in rapeseed. The module names (e.g., lightblue3, firebrick, indianred2) correspond to coexpression modules identified by the WGCNA algorithm.

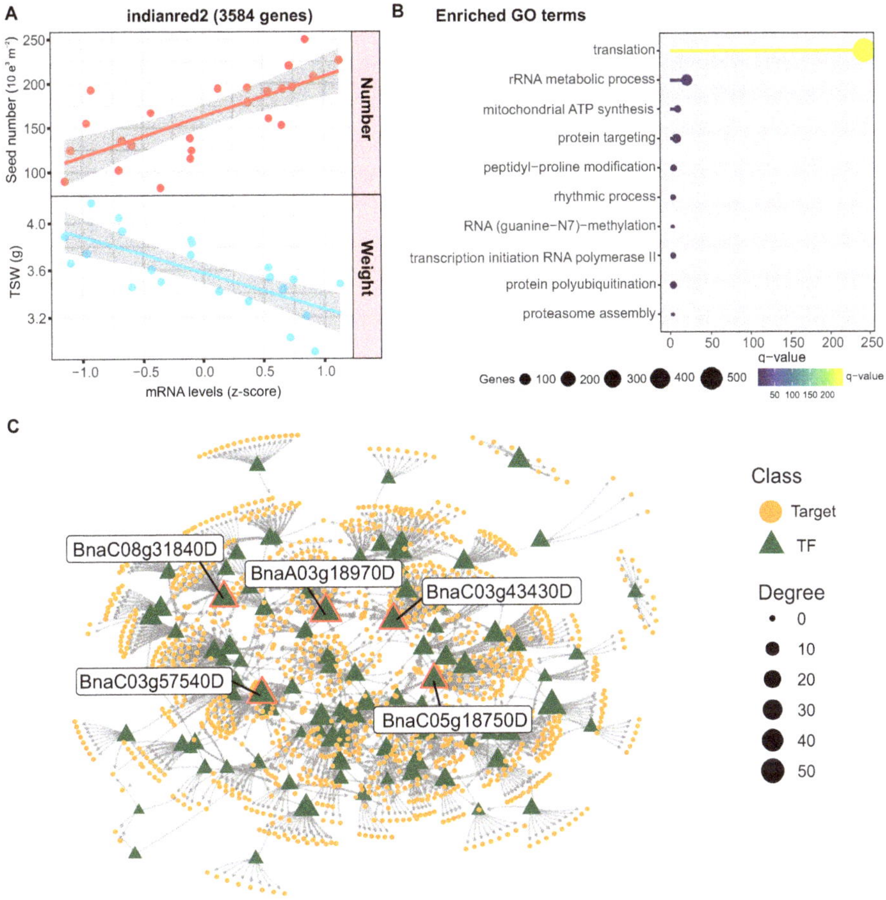

Figure 7. Co-expression network analysis reveals regulators of seed number and weight tradeoff. (**A**) Correlation plot showing the relationship between average expression of indianred2 module genes and seed number/weight across samples. Each point represents the mean expression level of module genes for a sample, with red indicating seed number and blue indicating seed weight. (**B**) Enriched Gene Ontology terms for the indianred2 module related to translation, protein targeting, and other functions. The x-axis shows statistical significance. GO term enrichment analysis was performed using g:Profiler with a q-value cutoff of 0.05 (**C**) Inferred gene regulatory network for the indianred2 module. The network was constructed using GENIE3, ARACNE, and CLR algorithms on gene expression data. Nodes represent transcription factors (triangles) and target genes (circles). The 5 transcription factors with highest connectivity (degree) are labeled. Node size corresponds to degree.

In this way, we inferred a network with 93 transcription factors and 1256 targets, being the five transcription factors with the highest connectivity *BnaC03g43430D* (*PHL2*), *BnaC05g18750D* (*INO1*), *BnaA03g18970D* (*ABI4*), *BnaC08g31840D* (*MYC67*), and *BnaC03g57540D* (*SEP3*) (Figure 7C). Interestingly, functional information available in Arabidopsis indicates that four of these five factors are related to floral architecture [31] or seed development [32–34].

In the case of the palevioletred module, the genes showed higher expression levels in samples with lower seed number and higher weight (Figure 8A). According to gene ontology analysis, the genes in this module are associated with biological functions related to intracellular transport, mRNA metabolism, cell signaling, and chromatin remodeling (Figure 8B). The regulatory network analysis revealed that this co-expression module is controlled by 118 TFs that potentially regulate 1483 targets. The five transcription factors with the highest connectivity were *BnaA06g29050D*, *BnaA04g24200D* (*OSX2*), *BnaC05g51140D*, *BnaC03g72370D* (*NGA1*) and *BnaA06g21460D* (*ARF2*) (Figure 8C). Importantly, the transcription factor ARF2 has been shown to increase seed size due to increased cell proliferation, suggesting that the regulators identified could be important for controlling seed weight [35].

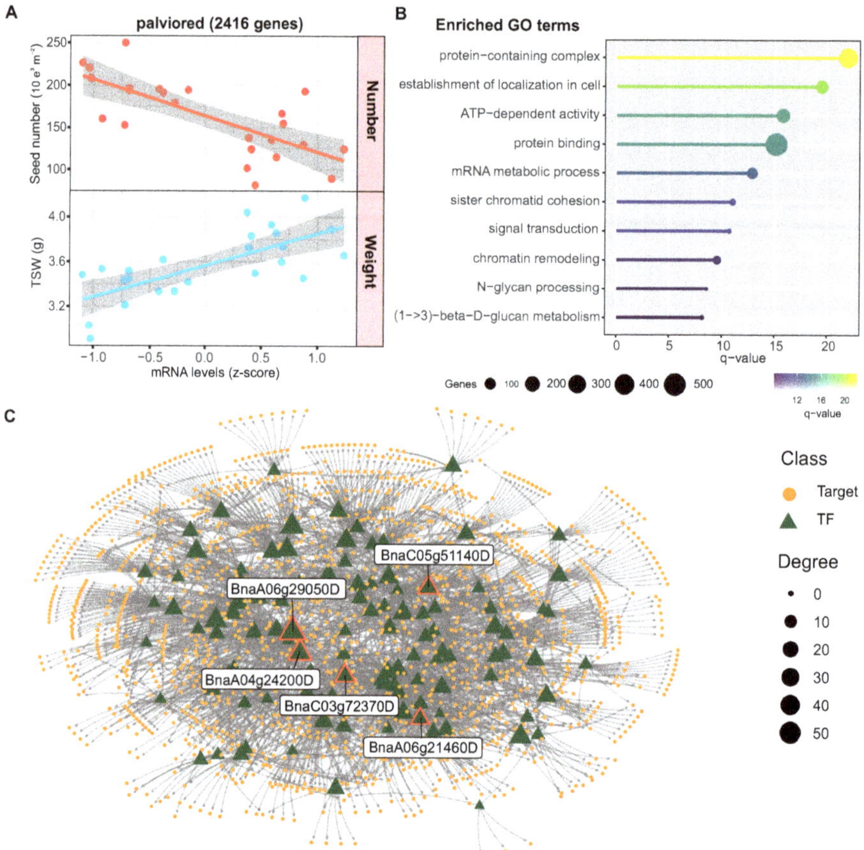

Figure 8. Co-expression module associated with seed weight in rapeseed. (**A**) Scatterplot showing the negative correlation between seed number and the expression pattern of genes in the palevioletred module ($r = -0.77$, $p < 0.0001$). Each dot represents a specific sample. (**B**) Enriched gene ontology (GO) terms for the palevioletred module related to intracellular transport, mRNA metabolism, signaling, and chromatin remodeling. The bar plot shows the number of genes annotated to each term out of the 2416 genes in the module. GO term enrichment analysis was performed using g:Profiler with a q-value cutoff of 0.05. (**C**) Regulatory network showing predicted transcription factor (TF) regulators of the palevioletred module. The network was constructed using GENIE3, ARACNE, and CLR algorithms on gene expression data. The five transcription factors with highest connectivity (degree) are labeled. Node size corresponds to degree.

3. Discussion

Brassica crops such as canola, mustard, and cabbage are cool-season plants that are susceptible to high temperatures, especially during reproductive development [14,36,37]. Even short episodes of heat stress during flowering can significantly reduce seed and oil yields in canola by impairing pollen viability, pollen tube growth, fertilization success, and embryo development [14,38]. Our study found genotype-specific responses to heat stress during flowering and early seed development. The Solar genotype showed greater reductions in seed yield and number compared to Lumen under heat stress. Seed oil and protein concentrations also showed genotype-dependent responses. Lumen maintained higher seed oil content whereas Solar increased protein content under heat stress. These results demonstrate differential heat tolerance between genotypes that impacts yield and quality; however, the underlying molecular mechanisms are unknown. This motivated a large-scale transcriptomic analysis (112 samples) which provides novel insights into the molecular mechanisms governing heat stress responses in rapeseed during the critical reproductive stages of flowering and early seed set. Leveraging field experiments and multi-level omics across source and sink tissues, we reveal both conserved and divergent strategies for thermal adaptation.

3.1. Divergent Molecular Responses to Heat Stress in Source and Sink Tissues

Our multivariate analysis of RNA-seq data revealed distinct effects of heat stress on siliques (source) versus seeds (sink). In the pod wall, heat stress triggered extensive transcriptional repression of genes regulating photosynthesis and carbon assimilation. This corroborates previous observations in model species like Arabidopsis, where high temperature inhibits RuBisCO activase [39], electron transport, and carbon fixation [40]. Exposure to high temperatures was found to negatively impact components of the photosynthetic machinery in *B. napus* seeds, including reduced oil accumulation, lowered photosynthetic and respiration rates, decreased maximum quantum yield of photosystem II, and reduced levels of light-harvesting pigments like chlorophyll [41]. Moreover, a previous transcriptomic analysis in *B. napus* seeds also revealed up-regulation of genes related to the response to high light intensity and genes associated with chloroplasts and photoinhibition in response to heat stress [41].

In maturing seeds, the predominant heat response entailed upregulation of protein homeostasis processes like folding and stabilization. Such activation of heat shock factors, chaperones, and related pathways enables survival under protein-denaturing conditions [42]. Intriguingly, Lumen exhibited fluctuating induction of these genes whereas Solar showed constitutively high expression. This hints at distinct regulatory logic where Lumen employs a reactive mechanism triggered by stress, whereas Solar utilizes an anticipatory strategy.

Our finding that photosynthesis-related genes were downregulated in siliques but not seeds under short heat waves suggests reproductive tissues are more sensitive than vegetative tissues to high temperature damage during flowering. This aligns with previous work showing decreased expression of sucrose transporters like OsSUT1 in rice grains exposed to heat stress, which limits photoassimilate supply and grain filling [43]. Reduced photosynthate translocation is a common consequence of heat stress and a major cause of yield loss [44–46].

3.2. Differential Transcriptome Rapeseed Genotype Responses to Heat Stress

We newly uncover genotype-dependent variation in the extent of photosynthetic gene repression. The heat-tolerant Lumen maintained higher expression of light reaction and Calvin cycle transcripts compared to the sensitive Solar. Sustaining pod photosynthesis likely maintains carbon supply to developing seeds under heat stress, as reflected by Lumen's yield stability (Figure 1). We found that an important gene of the Calvin cycle, *SBPase*, was downregulated under heat stress and consistently showed lower levels in the Solar genotype. Previous research has shown the importance of SBPase activity in photosynthesis and plant growth [47]. Specifically, transgenic plants with decreased SBPase activity have exhibited reduced rates of CO_2 assimilation and carbon fixation [48,49]. This provides evidence that SBPase plays a key role in regulating carbon flow in the Calvin cycle. Furthermore, overexpressing SBPase has been demonstrated to increase photosynthetic capacity, carbon fixation, biomass accumulation, and starch and sucrose levels in tobacco [50,51]. Therefore, the downregulation of SBPase we observed under heat stress likely contributes to decreases in photosynthesis and growth in heat sensitive genotypes. Our findings suggest *SBPase* could be a useful target for genetic modification to improve heat tolerance by maintaining carbon assimilation and plant productivity under high temperatures.

Our results are consistent with the study by Jedlickova et al., 2023 in which photosynthesis-related genes clustered in two co-expression modules showing higher expression in heat-sensitive cultivars DH12075 and Westar compared to heat-tolerant Topas after heat stress [17]. The authors suggest induction of these genes may indicate heat damage to photosystems in sensitive cultivars. Genes connected to photosynthesis and light-harvesting proteins were also up-regulated by heat specifically in seeds of sensitive cultivars [17]. The effects of heat stress on impairing photosynthesis and related gene expression likely contribute to reduced seed production and oil content in sensitive *B. napus* cultivars. Identifying and selecting for thermotolerance factors that protect photosynthesis may improve yield stability.

3.3. Distinct Regulators Mediate the Seed Number versus Weight Trade-Off

Beyond heat responses, our integrated omics approach discovered novel gene modules regulating the complex physiological trade-off between seed number and weight, an area of intense agronomic interest. The identification of the indianred2 and palevioletred modules that show an inverse correlation between seed number and weight suggests there are specific genetic networks controlling this trade-off in rapeseed. To further explore these networks, we analyzed the transcription factors with the highest connectivity in each module.

In the case of indianred2, the transcription factors *INO* (*INNER NO OUTER*) and ABI4 (ABA INSENSITIVE 4) stand out for their relationship with seed development [33,52]. The Arabidopsis *INO* gene encodes a YABBY family transcription factor that acts exclusively in regulating development of the ovule outer integument, with loss-of-function mutants lacking this structure [32,53]. We identified the ortholog of INO in *B. napus* as a hub of a gene coexpression module that showed a positive correlation with seed number per silique and a negative correlation with seed weight. This correlation is consistent with the specific role of INO in promoting outer integument growth. In wild-type Arabidopsis plants, the outer integument facilitates pollen tube guidance, fertilization, and supports embryo development [32]. Thus, loss of INO function leads to reduced seed number due to failed fertilization, while seeds that do form in heterozygous plants show increased weight and size, likely due to additional resources available from unfertilized ovules [54]. The central role of *INO* in regulating outer integument growth and its downstream effects on seed number and weight underlies its identification as a hub of the correlated gene coexpression module.

Our finding that *ABI4* expression positively correlates with seed number in *B. napus* is consistent with previous studies demonstrating a role for *ABI4* in promoting seed production. Specifically, Shu et al., 2016 [33] showed that constitutive overexpression of *ABI4* in *Arabidopsis thaliana* resulted in plants with a 20–26% reduction in total seed yield compared to wildtype. This decrease was attributed to a significant reduction in seed number, as the 1000-grain weight was not affected by ABI4 overexpression [33]. The reduction in seed number was likely due to the altered ABA/GA balance caused by ABI4-mediated upregulation of ABA biosynthesis genes like NCED6 and downregulation of GA catabolic genes like *GA2ox7* [33]. The resulting increase in ABA and decrease in GA probably impaired fertilization and seed development. Taken together, these results indicate that precise control of ABI4 expression is critical for proper seed yield, as both too little and too much ABI4 can negatively impact production by reducing seed number. Thus, our observed positive correlation between ABI4 and seed number in *B. napus* suggests this gene may play a similar role in regulating the ABA/GA balance and reproductive development in oilseed crop species. Further characterization of ABI4's effects on hormone signaling and the seed setting process in *B. napus* will clarify its utility as a target for improving seed yield.

Conversely, in the palevioletred module, *ARF2* (*AUXIN RESPONSE FACTOR 2*) emerges as a high-connectivity transcription factor. *ARF2* is a pivotal regulator in auxin signaling and is implicated in controlling seed size through the modulation of cell expansion and endoreduplication [35]. Our gene co-expression analysis shows a negative correlation with seed number but a positive correlation with seed weight in *Brassica napus*. This result aligns with previous findings in Arabidopsis thaliana that loss-of-function mutations in ARF2 lead to increased seed number but decreased individual seed weight, whereas *ARF2* overexpression has the opposite effect of decreasing seed number but increasing seed size [35]. The conserved role for *ARF2* in promoting seed growth across distinct *Brassicaceae* species is likely due to its activity in stimulating cell proliferation in the integuments of developing ovules, as demonstrated in Arabidopsis [55]. The enlarged integuments generated more cells in the seed coat, allowing increased embryo growth and final seed size. Taken together, these data establish *ARF2* as a key maternal regulator of seed size in multiple Brassicaceae species, functioning to promote integument cell proliferation and limit seed number in order to increase the biomass allocated per seed. Modulating *ARF2* expression or activity provides a promising route to altering seed size and yield in oilseed crops like canola. These transcription factors may represent important nodes in the co-expression modules that modulate the trade-off between seed number and weight in rapeseed.

The co-expression modules associated with the seed number versus weight trade-off differed between this work and our previous work focused on shading treatments effects on seed yield [20]. In the present study under heat stress, the indianred2 module linked to higher seed number showed enrichment for translation and protein functions, whereas the palevioletred module for higher seed weight was associated with transport and signaling. In contrast, in our previous work assessing light stress [20], the turquoise module correlated with seed number was enriched for cell cycle and DNA replication genes, whereas the blue module for seed weight showed lipid metabolism functions. These results indicate distinct molecular processes mediate the trade-off in response to heat versus shading. The hub transcription factors also differed between studies. This work identified floral, embryo and growth regulators as key for seed number or weight (*INO*, *ABI4*, *ARF2*) whereas Canales et al., 2021 [20] highlighted *NAC082*, *NGAL2*, and *IDD1*. The distinct functional enrichment and hub genes likely reflect the different environmental stimuli used in each study to modify source-sink balance during seed set. The low overlap in the modules and regulators identified between the two studies indicates there are mostly specific, rather than conserved, molecular mechanisms mediating the trade-off in response to heat versus shading. Further comparative analyses to find common genetic factors influencing plasticity in seed number and weight would be valuable, but the current results suggest the underlying networks are largely distinct.

4. Materials and Methods

4.1. Field Experiment, Treatments, and Crop Management

A detailed description of the experimental setup, crop management, and weather data is provided in the Supplementary Methods section. Briefly, two field experiments were set in the 2019–20 and 2020–21 growing seasons at the Austral Farming Experimental Station (EEAA) of the Universidad Austral de Chile in Valdivia, Chile (39°47′ S, 73°14′ W). In each experiment, two adapted spring rapeseed hybrids (Lumen and Solar from NPZ-Lembke, Germany) were assessed under two heat stress treatments (Figure S3): a control without manipulation and a thermal increase treatment where plots were heated by 5 °C on average above the environmental temperature from the beginning of flowering [BBCH 61] to 15 days after flowering. The hybrids Lumen and Solar were chosen for their similar phenology and confirmed adaptation to southern Chile. The experiment of the 2019–20 growing season was arranged in a split plot design with three replicates in which the heat stress treatments were assigned to the main plots and genotypes to sub-plots. In the experiment of the 2020–21 growing season, plots were arranged in a randomized block design with three replicates in which the heat stress treatments and genotypes were fully randomized in each block. In both experiments, rapeseed plants were sown at a plant density of 55 plants m^{-2}.

Air temperature and average solar radiation between emergence and the start of flowering recorded across the experiments were 10.9 and 11.4 °C and 17.2 to 18.9 MJ m^2 day^{-1}, respectively (Table S3). Crop phenology of control plants was similar between genotypes and seasons.

4.2. Phenology and Physiological Plant Sampling

Crop development was followed twice a week in both experiments according to the BBCH phenological scale for rapeseed [56]. In order to determine seed yield (SY), seed number (SN), thousand seed weight (TSW), and quality traits as grain oil and protein concentrations, the seed samples were harvested in one lineal meter from the central rows of each plot at maturity (BBCH 89), when the seeds inside the siliques were dark and hard. Seed yield and number was measured after oven drying the samples at 65 °C for 48 h. Seed number was counted by using a grain counter (Pfeuffer GmbH, Kitzengen, Germany). Then, SY was measured and TSW was estimated as the ratio between seed yield and SN.

Oil concentration of seeds was determined by near infrared reflectometry (NIR) (Foss Infratec 1241, Hilleroed, Denmark) and nitrogen concentration of seed by the Kjeldahl procedure [57]. Seed protein concentration was calculated with a conversion factor of 5.8 [58]. Seed concentrations of both oil and protein were expressed on a dry matter basis.

4.3. Seed Sampling and RNA Isolation

Seed samples for RNA-seq analysis were collected from 5 plants per plot (25 siliques per plant from the basal portion of the primary inflorescence) across treatments at two developmental stages: 7 and 14 days after flowering, corresponding to one and two weeks after initiation of the heat stress treatment, respectively. The precise collection dates for each genotype are delineated in Supplementary Table S4. Immediately after harvest, the seed samples were flash frozen in liquid nitrogen and stored at −80 °C until processing. The frozen seeds were ground into a fine powder using liquid nitrogen and a chilled mortar and pestle. Total RNA was extracted from 100 mg of each pulverized seed sample using a protocol adapted for Brassica seeds [59] and NucleoSpin Gel and PCR clean-up columns (Macherey-Nagel, Düren, Germany) as described previously [60].

4.4. RNA-Seq Analysis

The RNA-seq analysis incorporated biological replication, with samples collected from multiple field plots under each condition. In the first field trial season, 3 biological replicates were analyzed per genotype, tissue type, temperature treatment, and time point, for a total of 48 samples. In the second field season, 4 biological replicates were analyzed, yielding 64 samples. Each biological replicate represents plant material harvested from an independent field plot receiving the same conditions. The replicates reflect the inherent biological variability between plants grown under field conditions. RNA samples were transported to Novogene facilities located in Sacramento, CA, USA, under cryogenic conditions utilizing dry ice. Subsequent analyses for RNA quality, as well as library construction and high-throughput sequencing, were executed by Novogene's Beijing-based laboratory. For each RNA sample, a unique Illumina RNA-seq library was constructed using the Illumina TruSeq Stranded Total RNA Sample Prep Kit according to the manufacturer's protocol. Individual barcodes were incorporated during the library preparation for later demultiplexing. Sequencing was conducted employing a 2 × 150 bp paired-end sequencing kit on an Illumina NovaSeq 6000 platform, targeting a sequencing depth of 6 Gb per individual sample. Quality control metrics revealed an average Q20 score of 97.9% and a Q30 score of 94.0% across all samples, thereby attesting to the high fidelity of the sequencing process.

The RNA-seq reads were subjected to a comprehensive computational workflow utilizing pseudoalignment algorithms for accurate transcript quantification and downstream differential expression analysis. Specifically, the Kallisto software (version 0.46.1) was employed for the initial quantification steps [61]. A reference transcriptome index was generated using the "kallisto index" function and rapeseed annotation version AST_PRJEB5043_v1, subsequent to which the "kallisto quant" function was executed on each sample with 100 bootstraps with default parameters. The output provided transcript abundance estimates in terms of transcripts per million (TPM).

For the identification of differentially expressed genes (DEGs) between distinct experimental conditions, the R-based statistical package Sleuth (version 0.30.0) was employed [62]. A Sleuth object was instantiated via the "sleuth_prep" function, incorporating multiple variables into the model, namely genotype, heat treatment, temporal factors, and seasonal variations. Two statistical models were then fitted using the "sleuth_fit" function: a reduced model incorporating all main effects and their respective interactions, and a full model encompassing a four-way interaction term. Likelihood ratio tests (LRT) were executed between these models using the "sleuth_lrt" function, culminating in the identification of DEGs using an adjusted p-value < 0.01.

4.5. Hierarchical Clustering Analysis

For the purpose of uncovering the latent structure in gene expression profiles, hierarchical clustering was conducted. Initially, the gene expression matrix was subjected to normalization, centering, and scaling using native R functions. The Pearson correlation distance matrix was subsequently employed for hierarchical clustering through the "hclust" function, utilizing average linkage as the agglomeration method. The Dynamic Tree Cut algorithm, implemented in the "cutreeDynamic" function from the dynamicTreeCut package [63], was leveraged to partition the hierarchical tree into an optimal number of gene expression modules. Visualization of these modules was facilitated using the ComplexHeatmap package [26].

Functional enrichment analysis of the gene modules was performed using the g:Profiler toolkit [64]. The gene sets corresponding to each identified module were subjected to Gene Ontology (GO) enrichment analysis against *Brassica napus* gene databases using the "gost" function implemented in the R package gprofiler2 [65]. This enabled the elucidation of the biological pathways and molecular functions significantly associated with each gene module.

4.6. Gene Coexpression Network Analysis

To identify modules of coexpressed genes associated with seed yield and quality traits, we performed weighted gene coexpression network analysis (WGCNA) as described previously [29]. The analysis was conducted using the R package BioNERO [66] on normalized RNA-seq count data.

As a first step, we fitted a soft thresholding function to the count data to determine the power (β) for weighted network construction. Using the "SFT_fit function" with a signed hybrid network type, Pearson correlation, and a target R^2 of 0.9 for scale independence, a power of 8 was selected. Next, the exponentiated transformed count data was used to construct a signed coexpression network with the "exp2gcn" function of BioNERO R package [66], specifying the previously determined soft threshold power, a Pearson correlation, and a module merging threshold of 0.8.

The resulting gene coexpression network consisted of 35,432 genes grouped into 33 modules based on topological overlap. To identify modules associated with phenotypes, we calculated module-trait correlations using the module_trait_cor function. Eight modules showed significant correlations with at least one trait related to plant growth, seed yield, or quality ($|r| > 0.7$ and $p < 0.0001$).

To visualize the module-trait relationships, a customized heatmap was generated using the ComplexHeatmap package in R [26]. Modules were assigned to three broad categories (growth, yield, quality) and ordered accordingly. The heatmap depicts module–trait correlations, with the strength of positive correlations shown in red and negative correlations in blue. This systems-level visualization reveals major gene coexpression modules associated with different yield and quality traits.

To infer potential regulators of the yield and quality associated modules, we performed gene regulatory network analysis using the "exp2grn" function of BioNERO R package [66]. Transcription factors were specified based on PlantTFDB 4.0 annotations [30], and the expression data was used to predict regulatory interactions [66]. Hub regulators with the highest connectivity in each module were identified using the "get_hubs_grn" function of BioNERO R package [66], which implements three popular algorithms: GENIE3 [67], ARACNE [68], and CLR [69]. The transcriptomic network analysis provides a systems-level view of potential genetic factors influencing yield and quality traits in response to heat stress.

5. Conclusions

In summary, our multi-level study of rapeseed under field conditions provides new insights into heat stress responses during the critical reproductive phase. We reveal genotype-dependent variation in photosynthetic gene regulation that may confer thermotolerance through sustained carbon assimilation. Our integrated omics approach also newly identified candidate networks and transcriptional regulators governing the agronomically important trade-off between seed number and weight. These findings advance molecular understanding of heat adaptation and yield determination in oilseed crops.

Supplementary Materials: The supporting information can be downloaded at: https://www.mdpi.com/article/10.3390/ijms242115593/s1, references [70,71] are cited in Supplementary Materials.

Author Contributions: Conceptualization, J.C.; methodology, J.F.V.; software, J.C.; validation, J.C.; formal analysis, J.C.; investigation, J.F.V., J.C. and D.F.C.; resources, D.F.C. and J.C.; data curation, J.F.V. and J.C.; writing—original draft preparation J.C.; writing—review and editing, J.C., J.F.V. and D.F.C.; visualization, J.F.V. and J.C.; supervision, J.C. and D.F.C.; project administration J.C. and D.F.C.; funding acquisition, J.C. and D.F.C. All authors have read and agreed to the published version of the manuscript.

Funding: J.C. and D.F.C. were supported by the National Agency for Research and Development (ANID) Chile with Program FONDECYT Regular 1230833 and FONDECYT Regular 1170913. J.C. was also supported by ANID—Millennium Science Initiative Program—ICN17-022.

Institutional Review Board Statement: Not applicable.

Informed Consent Statement: Not applicable.

Data Availability Statement: The RNA-Seq datasets generated during this study are available in the NCBI Gene Expression Omnibus (GEO) repository, accession number GSE241954. All other data generated during this study are included in this published article and its Supplementary Materials.

Acknowledgments: We appreciate the technical assistance of María Beatriz Ugalde, Beatriz Shibar, Eusebio Miranda, and the staff at the Austral Farming Experimental Station (EEAA) of the Universidad Austral de Chile. Furthermore, we extend our heartfelt thanks to every member of the Plant Nutrition and Genomics Laboratory of the same University for their unwavering support and collaborative efforts throughout this study.

Conflicts of Interest: The authors declare no conflict of interest. The funders had no role in the design of the study; in the collection, analyses, or interpretation of data; in the writing of the manuscript; or in the decision to publish the results.

References

1. Cheng, W.; Sakai, H.; Yagi, K.; Hasegawa, T. Interactions of elevated [CO_2] and night temperature on rice growth and yield. *Agric. For. Meteorol.* **2009**, *149*, 51–58. [CrossRef]
2. Kim, H.R.; You, Y.H. Effects of elevated CO_2 concentration and increased temperature on leaf related-physiological responses of Phytolacca insularis (native species) and Phytolacca americana (invasive species). *J. Ecol. Environ.* **2010**, *33*, 195–204. [CrossRef]
3. Ray, D.K.; West, P.C.; Clark, M.; Gerber, J.S.; Prishchepov, A.V.; Chatterjee, S. Climate change has likely already affected global food production. *PLoS ONE* **2019**, *14*, e0217148. [CrossRef] [PubMed]
4. IPCC. *Climate Change 2021—The Physical Science Basis: Working Group I Contribution to the Sixth Assessment Report of the Intergovernmental Panel on Climate Change*; Cambridge University Press: Cambridge, UK, 2023. [CrossRef]
5. Rivelli, G.M.; Long, M.E.F.; Abeledo, L.G.; Calderini, D.F.; Miralles, D.J.; Rondanini, D.P. Assessment of heat stress and cloudiness probabilities in post-flowering of spring wheat and canola in the Southern Cone of South America. *Theor. Appl. Climatol.* **2021**, *145*, 1485–1502. [CrossRef]
6. Sandaña, P.A.; Harcha, C.I.; Calderini, D.F. Sensitivity of yield and grain nitrogen concentration of wheat lupin and pea to source reduction during grain filling. A comparative survey under high yielding conditions. *Field Crops Res.* **2009**, *114*, 233–243. [CrossRef]
7. Mera, M.; Lizana, X.C.; Calderini, D.F. Cropping systems in environments with high yield potential of southern Chile. In *Crop Physiology*; Elsevier: Amsterdam, The Netherlands, 2015; pp. 111–140. [CrossRef]
8. Verdejo, J.; Calderini, D.F. Plasticity of seed weight in winter and spring rapeseed is higher in a narrow but different window after flowering. *Field Crops Res.* **2020**, *250*, 107777. [CrossRef]
9. del Pozo, A.; Jobet, C.; Matus, I.; Méndez-Espinoza, A.M.; Garriga, M.; Castillo, D.; Elazab, A. Genetic Yield Gains and Changes in Morphophysiological-Related Traits of Winter Wheat in Southern Chilean High-Yielding Environments. *Front. Plant Sci.* **2022**, *12*, 732988. [CrossRef]
10. Slafer, G.; Rawson, H. Sensitivity of Wheat Phasic Development to Major Environmental Factors: A Re-Examination of Some Assumptions Made by Physiologists and Modellers. *Funct. Plant Biol.* **1994**, *21*, 393. [CrossRef]
11. Sadras, V.O.; Calderini, D.F. *Crop Physiology Case Histories for Major Crops*; Elsevier: Amsterdam, The Netherlands, 2021. [CrossRef]
12. Secchi, M.A.; Fernandez, J.A.; Stamm, M.J.; Durrett, T.; Prasad, P.V.; Messina, C.D.; Ciampitti, I.A. Effects of heat and drought on canola (*Brassica napus* L.) yield oil, and protein: A meta-analysis. *Field Crops Res.* **2023**, *293*, 108848. [CrossRef]
13. Kourani, M.; Mohareb, F.; Rezwan, F.I.; Anastasiadi, M.; Hammond, J.P. Genetic and Physiological Responses to Heat Stress in Brassica napus. *Front. Plant Sci.* **2022**, *13*, 832147. [CrossRef]
14. Young, L.W.; Wilen, R.W.; Bonham-Smith, P.C. High temperature stress of Brassica napus during flowering reduces micro- and megagametophyte fertility induces fruit abortion, and disrupts seed production. *J. Exp. Bot.* **2004**, *55*, 485–495. [CrossRef] [PubMed]
15. de Almeida, L.M.M.; Coquemont-Guyot, M.; Elie, N.; Morvan-Bertrand, A.; Avice, J.C.; Mollier, A.; Brunel-Muguet, S. Repeated heat stress events during the reproductive phase impact the dynamic development of seeds in *Brassica napus* L. *Plant Sci.* **2023**, *327*, 111559. [CrossRef]
16. Chen, S.; Stefanova, K.; Siddique, K.H.M.; Cowling, W.A. Transient daily heat stress during the early reproductive phase disrupts pod and seed development in *Brassica napus* L. *Food Energy Secur.* **2020**, *10*, e262. [CrossRef]
17. Jedličková, V.; Hejret, V.; Demko, M.; Jedlička, P.; Štefková, M.; Robert, H.S. Transcriptome analysis of thermomorphogenesis in ovules and during early seed development in Brassica napus. *BMC Genom.* **2023**, *24*, 236. [CrossRef]
18. Gao, G.; Hu, J.; Zhang, X.; Zhang, F.; Li, M.; Wu, X. Transcriptome analysis reveals genes expression pattern of seed response to heat stress in Brassica napus L. *Oil Crop Sci.* **2021**, *6*, 87–96. [CrossRef]
19. Lohani, N.; Singh, M.B.; Bhalla, P.L. RNA-Seq Highlights Molecular Events Associated With Impaired Pollen-Pistil Interactions Following Short-Term Heat Stress in Brassica napus. *Front. Plant Sci.* **2021**, *11*, 622748. [CrossRef] [PubMed]
20. Canales, J.; Verdejo, J.; Carrasco-Puga, G.; Castillo, F.M.; Arenas-M, A.; Calderini, D.F. Transcriptome Analysis of Seed Weight Plasticity in Brassica napus. *Int. J. Mol. Sci.* **2021**, *22*, 4449. [CrossRef]
21. Li, N.; Song, D.; Peng, W.; Zhan, J.; Shi, J.; Wang, X.; Liu, G.; Wang, H. Maternal control of seed weight in rapeseed (*Brassica napus* L.): The causal link between the size of pod (mother source) and seed (offspring, sink). *Plant Biotechnol. J.* **2018**, *17*, 736–749. [CrossRef]
22. Dong, H.; Tan, C.; Li, Y.; He, Y.; Wei, S.; Cui, Y.; Chen, Y.; Wei, D.; Fu, Y.; He, Y.; et al. Genome-Wide Association Study Reveals Both Overlapping and Independent Genetic Loci to Control Seed Weight and Silique Length in Brassica napus. *Front. Plant Sci.* **2018**, *9*, 921. [CrossRef]
23. Kirkegaard, J.A.; Lilley, J.M.; Brill, R.D.; Ware, A.H.; Walela, C.K. The critical period for yield and quality determination in canola (*Brassica napus* L.). *Field Crops Res.* **2018**, *222*, 180–188. [CrossRef]
24. Labra, M.H.; Struik, P.C.; Evers, J.B.; Calderini, D.F. Plasticity of seed weight compensates reductions in seed number of oilseed rape in response to shading at flowering. *Eur. J. Agron.* **2017**, *84*, 113–124. [CrossRef]
25. Rivelli, G.; Calderini, D.F.; Abeledo, L.G.; Miralles, D.J.; Rondanini, D.P. Yield and Quality Traits of Wheat and Rapeseed in Response to Source-Sink Ratio and Heat Stress in Post-Flowering. *Soc. Sci. Res. Netw.* **2023**. [CrossRef]
26. Gu, Z. Complex heatmap visualization. *iMeta* **2022**, *1*, e43. [CrossRef]

27. Liu, X.; Yu, H.; Guan, Y.; Li, J.; Guo, F. Carbonylation and loss-of-function analyses of SBPase reveal its metabolic interface role in oxidative stress, carbon assimilation, and multiple aspects of growth and development in Arabidopsis. *Mol. Plant* **2012**, *5*, 1082–1099. [CrossRef]
28. Hashida, S.; Miyagi, A.; Nishiyama, M.; Yoshida, K.; Hisabori, T.; Kawai-Yamada, M. Ferredoxin/thioredoxin system plays an important role in the chloroplastic NADP status of Arabidopsis. *Plant J.* **2018**, *95*, 947–960. [CrossRef]
29. Langfelder, P.; Horvath, S. WGCNA: An R package for weighted correlation network analysis. *BMC Bioinform.* **2008**, *9*, 559. [CrossRef]
30. Jin, J.; Tian, F.; Yang, D.C.; Meng, Y.Q.; Kong, L.; Luo, J.; Gao, G. PlantTFDB 4.0: Toward a central hub for transcription factors and regulatory interactions in plants. *Nucleic Acids Res.* **2016**, *45*, D1040–D1045. [CrossRef]
31. Yu, H.; Murchie, E.H.; González-Carranza, Z.H.; Pyke, K.A.; Roberts, J.A. Decreased photosynthesis in the erect panicle 3 (ep3) mutant of rice is associated with reduced stomatal conductance and attenuated guard cell development. *J. Exp. Bot.* **2015**, *66*, 1543–1552. [CrossRef]
32. Villanueva, J.M.; Broadhvest, J.; Hauser, B.A.; Meister, R.J.; Schneitz, K.; Gasser, C.S. INNER NO OUTER regulates abaxial-adaxial patterning in Arabidopsis ovules. *Genes Dev.* **1999**, *13*, 3160–3169. [CrossRef]
33. Shu, K.; Chen, Q.; Wu, Y.; Liu, R.; Zhang, H.; Wang, P.; Li, Y.; Wang, S.; Tang, S.; Liu, C.; et al. ABI4 mediates antagonistic effects of abscisic acid and gibberellins at transcript and protein levels. *Plant J.* **2016**, *85*, 348–361. [CrossRef]
34. Renard, J.; Niñoles, R.; Martínez-Almonacid, I.; Gayubas, B.; Mateos-Fernández, R.; Bissoli, G.; Bueso, E.; Serrano, R.; Gadea, J. Identification of novel seed longevity genes related to oxidative stress and seed coat by genome-wide association studies and reverse genetics. *Plant Cell Environ.* **2020**, *43*, 2523–2539. [CrossRef]
35. Schruff, M.C.; Spielman, M.; Tiwari, S.; Adams, S.; Fenby, N.; Scott, R.J. The *AUXIN RESPONSE FACTOR 2* gene of *Arabidopsis* links auxin signalling, cell division, and the size of seeds and other organs. *Development* **2006**, *133*, 251–261. [CrossRef]
36. Angadi, S.V.; Cutforth, H.W.; Miller, P.R.; McConkey, B.G.; Entz, M.H.; Brandt, S.A.; Volkmar, K.M. Response of three *Brassica* species to high temperature stress during reproductive growth. *Can. J. Plant Sci.* **2000**, *80*, 693–701. [CrossRef]
37. Lohani, N.; Singh, M.B.; Bhalla, P.L. High temperature susceptibility of sexual reproduction in crop plants. *J. Exp. Bot.* **2019**, *71*, 555–568. [CrossRef]
38. Mácová, K.; Prabhullachandran, U.; Štefková, M.; Spyroglou, I.; Pěnčík, A.; Endlová, L.; Novák, O.; Robert, H.S. Long-Term High-Temperature Stress Impacts on Embryo and Seed Development in Brassica napus. *Front. Plant Sci.* **2022**, *13*. [CrossRef]
39. Kim, K.; Portis, A.R. Temperature Dependence of Photosynthesis in Arabidopsis Plants with Modifications in Rubisco Activase and Membrane Fluidity. *Plant Cell Physiol.* **2005**, *46*, 522–530. [CrossRef] [PubMed]
40. Wang, L.; Ma, K.B.; Lu, Z.G.; Ren, S.X.; Jiang, H.R.; Cui, J.W.; Chen, G.; Teng, N.J.; Lam, H.M.; Jin, B. Differential physiological transcriptomic and metabolomic responses of Arabidopsis leaves under prolonged warming and heat shock. *BMC Plant Biol.* **2020**, *20*, 86. [CrossRef]
41. Huang, R.; Liu, Z.; Xing, M.; Yang, Y.; Wu, X.; Liu, H.; Liang, W. Heat Stress Suppresses Brassica napus Seed Oil Accumulation by Inhibition of Photosynthesis and BnWRI1 Pathway. *Plant Cell Physiol.* **2019**, *60*, 1457–1470. [CrossRef] [PubMed]
42. ul Haq, S.; Khan, A.; Ali, M.; Khattak, A.M.; Gai, W.X.; Zhang, H.X.; Wei, A.M.; Gong, Z.H. Heat Shock Proteins: Dynamic Biomolecules to Counter Plant Biotic and Abiotic Stresses. *Int. J. Mol. Sci.* **2019**, *20*, 5321. [CrossRef] [PubMed]
43. Miyazaki, M.; Araki, M.; Okamura, K.; Ishibashi, Y.; Yuasa, T.; Iwaya-Inoue, M. Assimilate translocation and expression of sucrose transporter OsSUT1, contribute to high-performance ripening under heat stress in the heat-tolerant rice cultivar Genkitsukushi. *J. tPlant Physiol.* **2013**, *170*, 1579–1584. [CrossRef]
44. Moore, C.E.; Meacham-Hensold, K.; Lemonnier, P.; Slattery, R.A.; Benjamin, C.; Bernacchi, C.J.; Lawson, T.; Cavanagh, A.P. The effect of increasing temperature on crop photosynthesis: From enzymes to ecosystems. *J. Exp. Bot.* **2021**, *72*, 2822–2844. [CrossRef]
45. Abdelrahman, M.; Burritt, D.J.; Gupta, A.; Tsujimoto, H.; Tran, L.S.P. Heat stress effects on source–sink relationships and metabolome dynamics in wheat. *J. Exp. Bot.* **2019**, *71*, 543–554. [CrossRef] [PubMed]
46. Kumar, S.; Bhushan, B.; Wakchaure, G.C.; Dutta, R.; Jat, B.S.; Meena, K.K.; Rakshit, S.; Pathak, H. Unveiling the impact of heat stress on seed biochemical composition of major cereal crops: Implications for crop resilience and nutritional value. *Plant Stress* **2023**, *9*, 100183. [CrossRef]
47. Harrison, E.P.; Olcer, H.; Lloyd, J.C.; Long, S.P.; Raines, C.A. Small decreases in SBPase cause a linear decline in the apparent RuBP regeneration rate but do not affect Rubisco carboxylation capacity. *J. Exp. Bot.* **2001**, *52*, 1779–1784. [CrossRef]
48. Harrison, E.P.; Willingham, N.M.; Lloyd, J.C.; Raines, C.A. Reduced sedoheptulose-1,7-bisphosphatase levels in transgenic tobacco lead to decreased photosynthetic capacity and altered carbohydrate accumulation. *Planta* **1997**, *204*, 27–36. [CrossRef]
49. Lawson, T.; Bryant, B.; Lefebvre, S.; Loyd, J.C.; Raines, C.A. Decreased SBPase activity alters growth and development in transgenic tobacco plants. *Plant Cell Environ.* **2006**, *29*, 48–58. [CrossRef] [PubMed]
50. Miyagawa, Y.; Tamoi, M.; Shigeoka, S. Overexpression of a cyanobacterial fructose-1,6-/sedoheptulose-1,7-bisphosphatase in tobacco enhances photosynthesis and growth. *Nat. Biotechnol.* **2001**, *19*, 965–969. [CrossRef]
51. Lefebvre, S.; Lawson, T.; Fryer, M.; Zakhleniuk, O.V.; Lloyd, J.C.; Raines, C.A. Increased Sedoheptulose-1,7-Bisphosphatase Activity in Transgenic Tobacco Plants Stimulates Photosynthesis and Growth from an Early Stage in Development. *Plant Physiol.* **2005**, *138*, 451–460. [CrossRef]
52. Skinner, D.J.; Dang, T.; Gasser, C.S. The Arabidopsis *INNER NO OUTER (INO)* gene acts exclusively and quantitatively in regulation of ovule outer integument development. *Plant Direct* **2023**, *7*, e485. [CrossRef]

53. Skinner, D.J. Regulation of Ovule Development. *Plant Cell Online* **2004**, *16*, S32–S45. [CrossRef]
54. Jia, D.; Chen, L.G.; Yin, G.; Yang, X.; Gao, Z.; Guo, Y.; Sun, Y.; Tang, W. Brassinosteroids regulate outer ovule integument growth in part via the control of *INNER NO OUTER* by BRASSINOZOLE-RESISTANT family transcription factors. *J. Integr. Plant Biol.* **2020**, *62*, 1093–1111. [CrossRef] [PubMed]
55. Adamski, N.M.; Anastasiou, E.; Eriksson, S.; O'Neill, C.M.; Lenhard, M. Local maternal control of seed size by *KLUH/CYP78A5*-dependent growth signaling. *Proc. Natl. Acad. Sci. USA* **2009**, *106*, 20115–20120. [CrossRef]
56. Meier, U. *Growth Stages of Mono-And Dicotyledonous Plants*; Julius Kühn-Institut: Quedlinburg, Germany, 2018.
57. Kirk, P.L. Kjeldahl Method for Total Nitrogen. *Anal. Chem.* **1950**, *22*, 354–358. [CrossRef]
58. Merrill, A.; Watt, B. *Energy Value of Foods: Basis and Derivation*; Scientific Research Publishing: Wuhan, China, 1973.
59. Graeber, K.; Linkies, A.; Wood, A.T.; Leubner-Metzger, G. A Guideline to Family-Wide Comparative State-of-the-Art Quantitative RT-PCR Analysis Exemplified with a Brassicaceae Cross-Species Seed Germination Case Study . *Plant Cell* **2011**, *23*, 2045–2063. [CrossRef] [PubMed]
60. Yaffe, H.; Buxdorf, K.; Shapira, I.; Ein-Gedi, S.; Zvi, M.M.B.; Fridman, E.; Moshelion, M.; Levy, M. LogSpin: A simple economical and fast method for RNA isolation from infected or healthy plants and other eukaryotic tissues. *BMC Res. Notes* **2012**, *5*, 45. [CrossRef] [PubMed]
61. Bray, N.L.; Pimentel, H.; Melsted, P.; Pachter, L. Near-optimal probabilistic RNA-seq quantification. *Nat. Biotechnol.* **2016**, *34*, 525–527. [CrossRef]
62. Pimentel, H.; Bray, N.L.; Puente, S.; Melsted, P.; Pachter, L. Differential analysis of RNA-seq incorporating quantification uncertainty. *Nat. Methods* **2017**, *14*, 687–690. [CrossRef]
63. Langfelder, P.; Zhang, B.; Horvath, S. Defining clusters from a hierarchical cluster tree: The Dynamic Tree Cut package for R. *Bioinformatics* **2007**, *24*, 719–720. [CrossRef] [PubMed]
64. Raudvere, U.; Kolberg, L.; Kuzmin, I.; Arak, T.; Adler, P.; Peterson, H.; Vilo, J. g:Profiler: A web server for functional enrichment analysis and conversions of gene lists (2019 update). *Nucleic Acids Res.* **2019**, *47*, W191–W198. [CrossRef]
65. Kolberg, L.; Raudvere, U.; Kuzmin, I.; Vilo, J.; Peterson, H. gprofiler2—An R package for gene list functional enrichment analysis and namespace conversion toolset g:Profiler. *F1000Research* **2020**, *9*, 709. [CrossRef]
66. Almeida-Silva, F.; Venancio, T.M. BioNERO: An all-in-one R/Bioconductor package for comprehensive and easy biological network reconstruction. *Funct. Integr. Genom.* **2021**, *22*, 131–136. [CrossRef] [PubMed]
67. Huynh-Thu, V.A.; Irrthum, A.; Wehenkel, L.; Geurts, P. Inferring Regulatory Networks from Expression Data Using Tree-Based Methods. *PLoS ONE* **2010**, *5*, e12776. [CrossRef]
68. Margolin, A.A.; Nemenman, I.; Basso, K.; Wiggins, C.; Stolovitzky, G.; Favera, R.D.; Califano, A. ARACNE: An Algorithm for the Reconstruction of Gene Regulatory Networks in a Mammalian Cellular Context. *BMC Bioinform.* **2006**, *7*, S7. [CrossRef]
69. Faith, J.J.; Hayete, B.; Thaden, J.T.; Mogno, I.; Wierzbowski, J.; Cottarel, G.; Kasif, S.; Collins, J.J.; Gardner, T.S. Large-Scale Mapping and Validation of Escherichia coli Transcriptional Regulation from a Compendium of Expression Profiles. *PLoS Biol.* **2007**, *5*, e8. [CrossRef] [PubMed]
70. Lizana, X.C.; Calderini, D.F. Yield and grain quality of wheat in response to increased temperatures at key periods for grain number and grain weight determination: Considerations for the climatic change scenarios of Chile. *J. Agric. Sci.* **2013**, *151*, 209–221. [CrossRef]
71. Kutner, M.; Nachtsheim, C.; Neter, J. *Applied Linear Regression Models*; McGraw-Hill Education: Boston, MA, USA, 2004; p. 701.

International Journal of
Molecular Sciences

MDPI

Article

Integrated Transcriptomic and Metabolomic Analysis Reveal the Underlying Mechanism of Anthocyanin Biosynthesis in *Toona sinensis* Leaves

Jing Xu [1,2], Yanru Fan [1,2], Xiaojiao Han [1,2], Huanhuan Pan [1,2], Jianhua Dai [1,2], Yi Wei [1,2], Renying Zhuo [1,2] and Jun Liu [1,2,*]

[1] State Key Laboratory of Tree Genetics and Breeding, Chinese Academy of Forestry, Beijing 100091, China
[2] Key Laboratory of Tree Breeding of Zhejiang Province, Research Institute of Subtropical Forestry, Chinese Academy of Forestry, Hangzhou 311400, China
* Correspondence: liujun@caf.ac.cn; Tel.: +86-571-6313-1172

Abstract: *Toona sinensis*, commonly known as Chinese Toon, is a plant species that possesses noteworthy value as a tree and vegetable. Its tender young buds exhibit a diverse range of colors, primarily determined by the presence and composition of anthocyanins and flavonoids. However, the underlying mechanisms of anthocyanin biosynthesis in *Toona sinensis* have been rarely reported. To explore the related genes and metabolites associated with composition of leaf color, we conducted an analysis of the transcriptome and metabolome of five distinct *Toona* clones. The results showed that differentially expressed genes and metabolites involved in anthocyanin biosynthesis pathway were mainly enriched. A conjoint analysis of transcripts and metabolites was carried out in JFC (red) and LFC (green), resulting in the identification of 510 genes and 23 anthocyanin-related metabolites with a positive correlation coefficient greater than 0.8. Among these genes and metabolites, 23 transcription factors and phytohormone-related genes showed strong coefficients with 13 anthocyanin derivates, which mainly belonged to the stable types of delphinidin, cyanidin, peonidin. The core derivative was found to be Cyanidin-3-O-arabinoside, which was present in JFC at 520.93 times the abundance compared to LFC. Additionally, the regulatory network and relative expression levels of genes revealed that the structural genes *DFR*, *ANS*, and *UFGT1* might be directly or indirectly regulated by the transcription factors SOC1 (MADS-box), CPC (MYB), and bHLH162 (bHLH) to control the accumulation of anthocyanin. The expression of these genes was significantly higher in red clones compared to green clones. Furthermore, RNA-seq results accurately reflected the true expression levels of genes. Overall, this study provides a foundation for future research aimed at manipulating anthocyanin biosynthesis to improve plant coloration or to derive human health benefits.

Keywords: *Toona sinensis*; transcriptome; metabolome; anthocyanin; transcription factor; plant hormones

1. Introduction

As the national economy grows rapidly, and living standards continue to improve, consumers today demand more than just food and clothing. They desire a healthy diet that also helps prevent diseases. Food that is high in nutrition and has antioxidant properties is particularly appealing [1]. Chinese Toon (*Toona sinensis*), a tree with both medicinal and edible uses, contains rich chemical substances with functions such as antifungal, antiglycation, or anti-tumor activities. Flavonoid compounds, especially anthocyanin, have been identified as the major bioactive constituents in *Toona* [2–4] which are considered to be of the most beneficial plant-derived antioxidants and have led to increased efforts to analyze the genetic controls of the mechanism of flavonoids synthesis [5,6].

Anthocyanins belong to the flavonoid class of phenolic compounds. Anthocyanins are water-soluble pigments present in plants that contribute to the colors of leaves, flowers, and fruit, which are also crucial for plant growth and beneficial for human health due to their ability to scavenge active oxygens [7]. More than 600 distinct anthocyanin types have been identified so far through previous research [8]. The most prevalent anthocyanin pigments, which generate purple, blue, and red colors, are cyanidin (Cy), delphinidin (Dp), pelargonidin (Pg), peonidin (Pn), petunidin (Pt), and malvidin (Mv) [9]. Anthocyanin biosynthesis and regulation is a complex process influenced by various internal and external factors [10,11].

The anthocyanin biosynthesis pathway is a well-characterized process in plants such as *Arabidopsis thaliana*, apple, pear and tomato [12–18]. The pathway is initiated by the phenylpropanoid and flavonoid pathways, which are controlled by both structural and regulatory genes. The expression of functional structural genes was related with the accumulation of anthocyanin [19]. Overexpression of enzymes such as dihydroflavonol-4-reductase (DRF), anthocyanidin synthase (ANS) and UDP-glucoside: flavonoid glucosyltransferase (UFGT) can increase color accumulation in fruits. Meanwhile, The MBW complex, which is formed by MYB-bHLH-WD40 transcription factors, regulates anthocyanin biosynthesis by promoting or inhibiting the expression of structural genes [18,20–25]. The MYB transcription factor TRANSPARENT TESTA 2 (TT2) in *Arabidopsis thaliana* interacts with TT8 (bHLH42) and TTG1 (WD40) to regulated the expression of late biosynthetic genes, resulting in pigmentation of the seed coat [26]. In addition, the transcription factor, such as WRKY, NAC, bZIP, and MADS, have also been proven to play a role in regulating anthocyanin biosynthesis. *LONG HYPOCOTYL 5 (HY5)*, a bZIP transcription factor, directly participates in photosensitive morphology, UV-B resistance, and anthocyanin biosynthesis [27,28]. Studies have revealed its interaction with B-Box proteins (BBX) to control anthocyanin biosynthesis [29,30]. In addition, plant hormones also interact with transcription factors to influence the regulation of anthocyanin accumulation [31,32]. Jasmonate ZIM-domain (JAZ) proteins interact with essential components of WD-repeat/bHLH/MYB transcriptional complexes, namely bHLH (Transparent Testa8, Glabra3 [GL3], and Enhancer of Glabra3 [EGL3]) and R2R3 MYB transcription factors (MYB75 and Glabra1), to repress JA (Jasmonic acid)-regulated anthocyanin accumulation and trichome initiation [33]. Identification of structural enzymes and regulatory factors is crucial for progress in understanding the underlying mechanism of anthocyanin biosynthesis regulation.

The investigation of anthocyanin-regulating genes and associated flavonoid metabolites is critical to elucidate the mechanism that governs the biological synthesis of anthocyanins. To study gene expression and metabolic changes in different cultivars, researchers mainly rely on transcriptomic and metabolic profiling, which complement each other [34,35]. In this study, we conducted a dual-level exploration of genes and metabolites that control the pigmentation of *Toona sinensis* leaves using transcriptome and metabolomics approaches. Specifically, we analyzed the transcriptome and metabolome data of five *Toona sinensis* clones exhibiting a range of leaf colors, including red, light red, pink, dark, and green, to clarify regulatory networks for biosynthesis-related genes and key metabolomes. Overall, our findings unveil the regulatory mechanisms driving the coloration of *Toona sinensis* leaves.

2. Results

2.1. Morphological Observation and Anthocyanin Content of Leaves in Toona sinensis

Five clones of *Toona sinensis* with distinct leaf colors were selected to analyze the regulatory mechanism of anthocyanin biosynthesis. Among the clones, JFC *Toona* is characterized by brilliant-red leaves and is commonly cultivated in Zhejiang province. The leaf color of other clones was pink buds (FHXN), light-red with a little green buds (BSH), dark-purple buds (HBG), and green buds (LFC) (Figure 1A). Then, the total anthocyanin was extracted and the resulting color were displayed in Figure 1B, which reflect the anthocyanin content of different clones. Total anthocyanin present in the leaves were quantified, with JFC *Toona sinensis* found to have the highest concentration of anthocyanin, followed by BSH, FHXN and HBG, and LFC (Figure 1C). Obviously, LFC exhibited green leaves due to a lack of anthocyanin accumulation.

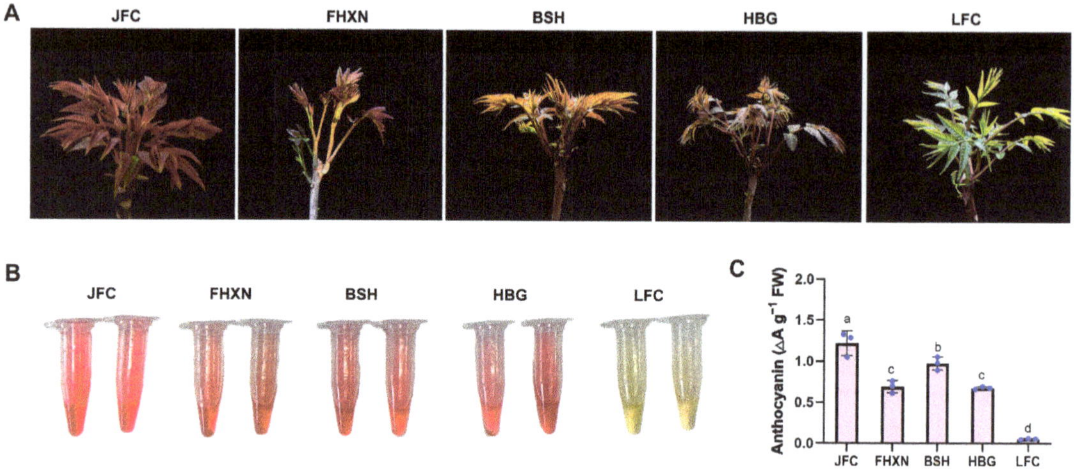

Figure 1. Morphological observation and anthocyanin content of leaves from different *Toona* clones (JFC, FHXN, BSH, HBG, and LYC). (**A**) The morphological characteristics of leaves from the five *Toona* clones. (**B**) The color of anthocyanin present in the leaves of these clones. (**C**) The quantification of anthocyanin content in each of the *Toona* clones. Data are shown as mean ± standard deviation, $n = 3$. Bars with different letters are significantly different at $p < 0.05$.

2.2. Metabolome Analysis of Anthocyanin in Five Toona Clones

To compare the flavonoids and anthocyanin compound composition among five *Toona* clones, three biological replicates of each clone were sampled to detect anthocyanin-related metabolites employing the UPLC-MS/MS platform. The resulting metabolite composition data sets underwent PCA analysis (Figure 2A). The PCA plot for the anthocyanin-related metabolites showed a clear separation between the red, pink, light red, green, dark-purple samples. A total of 43, 40, 44, 40, and 32 anthocyanin-related metabolites were identified in leaves of JFC, BSH, FHXN, HBG, and LFC, respectively (Table S1). A total of 31 anthocyanins were identified in all of the samples. Then, the $|\log2(\text{fold-change})| \geq 1$ and p value < 0.05 was used to determine significantly metabolites between groups (Figure 2B). In group LFC vs. JFC, 23 upregulated metabolites and 8 downregulated metabolites were found to be significantly different (Figure 2B). In LFC vs. BSH, there were 24 upregulated metabolites and 6 downregulated metabolites (Figure 2B). In BSH vs. JFC, there were 10 upregulated metabolites and 11 downregulated metabolites (Figure 2B). Counting the differential metabolites compared with both LFC and JFC for each individual plant, a Venn diagram was used to show that there were 6 differential metabolites in the comparison group compared with JFC and 24 common differential metabolites compared with LFC (Figure 2C,D). Thirty-one common detected metabolites were used to draw clustering heatmap by TBtools (https://bio.tools/tbtools, accessed on 13 September 2023). The contents of cyanidin and delphinidin could be mainly responsible for the leaves color of *Toona sinensis* (Figure 2E).

Metabolomic analyses revealed significantly higher levels of several anthocyanin pigments, namely Naringenin, Delphinidin-3-O-rutinoside, Delphinidin-3-O-5-O-(6-O-coumaroyl)-diglucoside, Cyanidin-3-O-arabinoside, Cyanidin-3-O-glucoside, Cyanidin-3-O-rutinoside, Cyanidin-3-O-galactoside, Cyanadin-3-O-xyloside, Cyanidin-3-O-sambubioside, Cyanidin-3,5-O-diglucoside, Peonidin-3-O-rutinoside, in the JFC variety compared to other varieties. Interesting, among these anthocyanins, the content of Cyanidin-3-O-arabinoside was found to be 520.93-fold higher in JFC than in LFC.

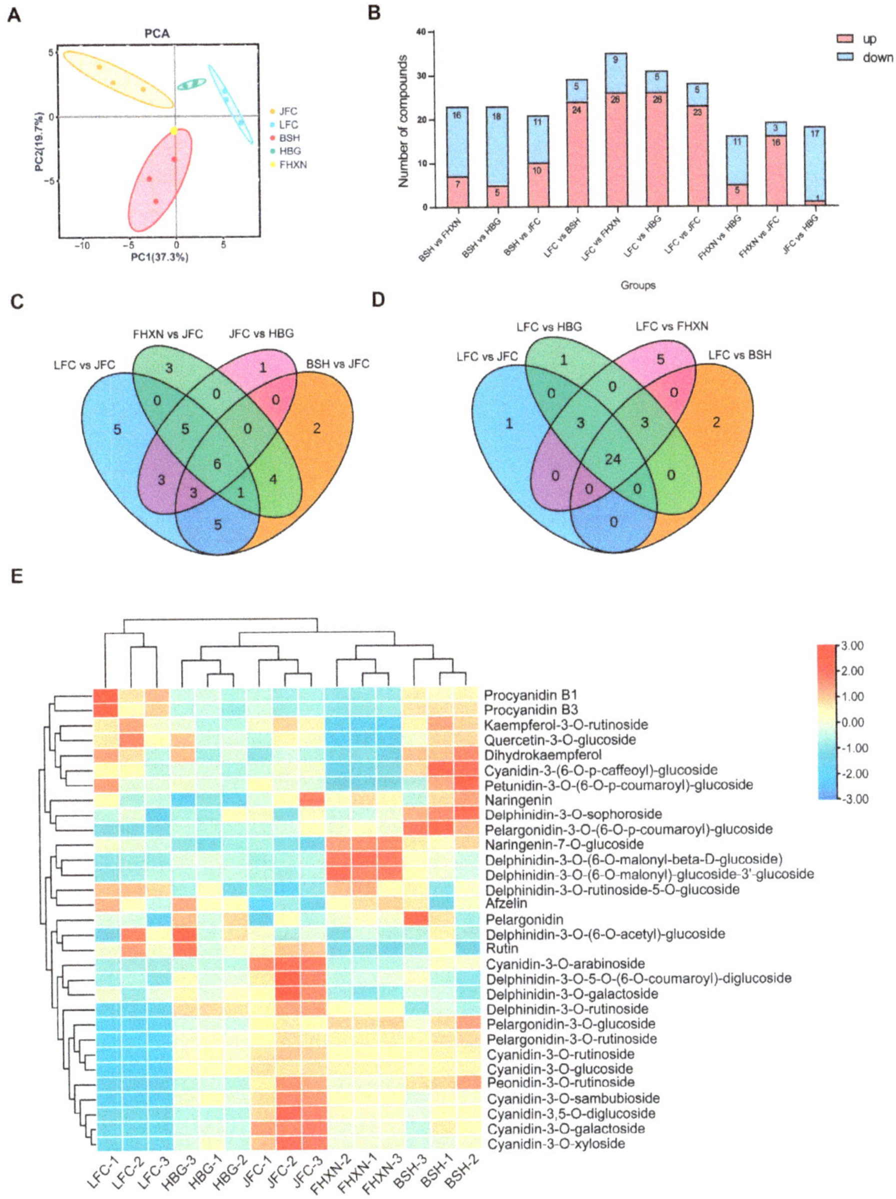

Figure 2. Targeted metabolome profiles of the anthocyanin biosynthesis pathway. (**A**) A PCA score plot demonstrates variations among different colored *Toona* clones. (**B**) The number of up- and down-regulated metabolites varies across different comparison groups. (**C**,**D**) Venn diagram illustrate the number of differential metabolites when compared with JFC and LFC. (**E**) Heatmaps reveals the differential metabolites present in diverse *Toona* clones. Three independent replicates of each clone are displayed in the heatmap.

2.3. *Transcriptome Analysis of Five Toona sinensis Clones*

Fifteen cDNA libraries (each clone with three biological replicates) were sequenced using the Illumina HiSeq platform. After removing low-quality reads, each sample yielded between 42.16 M to 52.03 M clean reads (Supplementary Figure S1). The Q30 base exceeded 92.56% and the guanine–cytosine (GC) content of each sample was greater than 42.27%, indicating acceptable sequencing quality. After filtration, the clean data reached 6.32–7.80 Gb for each sample. The genome of *Toona sinensis* var. Heiyouchun was utilized, which marked a significant was the first chromosome-level genome assembled of Toona, as our reference genome [36]. Furthermore, comparison of all clean reads with this reference genome and subsequently performed gene annotations, the results revealed a high rate of similarity, ranging from 92.26% to 94.03%. These results validate the reliability of our data. Novel genes were defined as unigenes found in the sequencing results but not included in the reference genome (or reference gene set) after reconstructing the transcripts using StringTie v1.3.4d software. Comparison identified 9127 novel genes. The PCA results of transcriptome showed significant differences among five clones (Figure 3A).

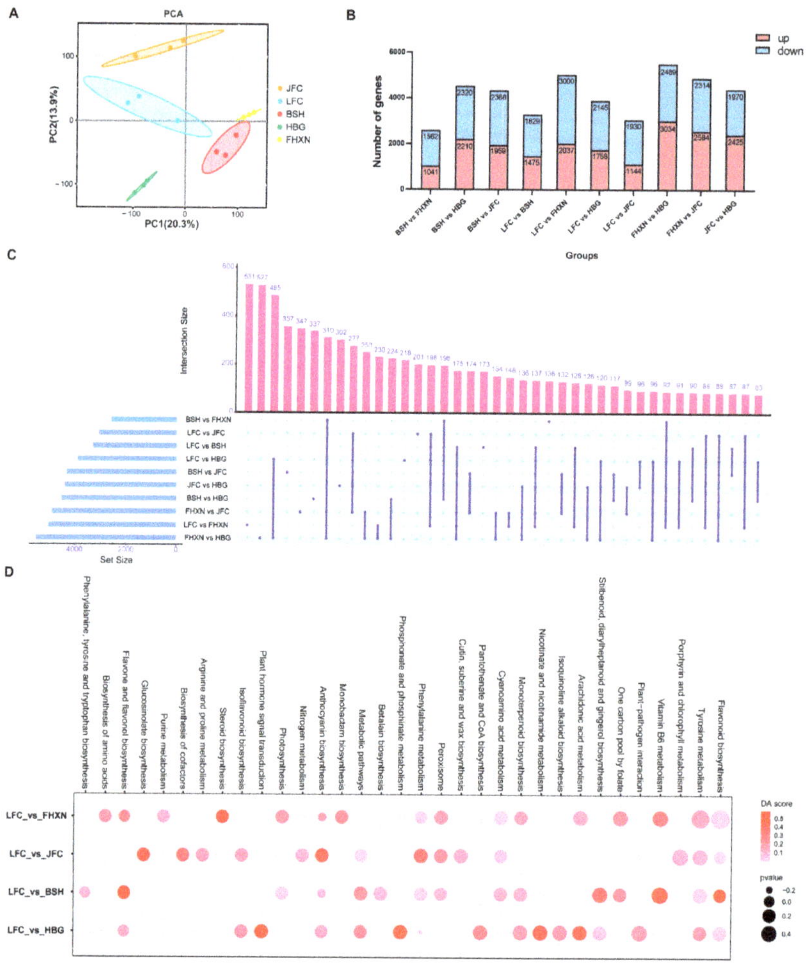

Figure 3. Transcriptome analysis of leaves among *Toona* clones. (**A**) A PCA score plot of the RNA-seq results of various colored *Toona*. (**B**) The number of up- and down-regulated genes differs across different comparison groups. (**C**) A shared number of DEGs across varying comparison groups. (**D**) A KEGG pathway enrichment bubble plot for DEGs between different comparison groups.

To investigate the function of unigenes in *Toona*, we conducted a comparative analysis with various databases including KEGG, NR, Swiss-Prot, GO, KOG, and Trembl using BLAST (Diamond version 2.0.9) software. To obtain annotation information on unigenes, amino acid sequences were derived from the unigene sequences and compared against the Pfam database using HMMER V3.2 (http://hmmer.org/, accessed on 10 June 2022) software. Interestingly, when compared to data in the NR database, it was observed that gene sequences from *Toona sinensis* exhibited a greater degree in similarity to *Citrus clementina* (30.91%), followed by *Citrus sinensis* (30.47%) as depicted in Supplementary Figure S1B. The function terms of unigenes were annotated in the KOG and GO database, and the function groups were showed in Supplementary Figure S1C,D.

Based on the reference transcriptome, 40,421 genes were assembled from clean reads, which included 31,647 genes annotation in reference genome and 9127 novel transcripts. In addition, 23,909, 16,532, 25,617, 15,566, 32,348, and 25,575 unigenes were annotated by KEGG, NR, SWISSPORT, THEM, GO, and KOG database. There were 14,153 unigenes annotated by all of the databases. A comparison with six public databases (KEGG, NR, SwissProt, Tremble, GO and KOG) annotated a total of 13,295 differentially expressed genes (DEGs) and predicated their potential functions. The DEGs were counted with an absolute $|\log2\text{fold-change}| \geq 1$ and a false discovery rate (FDR) ≤ 0.05 in each compared groups. According to five clones of *Toona* leaves, 10 pairs of comparisons were performed and the number of up-regulated and down-regulated differential genes was counted in all groups (Figure 3B). Across all compared groups, a total of 310 DEGs were co-detected (Figure 3C). The KEGG-based enrichment analysis revealed that all DEGs compared to LFC were enriched into 140 KEGG pathways. These pathways included flavonoid biosynthesis, phenylpropanoid biosynthesis, and anthocyanin biosynthesis, all of which were significantly enriched (Figure 3D). These terms have previously been associated with anthocyanin production.

2.4. *Analysis of Anthocyanidin Biosynthetic Pathway Genes and Anthocyanin Derivates in JFC, BSH, and LFC*

According to transcriptome and metabolites studies, anthocyanin biosynthesis may be responsible for the distinct red leaf colorations of *Toona*. Therefore, it is necessary to investigate the mechanisms that underlie this process. Genes encoding anthocyanin biosynthetic enzymes were found to be associated with different compositions of the compounds in the anthocyanin biosynthesis pathways. We constructed a pathway diagram displaying the expression heat map of structural genes and compounds for red, light-red, and green leaves (Figure 4). The *PAL* gene catalyzed the conversion of L-Phenylalanine to cinnamic acid and was highly expressed in JFC. Among the *4CL* genes, two exhibited a gradual decrease in expression from red to green leaves across JFC, BSH, and LFC. CHS, the first key enzyme in the flavonoid pathway, played a pivotal role in catalyzing the synthesis of chalcone. Three *CHS* genes were differentially expressed in these clones. Two of these genes were significantly up-regulated in BSH and JFC, compared with LFC, whereas one was more highly expressed in LFC than in BSH and JFC. Two *CHI* genes responsible for converting naringenin chalcone to naringenin exhibited higher expression levels in BSH and lower expression in LFC. The accumulation of anthocyanin was positively correlated with the expression level of these genes. The study found that there was a gradual decrease in the content of naringenin metabolite in JFC, BSH, and LFC. Moreover, the study found that the *FLS*, *DFR*, and *ANS* genes play a crucial role in transforming naringenin into delphinidin, cyanidin, and pelargonidin, respectively, with their expression levels being up-regulated in JFC and down-regulated in LFC (Figure 4). UFGT is a critical enzyme involved in the biosynthesis of stable anthocyanins from unstable anthocyanin glycosides. The study identified two *UFGT* genes whose expression levels were significantly higher in JFC than in BSH, with the lowest levels observed in LFC, resulting in the highest accumulation of different anthocyanin compounds in JFC compared to LFC. These results revealed that the expression of genes in anthocyanin biosynthesis was consistent with accumulation of anthocyanins.

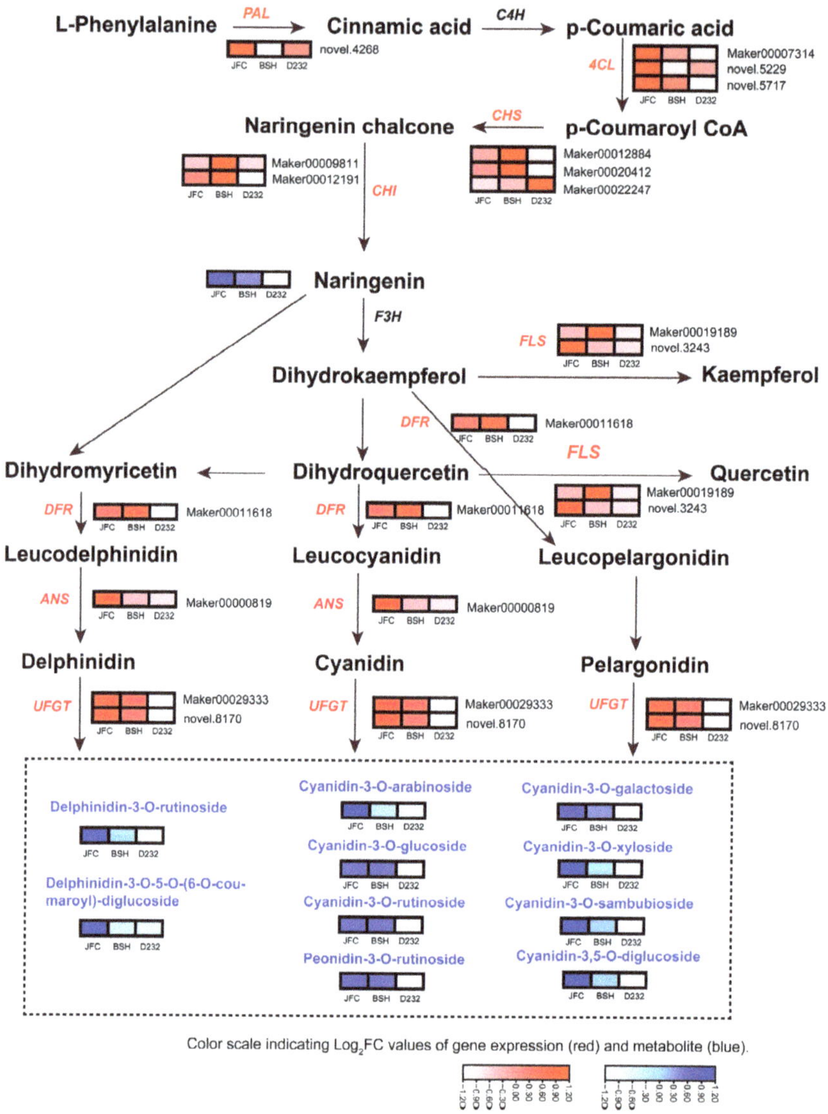

Figure 4. Biosynthetic pathway of anthocyanins. The construction of this pathway is based on the KEGG pathway and pertinent literature references. The red box indicated the expression of genes, and the blue box indicated the content of compounds in this pathway.

2.5. Correlation Analysis of Differentially Expressed Transcripts and Anthocyanin Compounds to Identify Key TFs and Plant Hormones Genes

To achieve the regulatory network between regulatory genes, including transcription factors and plant hormone-related genes, and anthocyanin metabolites, an association analysis was conducted between transcripts and metabolome (Supplementary Figure S2). The major transcription factors and plant hormone-related genes were then identified using differentially expressed genes and metabolites that showed significant differences between JFC (red) and LFC (green). A conjoint analysis was conducted to establish relationships between 510 genes and 23 metabolites with a positive correlation coefficient greater than

0.8 within JFC and LFC. Among these genes and metabolites, 23 TFs and phytohormone-related genes were identified and showed strong correlations with 13 anthocyanin derivates (Table 1). Using Cytoscape 3.10.0, a regulatory network was constructed between these 23 genes and 13 anthocyanin derivatives (Figure 5A). The top 3 TFs associated with the biosynthesis of anthocyanins belonged to *AGL9* (MADS-box), *SOC1* (MADS-box), *MYB* (CPC) (Table 1). Furthermore, among the top five compounds, Cyanidin-3-O-arabinoside and Cyanidin-3-O-galactoside emerged as the most significantly different anthocyanin derivatives in *Toona sinensis* (Table 1). Consequently, a regulatory network was constructed between these top transcription factors and structural regulatory genes involved in anthocyanin biosynthesis (Figure 5B). These findings suggest that MADS-box, MYB, and bHLH transcription factors may regulate *DFR, ANS, UFGT,* and other structural genes involved in the biosynthesis of anthocyanin in *Toona sinensis* by either affecting or participating in the regulation of structural gene.

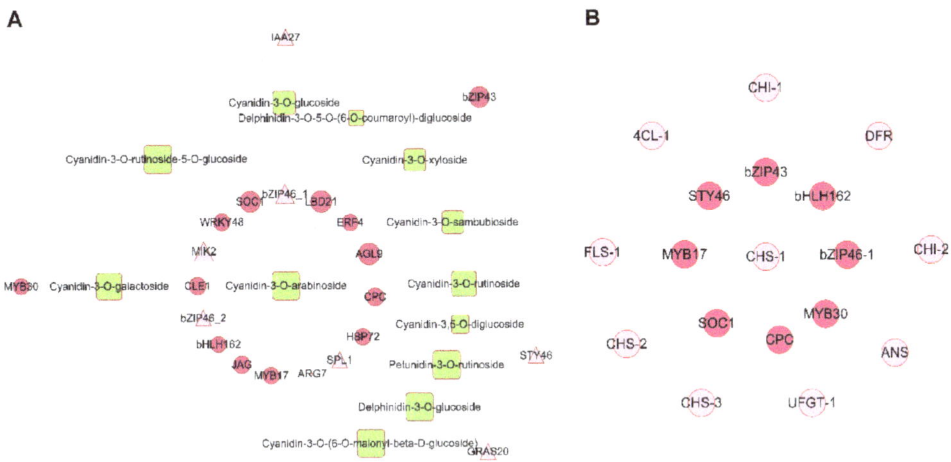

Figure 5. Regulatory network related to the anthocyanin biosynthesis pathway. (**A**) Connection network between regulatory genes and metabolites involved in anthocyanin biosynthesis. Purple circles represent transcription factors, light-purple triangle represent plant hormones related genes, green round rectangle represent the metabolites. The size of the cells represented the log2FC (gene or compounds). (**B**) Regulatory network between transcription factors and structural genes responsible for anthocyanin biosynthesis. Purple circles represent transcription factors, light-purple circles represent structural genes involved in the biosynthesis of anthocyanin pathway.

Table 1. Correlation analysis of transcription factors and plant hormone-related genes with anthocyanin-related metabolites.

Gene ID	Log2FC_gene (JFC: LFC)	Gene Name	AGI	Gene Annotation	Compounds	Log2FC_meta (JFC: LFC)	Coefficient	p Value
Maker00012886	1.5011	ERF4	At5g44210	AP2 domain	Cyanidin-3-O-galactoside	7.9568	0.8479	6.52×10^{-5}
					Cyanidin-3-O-sambubioside	5.4217	0.8049	0.0002953
					Cyanidin-3-O-arabinoside	9.025	0.8034	0.000309044
					Cyanidin-3-O-xyloside	5.2648	0.8327	0.000116859
					Delphinidin-3-O-glucoside	9.025	0.8197	0.00018361
Maker00009687	2.2469	CLE1	At2g22850	bZIP transcription factor	Cyanidin-3-O-rutinoside-5-O-glucoside	9.025	0.8066	0.000279669
					Cyanidin-3-O-galactoside	7.9568	0.8819	1.37×10^{-5}
					Cyanidin-3-O-arabinoside	9.025	0.8696	2.54×10^{-5}
Maker00020104	2.8632	bZIP46_1	At1g68640	bZIP transcription factor	Cyanidin-3-O-rutinoside-5-O-glucoside	9.025	0.8042	0.000301857
					Cyanidin-3-O-rutinoside	7.2774	0.8418	8.30×10^{-5}
					Cyanidin-3-O-galactoside	7.9568	0.8692	2.59×10^{-5}
					Cyanidin-3-O-sambubioside	5.4217	0.8169	0.000201339
					Cyanidin-3-O-glucoside	5.7951	0.8047	0.000296932
					Cyanidin-3-O-arabinoside	9.025	0.8256	0.000149917
					Cyanidin-3,5-O-diglucoside	2.6736	0.8462	6.98×10^{-5}
					Cyanidin-3-O-xyloside	5.2648	0.8548	4.91×10^{-5}
Maker00026824	4.2921	bZIP43	At2g22850	bZIP transcription factor	Cyanidin-3-O-rutinoside-5-O-glucoside	9.025	0.844	7.62×10^{-5}
					Cyanidin-3-O-rutinoside	7.2774	0.8201	0.00018127
					Cyanidin-3-O-galactoside	7.9568	0.8575	4.39×10^{-5}
					Cyanidin-3-O-sambubioside	5.4217	0.8221	0.000169159
					Cyanidin-3,5-O-diglucoside	2.6736	0.8355	0.000105219
					Cyanidin-3-O-xyloside	5.2648	0.8607	3.81×10^{-5}

Table 1. *Cont.*

Gene ID	Log2FC_gene (JFC: LFC)	Gene Name	AGI	Gene Annotation	Compounds	Log2FC_meta (JFC: LFC)	Coefficient	*p* Value
Maker00032838	1.3156	bZIP46_2	At1g68640	bZIP transcription factor	Cyanidin-3-O-galactoside	7.9568	0.8368	0.00010049
					Cyanidin-3-O-arabinoside	9.025	0.8284	0.000136002
novel.8684	2.8522	POLX	At3g45520	bZIP transcription factor	Pelargonidin-3-O-(6-O-p-coumaroyl)-glucoside	4.1336	0.8159	0.000208096
Maker00014746	2.4849	CPC	At2g46410	Myb-like DNA-binding domain	Cyanidin-3-O-(6-O-malonyl-beta-D-glucoside)	9.025	0.801	0.000332643
					Cyanidin-3-O-rutinoside	7.2774	0.8151	0.000213466
					Cyanidin-3-O-galactoside	7.9568	0.8915	8.09×10^{-6}
					Cyanidin-3-O-sambubioside	5.4217	0.811	0.000243476
					Cyanidin-3-O-arabinoside	9.025	0.9149	1.78×10^{-6}
					Cyanidin-3,5-O-diglucoside	2.6736	0.8358	0.000104028
					Cyanidin-3-O-xyloside	5.2648	0.8778	1.69×10^{-5}
					Delphinidin-3-O-glucoside	9.025	0.8114	0.000240963
					Delphinidin-3-O-5-O-(6-O-coumaroyl)-diglucoside	1.0838	0.8249	0.000153878
Maker00016969	1.3602	MYB30	At4g09460	Myb-like DNA-binding domain	Cyanidin-3-O-galactoside	7.9568	0.8189	0.000188654
Maker00023867	1.6363	MYB17	At3g61250	Myb-like DNA-binding domain	Cyanidin-3-O-galactoside	7.9568	0.8493	6.17×10^{-5}
					Cyanidin-3-O-arabinoside	9.025	0.8159	0.000207999
Maker00029632	4.4878	SOC1	At2g45660	MADS-box transcription factor	Cyanidin-3-O-galactoside	7.9568	0.9012	4.53×10^{-6}
					Cyanidin-3-O-arabinoside	9.025	0.9525	4.40×10^{-8}
					Cyanidin-3-O-xyloside	5.2648	0.8064	0.000281454
					Delphinidin-3-O-5-O-(6-O-coumaroyl)-diglucoside	1.0838	0.8166	0.000203571

Table 1. *Cont.*

Gene ID	Log2FC_gene (JFC: LFC)	Gene Name	AGI	Gene Annotation	Compounds	Log2FC_meta (JFC: LFC)	Coefficient	p Value
novel.3727	4.9012	AGL9	At1g24260	MADS-box transcription factor	Cyanidin-3-O-rutinoside-5-O-glucoside	9.025	0.8712	2.35×10^{-5}
					Cyanidin-3-O-(6-O-malonyl-beta-D-glucoside)	9.025	0.8	0.000341781
					Cyanidin-3-O-rutinoside	7.2774	0.8165	0.000204
					Cyanidin-3-O-galactoside	7.9568	0.9652	5.96×10^{-9}
					Cyanidin-3-O-sambubioside	5.4217	0.8078	0.000269708
					Cyanidin-3-O-arabinoside	9.025	0.9734	1.07×10^{-9}
					Cyanidin-3,5-O-diglucoside	2.6736	0.8416	8.35×10^{-5}
					Cyanidin-3-O-xyloside	5.2648	0.8968	5.93×10^{-6}
					Delphinidin-3-O-glucoside	9.025	0.8552	4.83×10^{-5}
					Delphinidin-3-O-5-O-(6-O-coumaroyl)-diglucoside	1.0838	0.8735	2.10×10^{-5}
					Petunidin-3-O-rutinoside	9.025	0.8018	0.000324031
Maker00006536	1.8404	WRKY48	At5g49520	WRKY DNA-binding protein 48	Cyanidin-3-O-rutinoside-5-O-glucoside	9.025	0.832	0.000119819
					Cyanidin-3-O-galactoside	7.9568	0.8644	3.22×10^{-5}
					Cyanidin-3-O-arabinoside	9.025	0.8435	7.77×10^{-5}
					Cyanidin-3-O-xyloside	5.2648	0.8093	0.000257445
novel.6183	3.5074	MIK2	At4g08850	Leucine-rich repeat receptor-like protein kinase	Cyanidin-3-O-rutinoside-5-O-glucoside	9.025	0.8304	0.000126884
					Cyanidin-3-O-galactoside	7.9568	0.9136	1.95×10^{-6}
					Cyanidin-3-O-arabinoside	9.025	0.9219	1.03×10^{-6}
					Delphinidin-3-O-5-O-(6-O-coumaroyl)-diglucoside	1.0838	0.8338	0.000112199
Maker00021798	1.1527	bHLH162	At4g20970	Helix-loop-helix DNA-binding domain	Cyanidin-3-O-galactoside	7.9568	0.8684	2.68×10^{-5}
					Cyanidin-3-O-arabinoside	9.025	0.8782	1.66×10^{-5}
Maker00003242	2.1058	JAG	At1g13400	C2H2 and C2HC zinc fingers superfamily protein	Cyanidin-3-O-galactoside	7.9568	0.8668	2.90×10^{-5}
					Cyanidin-3-O-arabinoside	9.025	0.8657	3.04×10^{-5}
Maker00011793	1.1426	GRAS20	At3g54220	GRAS domain family	Cyanidin-3-O-(6-O-malonyl-beta-D-glucoside)	9.025	0.8327	0.000116807
					Cyanidin-3,5-O-diglucoside	2.6736	0.8072	0.000275174

Table 1. *Cont.*

Gene ID	Log2FC_gene (JFC: LFC)	Gene Name	AGI	Gene Annotation	Compounds	Log2FC_meta (JFC: LFC)	Coefficient	*p* Value
novel.4188	1.6057	HSP72	At1g56410	heat shock protein 70 family protein	Cyanidin-3-O-(6-O-malonyl-beta-D-glucoside)	9.025	0.8302	0.000127565
					Cyanidin-3-O-galactoside	7.9568	0.9167	1.55×10^{-6}
					Cyanidin-3-O-arabinoside	9.025	0.9144	1.84×10^{-6}
					Cyanidin-3-O-xyloside	5.2648	0.807	0.00027619
					Delphinidin-3-O-glucoside	9.025	0.8147	0.000216435
novel.4220	4.8378	LED21	At3g11090	Lateral organ boundaries domain	Cyanidin-3-O-rutinoside-5-O-glucoside	9.025	0.8909	8.38×10^{-6}
					Cyanidin-3-O-rutinoside	7.2774	0.8734	2.11×10^{-5}
					Cyanidin-3-O-galactoside	7.9568	0.8584	4.21×10^{-5}
					Cyanidin-3-O-sambubioside	5.4217	0.8307	0.00012547
					Cyanidin-3-O-arabinoside	9.025	0.8884	9.68×10^{-6}
					Cyanidin-3,5-O-diglucoside	2.6736	0.8797	1.54×10^{-5}
					Cyanidin-3-O-xyloside	5.2648	0.8999	4.89×10^{-6}
					Delphinidin-3-O-5-O-(6-O-coumaroyl)-diglucoside	1.0838	0.9109	2.36×10^{-6}
					Petunidin-3-O-rutinoside	9.025	0.8911	8.31×10^{-6}
Maker00033233	1.0573	STY46	At4g35780	Protein tyrosine kinase	Cyanidin-3-O-(6-O-malonyl-beta-D-glucoside)	9.025	0.8235	0.000161377
					Cyanidin-3-O-galactoside	7.9568	0.8026	0.00031603
					Cyanidin-3,5-O-diglucoside	2.6736	0.8017	0.000324943
Maker00000350	1.9217	ARR9	At3g57040	Response regulator receiver domain	Cyanidin-3-O-rutinoside-5-O-glucoside	9.025	0.8952	6.53×10^{-6}
					Cyanidin-3-O-rutinoside	7.2774	0.8032	0.000310451
					Cyanidin-3-O-galactoside	7.9568	0.842	8.23×10^{-5}
					Cyanidin-3-O-arabinoside	9.025	0.8663	2.95×10^{-5}
					Cyanidin-3,5-O-diglucoside	2.6736	0.833	0.000115529
Maker00026791	1.2059	SPL1	At3g57920	SBP domain	Cyanidin-3-O-arabinoside	9.025	0.8222	0.000168825
Maker00008503	1.3275	IAA27	At4g29080	AUX/IAA family	Cyanidin-3-O-glucoside	5.7951	0.8337	0.000112681
Maker00022256	1.0026	ARG7	At1g75590	Auxin responsive protein	Cyanidin-3-O-arabinoside	9.025	0.8388	9.31×10^{-5}

2.6. Expression Profiles of TFs and Plant Hormone Genes Related to Anthocyanin Biosynthesis

Previous studies have demonstrated that the biosynthesis of anthocyanins requires the involvement of transcription factors, such as MADS-box, MYB, bHLH, and bZIP. In this study, we identified a total of 11 MADS-box genes, 33 MYBs, 42 bHLHs, 36 bZIPs, and 15 WD40 genes in *Toona*. The RNA expression levels in JFC, BSH, and LFC was compared, which revealed that 12 *bHLH*, 10 *bZIP*, 12 *MYB*, 2 *MADS-box*, and 1 WD40 were significantly highly expressed in JFC, lowest in LFC (Figure 6A). Furthermore, the heatmap showed that all highly correlated TFs with anthocyanin derivates belonged to these groups.

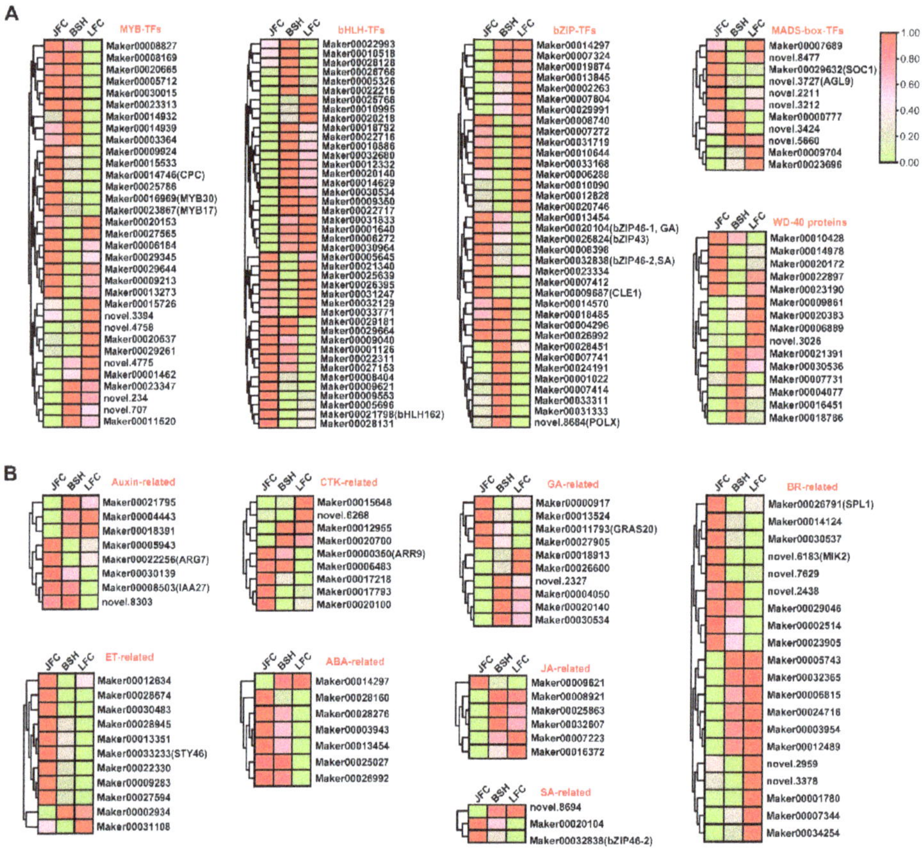

Figure 6. Analysis of potential transcriptional regulators linked to anthocyanins in JFC, BSH, and LFC. (**A**) The heatmap representation of RNA expression levels for various transcription factors that participated in anthocyanins production. These TFs included MYB, bHLH, bZIP, MADS-box, and WD-40 repeats. (**B**) The RNA expression level of genes responsible for plant hormones involved in anthocyanins accumulation. These genes comprise auxin, cytokinin (CTK), gibberellin (GA), brassinolide (BR), ethylene (ET), abscisic acid (ABA), jasmonic acid (JA), and salicylic acid (SA).

Studies suggest that plant hormones have a crucial function in regulating anthocyanin concentrations in leaves and fruits via direct or indirect regulation of transcription factors responsible for anthocyanin biosynthesis. The hormones capable of activating or inhibiting the structural genes involved in the biosynthesis pathway include auxin, gibberellins, cytokinin, brassinosteroids, jasmonic acid, salicylic acid, and abscisic acid. Furthermore, we identified several differentially expressed genes that participate in plant hormone signal transduction, including 8 Auxin-related genes, 9 CTK-related genes, 10 GA-related genes, 23 BR-related genes, 7 ABA-related genes, 11 ethlye-responsive genes, 6 JA-related genes, and 3SA-related genes. Among these genes, the following exhibited a positive correlation coefficient greater than 0.8 with anthocyanin compounds: 2 Auxin-related genes (*Maker00008503* and *Maker00022256*), 1 CTK-related gene (*Maker00000350*), 1 GA-related gene (*Maker000117930*), 2 BR-related genes (*Maker00026791* and *novel.6183*), 1 Ethylene-responsive genes (*Marker00033233*), and 2 SA-related genes (*Maker00020104* and *Maker00032838*) (Figure 6B). This indicates that these genes may positively regulate anthocyanin biosynthesis either directly or indirectly.

2.7. Expression Analysis of Genes through qRT-PCR

To validate the RNA-seq results, we utilized qRT-PCR to evaluate the expression of 8 structural genes and 4 TFs involved in the anthocyanin biosynthesis pathway. Our observations revealed that *bHLH162*, *CPC*, and *SOC1* TFs were significantly upregulated in JFC and exhibited the lowest expression levels in LFC (Figure 7A). Additionally, there was a strong correlation between the RNA-seq and the characteristic leaf color of these samples. The qRT-PCR outcomes of structural genes were also significantly correlated with RNA-seq (Figure 7B). Moreover, our data suggested that the expression levels of *bHLH162* (*Maker00021798*, 2.99 times higher in JFC than in LFC), *CPC* (*Maker00014746*, 10.48 times), and *SOC1* (*Maker00029632*, 6.67 times) were synchronized with *DFR* (*Marker00011618*, 1.84 times), *ANS* (*Marker00000819*, 2.95 times), and *UFGT-1* (*Marker00029333*, 3.07 times), indicating that they may be involved in regulating anthocyanin production. These findings demonstrated that transcription data accurately reflect gene expression in the anthocyanin biosynthesis of *Toona sinesis*.

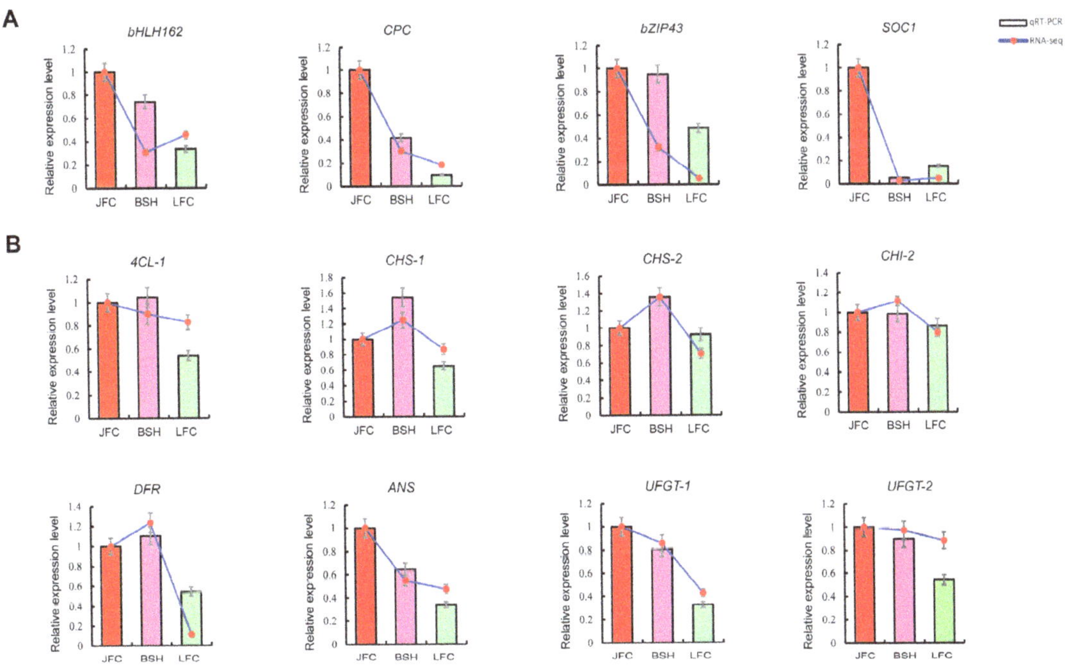

Figure 7. Quantitative real-time RT-PCR (qRT-PCR) and RNA-seq analysis of genes involved in anthocyanin biosynthesis pathway and putative transcription regulators. (**A**) Relative expression level of transcription factors. (**B**) Relative expression level of structural genes. Data are shown as mean ± standard deviation of three biological replicates.

3. Discussion

Anthocyanins present in *Toona sinensis* leaves possess various health benefits, including antioxidant and anti-inflammatory properties, and potential cancer-preventing properties. Understanding the regulatory mechanisms involved in anthocyanin biosynthesis in *Toona sinensis* leaves has significant implications for human health and agriculture. In this study, we utilized metabolomic and transcriptomic approaches to elucidate the molecular mechanisms and regulatory network involved in anthocyanin biosynthesis in *Toona*. A total of 32–44 metabolites related to anthocyanins were identified in five different clones of *Toona* leaves using UPLC-MS/MS. We found that JFC had the highest total anthocyanin content, with Delphinidin, Cyanidin, Peonidin being the most abundant pigments in JFC. Specifically, Cyanidin-3-O-arabinoside was found to be 520.93 times more prevalent in JFC than in LFC, indicating that cyanidin compounds are the dominant anthocyanin in JFC. These findings are consistent with those of previous studies on strawberry petals, which have shown that cyanidins are the primary anthocyanin compound present [37].

Transcriptome sequencing of *Toona* leaves with different colors can provide valuable insights into the regulating genes involved in anthocyanin biosynthesis. The KEGG pathway enrichment analysis of differentially expressed genes (DEGs) in colored leaves of different clones compared to green leaves revealed significant enrichment of genes involved in flavonoid biosynthesis, phenylpropanoid biosynthesis, and the anthocyanin biosynthesis pathway. This finding supports previous studies demonstrating the involvement of multiple enzymes encoded by early biosynthesis genes and anthocyanin biosynthesis genes in anthocyanin biosynthesis. The study indicates that the biosynthesis of naringenin is a crucial step in the flavonoids metabolic pathway and plays a decisive role in the synthesis of anthocyanin compounds. The *PAL, C4H, 4CL, CHS,* and *CHI* are key genes in the flavonoids metabolic pathway and play a decisive role in determining naringenin production. The study found that JFC had a higher naringenin content than BSH and LFC, but only two *4CL* genes (*Maker00007314, novel.5717*) were expressed at high levels in JFC and low levels in LFC. The expression of *4CL* genes was highly positively correlated with naringenin content, indicating that the number and function of genes in the naringenin pathway may affect naringenin content. During the biosynthesis of anthocyanins, DFR catalyzes the reduction reaction of flavonoid-3′,5′-hydroxylase to form anthocyanins, which was confirmed as a key step for regulating anthocyanin types. ANS converts anthocyanin precursors to anthocyanin glycosides. UFGT can catalyze the conversion of unstable anthocyanidins into stable anthocyanins. The absence of these genes directly affects anthocyanins biosynthesis, leading to pigment loss [38,39]. Down-regulation of the apple *DFR* gene can inhibit both cyanidin and procyanidin accumulation. In *Vitis vinifera*, the silencing of the DFR gene leads to the absence of anthocyanins [40]. In *Duchesnea indica*, a plant belonging to the Rosaceae family, decreased expression of the *ANS* gene results in white-colored fruits [41]. Conversely, overexpressing the *SmANS* gene in *Salvia miltiorrhiza* can increase anthocyanin content but decrease the biosynthesis of salvianolic acid [42]. In our study, we found significantly increased expression of the *DFR* gene (*Marker00011618*), *ANS* gene (*Marker00000819*), *UFGT* (*Marker 00029333* and *novel.8170*) in JFC and BSH compared to LFC, which may explain the red color leaves in JFC and the green color leaves in LFC. Difference in gene sequence and promoter regions of these structural genes may be the primary factors contributing to difference in gene function and expression, ultimately impacting anthocyanin biosynthesis in different clones.

The MBW protein complex was proven to activate the expression of structural genes involved in anthocyanin production. In addition to the MBW complex, other TFs such as bZIP [32,43], NAC [44], WRKY [45,46], MADS-box [47,48], and zinc finger [49,50] proteins have also shown to regulate anthocyanin biosynthesis in plants. In *Toona*, we found that MADS-box, MYB, and C2H2 possessed top five TFs related to anthocyanin derivates according to the joint analysis between transcriptome and metabolome. The gene-gene regulatory network also showed that the *SOC1, CPC,* and *bHLH162* were the major TFs regulating the anthocyanin biosynthesis structural genes. *SOC1* encodes a MADS box transcription factor and is involved in the regulation of flowering in response to temperature or light in various plants species, such as *Arabidopsis* [51], *Oryza sativa* L. [52], *Gossypium hissytum* [53]. In *Toona sinensis*, the young leaf color is highly sensitive to temperature, indicating a critical role of *SOC1* in temperature or light-induced anthocyanin biosynthesis. These transcription factors might regulate anthocyanin biosynthesis process by directly active or repress gene expression by binding to the promoter regions, or indirectly though protein-protein interactions in model plants and other fruits.

Plant hormones can affect anthocyanin biosynthesis through various mechanisms. Previous studies have reported that overexpression of *MdIAA26* in apple calli and *Arabidopsis* promotes the accumulation of anthocyanin, while auxin inhibit it by degrading the MdIAA26 protein [54]. Exogenous ethylene treatment increased anthocyanin accumulation in grape skins and induced the expression of structural genes (*VvPAL, Vv4CH, VvCHS, VvCHI, VvF3H,* and *VvUFGT*) and regulatory genes (*VvMYBA1, VvMYBA2,* and *VvMYBA3*) related to anthocyanin biosynthesis [55]. Our study identified nine hormone-related genes with a high correlation to regulate anthocyanins biosynthesis, which are highly expressed in JFC and play a significant role in the controlling leaf color variation in *Toona*. Leaf color is an important trait that is susceptible to both endogenous and exogenous influences. Variations in hormone signal transduction gene expression reflect the essential mechanisms by which plant hormones affect the anthocyanin biosynthesis pathway. These findings enhance our understanding of how anthocyanin biosynthesis works in *Toona* and how it can be controlled, with implications for breeding new *Toona* cultivars. Identifying and understanding the beneficial metabolic constituents and regulatory networks of *Toona* can facilitate the development and utilization of anthocyanin-enriched varieties.

4. Materials and Methods

4.1. Plant Materials

The study selected five Chinese toon clones, JinFuChun (JFC), BaShanHong (BSH), LvFuChun (LFC), FenHongXinNiang (FHXN), and HeBeiGu (HBG), which cultivated in Research Institution of Subtropical Forestry, Chinese Academy of Forestry. The colors of these varieties are: brilliant-red buds (JFC), pink buds (FHXN), light-red with a little green buds (BSH), dark-purple buds (HBG), and green buds (LFC). Fresh buds from five different clones were collected, with three replicates taken for each clone, each replicate sourced from different trees. Buds measuring 10–15 cm in length and weighing approximately 5–10 g were collected for each sample and subsequently underwent freeze-drying for RNA-seq and metabolites measures.

4.2. Sample Preparation and Extraction

The sample underwent freeze-drying, followed by grinding into a powder (30 Hz, 1.5 min), and subsequently, it was stored at −80 °C until required. Subsequently, 50 mg of the powder was precisely weighed and subjected to extraction using 0.5 mL of methanol/water/hydrochloric acid (500:500:1, *v*/*v*/*v*). The resulting extract was vigorously vortexed for 5 min, followed by ultrasonication for another 5 min, and then centrifuged at 12,000× *g* at 4 °C for 3 min. The residue underwent a repeat extraction using the same procedure and conditions. The supernatants obtained were collected and filtered through a 0.22 μm membrane filter (Anpel) prior to LC-MS/MS analysis.

4.3. UPLC Conditions

The sample extracts underwent analysis utilizing a UPLC-ESI-MS/MS system, comprising the UPLC (ExionLC™ AD, available at https://sciex.com.cn/, accessed on 17 June 2022) and the MS (Applied Biosystems 6500 Triple Quadrupole (Waltham, MA, USA), also accessible at https://sciex.com.cn/, accessed on 17 June 2022). The analytical parameters were configured as follows: UPLC:Column: Waters ACQUITY BEH C18 (1.7 μm, 2.1 mm × 100 mm); Solvent System: Water (0.1% formic acid) and Methanol (0.1% formic acid); Gradient Program: 95:5 *v*/*v* at 0 min, 50:50 *v*/*v* at 6 min, 5:95 *v*/*v* at 12 min (maintained for 2 min), 95:5 v/v at 14 min (maintained for 2 min); Flow Rate: 0.35 mL/min; Temperature: 40 °C; Injection Volume: 2 µL.

4.4. ESI-MS/MS Conditions

Linear ion trap (LIT) and triple quadrupole (QQQ) scans were obtained using a triple quadrupole-linear ion trap mass spectrometer, the AB Scoex Qtrap® 6500+ LC-MS/MS System (Sciex, Framingham, MA, USA), equipped with an ESI Turbo Ion-Spray interface. The instrument operated in positive ion mode and was controlled by Analyst 1.6.3 software (Sciex). The ESI source operated with the following parameters: Ion Source: ESI+, Source Temperature: 550 °C, Ion Spray Voltage (IS): 5500 V, Curtain Gas (CUR): Set at 35 psi. Anthocyanins were subjected to analysis using scheduled multiple reaction monitoring (MRM). Data acquisition was conducted through Analyst 1.6.3 software (Sciex). Quantification of all metabolites was performed using Multiquant 3.0.3 software (Sciex). Further optimization of declustering potentials (DP) and collision energies (CE) for individual MRM transitions was carried out. A specific set of MRM transitions was monitored for each period, corresponding to the elution of metabolites during that timeframe.

4.5. Identification of Differentially Accumulated Metabolites

Significantly regulated metabolites between groups using the identification criterion of the absolute $|\log_2(\text{fold-change})| \geq 1$ and p value < 0.05, based on the Student's t-test. Identified metabolites using the Kyoto Encyclopedia of Gene and Genomes (KEGG) compound database available at http://www.kegg.jp/kegg/compound/, accessed on 20 June 2022. Next, we mapped the annotated metabolites to KEGG Pathway database found at http://www.kegg.jp/kegg/pathway.html, accessed on 20 June 2022. The pathways that significantly regulated metabolites mapped to were integrated into MSEA (metabolite sets enrichment analysis) and evaluated for significance by checking the hypergeometric test's p-values.

4.6. RNA Sequencing

Total RNA was extracted from frozen leaves utilizing the RNAprep Pure Plant Kit (Tiangen Biotech, Beijing, China). The Agilent 2100 Bioanalyzer (Agilent Technologies, Santa Clara, CA, USA) was used to evalute the quality of the gathered RNAs by examining their integrity. Subsequently, Poly (A) mRNA was enriched from total RNA using Oligo (dT) magnetic beads. To facilitate sequencing, Poly (A) mRNA was fragmented by an RNA fragmentation kit (Ambion, Austin, TX, USA). First-strand cDNA was produced via transcriptase reaction with random hexamer primers. Next, DNA polymerase I and RNase H enzymes were employed to create the second-strand cDNA (Invitrogen, Carlsbad, CA, USA). Following this, DNA fragments of suitable lengths were obtained, end-repaired, poly(A)-tailed, and connected with sequencing adaptors. Eventually, these fragments underwent Illumina HiSeq™ 2500 platform sequencing.

4.7. Transcript Profiles and Annotation

High-quality reads were obtained by processing the raw reads in fastq format using Perl scripts developed in-house. Clean reads were obtained from the raw data by eliminating adaptor sequences, low-quality reads, and reads containing polyN. All downstream analyses were based on clean, highquality data. Gene function was annotated employing several databases: KEGG pathway database, the NCBI non-redundant (Nr) database, the Swiss-Prot protein database, the euKaryotic Clusters of Orthologous Groups (KOG) database, the Gene Ontology (GO) database, and the Pfam database. We analyzed the differentially expressed genes of both groups by utilizing the DESeq R package (version 1.10.1). The DESeq R package utilizes a statistical model based on the negative binomial distribution to identify differentially expressed genes. We corrected the outcomes of all statistical tests by employing the false discovery rate of Benjamini and Hochberg to account for multiple testing. According to DESeq, genes were considered substantially differentially expressed if their adjusted p-value was less than 0.05. We used the top GO R package v.1, which is based on the Kolmogorov-Smirnov test, to carry out GO enrichment analysis of the differentially expressed genes. We performed pathway analysis utilizing the KEGG database (http://www.genome.jp/kegg/, accessed on 10 June 2022) to investigate relevant pathways of substantially differentially expressed genes [56–58].

4.8. Correlation Analysis of Transcriptome and Metabolome

To integrate transcriptome and metabolome datasets, Pearson correlation coefficients were utilized. Gene-metabolite coefficients were obtained by calculating the average expression levels of transcripts and metabolite contents. The fold changes for both transcriptome and metabolome data in each group were also computed. A correlation was considered significant if the Pearson correlation coefficients exceed 0.8 and p-values were less than 0.05 (Table 1). Significant positive correlations between transcription factors (TFs) and anthocyanin derivatives in groups JFC and LFC were detected and visually presented utilizing Cytoscape 3.10.0.

4.9. Verification of RNA-Seq Data by qRT-PCR

Total RNA was extracted from *Toona sinesis* leaves and reverse-transcribed using the Quantscript Reverse Transcriptase Kit. Each clone had three biological replicates, and each sample had three technical replicates. The obtained cDNA served as a template for determining gene expression levels, employing specific primers for genes linked to anthocyanin biosynthesis as well as *ACTIN* gene (used as an internal control). The primers used in the qRT-PCR analysis were shown in Table 2. The melt curve of *ACTIN* in the qPCR products was presented in Supplementary Figure S3. The reaction system contained 5 μL 2 × Q3 SYBR qPCR Master Mix-Universal (TOLOBIO), 1 μL cDNA template, 0.5 μL of each forward and reverse primer, and 3 μL of RNase-free water. qRT-PCR was performed using Applied Biosystems 7500 Fast Real-Time PCR System.

Table 2. The primer sequences used in this study.

Primer Name	Primer Sequence F (5′-3′)	Primer Sequence R (5′-3′)
RT-bHLH162 (Maker00021798)	CAGGTTAATGGGAATCGAAAAG	CCAACCCAGTTACCAAAACAAT
RT-CPC (Maker00014746)	AAACTCGTTGGAGACAGGTG	GCAAACCCTTCTCCATGTTT
RT-bZIP43 (Maker00026824)	TCGGAGACTATGATCGGAAATC	ACTTGAAACGACAGGTTTTGGT
RT-SOC1 (Maker00029632)	CAATTGAAGCACGAAATAGCAA	TGTACCGTTTAGAGGGAAGGAA
RT-4CL-1 (Maker00007314)	AGGTTTGAGGCTTCGCAGTA	GTCCCCAGTGACAACAATCC
RT-CHS-1 (Maker00012884)	TGCAATTCTTGACCAAGTCG	GATGAACAAAACGCATGCAC
RT-CHS-2 (Maker00020412)	TGGTGCAATTGATGGACACT	TGCTTCGACCAAGCTTTTCT
RT-CHI-2 (Maker00012191)	GGGTGTTGGAAGAAAATCCA	GGTCGACTCCATCACCCTTA
RT-DFR (Maker00011618)	CCATTCCTGGCTTCAAGTGT	AAGCAATTGACCCCATTCTG
RT-ANS (Maker00000819)	GCTTTCAGGCAACTGGTCAT	TGCTCGGGAATATTTCTGCT
RT-UFGT-1 (Maker00029333)	GTGGTCGAGACTGGGAGAAG	AGGACCTGCAGTCCAAAGTG
RT-UFGT-2 (novel.8170)	GTGACAAAACCCTGGAAGGA	CAATTTTGGGAAAGCTGCAT
RT-Actin	TGATTGGGATGGAAGCAGCA	GAACATGGTTGAACCGCCAC

5. Conclusions

In this study, an integrative analysis of the transcriptome and metabolome of five distinct *Toona sinensis* clones were performed to explore the related genes and metabolites associated with the anthocyanin biosynthesis. Furthermore, our analyses found that the red leaves 'JFC' contained the highest content of stable cyanidin, delphinidin, and peonidin, especially Cyanidin-3-O-arabinoside. The integrated analysis also identified the major transcription factor SOC1, CPC, and bHLH162. Moreover, the regulatory network construction of TFs, metabolic, structural genes reveals the underlying mechanisms of the anthocyanin biosynthesis pathway in *Toona sinensis*. Overall, this study provides a foundation for future research aimed at manipulating anthocyanin biosynthesis to improve plant coloration or to derive human health benefits.

Supplementary Materials: The supporting information can be downloaded at: https://www.mdpi.com/article/10.3390/ijms242015459/s1.

Author Contributions: Conceptualization, J.L.; Data curation, J.X. and Y.F.; Formal analysis, J.X., Y.F., X.H., H.P., J.D., Y.W. and R.Z.; Funding acquisition, J.L.; Investigation, J.X., X.H., H.P., J.D., Y.W. and R.Z.; Methodology, J.X.; Project administration, J.L.; Writing—original draft, J.X.; Writing—review & editing, J.X. All authors have read and agreed to the published version of the manuscript.

Funding: This research was supported by the National Nonprofit Institute Research Grant of Chinese Academy of Forestry (CAFYBB2020SZ004) to J.L. and the Zhejiang Science and Technology Major Program on Agricultural New Variety Breeding" (2021C02070-4) to J.L.

Data Availability Statement: The raw sequence data reported in this paper have been deposited in the Genome Sequence Archive (Genomics, Proteomics & Bioinformatics 2021) in National Genomics Data Center (Nucleic Acids Res 2022), China National Center for Bioinformation/Beijing Institute of Genomics, Chinese Academy of Sciences (GSA: CRA011954) that are publicly accessible at https://ngdc.cncb.ac.cn/gsa, accessed on 26 July 2023.

Conflicts of Interest: The authors declare no conflict of interest.

References

1. Parada, J.; Aguilera, J.M. Food microstructure affects the bioavailability of several nutrients. *J. Food Sci.* **2007**, *72*, R21–R32. [CrossRef] [PubMed]
2. Shen, Y.; Xu, M.; Deng, P.; Gu, Q.; Yin, H.; Xia, G.; Jia, X.; Yang, H.; Tam, J. Establishment of a rapid method to quantify eight flavonol glycosides for quality assessment of red toon using UPLC. *Acta Chromatogr.* **2018**, *30*, 31–37. [CrossRef]
3. Sun, X.; Zhang, L.; Cao, Y.; Gu, Q.; Yang, H.; Tam, J.P. Quantitative Analysis and Comparison of Four Major Flavonol Glycosides in the Leaves of *Toona sinensis* (A. Juss.) Roemer (Chinese Toon) from Various Origins by High-Performance Liquid Chromatography-Diode Array Detector and Hierarchical Clustering Analysis. *Pharmacogn. Mag.* **2016**, *12* (Suppl. S2), S270–S276. [PubMed]
4. Wang, K.J.; Yang, C.R.; Zhang, Y.J. Phenolic antioxidants from Chinese toon (fresh young leaves and shoots of *Toona sinensis*). *Food Chem.* **2007**, *101*, 365–371. [CrossRef]
5. Falcone Ferreyra, M.L.; Rius, S.P.; Casati, P. Flavonoids: Biosynthesis, biological functions, and biotechnological applications. *Front. Plant Sci.* **2012**, *3*, 222. [CrossRef]
6. Winkel-Shirley, B. Flavonoid Biosynthesis. A Colorful Model for Genetics, Biochemistry, Cell Biology, and Biotechnology. *Plant Physiol.* **2001**, *126*, 485–493. [CrossRef]
7. He, J.; Giusti, M.M. Anthocyanins: Natural colorants with health-promoting properties. *Annu. Rev. Food Sci. Technol.* **2010**, *1*, 163–187. [CrossRef]
8. Yousuf, B.; Gul, K.; Wani, A.A.; Singh, P. Health Benefits of Anthocyanins and Their Encapsulation for Potential Use in Food Systems: A Review. *Crit. Rev. Food Sci. Nutr.* **2016**, *56*, 2223–2230. [CrossRef]
9. Jaakola, L. New insights into the regulation of anthocyanin biosynthesis in fruits. *Trends Plant Sci.* **2013**, *18*, 477–483. [CrossRef]
10. Grotewold, E. The genetics and biochemistry of floral pigments. *Annu. Rev. Plant Biol.* **2006**, *57*, 761–780. [CrossRef]
11. Jaakola, L.; Hohtola, A. Effect of latitude on flavonoid biosynthesis in plants. *Plant Cell Environ.* **2010**, *33*, 1239–1247. [CrossRef] [PubMed]
12. Feng, F.; Li, M.; Ma, F.; Cheng, L. Phenylpropanoid metabolites and expression of key genes involved in anthocyanin biosynthesis in the shaded peel of apple fruit in response to sun exposure. *Plant Physiol. Bioch.* **2013**, *69*, 54–61. [CrossRef] [PubMed]
13. Gonzalez, A.; Zhao, M.; Leavitt, J.M.; Lloyd, A.M. Regulation of the anthocyanin biosynthetic pathway by the TTG1/bHLH/Myb transcriptional complex in *Arabidopsis* seedlings. *Plant J.* **2008**, *53*, 814–827. [CrossRef] [PubMed]
14. Liu, H.; Liu, Z.; Wu, Y.; Zheng, L.; Zhang, G. Regulatory Mechanisms of Anthocyanin Biosynthesis in Apple and Pear. *Int. J. Mol. Sci.* **2021**, *22*, 8441. [CrossRef]
15. Mattoo, A.K.; Dwivedi, S.L.; Dutt, S.; Singh, B.; Garg, M.; Ortiz, R. Anthocyanin-Rich Vegetables for Human Consumption-Focus on Potato, Sweetpotato and Tomato. *Int. J. Mol. Sci.* **2022**, *23*, 2634. [CrossRef]
16. Sun, C.; Deng, L.; Du, M.; Zhao, J.; Chen, Q.; Huang, T.; Jiang, H.; Li, C.B.; Li, C. A Transcriptional Network Promotes Anthocyanin Biosynthesis in Tomato Flesh. *Mol. Plant.* **2020**, *13*, 42–58. [CrossRef]
17. Wang, Y.; Wang, Y.; Song, Z.; Zhang, H. Repression of *MYBL2* by Both microRNA858a and HY5 Leads to the Activation of Anthocyanin Biosynthetic Pathway in *Arabidopsis*. *Mol. Plant.* **2016**, *9*, 1395–1405. [CrossRef]
18. Xu, W.; Grain, D.; Bobet, S.; Le Gourrierec, J.; Thévenin, J.; Kelemen, Z.; Lepiniec, L.; Dubos, C. Complexity and robustness of the flavonoid transcriptional regulatory network revealed by comprehensive analyses of MYB-bHLH-WDR complexes and their targets in *Arabidopsis* seed. *New Phytol.* **2014**, *202*, 132–144. [CrossRef]
19. Li, L.; Ban, Z.J.; Li, X.H.; Wu, M.Y.; Wang, A.L.; Jiang, Y.Q.; Jiang, Y.H. Differential Expression of Anthocyanin Biosynthetic Genes and Transcription Factor PcMYB10 in Pears (*Pyrus communis* L.). *PLoS ONE* **2012**, *7*, e46070. [CrossRef]

20. An, X.H.; Tian, Y.; Chen, K.Q.; Wang, X.F.; Hao, Y.J. The apple WD40 protein MdTTG1 interacts with bHLH but not MYB proteins to regulate anthocyanin accumulation. *J. Plant Physiol.* **2012**, *169*, 710–717. [CrossRef]
21. Chen, L.; Hu, B.; Qin, Y.; Hu, G.; Zhao, J. Advance of the negative regulation of anthocyanin biosynthesis by MYB transcription factors. *Plant Physiol. Bioch.* **2019**, *136*, 178–187. [CrossRef] [PubMed]
22. Das, P.K.; Shin, D.H.; Choi, S.-B.; Park, Y.-I. Sugar-hormone cross-talk in anthocyanin biosynthesis. *Mol. Cells* **2012**, *34*, 501–507. [CrossRef] [PubMed]
23. Dubos, C.; Stracke, R.; Grotewold, E.; Weisshaar, B.; Martin, C.; Lepiniec, L. MYB transcription factors in *Arabidopsis*. *Trends Plant Sci.* **2010**, *15*, 573–581. [CrossRef]
24. Lei, Z.; Zhou, C.; Ji, X.; Wei, G.; Huang, Y.; Yu, W.; Luo, Y.; Qiu, Y. Transcriptome Analysis Reveals genes involved in flavonoid biosynthesis and accumulation in *Dendrobium catenatum* from different locations. *Sci. Rep.* **2018**, *8*, 6373. [CrossRef]
25. Li, H.; Yang, Z.; Zeng, Q.; Wang, S.; Luo, Y.; Huang, Y.; Xin, Y.; He, N. Abnormal expression of *bHLH3* disrupts a flavonoid homeostasis network, causing differences in pigment composition among mulberry fruits. *Hortic. Res.* **2020**, *7*, 83. [CrossRef] [PubMed]
26. Baudry, A.; Heim, M.A.; Dubreucq, B.; Caboche, M.; Weisshaar, B.; Lepiniec, L. TT2, TT8, and TTG1 synergistically specify the expression of *BANYULS* and proanthocyanidin biosynthesis in *Arabidopsis thaliana*. *Plant J.* **2004**, *39*, 366–380. [CrossRef]
27. Lau, O.S.; Deng, X.W. The photomorphogenic repressors COP1 and DET1: 20 years later. *Trends Plant Sci.* **2012**, *17*, 584–593. [CrossRef]
28. Stracke, R.; Favory, J.J.; Gruber, H.; Bartelniewoehner, L.; Bartels, S.; Binkert, M.; Funk, M.; Weisshaar, B.; Ulm, R. The Arabidopsis bZIP transcription factor HY5 regulates expression of the *PFG1/MYB12* gene in response to light and ultraviolet-B radiation. *Plant Cell Environ.* **2010**, *33*, 88–103.
29. Lee, J.; He, K.; Stolc, V.; Lee, H.; Figueroa, P.; Gao, Y.; Tongprasit, W.; Zhao, H.; Lee, I.; Deng, X.W. Analysis of transcription factor HY5 genomic binding sites revealed its hierarchical role in light regulation of development. *Plant Cell* **2007**, *19*, 731–749. [CrossRef]
30. Zhao, Y.; Min, T.; Chen, M.; Wang, H.; Zhu, C.; Jin, R.; Allan, A.C.; Lin-Wang, K.; Xu, C. The Photomorphogenic Transcription Factor PpHY5 Regulates Anthocyanin Accumulation in Response to UVA and UVB Irradiation. *Front. Plant Sci.* **2020**, *11*, 603178. [CrossRef]
31. Gao, H.-N.; Jiang, H.; Cui, J.-Y.; You, C.-X.; Li, Y.-Y. Review: The effects of hormones and environmental factors on anthocyanin biosynthesis in apple. *Plant Sci.* **2021**, *312*, 111024. [CrossRef] [PubMed]
32. An, J.-P.; Yao, J.-F.; Xu, R.-R.; You, C.-X.; Wang, X.-F.; Hao, Y.-J. Apple bZIP transcription factor MdbZIP44 regulates abscisic acid-promoted anthocyanin accumulation. *Plant Cell Environ.* **2018**, *41*, 2678–2692. [CrossRef] [PubMed]
33. Qi, T.; Song, S.; Ren, Q.; Wu, D.; Huang, H.; Chen, Y.; Fan, M.; Peng, W.; Ren, C.; Xie, D. The Jasmonate-ZIM-domain proteins interact with the WD-Repeat/bHLH/MYB complexes to regulate Jasmonate-mediated anthocyanin accumulation and trichome initiation in *Arabidopsis thaliana*. *Plant Cell* **2011**, *23*, 1795–1814. [CrossRef] [PubMed]
34. Jiang, T.; Zhang, M.; Wen, C.; Xie, X.; Tian, W.; Wen, S.; Lu, R.; Liu, L. Integrated metabolomic and transcriptomic analysis of the anthocyanin regulatory networks in *Salvia miltiorrhiza* Bge. flowers. *BMC Plant Biol.* **2020**, *20*, 349. [CrossRef]
35. Xie, Z.; Wang, J.; Wang, W.; Wang, Y.; Xu, J.; Li, Z.; Zhao, X.; Fu, B. Integrated Analysis of the Transcriptome and Metabolome Revealed the Molecular Mechanisms Underlying the Enhanced Salt Tolerance of Rice Due to the Application of Exogenous Melatonin. *Front. Plant Sci.* **2020**, *11*, 618680. [CrossRef] [PubMed]
36. Ji, Y.T.; Xiu, Z.; Chen, C.H.; Wang, Y.; Yang, J.X.; Sui, J.J.; Jiang, S.J.; Wang, P.; Yue, S.Y.; Zhang, Q.Q.; et al. Long read sequencing of *Toona sinensis* (A. Juss) Roem: A chromosome-level reference genome for the family Meliaceae. *Mol. Ecol. Resour.* **2021**, *21*, 1243–1255. [CrossRef]
37. Xue, L.; Wang, J.; Zhao, J.; Zheng, Y.; Wang, H.F.; Wu, X.; Xian, C.; Lei, J.J.; Zhong, C.F.; Zhang, Y.T. Study on cyanidin metabolism in petals of pink-flowered strawberry based on transcriptome sequencing and metabolite analysis. *BMC Plant Biol.* **2019**, *19*, 423. [CrossRef]
38. Bogs, J.; Jaffé, F.W.; Takos, A.M.; Walker, A.R.; Robinson, S.P. The grapevine transcription factor VvMYBPA1 regulates proanthocyanidin synthesis during fruit development. *Plant Physiol.* **2007**, *143*, 1347–1361. [CrossRef]
39. Clark, S.T.; Verwoerd, W.S. A systems approach to identifying correlated gene targets for the loss of colour pigmentation in plants. *BMC Bioinform.* **2011**, *12*, 343. [CrossRef]
40. Terrier, N.; Torregrosa, L.; Ageorges, A.; Vialet, S.; Verries, C.; Cheynier, V.; Romieu, C. Ectopic expression of VvMybPA2 promotes proanthocyanidin biosynthesis in grapevine and suggests additional targets in the pathway. *Plant Physiol.* **2009**, *149*, 1028–1041. [CrossRef]
41. Debes, M.A.; Arias, M.E.; Grellet-Bournonville, C.F.; Wulff, A.F.; Martínez-Zamora, M.G.; Castagnaro, A.P.; Díaz-Ricci, J.C. White-fruited *Duchesnea indica* (Rosaceae) is impaired in *ANS* gene expression. *Am. J. Bot.* **2011**, *98*, 2077–2083. [CrossRef] [PubMed]
42. Li, H.; Liu, J.; Pei, T.; Bai, Z.; Han, R.; Liang, Z. Overexpression of *SmANS* Enhances Anthocyanin Accumulation and Alters Phenolic Acids Content in *Salvia miltiorrhiza* and *Salvia miltiorrhiza* Bge f. alba Plantlets. *Int. J. Mol. Sci.* **2019**, *20*, 2225. [CrossRef]
43. Liu, H.; Su, J.; Zhu, Y.; Yao, G.; Allan, A.C.; Ampomah-Dwamena, C.; Shu, Q.; Lin-Wang, K.; Zhang, S.; Wu, J. The involvement of PybZIPa in light-induced anthocyanin accumulation via the activation of *PyUFGT* through binding to tandem G-boxes in its promoter. *Hortic. Res.* **2019**, *6*, 134. [CrossRef] [PubMed]
44. Sun, Q.; Jiang, S.; Zhang, T.; Xu, H.; Fang, H.; Zhang, J.; Su, M.; Wang, Y.; Zhang, Z.; Wang, N.; et al. Apple NAC transcription factor MdNAC52 regulates biosynthesis of anthocyanin and proanthocyanidin through MdMYB9 and MdMYB11. *Plant Sci.* **2019**, *289*, 110286. [CrossRef] [PubMed]

45. Alabd, A.; Ahmad, M.; Zhang, X.; Gao, Y.; Peng, L.; Zhang, L.; Ni, J.; Bai, S.; Teng, Y. Light-responsive transcription factor PpWRKY44 induces anthocyanin accumulation by regulating PpMYB10 expression in pear. *Hortic. Res.* **2022**, *9*, uhac199. [CrossRef] [PubMed]
46. Zhang, H.; Zhang, Z.; Zhao, Y.; Guo, D.; Zhao, X.; Gao, W.; Zhang, J.; Song, B.; Cahill, D. StWRKY13 promotes anthocyanin biosynthesis in potato (*Solanum tuberosum*) tubers. *Funct. Plant Biol.* **2021**, *49*, 102–114. [CrossRef]
47. Qi, F.; Liu, Y.; Luo, Y.; Cui, Y.; Lu, C.; Li, H.; Huang, H.; Dai, S. Functional analysis of the ScAG and ScAGL11 MADS-box transcription factors for anthocyanin biosynthesis and bicolour pattern formation in *Senecio cruentus* ray florets. *Hortic. Res.* **2022**, *9*, uhac071. [CrossRef]
48. Qi, X.; Liu, C.; Song, L.; Li, M. PaMADS7, a MADS-box transcription factor, regulates sweet cherry fruit ripening and softening. *Plant Sci.* **2020**, *301*, 110634. [CrossRef]
49. Shi, H.; Liu, G.; Wei, Y.; Chan, Z. The zinc-finger transcription factor ZAT6 is essential for hydrogen peroxide induction of anthocyanin synthesis in *Arabidopsis*. *Plant Mol. Biol.* **2018**, *97*, 165–176. [CrossRef]
50. Wang, D.R.; Yang, K.; Wang, X.; Lin, X.L.; Rui, L.; Liu, H.F.; Liu, D.D.; You, C.X. Overexpression of MdZAT5, an C2H2-Type Zinc Finger Protein, Regulates Anthocyanin Accumulation and Salt Stress Response in Apple Calli and *Arabidopsis*. *Int. J. Mol. Sci.* **2022**, *23*, 1897. [CrossRef]
51. Kimura, Y.; Aoki, S.; Ando, E.; Kitatsuji, A.; Watanabe, A.; Ohnishi, M.; Takahashi, K.; Inoue, S.I.; Nakamichi, N.; Tamada, Y.; et al. A flowering integrator, SOC1, affects stomatal opening in *Arabidopsis thaliana*. *Plant Cell Physiol.* **2015**, *56*, 640–649. [CrossRef] [PubMed]
52. Ryu, C.H.; Lee, S.; Cho, L.H.; Kim, S.L.; Lee, Y.S.; Choi, S.C.; Jeong, H.J.; Yi, J.; Park, S.J.; Han, C.D.; et al. OsMADS50 and OsMADS56 function antagonistically in regulating long day (LD)-dependent flowering in rice. *Plant Cell Environ.* **2009**, *32*, 1412–1427. [CrossRef]
53. Ma, L.; Yan, Y. GhSOC1s Evolve to Respond Differently to the Environmental Cues and Promote Flowering in Partially Independent Ways. *Front. Plant Sci.* **2022**, *13*, 882946. [CrossRef]
54. Wang, C.K.; Han, P.L.; Zhao, Y.W.; Ji, X.L.; Yu, J.Q.; You, C.X.; Hu, D.G.; Hao, Y.J. Auxin regulates anthocyanin biosynthesis through the auxin repressor protein MdIAA26. *Biochem. Biophys. Res. Commun.* **2020**, *533*, 717–722. [CrossRef]
55. Wang, P.; Ge, M.; Yu, A.; Song, W.; Fang, J.; Leng, X. Effects of ethylene on berry ripening and anthocyanin accumulation of "Fujiminori" grape in protected cultivation. *J. Sci. Food Agric.* **2022**, *102*, 1124–1136. [CrossRef]
56. Kanehisa, M. Toward understanding the origin and evolution of cellular organisms. *Protein Sci.* **2019**, *28*, 1947–1951. [CrossRef]
57. Kanehisa, M.; Goto, S. KEGG: Kyoto encyclopedia of genes and genomes. *Nucleic Acids Res.* **2000**, *28*, 27–30. [CrossRef] [PubMed]
58. Kanehisa, M.; Furumichi, M.; Sato, Y.; Kawashima, M.; Ishiguro-Watanabe, M. KEGG for taxonomy-based analysis of pathways and genomes. *Nucleic Acids Res.* **2023**, *51*, D587–D592. [CrossRef] [PubMed]

International Journal of
Molecular Sciences

MDPI

Article

Screening and Validation of Appropriate Reference Genes for Real-Time Quantitative PCR under PEG, NaCl and $ZnSO_4$ Treatments in *Broussonetia papyrifera*

Mengdi Chen [1], Zhengbo Wang [1], Ziyuan Hao [1], Hongying Li [1,*], Qi Feng [1], Xue Yang [1], Xiaojiao Han [2,*] and Xiping Zhao [1]

1 College of Horticulture and Plant Protection, Henan University of Science and Technology, Luoyang 471023, China; a1442388073a@163.com (M.C.)

2 Key Laboratory of Tree Breeding of Zhejiang Province, Research Institute of Subtropical Forestry, Chinese Academy of Forestry, Hangzhou 311400, China

* Correspondence: lihy@haust.edu.cn (H.L.); hanxj@caf.ac.cn (X.H.)

Abstract: Real-time quantitative PCR (RT-qPCR) has a high sensitivity and strong specificity, and is widely used in the analysis of gene expression. Selecting appropriate internal reference genes is the key to accurately analyzing the expression changes of target genes by RT-qPCR. To find out the most suitable internal reference genes for studying the gene expression in *Broussonetia papyrifera* under abiotic stresses (including drought, salt, and $ZnSO_4$ treatments), seven different tissues of *B. papyrifera*, as well as the roots, stems, and leaves of *B. papyrifera* under the abiotic stresses were used as test materials, and 15 candidate internal reference genes were screened based on the transcriptome data via RT-qPCR. Then, the expression stability of the candidate genes was comprehensively evaluated through the software geNorm (v3.5), NormFinder (v0.953), BestKeeper (v1.0), and RefFinder. The best internal reference genes and their combinations were screened out according to the analysis results. *rRNA* and *Actin* were the best reference genes under drought stress. Under salt stress, *DOUB*, *HSP*, *NADH*, and *rRNA* were the most stable reference genes. Under heavy metal stress, *HSP* and *NADH* were the most suitable reference genes. *EIF3* and *Actin* were the most suitable internal reference genes in the different tissues of *B. papyrifera*. In addition, *HSP*, *rRNA*, *NADH*, and *UBC* were the most suitable internal reference genes for the abiotic stresses and the different tissues of *B. papyrifera*. The expression patterns of *DREB* and *POD* were analyzed by using the selected stable and unstable reference genes. This further verified the reliability of the screened internal reference genes. This study lays the foundation for the functional analysis and regulatory mechanism research of genes in *B. papyrifera*.

Keywords: real-time quantitative PCR; internal reference genes; *Broussonetia papyrifera*; abiotic stresses; gene expression

1. Introduction

Broussonetia papyrifera is a deciduous tree of the genus Broussonetia in the Moraceae family. It is distributed in most parts of China and Southeast Asia. It is a typical native tree species and a pioneer plant [1]. *B. papyrifera* has the advantages of easy reproduction, strong stress resistance, and fast growth, and is widely used in the fields of feed, papermaking, and vegetation restoration [2,3]. Moreover, *B. papyrifera* has medicinal values, as well as flavonoids, polyphenols, and fructose contents that are much higher than those of other plants [4]. Flavonoid derivatives in *B. papyrifera* have inhibitory effects on cancer cells [5], and polyphenols can inhibit coronavirus proteases [6]. Generally speaking, *B. papyrifera* is a woody plant with great potential for development, combining economic value and excellent resistance. At the same time, due to the characteristics of its growing environment, the *B. papyrifera* also has a strong ability to tolerate a variety of unfavorable environments, such as drought, salt, and heavy metals [7–9]. Currently, research on *B. papyrifera* focuses on their breeding [10], physiological characteristics [11], medicinal [12], and pasture values [13], but less research has been performed on the molecular mechanisms of their stress tolerance. As a pioneer tree species widely cultivated in harsh environments, stress resistance is a hotspot in molecular biology research in *B. papyrifera* [14]. Therefore, it is important to carry out research on the molecular mechanism of *B. papyrifera* for resistance breeding and the genetic improvement of *B. papyrifera*.

Real-time quantitative PCR (RT-qPCR) has many advantages, such as convenience, strong specificity, and high sensitivity, and is an effective means to study gene functions [15]. However, the accuracy of RT-qPCR is affected by various factors, such as RNA integrity, cDNA quality, sample dilution factor, and experimental operations [16]. In the study of the expression levels and regulation mechanisms of plant functional genes, the optimization of internal reference genes is the key and the basis for correcting and normalizing the expression of the functional genes [17]. Therefore, the introduction of appropriate internal reference genes is crucial for the normalized analysis of target gene expression [18]. An ideal internal reference gene should be the gene that can be stably expressed in cells and whose expression level is almost not disturbed by the external environment. It is generally the housekeeping gene that maintains the basic life activities of cells, such as *Actin, 18s Ribosome RNA, Tublin* and *Ubiquitin*, etc. [19,20]. However, many studies have proved that the transcription levels of the housekeeping genes may change with different species, tissues, and organs [21–23]. Therefore, appropriate internal reference genes should be selected according to specific experimental conditions to reduce experimental errors. However, there has been no report on the screening of the internal reference genes in *B. papyrifera*. This limits the research on the regulation mechanisms of gene expression in *B. papyrifera* under adversity stresses.

In this study, 15 candidate internal reference genes (*NADH, L13, EIF3, HIS, Actin, PP2A, DOUB, UBE2, UBC, PTB, rRNA, GAPDH, HSP, RPL8,* and *TUA*) were selected based on the transcriptome data in *B. papyrifera*. The RT-qPCR technology, and the geNorm (Version 3.5) [24], NormFinder (version 0.953) [25], and BestKeeper (version 1.0) [26] software were used to analyze the expression stability of the candidate internal reference genes under the abiotic stresses (i.e., drought stress, salt stress, and heavy metal stress) and in different tissues. In addition, the online analysis tool RefFinder [27] was used to comprehensively evaluate the results obtained by the above software. Then, the selected internal reference genes were verified with target genes *DREB* and *POD*. This study is the first to screen and verify the internal reference genes used for RT-qPCR normalization in *B. papyrifera* under the abiotic stresses and the different tissues, which lays the foundation for gene expression analysis in *B. papyrifera*.

2. Results

2.1. Determination of Primer Specificity and Amplification Efficiency

The PCR amplifications were performed using equal amounts of the mixed cDNA as templates (Figure 1). The target fragments were unique and bright for all the primers, without primer dimers and non-specific amplifications. The band sizes were in line with the expected values. The primer specificity was verified by the RT-qPCR technology (Figure 2). The melting curves of individual genes showed a single melting peak. This indicates that those primers can perform specific amplifications. After calculation, the amplification efficiency of each candidate internal reference gene was between 90.26–117.99%, and the correlation coefficient (R^2) was between 0.987–0.999 (Table 1). Therefore, those primers achieved good specificity and efficiency for amplifying the candidate genes, suggesting that the candidate genes can be used in subsequent experiments.

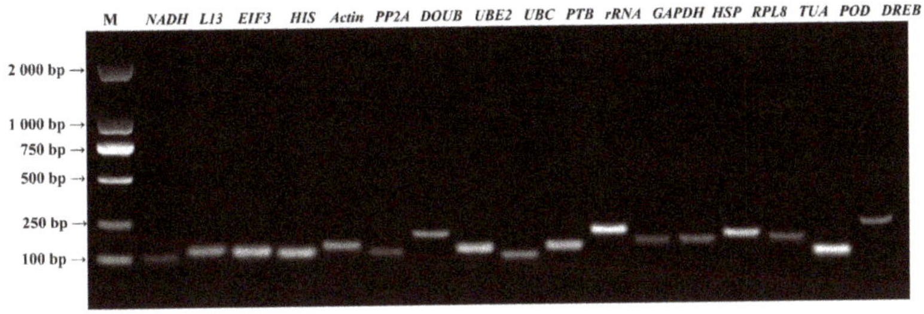

Figure 1. Agarose gel electrophoresis of the conventional PCR products of the candidate internal reference genes in *B. papyrifera*.

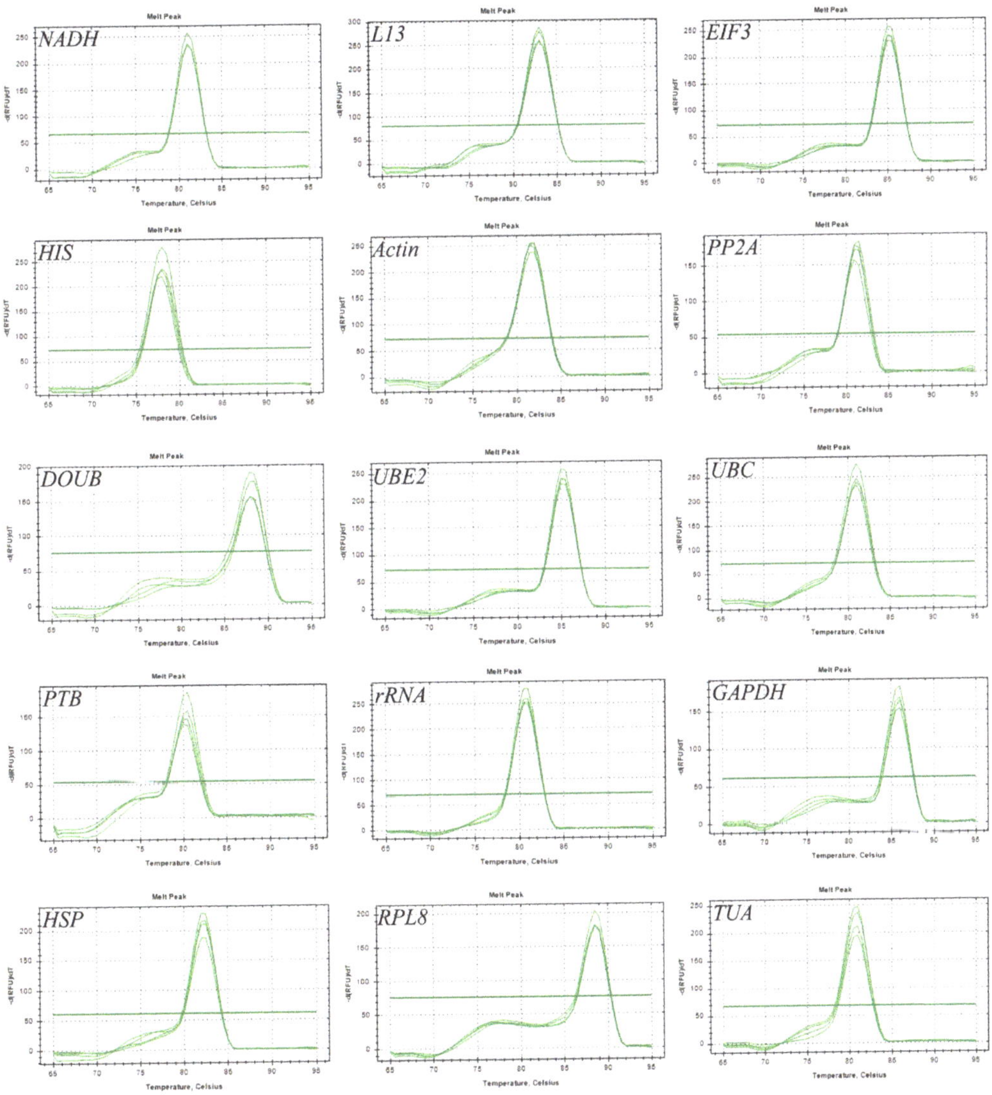

Figure 2. Melting curves of the candidate reference genes in *B. papyrifera*.

Table 1. Primer sequence and amplification efficiency of the 15 candidate reference genes.

Gene	Primer ID	Primer sequence (5′-3′)	Amplicon Size (Bp)	Efficiency (%)	R^2
NADH	*NADH*-F *NADH*-R	GGACAGGTGGAAGATCGTCTG GGAATCTTCAGAACCCCGGAA	111	97.18	0.987
L13	*L13*-F *L13*-R	TGCCAGCCCTAACTTTCATGT AGACCCGGAGAAGAATTGCTC	126	92.17	0.999
EIF3	*EIF3*-F *EIF3*-R	GTCCACATCATTCGAAGCAGC GATCTATGAAGTGCCTGCGGA	130	106.31	0.997
HIS	*HIS*-F *HIS*-R	TGGCCTTGCATTCTCCAGTAG GACAAGCTGCGAGAGTGGTAT	118	98.87	0.996
Actin	*Actin*-F *Actin*-R	TACGCATTGAAGACCCTCCAC TGGCCACACTTGCTTAGACAA	148	90.26	0.998
PP2A	*PP2A*-F *PP2A*-R	TCCTTTTGCGAGTCGATGGAA CTTTGACGTTTGAAGCGAGCA	119	117.99	0.988
DOUB	*DOUB*-F *DOUB*-R	CCTGATCTTCGCCGGAAAACA TGGAGAGGGTTGAAGAGAGCT	194	97.48	0.999
UBE2	*UBE2*-F *UBE2*-R	TCTCTGCTTACGGACCCAAAC GAGGAGGAGCTATTGGGCCTA	144	92.29	0.998
UBC	*UBC*-F *UBC*-R	AGCATTACTTTCCGCTCCACA TGGCGAAAGTTTCTGTCCAGT	119	91.44	0.995
PTB	*PTB*-F *PTB*-R	CTGGAAACCTGCTGCCTTTTC ATTGAGGGTGTAGAAGCTGGC	151	96.29	0.999
rRNA	*rRNA*-F *rRNA*-R	CAGGTTTCGATGTTGGGGAGA CCAGCTTCCGAGAACATTCCT	196	95.56	0.999
GAPDH	*GAPDH*-F *GAPDH*-R	CCATGGAAGGACTTGGGGATC GTTCACTCCCACCACGTATGT	156	90.43	0.995
HSP	*HSP*-F *HSP*-R	CCAGCGCTGATGTTAGATTGC TTGCCATCAGAGCCTTTTCCT	174	92.66	0.993
RPL8	*RPL8*-F *RPL8*-R	TGATCACCGACATCATCCACG TCTGATCGGAAGGACATTGCC	185	90.55	0.992
TUA	*TUA*-F *TUA*-R	TCGAAAGGCCAACATACACCA GAGATGACAGGGGCATACGAG	175	96.59	0.997
POD	*POD*-F *POD*-R	CTCCTGTGACCTCAACTGCAA GAGTTGAACCATGGCGCAAAT	136	91.71	0.987
DREB	*DREB*-F *DREB*-R	TAAACCAGCTCACCCAATCCC CGGTTCTTGGGGAGTCTGATC	274	90.99	0.989

2.2. C_t Values of the Internal Reference Genes

The cycle threshold (C_t) value is inversely proportional to the gene expression level. A low C_t value reflects a high gene expression level. In the analysis of the box plot (Figure 3), the C_t values of most of the genes were between 22 and 28, indicating moderate expression abundances. As it can be seen in the box plot, the *GAPDH* gene has a broad range of C_t values (20.30–33.24), indicating a low gene expression stability. However, the *Actin* gene has a narrow range of C_t values (23.90–26.59), indicating a high gene expression stability. Therefore, different internal reference genes have different expression levels under the abiotic stresses and in the different tissues of *B. papyrifera*. According to the range of C_t values, the expression level of the *Actin* gene was stable. Therefore, the *Actin* gene was the best candidate internal reference gene.

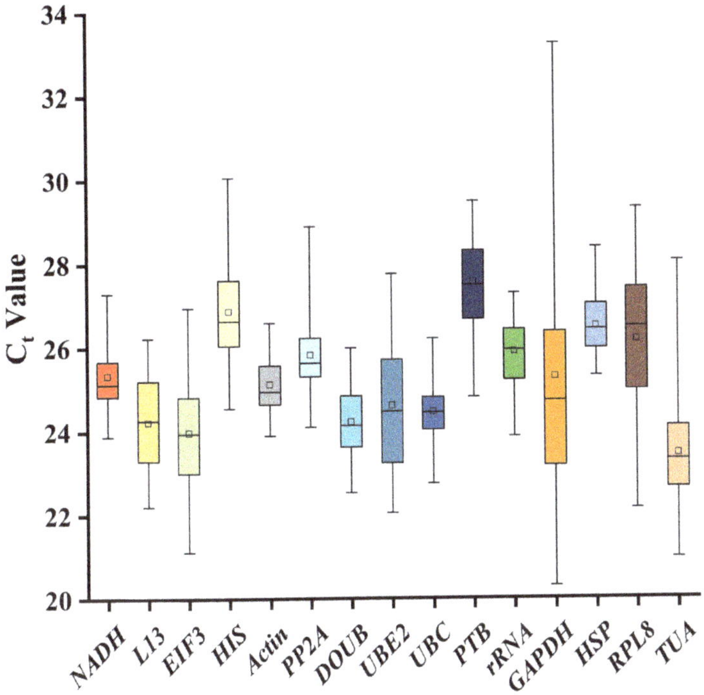

Figure 3. Box plot of the C_t values of the 15 candidate reference genes in *B. papyrifera*. The box represents the concentrated range of C_t values. The horizontal line in the middle of the box represents the median value, and the black square represents the average value. The upper and lower edges of the box represent the upper and lower quartiles, respectively. The upper and lower ends of the box represent the maximum and minimum values of the gene, respectively.

2.3. geNorm Analysis

The M value of the expression stability of each candidate internal reference gene under the drought stress and the different tissues was calculated by the geNorm program. The program takes M = 1.5 as the critical value. The smaller the value of M, the more stable the internal reference gene [24]. The geNorm analyses are shown in Figure 4. The M values of each candidate internal reference gene under the abiotic stresses and in the different tissues were lower than 1.5, suggesting that the expression level of each candidate internal reference gene was stable. The expression levels of the *Actin* and *EIF3* were stable under the drought stress. The *DOUB* and *HSP* were the most stable reference genes under the salt stress. The expression levels of the *NADH* and *HSP* were the most stable under the heavy metal stress. However, among the different tissues of *B. papyrifera*, the *PP2A* and *EIF3* were the most stable genes among the 15 reference genes. Comprehensive analysis of the expression levels of the 15 candidate internal reference genes in all the samples showed that the M values of the other genes were less than 1.5, except for the *GAPDH*, whose M values were greater than 1.5. The order of expression stability from high to low is *HSP* = *rRNA* > *NADH* > *PP2A* > *Actin* > *UBC* > *DOUB* > *L13* > *PTB* > *HIS* > *TUA* > *EIF3* > *RPL8* > *UBE2* > *GAPDH*. Therefore, the *HSP* and *rRNA* genes are the most stable internal reference genes in all the samples, and the *GAPDH* was the least stable one.

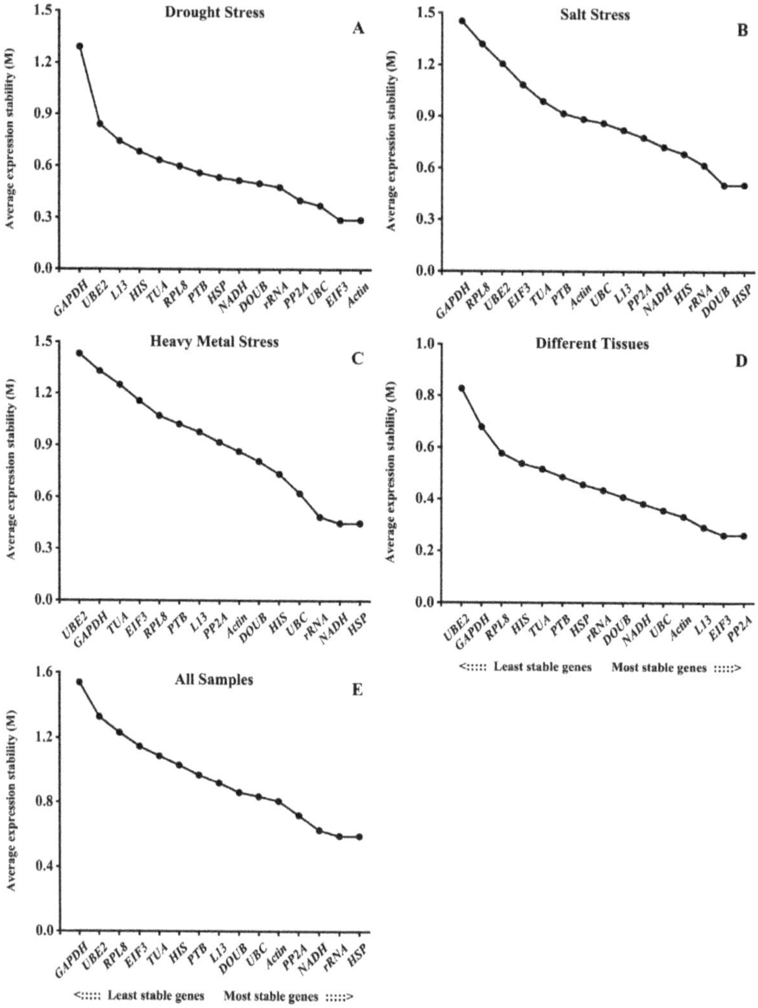

Figure 4. Average expression stability of the 15 candidate reference genes in *B. papyrifera* analyzed by the geNorm program. (**A**) Drought stress, (**B**) salt stress, (**C**) heavy metal stress, (**D**) different tissues, (**E**) all samples.

Determining the optimal number of internal reference genes can reduce the bias and fluctuation caused by a single internal reference gene. In Figure 5, the $V_{2/3}$ of *B. papyrifera* were 0.130, 0148, and 0.094 under the drought stress, under the heavy metal stress, and in the different tissues, respectively, all of which were less than 0.15. This shows that the results of selecting two internal reference genes were stable and reliable, and thus there was no need to select more than two reference genes. Under the salt stress and all samples, the coefficients of variation were $V_{2/3}$ (0.208) and $V_{2/3}$ (0.191), both of which were higher than the critical value 0.15. Both $V_{4/5}$ (0.135) and $V_{4/5}$ (0.148) were less than 0.15. This indicates that the samples under the salt stress and all samples need to introduce four internal reference genes for correction to keep the results stable and reliable.

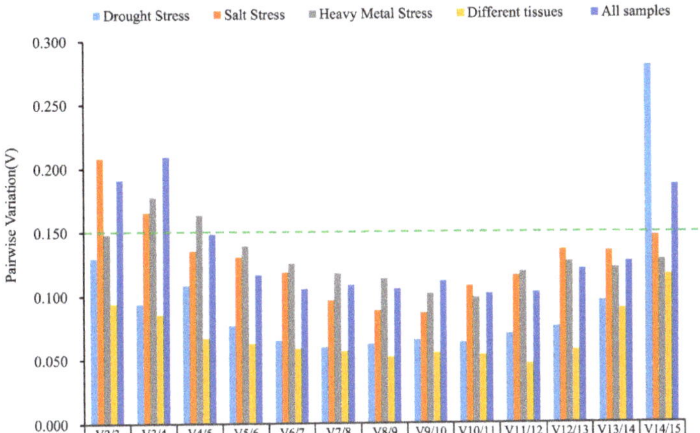

	V2/3	V3/4	V4/5	V5/6	V6/7	V7/8	V8/9	V9/10	V10/11	V11/12	V12/13	V13/14	V14/15
Drought Stress	0.130	0.094	0.108	0.077	0.065	0.060	0.062	0.065	0.063	0.070	0.075	0.095	0.279
Salt Stress	0.208	0.165	0.135	0.130	0.118	0.096	0.088	0.086	0.107	0.115	0.135	0.134	0.146
Heavy Metal Stress	0.148	0.177	0.163	0.139	0.125	0.117	0.113	0.101	0.098	0.118	0.126	0.121	0.127
Different tissues	0.094	0.085	0.067	0.062	0.058	0.056	0.051	0.055	0.053	0.046	0.057	0.089	0.115
All samples	0.191	0.209	0.148	0.116	0.105	0.108	0.105	0.111	0.101	0.102	0.120	0.126	0.186

Figure 5. Analysis of the paired variation of the 15 candidate internal reference genes in *B. papyrifera* by geNorm program. The optimal number of candidate reference genes required for accurate normalization was determined by paired variation $V_{n/(n+1)}$. The threshold value of $V_{n/(n+1)}$ was 0.15. When $V_{n/(n+1)}$ is less than 0.15, n genes can ensure stable and reliable results.

2.4. NormFinder Analysis

NormFinder needs to convert the C_t values of the genes into relative expressions before analyzing the data. Then, the stability of the candidate internal reference genes was sorted based on variance analysis. A smaller expression stability value of the candidate internal reference gene indicates a more stable expression [25]. The NormFinder analysis results are shown in Table 2. Under the drought stress and the salt stress, the most stable internal reference gene was the *DOUB*, and the least stable gene was the *GAPDH*. Under the heavy metal stress, the most stable internal reference gene was the *HSP*. In the different tissues, the *rRNA* has a relatively stable expression. The ranking of the gene expression stability in all samples from high to low was: *rRNA* (0.338) > *HSP* (0.383) > *NADH* (0.495) > *PP2A* (0.691) > *UBC* (0.738) > *Actin* (0.751) > *DOUB* (0.753) > *PTB* (0.792) > *HIS* (0.966) > *L13* (1.028) > *TUA* (1.090) > *EIF3* (1.277) > *RPL8* (1.598) > *UBE2* (1.652) > *GAPDH* (2.786). Therefore, the *rRNA* (0.338) was the most stable gene in all samples, and the *GAPDH* (2.786) was the least stable gene.

Table 2. Expression stability of the reference genes in *B. papyrifera* calculated by NormFinder.

Rank	Drought Stress		Salt Stress		Heavy Metal Stress		Different Tissues		All Samples	
	Gene	Stability	Gene	Stability	Gene	Stability	Gene	Stability	Gene	Stability
1	*DOUB*	0.152	*DOUB*	0.172	*HSP*	0.359	*rRNA*	0.051	*rRNA*	0.338
2	*rRNA*	0.162	*HSP*	0.396	*rRNA*	0.362	*Actin*	0.222	*HSP*	0.383
3	*Actin*	0.394	*NADH*	0.540	*NADH*	0.401	*EIF3*	0.229	*NADH*	0.495
4	*HSP*	0.429	*rRNA*	0.569	*UBC*	0.513	*TUA*	0.343	*PP2A*	0.691
5	*UBC*	0.436	*PP2A*	0.630	*DOUB*	0.707	*DOUB*	0.346	*UBC*	0.738
6	*PTB*	0.520	*HIS*	0.635	*HIS*	0.745	*PP2A*	0.383	*Actin*	0.751
7	*EIF3*	0.547	*UBC*	0.654	*Actin*	0.805	*PTB*	0.383	*DOUB*	0.753
8	*TUA*	0.586	*L13*	0.720	*PP2A*	0.907	*HIS*	0.456	*PTB*	0.792
9	*PP2A*	0.596	*Actin*	0.742	*PTB*	0.969	*NADH*	0.484	*HIS*	0.966

Table 2. *Cont.*

Rank	Drought Stress		Salt Stress		Heavy Metal Stress		Different Tissues		All Samples	
	Gene	Stability	Gene	Stability	Gene	Stability	Gene	Stability	Gene	Stability
10	*NADH*	0.616	*PTB*	0.817	*L13*	0.992	*HSP*	0.493	*L13*	1.028
11	*RPL8*	0.729	*TUA*	1.133	*RPL8*	1.265	*UBC*	0.503	*TUA*	1.090
12	*HIS*	0.917	*EIF3*	1.461	*TUA*	1.400	*L13*	0.557	*EIF3*	1.277
13	*L13*	0.998	*UBE2*	1.791	*GAPDH*	1.511	*RPL8*	0.836	*RPL8*	1.598
14	*UBE2*	1.090	*RPL8*	1.924	*EIF3*	1.672	*GAPDH*	1.178	*UBE2*	1.652
15	*GAPDH*	4.183	*GAPDH*	2.216	*UBE2*	1.921	*UBE2*	1.727	*GAPDH*	2.786

2.5. BestKeeper Analysis

The Bestkeeper mainly evaluates the stability of genes by comparing the SD values among C_t values of the candidate internal reference genes [26]. The analysis results of the Bestkeeper are shown in Table 3. Under the drought stress, the *UBC* was the most stable reference gene, and the *GAPDH* was the least stable gene. Under the salt stress, the *HSP* was the most stable reference gene, and the *GAPDH* was the least stable gene. Under the heavy metal stress, the *UBC* was the most stable reference gene, and the *UBE2* was the least stable gene. In addition, the *HSP* was the most stable reference gene, and the *UBE2* was the least stable gene in the different tissues. In all samples, the ranking of the gene expression stability from high to low was: *HSP* (0.518) > *UBC* (0.535) > *NADH* (0.607) > *DOUB* (0.615) > *Actin* (0.643) > *PP2A* (0.647) > *rRNA* (0.672) > *TUA* (0.886) > *L13* (0.913) > *PTB* (0.917) > *HIS* (1.018) > *EIF3* (1.023) > *RPL8* (1.353) > *UBE2* (1.477) > *GAPDH* (2.123). It indicates that the *HSP* (0.518) was the most stable gene in all samples, and the *GAPDH* (2.123) was the least stable gene.

Table 3. Analysis of the expression stability by Bestkeeper and the ranking of the internal reference genes in *B. papyrifera*.

Rank	Drought Stress		Salt Stress		Heavy Metal Stress		Different Tissues		All Samples	
	Gene	Stability	Gene	Stability	Gene	Stability	Gene	Stability	Gene	Stability
1	*UBC*	0.398	*HSP*	0.453	*UBC*	0.382	*HSP*	0.215	*HSP*	0.518
2	*rRNA*	0.465	*NADH*	0.49	*NADH*	0.416	*RPL8*	0.216	*UBC*	0.535
3	*HSP*	0.481	*DOUB*	0.521	*HSP*	0.439	*DOUB*	0.256	*NADH*	0.607
4	*DOUB*	0.513	*UBC*	0.561	*rRNA*	0.493	*UBC*	0.475	*DOUB*	0.615
5	*PP2A*	0.515	*HIS*	0.596	*DOUB*	0.524	*NADH*	0.489	*Actin*	0.643
6	*Actin*	0.530	*PP2A*	0.609	*Actin*	0.607	*Actin*	0.514	*PP2A*	0.647
7	*EIF3*	0.557	*rRNA*	0.723	*PP2A*	0.645	*PP2A*	0.566	*rRNA*	0.672
8	*PTB*	0.634	*L13*	0.729	*HIS*	0.741	*L13*	0.588	*TUA*	0.886
9	*RPL8*	0.639	*Actin*	0.772	*L13*	0.768	*EIF3*	0.622	*L13*	0.913
10	*TUA*	0.671	*TUA*	0.784	*PTB*	0.866	*TUA*	0.641	*PTB*	0.917
11	*NADH*	0.727	*PTB*	0.871	*RPL8*	0.972	*rRNA*	0.650	*HIS*	1.018
12	*UBE2*	0.844	*EIF3*	1.336	*TUA*	1.088	*HIS*	0.651	*EIF3*	1.023
13	*L13*	0.928	*UBE2*	1.356	*GAPDH*	1.137	*PTB*	0.819	*RPL8*	1.353
14	*HIS*	0.982	*RPL8*	1.551	*EIF3*	1.303	*GAPDH*	1.213	*UBE2*	1.477
15	*GAPDH*	3.496	*GAPDH*	1.554	*UBE2*	1.593	*UBE2*	1.619	*GAPDH*	2.123

2.6. RefFinder Analysis

To avoid the error caused by the evaluation program of a single internal reference gene, the online analysis tool RefFinder was used to calculate the geometric mean of the gene expression stability rankings of the above three programs (geNorm, NormFinder, and Bestkeeper). The smaller the geometric mean, the more stable the gene expression [27]. From Table 4, under the drought stress, the *rRNA* and *Actin* were the most stable internal reference genes, and the *GAPDH* was the least stable gene. Under the salt stress, the *DOUB* and *HSP* were the most stable internal reference genes, and the *GAPDH* was the least stable gene. Under the heavy metal stress, the *HSP* and *NADH* were the most stable internal reference genes, and the *UBE2* was the least stable gene. In the different tissues, the *EIF3* and *Actin* were the most stable internal reference genes, and the *UBE2* was the least stable gene. The ranking of the gene expression stability in all samples was: *HSP* (1.414) > *rRNA* (1.627) > *NADH* (3) > *UBC* (3.761) > *PP2A* (5.091) > *Actin* (5.477) > *DOUB* (5.856) > *PTB* (8.459) > *L13* (9.24) > *HIS* (9.975) > *TUA* (10.158) > *EIF3* (12) > *RPL8* (13) > *UBE2* (14) > *GAPDH* (15). Among them, the *HSP* and *rRNA* were identified as the most stable internal reference genes in all samples, and the *GAPDH* was the least stable gene in all samples.

Table 4. Ranking of the expression stability of the reference genes in *B. papyrifera* by RefFinder.

Rank	Drought Stress		Salt Stress		Heavy Metal Stress		Different Tissues		All Samples	
	Gene	Stability	Gene	Stability	Gene	Stability	Gene	Stability	Gene	Stability
1	*rRNA*	2.115	*DOUB*	1.316	*HSP*	1.316	*EIF3*	2.28	*HSP*	1.414
2	*Actin*	2.449	*HSP*	1.414	*NADH*	2.06	*Actin*	3.13	*rRNA*	1.627
3	*UBC*	2.590	*NADH*	3.31	*rRNA*	2.632	*PP2A*	3.742	*NADH*	3
4	*DOUB*	3.130	*rRNA*	3.984	*UBC*	2.828	*rRNA*	4.031	*UBC*	3.761
5	*EIF3*	3.956	*HIS*	4.949	*DOUB*	5.477	*DOUB*	4.787	*PP2A*	5.091
6	*HSP*	4.899	*PP2A*	5.733	*HIS*	5.886	*HSP*	5.477	*Actin*	5.477
7	*PP2A*	6.160	*UBC*	6.293	*Actin*	6.735	*NADH*	6.344	*DOUB*	5.856
8	*PTB*	7.416	*L13*	7.737	*PP2A*	7.737	*UBC*	6.477	*PTB*	8.459
9	*NADH*	9.124	*Actin*	9	*PTB*	9.487	*L13*	7.502	*L13*	9.24
10	*TUA*	9.685	*PTB*	10.241	*L13*	9.487	*TUA*	7.933	*HIS*	9.975
11	*RPL8*	10.215	*TUA*	10.741	*RPL8*	11	*RPL8*	8.142	*TUA*	10.158
12	*HIS*	12.471	*EIF3*	12	*TUA*	12.243	*PTB*	8.596	*EIF3*	12
13	*L13*	13	*UBE2*	13	*GAPDH*	13.243	*HIS*	10.843	*RPL8*	13
14	*UBE2*	13.471	*RPL8*	14	*EIF3*	13.741	*GAPDH*	14	*UBE2*	14
15	*GAPDH*	15	*GAPDH*	15	*UBE2*	15	*UBE2*	15	*GAPDH*	15

2.7. Verification of the Expression Stability of the Internal Reference Genes

To verify the reliability of the selected internal reference genes, the most stable and least stable internal reference genes screened by the ReFinder program were used as normalization factors. Then, the *DREB* and *POD* gene expression levels were independently validated. As shown in Figure 6, there were great differences in the *DREB* and *POD* expression levels obtained by using different internal reference genes. When the selected stable genes were used alone or in combination as normalized internal reference genes, the relative expression patterns of the *DREB* and *POD* genes showed a similar trend. On the contrary, when relatively unstable genes were used for relative quantification, the relative expression levels of the *DREB* and *POD* were quite different.

Under the drought stress, when the most stable internal reference genes *rRNA*, *Actin*, and their combination (*rRNA* + *Actin*) were used, the expression levels of the *DREB* and *POD* generally showed a trend of increasing first and then decreasing. The *DREB* and *POD* expression level reached the peak at 24 h, and 12 h, respectively. However, when the least stable gene *GAPDH* was used for calculation, the expression levels of the *DREB* and *POD* showed an overall trend of first increasing, then decreasing, followed by increasing. Under the salt stress, when the most stable internal reference genes *DOUB*, *HSP*, *NADH*, *rRNA*, and their combination (*DOUB* + *HSP* + *NADH* + *rRNA*) were used as candidate internal reference genes, the relative expression of the *DREB* showed an overall trend of first increasing, and reached the peak at 12 h. The relative expression of the *POD* decreased first and then increased, and reached a peak at 72 h, with its expression remaining at a low level. When calculated with the unstable gene *GAPDH*, the *DREB* reached the peak at 72 h. Although the *POD* gene expression also reached its peak at 72 h, its expression remained at a high level. Under the heavy metal stress, when the stable internal reference genes *HSP*, *NADH*, and their combination (*HSP* + *NADH*) were used, the relative expression levels of the *DREB* and *POD* both reached their peaks at 72 h. When using the least stable internal reference gene *UBE2*, the *DREB* and *POD* reached their peaks at 48 h. Therefore, the selection of internal reference genes has a great impact on the expression levels of the target genes. Appropriate internal reference genes are conducive to obtaining accurate RT-qPCR results. Using unstable reference genes can lead to unreliable results.

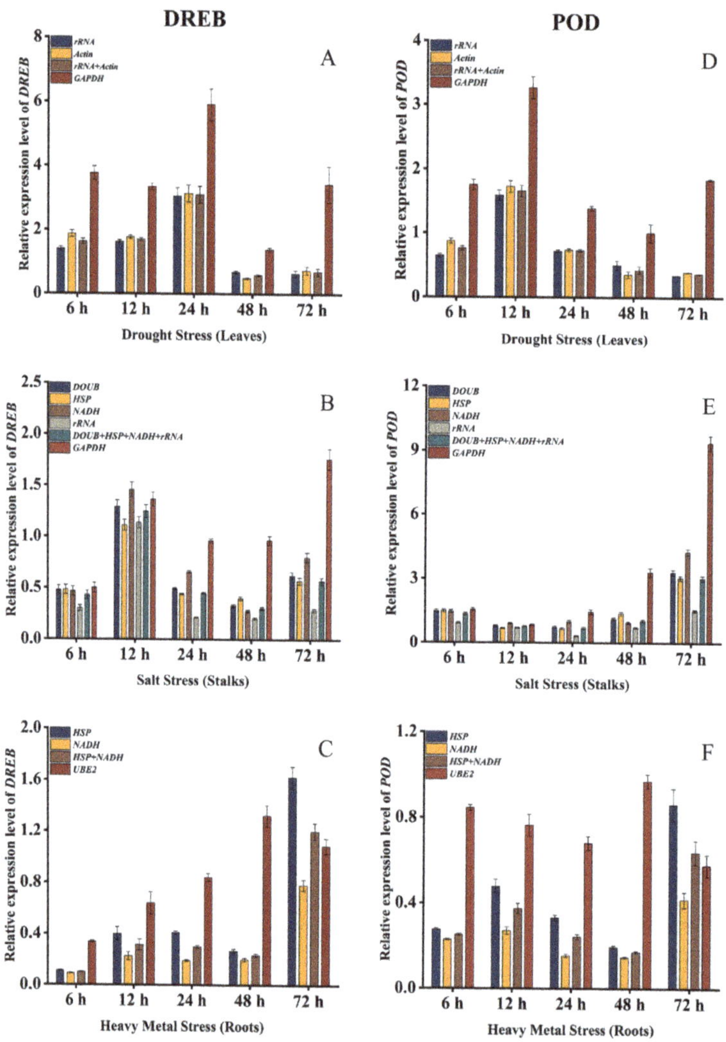

Figure 6. Using *DREB* and *POD* to verify the expression stability of internal reference genes screened under the abiotic stress in *B. papyrifera*. (**A**,**D**): Drought stress (leaves); (**B**,**E**): Salt stress (stalks); (**C**,**F**): Heavy metal stress (roots). The error bars indicate the standard deviations (SDs).

3. Discussion

With the development of molecular biology research, the study of key genes controlling plant stress tolerance and its molecular stress tolerance mechanism will provide important information for plant breeding [28]. The RT-qPCR is one of the main methods for analyzing gene expression levels and regulatory patterns [29]. Selection of internal reference genes with stable expression levels is the key to accurately analyzing the target gene expression [30]. The screening of the internal reference genes in combination with the plant transcriptome database is one of the most effective methods for the research in non-model plants [31]. It has been applied in various plants such as *Malpighia emarginata* [32], *Sinocalycanthus chinensisb* [33], and *Oryza sativab* [34]. Therefore, through the transcriptome database, this study screened a batch of stable candidate internal reference genes, *NADH, L13, EIF3, HIS, Actin, PP2A, DOUB, UBE2, UBC, PTB, rRNA, GAPDH, HSP, RPL8,* and *TUA*. Then, their expression stability under abiotic stresses (i.e., drought stress, salt stress, and heavy metal stress) and in seven different tissues were studied.

Due to the differences in the operational logic and statistical methods used in each program, the ranking of the expression stability of the internal reference genes were slightly different among the three programs. For example, under the drought stress, the *Actin* and *EIF3* genes are the most suitable reference genes verified by geNorm software. The *DOUB* and *rRNA* genes are the most stable internal reference genes analyzed by the Normfinder software. The *UBC* and *rRNA* genes are the most stable reference genes verified by the Bestkeeper. This phenomenon also appeared in *Carya illinoinensis* [35], *Passiflora edulis* [36], *Forsythia suspensa* [37], etc. Therefore, in order to avoid the one-sidedness of the analysis caused by a single piece of software, scholars usually choose the RefFinder as the comprehensive analysis program for internal reference gene stability. It is widely used in reference gene screening studies [38,39]. In this study, RefFinder was used to comprehensively evaluate the results of the above three kinds of software to determine the ranking of the expression stability of the candidate internal reference genes. However, accurate RT-qPCR analysis results cannot be obtained with only one single internal reference gene. Therefore, the geNorm is often used to determine the optimal number of reference genes under abiotic stresses and in different tissues. This can determine the most appropriate combination of internal reference genes for different experimental samples.

The *DREB* is a transcription factor unique to plants. Under adversity stresses, the *DREB* interacts with the DRE/CRT (dehydration response element) cis-element in the promoter region of the stress resistance genes, regulating the expression of a series of downstream genes (including DRE/CRT elements), and enhancing the resistance of plants to stresses [40,41]. The *DREB* gene can be induced to up-regulate its expression under the adversity stresses in *Musa acuminata* [42], *Glycine max* [43], and *Ricinus communis* [44]. The *POD* gene is a functional gene of antioxidant enzymes, and up-regulating its expression can help plants resist external damages when they encounter abiotic stresses [45]. This has been verified in *Phytophtora capsici* [46], *Ipomoea batatas* [47], *Tamarix hispida* [48], etc. Therefore, the *DREB* and *POD* genes can be used to verify the reliability of the screened reference genes. This study screened the most stable internal reference genes and their combinations under drought stress, salt stress, and heavy metal stress. Furthermore, the expression patterns of the *DREB* and *POD* genes in *B. papyrifera* under abiotic stresses were analyzed with the least stable genes as reference genes. When normalizing the gene expression levels with genes of stable expression, the expression patterns of the stress-responsive genes *DREB* and *POD* were consistent. However, when the unstable gene was used as a reference gene, the *DREB* and *POD* gene expression levels were significantly different. This further verified the accuracy of the screened internal reference genes. In summary, the selection of suitable internal reference genes is the key to analyzing the expression changes of target genes.

4. Materials and Methods

4.1. Materials

The seeds of *B. papyrifera* were collected from Yanji Town, Shuyang County, Suqian City, Jiangsu Province (N34.16560, E118.58537) in China. The seeds were soaked in 1600 mg L^{-1} Gibberellin A3 (GA3) solution (Coolaber, Beijing, China) for 24 h, rinsed with distilled water 2–3 times, and then sowed in a mixed substrate of peat soil (Pindstrup Mosebrug A/S, Ryomgaard, Denmark) and vermiculite (Guangdong Chenxing Agriculture Co., Ltd., Guangzhou, China). Then, they were cultured in a light incubator (LHP-300H, Changzhou

Putian Instrument Manufacturing Co., Ltd., Changzhou, China) (at a temperature of 30 °C, with a humidity between 60% and 70%, a light intensity of 800 μmol m^{-2} s^{-1}, and a photoperiod of 12 h light/12 h dark). After six months of culture, samples were collected from seven different tissues (i.e., terminal bud, young leaf, petiole, old leaf, phloem, xylem, and root). We selected seedlings of *B. papyrifera* with good growth and uniform size, carefully peeled off the nutrient matrix, placed them in the 1/2 Hoagland nutrient solution (Phygene Biotechnology Co., Ltd., Fuzhou, China) for one week, and then applied the abiotic stresses on the seedlings. The drought stress was simulated by a 30 g/L PEG-6000 solution (Tianjin Kermel Chemical Reagent Co., Ltd., Tianjin, China). The salt stress was carried out with a 300 mmol L^{-1} NaCl solution (Tianjin Kermel Chemical Reagent Co., Ltd., Tianjin, China). The heavy metal stress was applied by a 500 μmol L^{-1} $ZnSO_4{\cdot}7H_2O$ solution (Tianjin Bodi Chemicals Co., Ltd., Tianjin, China). Then, at 0 h (CK), 6 h, 12 h, 24 h, 48 h, and 72 h after the treatment, the roots, stems, and leaves were cut off and used as samples. Finally, the samples were frozen in liquid nitrogen immediately, and stored at −80 °C in an ultra-low temperature freezer (DW-HL668, Zhongke Meiling Cryogenics Co., Ltd., Hefei, China) until use. All the experiments were repeated three times.

4.2. Extraction and Detection of RNA

The RNA prep Pure Polysaccharide Polyphenol Plant Total RNA Extraction Kit (Cat. #DP441, Tiangen Biotech (Beijing) Co., Ltd., Beijing, China) was used to extract RNA according to the manufacturer's instructions. In addition, 1% agarose gel (Biosharp Life Sciences, Anhui, China) electrophoresis was used to test the integrity of the RNA. Then, the concentration and purity of RNA were tested using an ultra-micro ultraviolet-visible spectrophotometer (NanoDrop2000) (NanoDrop2000, Thermo Fisher Scientific, Waltham, MA, USA).

4.3. Synthesis of cDNA

The RNase-free water was used to dilute the RNA to a concentration of 200 ng/μL, and the concentrations of the RNA were the same for all samples. To achieve a higher efficiency of synthesis, the RNA templates were incubated at 65 °C for 5 min, and then the samples were placed on ice for 2 min. According to the instructions of the M5 Super qPCR RT Kit (Cat. #MF012, Mei5 Biotechnology Co., Ltd., Beijing, China), the PCR reaction system was configured as follows: 4 μL 5 × M5 RT Super Mix and 2 μg RNA template were blended to a total volume of 20 μL with RNase-free water (Cat. #CD4381, Phygene, Biotechnology Co., Ltd., Fuzhou, China). The operations were performed on ice. Samples were reverse-transcribed using a gradient PCR amplification instrument from Bio-Rad ($T100^{TM}$ Thermal Cycler, Bio-Rad, Hercules, CA, USA). The PCR reaction process was as follows: first incubated at 37 °C for 15 min, then incubated at 50 °C for 5 min, and finally heated at 96 °C for 5 min to deactivate the enzyme. After the reaction, the reverse-transcribed cDNA was stored at −20 °C for subsequent experiments.

4.4. Screening of Candidate Internal Reference Genes and Designing of Primers

According to the common internal reference genes in other plants in the existing literature, 15 candidate internal reference genes were selected according to the transcriptome database of *B. papyrifera*, namely: *NADH*, *L13*, *EIF3*, *HIS*, *Actin*, *PP2A*, *DOUB*, *UBE2*, *UBC*, *PTB*, *rRNA*, *GAPDH*, *HSP*, *RPL8*, and *TUA*. Primer 3web (http://primer3.ut.ee/, accessed on 1 June 2022) was used to design primers. The principles of primer design included: the length of the PCR amplification product between 100 bp and 300 bp, the primer length between 18 bp and 25 bp, the annealing temperatures between 50 °C and 60 °C, and the GC base content between 45% and 55%. It is necessary to avoid the occurrence of hairpin structures and primer-dimer mismatches as much as possible. In addition, the NCBI Primer-BLAST (https://www.ncbi.nlm.nih.gov/tools/primer-blast/index.cgi?LINK_LOC=BlastHome, accessed on 1 June 2022) was utilized to test the primer specificity. Primers were synthesized by General Biology (Anhui) Co., Ltd., Chuzhou, China.

4.5. RT-qPCR Reaction Conditions

The RT-qPCR used the SYBR green dye method, and the following PCR reaction system was created according to the instructions of the 2 × M5 HiPer SYBR Premix EsTaq kit (Cat. #MF787, Mei5 Biotechnology Co., Ltd., Beijing, China): 1 μL cDNA template, 0.2 μL both forward and reverse primers (10 μmol L^{-1}), 3.6 μL ddH_2O and 5 μL 2 × M5 HiPer SYBR Premix EsTaq. These operations were performed three times for all samples. The samples were amplified using the CFX96 RT-qPCR instrument from Bio-Rad (CFX96 Real-time System, Bio-Rad, Hercules, CA, USA). The PCR reaction programs consisted of pre-denaturation at 95 °C for 30 s, denaturation at 95 °C for 5 s, then annealing at 60 °C for 30 s, for 39 cycles (the melting curve was from 65 °C to 95 °C, increasing by 0.5 °C for each cycle, and lasting for 0.05 s to reach the melting temperature). The fluorescence signals were collected.

4.6. Detection of Primer Specificity and Amplification Efficiency

The cDNA samples were mixed in equal amounts and diluted three times with ddH_2O as a template for the ordinary PCR amplification to test the specificity of the primers. Normal PCR amplification was performed according to the TaKaRa Taq (Cat. #R001A, TaKaRa, Kyoto, Japan) kit. The reaction system comprised 14.3 μL ddH_2O, 2 μL 10 × PCR Buffer (Mg^{2+} plus), 0.1 μL TaKaRa Taq (5 U μL^{-1}), 1.6 μL dNTP Mixture (2.5 mmol L^{-1}), 1.5 μL cDNA, 0.25 μL upstream primers, and 0.25 μL downstream primers (10 μmol L^{-1}). The reaction procedures consisted of 95 °C for 2 min; 98 °C for 10 s, 60 °C for 30 s, 72 °C for 30 s, 30 cycles, and 72 °C for 5 min at the end. After the reaction, the primer specificity was tested with a 1% agarose gel. The cDNA of all the samples were mixed in appropriate amounts and diluted into six gradients (1/3, 1/9, 1/27, 1/81, 1/243, and 1/729). These were then used as the templates to perform the RT-qPCR amplifications for a standard curve. In addition, the primer amplification efficiency was calculated by the formula:

$$E\% = (3^{-1/\text{slope}} - 1) \times 100\% \quad (1)$$

4.7. The Stability of Candidate Internal Reference Genes

The C_t of the 15 candidate genes in *B. papyrifera* was obtained via the RT-qPCR. The original C_t values were sorted out by using the software Microsoft Excel 2016, and three programs (geNorm [24], NormFinder [25], BestKeeper [26]) and the online analysis tool RefFinder [27] were operated to comprehensively evaluate the expression stability of the 15 candidate internal reference genes. Finally, the best internal reference genes in *B. papyrifera* under the abiotic stresses and in the various tissues were screened out.

4.8. Verification of the Expression Stability of the Internal Reference Genes

The genes in response to adversity stresses, i.e., *DREB* and *POD*, were selected to verify the stability of the screened internal reference genes. The cDNAs of the leaves of *B. papyrifera* under the drought stress, the stems of *B. papyrifera* under the salt stress, and the roots of *B. papyrifera* under the heavy metal stress were used as templates. By the RT-qPCR technology, the best candidate internal reference genes and their combinations were used as normalization factors, and the unstable internal reference genes were used for comparisons. Then, the relative expressions of the *DREB* and *POD* genes of *B. papyrifera* under the abiotic stresses were analyzed using the $2^{-\Delta\Delta CT}$ method. The experiments were repeated three times. The reaction system and procedures were as described in Section 4.5.

5. Conclusions

In this study, 15 candidate internal reference genes were selected based on the transcriptome database of *B. papyrifera*, and their expression levels under abiotic stresses and in seven different tissues were studied. We used the programs geNorm, NormFinder, BestKeeper, and RefFinder to evaluate the expression stability of the candidate internal reference genes. Then, the accuracy of the screened reference genes was verified by the stress-responsive genes *DREB* and *POD*. This study provides a few reliable internal reference genes for the analysis of target gene expression in *B. papyrifera* under abiotic stresses and in the different tissues. This research lays the foundation for the study of the stress resistance and regulatory mechanisms in *B. papyrifera*, and the discovery of its important functional genes.

Author Contributions: Conceptualization, H.L. and X.H.; methodology, H.L. and M.C.; software, Z.W.; validation, Q.F. and X.Y.; formal analysis, Z.W., Z.H. and M.C.; investigation, Q.F.; resources, H.L.; data curation, M.C.; writing—original draft preparation, M.C.; writing—review and editing, M.C.; visualization, Z.H.; supervision, X.H.; project administration, H.L. and X.H.; funding acquisition, X.Z. All authors have read and agreed to the published version of the manuscript.

Funding: This research was funded by the National Natural Science Foundation of China (No. 32171701).

Institutional Review Board Statement: Not applicable.

Informed Consent Statement: Not applicable.

Data Availability Statement: Not applicable.

Acknowledgments: We thank Jianwei Ni for his generosity in sharing the transcriptome data and providing his experience in the cultivation of *B. papyrifera* seedlings. We also thank Enying Liu for language edit.

Conflicts of Interest: The authors declare no conflict of interest.

References

1. Peng, X.J.; Shen, S.H. The Paper Mulberry: A Novel Model System for Woody Plant Research. *Chin. Bull. Bot.* **2018**, *53*, 372–381. [CrossRef]
2. Wang, G.W.; Huang, B.K.; Qin, L.P. The Genus Broussonetia: A Review of its Phytochemistry and Pharmacology. *Phytother. Res.* **2012**, *26*, 1–10. [CrossRef] [PubMed]
3. Si, B.W.; Tao, H.; Zhang, X.L.; Guo, J.P.; Cui, K.; Tu, Y.; Diao, Q.Y. Effect of *Broussonetia papyrifera* L. (Paper Mulberry) Silage on Dry Matter Intake, Milk Composition, Antioxidant Capacity and Milk Fatty Acid Profile in Dairy Cows. *Asian. Austral. J. Anim.* **2018**, *31*, 1259–1266. [CrossRef] [PubMed]
4. Han, Q.H.; Wu, Z.l.; Huang, B.; Sun, L.Q.; Ding, C.B.; Yuan, S.; Zhang, Z.W.; Chen, Y.E.; Hu, C.; Zhou, L.J.; et al. Extraction, Antioxidant and Antibacterial Activities of *Broussonetia papyrifera* Fruits Polysaccharides. *Int. J. Biol. Macromol.* **2016**, *92*, 116–124. [CrossRef] [PubMed]
5. Guo, F.J.; Feng, L.; Huang, C.; Ding, H.X.; Zhang, X.T.; Wang, Z.Y.; Li, Y.M. Prenylflavone Derivatives from *Broussonetia papyrifera*, Inhibit the Growth of Breast Cancer Cells in Vitro and in Vivo. *Phytochem. Lett.* **2013**, *6*, 331–336. [CrossRef]
6. Park, J.Y.; Yuk, H.J.; Ryu, H.W.; Lim, S.H.; Kim, K.S.; Park, K.H.; Ryu, Y.B.; Lee, W.S. Evaluation of Polyphenols from *Broussonetia papyrifera* as Coronavirus Protease Inhibitors. *J. Enzyme Inhib. Med. Chem.* **2017**, *32*, 504–515. [CrossRef] [PubMed]
7. Zeng, P.; Guo, Z.H.; Xiao, X.Y.; Zhou, H.; Gu, J.; Liao, B. Tolerance capacities of *Broussonetia papyrifera* to heavy metal(loid)s and its phytoremediation potential of the contaminated soil. *Int. J. Phytorem.* **2022**, *24*, 580–589. [CrossRef] [PubMed]
8. Zhang, M.; Fang, Y.M.; Ji, Y.H.; Jiang, Z.P.; Wang, L. Effects of salt stress on ion content, antioxidant enzymes and protein profile in different tissues of *Broussonetia papyrifera*. *S. Afr. J. Bot.* **2013**, *85*, 1–9. [CrossRef]
9. Ding, F.; Yang, F.; Li, D.L.; Du, T.Z. Studies on the anatomical structure characteristic and drought resistance of *Broussonetia papyrifera*. *J. Anhui Agric. Sci.* **2010**, *38*, 20949–20952. [CrossRef]
10. Jin, Q.; Wu, L.; Zhao, Y.L.; Yang, G.Y.; Tang, T.C.; Ma, Y.Z.; Xu, Z.G. Methods for Rapid Seed Germination of *Broussonetia papyrifera*. *Pak. J. Bot.* **2023**, *55*, 941–948. [CrossRef]
11. Zhang, W.; Zhao, Y.L.; Xu, Z.G.; Huang, H.M.; Zhou, J.K.; Yang, G.Y. Morphological and Physiological Changes of *Broussonetia papyrifera* Seedlings in Cadmium Contaminated Soil. *Plants* **2020**, *9*, 1698. [CrossRef] [PubMed]
12. Xu, B.C.; Hao, K.Y.; Chen, X.G.; Wu, E.Y.; Nie, D.Y.; Zhang, G.Y.; Si, H.B. *Broussonetia papyrifera* Polysaccharide Alleviated Acetaminophen-Induced Liver Injury by Regulating the Intestinal Flora. *Nutrients* **2022**, *14*, 2636. [CrossRef] [PubMed]
13. Zhao, M.J.; Lv, D.L.; Hu, J.C.; He, Y.L.; Wang, Z.; Liu, X.Y.; Ran, B.K.; Hu, J.H. Hybrid *Broussonetia papyrifera* Fermented Feed Can Play a Role Through Flavonoid Extracts to Increase Milk Production and Milk Fatty Acid Synthesis in Dairy Goats. *Front. Vet. Sci.* **2022**, *9*, 794443. [CrossRef] [PubMed]
14. Huang, H.M.; Zhao, Y.L.; Xu, Z.G.; Zhang, W.; Jiang, K.K. Physiological Responses of *Broussonetia papyrifera* to Manganese Stress, a Candidate Plant for Phytoremediation. *Ecotox. Environ. Saf.* **2019**, *181*, 18–25. [CrossRef] [PubMed]
15. Abdallah, H.B.; Bauer, P. Quantitative Reverse Transcription-qPCR-Based Gene Expression Analysis in Plants. *Methods Mol. Biol.* **2016**, *1363*, 9–24. [CrossRef] [PubMed]

16. Die, J.V.; Román, B. RNA Quality Assessment: A View from Plant qPCR Studies. *J. Exp. Bot.* **2012**, *63*, 6069–6077. [CrossRef] [PubMed]
17. Zhan, X.Y.; Cui, H.L.; Ji, X.J.; Xue, J.A.; Jia, X.Y.; Li, R.Z. Selection of the optimal reference genes for transcript expression analysis of lipid biosynthesis-related genes in Okra (*Abelmoschus esculentus*). *Sci. Hortic.* **2021**, *282*, 110044. [CrossRef]
18. Soni, P.; Shivhare, R.; Kaur, A.; Bansal, S.; Ram, H. Reference Gene Identification for Gene Expression Analysis in Rice under Different Metal Stress. *J. Biotechnol.* **2021**, *332*, 83–93. [CrossRef]
19. Wei, T.L.; Wang, H.; Pei, M.S.; Liu, H.N.; Yu, Y.H.; Jiang, J.F.; Guo, D.L. Identification of Optimal and Novel Reference Genes for Quantitative Real-Time Polymerase Chain Reaction Analysis in Grapevine. *Aust. J. Grape Wine Res.* **2021**, *27*, 325–333. [CrossRef]
20. Lin, S.k.; Xu, S.c.; Huang, L.y.; Qiu, F.x.; Zheng, Y.h.; Liu, Q.h.; Ma, S.w.; Wu, B.; Wu, J.c. Selection and Validation of Reference Genes for Normalization of RT-qPCR Analysis in Developing or Abiotic-Stressed Tissues of Loquat (*Eriobotrya japonica*). *Phyton-Int. J. Exp. Bot.* **2023**, *92*, 1185–1201. [CrossRef]
21. Chauhan, A.S.; Tiwari, M.; Indoliya, Y.; Mishra, S.K.; Lavania, U.C.; Chauhan, P.S.; Chakrabarty, D.; Tripathi, R.D. Identification and Validation of Reference Genes in Vetiver (*Chrysopogon zizanioides*) Root Transcriptome. *Physiol. Mol. Biol. Plants.* **2023**, *29*, 613–627. [CrossRef]
22. Škiljaica, A.; Jagić, M.; Vuk, T.; Leljak Levanić, D.; Bauer, N.; Markulin, L. Evaluation of Reference Genes for RT-qPCR Gene Expression Analysis in *Arabidopsis thaliana* Exposed to Elevated Temperatures. *Plant Biol.* **2022**, *24*, 367–379. [CrossRef] [PubMed]
23. Zhao, G.J.; Wang, M.; Gan, Y.Q.; Gong, H.; Li, J.X.; Zheng, X.M.; Liu, X.X.; Zhao, S.Y.; Luo, J.N.; Wu, H.B. Identification of suitable reference genes for quantitative reverse transcription PCR in Luffa (*Luffa cylindrica*). *Physiol. Mol. Biol. Plants* **2022**, *28*, 737–747. [CrossRef] [PubMed]
24. Vandesompele, J.; De Preter, K.; Pattyn, F.; Poppe, B.; Van Roy, N.; De Paepe, A.; Speleman, F. Accurate Normalization of Real-Time Quantitative RT-PCR Data by Geometric Averaging of Multiple Internal Control Genes. *Genome Biol.* **2002**, *3*, research0034.1. [CrossRef] [PubMed]
25. Andersen, C.L.; Jensen, J.L.; Orntoft, T.F. Normalization of Real-Time Quantitative Reverse Transcription-PCR Data: A Model-Based Variance Estimation Approach to Identify Genes Suited for Normalization, Applied to Bladder and Colon Cancer Data Sets. *Cancer Res.* **2004**, *64*, 5245–5250. [CrossRef] [PubMed]
26. Pfaffl, M.W.; Tichopad, A.; Prgomet, C.; Neuvians, T.P. Determination of Stable Housekeeping Genes, Differentially Regulated Target Genes and Sample Integrity: Bestkeeper-Excel-Based Tool Using Pair-Wise Correlations. *Biotechnol. Lett.* **2004**, *26*, 509–515. [CrossRef] [PubMed]
27. Xie, F.; Wang, J.; Zhang, B. RefFinder: A Web-Based Tool for Comprehensively Analyzing and Identifying Reference Genes. *Funct. Integr. Genomics.* **2023**, *23*, 125. [CrossRef]
28. Zandalinas, S.I.; Fichman, Y.; Devireddy, A.R.; Sengupta, S.; Azad, R.K.; Mittler, R. Systemic Signaling During Abiotic Stress Combination in Plants. *Proc. Natl. Acad. Sci. USA* **2020**, *117*, 13810–13820. [CrossRef]
29. Bustin, S.A.; Benes, V.; Nolan, T.; Pfaffl, M.W. Quantitative Real-Time RT-PCR - a Perspective. *J. Mol. Endocrinol.* **2005**, *34*, 597–601. [CrossRef]
30. Kozera, B.; Rapacz, M. Reference genes in real-time PCR. *J. Appl. Genet.* **2013**, *54*, 391–406. [CrossRef]
31. Ahmed, U.; Xie, Q.; Shi, X.P.; Zheng, B. Development of Reference Genes for Horticultural Plants. *Crit. Rev. Plant Sci.* **2022**, *41*, 190–208. [CrossRef]
32. dos Santos, C.P.; da Cruz Saraiva, K.D.; Batista, M.C.; Germano, T.A.; Costa, J.H. Identification and Evaluation of Reference Genes for Reliable Normalization of Real-Time Quantitative PCR Data in Acerola Fruit, Leaf, and Flower. *Mol. Biol. Rep.* **2020**, *47*, 953–965. [CrossRef] [PubMed]
33. Zhang, C.; Jiang, Q.N.; Wang, Y.G.; Fu, J.X.; Dong, B.; Zhou, L.H.; Zhao, H.B. Transcriptome-Based Validation of Proper Reference Genes for Reverse Transcription Quantitative PCR Analysis of *Sinocalycanthus chinensis*. *Biol. Plant.* **2020**, *64*, 253–257. [CrossRef]
34. Zhao, Z.; Zhang, Z.; Ding, Z.; Meng, H.; Shen, R.; Tang, H.; Liu, Y.G.; Chen, L. Public-Transcriptome-Database-Assisted Selection and Validation of Reliable Reference Genes for qRT-PCR in Rice. *Sci. China-Life Sci.* **2020**, *63*, 92–101. [CrossRef] [PubMed]
35. Mo, Z.H.; Chen, Y.Q.; Lou, W.R.; Jia, X.D.; Zhai, M.; Xuan, J.P.; Guo, Z.R.; Li, Y.R. Identification of Suitable Reference Genes for Normalization of Real-Time Quantitative PCR Data in Pecan (*Carya illinoinensis*). *Trees* **2020**, *34*, 1233–1241. [CrossRef]
36. Zhao, M.Q.; Fan, H.; Tu, Z.H.; Cai, G.J.; Zhang, L.M.; Li, A.D.; Xu, M. Stable Reference Gene Selection for Quantitative Real-Time PCR Normalization in Passion Fruit (*Passiflora edulis* Sims.). *Mol. Biol. Rep.* **2022**, *49*, 5985–5995. [CrossRef] [PubMed]
37. Shen, J.S.; Wu, Y.T.; Jiang, Z.Y.; Xu, Y.; Zheng, T.C.; Wang, J.; Cheng, T.R.; Zhang, Q.X.; Pan, H.T. Selection and Validation of Appropriate Reference Genes for Gene Expression Studies in *Forsythia*. *Physiol. Mol. Biol. Plants* **2020**, *26*, 173–188. [CrossRef]
38. Zhu, X.L.; Wang, B.Q.; Wang, X.; Wei, X.H. Screening of Stable Internal Reference Gene of Quinoa under Hormone Treatment and Abiotic Stress. *Physiol. Mol. Biol. Plants* **2021**, *27*, 2459–2470. [CrossRef]
39. Amiruddin, N.; Chan, P.L.; Chan, K.L.; Ong, P.W.; Morris, P.E.; Ong-Abdullaw, M.; Masura, S.S.; Low, E.T.L. Determination of Reliable Reference Genes for Reverse Transcription Quantitative Real-Time PCR from Oil Palm Transcriptomes. *J. Oil Palm Res.* **2022**, *34*, 439–452. [CrossRef]
40. Agarwal, P.K.; Gupta, K.; Lopato, S.; Agarwal, P. Dehydration Responsive Element Binding Transcription Factors and Their Applications for the Engineering of Stress Tolerance. *J. Exp. Bot.* **2017**, *68*, 2135–2148. [CrossRef]
41. Erpen, L.; Devi, H.S.; Grosser, J.W.; Dutt, M. Potential Use of the DREB/ERF, MYB, NAC and WRKY Transcription Factors to Improve Abiotic and Biotic Stress in Transgenic Plants. *Plant Cell Tissue Organ Cult.* **2018**, *132*, 1–25. [CrossRef]

42. Jangale, B.L.; Chaudhari, R.S.; Azeez, A.; Sane, P.V.; Sane, A.P.; Krishna, B. Independent and Combined Abiotic Stresses Affect the Physiology and Expression Patterns of Dreb Genes Differently in Stress-Susceptible and Resistant Genotypes of Banana. *Physiol. Plant.* **2019**, *165*, 303–318. [CrossRef] [PubMed]
43. Zhou, Y.X.; Zhou, W.; Liu, H.; Liu, P.; Li, Z.G. Genome-Wide Analysis of the Soybean *DREB* Gene Family: Identification, Genomic Organization and Expression Profiles in Response to Drought Stress. *Plant Breed.* **2020**, *139*, 1158–1167. [CrossRef]
44. Ali, N.; Hadi, F. CBF/DREB Transcription Factor Genes Play Role in Cadmium Tolerance and Phytoaccumulation in *Ricinus communis* under Molybdenum Treatments. *Chemosphere* **2018**, *208*, 425–432. [CrossRef] [PubMed]
45. Onele, A.; Mazina, A.; Leksin, I.; Chasov, A.; Minibayeva, F.; Beckett, R. Class III Peroxidase Genes in the Moss *Dicranum scoparium*: Identification and Abiotic Stress Induced Expression Analysis. *S. Afr. J. Bot.* **2023**, *159*, 72–84. [CrossRef]
46. Wang, J.E.; Liu, K.K.; Li, D.W.; Zhang, Y.L.; Zhao, Q.; He, Y.M.; Gong, Z.H. A Novel Peroxidase *CanPOD* Gene of Pepper Is Involved in Defense Responses to *Phytophthora capsici* Infection as Well as Abiotic Stress Tolerance. *Int. J. Mol. Sci.* **2013**, *14*, 3158–3177. [CrossRef]
47. Lee, C.J.; Park, S.U.; Kim, S.E.; Lim, Y.H.; Ji, C.Y.; Kim, Y.H.; Kim, H.S.; Kwak, S.S. Overexpression of *IbLfp* in Sweetpotato Enhances the Low-Temperature Storage Ability of Tuberous Roots. *Plant Physiol. Biochem.* **2021**, *167*, 577–585. [CrossRef]
48. Gao, C.Q.; Wang, Y.C.; Liu, G.F.; Wang, C.; Jiang, J.; Yang, C.P. Cloning of Ten Peroxidase (*POD*) Genes from *Tamarix hispida* and Characterization of Their Responses to Abiotic Stress. *Plant Mol. Biol. Rep.* **2010**, *28*, 77–89. [CrossRef]

MDPI AG
Grosspeteranlage 5
4052 Basel
Switzerland
Tel.: +41 61 683 77 34

International Journal of Molecular Sciences Editorial Office
E-mail: ijms@mdpi.com
www.mdpi.com/journal/ijms

www.ingramcontent.com/pod-product-compliance
Lightning Source LLC
Chambersburg PA
CBHW061552220926
51850CB00005B/314
9783725856855